The Real Number System

Natural numbers	$N = \{1, 2, 3, \ldots\}$
Whole numbers	$W = \{0, 1, 2, 3, \ldots\}$
Integers	$J = \{\ldots, -3, -2, -1, 0, 1, 2, 3, \ldots\}$
Rational numbers	$Q = \left\{ \dfrac{a}{b} \,\middle|\, a \in J, b \in J, b \neq 0 \right\}$
Irrational numbers	$H = \{\text{nonterminating, nonrepeating decimals}\}$
Real numbers	$R = H \cup Q$
Absolute value	$\lvert x \rvert = \begin{cases} x & \text{if } x \geq 0 \\ -x & \text{if } x < 0 \end{cases}$

Properties of Real Numbers

If $a, b, c \in \mathbf{R}$, the following properties are assumed to be true:

	Addition	*Multiplication*
Closure	$a + b \in R$	$a \cdot b \in R$
Commutative	$a + b = b + a$	$a \cdot b = b \cdot a$
Associative	$a + (b + c) = (a + b) + c$	$a(b \cdot c) = (a \cdot b)c$
Identity	$a + 0 = 0 + a = a$	$a \cdot 1 = 1 \cdot a = a$
Inverse	$a + (-a) = (-a) + a = 0$	$a \cdot \dfrac{1}{a} = \dfrac{1}{a} \cdot a = 1,\ a \neq 0$
Distributive		$a(b + c) = ab + ac$

Theorems and Definitions

Zero-factor theorem	$a \cdot 0 = 0 \cdot a = 0$
Definition of subtraction	$a - b = a + (-b)$
Double negative	$-(-a) = a$

Properties of Equality

Reflexive	$a = a$
Symmetric	If $a = b$, then $b = a$.
Transitive	If $a = b$ and $b = c$, then $a = c$.
Substitution	If $a = b$, then a can be replaced by b or b by a without changing the truth of the statement.

Operations with Fractions

$$\frac{a}{b} \cdot \frac{c}{d} = \frac{ac}{bd},\ \ b, d \neq 0 \qquad \frac{a}{b} \div \frac{c}{d} = \frac{a}{b} \cdot \frac{d}{c} = \frac{ad}{bc},\ \ b, c, d \neq 0$$

$$\frac{a}{b} = \frac{ax}{bx},\ \ b, x \neq 0 \qquad \frac{a}{c} + \frac{b}{c} = \frac{a + b}{c},\ \ c \neq 0$$

$$-\frac{a}{b} = \frac{-a}{b} = \frac{a}{-b} = -\frac{-a}{-b},\ \ b \neq 0$$

Operations with Equations

Addition property of equality   If $a = b$, then $a + c = b + c$.
Subtraction property of equality   If $a = b$, then $a - c = b - c$.
Multiplication property of equality   If $a = b$, then $ac = bc$.
Division property of equality   If $a = b$, then $\dfrac{a}{c} = \dfrac{b}{c}$, $c \neq 0$.

Operations with Inequalities

If $a < b$, then $a + c < b + c$.

If $a < b$, then $a - c < b - c$

If $a < b$, then $\begin{cases} \dfrac{a}{c} < \dfrac{b}{c} \text{ if } c > 0; \\ \dfrac{a}{c} > \dfrac{b}{c} \text{ if } c < 0. \end{cases}$

If $a < b$, then $\begin{cases} ac < bc \text{ if } c > 0; \\ ac > bc \text{ if } c < 0. \end{cases}$

If $p > 0$, then $\begin{cases} |x| < p \text{ is equivalent to } -p < x < p; \\ |x| > p \text{ is equivalent to } x < -p \text{ or } x > p. \end{cases}$

Determinants

$\begin{vmatrix} a & b \\ c & d \end{vmatrix} = ad - bc$

$\begin{vmatrix} a_1 & b_1 & c_1 \\ a_2 & b_2 & c_2 \\ a_3 & b_3 & c_3 \end{vmatrix} = a_1 b_2 c_3 + a_3 b_1 c_2 + a_2 b_3 c_1 - a_3 b_2 c_1 - a_1 b_3 c_2 - a_2 b_1 c_3$

Properties of Exponents

If $a, b \in R$ and $m, n \in J$, then the following rules hold:

Product rule $b^m \cdot b^n = b^{m+n}$

Power rules $(b^m)^n = b^{mn}$; $(ab)^m = a^m b^m$; $\left(\dfrac{a}{b}\right)^m = \dfrac{a^m}{b^m}$, $b \neq 0$

Quotient rule $\dfrac{b^m}{b^n} = b^{m-n}$, $b \neq 0$

Negative exponents $b^{-m} = \dfrac{1}{b^m}$, $b \neq 0$

Zero as an exponent $b^0 = 1$, $b \neq 0$

Special Products and Factors

$(a + b)^2 = a^2 + 2ab + b^2$

$(a - b)^2 = a^2 - 2ab + b^2$

$(a + b)(a - b) = a^2 - b^2$

$a^3 + b^3 = (a + b)(a^2 - ab + b^2)$

$a^3 - b^3 = (a - b)(a^2 + ab + b^2)$

Intermediate Algebra

Intermediate Algebra

Gilbert M. Peter
Cuesta College

C. Lee Welch
Cuesta College

WEST PUBLISHING COMPANY
Minneapolis/St. Paul   New York   San Francisco   Los Angeles

PRODUCTION CREDITS

Copyediting	Susan Gerstein
Text Design	Geri Davis, Quadrata, Inc.
Chapter Opening Art	Nancy Wirsig McClure, Hand to Mouse Arts
Text Art	Ed Rose, VGS
Answers Art	Ross Rueger
Proofreading	Sylvia Dovner, Technical Texts
Indexer	Bernice Eisen
Composition	Bi-Comp Incorporated
Problem Checking	Chuck Heuer

WEST'S COMMITMENT TO THE ENVIRONMENT

In 1906, West Publishing Company began recycling materials left over from the production of books. This began a tradition of efficient and responsible use of resources. Today, up to 95 percent of our legal books and 70 percent of our college and school texts are printed on recycled, acid-free stock. West also recycles nearly 22 million pounds of scrap paper annually—the equivalent of 181,717 trees. Since the 1960s, West has devised ways to capture and recycle waste inks, solvents, oils, and vapors created in the printing process. We also recycle plastics of all kinds, wood, glass, corrugated cardboard, and batteries, and have eliminated the use of Styrofoam book packaging. We at West are proud of the longevity and the scope of our commitment to the environment.

Production, Prepress, Printing and Binding by West Publishing Company.

British Library Cataloguing-in-Publication Data. A catalogue record for this book is available from the British Library.

COPYRIGHT ©1996   By WEST PUBLISHING COMPANY
610 Opperman Drive
P.O. Box 64526
St. Paul, MN 55164-0526

All rights reserved
Printed in the United States of America
03 02 01 00 99 98 97 96   8 7 6 5 4 3 2 1 0

Library of Congress Cataloging-in-Publication Data

Peter, Gilbert M.
    Intermediate algebra / Gilbert M. Peter, C. Lee Welch.
      p. cm.
    Welch's name appears first on the earlier ed.
    Includes index.
    ISBN 0-314-04158-3
    1. Algebra.   I. Welch, C Lee.   II. Title.
QA152.2.W433   1995
512.9—dc20   94-10481
CIP

DEDICATION

We dedicate this book to our wives Marlene and Pauline and to the re-entry students whose desire to change their lives and careers through education has been an inspiration to both of us.

Brief Contents

1	The Real-Number System	2
2	Linear Equations and Inequalities	36
3	Linear Equations and Functions in Two Variables	102
4	Systems of Linear Equations and Inequalities	166
5	Laws of Exponents, Polynomials, and Polynomial Functions	246
6	Factoring Polynomials; Factoring by Grouping	294
7	Rational Expressions, Equations, and Functions	346
8	Radicals and Rational Exponents	392
9	Quadratic Equations, Functions, and Inequalities	444
10	Conic Sections	500
11	Exponential and Logarithmic Functions	554
12	Sequences, Series, and the Binomial Theorem	592

Appendices

A	Graphics Calculators	A-2
B	Tables	A-14
C	Answers	A-20

Index

I-1

Contents

Preface   xv

CHAPTER 1

The Real-Number System   2

1.1 Sets and Real Numbers   3
1.2 Properties of Real Numbers   12
1.3 Operations with Real Numbers   21
  Chapter Summary   30
  Cooperative Exercise   32
  Review Exercises   32
  Chapter Test   33

CHAPTER 2

Linear Equations and Inequalities   36

2.1 Solving Linear Equations in One Variable   37
2.2 From Words to Symbols: An Introduction to Applied Problems   47
2.3 Literal Equations and Formulas   54
2.4 Solving Linear Equations Involving Absolute Value   62
2.5 Solving Linear Inequalities   68
2.6 Solving Linear Inequalities Involving Absolute Value   79
2.7 More Applications   84
  Chapter Summary   96
  Cooperative Exercise   97
  Review Exercises   98
  Chapter Test   99

CHAPTER 3

Linear Equations and Functions in Two Variables 102

- **3.1** The Rectangular Coordinate System; Graphing Linear Equations in Two Variables 103
- **3.2** The Slope of a Line; Distance and Midpoint Formulas 113
- **3.3** Point–Slope, Slope–Intercept, and the General Form of the Equation of a Line 127
- **3.4** Relations, Functions, and Function Notation 135
- **3.5** The Algebra of Functions; Inverse Functions 147
  - CHAPTER SUMMARY 155
  - COOPERATIVE EXERCISE 156
  - REVIEW EXERCISES 157
  - CHAPTER TEST 160
  - CUMULATIVE REVIEW FOR CHAPTERS 1–3 161

CHAPTER 4

Systems of Linear Equations and Inequalities 166

- **4.1** Solving Systems of Linear Equations by Graphing, Elimination, and Substitution 167
- **4.2** Solving Systems of Linear Equations in Three or More Variables 183
- **4.3** Solving Systems of Linear Equations Using Matrices 195
- **4.4** Determinants and Cramer's Rule 205
- **4.5** Graphing Linear Inequalities in Two Variables 215
- **4.6** Linear Programming 224
- **4.7** More Applications 231
  - CHAPTER SUMMARY 240
  - COOPERATIVE EXERCISE 241
  - REVIEW EXERCISES 242
  - CHAPTER TEST 245

CHAPTER 5

Laws of Exponents, Polynomials, and Polynomial Functions 246

- **5.1** Exponents and Scientific Notation 247
- **5.2** Adding and Subtracting Polynomials 258
- **5.3** Multiplying Polynomials 265
- **5.4** Dividing Polynomials 274

5.5 Synthetic Division   284
  CHAPTER SUMMARY   290
  COOPERATIVE EXERCISE   291
  REVIEW EXERCISES   292
  CHAPTER TEST   293

CHAPTER 6

Factoring Polynomials; Factoring by Grouping   294

6.1 Monomial Factors; Factoring by Grouping   295
6.2 Factoring Type I Trinomials   302
6.3 Factoring Type II Trinomials; A Test for Factorability   307
6.4 Factoring Special Types of Polynomials; More Factoring by Grouping   315
6.5 General Methods of Factoring   322
6.6 Solving Nonlinear Equations by Factoring   326
6.7 More Applications; Solving Applied Problems by Factoring   334
  CHAPTER SUMMARY   340
  COOPERATIVE EXERCISE   342
  REVIEW EXERCISES   342
  CHAPTER TEST   343
  CUMULATIVE REVIEW FOR CHAPTERS 4–6   343

CHAPTER 7

Rational Expressions, Equations, and Functions   346

7.1 Reducing and Building Rational Expressions   347
7.2 Multiplying and Dividing Rational Expressions   353
7.3 Adding and Subtracting Rational Expressions   359
7.4 Simplifying Complex Rational Expressions   366
7.5 Solving Rational Equations   373
7.6 More Applications   378
  CHAPTER SUMMARY   387
  COOPERATIVE EXERCISE   389
  REVIEW EXERCISES   389
  CHAPTER TEST   391

CHAPTER 8

Radicals and Rational Exponents — 392

- **8.1** Rational Exponents   393
- **8.2** Radicals and Rational Exponents   400
- **8.3** Arithmetic Operations Involving Radicals   408
- **8.4** Division by Radicals; Rationalizing   414
- **8.5** Solving Equations Involving Radicals   422
- **8.6** Complex Numbers   428
- **8.7** More Applications   434
   - CHAPTER SUMMARY   439
   - COOPERATIVE EXERCISE   440
   - REVIEW EXERCISES   440
   - CHAPTER TEST   441

CHAPTER 9

Quadratic Equations, Functions, and Inequalities — 444

- **9.1** Solving Polynomial Equations by Factoring and the Square Root Property   445
- **9.2** Solving Quadratic Equations by Completing the Square   452
- **9.3** Solving Quadratic Equations Using the Quadratic Formula   460
- **9.4** Solving Equations Involving Quadratics   469
- **9.5** Solving Inequalities Involving Quadratics   476
- **9.6** Variation   483
- **9.7** More Applications   488
   - CHAPTER SUMMARY   495
   - COOPERATIVE EXERCISE   496
   - REVIEW EXERCISES   496
   - CHAPTER TEST   497
   - CUMULATIVE REVIEW FOR CHAPTERS 7–9   498

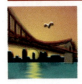

CHAPTER 10

Conic Sections — 500

- **10.1** Circles   501
- **10.2** Parabolas   509
- **10.3** Ellipses   525
- **10.4** Hyperbolas   533

10.5	Solving Nonlinear Systems of Equations	540
10.6	Solving Nonlinear Systems of Inequalities	546
	CHAPTER SUMMARY	550
	COOPERATIVE EXERCISE	552
	REVIEW EXERCISES	552
	CHAPTER TEST	553

CHAPTER 11

Exponential and Logarithmic Functions 554

11.1	Exponential Functions	555
11.2	Logarithmic Functions	565
11.3	Properties of Logarithms	572
11.4	Solving Exponential and Logarithmic Equations	578
11.5	More Applications	583
	CHAPTER SUMMARY	587
	COOPERATIVE EXERCISE	588
	REVIEW EXERCISES	589
	CHAPTER TEST	590

CHAPTER 12

Sequences, Series, and the Binomial Theorem 592

12.1	Sequences, Series, and Summation Notation	593
12.2	Arithmetic Sequences and Series	599
12.3	Geometric Sequences and Series	606
12.4	The Binomial Theorem	615
	CHAPTER SUMMARY	621
	COOPERATIVE EXERCISE	623
	REVIEW EXERCISES	623
	CHAPTER TEST	624
	CUMULATIVE REVIEW CHAPTERS 10–12	625

APPENDICES

A Graphics Calculators   A-2
B Tables   A-14
C Answers   A-20

INDEX   I-1

PREFACE

This text is intended for a one-term course in intermediate algebra and is the companion text to *Elementary Algebra* by the same authors. A primary goal is to help to prepare students for success in college algebra, precalculus, or statistics. The last three chapters—Conic Sections; Exponential and Logarithmic Functions; and Sequences, Series, and the Binomial Theorem—build students' skills and understanding beyond the typical coverage of intermediate algebra and are appropriate for expanded intermediate algebra courses. For briefer courses, this text is designed to allow the exclusion of some material. Chapter 1 (The Real Number System) can be omitted if students are comfortable working with fractions and are familiar with the set of real numbers and its subsets. In addition, Sections 2.4 (Solving Linear Equations Involving Absolute Value), 2.6 (Solving Linear Inequalities Involving Absolute Value), 4.3 (Solving Linear Systems of Equations Using Matrices), and 4.6 (Linear Programming) can also be eliminated in a briefer course of instruction.

It has been our experience, and the experience of many instructors we know, that students often have a difficult time making the transition from the more "skills-oriented" intermediate algebra course to the more "concept-oriented" presentation typical of college algebra, precalculus, and statistics courses. Instructors in these courses often find that students are simply not adequately prepared to take college algebra (for example) in spite of evidence indicating that they had performed adequately (or better) in the prerequisite courses. The texts used in these preceding courses are often a key factor in this lack of preparation. This is particularly true when the intermediate text duplicates much of the content and the approach of an elementary algebra text. It isn't until the college algebra course that the student (and instructor) discovers to what extent the student is underprepared for success. This has sometimes been our experience with texts we have used, and some of the reviewers of this text volunteered their own experiences and concerns of this kind. Therefore, we have written an intermediate algebra text that makes reasonable demands of the students while preparing them for success at the college algebra level. We've accomplished this goal by organizing and presenting the traditional contents of intermediate algebra in a way that we've found to be effective in guiding students to success in this and any subsequent mathematics courses.

To master skills and apply them at a more conceptual level in later courses, students must use the key skills and concepts in a number of different circumstances.

In this way they will develop a more intuitive and reasoning-oriented ability to use mathematics. Understanding the concept of a function and the correspondence between the algebraic expression of a function and the graphic one is essential for success in college algebra. We believe that too often functions and graphing are covered late in intermediate algebra, where they are not given proper emphasis and when students have too few opportunities to practice the skills and understand the power of the function concept. Therefore we have introduced graphing and the concept of a function in Chapter 3. Once introduced, the function concept is then used in the development of other topics to help students gain a more powerful understanding of different types of equations and their solutions. This gives students many opportunities to use functions during the semester and to obtain insight into analyzing mathematical relationships. In addition, the early introduction of functions and graphing provides opportunities for students to use a graphing calculator or computer software. These powerful graphing technologies help students to visualize concepts and to develop a deeper understanding of various algebraic equations and functions. Although the text encourages and helps the students to explore with graphing technology, reference to this technology is not invasive and its use is completely optional.

A major theme of this book is problem solving. This emphasis appears in a variety of ways, but it is most evident in the number of applications and applied problems throughout the book, and in the "More Applications" sections at the ends of Chapter 2, 4, 6–9, and 11. While the problem-solving theme is not limited to these sections, they do provide major examples of the way in which the mathematical concepts and skills being presented can be used to model and solve real-life applied problems. In addition, the introduction of graphing and the function concept in Chapter 3 and especially of systems of equations in Chapter 4 provides students with powerful tools that they can use for solving applied problems.

It has been said that a good intermediate algebra text can be measured by its ability to use examples to show the "how and why" of an algebraic concept or skill, rather than by presenting lengthy discussions of the theory. We believe that both theory and practice are equally important and we have tried to strike a balance between the two. We make an effort to show students how to apply the skills and concepts with an abundance of examples, thereby encouraging their confidence and providing them with a substantial opportunity to see mathematics in action. Corresponding to the nearly 900 in-text examples is a wealth of exercises, over 6,200 of them. Any experienced instructor knows that exercise sets are vital to a good text. This is where the students have the opportunity to practice the mathematics that they have just learned and to discover where additional work is needed. The exercises are structured to give instructors many options in assigning problems, including group problems and writing exercises.

Key Features

APPLIED PROBLEMS

Applied problems appear early in Chapter 1 and are included in most exercise sets as well as in separate sections called "More Applications" at the end of most chapters. Experience has shown that students often learn how to manipulate algebraic expressions without knowing how to use them as a tool for solving applied problems. The introduction of such problems early in the text helps the student to see how mathematics is used while helping them to develop the logical thinking necessary for problem-solving. Applied problems present a challenge that must be met to be successful in working with the more difficult material found in subsequent courses. To make the problems meaningful and not contrived, we have selected problems from physics, chemistry, business, finance, and areas of student aspirations or experience. Some of these problems may use units of measurement that are not familiar to the student, but such familiarity is not necessary to solve the problems successfully.

Systems of Equations and Graphing

Linear equations and inequalities are introduced in Chapter 2, expanded upon in Chapter 3, where graphing first occurs, and then reinforced in Chapter 4, when systems of linear equations and inequalities are developed. This organization provides the student with the option of using systems of equations or inequalities to solve applied problems throughout the remainder of the text. The introduction of graphing techniques early in the course gives the student a means to visualize the mathematics that they are learning and to use this technique of visualization to help them solve later problems.

Early Functions

The concept of a function is central to the development of mathematics and as such is introduced in Chapter 3, earlier than many intermediate algebra texts. Functions are then used throughout the remainder of the text where appropriate. For example, they are used to discuss polynomial functions and their properties in Chapter 5, rational functions in Chapter 7, quadratic functions in Chapter 8, and exponential and logarithmic functions in Chapter 11.

Applications of Geometry

Many students enroll directly in intermediate algebra after completing elementary algebra, without experiencing the benefit of a course in plane geometry. They are, however, aware of the geometry of the world about them, so we have included applications of algebra to geometry together with sufficient explanation so that they will understand. Common geometric figures together with the formulas for finding their perimeters, areas, and volumes are provided in the Appendix. For students who have completed a course in geometry, the problems will give them a chance to refresh and renew concepts already learned.

Calculators

Although we have chosen not to focus on the calculator throughout the book, we are certainly aware of its utility and we have included problems where its use may be helpful. Such problems are identified by words or by a graphics calculator icon beside the particular exercise or group of exercises. We chose to use a graphics calculator logo because a graphics calculator can be used to carry out the same operations that a scientific calculator can in addition to having graphics capabilities. Periodically, instructions are given showing how to use a calculator to carry out an operation or a specific order of operations. Basic instructions for using the TI-81 and the TI-82 are contained in the Appendix.

Theory verses Practice

A careful balance is maintained between the "how" and the "why" of each mathematical concept. The discussion is generally intuitive but it is based on carefully stated mathematical definitions and theorems that are supported by a large number of examples. Proofs are included in the areas where they enhance the "why" of a concept.

Set Notation

Set notation is introduced in Chapter 1 and is used to show the relationship between the components of the number system. It is also used to provide a consistent means of identifying solution sets of equations or systems of equations throughout the text. Solution sets of inequalities are described using interval notation to prepare the student for more advanced courses.

Fractions

Chapter 1 contains a review of fractions. The review can be either covered in class or used as a reference by students who feel they need additional work in this area.

AMATYC and NCTM Guidelines

The text was written with the NCTM and AMATYC guidelines in mind. There is an abundance of exercises that require critical thinking, writing, and the use of modern technology, as well as cooperative group projects.

Key Chapter Pedagogy

Chapter Openers

Every chapter opens with an applied problem that appears later in the chapter. This gives the student an idea of the kind of problems they will learn to solve by using the concepts covered in the chapter.

Section Objectives

The objectives to be accomplished in each section are clearly stated at its beginning. They are written in a format that tells the student exactly what they are expected to learn by the time the section is completed.

Historical Comments

All of us are indebted to those who preceded us in mathematics, so we have included many historical comments throughout the text. Some are brief glimpses of the lives of early mathematicians and describe the part they played in developing the topic at hand. Others trace the paths that have led to the mathematical symbolism that we use today. Still others present problems from ancient times or from texts that are hundreds of years old and that, except for the language, are still used today to teach important algebraic concepts. Complete solutions to the historical problems are presented at the end of the section in which the problem occurs. In every case, care has been taken to keep the historical comments brief enough so that they will be read by the students and interesting enough so that some students will want to learn more.

Key Concepts, Definitions, and Theorems
Important theorems and other key concepts are boxed to emphasize their importance and to make them easier to find for future reference and review.

Caution Warnings
Students are warned about common misconceptions with special caution notations in the left margins of pages where such mistakes may occur. These warnings help the student avoid the frustration that naturally occurs when these errors are made and the student is unable to discover what is wrong.

Chapter Summaries
A summary of the important concepts in each chapter is included at its end. The items discussed in the summaries are keyed to the chapter sections for quick reference.

Can You Match These?
A short matching quiz concludes the text discussion in most sections. These self tests give students the opportunity to check whether they are ready to begin the exercise set for that section. Answers are supplied after the last question, providing them with immediate feedback about their level of understanding.

Review Problems
Since algebraic concepts are best mastered and retained by constant practice, most exercise sets contain problems from previous sections and chapters. At the end of each chapter there is a complete set of review problems covering the topics in that chapter, and a cumulative review is included at the end of every third chapter. Complete solutions to all problems in the chapter reviews and the cumulative reviews are included in the Student's Solutions Manual.

Exercise Sets
Thousands of class-tested exercises appear at section ends and chapter ends. Odd- and even-numbered exercises are paired so that an instructor can assign just odd-numbered problems, with answers in the text and in the Student's Solutions Manual, or just even-numbered problems, with answers only in the Instructor's Solutions Manual.

Student Writing Assignments
We believe that students have not fully mastered a mathematical concept until they can discuss it and write about what they have learned. Where appropriate, the exercise sets contain a "Write in Your Own Words" section that directs the student to write about a particular concept. Responses can range from a single sentence to a paragraph or more, depending on the question.

For Extra Thought
Many of the exercise sets contain problems that are designed to challenge the student by requiring them to think critically. These problems use the concepts of the section and require that the student extend the concept a step or two further to arrive at a solution.

Cooperative Exercises

Each chapter contains a cooperative exercise after the chapter summary. Care has been taken, as much as possible, to coordinate the exercise with the concepts learned in the chapter. The contents of these exercises reflect situations relevant to everyday life.

Ancillaries

Instructor's Solutions Manual

An instructor's manual is available free of charge to all qualified adopters of the text. It contains the solutions to all problems not solved in the Student's Solutions Manual.

Test Bank

A test bank is available free of charge to qualified adopters of the text. It contains 80 questions for each chapter, half of which are multiple choice and half problems that the student is required to solve. The Test Bank Manual is keyed to West Test, a comptuer test generator that can create tests based on the instructor's selections from the manual.

Group Learning Resource Manual

A Group Learning Resource Manual for Elementary and Intermediate Algebra, by ViAnn Olson of Rochester Community College, contains 20 group learning exercises for both in-class and homework assignments. Each exercise is accompanied by suggestions for the instructor on how best to use that exercise. A section at the beginning of the manual contains guidelines on how to use group learning exercises effectively. The exercises are designed to be photocopied by instructors for distribution to the students.

Student's Solutions Manual

A student's manual, prepared by Ross Rueger of the College of the Sequoias, is available for purchase. It contains complete solutions to the odd-numbered exercises in the exercise sets, as well as all of the answers to the in-text chapter tests, chapter reviews, and cumulative reviews.

West Math Tutor

The West Math Tutor by Mathens is available for IBM and IBM-compatible PCs, as well as for Apple Macintosh computers. This software contains algorithmically based tutorials for most of the topics in the text. The tutorials are interactive in that they give feedback to the students when they make errors and provide hints that help the students when they try the questions again. The student is shown the correct solution after two incorrect attempts.

West Math Videos

The West Math Videos are designed to provide additional instruction for the students in the areas in which they may be experiencing difficulty. Worksheets accompany the videos so that the student can be actively involved in the lesson.

Acknowledgements

We wish to thank the many reviewers of this text whose invaluable comments and suggestions greatly improved the final revision.

Carol Achs, Mesa Community College
Judy Barclay, Cuesta College
Harold Bennett, Texas Tech University
Kathleen Bavelas, Manchester Community College
Mitzi Chaffer, Central Michigan University
Rodney Chase, Oakland Community College
Dharam Chopra, Wichita State University
John Coburn, St. Louis Community College at Florissant Valley
Sally Copeland, Johnson County Community College
Gregory J. Davis, University of Wisconsin—Green Bay
Sharon Edgmon, Bakersfield College
Rebecca Eller, Blue Ridge Community College
Robert Farinelli, Community College of Allegheny County—South Campus
Lenore Frank, State University of New York at Stony Brook
Jay Graening, University of Arkansas
Louise Hasty, Austin Community College
John Hill, Illinois State University
Elizabeth Hodes, Santa Barbara Community College
Norma James, New Mexico State University
Joyce Lindstrom, St. Charles County Community College
Peter Lindstrom, North Lake College
Giles Maloof, Boise State University
John Martin, Santa Rosa Junior College
Vince McGarry, Austin Community College
Sheila McNicholas, University of Illinois at Chicago
Ernest Multhaup, College of San Mateo
ViAnn Olson, Rochester Community College
James Peake, Iowa State University
Dennis Reissig, Suffolk Community College
Deborah Ritchie, Moorpark College
Robert Russell, West Valley College
Barbara Sausen, Fresno City College
Larry Small, Los Angeles Pierce College
Ara Sullenberger, Tarrant County Junior College
George Szoke, University of Akron
Bruce Teague, Santa Fe Community College

Anthony Vance, Austin Community College
Gerry Vidrine, Louisiana State University
George Wales, Ferris State University
Richard Werner, Santa Rosa Junior College
George Witt, Glendale Community College

In particular, we want to thank Marvin Johns of the Cuesta College Mathematics Department for his line-by-line review of the manuscript and his insistence that we add additional clarifying comments where his teaching experience has shown that students have difficulties. We would also like to thank Richard Mixter and Keith Dodson of West Publishing, who have been involved with this project since its inception, for their helpful suggestions and encouragement; Susan Gerstein, whose superb copyediting skills greatly improved the readability of the text; and Christine Hurney, our production editor, who smoothly directed the transition from manuscript to textbook.

Preface to the Student

Learning mathematics requires dedication on your part. Try to adhere to the 2-for-1 rule: for every hour in class, follow it with two hours of quality study outside of class.

When you registered for the class that uses this text, you probably did not plan any conflicts. It is worth your effort to try to ensure that none develops. In order to be successful, we offer the following five suggestions:

1. Except in an emergency, attend every class session. No text is designed to cover all your questions. Your instructor can. If you must miss a class, get the notes as soon as possible.
2. Ask questions in class. There is no such thing as a stupid question! What does qualify as stupid is the failure to ask about something you don't understand when the opportunity presents itself.
3. Take advantage of your instructor's posted office hours. Failure to understand the contents of one section will often hinder your understanding of the sections that follow. College classes can be less personal than those you experienced in high school and it is up to you to get to know your instructor well.
4. Do the assigned homework as soon as possible after class. Time is your enemy. If you have a small amount of time and can do part of the assignment before your next class, do it. This will solidify the concepts you have learned, making it much easier to complete the rest of the assignment when your schedule permits.
5. Rework all problems that you miss on any test until you understand what you did wrong. This will keep you from making the same mistakes again.
6. The greatest threat to success in a mathematics class is procrastination. Many college students hesitate to admit they don't understand a concept, hoping that something magic will happen. What *will* happen is that insurmountable problems will develop, causing you to drop the class or, worse yet, to fail the class. Get help as soon as you need it.

Intermediate Algebra

CHAPTER 1
THE REAL-NUMBER SYSTEM

1.1 Sets and Real Numbers
1.2 Properties of Real Numbers
1.3 Operations with Real Numbers

*A*PPLICATION

Find the vertical distance between the top of Mt. Whitney at 14,494 feet above sea level and the Salton Sea at 235 feet below sea level.

See Exercise Set 1.3, Number 30.

Mathematics, rightly viewed, possesses not only truth but supreme beauty, a beauty cold and austere, like that of a sculpture. It has no appeal to our weaker nature, nor gorgeous like a painting or sculpture, yet sublimely pure, and capable of requiring a stern perfection as only a great artist can show.

BERTRAND RUSSELL (1935)

1.1 Sets and the Real Numbers

HISTORICAL COMMENT   Georg Cantor (1845–1918) wrote the first formal publication on *set theory*. Since then the ideas and uses of sets have spread to almost all of mathematics, logic, and other areas. We will be introducing *sets* in this section, and we will be using them throughout most of the book.

OBJECTIVES   In this section we will learn to identify
1. sets, set notation, and symbols;
2. finite, infinite, and empty sets;
3. unions and intersections of sets;
4. some subsets of the set of real numbers; and
5. the set of real numbers.

We will begin by getting acquainted with some of the symbols, definitions, and properties of sets. The concept of a set is fundamental in helping us to understand our number system and how it is organized.

DEFINITION
Set
Element, or Member

A **set** is a well-defined collection of objects.
An **element**, or **member**, is any object within the set.

Sets are usually designated by a capital letter and by enclosing the elements within *braces*. For example,

$$N = \{1, 2, 3, 4, \ldots\}$$

describes the set of **natural, or counting numbers.** The three dots are used to indicate that the set continues on in the same manner without end.

To indicate that an object is an **element of** a set, the symbol "$\in$" is used. To indicate that an object is **not an element of** a set, the symbol "$\notin$" is used. Each of the following is true for the set of natural numbers, N.

NOTE A "/" (slash) through a symbol always negates it. For example, $\neq$ means "not equal."

$1 \in N$ One is an element of N.
$3 \in N$ Three is an element of N.
$\frac{1}{2} \notin N$ One-half is not an element of N.

3

$0 \notin N$ Zero is not an element of N.

Sets may be classified according to the number of elements they contain. A set without any elements is said to be an **empty,** or **null set.** The empty set is indicated by the symbol $\emptyset$. A set is said to be a **finite set** if there is a number that describes exactly how many elements it contains. A set that is not a finite set is said to be an **infinite set.** The empty set is finite because it contains *zero* (0) elements.

EXAMPLE 1 State whether the following sets are finite or infinite. If a set is finite, state whether it is also the empty set.

(a) $A = \{2, 4, 5\}$
(b) $B = \{1, 2, 3, \ldots\}$
(c) The set of cards in a standard deck of playing cards.
(d) The set of women over 10 feet tall.

SOLUTION
(a) A is a finite set: it has exactly three members.
(b) B is an infinite set, as indicated by the three dots.
(c) This is a finite set; it has 52 members.
(d) This is the empty set. It is also a finite set, with zero elements.

> **CAUTION**
> The empty set symbol, $\emptyset$, should not be confused with the symbol for zero, 0, on the computer. Also $\{\emptyset\}$ is not the empty set: it is a set with one member, the empty set.

We will be considering two ways to describe sets, the *listing method* and *set-builder notation*. With the **listing method,** a pattern is established, such as $N = \{1, 2, 3, 4, \ldots\}$, or, for a finite set, each element in the set is named, such as

$T = \{$Mercury, Venus, Earth, Mars, Jupiter, Saturn, Uranus, Neptune, Pluto$\}$

With **set-builder notation,** we describe the common characteristic of the elements in the set. For example,

$$T = \{\underbrace{x}_{\text{set of all } x} \mid \underbrace{}_{\text{``such that''}} \underbrace{x \text{ is a known planet in our solar system}}_{\text{characteristic}}\}$$

This expression is read, "T is the set of all x such that x is a known planet in our solar system." The vertical bar is read "such that." Notice that x can represent any one of the elements in the set. For this reason we call x a **variable.** A symbol that represents a fixed value (one value) is called a **constant.**

EXAMPLE 2 Write $A = \{x \mid x$ is a natural number less than 10$\}$ using the listing method.

SOLUTION $A = \{1, 2, 3, 4, 5, 6, 7, 8, 9\}$

EXAMPLE 3 Write $V = \{a, e, i, o, u\}$ using set-builder notation.

SOLUTION $V = \{x \mid x$ is a vowel in the English alphabet$\}$

Now let's consider two operations using sets.

DEFINITION

Union

Intersection

The **union** of two sets A and B, written $A \cup B$, is the set of all elements that are only in set A, only in set B, or in both sets A and B.

$$A \cup B = \{x \mid x \in A \text{ or } x \in B\}$$

The **intersection** of two sets A and B, written $A \cap B$, is the set of all elements that are common to both sets A and B.

$$A \cap B = \{x \mid x \in A \text{ and } x \in B\}$$

CAUTION

In the union of sets, it is not necessary to repeat the common elements.

$A \cup B = \{1, 2, 3, 4, 5\}$
not
$A \cup B = \{1, 2, 3, 3, 4, 5\}$

These ideas are illustrated in Figure 1.1 for the sets $A = \{1, 2, 3\}$ and $B = \{3, 4, 5\}$. Notice that the element 3 is located in the region shared by the two sets. The overlapping part indicates the intersection.

The intersection of the sets is $A \cap B = \{3\}$.
The union of the sets is $A \cup B = \{1, 2, 3, 4, 5\}$.

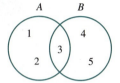

FIGURE 1.1

EXAMPLE 4 Let $A = \{1, 2, 3\}$, $B = \{3, 4, 5\}$, and $C = \{5, 6\}$ (see Figure 1.2). Find the following:

(a) $A \cup B$ (b) $A \cup C$ (c) $A \cap B$ (d) $A \cap C$

SOLUTION
(a) $A \cup B = \{1, 2, 3, 4, 5\}$
(b) $A \cup C = \{1, 2, 3, 5, 6\}$
(c) $A \cap B = \{3\}$
(d) $A \cap C = \emptyset$ since A and C have no elements in common.

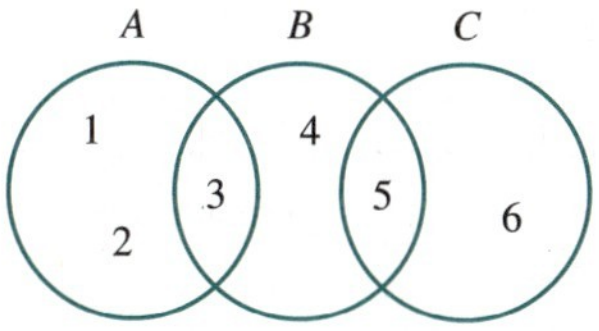

FIGURE 1.2

When all of the elements of one set are contained within a second set, the first set is said to be a *subset* of the second set.

DEFINITION

Subset

Set A is a **subset** of set B, written $A \subseteq B$, if and only if every element of A is also an element of B. In symbols this is written as $A \subseteq B$ if, whenever $x \in A$, then $x \in B$.

EXAMPLE 5 Let $A = \{2, 3, 4\}$ and $B = \{1, 2, 3, 4\}$ (see Figure 1.3).

(a) Is set A a subset of set B?
(b) Is set B a subset of set A?
(c) Does set A equal set B?

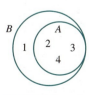

FIGURE 1.3

SOLUTION
(a) Yes, $A \subseteq B$ because every element of A is also an element of B.
(b) No, $B \not\subseteq A$ because $1 \in B$ but $1 \notin A$.
(c) No, $A \neq B$ because B has one element not in A.

> **Note** The symbol $\not\subseteq$ means "is not a subset of."

> Two sets A and B are said to be **equal**, written $A = B$, if and only if they contain exactly the same elements.

EXAMPLE 6 If $A = \{4, 6, 8\}$ and $B = \{x \mid x$ is an even natural number between 2 and 9$\}$, does set A equal set B?

SOLUTION Yes, set A equals set B because $A = \{4, 6, 8\}$ and $B = \{4, 6, 8\}$. They have exactly the same elements.

HISTORICAL COMMENT It took many centuries before *negative numbers* were considered to be anything but absurd. The Greeks, in the third century, refused to accept negative solutions to equations. The Hindus, however, discussed negative quantities as early as the seventh century and gave rules for their operation. In thirteenth-century Europe, Fibonacci (1170–?) wrote about a man's "profit" by saying, "I will show that this question cannot be solved unless it is conceded that the man be in debt." It was not until the close of the sixteenth century that negative numbers were readily accepted everywhere.

Now that we have some understanding of sets, let's get acquainted with our number system. The preceding Historical Comment refers to the Hindu–Arabic numeration system. It originally consisted of nine symbols to represent the numbers 1, 2, 3, 4, 5, 6, 7, 8, and 9. Later zero was added to the set. These ten symbols, along with fraction bars and decimal points, allow us to write almost any number.

A nice way to visualize sets of numbers is to draw a *number line*. To do this, we draw a horizontal line, choose a point on it, and label it "0"; then we choose another point to the right of 0 and label it "1." The distance from 0 to 1 will be our unit of measure that we use to locate more points. All numbers to the right of 0 we will call **positive numbers**, and those to the left of 0 we will call **negative numbers**. See Figure 1.4.

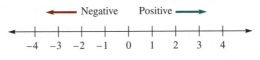

FIGURE 1.4

Number lines have the following characteristics:

1. The arrows at the ends of the number line indicate that the pattern continues without end.
2. The numbers increase in size as we move to the right. For example, 2 is greater than -3.
3. The numbers decrease in size as we move to the left. For example, -4 is less than -1.
4. Numbers associated with points on a number line are called the **coordinates** of the points. For example, in Figure 1.5, the coordinate of A is 0; that of B is 2; that of F is $-3\frac{1}{4}$; and the coordinate of D is $-\frac{1}{2}$.

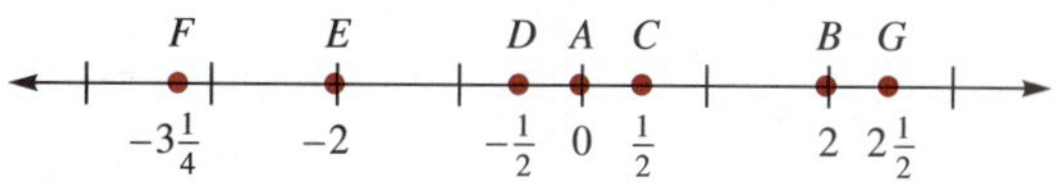

FIGURE 1.5

5. Points on a number line are called the **graphs** of the numbers. For example, in Figure 1.5, the point at -2 is the graph of E; the point at $\frac{1}{2}$ is the graph of C; and the point at $2\frac{1}{2}$ is the graph of G.
6. On a number line, each number to the left of 0 is known as the **additive inverse** of a corresponding number to the right of 0. Each number to right of 0 is known as the additive inverse of a corresponding number to the left of 0. For example, -3 is the additive inverse of 3, while $-\frac{1}{2}$ is the additive inverse of $\frac{1}{2}$. Also, 3 is the additive inverse of -3 and $\frac{1}{2}$ is the additive inverse of $-\frac{1}{2}$.

Now let's return to the discussion of our number system. Earlier we introduced the set of natural numbers, N. The graph of this set appears in Figure 1.6.

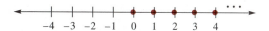

FIGURE 1.6

DEFINITION	$N = \{\text{natural numbers}\} = \{1, 2, 3, 4, \ldots\}$
The Set of Natural Numbers	

When zero is included in the set of natural numbers, the new set formed is called the set of **whole numbers,** denoted W. The graph of W is shown in Figure 1.7.

FIGURE 1.7

DEFINITION	$W = \{\text{whole numbers}\} = \{0, 1, 2, 3, 4, \ldots\}$
The Set of Whole Numbers	

Notice that N is a subset of W, written $N \subseteq W$.

When we form the *union* of the whole numbers and the *additive inverses of the natural numbers,* a new set is formed, called the set of **integers,** J. See Figure 1.8.

FIGURE 1.8

DEFINITION	$J = \{\text{integers}\} = \{\ldots, -4, -3, -2, -1, 0, 1, 2, 3, 4, \ldots\}$
The Set of Integers	

Notice that *W* is a subset of *J*, written $W \subseteq J$.

Next let's consider the set of **rational numbers,** *Q*. A rational number is the quotient of two integers. (Division by zero is excluded.)

DEFINITION	$Q = \{\text{rational numbers}\} = \left\{ \dfrac{a}{b} \,\middle|\, a, b \in J,\ b \neq 0 \right\}$
The Set of Rational Numbers	

Some examples of rational numbers are

$$\frac{1}{2} \qquad 5 = \frac{5}{1} \qquad 0 = \frac{0}{4} \qquad -2\frac{1}{3} = \frac{-7}{3} \qquad -6 = \frac{-6}{1} \qquad 0.75 = \frac{75}{100}$$

These examples illustrate that all integers are also rational numbers, so $J \subseteq Q$.

When rational numbers are written in *decimal form,* they either **terminate** (end) or, if they do not terminate, they **repeat.** The decimal form of a rational number is found by dividing the *numerator* by the *denominator*. Following are some examples of rational numbers that, when written in decimal form, terminate (end).

$$\frac{7}{10} = 0.7 \qquad \frac{3}{8} = 0.375 \qquad \frac{1}{4} = 0.25 \qquad \frac{16}{1000} = 0.016$$

Following are some examples of rational numbers in decimal form that do not terminate but do repeat.

$$\frac{1}{3} = 0.333\ldots \qquad \text{The 3 repeats.}$$

We usually write this as $\frac{1}{3} = 0.33\overline{3}$, where the bar indicates that the 3 repeats. Also,

$$\frac{2}{7} = 0.285714\overline{285714}$$

Again, the bar over 285714 indicates that it repeats.

If we were to *plot* (graph) all of the rational numbers on a number line, not all of the places on the line would be filled. There would still be many places on the line left unfilled. The numbers represented by the unfilled places *cannot be expressed as the quotient of two integers.* They are called **irrational numbers,** *H*.

DEFINITION	$H = \{\text{irrational numbers}\}$
The Set of Irrational Numbers	

Some examples of irrational numbers are

$$\pi \text{ (pi)} \qquad \sqrt{2} \qquad \sqrt{3} \qquad \sqrt[3]{2} \qquad \pi - 2 \qquad \sqrt[5]{7}$$

When written in decimal form, irrational numbers neither terminate nor repeat. For example,

$$\pi \approx 3.141592654 \quad \text{(rounded to 9 decimal places)}$$

The decimal form of the number π never repeats.

$$\sqrt{2} \approx 1.414213562 \quad \text{(rounded to 9 decimal places)}$$

The decimal form of $\sqrt{2}$ does not repeat.

If we graph (plot) all of the rational and irrational numbers on one number line, every place on the line will be filled. This new set of numbers is the *union* of the rational and irrational numbers and is called the set of **real numbers,** denoted R. See Figure 1.9.

> **N**OTE The symbol "$\approx$" means is approximately equal to.

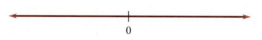

FIGURE 1.9

DEFINITION
The Set of Real Numbers

$$R = \{\text{real numbers}\} = Q \cup H$$

Every real number corresponds to exactly one point on the number line and every point on the number line corresponds to exactly one real number.

In summary:

$$N = \{\text{natural numbers}\} = \{1, 2, 3, \ldots\}$$
$$W = \{\text{whole numbers}\} = \{0, 1, 2, 3, \ldots\}$$
$$J = \{\text{integers}\} = \{\ldots, -3, -2, -1, 0, 1, 2, 3, \ldots\}$$
$$Q = \{\text{rational numbers}\} = \left\{\frac{a}{b} \,\middle|\, a, b \in J,\ b \neq 0\right\}$$
$$H = \{\text{irrational numbers}\} = \{x \mid x \in R,\ x \notin Q\}$$
$$R = \{\text{real numbers}\} = Q \cup H$$

The relationship between the subsets of the real numbers is shown in Figure 1.10 on the next page.

The natural numbers are a subset of the whole numbers, $N \subseteq W$.
The whole numbers are a subset of the integers, $W \subseteq J$.
The integers are a subset of the rational numbers, $J \subseteq Q$.
The rational numbers are a subset of the real numbers, $Q \subseteq R$.
The irrational numbers are a subset of the real numbers, $H \subseteq R$.
The rational and irrational numbers have no elements in common, $Q \cap H = \emptyset$.

The numbers shown in the figure are typical examples of the elements of each set.

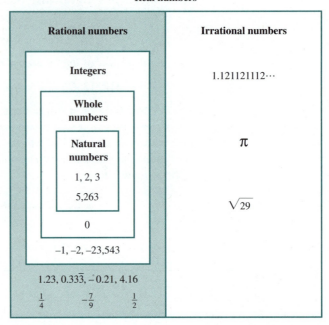

FIGURE 1.10
The real numbers and their subsets.

Do You Remember?

Can you match these?

_____ 1. Set symbol a) R
_____ 2. The set of irrational numbers b) {1, 2, 3, 4, 5}
_____ 3. Element of a set c) $0.11\overline{1}$
_____ 4. The set of real numbers d) { }
_____ 5. Empty set e) x
_____ 6. Example of a repeating decimal f) $\cap$
_____ 7. Example of a finite set g) $\in$
_____ 8. Example of a terminating decimal h) 7
_____ 9. Subset symbol i) $\cup$
_____ 10. Example of a variable j) N
_____ 11. Union of sets k) $\varnothing$
_____ 12. Example of a whole-number constant l) 0.75
_____ 13. Intersection of sets m) $\approx$
_____ 14. Additive inverse of x n) W
_____ 15. The set of natural numbers o) H
_____ 16. Approximately equal p) Q
_____ 17. The set of whole numbers q) $\subseteq$
_____ 18. The set of rational numbers r) $-x$

Answers: 1. d, 2. o, 3. g, 4. a, 5. k, 6. c, 7. b, 8. l, 9. q, 10. e, 11. i, 12. h, 13. f, 14. r, 15. j, 16. m, 17. n, 18. p

Exercise Set 1.1

Identify as true *or* false.

1. Zero is a natural number.
2. Every natural number is a whole number.
3. Every whole number is an integer.
4. $0.11\overline{1}$ is an example of an irrational number.
5. $\sqrt{3}$ is a real number.
6. Every integer is a rational number.
7. The union of the set of rational and irrational numbers is an empty set.
8. The set of irrational numbers is a subset of the set of real numbers.
9. The intersection of the set of rational and irrational numbers is an empty set.
10. The set of integers is a subset of the set of rational numbers.
11. The set of integers is a subset of the set of irrational numbers.
12. The union of the sets of natural and whole numbers is the set of whole numbers.
13. The union of the sets of rational and irrational numbers is the set of real numbers.
14. Every element of the set of integers is also an element of the set of real numbers.

Indicate which of the numbers -18, -5.25, $-\sqrt{3}$, $-0.22\overline{2}$, 0, 1, 4.875, $\sqrt{8}$, *and* $6.83\overline{33}$ *are in each of the following sets. Some may be in more than one set.*

15. The natural numbers, N
16. The whole numbers, W
17. The integers, J
18. The rational numbers, Q
19. The irrational numbers, H
20. The real numbers, R
21. Both the natural and the whole numbers
22. Either the rational or the irrational numbers
23. The negative integers
24. The nonnegative integers

Write each of the following sets using the listing method. If the set is empty, write $\emptyset$.

25. $\{x \mid x$ is a letter in the word *add*$\}$
26. $\{t \mid t$ is a letter in the word *calculus*$\}$
27. $\{y \mid y$ is a day of the week starting with the letter $A\}$
28. $\{p \mid p$ is or was a woman president of the United States$\}$
29. $\{m \mid m = 4\}$
30. $\{x \mid x = 2\}$
31. $\{x \mid x$ is a natural number less than $5\}$
32. $\{y \mid y$ is a natural number between 7 and 11$\}$
33. $\{2, 3, 4\} \cup \{1, 3, 5\}$
34. $\{2, 3, 4\} \cap \{1, 3, 5\}$

Indicate whether each of the following statements is true *or* false.

35. $2 \in \{2, 3, 5\}$
36. $6 \in \{x \mid x \in N$ and x is less than $9\}$
37. $5 \subseteq \{4, 5, 9\}$
38. $-5 \in \{3, 4, 5\}$
39. $\{1\} \subseteq \{1\}$
40. $\{4, 1, 5\} \subseteq \{1, 4, 5\}$
41. If $C \cup D = D$, then $C \subseteq D$
42. $\emptyset = \{0\}$
43. If C and D are any two sets, then $C \cap D \subseteq D$
44. If $A = \{n \mid n$ is a natural number greater than 100$\}$, then A is an example of an infinite set.
45. $3 \in N$
46. $5 \in W$
47. $\frac{1}{4} \in Q$
48. $0.325 \in H$

List the elements of the given set if $A = \{1, 3, 5, 9\}$, $B = \{2, 4, 6, 8\}$, $C = \{1, 2, 3, 6\}$, $D = \emptyset$, $E = \{0\}$, *and* $F = \{1, 2, 3, 4, \ldots\}$.

49. $E \cup F$
50. $C \cup D$
51. $B \cap F$
52. $A \cap F$
53. $B \cup E$
54. $D \cap F$
55. $D \cup E$

Indicate whether each of the following sets is finite, infinite, *or the* empty *set.*

56. $\{n \mid n$ is an even whole number greater than 1000$\}$
57. $\{n \mid n$ is a natural number and a multiple of 2$\}$
58. $\{n \mid n$ is an odd natural number and divisible by 4$\}$
59. $\{n \mid n$ is a whole number that is a multiple of 3 and divisible by 6$\}$

List the elements in the following sets.

60. The names of the months of the year that have four letters.
61. The names of the countries on the North American continent.

62. The last six letters of the English alphabet.
63. The names of the states in the United States that have four letters.
64. The union of the set of days of the week that start with the letter *T* and the set of those days that start with the letter *S*.
65. The intersection of the set of months of the year that have 30 days and the set of months of the year that begin with the letter *N*.

Write each of the following sets using set-builder notation

66. {Sunday, Monday, ..., Saturday}
67. {January, February, ..., December}
68. {1, 2, 3, ...}
69. {0, 1, 2, 3, ...}
70. {March, May}
71. {November}

The Eggleston family consists of Verle, Pam, Sherry, and Lisa; we will designate the family by {V, P, S, L}. In order to develop a cooperative work schedule, the family has agreed to set up work groups for house cleaning, yard cleanup, and other chores.

72. List all of the different possible sets of one-person work groups.
73. List all of the different possible sets of two-person work groups.
74. List all of the different possible three-person work groups.
75. List all of the different possible four-person work groups.

Fill in the blank with "∈" or "∉" to make each statement correct.

76. 3 __ N
77. $\frac{1}{4}$ __ J
78. $0.33\overline{3}$ __ H
79. 0.675 __ Q

Fill in the blank with "⊆" or "⊄" to make each statement correct.

80. N __ W
81. Q __ R
82. H __ R
83. Q __ H

Change each of the following fractions to decimal form. State whether each is a rational or an irrational number. A calculator may be helpful.

84. $\frac{1}{2}$
85. $\frac{6}{8}$
86. $\frac{7}{5}$
87. $\frac{3}{11}$
88. $\frac{1}{9}$
89. $\frac{2}{11}$
90. $\frac{5}{6}$

Let A and B be any two sets. Answer the following with yes or no.

91. If $x \in A \cup B$, must x be an element of B?
92. If $x \in A \cap B$, must x be an element of B?
93. If $x \in A \cup B$, must x be an element of both A and B?
94. If $x \in A$ and $A \subseteq B$, must x be an element of B?
95. If $x \in A$ and $A \not\subseteq B$, must x be an element of B?
96. If $x \in A \cap B$, must x be an element of both A and B?
97. If x is an odd natural number, can x ever be an element of the set of even natural numbers?
98. If x is an odd natural number and a muptiple of 3, can x ever be an element of the set of even natural numbers?

Write a sentence (outside the area of mathematics) using the following words. Use a dictionary if needed.

99. set
100. intersection
101. union
102. element
103. finite
104. infinite

1.2 Properties of Real Numbers

HISTORICAL COMMENT   Girolamo Cardano (1501–1576) was the first mathematician to make extensive use of the idea of a *negative number*. He determined that the product of two negative numbers was a positive number, among many other things. See illustrations and uses of positive and negative numbers in this section.

1.2 · PROPERTIES OF REAL NUMBERS

OBJECTIVES

In this section we will learn to
1. recognize and use symbols of equality and inequality;
2. understand and use symbols of operation;
3. understand and work with absolute value;
4. identify and use the properties of equality;
5. identify and use the properties of real numbers.

We will begin by reviewing some symbols and definitions that we will need in the rest of the book. Several symbols are used to *compare* two numbers a and b. These are listed in the following table.

Symbol	Interpretation	Example
$a = b$	a is **equal to** b	$7 = 7$
$a < b$	a is **less than** b	$-3 < 8$
$a > b$	a is **greater than** b	$4 > -10$
$a \leq b$	a is **less than or equal to** b	$8 \leq 8$
$a \geq b$	a is **greater than or equal to** b	$5 \geq 0$

EXAMPLE 1 Fill in the blank with "$<$" or "$>$" to make each statement true.

(a) $3 \_ 8$ (b) $-5 \_ -1$ (c) $2 \_ -1$ (d) $1 \_ -6$ (e) $-5 \_ 0$

SOLUTION
(a) $3 < 8$ 3 is to the left of 8 on the number line.
(b) $-5 < -1$
(c) $2 > -1$ 2 is to the right of -1 on the number line.
(d) $1 > -6$
(e) $-5 < 0$ -5 is to the left of 0 on the number line.

Recall from Section 1.1 that a slash, "/", through any symbol negates its meaning. For example, $a \not< b$ means "a is **not** less than b."

EXAMPLE 2 State each of the following inequalities in words.

(a) $5 \leq 8$ (b) $6 \not\geq 10$ (c) $15 \not> 15$ (d) $-11.3 \leq -11.3$
(e) $5\frac{1}{2} \not\leq 5\frac{1}{4}$

SOLUTION
(a) 5 is less than or equal to 8.
(b) 6 is not greater than or equal to 10.
(c) 15 is not greater than 15.
(d) -11.3 is less than or equal to -11.3.
(e) $5\frac{1}{2}$ is not less than or equal to $5\frac{1}{4}$.

Operation Symbols

Symbol	Operation	Interpretation
$a + b$	Addition	The *sum* of a and b
$a - b$	Subtraction	The *difference* of a and b
ab or $a \cdot b$	Multiplication	The *product* of a and b
$a \div b$ or $\dfrac{a}{b}$	Division	The *quotient* of a and b
a^n	Powers	a raised to the nth *power*
$\sqrt[n]{b}$	Roots	The nth root of b

EXAMPLE 3 Using symbols, complete each of the following.

(a) Find the sum of 17 and 14.
(b) Find the difference of 64 and 12.
(c) Find the product of 9 and 7.
(d) Find the quotient of 56 and 8.
(e) Find 2 raised to the third power.
(f) Find the square root of 36.

SOLUTION

(a) $17 + 14 = 31$
(b) $64 - 12 = 52$
(c) $9 \cdot 7 = 63$
(d) $56 \div 8 = 7$
(e) $2^3 = 2 \cdot 2 \cdot 2 = 8$
(f) $\sqrt{36} = 6$

In Section 1.1 we learned that every point on one side of zero corresponds to a point on the other side of zero and the same distance away. The numbers represented by these pairs of points are called *additive inverses*. For example,

5 and -5 are additive inverses;

-4 and 4 are additive inverses;

$-\frac{1}{4}$ and $\frac{1}{4}$ are additive inverses.

The additive inverse of any real number can be found by placing a negative sign in front of that real number. Thus the additive inverse of 4 is -4 and the additive inverse of -5 is $-(-5) = 5$.

The Double Negative Property

For any real number a,

$$-(-a) = a$$

Two numbers that are additive inverses are said to have the same **absolute value.** The absolute value of a real number is indicated by placing the number between vertical bars.

$|a|$ is read "the **absolute value** of a."

The distance between a real number and the origin (0) on a number line is its absolute value. Thus the absolute value of 4 is 4, or $|4| = 4$; the absolute value of -4 is 4, or $|-4| = 4$. Both 4 and -4 are 4 units from the origin; see Figure 1.11.

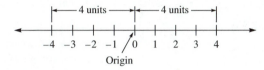

FIGURE 1.11

EXAMPLE 4   Find: $|6|$; $|-3.7|$; $\left|\dfrac{1}{2}\right|$; $\left|-\dfrac{1}{4}\right|$; and $|0|$.

SOLUTION
$|6| = 6$ 6 is 6 units from the origin.
$|-3.7| = 3.7$ -3.7 is 3.7 units from the origin.
$\left|\dfrac{1}{2}\right| = \dfrac{1}{2}$ $\dfrac{1}{2}$ is $\dfrac{1}{2}$ unit from the origin.
$\left|-\dfrac{1}{4}\right| = \dfrac{1}{4}$ $-\dfrac{1}{4}$ is $\dfrac{1}{4}$ unit from the origin.
$|0| = 0$ 0 is 0 units from the origin.

More generally, we define the absolute value of any real number as follows.

DEFINITION

Absolute Value

$$|a| = \begin{cases} a \text{ if } a \text{ is positive or zero, } a \geq 0 \\ -a \text{ if } a \text{ is less than zero (negative), } a < 0 \end{cases}$$

AUTION

The absolute value of a number is always nonnegative.

At first glance, the second part of the definition seems to say that the absolute value of a number is negative. But since a is negative ($a < 0$), the negative of a is a positive number. For example, $|-3| = -(-3) = 3$.

EXAMPLE 5   Find: $|-81|$; $|-1.5|$; $-|6|$; $-|-13|$; and $-[-|-8|]$.

SOLUTION
$|-81| = 81$
$|-1.5| = 1.5$

$-|6| = -6$ The negative sign is outside the absolute value sign.

$-|-13| = -(13) = -13$ The absolute value of -13 is 13.

$-[-|-8|] = -[-(8)] = 8$ The absolute value of -8 is 8.

In Section 1.1 we discussed the set of *real numbers*. Now we will learn about the properties that govern the ways that real numbers can be added, subtracted, multiplied, and divided. The first four are called the *properties of equality*.

DEFINITION
Properties of Equality

For all real numbers a, b, and c, we have

Reflexive property	$a = a$
Symmetric property	If $a = b$, then $b = a$.
Transitive property	If $a = b$, and $b = c$ then $a = c$.
Substitution property	If $a = b$, then a can be replaced by b or b can be replaced by a without changing the truth of a statement.

EXAMPLE 6   State which property of equality is illustrated by each of the following statements.

(a) $y + 3 = y + 3$
(b) If $2x - 1 = p$, then $p = 2x - 1$.
(c) If $2m + 3 = 5$ and $5 = z$, then $2m + 3 = z$.
(d) If $3x + 6 = 12$ and $x = 2$, then $3(2) + 6 = 12$.

SOLUTION
(a) Reflexive property
(b) Symmetric property
(c) Transitive property
(d) Substitution property

Now let's consider some properties of the real numbers as they apply to the operations of *addition* and *multiplication*. In the following properties, when we say a number is *unique,* we mean that it is the only number that possesses a particular property.

DEFINITION
Properties of Real Numbers

For all real numbers a, b, and c:

Closure properties	$a + b$ is again a real number.
	$a \cdot b$ is again a real number.
Commutative properties	$a + b = b + a$
	$a \cdot b = b \cdot a$
Associative properties	$a + (b + c) = (a + b) + c$
	$a(b \cdot c) = (a \cdot b)c$

continues

DEFINITION
Properties of Real Numbers—*Continued*

For all real numbers a, b, and c:

Identity properties There is a unique number "0" such that $a + 0 = 0 + a = a$. Zero is called the **additive identity.**
There is a unique number "1" such that $a \cdot 1 = 1 \cdot a = a$. One is called the **multiplicative identity.**

Inverse properties There are unique numbers $-a$ and $\frac{1}{a}$ such that $a + (-a) = 0$ and $a \cdot \frac{1}{a} = 1$. The **additive inverse** of a is $-a$ and the **multiplicative inverse** of a is $\frac{1}{a}$, $a \neq 0$.

Distributive properties $a(b + c) = a \cdot b + a \cdot c$
$a(b - c) = a \cdot b - a \cdot c$

The following examples illustrate the properties:

CLOSURE PROPERTIES
$11 + 34 = 45$ 11 and 34 are real numbers and their sum, 45, is a real number.
$8(1.2) = 9.6$ 8 and 1.2 are real numbers and their product, 9.6, is a real number.

COMMUTATIVE PROPERTIES
$6 + 15 = 15 + 6 = 21$
$13 \cdot 5 = 5 \cdot 13 = 65$

The commutative properties state that the order in which numbers are added or multiplied does not change the sum or product.

ASSOCIATIVE PROPERTIES
$2 + (8 + 13) = 2 + (21) = 23$ or $(2 + 8) + 13 = 10 + 13 = 23$
$5(6 \cdot 12) = 5(72) = 360$ or $(5 \cdot 6)12 = (30)12 = 360$

The associative properties state that the way in which three or more numbers are associated (grouped) does not change the sum or product.

INVERSE PROPERTIES
$63 + (-63) = 0$

$\left(-\frac{3}{7}\right) + \frac{3}{7} = 0$

$\sqrt{3} \cdot \frac{1}{\sqrt{3}} = 1$

$6\left(\frac{1}{6}\right) = 1$

DISTRIBUTIVE PROPERTY
$2(11 + 18) = 2 \cdot 11 + 2 \cdot 18 = 22 + 36 = 58$
$5x + 7x = (5 + 7)x = 12x$

CAUTION
The commutative and associative properties do not hold for the operations of subtraction or division.

$$6(x - 2y) = 6x - 12y$$
$$15(2a - 3b) = 30a - 45b$$

EXAMPLE 7 State the *property of real numbers* that makes each of the following true.

(a) $5 + (-5) = 0$
(b) $3.5 + 1 = 1 + 3.5$
(c) $6 \cdot \dfrac{1}{6} = 1$
(d) $2 + 4$ is a real number.
(e) $7(3 + 9) = 7 \cdot 3 + 7 \cdot 9$
(f) If $a = 3$ and $3 = b$, then $a = b$
(g) $2(4 \cdot 6) = (2 \cdot 4)6$
(h) $7 + 0 = 7$
(i) $2 + (4 + 7) = (2 + 4) + 7$
(j) $5 \cdot 1 = 5$
(k) $4 \cdot 9$ is a real number
(l) $26 = 26$
(m) If $5 + 17 = 22$, then $22 = 5 + 17$.
(n) $7(6 \cdot 2) = 7(2 \cdot 6)$

SOLUTION

(a) Additive inverse property
(b) Commutative property for addition
(c) Multiplicative inverse property
(d) Closure property for addition
(e) Distributive property
(f) Transitive property of equality
(g) Associative property for multiplication
(h) Identity property for addition
(i) Associative property for addition
(j) Identity property for multiplication
(k) Closure property for multiplication
(l) Reflexive property of equality
(m) Symmetric property of equality
(n) Commutative property for multiplication. Notice that the *order* has been changed, but not the *grouping*.

Although we shall not generally do so, it is instructive to see how the properties of real numbers can be used to *prove* further properties. A property that has been proven is called a **theorem.** Consider the following.

THEOREM 1.1

The Multiplication Property of Zero

If $a \in R$, then $a \cdot 0 = 0 \cdot a = 0$.

To prove this theorem, we will start with a statement that is known to be true and change it through a series of steps, each of which is justified by one of the properties of real numbers.

Proof

Statement	Justification
1. $a = a \cdot 1$	Multiplicative identity property
2. $a = a(1 + 0)$	$1 = 1 + 0$ by the additive identity property
3. $a = a \cdot 1 + a \cdot 0$	Distributive property
4. $a = a + a \cdot 0$	Multiplicative identity property
5. $a = a + 0$	Additive identity property
6. $a \cdot 0 = 0$	By the uniqueness of the additive identity

This completes our proof that $a \cdot 0 = 0$.

Do You Remember?

Can you match these?

_____ 1. Double negative property
_____ 2. Real numbers
_____ 3. Absolute value
_____ 4. Reflexive property
_____ 5. Symmetric property
_____ 6. Transitive property
_____ 7. Substitution property
_____ 8. Closure property
_____ 9. Commutative property
_____ 10. Associative property
_____ 11. Additive identity
_____ 12. Multiplicative inverse
_____ 13. Additive inverse
_____ 14. Distributive property
_____ 15. Multiplication property of zero
_____ 16. Multiplicative identity

a) $13 \cdot 1 = 13$
b) $6 + 8 = 8 + 6$
c) $6 + 3 \in R$
d) $6 \cdot 0 = 0$
e) $5(a + 3) = 5a + 15$
f) $-(-5) = 5$
g) $a + b = a + b$
h) If $a = b$ and $b = c$, then $a = c$.
i) $2 \cdot \frac{1}{2} = 1$
j) $9 + 0 = 9$
k) If $a = b$, then $b = a$.
l) $7 + (-7) = 0$
m) R
n) $7 + (9 + 11) = (7 + 9) + 11$
o) If $a + b = 3$ and $b = 2$, then $a + 2 = 3$.
p) $|-6| = 6$

Answers: 1. f, 2. m, 3. p, 4. g, 5. k, 6. h, 7. o, 8. c, 9. b, 10. n, 11. j, 12. i, 13. l, 14. e, 15. d, 16. a

Exercise Set 1.2

Fill in the blank with "<" or ">" to make each statement true.

1. 4 __ 9
2. 3 __ 1
3. 5 __ −8
4. −11 __ 1
5. −8 __ −6
6. −11 __ −18
7. −(−5) __ 0
8. 3 __ −(−12)
9. |6| __ 5
10. |3| __ 13
11. |−7| __ −15
12. |−8| __ |−1|
13. 0 __ −|−3|
14. −|−4| __ 4
15. −|9| __ −|−10|
16. −|−3| __ 0

Write the additive inverse for each of the following quantities.

17. 8
18. 11
19. −5
20. −1
21. a
22. b
23. −a
24. −b
25. −(−b)
26. −(−a)
27. $|x|$
28. $|y|$
29. −$|x|$
30. −$|y|$
31. −$|-a|$
32. −$|-b|$

Rewrite each of the following using $<$, $>$, $\leq$, $\geq$, or $=$ in place of the words.

33. 7 is less than 12.
34. −4 is greater than −7.
35. $x + 1$ is positive.
36. $x + 7$ is negative.
37. $m + 2$ is nonnegative (zero or positive).
38. $6x - 5$ is greater than or equal to 0.
39. $3y - 2$ is less than or equal to 0.
40. $x - 4$ is not positive.
41. $9x + 7$ is equal to −9.
42. $5y + 4$ is equal to 13.

Use the indicated property of real numbers to complete each of the following.

43. $7 + 2 = $ ____; closure property
44. $8 \cdot 1 = $ ____; identity property
45. $x + (-x) = $ ____; additive inverse property
46. $x + y = $ ____; commutative property for addition
47. $7(8y) = $ ____; associative property for multiplication
48. $5m + 5n = $ ____; distributive property

Use the indicated property of equality to complete each of the following.

49. If $x = 5 + b$, then ____ $= x$; symmetric property
50. If $a + b = p$ and $a = 5$, then ____ $+ b = p$; substitution property
51. $p = $ ____; reflexive property
52. If $a = t$ and $t = b$, then ____; transitive property

Use the distributive property to rewrite each of the following.

53. $2(a + 3) = $ ____
54. $6(x + 1) = $ ____
55. $3t + 3 \cdot 2 = $ ____
56. $5m + 5 \cdot 3 = $ ____
57. $13(r - 4t) = $ ____
58. $11(2a - 3b) = $ ____

State which property or properties are illustrated by each of the following.

59. $15 + 31 = 31 + 15$
60. $(6 + 2) + 9 = 6 + (9 + 2)$
61. $0 + m = m$
62. $x \cdot 3 = 3 \cdot x$
63. $m \cdot 1 = m$
64. $p + (-p) = 0$
65. $a + b = a + b$
66. $6w + 7w = (6 + 7)w$
67. $(6x)y = 6(xy)$
68. $y \cdot \dfrac{1}{y} = 1$
69. $15r \cdot \dfrac{1}{15r} = 1$
70. $-(t - s) + (t - s) = 0$

Use the distributive property to rewrite Exercises 71–76.

71. $2a + 3a$
72. $5t + 6t$
73. $11m - 5m$
74. $13b - 8b$
75. $7x + x$
76. $9y + y$

Write an example to illustrate the use of each of the following concepts.

77. $\neq$
78. $<$
79. Identity property for addition
80. The multiplicative inverse property
81. $\in$
82. $\nsubseteq$
83. The distributive property
84. The commutative property for multiplication
85. Subtraction is not a commutative operation.
86. Division is not a commutative operation.

REVIEW EXERCISES

Fill in the blank in Exercises 87–92 with "$\in$" or "$\subseteq$" to make each statement correct. Reminder: N represents the set of natural numbers, etc.

87. 5 __ N
88. $\dfrac{1}{4}$ __ Q
89. W __ J
90. H __ R
91. $\sqrt{7}$ __ H
92. $0.33\overline{3}$ __ R

Explain (in words) how the commutative, associative, distributive, identity, and inverse properties could be used to make the following problems easier to simplify.

93. $67 + \dfrac{1}{5}(5) + [7 + (-7)] + 33$
94. $39 + (-15) + \dfrac{1}{2}(2) + 15 + 5(0)$
95. $(-58) + (-13) + 58 + \dfrac{1}{2}(29 - 3)$
96. $\dfrac{2}{3}\left(\dfrac{3}{2}\right) + [5 + (-5)] - \dfrac{1}{4}(9 - 5)$

1.3 Operations with Real Numbers

HISTORICAL COMMENT Johannes de Lineriis (French), in about the year 1340, described what we now call *common fractions*. He introduced the idea of placing the numerator over the denominator separated by a horizontal line, such as $\frac{3}{4}$. In the mid-sixteenth century, one scheme for adding fractions such as $\frac{7}{6} + \frac{3}{4}$ was as follows:

$$\frac{\overset{14}{\frac{7}{6}} + \overset{9}{\frac{3}{4}}}{12} = \frac{23}{12}$$

Can you discover the pattern?

OBJECTIVES In this section we will learn to
1. add, subtract, multiply, divide, and simplify fractions;
2. add, subtract, multiply, and divide real numbers;
3. understand and use order of operations;
4. recognize and work with exponents.

Now that we have an understanding of real numbers, their subsets, and their properties, we'll learn how to work with them. We will examine the rules governing addition, subtraction, multiplication, division, and powers of real numbers.

First let's review the operations as they apply to fractions.

Multiplication of Fractions

$$\frac{a}{b} \cdot \frac{c}{d} = \frac{a \cdot c}{b \cdot d} = \frac{ac}{bd}, \quad b \neq 0, d \neq 0$$

The product of two fractions is obtained by finding the product of the numerators and dividing by the product of the denominators.

Division of Fractions

$$\frac{a}{b} \div \frac{c}{d} = \frac{a}{b} \cdot \frac{d}{c} = \frac{ad}{bc}, \quad b, c, \text{ and } d \neq 0$$

The quotient of two fractions is found by inverting the divisor and following the rule for multiplication.

Fundamental Property of Fractions

$$\frac{a \cdot x}{b \cdot x} = \frac{a}{b} \quad \text{and} \quad \frac{a}{b} = \frac{a \cdot x}{b \cdot x}, \quad b, x \neq 0$$

If the numerator and denominator of a fraction are multiplied by the same nonzero real number, the new fraction is equivalent to the original fraction.

For example,

$$\frac{3}{4} = \frac{3}{4} \cdot \frac{5}{5} = \frac{3 \cdot 5}{4 \cdot 5} = \frac{15}{20}$$

When these steps are reversed, the process is called **reducing to lowest terms,** or **simplifying,** as follows:

$$\frac{15}{20} = \frac{3 \cdot 5}{4 \cdot 5} = \frac{3}{4}$$

Addition and Subtraction of Fractions

$$\frac{a}{c} + \frac{b}{c} = \frac{a+b}{c} \quad \text{and} \quad \frac{a}{c} - \frac{b}{c} = \frac{a-b}{c}, \quad c \neq 0$$

---

*To add or subtract fractions with a **common denominator**, add or subtract the numerators and write the result over the common denominator.*

When adding or subtracting fractions that do not have a common denominator, we use the fundamental property of fractions to find a **least common denominator (LCD)**. For example,

$$\frac{1}{2} + \frac{3}{5} = \frac{1 \cdot 5}{2 \cdot 5} + \frac{3 \cdot 2}{5 \cdot 2} \quad \text{Fundamental property of fractions}$$

$$= \frac{5}{10} + \frac{6}{10}$$

$$= \frac{11}{10}$$

EXAMPLE 1 Carry out the indicated operations.

(a) $\frac{1}{7} + \frac{3}{7}$ (b) $\frac{3}{4} - \frac{1}{2}$ (c) $2\frac{1}{2} \cdot 5\frac{1}{4} \cdot 3$ (d) $\frac{3}{16} \div \frac{1}{8}$

SOLUTION (a) $\frac{1}{7} + \frac{3}{7} = \frac{4}{7}$

(b) $\frac{3}{4} - \frac{1}{2} = \frac{3}{4} - \frac{1 \cdot 2}{2 \cdot 2} = \frac{3}{4} - \frac{2}{4} = \frac{1}{4}$

(c) $2\frac{1}{2} \cdot 5\frac{1}{4} \cdot 3 = \frac{5}{2} \cdot \frac{21}{4} \cdot 3 = \frac{5}{2} \cdot \frac{21}{4} \cdot \frac{3}{1} = \frac{315}{8}$

(d) $\frac{3}{16} \div \frac{1}{8} = \frac{3}{16} \cdot \frac{8}{1} = \frac{3 \cdot 8}{16 \cdot 1} = \frac{3 \cdot 8}{16} = \frac{3 \cdot 8}{2 \cdot 8} = \frac{3}{2}$

Now let's consider how to perform the operations of addition, subtraction, multiplication, and division using *signed numbers*. A **signed number** is either a positive number, such as +5, or a negative number, such as −5. We will begin with **addition of signed numbers.**

> ### To Add Signed Numbers
> 1. To add two numbers with *like* signs, both positive or both negative, find the sum of their absolute values and keep the same sign.
> 2. To add two numbers with *unlike* signs, one positive and one negative, find the difference of their absolute values by subtracting the smaller from the larger, and attach the sign of the number with the greater absolute value.

EXAMPLE 2 Find the sum of −8 and −5.

SOLUTION First we find the absolute value of each number:

$$|-8| = 8 \quad \text{and} \quad |-5| = 5$$

Since both numbers have the same sign, we add: $8 + 5 = 13$. Then, since both numbers are negative, the sum will be negative. Thus, $-8 + (-5) = -13$.

EXAMPLE 3 Find the sum of $-\dfrac{7}{13}$ and $\dfrac{2}{13}$.

SOLUTION $\left|-\dfrac{7}{13}\right| = \dfrac{7}{13} \quad \text{and} \quad \left|\dfrac{2}{13}\right| = \dfrac{2}{13}$

Since the two numbers have unlike signs, we subtract:

$$\frac{7}{13} - \frac{2}{13} = \frac{5}{13}$$

Since $\left|-\dfrac{7}{13}\right|$ is greater than $\left|\dfrac{2}{13}\right|$, the sum will be negative. Thus

$$-\frac{7}{13} + \frac{2}{13} = -\frac{5}{13}$$

EXAMPLE 4 Add: $6 + (-9) + (-3)$.

SOLUTION $6 + (-9) + (-3) = -3 + (-3) = -6$

Now let's turn our attention to **subtraction of signed numbers.**

> **To Subtract Signed Numbers**
>
> To subtract one number from another, change the sign of the number being subtracted, then follow the rules for addition of signed numbers:
>
> $a - b = a + (-b)$

EXAMPLE 5   Subtract each of the following.

(a) $11 - 3$ (b) $-15 - 6$ (c) $-6.4 - (-3.1)$ d. $-\dfrac{3}{8} - \left(-\dfrac{5}{8}\right)$

SOLUTION
(a) $11 - 3 = 11 + (-3) = 8$
(b) $-15 - 6 = -15 + (-6) = -21$
(c) $-6.4 - (-3.1) = -6.4 + (3.1) = -3.3$
(d) $-\dfrac{3}{8} - \left(-\dfrac{5}{8}\right) = -\dfrac{3}{8} + \dfrac{5}{8} = \dfrac{2}{8} = \dfrac{1}{4}$

EXAMPLE 6   Simplify each of the following.

(a) $6 - (-3) + 9$ (b) $1.3 - |-2.7|$

SOLUTION
(a) $6 - (-3) + 9 = 6 + 3 + 9 = 18$
(b) $1.3 - |-2.7| = 1.3 - 2.7$ $|-2.7| = 2.7$
    $= 1.3 + (-2.7) = -1.4$

EXAMPLE 7   Evaluate $x - y + z$ for $x = 3$, $y = -4$, and $z = 0$.

SOLUTION   We replace x with 3, y with -4, and z with 0:

$3 - (-4) + 0 = 3 + 4 + 0 = 7$

> **To Multiply Signed Numbers**
>
> 1. The product of two numbers with *like* signs is a *positive* number.
> 2. The product of two numbers with *unlike* signs is a *negative* number.

EXAMPLE 8   Multiply each of the following.

(a) $3 \cdot 7$ (b) $(-5)(-9)$ (c) $(-8)(6)$ (d) $11(-15)$ (e) $4\left[\left(-\dfrac{2}{3}\right)\left(\dfrac{6}{5}\right)\right]$

SOLUTION (a) $3 \cdot 7 = 21$
(b) $(-5)(-9) = 45$ -5 and -9 have like signs.
(c) $(-8)(6) = -48$ -8 and 6 have unlike signs.
(d) $11(-15) = -165$
(e) $4 \cdot \left[\left(-\dfrac{2}{3}\right)\left(\dfrac{6}{5}\right)\right] = 4 \cdot \left(-\dfrac{4}{5}\right) = -\dfrac{16}{5}$

To Divide Signed Numbers

1. The quotient of two numbers with *like* signs is a *positive* number.
2. The quotient of two numbers with *unlike* signs is a *negative* number.

EXAMPLE 9   Divide each of the following quantities.

(a) $\dfrac{-36}{-18}$ (b) $\dfrac{34}{-2}$ (c) $\dfrac{-63}{21}$

SOLUTION   (a) $\dfrac{-36}{-18} = 2$ (b) $\dfrac{34}{-2} = -17$ (c) $\dfrac{-63}{21} = -3$

Using the rules for dividing signed numbers, it can be shown that

$$\dfrac{-a}{b} = \dfrac{a}{-b} = -\dfrac{a}{b}, \quad \text{where } b \neq 0$$

For example,

$$\dfrac{-6}{3} = \dfrac{6}{-3} = -\dfrac{6}{3} = -2$$

These are called **equivalent fractions.** *In this book, all negative numbers in fraction form* will be written

$$\dfrac{-a}{b} \quad \text{or} \quad -\dfrac{a}{b}$$

EXAMPLE 10   Write $\dfrac{5}{-6}$ in two equivalent fraction forms.

SOLUTION $\dfrac{5}{-6} = \dfrac{-5}{6} = -\dfrac{5}{6}$

Let's consider what is meant by division. For example,

$$\frac{6}{2} = 3 \quad \text{if and only if} \quad 2 \cdot 3 = 6$$

and

$$\frac{-15}{-3} = 5 \quad \text{if and only if} \quad -3 \cdot 5 = -15$$

Now let's consider the quotient q when a nonzero real number a is divided by **zero**:

$$\frac{a}{0} = q, \quad a \neq 0$$

We know that

$$\frac{a}{0} = q \quad \text{if and only if} \quad 0 \cdot q = a$$

But $0 \cdot q = 0$, and we said $a \neq 0$. Therefore no such quotient "q" exists. Thus we say that *division by zero is undefined.*

Now, if $a = 0$, then $\frac{0}{0} = q$ and $0 \cdot q = 0$. This is true for every real number q, so no *unique* result can be found. (Try dividing by zero on your calculator.)

We now turn our attention to the operation of **raising to powers.** Recall that exponents can be used to indicate repeated multiplication. For example, $2 \cdot 2 \cdot 2 \cdot 2 = 2^4$, where the **exponent** 4 indicates that the **base** 2 is used as a **factor** (multiplier) 4 times.

> **NOTE** When two real numbers are divided, the divisor can **never** be zero.

> **NOTE** $\frac{0}{a} = 0$ for all $a \neq 0$.

> **CAUTION**
> Division by zero is undefined

DEFINITION
Powers

If n is a positive integer and b is a real number, then

$$b^n = \underbrace{b \cdot b \cdots b}_{n \text{ factors}}$$

Here b is the base and n the exponent.

We read b^n as "b raised to the nth power" or simply as "b to the nth."

EXAMPLE 11 Rewrite each of the following without exponents and evaluate.

(a) 2^3 (b) $(-5)^2$ (c) $(-5)^3$ (d) -7^2 (e) $\left(\frac{3}{4}\right)^4$

SOLUTION
(a) $2^3 = 2 \cdot 2 \cdot 2 = 8$
(b) $(-5)^2 = (-5)(-5) = 25$
(c) $(-5)^3 = (-5)(-5)(-5) = -125$
(d) $-7^2 = -(7 \cdot 7) = -49$ The negative sign is not part of the base.
(e) $\left(\frac{3}{4}\right)^4 = \frac{3}{4} \cdot \frac{3}{4} \cdot \frac{3}{4} \cdot \frac{3}{4} = \frac{81}{256}$

> **CAUTION**
> In Example 11(b), the base, -5, is used as a factor twice. In part (d), the base is 7, also used as a factor twice. Unless a negative sign is enclosed in grouping symbols, as in parts (b) and (c), it is not part of the base.

Many times an arithmetic or algebra problem involves more than one operation. When this occurs, a *definite order* must be followed. For example, $2 \cdot 8 + 3 = 16 + 3 = 19$ if the multiplication is completed first and $2 \cdot 8 + 3 = 2 \cdot 11 = 22$

if the addition is completed first. The first result is the accepted order. An outline specifying the accepted order of operations is given in the box that follows.

The most common grouping symbols are ( ) parentheses, [ ] brackets, { } braces, — the fraction bar, and $\sqrt{}$ the radical symbol.

> **To Perform Operations in Order**
> 1. Perform all operations within grouping symbols, working from the innermost symbol outward. When the grouping symbol is a fraction bar, carry out the operations in the numerator and in the denominator independently.
> 2. Perform all operations indicated by exponents.
> 3. Perform all multiplication and division, working from left to right.
> 4. Perform all addition and subtraction, working from left to right.

EXAMPLE 12 Simplify: $4(8 + 3) + 9$.

SOLUTION

$4(8 + 3) + 9$
$= 4(11) + 9$ Work within grouping symbols first.
$= 44 + 9$ Multiply.
$= 53$ Add.

EXAMPLE 13 Simplify: $\dfrac{5(-3) + (-5)}{2(-8) + 12}$.

SOLUTION

$\dfrac{5(-3) + (-5)}{2(-8) + 12}$

$= \dfrac{-15 + (-5)}{-16 + 12}$ Multiply first.

$= \dfrac{-20}{-4}$ Add.

$= 5$

EXAMPLE 14 Simplify: $(-2)^2 + 2 \cdot 3^3$.

SOLUTION

$(-2)^2 + 2 \cdot 3^3$
$= 4 + 2 \cdot 27$ Exponents first.
$= 4 + 54$ Multiply.
$= 58$ Add.

EXAMPLE 15 Simplify: $4 + [3(-2) + 5(-2)^3]^2$.

SOLUTION

$4 + [3(-2) + 5(-2)^3]^2 = 4 + [3(-2) + 5(-8)]^2$
$ = 4 + [-6 + (-40)]^2$
$ = 4 + [-46]^2$

$$= 4 + 2116$$
$$= 2120$$

EXAMPLE 16 Simplify: $2\{8 + 6[4 - 2] - 5\} - 9$.

SOLUTION
$$2\{8 + 6[4 - 2] - 5\} - 9 = 2\{8 + 6[2] - 5\} - 9$$
$$= 2\{8 + 12 - 5\} - 9$$
$$= 2\{15\} - 9$$
$$= 30 - 9$$
$$= 21$$

EXAMPLE 17 $\dfrac{\frac{1}{2}[15 - (-7)] + \frac{1}{2}(10)}{\frac{1}{4}(-7 + 3)} = ?$

SOLUTION
$$\frac{\frac{1}{2}[15 - (-7)] + \frac{1}{2}(10)}{\frac{1}{4}(-7 + 3)} = \frac{\frac{1}{2}[22] + \frac{1}{2}(10)}{\frac{1}{4}(-4)}$$
$$= \frac{11 + 5}{-1}$$
$$= \frac{16}{-1}$$
$$= -16$$

Do You Remember?

Can you match these?

____ 1. Addition and subtraction of fractions
____ 2. Multiplication and division of fractions
____ 3. Fundamental property of fractions
____ 4. Addition of signed numbers
____ 5. Subtraction of signed numbers
____ 6. Multiplication of signed numbers
____ 7. Division of signed numbers
____ 8. Division by zero
____ 9. Raising to powers
____ 10. Order of operation

a) $3^4 = 81$
b) $-6 + (-13) = -19$
c) $\dfrac{56}{0}$ is undefined.
d) $6 + 8 \cdot 4 = 6 + 32 = 38$
e) $\dfrac{5}{8} + \dfrac{3}{8} - \dfrac{1}{8} = \dfrac{7}{8}$
f) $\dfrac{-27}{3} = -9$
g) $\dfrac{8}{11} = \dfrac{8 \cdot 2}{11 \cdot 2}$
h) $(5)(-14) = -70$
i) $\dfrac{1}{3} \cdot \dfrac{2}{5} \div \dfrac{2}{3} = \dfrac{1}{5}$
j) $(-16) - (-11) = -5$

Answers: 1. e, 2. i, 3. g, 4. b, 5. j, 6. h, 7. f, 8. c, 9. a, 10. d

Exercise Set 1.3

Simplify each of the following quantities as indicated.

1. $10 + (-4)$
2. $18 + (-10)$
3. $-15 + (-17)$
4. $-27 + (-31)$
5. $-1 + 17 + (-16)$
6. $(-2) + (-10) + (-3)$
7. $-8 - 15$
8. $-1 - 35$
9. $2^2 + 5^2$
10. $3^2 + 4^2$
11. $4^2 - 34$
12. $2^3 - 52$
13. $-3 - (-5) - 2^3$
14. $-2 - (-3^2) - 6$
15. $\frac{7}{11} - \frac{3}{11} + 1$
16. $\frac{4}{9} - \frac{2}{9} + 1$
17. $\frac{5}{7} - \left(\frac{1}{7} + \frac{4}{4}\right)$
18. $1 - \left(\frac{4}{11} - \frac{1}{6}\right)$
19. $1\frac{3}{8} - 2\frac{1}{4} - \frac{1}{2}$
20. $5\frac{3}{7} + 2\frac{1}{4} - \frac{3}{2}$
21. $-18 - (-2) - (-13)$
22. $-41 - (-9) - (-6)$
23. $-|-3| + |6|$
24. $|23| - |-5|$
25. $-16p - (-7p)$
26. $-28k - (-7k)$
27. $-16a + 18b - (-11a) + 2b$
28. $-60r - (-6t) + 75r - 15t$

29. Find the vertical distance between the top of Mount McKinley (20,320 feet above sea level) and Death Valley (282 feet below sea level).
30. Find the vertical distance between the top of Mount Whitney (14,494 feet above sea level) and the Salton Sea (235 feet below sea level).

Carry out the indicated operations.

31. $-2^3(15)$
32. $-3^2(-2)^3$
33. $-10(-7)$
34. $6(-13)$
35. $-7(-13)(-6)$
36. $-9(-7)(-3)(-5)$
37. $-36 \div 12$
38. $-81 \div (-27)$
39. $\frac{-256}{-8}$
40. $\frac{-8 - 4(-2)}{15}$
41. $\frac{28 - 7(4)}{13(-2)}$
42. $\left(\frac{1}{2} \cdot \frac{3}{3}\right) \div \left(-\frac{1}{6}\right)$
43. $-\frac{5}{8}\left(\frac{3}{4} \cdot \frac{4}{1}\right)$
44. $\frac{7}{15} \div \left(\frac{5}{21} \div \frac{10}{6}\right)$

Simplify, using the correct order of operations. If the division is undefined, say so.

45. $-6(2^3 - 2^2) \div 2^3$
46. $-5(13 - 2^3) \div (-5^2)$
47. $-33(18 - 15)(0)$
48. $-54(17 - 14)(0)$
49. $-13 + (-5)^2(6)$
50. $-18 + (-2)(-5)^2$
51. $-2[-6 + (-3)(-2)] \div 15$
52. $-[-5 + 2(4 - 1) - 1] \div (-17)$
53. $[-3 + (-2)][-6 - (-4)]$
54. $[6 + (-4)][-8 - (-7)]$
55. $-11 - 2^2 \cdot 8 + 30$
56. $-4^2 - 3 \cdot 3^2 + 26$
57. $\left(\frac{3}{7} - \frac{2}{5}\right)\left(\frac{7}{5} - \frac{1}{7}\right)$
58. $\left(\frac{8}{5} - \frac{5}{4}\right)\left(\frac{27}{10} - \frac{3}{20}\right)$
59. $\frac{8 + (-2)(5 - 6)}{-5}$
60. $\frac{11 + (-3)(1 - 3)}{-17}$
61. $15 \div 3 \div 4 \div 0$
62. $26 \div 13 \div (-2) \div 0$
63. $(-3 - 2)(1 - 5)(3 - 7)$
64. $(4 - 1)(2 - 5)(7 - 10)$
65. $\frac{\frac{1}{2}[17 - (8 - 5)][\frac{1}{2}(10 - 2) \div \frac{1}{2}](100)}{\frac{1}{4}[18 - \frac{1}{2}(32 - 4)]}$
66. $\frac{\frac{1}{4}[36 + 5 + \frac{1}{2}(17 - 3)][\frac{1}{2}(31 - 7)]}{\frac{1}{2}[15.5 - 7.3 + 1.8] \div 5}$
67. $\left(\frac{1}{2} + \frac{1}{3}\right) \div \left(\frac{3}{5} - \frac{1}{2}\right)$
68. $\frac{3}{7} + \frac{2}{9}\left(\frac{3}{4} \div 2\right) - 1$
69. $\frac{-2[-15 - 3(8 - 3) + 5]}{-5[-31 - 15(-2) + 1]}$
70. $\frac{-3[-24 - 2(4 - 8) + 8]}{-8[-18 + 5(10 - 13) + 30]}$
71. $\frac{51t}{50} \div \frac{-17t}{30}$
72. $\frac{7y}{13} \div \frac{-49y}{91}$
73. $(0.5 + 1.6 - 0.12)(7.2 \div 6)$
74. $(1.6)(0.1 - 2.3 + 4.56) \div 0.8$
75. $(-1.5)[3.02 - (6.43 - 4.41)] \div 0.05$
76. $(-3.05)[2.17 - (8.83 - 6.66)] \div (-0.5)$
77. $(0.3^2)[(2.5)^2 - 5^2(-0.5)^2] \div 0.7^3$
78. $(0.8^3)[(3.6)^2 - 6^2(-0.6)^2] \div 0.9^3$

Evaluate each of the following.

79. $2x - y + z$ for $x = -2$, $y = 3$, $z = 9$

80. $\frac{1}{2}bh$ for $b = 2\frac{1}{4}, h = 3\frac{1}{2}$

81. $2L + 2W + 2H$ for $L = \frac{1}{2}, W = \frac{1}{5}, H = \frac{3}{4}$

82. $2xy + 2xz + 2yz$ for $x = 2, y = 1, z = 3$

REVIEW EXERCISES

State the property illustrated by each of the following.

83. $x + y = y + x$
84. $4 \cdot \frac{1}{4} = 1$
85. $x + (y + z) = (x + y) + z$
86. $16 + (-16) = 0$
87. $6(x - y) = 6x - 6y$
88. $15 \cdot 4 = 4 \cdot 15$
89. If $x = y$ and $y = z$, then $x = z$.
90. $7x + 7y = 7(x + y)$

State whether Exercises 91–100 are true *or* false.

91. $6 \cdot 0 = 0$
92. $9 \div 0 = 0$
93. $9 + (-9) = 0$
94. $\frac{64 - 8^2}{7 - 4} = 0$
95. $\frac{23 - 4^2}{9 - 3 \cdot 3} = 0$
96. $54 \cdot \frac{1}{54} = 1$
97. $|-8| < |-5|$
98. $-|(-3)^2| < |-3^2|$
99. $-3 \in N$
100. $\{2, 3\} \in W$

WRITE IN YOUR OWN WORDS

101. In which order would you calculate $\{3 - 2^4 \cdot 3\}(-5)^2$?
102. Why is $(-5)^2 = 25$ but $-5^2 = -25$?
103. Why is $|-7|$ greater than -7?
104. How do you add two fractions with the same denominator?
105. How do you multiply two fractions?
106. Why is $a \div 0 \neq 0$ but $0 \div a = 0$, where $a \neq 0$?
107. How do you divide two fractions?
108. How do you evaluate $x - y^2$ when $x = 3$ and $y = -2$?
109. Why is $\{3\} \in \{1, 2, 3\}$ not a correct statement?

SUMMARY

1.1 Sets and Real Numbers

A set is a well-defined collection of objects. Each object is an **element,** or **member,** of the set. Sets are generally designated by capital letters, and the members enclosed within **braces.** The **listing method** lists each element of a set individually or indicates that a pattern occurs. **Set-builder notation** (also known as the **rule method**) describes the elements.

A set without any elements is called the **empty,** or **null set.** A set is said to be **finite** if there is a number that describes exactly how many elements it contains. A set that is not a finite set is said to be an **infinite set.** The **union** of two sets A and B, written $A \cup B$, includes the elements just in A, just in B, or in both A and B. The **intersection** of two sets A and B, written $A \cap B$, consists only of elements common to both A and B.

Set A is a **subset** of set B, written $A \subseteq B$, if every element in A is an element of B.

Natural numbers $N = \{1, 2, 3, \ldots\}$
Whole numbers $W = \{0, 1, 2, 3, \ldots\}$
Integers $J = \{\ldots, -3, -2, -1, 0, 1, 2, 3, \ldots\}$
Rational numbers $Q = \left\{\dfrac{a}{b} \mid a \in J, b \in J, b \neq 0\right\}$

Irrational numbers $H = \{\text{nonrepeating, nonterminating decimals}\}$
Real numbers $R = Q \cup H$

1.2 Properties of Real Numbers

The **absolute value** of a number is its distance from the origin on the number line. More generally

$$|a| = \begin{cases} a \text{ if } a \geq 0 \\ -a \text{ if } a < 0 \end{cases}$$

Properties of Equality
Reflexive $a = a$
Symmetric If $a = b$, then $b = a$.
Transitive If $a = b$ and $b = c$, then $a = c$.
Substitution If $a = b$, then a can be replaced by b or b by a in any mathematical statement.
Multiplication property of zero $a \cdot 0 = 0 \cdot a = 0$
Double negative property $-(-a) = a$

Properties of Real Numbers
If a, b, and $c \in R$, the following properties are true.
Closure $a + b \in R$ and $a \cdot b \in R$
Commutative $a + b = b + a$ and $a \cdot b = b \cdot a$
Associative $a + (b + c) = (a + b) + c$ and $a(b \cdot c) = (a \cdot b)c$
Identity $a + 0 = 0 + a = a$ and $a \cdot 1 = 1 \cdot a = a$
Inverse $a + (-a) = 0$ and $a \cdot \dfrac{1}{a} = 1, a \neq 0$
Distributive $a(b + c) = ab + ac$

1.3 Operations with Real Numbers

Fractions are multiplied, divided, added, and subtracted as follows.

$$\frac{a}{b} \cdot \frac{c}{d} = \frac{a \cdot c}{b \cdot d}, \quad b, d \neq 0$$

$$\frac{a}{b} \div \frac{c}{d} = \frac{a}{b} \cdot \frac{d}{c}, \quad b, c, d \neq 0$$

$$\frac{a}{c} + \frac{b}{c} = \frac{a + b}{c}, \quad c \neq 0$$

$$\frac{a}{c} - \frac{b}{c} = \frac{a - b}{c}, \quad c \neq 0$$

Fractions are simplified or reduced by using the **fundamental property of fractions:**

$$\frac{a \cdot x}{b \cdot x} = \frac{a}{b} \quad \text{or} \quad \frac{a}{b} = \frac{a \cdot x}{b \cdot x}, \quad b, x \neq 0$$

To add fractions that do not have a common denominator, use the fundamental property of fractions to generate equivalent fractions having a **least common denominator (LCD).**

Signed numbers are added, subtracted, multiplied, and divided by the following rules.

Addition To add two numbers with **like signs,** find the sum of their absolute values and prefix the common sign.
To add two numbers with **unlike signs,** find the difference of their absolute values

by subtracting the smaller from the larger, and prefix the sign of the number with the greater absolute value.

Subtraction To subtract two numbers, change the sign (find the additive inverse) of the number being subtracted and follow the rules for addition.

Multiplication The product of two numbers with **like signs** is *positive*. The product of two numbers with **unlike signs** is *negative*.

Division The quotient of two numbers with **like signs** is *positive*. The quotient of two numbers with **unlike signs** is *negative*. Division by zero is **undefined.**

Order of Operations

1. Perform all operations within grouping symbols, working from the innermost symbol outward. When a grouping symbol is a fraction bar, carry out the operations in the numerator and in the denominator independently.
2. Perform all operations indicated by exponents.
3. Perform all division and multiplication, working from left to right.
4. Perform all additions and subtractions, working from left to right.

Cooperative Exercise

At Cuesta College, 720 new students were asked to answer the following two questions:

A. Do you plan to enroll in an elementary algebra class?

B. Do you plan to enroll in an intermediate algebra class?

The resulting information is illustrated in the accompanying diagram. Use this information to answer the following questions. The numbers indicate the number of students who plan to enroll.

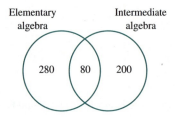

1. How many students did not answer either question?
2. How many students plan to enroll in an elementary algebra class?
3. How many students plan to enroll in an intermediate algebra class?
4. How many students plan to enroll in an elementary algebra or an intermediate algebra class, but not both?
5. How many students plan to enroll in an elementary algebra class but not in an intermediate algebra class?
6. How many students plan to enroll in an intermediate algebra class but not in an elementary algebra class?
7. How many students plan to enroll in both an elementary algebra class and an intermediate algebra class?
8. How many students do not plan to enroll in either kind of class?
9. How many sections of elementary algebra (40 students per section) should be made available to accomodate the indicated need?
10. How many sections of intermediate algebra (40 students per section) should be made available to accomodate the indicated need?

Review Exercises

Complete the following statements.

1. A well-defined collection of objects is called a _____.
2. The set N is called the set of _____ numbers.
3. The set of letters in the word "irrational" is written _____.

4. Two subsets of the set *J* are _____ and _____.
5. List $B = \{x \mid x \in J, x > -4, x < 10, \text{ and } x \text{ an even number}\}$.
6. $\{x \mid x \in A \text{ and } x \in B\}$ represents the _____ of *A* and *B*.
7. If $A \cap B = \emptyset$, then *A* has _____ elements that are also in *B*.
8. *A* is said to be a _____ of *B* if whenever $x \in A$ then $x \in B$.
9. $a \not\equiv b$ means *a* _____ *b*.
10. If $a > b$ then *b* is to the _____ of *a* on the number line.
11. *x* and $-x$ are called _____ of each other.
12. The set $\{\ldots, -3, -2, -1, 0, 1, 2, 3, \ldots\}$ is known as the set of _____.
13. The set $\left\{ \dfrac{a}{b} \mid a \in J, b \in J, b \neq 0 \right\}$ is known as the set of _____ numbers.
14. The number π is an element of the set of _____ numbers.
15. In decimal form, the fraction $\dfrac{3}{11}$ is _____.
16. The statement, "If $x, y \in R$, then $x \cdot y$ is a real number" illustrates the _____ property of multiplication.
17. The equation $6 + y = y + 6$ is an illustration of the _____ property of addition.
18. The equation $6x - 6y = 6(x - y)$ is an illustration of the _____ property.
19. The statement, "If $a, b \in R$ and $a = b$, then $b = a$," illustrates the _____ property of equality.

Simplify each of the following.

20. $-\dfrac{3}{5} + \dfrac{2}{7} - \dfrac{1}{2}$
21. $-\dfrac{5}{8} \cdot \dfrac{1}{4} \div \dfrac{1}{16}$
22. $-5\dfrac{7}{9} \div 2\dfrac{1}{8}$
23. $-6 + \dfrac{2}{3}\left(2 - \dfrac{7}{8}\right)$
24. $\dfrac{49}{5} - 2 + \dfrac{1}{3} \div \dfrac{3}{2}$
25. $-4\dfrac{1}{9} + \dfrac{1}{2}\left(6\dfrac{1}{2} + \dfrac{1}{3}\right) - 1$
26. $\dfrac{156}{72} \div \left(\dfrac{-90}{48}\right)$
27. $-1\dfrac{1}{5}\left(-3\dfrac{1}{8}\right) \div \left(-\dfrac{1}{4}\right)$

Complete each of the following.

28. The sum of -11 and 7 is _____.
29. The product of -5 and 9 is _____.
30. $-16 - (-3)$ equals _____.
31. The value of $x + y - z$ for $x = -5, y = 7,$ and $z = -2$ is _____.
32. The sign of the product of two numbers with unlike signs is _____.
33. $\dfrac{-8(6)}{-12}$ equals _____.
34. Division by zero is _____.
35. $\dfrac{6(-8 + 4^2) + (-3)^3}{15(-5 + 4) + 6}$ equals _____.
36. $\dfrac{\frac{1}{4}[16 - (-2)] + \frac{1}{2}(-8)}{\frac{1}{2}[18 + (-4)] - \frac{1}{4}(-12)} =$ _____
37. $-4(x - 2y) + 5(-3x + y)$ equals _____.
38. $|-5|$ equals _____.
39. If $x < 0$ then $|x|$ equals _____.
40. $-|-3^2|$ equals _____.

Simplify each of the following.

41. $|-23| - |-17|$
42. $|3\frac{1}{4}| + |-1\frac{1}{2}| - |-5.5|$
43. $(0.8)^2[0.19 - (2.30 - 0.11)] \div 2.0$
44. $-18x - (-5x)$
45. $-4(3^3 - 3^2) \div 3$
46. $\dfrac{65y}{40} \div \dfrac{13y}{50}$
47. $(-61)(49 - 17)(0)$
48. $\dfrac{8}{17} \div \left(\dfrac{3}{7} \div \dfrac{18}{21}\right)$
49. $(1.2)^2[6.36 - (0.6^2 - 0.3^2)] \div 0$
50. $(7.21)^2[15.1 - (6.9 + 8.2)] \div 2.15$

Chapter Test

1. Rewrite the set $\{x \mid x < 5, x \in N\}$ using the *listing method*.
2. Give an example of an empty set.
3. Write $\dfrac{5}{9}$ in decimal form.

4. Find $A \cap B$ when $A = \{1, 3, 5, 7\}$ and $B = \{1, 2, 3, 4\}$.
5. Simplify $\dfrac{7}{8}\left(\dfrac{5}{3} \div \dfrac{4}{3}\right)$.
6. Simplify $\dfrac{1}{3}\left(\dfrac{1}{2} + \dfrac{3}{5}\right)$.
7. Evaluate $-x - [-(-y) + 2x]$ for $x = -1$ and $y = 0$.
8. Simplify $16 - (6 - 2^2) - (-4) \cdot (-7)$.

Name the property illustrated by each of the following statements.

9. $15a - 6b = 3(5a - 2b)$
10. $(17 + 3) + 2 = 17 + (3 + 2)$
11. $13 \cdot \dfrac{1}{13} = 1$
12. $19 + (-19) = 0$
13. $6(7 \cdot 3) = (6 \cdot 7)3$
14. $a = a$
15. If $x = y$ and $y = 3$, then $x = 3$.

Name the subset of the real numbers illustrated by each of the following sets.

16. $\{1, 2, 3, \ldots\}$
17. $J \cup Q$
18. $\{\ldots, -3, -2, -1, 0, 1, 2, 3, \ldots\}$
19. $\left\{\dfrac{a}{b} \,\middle|\, a \in J, b \in J, b \neq 0\right\}$
20. $\{0, 1, 2, 3, \ldots\}$
21. $N \cup W$
22. $J \cap W$
23. Write the set of letters in the word "January" using set-builder notation.
24. Give a numerical example to show that division of numbers is not commutative.

Complete each of the following.

25. The set of real numbers is commutative for the operations of _____ and _____.
26. The number $-|-24|$ is equal to _____.
27. $a \cdot 0 = 0$ is an example of the _____ property of zero.
28. The number π is an element of the set of _____ numbers.
29. $H \cap Q$ equals the set _____.
30. $27 + 0 = 27$ illustrates the _____ property.

Simplify each of the following quantities.

31. $\dfrac{3[-7 + (-5)] + 2(8 + 4)}{6[|-5| - (3)]}$
32. $\dfrac{168}{72} \div \dfrac{21}{36}$
33. $|-3.5| \cdot [2.1 - (3.4)] \div 1.1$
34. $2\tfrac{1}{2}(3\tfrac{1}{4} - 1\tfrac{1}{2})$
35. $\dfrac{\tfrac{1}{2}[(0.5 + 3.5) \div 2.0]}{\tfrac{1}{4}[(3 + \tfrac{1}{2}) - (-\tfrac{1}{2})] \div \tfrac{1}{2}}$
36. $3^3[2^4 - 3^2 \cdot 6 + 2^2] \div |-0.5|$
37. $-1\tfrac{1}{4}\left(-3\tfrac{1}{2}\right) \div \left(-\tfrac{1}{4}\right)$
38. $5\tfrac{1}{3} - \tfrac{1}{5}\left(6\tfrac{1}{4} - 1\tfrac{1}{5}\right)$
39. $\dfrac{560}{270} \cdot \dfrac{90}{80} \div \dfrac{3}{7}$
40. $-(-5 + 3)(7 - 2)(-6 + 2) \cdot |-5 + 2|$

CHAPTER 2
LINEAR EQUATIONS AND INEQUALITIES

2.1 Solving Linear Equations in One Variable
2.2 From Words to Symbols: An Introduction to Applied Problems
2.3 Literal Equations and Formulas
2.4 Solving Linear Equations Involving Absolute Value
2.5 Solving Linear Inequalities
2.6 Solving Linear Inequalities Involving Absolute Value
2.7 More Applications

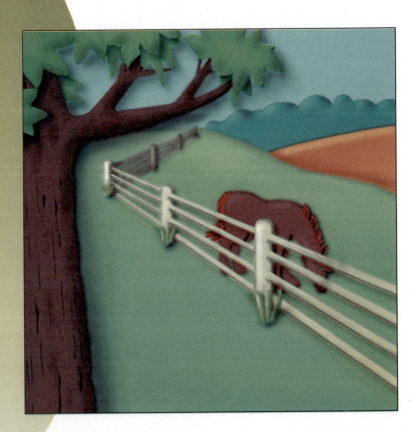

APPLICATION
A rancher has 2000 yards of fencing to enclose a rectangular pasture. The length of the pasture is 500 yards longer than the width. Find the dimensions of the pasture.

See Section 2.7, Example 4.

The higher arithmetic gives us an almost inexhaustible store of exciting truths, which stand in close internal connection, and from which we constantly find new usually unexpected ties. They are usually very simple, and can often be discovered by induction, and yet so profound that we cannot find how to use them until after many vain tries. Even if we do succeed the simpler methods may long remain concealed.

ALBERT EINSTEIN (1879–1955)

2.1 Solving Linear Equations in One Variable

HISTORICAL COMMENT Francois Vietta (1540–1603) was a French lawyer who studied mathematics for recreation. His works on mathematics, published privately for distribution among friends, were the first known to use letters to represent numbers. Vietta used consonants to represent known quantities and vowels to represent unknown quantities.

OBJECTIVES In this section we will learn to
1. solve linear equations;
2. use the addition, subtraction, multiplication, and division properties of equality; and
3. recognize and solve conditional equations and identities.

DEFINITION **Linear equation**	A **linear (first-degree) equation** in one variable is one that can be written in the form $ax + b = c$ where a, b, and c represent real numbers, with $a \neq 0$.

For example, the equation $2x + 3 = 13$ is linear, where $a = 2$, $b = 3$, and $c = 13$. The equation $3x - 5 = 0$ is also linear, with $a = 3$, $b = -5$, and $c = 0$.

DEFINITION **A Solution, or Root, of an Equation**	A **solution**, or **root, of an equation** involving a single varible is a number that makes the original equation true when substituted for the variable.

In the equation $2x + 7 = 13$, the solution is 3. To verify the solution, we replace x by 3:

$$2 \cdot 3 + 7 \stackrel{?}{=} 13$$
$$6 + 7 \stackrel{?}{=} 13$$
$$13 \stackrel{\checkmark}{=} 13 \quad \text{The check mark means that the solution is correct.}$$

The set of all solutions to an equation is called its **solution set.** The solution set of the equation $2x + 7 = 13$ is $\{3\}$. In this text we will write the solutions to all equations in set notation.

Linear equations are solved by writing them in simpler forms, called **equivalent equations,** until the solution set can be determined by inspection. *Equivalent equations are equations with exactly the same solution set.*

To generate equivalent equations, four theorems are used. They are the addition, subtraction, multiplication, and division properties of equality, which we will state without proof. We begin with the **addition property of equality.**

Theorem 2.1
The Addition Property of Equality

If a, b, and c represent real numbers and

$$a = b$$

then

$$a + c = b + c$$

In simplest terms, the addition property states that *if the same real number is added to each side of an equation, the result is an equation equivalent to the original equation.*

Example 1 shows how the addition property is used to solve an equation.

EXAMPLE 1 Solve the equation $x - 7 = 6$.

SOLUTION To solve the equation, we need to find an equivalent equation in which the variable x stands alone.

$$x - 7 = 6$$
$$x - 7 + 7 = 6 + 7 \quad \text{Add 7 to each side.}$$
$$x = 13$$

To check the solution, we substitute 13 for x in the original equation.

CHECK
$$x - 7 = 6$$
$$13 - 7 \stackrel{?}{=} 6$$
$$6 \stackrel{\checkmark}{=} 6$$

The solution set is $\{13\}$.

Recall from Chapter 1 that $a - c = a + (-c)$. Adding the negative of a number is the same as subtracting the number. This leads to the **subtraction property of equality.**

Theorem 2.2

The Subtraction Property of Equality

If a, b, and c, represent real numbers and

$$a = b$$

then

$$a - c = b - c$$

In simplest terms, the subtraction property of equality states that *if the same real number is subtracted from each side of an equation, the result is an equation equivalent to the original equation.*

The next example shows how the subtraction property is used to solve an equation.

EXAMPLE 2 Solve the equation $x + 5 = 7$.

SOLUTION

$$x + 5 = 7$$
$$x + 5 - 5 = 7 - 5 \quad \text{Subtract 5 from each side.}$$
$$x = 2$$

To check the solution, we substitute 2 for x in the original equation.

CHECK $x + 5 = 7$
$2 + 5 \stackrel{?}{=} 7$
$7 \stackrel{\checkmark}{=} 7$

The solution set is $\{2\}$.

Two terms that are exactly alike except for their numerical coefficients are called **like terms.** For example, $3y$ and $5y$ are like terms because the variable part, y, is the same in each. Their *numerical coefficients* are 3 and 5, respectively. Like terms can be combined (added or subtracted) by using the distributive property.

EXAMPLE 3 Combine like terms: **(a)** $3y + 5y$ **(b)** $7x - 2x$ **(c)** $17m - 5c + 2m$

SOLUTION
(a) $3y + 5y = (3 + 5)y = 8y$ Distributive property.
(b) $7x - 2x = (7 - 2)x = 5x$
(c) $17m - 5c + 2m = 17m + 2m - 5c = 19m - 5c$

EXAMPLE 4 Solve: $3m - m + 3 = m + 8$.

SOLUTION

$$3m - m + 3 = m + 8$$
$$2m + 3 = m + 8 \quad \text{Combine like terms.}$$
$$2m - m + 3 = m - m + 8 \quad \text{Subtract } m \text{ from each side.}$$
$$m + 3 = 8$$
$$m + 3 - 3 = 8 - 3 \quad \text{Subtract 3 from each side.}$$
$$m = 5$$

CHECK $\quad 3m - m + 3 = m + 8$
$3 \cdot 5 - 5 + 3 \stackrel{?}{=} 5 + 8$
$15 - 5 + 3 \stackrel{?}{=} 5 + 8$
$13 \stackrel{\checkmark}{=} 13$

The solution set is {5}.

The equations in Examples 1, 2, and 4 were solved by using the addition and/or subtraction properties of equality to generate simpler equations. Some equations cannot be solved using only these properties. Two such equations are

$$6x + 5 = 8 \quad \text{and} \quad \frac{x}{2} - \frac{3x}{2} = 9$$

Their solutions require the **multiplication** and **division properties of equality.**

Theorem 2.3
The Multiplication Property of Equality

If a, b, and c represent real numbers and

$a = b$

then

$a \cdot c = b \cdot c, \quad c \neq 0$

In simplest terms, the multiplication property of equality states that *if each side of an equation is multiplied by the same* **nonzero** *real number, the result is an equation equivalent to the original equation.*

Theorem 2.4
The Division Property of Equality

If a, b, and c represent real numbers and

$a = b$

then

$\dfrac{a}{c} = \dfrac{b}{c}, \quad c \neq 0$

In simplest terms, the division property of equality states that *if each side of an equation is divided by the same* **nonzero** *real number, the result is an equation equivalent to the original equation.*

In the next example, we will show how the division property of equality is used to solve an equation.

EXAMPLE 5   Solve the equation $6x + 5 = 8$.

SOLUTION   First we find an equivalent equation in which the term involving the variable stands alone on one side of the equation:

$$6x + 5 = 8$$
$$6x + 5 - 5 = 8 - 5 \quad \text{Subtract 5 from each side.}$$
$$6x = 3$$

Now to solve for x, we use the division property of equality.

$$\frac{6x}{6} = \frac{3}{6}$$
$$x = \frac{1}{2}$$

The solution set is $\left\{\frac{1}{2}\right\}$. The check is left to the reader.

EXAMPLE 6   Solve the equation $2(w + 1) - 4(w - 6) = 6$.

SOLUTION   First we use the distributive property to remove grouping symbols:

$$2(w + 1) - 4(w - 6) = 6$$
$$2w + 2 - 4w + 24 = 6 \quad \text{Distributive property.}$$
$$-2w + 26 = 6 \quad \text{Combine like terms.}$$

Now we use the subtraction and division properties to isolate w:

$$-2w + 26 - 26 = 6 - 26 \quad \text{Subtract 26 from each side.}$$
$$-2w = -20$$
$$\frac{-2w}{-2} = \frac{-20}{-2} \quad \text{Divide each side by } -2.$$
$$w = 10$$

CHECK
$$2(w + 1) - 4(w - 6) = 6$$
$$2(10 + 1) - 4(10 - 6) \stackrel{?}{=} 6$$
$$2(11) - 4(4) \stackrel{?}{=} 6$$
$$22 - 16 \stackrel{?}{=} 6$$
$$6 \stackrel{\checkmark}{=} 6$$

The solution set is $\{10\}$.

EXAMPLE 7   Solve the equation $\dfrac{x}{2} - \dfrac{3x}{7} = 9$.

SOLUTION   To solve this equation, it will be necessary to use the multiplication property of equality. We begin by multiplying each side by the least common denominator (LCD). The least common denominator in this case is $2 \times 7 = 14$.

$$\frac{x}{2} - \frac{3x}{7} = 9$$

$$14\left(\frac{x}{2} - \frac{3x}{7}\right) = 14(9) \qquad \text{Multiply each side by 14, the LCD.}$$

$$14\left(\frac{x}{2}\right) - 14\left(\frac{3x}{7}\right) = 14(9) \qquad \text{Distributive property.}$$

$$7x - 6x = 126 \qquad \text{Simplify.}$$

$$x = 126$$

CHECK
$$\frac{x}{2} - \frac{3x}{7} = 9$$

$$\frac{126}{2} - \frac{3(126)}{7} \stackrel{?}{=} 9$$

$$63 - 54 \stackrel{?}{=} 9$$

$$9 \stackrel{\checkmark}{=} 9$$

The solution set is {126}.

An equation that is true when the variable is replaced by any number is called an **identity.** *The solution set for an identity is the set of all real numbers.* The next example illustrates an identity.

EXAMPLE 8 Solve the equation $x + (3x + 2) - 4 = 2x + 2(x - 1)$.

SOLUTION
$$x + 3x + 2 - 4 = 2x + 2x - 2$$
$$4x - 2 = 4x - 2 \qquad \text{Combine like terms.}$$

Since $4x - 2 = 4x - 2$ for any real-number replacement for x, the equation is an identity. The solution set is the set of all real numbers, R.

In Example 8, we could have subtracted $4x$ from each side to obtain

$$4x - 2 = 4x - 2$$
$$-2 = -2$$

If all of the variables are eliminated when solving an equation and the result is a true statement, the equation is an identity. An equation that is true only for specified values of the variable is a **conditional equation.** The equations in Examples 1, 2, 4, 5, 6, and 7 are conditional equations.

As the examples of this section show, it is helpful to follow a series of steps when solving linear equations in one variable. The steps are summarized in the following box. Not all of the steps are necessary to solve every linear equation.

To Solve a Linear Equation

Step 1 Clear the equation of fractions by using the multiplication property, with the LCD (least common denominator) as the multiplier.
Step 2 Remove grouping symbols by using the distributive property.
Step 3 Combine like terms on each side of the equation.
Step 4 Isolate the variable by using the addition or subtraction property. Write the equation in the form $ax = b$ or $b = ax$.
Step 5 Solve for the variable by using the multiplication or division property.
Step 6 Check the solution.

EXAMPLE 9 Solve the equation $\dfrac{2(x+5)}{3} = \dfrac{4(x-2)}{5} - 2$.

SOLUTION First we clear of fractions by multiplying each side by 15, the LCD.

Be sure to multiply **each** term on both sides by the LCD.

$$\frac{2(x+5)}{3} = \frac{4(x-2)}{5} - 2$$

$$15\left[\frac{2(x+5)}{3}\right] = 15\left[\frac{4(x-2)}{5}\right] - 15(2)$$

$5[2(x+5)] = 3[4(x-2)] - 15(2)$ Simplify.
$5[2x+10] = 3[4x-8] - 15(2)$ Distributive property.
$10x + 50 = 12x - 24 - 30$ Distributive property.
$10x + 50 = 12x - 54$ Combine like terms.
$10x + 50 - 10x = 12x - 54 - 10x$ Subtract $10x$ from each side.
$50 = 2x - 54$
$50 + 54 = 2x - 54 + 54$ Add 54 to each side to isolate the variable.
$104 = 2x$
$52 = x$ Solve for the variable.

CHECK $\quad \dfrac{2(x+5)}{3} = \dfrac{4(x-2)}{5} - 2$

$\dfrac{2(52+5)}{3} \stackrel{?}{=} \dfrac{4(52-2)}{5} - 2$ Substitute 52 for x.

$\dfrac{2(57)}{3} \stackrel{?}{=} \dfrac{4(50)}{5} - 2$

$\dfrac{114}{3} \stackrel{?}{=} \dfrac{200}{5} - 2$

$38 \stackrel{?}{=} 40 - 2$

$38 \stackrel{\checkmark}{=} 38$

The solution set is {52}.

EXAMPLE 10 Solve the equation $2\left(\dfrac{x}{3} + \dfrac{1}{6}\right) = 5$.

SOLUTION

$$2\left(\dfrac{x}{3} + \dfrac{1}{6}\right) = 5$$

$6 \cdot 2\left(\dfrac{x}{3} + \dfrac{1}{6}\right) = 6 \cdot 5$ Multiply each side by 6, the LCD.

$12\left(\dfrac{x}{3} + \dfrac{1}{6}\right) = 30$ Simplify.

$4x + 2 = 30$ Distributive property.

$4x + 2 - 2 = 30 - 2$ Subtract 2 from each side.

$4x = 28$

$x = 7$ Divide each side by 4.

CHECK

$2\left(\dfrac{x}{3} + \dfrac{1}{6}\right) = 5$

$2\left(\dfrac{7}{3} + \dfrac{1}{6}\right) \stackrel{?}{=} 5$ Substitute 7 for x.

$2\left(\dfrac{14}{6} + \dfrac{1}{6}\right) \stackrel{?}{=} 5$

$2\left(\dfrac{15}{6}\right) \stackrel{?}{=} 5$

$5 \stackrel{\checkmark}{=} 5$

The solution set is {7}.

EXAMPLE 11 Solve the equation $-3\{2x - 3[x + 5(2x - 7)] - 9\} = 4\{-2(x - 9) + 1\}$.

SOLUTION We begin by carrying out the operations within the grouping symbols. Recall that the proper order is to start with the innermost symbol and work your way out until all symbols are removed.

$-3\{2x - 3[x + 5(2x - 7)] - 9\} = 4\{-2(x - 9) + 1\}$

$-3\{2x - 3[x + 10x - 35] - 9\} = 4\{-2x + 18 + 1\}$

$-3\{2x - 3x - 30x + 105 - 9\} = -8x + 72 + 4$

$-6x + 9x + 90x - 315 + 27 = -8x + 72 + 4$

$93x - 288 = -8x + 76$

$93x + 8x = 76 + 288$

$$101x = 364$$
$$x = \frac{364}{101}$$

The solution set is $\left\{\frac{364}{101}\right\}$. The check is left to the reader.

It should be noted in Example 11 that like terms can be combined whenever they occur upon removing grouping symbols. It is largely a matter of personal preference.

EXAMPLE 12 Solve $1.2(3.5 - y) - 1.7(y - 4) = 2y + 12$. Write the solution to the nearest tenth.

SOLUTION

$1.2(3.5 - y) - 1.7(y - 4) = 2y + 12$	
$4.2 - 1.2y - 1.7y + 6.8 = 2y + 12$	Remove grouping symbols.
$11 - 2.9y = 2y + 12$	Combine like terms.
$-4.9y = 1$	Subtract 11 and $2y$ from each side.
$y = \dfrac{1}{-4.9}$	Divide each side by -4.9.
$y \approx -0.2$	

The solution set is $\{-0.2\}$. A check using -0.2 can be only an approximation, since it is an approximation to the exact solution.

Do You Remember?

Can you match these?

_____ 1. Like terms
_____ 2. Equivalent equations
_____ 3. Conditional equation
_____ 4. Division property
_____ 5. Addition property
_____ 6. Identity
_____ 7. Solution set
_____ 8. Multiplication property
_____ 9. Subtraction property

a) Equations with the same solution set
b) An equation with a solution set that is the set of all permissible real numbers
c) The set of all solutions to an equation
d) If $a = b$, then $a + c = b + c$.
e) If $a = b$, then $a \cdot c = b \cdot c, c \neq 0$.
f) An equation that is true for only specified values of the variable
g) If $a = b$, then $a - c = b - c$.
h) If $a = b$, then $a \div c = b \div c, c \neq 0$.
i) Two terms that are exactly alike except perhaps for their numerical coefficients

Answers: 1. i 2. a 3. f 4. h 5. d 6. b 7. c 8. e 9. g

Exercise Set 2.1

Solve each equation. Check each solution.

1. $x - 3 = 5$
2. $y + 4 = 7$
3. $m + 8 = 3$
4. $p - 7 = 2$
5. $3a - 2 = 7$
6. $4s - 1 = 3$
7. $4y - 9 = -3$
8. $5x + 7 = -4$
9. $3p + 5 = -6$
10. $6a - 4 = -1$
11. $-2s + 1 = 9$
12. $-3m - 2 = -5$
13. $0.2r - 0.6 = 1.2r + 3.4$
14. $0.5q + 1.3 = -2q + 3.8$
15. $1.5k - 2.8 + 1.4k = -0.8 - 0.1k$
16. $6.2y - 3.7 = 4.1y + 1.3 - 2.9y$
17. $11x - 5 - 4x = 7 - x$
18. $2c - 11 - 8c = 5c - 11$
19. $16a - 5 + 7a = 6a - 5$
20. $-s + 3 - 2s = -3s + 3 - s$
21. $7r - 5 + 2r = -6r + 5 + 14r$
22. $-8m + 8 + 9m = 5 + 2m + 3$

Solve and then state whether each equation is a conditional equation or an identity.

23. $x + 4 = 3$
24. $3x + 4 = 2x - 1$
25. $1.93a + 7.28 = 1.93a + 7.28$
26. $2.85m - 3.2 = -3.2 + 2.85m$
27. $8k - 3 = 6k + 3 + k$
28. $7r - 5 = 7r + 5 - r$
29. $3(y - 1) = 3y - 3$
30. $-2(3 - s) = 2s - 6$

Use the distributive property to remove the grouping symbols, then combine like terms and solve.

31. $2(a + 3) - 3(a - 4) = 8$
32. $3t + 4(t - 1) = 5$
33. $-2(x - 5) + 4(2x - 1) = x + 3$
34. $-5(y + 1) + 2(3 - 2y) = y - 9$
35. $2(3x - 4) + 3(2 - x) = -(x - 1)$
36. $3(s - 5) + 4(-2s + 1) = 2(s - 4)$
37. $6 + 3(k - 1) - 2(2k + 4) = 5 - 3(2k + 5)$
38. $6m - (3m - 4) - 2 = 11 - (4m + 1) - 3m$
39. $5r + (6 - 7r) - 5 = 8 - (2 - 5r) + 8r$
40. $t + (t + 2) + 3(t + 4) = 50$
41. $-3\{2 - [p - (3p + 2)]\} = 4[-5 - (-3p + 7)] + 2$
42. $-2\{-5 + [-4 - 3(4 - 2r)]\}$
    $= 7[-2 - (-4r + 3)] - 5$
43. $-\{-6 - 2[3 - (2y - 4)] + 3\}$
    $= -[-5 - 3(4 - y) + 2] + 1$
44. $-\{1 - [2 - 3(4 - x) + 3x] - 5\}$
    $= -[2 + 7(6 - x) - 4x] - 8$

Solve each equation. Check each solution.

45. $\dfrac{x}{2} = 4$
46. $\dfrac{a}{3} = 1$
47. $\dfrac{2m}{3} = -5$
48. $\dfrac{4s}{7} = -2$
49. $\left(\dfrac{0.2}{5}\right)k = \dfrac{0.3}{7}$
50. $\left(\dfrac{0.01}{3}\right)r = \dfrac{0.01}{2}$
51. $-\dfrac{2}{9}y = \dfrac{5}{7}$
52. $-\dfrac{1}{4}b = \dfrac{2}{3}$
53. $4a = \dfrac{3}{2}$
54. $7x = -\dfrac{2}{3}$
55. $-4r = -\dfrac{7}{4}$
56. $-9b = -\dfrac{4}{3}$
57. $-\dfrac{1}{2}r - \dfrac{1}{3} = \dfrac{2}{3}$
58. $\dfrac{5m}{7} + \dfrac{2}{5} = -\dfrac{3}{5}$
59. $\dfrac{7m}{3} - \dfrac{4m}{2} = 5$
60. $\dfrac{1}{3}y + \dfrac{1}{4} = \dfrac{1}{2}y - \dfrac{2}{3}$
61. $\dfrac{3}{5}k - \dfrac{2}{3} = \dfrac{1}{4}k + \dfrac{1}{2}$
62. $\dfrac{1}{7}a - \dfrac{5}{3} = \dfrac{3}{8}a + \dfrac{2}{5}$
63. $\dfrac{2}{3}\left(4x + \dfrac{5}{4}\right) = x - 2$
64. $\dfrac{1}{2}\left(3y + \dfrac{1}{3}\right) = y + 1$
65. $\dfrac{1}{4}\left(3r - \dfrac{1}{2}\right) = \dfrac{1}{3}r + 3$
66. $\dfrac{1}{5}\left(2m - \dfrac{4}{3}\right) = -2m$
67. $\dfrac{z}{7} = \dfrac{3z}{8}$
68. $\dfrac{5t}{2} = \dfrac{t}{3}$
69. $\dfrac{7y - 1}{5} = \dfrac{11y + 1}{2}$
70. $\dfrac{5p - 3}{2} = \dfrac{2p - 1}{3}$
71. $-\dfrac{1}{2}\left(3x - \dfrac{1}{2}\right) = \dfrac{1}{3}(2x + 3)$
72. $-\dfrac{2}{5}(x + 7) = -\dfrac{1}{3}(x - 4)$

Review Exercises

In Exercises 73–90, indicate whether each statement is true or false. If false, then rewrite it as a true statement.

73. $-5 \in N$
74. $\{2, 3\} \subseteq \{2, 3, 4\}$
75. $N \subseteq J$
76. $6 \in \{3, 6, 9\}$

77. If $\{x \mid x + 2 = 4\}$, then $x = \{2\}$.
78. $\{1, 2, 3\} \cap \{2, 3, 4\} = \{1, 2, 3, 4\}$
79. $\{1, 2\} \cup \{2, 3\} = \{2\}$
80. $\{x \mid x \in N, x < 3\} = \{1, 2\}$
81. $|x| = x$ if and only if $x \geq 0$.
82. $\frac{1}{3} = 0.33\overline{3}$
83. $a + 0 = a$ is an illustration of the additive inverse property.
84. $a \cdot b = b \cdot a$ is an illustration of the associative property of multiplication.
85. $x \not> 11$ means that x is greater than 11.
86. $-60 - (-10) = -50$
87. $-(-|-3|) = 3$
88. $-15 + (-5)^2(2) \div (-10) = -40$
89. $(-8)(-5 - 1^3) - (3^2 - 4) = 43$
90. $12(15 - 90 \div 3^2) = 60$

WRITE IN YOUR OWN WORDS

Give an example of each.

91. Equivalent equations
92. Like terms
93. Linear equations
94. Addition property of equality
95. Multiplication property of equality
96. Conditional equation
97. Identity

2.2 From Words to Symbols: An Introduction to Applied Problems

HISTORICAL COMMENT   Diophantus (c. 250 A.D.) was probably the first mathematician (and philosopher) whose interest centered on what we call algebra today. He added much to the mathematical notation of his day. Little is known of his life, but it is said that it can be determined from the solution to the following problem.

Diophantus passed $\frac{1}{6}$ of his life in childhood, $\frac{1}{12}$ in youth, and $\frac{1}{7}$ in bachelorhood. Five years after his marriage, a son was born. This son died four years before Diophantus while only half as old as his father.

OBJECTIVES   In this section we will learn to
1. translate verbal statements to algebraic symbols; and
2. solve applied problems.

To use algebra as a problem-solving tool, we must learn how to translate a verbal statement into one involving symbols. The first step is to learn to look for key words and phrases. The following table indicates many common phrases and their interpretations. Notice that several phrases have exactly the same interpretation.

Common Algebraic Expressions

OPERATION	EXPRESSION IN WORDS	EXPRESSION IN SYMBOLS
Addition	a plus b a increased by b b more than a	$a + b$

OPERATION	EXPRESSION IN WORDS	EXPRESSION IN SYMBOLS
	the sum of a and b a more than b	$b + a$
Subtraction	the difference of a and b b subtracted from a b less than a a diminished by b a decreased by b a minus b b fewer than a a less b	$a - b$
Multiplication	a multiplied by b the product of a and b a times b	$a \cdot b,\ a(b),\ (a)(b),\ a \times b$
Division	the quotient of a and b a divided by b b divided into a	$\dfrac{a}{b}$ or $a \div b$

We now consider a variety of examples that illustrate these same four operations.

VERBAL STATEMENT	ALGEBRAIC EXPRESSION
5 more than a number x	$x + 5$
A number x decreased by 2	$x - 2$
The quotient of 7 and a number x	$7 \div x$ or $\dfrac{7}{x}$
The quotient of two numbers x and y	$x \div y$ or $\dfrac{x}{y}$
The product of two numbers x and y	$x \cdot y$ or xy
Twice a number x, increased by 5	$2x + 5$
Twice the sum of a number x and 5	$2(x + 5)$
8 less than the quotient of a number n and 6	$\dfrac{n}{6} - 8$
The sum of a number N and 8, divided by 12	$\dfrac{N + 8}{12}$
6 more than the quotient of a number n and 15	$\dfrac{n}{15} + 6$

Punctuation is often the key to proper interpretation of verbal statements. For example,

"The product of 4 and x, plus 9" means $4x + 9$

↑ Comma after the x

whereas

"The product of 4, and x plus 9" means $4(x + 9)$

↑ Comma after the 4.

In the table of expressions, many different letters were used to represent the unknown number. When translating from a verbal statement to one using symbols, it's a good idea to choose letters that symbolize what they represent. For example,

d = number of *d*ollars

m = *M*ary's age

t = number of *t*ickets sold

w = *w*idth of a rectangle

More examples of translation follow.

EXAMPLE 1 Change each verbal statement to one using variables.

(a) Seven times Cynthia's age
(b) The value of three identical shirts

SOLUTION (a) Let c = Cynthia's age. Then the quantity we want is

$7c$

(b) Let s = the value of one shirt. Then the required quantity is

$3s$

EXAMPLE 2 Write a symbolic expression to illustrate each of the following.

(a) The sum of two consecutive odd integers
(b) The sum of two consecutive multiples of 5

SOLUTION (a) Odd integers differ by 2, so we let

x = the first odd integer

$x + 2$ = the next consecutive odd integer

The sum is $x + (x + 2) = 2x + 2$.

(b) Consecutive multiples of 5 differ by 5, such as 5, 10, and 15. So we let

x = the first multiple of 5

$x + 5$ = the next consecutive multiple of 5

The sum is $x + (x + 5) = 2x + 5$.

EXAMPLE 3 Find an expression for the number of theater tickets that remain unsold if the theater seats 600 and t tickets have been sold.

SOLUTION $600 - t$

Translation is usually needed to solve **applied problems,** which are problems stated in a verbal format. Many translations involve the equal sign or a symbol of inequality, such as "$<$" or "$>$". Such expressions are called **mathematical**

sentences. Notice in the following examples that the verb "is" signals the point where either an equal sign or an inequality symbol is to be used.

EXAMPLE 4   Translate each of the following English sentences into a mathematical one.

(a) The sum of a number x and 6 is 5.
(b) A student's average of two test grades of x and 82 is between 70 and 80.

SOLUTION   (a) $x + 6 = 5$

(b) $70 < \dfrac{x + 82}{2} < 80$

We are now ready to use algebraic equations to solve problems.

Number Problems

EXAMPLE 5   The sum of two numbers is 37. The larger one is 7 more than twice the smaller. Find the smaller one.

SOLUTION   Let $x =$ the smaller number. Then $2x + 7 =$ the larger number, and

$x + (2x + 7) = 37$
$3x + 7 = 37$ Combine like terms.
$3x = 30$ Subtract 7 from each side.
$x = 10$ Divide each side by 3.

Since the smaller number is 10, the larger one is $2x + 7 = 2(10) + 7 = 20 + 7 = 27$. The larger plus the smaller is $27 + 10 = 37$ and the larger is 7 more than twice the smaller. The solution checks.

EXAMPLE 6   The sum of two consecutive integers is 167. Find the integers.

SOLUTION   Consecutive integers differ by 1, such as 13 and 14, and -6 and -5. So we let

$x =$ the smaller integer
$x + 1 =$ the next consecutive integer

THE SUM OF THE TWO CONSECUTIVE INTEGERS IS 167.
$x + (x + 1) = 167$
$2x + 1 = 167$ Combine like terms.
$2x = 166$ Subtract 1 from each side.
$x = 83$ Divide each side by 2.

Thus $x = 83$ and $x + 1 = 84$. The check of an applied problem must always be done using the words of the original problem. Here the two integers, 83 and 84, are consecutive and their sum is 167, as required. The solution checks.

Geometry Problems

EXAMPLE 7 The sum of the interior angles of a parallelogram (opposite sides parallel and opposite angles equal) is 360°. Find the measure of the four interior angles of the parallelogram shown in Figure 2.1.

SOLUTION The sum of the angles is 360°, so

$x + x + 2x + 2x = 360$ The sum of the angles is 360°.
$6x = 360$ Combine like terms.
$x = 60$ Divide each side by 6.

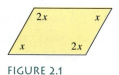

FIGURE 2.1

Therefore the angles are 60° and 120°.

CHECK The sum of the angles is 360°:
$60 + 60 + 120 + 120 = 360$
The solution checks.

EXAMPLE 8 The perimeter of a triangle is 24 feet. The shortest side is 4 feet less than the longest side. The third side is 2 feet shorter than the longest side. The perimeter of a triangle is given by the formula $p = a + b + c$, where a, b, and c are the length of the three sides. See Figure 2.2.

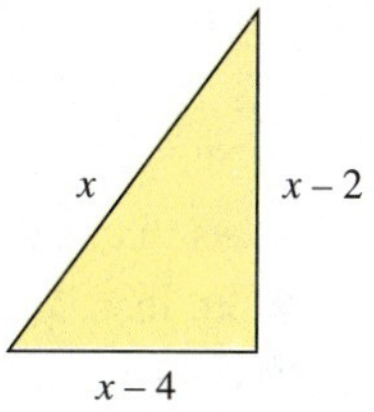

SOLUTION Let

$x =$ the length of the longest side
$x - 4 =$ the length of the shortest side
$x - 2 =$ the length of the third side
$x + (x - 4) + (x - 2) = 24$
$3x - 6 = 24$ Combine like terms.
$3x = 30$ Add 6 to each side.
$x = 10$ Divide each side by 3.

The three sides are $x = 10$ feet, $x - 4 = 6$ feet, and $x - 2 = 8$ feet.

CHECK $10 + 6 + 8 = 24$, so the perimeter is 24. Also, the shortest side, 6 feet, is 4 feet shorter than the longest side, 10 feet, and the third side, 8 feet, is 2 feet shorter, as required.

FIGURE 2.2

Exercise Set 2.2

Name the variable to be used, then translate the following into algebraic expressions.

1. The cost of six hot dogs
2. Wages earned for 5 hours of work
3. The cost of 10 gallons of gas
4. The value of x nickels (in pennies)
5. The value of x dimes (in pennies)
6. The value of x quarters (in pennies)

7. Add a number to 7, then multiply the sum by 3.
8. 5 times the sum of a number and 8
9. 3 times Mario's age
10. 5 less than Gina's age
11. The quotient of a number and 4, less 8
12. The sum of a number and 6, divided by 5
13. The average of a number and 50
14. The average of 34, 56, and a number
15. The average of the test scores 86, 92, and a third score
16. The sum of three consecutive even integers
17. The sum of two consecutive odd integers
18. The sum of two consecutive multiples of 3
19. The sum of two consecutive multiples of 4
20. 25 dollars subtracted from Rico's salary
21. 10 points added to Luis's grade
22. The sum of two integers, if the second integer is 5 less than the first integer
23. The sum of two integers, if the second integer is 4 more than the first integer
24. The sum of two integers, if the second integer is 1 less than twice the first integer

Translate each of the following into an algebraic expression.

25. 26 less than $3h$
26. 7 more than $5m$
27. The quotient of y and 11
28. The quotient of 15 and x
29. 8 more than one-fourth of a
30. 37 less than one-fifth of b
31. 51 subtracted from 6 times r
32. m subtracted from 29

Name the variable to be used and translate each sentence into a mathematical one.

33. The sum of two numbers, if one is 1 more than twice the other, is 35.
34. The sum of two numbers, if one is 6 less than one-third the other, is 12.
35. A number added to 8 is 10.
36. A number less 5 is 6.
37. Eleven more than 3 times a number is 24.
38. Fifteen less than 8 times a number is 41.
39. The sum of three consecutive odd integers is 21.
40. The sum of three consecutive even integers is 30.
41. Thirty less than the sum of three consecutive natural numbers is 18.
42. Seventeen more than the sum of three consecutive whole numbers is 23.

Write a mathematical sentence for each of the following, solve it, and check your answer.

NUMBER PROBLEMS

43. One number is 6 more than another number. The sum of the two numbers is 22. Find the numbers.
44. Five times the sum of a number and 3 is 25. Find the number.
45. The sum of two consecutive multiples of 4 is 28. Find the numbers.
46. The sum of a number and 7, divided by 3, is equal to 5. Find the number.
47. Three times the difference of a number and 6 is 3. Find the number.
48. The sum of two consecutive multiples of 3 is 33. Find the numbers.
49. The sum of two numbers is 22. One number is 1 more than 6 times the other number. Find the numbers.
50. One number is 3 more than twice another number. Thirteen times the smaller number equals 5 times the larger. Find the numbers.

GEOMETRY PROBLEMS

See the formulas in Appendix A as needed.

51. The perimeter of an *equilateral triangle* (all sides equal) is 81 feet. Find the length of a side.

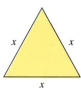

52. The perimeter of a square is 64 meters. Find the length of a side.
53. The sum of the measures of the interior angles of a square is 360°. Find the measure of an angle.

54. The sum of the measures of the interior angles of an equilateral triangle is 180°. Find the measure of an angle. (*Note:* All angles are equal.)
55. The perimeter of a rectangle is 90 centimeters. The length is twice the width. Find the dimensions.
56. The perimeter of a rectangle is 120 feet. The width is one-third of the length. Find the dimensions.
57. The perimeter of a parallelogram (opposite sides equal) is 30 centimeters. A short side is one-half the length of a long side. Find the dimensions.
58. The sum of the measures of an *isosceles triangle* (two sides and two angles equal) is 180°. One of the equal angles has a measure of 45°. Find the measure of the unequal angle.
59. The sum of the measures of the interior angles of a parallelogram (opposite sides and opposite angles equal) is 360°. If one angle has a measure of 80°, find the measure of the adjacent angle.
60. The perimeter of a trapezoid (see the accompanying figure) is 37.5 feet. Find its dimensions.

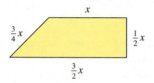

61. The perimeter of a *regular pentagon* (all five sides equal) is 85 meters. Find the length of a side.
62. The sum of the measures of the interior angles of a regular pentagon is 540°. Find the measure of an angle.

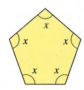

63. The sum of the measures of the interior angles of a trapezoid is 360° (see the accompanying figure). Find the measures of the other two angles.

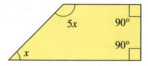

64. The area of a circle (see the figure) is 16π square centimeters. Find the length of the radius x. (Recall that the formula for the area of a circle is $A = \pi x^2$.)

65. The *circumference* (perimeter) of a circle (see the figure) is 8π centimeters. Find the length of the radius x if the circumference is $C = 2\pi x$.

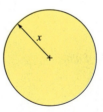

66. One square has a side of length x inches, and a second square has a side of length $2x + 3$ inches. The difference of their perimeters is 16 inches. Find the length of a side of each square.
67. The width of a rectangle is 5 inches less than the length. (See the figure.) The perimeter is 40 inches. Find the dimensions.

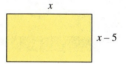

68. The length of a rectangle is twice the width, and the perimeter is 60 feet. Find the dimensions.
69. The perimeter of a triangle is 45 centimeters. The longest side is 6 centimeters longer than the shortest side. The third side is 3 centimeters longer than the shortest side. (See the figure.) Find the dimensions.

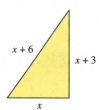

70. The longest side of a triangle is 5 feet more than the shortest side. The third side is 2 feet less than the longest side. If the perimeter is 38 feet, find the dimensions.
71. The perimeter of a parallelogram is 56 meters. Use the given figure to determine the dimensions. *Note:* The opposite sides in a parallelogram are equal in length.

72. The perimeter of a parallelogram is 62 feet. If the shorter side is 13 feet, find the length of the longer side.

REVIEW EXERCISES

Simplify each of the following.

73. $-7(3 - 6) \div 7$

74. $[18 - (15 - 2)] \div 5$

75. $\left(\dfrac{1}{2} + \dfrac{1}{4}\right)\left(\dfrac{1}{4} - \dfrac{1}{2}\right)$

76. $|-26| + |-5|$

77. $4(2x + 3) + 5(x + 4)$

78. $16 - (3 - 8) - 4$

79. $-2[-3 - 2(5 - 8)] \div 2$

80. $\dfrac{-17(3 - 2) + 17}{5 - 1}$

81. $22 + 3 \cdot 6 \div 9$

Solve each of the following.

82. $\dfrac{x}{5} = \dfrac{3}{4}$

83. $\dfrac{3x}{4} + 1 = \dfrac{5}{2}$

84. $3x - 4(x - 1) = 5$

85. $3x - 5 = 2x + 1$

86. $\dfrac{1}{2}(4x - 3) = \dfrac{1}{3}$

87. $7 - 2(3x - 5) = 4(x + 2)$

WRITE IN YOUR OWN WORDS

88. What is the difference between an algebraic *expression* and an algebraic *equation*?

89. What is a parallelogram? What can you say about the interior angles?

90. What is the relationship between a square and a rectangle?

2.3 Literal Equations and Formulas

HISTORICAL COMMENT

Hippocrates (c. 460 B.C.) was very interested in finding a way to determine the area of a circle. He supposed that he would eventually find a regular polygon (a many-sided geometric figures with sides that are straight lines of equal length, such as an equilateral triangle or a square) that had an area that was exactly the same as a circle. Today we know that this is impossible and that the area of a circle is given by the formula $A = \pi r^2$. Although Hippocrates did not know the actual value of π, he did show that such a number had to exist. Finding the actual value of π eluded mathematicians for centuries, and it was not until 1882 that it was shown to be a nonterminating, nonrepeating decimal. We will need an approximation for π to solve many of the problems and examples in this section.

OBJECTIVES

In this section we will learn to
1. solve literal equations for a specified variable; and
2. recognize and use many common formulas.

In subjects such as statistics, physics, and business, it is often useful to be able to solve a given formula for a particular variable. This is done by using the addition, subtraction, multiplication, and division properties of equality that were developed in Section 2.1. For example, consider the common formula

$$d = rt$$
↑   ↑↑
Distance Rate Time

As it stands, the formula allows us to consider the distance traveled if the rate (speed) and the time are known. Sometimes, however, we want to solve for one of the other variables, such as *t* in terms of *r* and *d*. We start with the original formula:

$$d = rt$$

$$\frac{d}{r} = \frac{rt}{r} \qquad \text{Division property of equality: divide each side by } r, \text{ since } r \neq 0.$$

$$\frac{d}{r} = t$$

Thus

$$\text{time} = \frac{\text{distance}}{\text{rate}}$$

Equations such as $d = rt$ are called **literal equations.** In addition to the subjects mentioned at the beginning of this section, they are also found in many other fields, such as chemistry, auto mechanics, electricity, the life sciences, and economics. The next examples show how to work with literal equations.

EXAMPLE 1 The volume of a rectangular solid is $V = lwh$. (See Figure 2.3.) Solve for *h*.

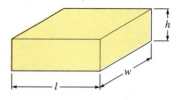

FIGURE 2.3

SOLUTION To solve for *h*, we divide both sides by *lw*.

$$V = lwh$$

$$\frac{V}{lw} = \frac{lwh}{lw}$$

$$\frac{V}{lw} = h$$

The height *h* of the rectangular solid is equal to the volume *V* divided by the product of the length and the width.

EXAMPLE 2 The volume of a cone is

$$V = \frac{1}{3}\pi r^2 h$$

Find a formula for the height of the cone in terms of the radius and the volume. (See Figure 2.4.)

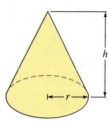

FIGURE 2.4

SOLUTION First we clear the equations of fractions by multiplying both sides by 3:

$$V = \frac{1}{3}\pi r^2 h$$

$$3V = 3\left(\frac{1}{3}\right)\pi r^2 h \qquad \text{Multiply each side by 3.}$$

$$3V = \pi r^2 h$$

Now we divide each side by πr^2 to solve for h:

$$\frac{3V}{\pi r^2} = \frac{\pi r^2 h}{\pi r^2}$$

$$\frac{3V}{\pi r^2} = h$$

The height h of a cone is equal to three times the volume $3V$ divided by π times the radius squared, πr^2.

EXAMPLE 3 The formula

$$C = \frac{5}{9}(F - 32)$$

is used to convert degrees Fahrenheit (°F) to degrees Celsius (°C). Use the properties of equality to find a formula for converting degrees Celsius to degrees Fahrenheit. Use the new formula to convert 100°C to degrees Fahrenheit.

SOLUTION

$$C = \frac{5}{9}(F - 32)$$

$$\frac{9}{5}C = \frac{9}{5} \cdot \frac{5}{9}(F - 32) \qquad \text{Multiply each side by } \frac{9}{5}.$$

$$\frac{9}{5}C = F - 32$$

$$\frac{9}{5}C + 32 = F - 32 + 32 \qquad \text{Add 32 to each side.}$$

$$\frac{9}{5}C + 32 = F$$

or

$$F = \frac{9}{5}C + 32$$

To convert 100°C to °F, we substitute 100 for C in the formula.

$$F = \frac{9}{5}(C) + 32$$

$$F = \frac{9}{5}(100) + 32$$

$$F = 180 + 32$$
$$F = 212$$

This temperature, 100°C or 212°F, is the boiling point of water at sea level.

EXAMPLE 4 A person's intelligence quotient, or IQ, is 100 times the person's mental age, M (how old they are mentally), divided by his or her chronological age, C (how old they are in years). Thus

$$IQ = \frac{100M}{C}$$

Solve for C.

SOLUTION

$$IQ = \frac{100M}{C}$$

$$C(IQ) = C\frac{100M}{C} \quad \text{Multiply each side by } C.$$

$$C(IQ) = 100M$$

$$\frac{C(IQ)}{IQ} = \frac{100M}{IQ} \quad \text{Divide each side by IQ.}$$

$$C = \frac{100M}{IQ}$$

Thus a girl with an IQ of 140 and a mental age of 14 years must be 10 years old. To see this, substitute 140 for IQ and 14 for M in the formula we calculated:

$$C = \frac{100M}{IQ} = \frac{100(14)}{140} = \frac{1400}{140} = 10 \text{ years old}$$

EXAMPLE 5 The formula to determine the area of a trapezoid is

$$A = \frac{h}{2}(B + b)$$

The letters h, B, and b represent the dimensions shown in Figure 2.5. Solve the formula for b.

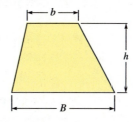

FIGURE 2.5

SOLUTION

$$A = \frac{h}{2}(B+b)$$

$$2A = 2 \cdot \frac{h}{2}(B+b) \quad \text{Multiply each side by 2.}$$

$$2A = h(B+b)$$

$$\frac{2A}{h} = \frac{h(B+b)}{h} \quad \text{Divide each side by } h.$$

$$\frac{2A}{h} = B + b$$

$$\frac{2A}{h} - B = B + b - B \quad \text{Subtract } B \text{ from each side.}$$

$$\frac{2A}{h} - B = b$$

An equivalent form is

$$\frac{2A - Bh}{h} = b$$

which is obtained by combining the terms on the left side of the equation over h, the LCD.

EXAMPLE 6 Clint found that on the average, it took him 50 minutes, or $\frac{5}{6}$ of an hour, to drive the distance of 25 miles to work during the morning rush hour. What was his average speed in miles per hour?

SOLUTION **Average speed** is the distance covered in a given period of time divided by the time it takes to cover that distance. If we divide each side of the distance formula, $d = rt$, by t, we will have

$$r = \frac{d}{t}$$

Since Clint's speed is in miles per hour, his time must be in hours. To determine his average speed, we substitute 25 miles for d and $\frac{5}{6}$ of an hour for t:

$$r = \frac{25}{\frac{5}{6}}$$

$$= 25\left(\frac{6}{5}\right)$$

$$= 30$$

His average speed is 30 miles per hour.

EXAMPLE 7 Solve the formula $A = p(1 + rt)$ for t if $r = 0.08$, $p = 1000$, and $A = 1160$.

SOLUTION   This formula describes how much money (A) an investor will have at the end of t years if $p = \$1000$ is invested at $r = 8\%$ interest. To find t, we begin by replacing each of the other quantities by their given values:

$$A = p(1 + rt)$$
$$1160 = 1000(1 + 0.08t) \quad A = 1160, p = 1000, \text{ and } r = 8\% = 0.08$$
$$1160 = 1000 + 80t \quad \text{Distributive property.}$$
$$160 = 80t \quad \text{Subtract 1000 from each side.}$$
$$2 = t \quad \text{Divide each side by 80.}$$

Therefore $t = 2$ years.

EXAMPLE 8   Solve $\dfrac{1}{t} = \dfrac{1}{a} + \dfrac{1}{b}$ for t.

SOLUTION   First we clear the equation of fractions by multiplying each side by the least common denominator, abt.

$$\frac{1}{t} = \frac{1}{a} + \frac{1}{b}$$
$$abt \cdot \frac{1}{t} = abt \cdot \frac{1}{a} + abt \cdot \frac{1}{b}$$
$$ab = bt + at \quad \text{Simplify each term.}$$
$$ab = (b + a)t \quad \text{Distributive property.}$$
$$\frac{ab}{b + a} = t \quad \text{Divide each side by } b + a.$$

EXAMPLE 9   The formula for finding the volume of a sphere with radius r is

$$V = \frac{4}{3}\pi r^3$$

Use a calculator to find the volume of the sphere shown in Figure 2.6, with radius 3.5 feet.

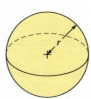

FIGURE 2.6

SOLUTION   We let $\pi = 3.14$ for an approximation. Since we are computing volume, the answer will be in *cubic* feet.

$$V = \frac{4}{3}\pi r^3$$

$$= \frac{4}{3}(3.14)(3.5)^3$$

$$= \frac{4}{3}(3.14)(42.875) \quad (3.5)^3 \text{ means } (3.5)(3.5)(3.5) = 42.875$$

$$V = 179.5 \text{ cubic feet} \quad \text{(rounded to the nearest tenth of a foot)}$$

The next exercise set contains many formulas that will be helpful to you in the remainder of the text. Make a mental note of their location so that you can refer to them when necessary.

Exercise Set 2.3

Solve for the specified variable in each literal equation and simplify.

1. $I = prt$ for r (interest = principal · rate · time)
2. $ax + b = c$ for a (a linear equation in one variable, x)
3. $C = 2\pi r$ for r (the circumference of a circle = 2π · radius)
4. $P = 2(l + w)$ for l (the perimeter of a rectangle = 2[length + width])
5. $ax + by = c$ for y (a linear equation in *two* variables, x and y)
6. $s = r\theta$ for θ (arc length = radius · measure of the angle in radians)
7. $S = 2(lw + lh + wh)$ for w (the surface of a rectangular box of length l, width w, and height h)
8. $y = mx + b$ for m (the slope–intercept form of the equation of a line)
9. $S = 2\pi rh$ for h (the lateral surface area of a right circular cylinder = 2π · radius of base · height)
10. $A = \frac{1}{2} sr$ for s (the area of a circular sector
   $= \frac{1}{2}$ arc length · radius)
11. $A = \frac{1}{2} h(B + b)$ for B (the area of a trapezoid
   $= \frac{1}{2}$ height · $(b + B)$
12. $d = \frac{C}{\pi}$ for π (the diameter of a circle of circumference C)
13. $A = \frac{1}{2} bh$ for b (the area of a triangle
   $= \frac{1}{2}$ · base · height)
14. $d = \frac{1}{2} gt^2$ for g (the distance a body has fallen at time t under the influence of gravity)
15. $A = \frac{1}{2} bh$ for h (the area of a triangle)
16. $E = mc^2$ for m (energy = mass · speed of light squared)
17. $A = p(1 + rt)$ for t (the amount of money A if p dollars are invested at a rate r for t years)
18. $h = vt - 16t^2$ for v (the height of a projectile with velocity v at time t)
19. $V = \pi r^2 h$ for h (the volume of a right circular cylinder of radius r and height h)
20. $\frac{1}{f} = \frac{1}{a} + \frac{1}{b}$ for f (the focal length of a lens)
21. $\frac{P_1 V_1}{T_1} = \frac{P_2 V_2}{T_2}$ for T_1 (Boyle's gas law relating the pressure, volume, and temperature of two gasses)
22. $E = IR$ for I (Ohm's law for electrical circuits)
23. $C = \frac{100B}{L}$ for L (anthropology: the cephalic index)
24. $F = \frac{gm_1 m_2}{d^2}$ for g (the force of the gravitational pull between two masses, m_1 and m_2)
25. $\frac{1}{R} = \frac{1}{R_1} + \frac{1}{R_2}$ for R (the resistance for parallel electrical circuits with resistance R_1 and R_2)
26. $\frac{12ds}{w} = CD$ for w (the cutting speed of a saw)

27. $D = \frac{11}{5}(P - 15)$ for P (pressure P is related to the depth D in feet below the surface of the ocean)

28. $H = \frac{62.4NS}{33,000}$ for N (the horsepower H generated by N cubic feet of water flowing over a dam S feet high)

29. $M = \frac{f_o}{f_e}$ for f_e (the magnifying power of a telescope)

30. $K = \frac{1}{2}mv^2$ for m (kinetic energy of a particle of mass m moving at velocity v)

31. $F = \frac{9}{5}C + 32$ for C (temperature conversion)

32. $A = 2\pi r^2 + 2\pi rh$ for h (the surface area of a right circular cylinder of height h and with base of radius r)

33. $A = S(1 - DN)$ for N (the proceeds of a discounted loan)

34. $y - y_1 = m(x - x_1)$ for x (the point–slope form of the equation of a line)

35. $y^2 = 2px$ for p (the equation of a parabola)

36. $A = \frac{1}{2}nal$ for n (the lateral surface area of a regular pyramid with slant height l and base with n sides of length a)

37. $V = \frac{1}{3}\pi h^2(3r - h)$ for r (the volume of a spherical segment of radius r)

38. $V = \frac{1}{3}\pi r^2 h$ for r^2 (the volume of a right circular cone of radius r and height h)

39. $V = \frac{1}{6}H(S_0 + 4S_1 + S_2)$ for S_2 (prismoidal formula)

40. $A = \frac{1}{3}h(y_0 + 4y_1 + y_2)$ for h (Simpson's rule)

Solve as indicated.

41. The area of a circle of radius r is calculated by $A = \pi r^2$; the circumference by $C = 2\pi r$. Solve for the area A in terms of the circumference C.

42. The area of a circle is calculated by $A = \pi r^2$ and the radius by $r = \frac{1}{2}d$. Solve for the area A in terms of the diameter d.

43. The area of a square of side s is calculated by $A = s^2$; the perimeter by $P = 4s$. Solve for the area A in terms of the perimeter P.

44. The area of a rectangle of length l and width w is calculated by $A = lw$; the perimeter by $P = 2l + 2w$. Solve for the area A in terms of the perimeter P and the width w.

Evaluate each of the following by substituting the given values into the equation, then solve for the remaining variable.

45. $C = 2\pi r$; $C = 4\pi$

46. $A = \frac{1}{2}bh$; $A = 8$ and $b = 3$

47. $V = \pi r^2 h$; $V = 64\pi$ and $r = 2$

48. $I = prt$; $I = 240$, $p = 1000$, and $t = 2$

49. $V = lwh$; $V = 800$, $l = 20$, and $h = 10$

50. $F = \frac{9}{5}C + 32$; $F = 63°$

51. $A = \frac{1}{2}h(B + b)$; $A = 100$, $h = 2$, and $B = 5$

52. $I = \frac{PN}{RN + A}$; $P = 400$, $N = 10$, $R = 20$, and $A = 5$

53. $\frac{1}{R} = \frac{1}{R_1} + \frac{1}{R_2}$; $R_1 = 40$ and $R_2 = 20$

54. $a_n = a_1 + (n - 1)d$; $d = 2$, $n = 15$, and $a_n = 60$

55. $\frac{1}{f} = \frac{1}{a} + \frac{1}{b}$; $a = 20$ and $b = 60$

56. $s = \frac{a}{1 - r}$; $a = 10$ and $s = 20$

57. $y = mx + b$; $y = 15$, $b = 1$, and $m = 7$

58. $h = vt - 16t^2$; $t = 1$ and $v = 64$

59. $S = 2(lw + lh + wh)$; $S = 128$, $l = 3$, and $w = 5$

60. $\frac{1}{f} = \frac{1}{a} + \frac{1}{b}$; $f = 4$ and $b = \frac{1}{16}$

Use a calculator to evaluate each of the following. Find an appropriate formula from Exercises 1–40. Round your answers to one decimal place, if necessary.

61. Find the temperature in degrees Fahrenheit (°F) if the temperature is 48.6°C.

62. Find the force F of the gravitational pull between two masses of 28.3 and 51.9 if the distance between them is 6.75 and G is 62.4.

63. Find the pressure P on a person at a depth D of 111.5 feet below the surface of the ocean.

64. Find the amount A of money accumulated if $1000 is invested at 13% for two and one-half years.

65. Find the height h of a projectile if its velocity v is 60 feet/second at time $t = 2$ seconds after launching.

66. Find the number n of sides of a regular pyramid with a lateral surface area A of 31.11 cm² if the length a of one side of its base is 6.1 cm and the slant height l is 1.7 cm.
67. Find the volume V of a right circular cone with a height h of 4 feet and a radius r of 3 feet. Use $\pi = 3.14$.

REVIEW EXERCISES

Translate each into a mathematical sentence and solve.

68. Find two consecutive integers with a sum of 107.
69. Find two consecutive odd natural numbers with a sum of 136.
70. One number is 6 more than another, and their sum is 108. Find the numbers.
71. One integer is 5 more than another, and their sum is -13. Find the integers.
72. The length of a rectangle is twice the width. Find the dimensions if the perimeter is 120 meters. (See the figure.)

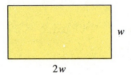

73. The perimeter of a rectangular pasture is 80 kilometers. If the width is 10 kilometers less than the length, find the dimensions.
74. The cost of mailing a package by first-class mail is given by $C = 0.29 + 0.23(x - 1)$, where C is the cost in dollars and x is the number of ounces. If a package costs $1.21 to mail, what does it weigh?
75. The cost of mailing a package by second-class mail is given by $C = 0.22 + 0.20(w - 1)$, where C is the cost in dollars and w is the weight in ounces. If a package costs $3.62 to mail, what does it weigh?

WRITE IN YOUR OWN WORDS

76. What is the literal equation?
77. What is a formula?
78. How would you find the total cost T of a new car if the car costs C dollars, the tax is 6% of the cost C, and the license is L dollars?
79. How you would find the total cost C that Donna paid if she had 18 square yards of linoleum installed at a cost of D dollars per square yard plus a sales tax of $7\frac{1}{4}\%$ of the total cost?

2.4 Solving Linear Equations Involving Absolute Value

HISTORICAL COMMENT The equal sign is said to be the invention of Robert Recorde (1510–1575), since it first appeared in an English text in 1557. He said that he would use twin parallel lines to represent two equal quantities because no two things could be more equal. Despite the simplicity of this symbol, it was not readily accepted and did not appear again in print until 1618. Some of the other early symbols for equality and the year when they appeared were:

| 1559 | [| 1575 | ‖ | | |
| 1634 | ⌒ | 1637 | ∞ | 1680 | ⌐ |

OBJECTIVE In this section we will learn to solve equations involving the absolute value.

In Chapter 1 the *absolute value of a number* was defined to be its distance from the origin on the number line. The solution to an equation such as

$$|x| = 9$$

is found by locating the points on the number line that are 9 units from the origin. The only such numbers are −9 and 9. (See Figure 2.7.)

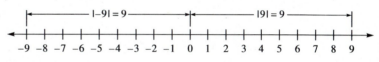

FIGURE 2.7

The solution set therefore is {−9, 9}.

The equation $|x| = 9$ can also be solved by using the more formal definition from Section 1.2.

$|a| = a$, if $a \geq 0$ or $|a| = -a$, if $a < 0$

When the latter definition is used, $|x| = 9$ becomes

$|x| = x = 9$, if $x \geq 0$
$\quad x = 9$

or

$|x| = -x = 9$, if $x < 0$
$\quad x = -9$ Multiply each side by −1.

The solution set is {−9, 9}, which is the same as we found by the distance formula.

Solutions to absolute value equations are found by using Theorem 2.5, as follows.

THEOREM 2.5

If p is a positive real number, then

$|ax + b| = p$

is equivalent to

$ax + b = p$ or $ax + b = -p$

Here are some examples of how the theorem works.

EXAMPLE 1 Solve the equation $|y + 2| = 6$.

SOLUTION Since 6 is a positive number, $|y + 2| = 6$ means

$y + 2 = 6$ or $y + 2 = -6$ $\quad|a| = p$ means $a = p$ or $a = -p$.
$y = 4$ or $y = -8$ $\quad$Subtract 2 from each side.

CHECK $|4 + 2| \stackrel{?}{=} 6$ or $|-8 + 2| \stackrel{?}{=} 6$ $\quad$Substitute 4 or −8 for y.
$\quad\quad\quad |6| \stackrel{?}{=} 6$ or $|-6| \stackrel{?}{=} 6$
$\quad\quad\quad 6 \stackrel{\checkmark}{=} 6$ $\quad\quad\quad\quad 6 \stackrel{\checkmark}{=} 6$

The solution set is {−8, 4}.

EXAMPLE 2 Solve $|2x + 4| = 9$.

SOLUTION

$2x + 4 = 9 \quad$ or $\quad 2x + 4 = -9 \quad\quad$ $|a| = p$ means $a = p$ or $a = -p$.

$2x = 5 \quad$ or $\quad 2x = -13$

$x = \dfrac{5}{2} \quad$ or $\quad x = \dfrac{-13}{2} = -\dfrac{13}{2}$

CHECK

$\left|2\left(\dfrac{5}{2}\right) + 4\right| \overset{?}{=} 9 \quad$ or $\quad \left|2\left(\dfrac{-13}{2}\right) + 4\right| \overset{?}{=} 9 \quad\quad$ Substitute $\dfrac{5}{2}$ or $-\dfrac{13}{2}$ for x.

$|5 + 4| \overset{?}{=} 9 \quad$ or $\quad |-13 + 4| \overset{?}{=} 9$

$|9| \overset{?}{=} 9 \quad$ or $\quad |-9| \overset{?}{=} 9$

$9 \overset{\checkmark}{=} 9 \quad\quad\quad\quad\quad 9 \overset{\checkmark}{=} 9$

The solution is $\{\frac{5}{2}, -\frac{13}{2}\}$.

EXAMPLE 3 Solve $|x + 3| + 2 = 8$.

SOLUTION First we isolate $|x + 3|$ by subtracting 2 from each side of the equation:

$|x + 3| + 2 - 2 = 8 - 2$

$|x + 3| = 6$

> **CAUTION**
> To solve an absolute value equation, the term involving the absolute value should be isolated on one side of the equation if possible.

Now we have

$x + 3 = 6 \quad$ or $\quad x + 3 = -6$

$x = 3 \quad$ or $\quad x = -9 \quad\quad$ Subtract 3 from each side.

CHECK

$|x + 3| + 2 = 8 \quad$ or $\quad |x + 3| + 2 = 8$

$|3 + 3| + 2 \overset{?}{=} 8 \quad$ or $\quad |-9 + 3| + 2 \overset{?}{=} 8 \quad\quad$ Substitute 3 or -9 for x.

$|6| + 2 \overset{?}{=} 8 \quad$ or $\quad |-6| + 2 \overset{?}{=} 8$

$6 + 2 \overset{?}{=} 8 \quad$ or $\quad 6 + 2 \overset{?}{=} 8$

$8 \overset{\checkmark}{=} 8 \quad\quad\quad\quad\quad 8 \overset{\checkmark}{=} 8$

The solution set is $\{3, -9\}$.

EXAMPLE 4 Solve the equation $|x + 6| = -2$.

SOLUTION The absolute value of a number can *never* be negative, so this equation has no solution. If we were to ignore this fact and try to solve it, then we would get

$x + 6 = -2 \quad$ or $\quad x + 6 = -(-2)$

$x = -8 \quad$ or $\quad x + 6 = 2$

$x = -4$

CHECK $|-8 + 6| \stackrel{?}{=} -2$ or $|-4 + 6| \stackrel{?}{=} -2$ Substitute -8 or -4 for x.
$|-2| \stackrel{?}{=} -2$ or $|2| \stackrel{?}{=} -2$
$2 \neq -2$ $2 \neq -2$

Neither solution checks. The solution set is the empty set, $\emptyset$.

When both sides of an equation involve an absolute value, such as $|a| = |b|$, we use the fact that the two quantities must be the same distance from the origin. If they are the same distance from the origin they must either be the same number or be additive inverses of each other. Thus,

$$|a| = |b| \quad \text{means that} \quad a = b \text{ or } a = -b$$

EXAMPLE 5 Solve $|2y + 3| = |y + 5|$.

SOLUTION

$2y + 3 = y + 5$ or $2y + 3 = -(y + 5)$
$2y - y = 5 - 3$ or $2y + 3 = -y - 5$
$y = 2$ or $2y + y = -5 - 3$
$3y = -8$
$y = \dfrac{-8}{3}$

CHECK $|2 \cdot 2 + 3| \stackrel{?}{=} |2 + 5|$ or $\left|2\left(\dfrac{-8}{3}\right) + 3\right| \stackrel{?}{=} \left|\dfrac{-8}{3} + 5\right|$

$|4 + 3| \stackrel{?}{=} |2 + 5|$ or $\left|\dfrac{-16}{3} + \dfrac{9}{3}\right| \stackrel{?}{=} \left|\dfrac{-8}{3} + \dfrac{15}{3}\right|$

$|7| \stackrel{?}{=} |7|$ $\left|\dfrac{-7}{3}\right| \stackrel{?}{=} \left|\dfrac{7}{3}\right|$

$7 \stackrel{\checkmark}{=} 7$ $\dfrac{7}{3} \stackrel{\checkmark}{=} \dfrac{7}{3}$

The solution set is $\left\{-\dfrac{8}{3}, 2\right\}$.

EXAMPLE 6 Solve $-|2a + 7| = -6$.

SOLUTION First we multiply each side of the equation by -1 to eliminate the negative signs.

$(-1)(-|2a + 7|) = (-1)(-6)$
$|2a + 7| = 6$

The equation is now similar to the one in Example 2 and can be solved in the same manner.

EXAMPLE 7 Solve $|2x - 3| = x - 4$.

SOLUTION Since an absolute value is always positive or zero, $x - 4$ must be greater than or equal to zero. If $x - 4 \geq 0$, then $x \geq 4$. Using the definition of absolute value, we have

$$2x - 3 = x - 4 \quad \text{or} \quad 2x - 3 = -(x - 4)$$
$$x = -1 \quad \text{or} \quad 2x - 3 = -x + 4$$
$$3x = 7$$
$$x = \frac{7}{3}$$

Since neither -1 nor $\frac{7}{3}$ is greater than 4, the equation has no solution. In other words, the solution set is the empty set, $\varnothing$.

EXAMPLE 8 Determine whether $|-6x| = 6|x|$ is true for $x \in \{-5, 5\}$.

SOLUTION

For $x = -5$:
$|-6(-5)| = 6|-5|$
$|30| = 6(5)$
$30 = 30$ True.

For $x = 5$:
$|-6(5)| = 6|5|$
$|-30| = 6(5)$
$30 = 30$ True.

EXAMPLE 9 Use an absolute value equation to describe the set of numbers that are six units from the origin.

SOLUTION The numbers that are six units from the origin are 6 and -6. Since $|x| = 6$ is equivalent to saying $x = 6$ or $x = -6$, the equation is $|x| = 6$.

EXAMPLE 10 If 10 is subtracted from twice a number, the absolute value of the difference is 4. Find the number.

SOLUTION Let x represent the number. Then $|2x - 10| = 4$. Thus

$$2x - 10 = 4 \quad \text{or} \quad 2x - 10 = -4$$

The reader should verify in the words of the problem that the number is 3 or 7. Therefore the solution set is $\{3, 7\}$.

Exercise Set 2.4

Solve each of the following equations. If no solution exists, say so. Check your answers.

1. $|x| = 4$
2. $|y| = 25$
3. $|p| = -1$
4. $|k| = -9$
5. $|t| = 169$
6. $|m| = 121$
7. $-|2x| = 16$
8. $-|3p| = 81$
9. $|x| = \frac{7}{36}$
10. $|y| = \frac{5}{9}$

11. $|r| = \dfrac{1}{4}$
12. $|n| = \dfrac{3}{25}$
13. $|t + 2| = 9$
14. $|m - 1| = 3$
15. $|k + 5| = -5$
16. $|2r + 3| = -11$
17. $-|3 - y| = -5$
18. $-|2x - 4| = -3$
19. $|m - 3| + 2 = 5$
20. $|4 + p| - 3 = 7$
21. $|3 + 6t| = 0$
22. $|k - 3| = 0$
23. $-|5 - 3x| = 1$
24. $-|2 - 5y| = 2$
25. $\left|\dfrac{r - 2}{3}\right| = 4$
26. $\left|\dfrac{p + 5}{7}\right| = 1$
27. $\left|\dfrac{2x - 3}{5}\right| = 6$
28. $\left|\dfrac{6m - 7}{8}\right| = \dfrac{3}{8}$
29. $\left|\dfrac{5k - 3}{7}\right| = \dfrac{5}{7}$
30. $\left|\dfrac{11p - 9}{26}\right| = \dfrac{3}{13}$
31. $\left|\dfrac{2y + 9}{18}\right| = \dfrac{5}{6}$

Solve each of the following equations. If no solution exists, say so.

32. $|x + 1| = |2x - 3|$
33. $|p - 3| = |3p|$
34. $|2t + 3| = 2|t - 1|$
35. $|3m + 4| = 3|m + 1|$
36. $|r + 3| = |r + 3|$
37. $|2y - 3| = |2y - 3|$
38. $|x - 1| = x$
39. $|k + 3| = 3k$
40. $|m| = -m$
41. $|p| = -p$
42. $|2x + 3| = 8 - x$
43. $|6a - 1| = |9 - 3a|$
44. $|3x - 2| = 4 + x$
45. $|2 + 3y| = -(7 - 2y)$
46. $|2k + 7| = |7 + 2k|$
47. $|p - 4| = p - 4$
48. $|5r - 1| = |1 - 5r|$
49. $|3y - 5| = |5 - 3y|$
50. $|s - 3| = |2s + 3|$
51. $|2x - 1| = |3x + 2|$
52. $|5x - 8| = -(8 - 5x)$
53. $|4a - 3| = -|3 - 4a|$
54. $|x + 3| = |x + 2|$
55. $|y - 5| = |6 - y|$
56. $|2k + 5| = |6 - 2k|$
57. $|2x - 1| = |5 - 3x|$

Determine whether each equation is true for the given value of the variable.

58. $|5x| = 5|x|$; $x = -4$
59. $|ab| = ab$; $a = -3$, $b = 2$
60. $\left|\dfrac{x}{y}\right| = \dfrac{x}{y}$; $x = -8$, $y = 3$
61. $\left|\dfrac{k}{10}\right| = \dfrac{|k|}{10}$; $k = 0$
62. $|-6x| = 6|x|$; $x = 7$
63. $|(-a)(-b)| = |ab|$; $a = 1$, $b = -9$
64. $\left|\dfrac{x}{y}\right| = \dfrac{|x|}{|y|}$; $x = 0$, $y = -6$
65. $|xy| = |x| \cdot |y|$; $x = 0$, $y = -11$

Use a calculator to solve each of the following equations. Round your answers to one decimal place.

66. $|-6.85x| = 3.27$
67. $|3.01k| = 9.03$
68. $|2.7y - 5.8| = 11.64$
69. $|3.91p + 2.01| = 1.95$
70. $|0.08x - 1.89| = |6.01x + 3.25|$
71. $|11.32 - 9.63y| = |2.79y - 5.26|$

Rewrite each of the following sentences as an absolute value equation and solve it. If no solution exists, say so.

72. The absolute value of a number is 1.
73. The absolute value of a number is 8.
74. If 6 is subtracted from a number, the absolute value of the result is 10.
75. If 3 is subtracted from twice a number and this quantity is divided by 4, the absolute value of the result is 0.
76. If 5 is added to one-third of a number, the absolute value of the result is -8.
77. If 1 is added to three-fourths of a number, the absolute value of the result is -10.

Use an absolute value equation to describe each of the following sets.

78. The numbers that are 2 units from the origin.
79. The numbers that are 4 units from the origin.
80. The numbers that are 6 units from zero.
81. The numbers that are 3 units from zero.

REVIEW EXERCISES

Solve for the indicated variable.

82. $y - y_1 = m(x - x_1)$ for y
83. $\dfrac{y - y_1}{x - x_1} = \dfrac{y_2 - y_1}{x_2 - x_1}$ for x
84. $V = \dfrac{4}{3}\pi r^3$ for r^3
85. $S = 2\pi r^2 + 2\pi rh$ for h
86. $S = 2(lw + lh + wh)$ for l

Write In Your Own Words

87. What is meant by the statement $|x| = p$, $p \geq 0$?
88. When is $|ax + b| = -|cx + d|$ a false statement? When is it a true statement?
89. Why it is possible for $|x|$ to be equal to $-x$?
90. In Chapter 1, we described $|a| = p$, $p \geq 0$, to be the set of numbers p units from the origin. Describe $|x + c| = p$, $p \geq 0$, in a similar manner.

2.5 Solving Linear Inequalities

HISTORICAL COMMENT The symbols for inequality, "<" and ">," date back to the early 1600s and were known to be used by an English mathematician, Thomas Harriot (1560–1621). Another English mathematician, the inventor of the slide rule, William Oughtred (1574–1660), used the symbol "⊏" to mean "is greater than" and the symbol "⊐" to mean "is less than." These symbols were known to be still in use more than a century later at Harvard.

OBJECTIVES In this section we will learn to
1. solve linear inequalities; and
2. use interval notation.

DEFINITION

Linear Inequality

A **linear inequality in one variable** is one that can be written in the form

$$ax + b < c$$

where a, b, and c, are real numbers and $a \neq 0$.

The expressions $ax + b > c$, $ax + b \leq c$, and $ax + b \geq c$ are also examples of linear inequalities. Throughout this section all rules and theorems will be stated in terms of $<$, but they are also true for the other inequality symbols, $>$, $\geq$, and $\leq$.

Linear inequalities are solved in a manner similar to that used to solve linear equations. That is, a series of simpler *equivalent inequalities* are generated until the solution set can be determined by inspection. **Equivalent inequalities** are inequalities that have the same solution set.

THEOREM 2.6

The Addition and Subtraction Properties of Inequality

For any real numbers a, b, and c:

If $a < b$, then $a + c < b + c$, by the **addition property**.
If $a < b$, then $a - c < b - c$, by the **subtraction property**.

It is not difficult to illustrate the truth of these properties. Consider the following examples.

(a) $\quad -2 < 3$
$\quad -2 + 2 < 3 + 2 \quad$ Add 2 to each side.
$\quad 0 < 5 \quad$ True.

(b) $\quad -2 < 3$
$\quad -2 - 2 < 3 - 2 \quad$ Subtract 2 from each side.
$\quad -4 < 1 \quad$ True.

EXAMPLE 1 Solve the inequality $x + 7 < 9$.

SOLUTION
$$x + 7 < 9$$
$$x + 7 - 7 < 9 - 7 \quad \text{Subtract 7 from each side.}$$
$$x < 2$$

It is frequently desirable to be able to show the *graph* of an inequality on a number line. To do this, we must first consider the concept of an **interval.** Suppose that x represents a number that lies between two numbers c and d on a number line, as shown in Figure 2.8. To say that x lies between c and d means that $c < x < d$. This is indicated, in what we call **interval notation,** by (c, d), which is not to be confused with the notation for an ordered pair. An interval that does *not* include its endpoints is called an **open interval.** The parentheses indicate that the endpoints of the interval are not a part of the set. If the endpoints c and d *are* included in the set, as in the inequality $c \leq x \leq d$, the interval is called a **closed interval** and is written $[c, d]$. A **half-open interval** is shown by using both a bracket and a parenthesis. For example, the interval $[c, d)$ includes the endpoint c but excludes the endpoint d. Various types of intervals are shown in the following table. The symbols ∞ (*infinity*) and $-\infty$ (*negative infinity*) indicate that an interval has no upper or lower boundary. These symbols do not represent real numbers. Unless otherwise indicated, the variable x represents a real number.

FIGURE 2.8

INTERVAL NOTATION	SET-BUILDER NOTATION	INTERPRETATION
(c, d)	$\{x \mid c < x < d\}$	Open interval from c to d
$(c, d]$	$\{x \mid c < x \leq d\}$	Half-open interval from c to d, excluding c
$[c, d)$	$\{x \mid c \leq x < d\}$	Half-open interval from c to d, excluding d
$[c, d]$	$\{x \mid c \leq x \leq d\}$	Closed interval from c to d
$[d, \infty)$	$\{x \mid x \geq d\}$	Right half-open interval
(d, ∞)	$\{x \mid x > d\}$	Right open interval
$(-\infty, c]$	$\{x \mid x \leq c\}$	Left half-open interval
$(-\infty, c)$	$\{x \mid x < c\}$	Left open interval
$(-\infty, \infty)$	$\{x \mid x \in R\}$	The real-number line

Returning to Example 1, the solution $x < 2$ in interval notation is $(-\infty, 2)$. The parenthesis at $x = 2$ indicates that 2 is *not* in the solution set. The graph is shown in Figure 2.9. The thick arrow pointing to the left indicates that all real

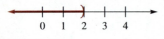

FIGURE 2.9

numbers less than 2 are in the solution set. The symbols ∞ and $-\infty$ will always appear next to a parenthesis (not a bracket) in interval notation, since they do not represent real numbers and so cannot be part of the solution set.

EXAMPLE 2   Graph the set $(2, 6) = \{x \mid 2 < x < 6\}$.

SOLUTION   The graph is shown in Figure 2.10. The heavy line indicates that any number between 2 and 6 lies in the interval and is therefore in the solution set.

FIGURE 2.10

EXAMPLE 3   Graph the set $[2, 6) = \{x \mid 2 \leq x < 6\}$.

SOLUTION   The graph is shown in Figure 2.11. The bracket at 2 shows that 2 *is* in the solution set, while the parenthesis at 6 shows that 6 is *not*.

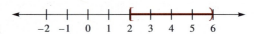

FIGURE 2.11

EXAMPLE 4   Graph $[3, \infty) = \{x \mid x \geq 3\}$.

SOLUTION   The graph is shown in Figure 2.12. The bracket indicates that 3 is a part of the solution set, and the thick arrow pointing to the right indicates that all numbers to the right of 3 are also in the solution set.

FIGURE 2.12

EXAMPLE 5   Convert each of the following sets to interval notation and graph them.

(a) $\{x \mid x < -2\}$ (b) $\{x \mid -4 \leq x \leq 2\}$

SOLUTION   (a) $\{x \mid x < -2\} = (-\infty, -2)$; the graph is shown in Figure 2.13(a).
(b) $\{x \mid -4 \leq x \leq 2\} = [-4, 2]$; the graph is shown in Figure 2.13(b).

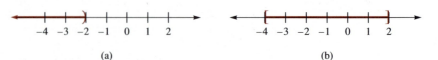

(a)   (b)

FIGURE 2.13

EXAMPLE 6 Graph $(-3, 2] \cap [-1, 5]$.

SOLUTION The graph will include all the points that lie in both intervals. The real numbers from -1 to 2 are in both sets. The graph is the intersection, $[-1, 2]$ shown in Figure 2.14.

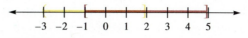

FIGURE 2.14

EXAMPLE 7 Solve and graph the inequality $x - 3 \geq 13$.

SOLUTION
$$x - 3 \geq 13$$
$$x - 3 + 3 \geq 13 + 3 \quad \text{Addition property.}$$
$$x \geq 16$$

The solution set is $[16, \infty)$. The graph is shown in Figure 2.15. The bracket at 16 indicates that 16 is part of the solution set. The arrow indicates that all real numbers greater than 16 are also in the solution set.

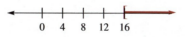

FIGURE 2.15

EXAMPLE 8 Solve $2t + 2(t + 3) \leq 3(t - 1)$, where t is an integer, and graph the solution. Recall that J represents the set of integers.

SOLUTION To solve this inequality, we follow the steps that were appropriate for solving equations.

$$2t + 2(t + 3) \leq 3(t - 1)$$
$$2t + 2t + 6 \leq 3t - 3 \quad \text{Distributive property.}$$
$$4t + 6 \leq 3t - 3 \quad \text{Combine like terms.}$$
$$4t - 3t \leq -3 - 6 \quad \text{Subtraction property.}$$
$$t \leq -9 \quad \text{Combine like terms.}$$

Since t is an integer, the solution set is $\{\ldots, -12, -11, -10, -9\}$, where the three dots indicates that the pattern continues to the left without end. The solution could also have been written $\{t \mid t \leq -9, t \in J\}$. See Figure 2.16.

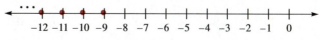

FIGURE 2.16

Inequalities such as

$$\frac{x}{2} \geq 5 \quad \text{and} \quad -3x < 12$$

cannot be solved by use of the addition or subtraction property. Their solution requires the **multiplication** and **division properties of inequality.**

THEOREM 2.7

The Multiplication Property of Inequality

For any nonzero real numbers a, b, and c:

If $a < b$, then $a \cdot c < b \cdot c$, if c is *positive*.
If $a < b$, then $a \cdot c > b \cdot c$, if c is *negative*.

Here are two numerical examples illustrating Theorem 2.7.

(a) $\quad -1 < 3$
$\quad\quad 2(-1) < 2(3) \quad$ Multiply each side by 2.
$\quad\quad -2 < 6 \quad$ True.

(b) $\quad -1 < 3$
$\quad\quad (-2)(-1) > (-2)(3) \quad$ Multiply each side by -2.
$\quad\quad 2 > -6 \quad$ True.

Example (a) illustrates that when each side of an inequality is multiplied by a *positive* number, the direction that the inequality symbol points remains the same. Example (b) illustrates that when each side of an inequality is multiplied by a *negative* number, the direction the inequality symbol points is reversed.

EXAMPLE 9 Solve $\frac{1}{3}x + 2 > 3$ and graph the solution set.

SOLUTION

$\quad\quad \frac{1}{3}x + 2 > 3$

$\quad\quad 3\left(\frac{1}{3}x + 2\right) > 3(3) \quad$ Multiply each side by 3.

$\quad\quad x + 6 > 9 \quad$ Distributive property.

$\quad\quad x > 3 \quad$ Subtract 6 from each side.

Let's pick a value in the proposed solution set, such as 4, and see if it makes the original inequality true:

CHECK $\quad \frac{1}{3}(4) + 2 \stackrel{?}{>} 3$

$\quad\quad\quad\quad \frac{4}{3} + 2 \stackrel{?}{>} 3$

$\quad\quad\quad\quad \frac{10}{3} \stackrel{\checkmark}{>} 3 \quad$ True.

While this method of checking is not a guarantee that the solution is correct, it is reassuring. Really, the only guarantee is to check each step of your work carefully. The solution set is $(3, \infty)$. The graph is shown in Figure 2.17.

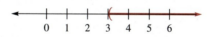

FIGURE 2.17

EXAMPLE 10   Solve $2(x + 5) - 6 < 3(x + 4)$ and graph the solution set.

SOLUTION

$2(x + 5) - 6 < 3(x + 4)$	
$2x + 10 - 6 < 3x + 12$	Distributive property.
$2x + 4 < 3x + 12$	Combine like terms.
$-x < 8$	Subtract $3x$ and 4 from each side.

To solve for x we multiply each side by -1. Remember that the direction of the inequality will change.

$$-1(-x) > -1(8)$$
$$x > -8$$

The solution set is $(-8, \infty)$. The graph is shown in Figure 2.18.

FIGURE 2.18

THEOREM 2.8

The Division Property of Equality

For any nonzero real numbers a, b, and c:

If $a < b$, then $\dfrac{a}{c} < \dfrac{b}{c}$, if c is *positive*.

If $a < b$, then $\dfrac{a}{c} > \dfrac{b}{c}$, if c is *negative*.

The following numerical examples illustrate Theorem 2.8. Recall that the theorem is true whether the inequality symbol is $<$, $>$, $\leq$, or $\geq$.

(a) $6 > -3$

$\dfrac{6}{3} > \dfrac{-3}{3}$	Divide each side by 3.
$2 > -1$	True.

(b) $6 > -3$

$\dfrac{6}{-3} < \dfrac{-3}{-3}$	Divide each side by -3.
$-2 < 1$	True.

Example (a) illustrates that when each side of an inequality is divided by a *positive* number, the direction of the inequality remains the same. Example (b) illustrates that when each side of an inequality is divided by a *negative* number the direction of the inequality is reversed.

EXAMPLE 11 Solve the inequality $3(x + 2) \geq 5(x - 9) + 4$.

SOLUTION

$3(x + 2) \geq 5(x - 9) + 4$

$3x + 6 \geq 5x - 45 + 4$ Distributive property.

$3x + 6 \geq 5x - 41$ Combine like terms.

$3x - 5x \geq -41 - 6$ Subtract $5x$ and 6 from each side.

$-2x \geq -47$

$\dfrac{-2x}{-2} \leq \dfrac{-47}{-2}$ Divide each side by -2, which changes $\geq$ to $\leq$.

$x \leq \dfrac{47}{2}$

The solution set is $\left(-\infty, \dfrac{47}{2}\right]$.

We frequently encounter inequalities in which the value of the variable lies *between* two numbers, such as $-2 \leq x \leq 3$. Such an inequality is called a **compound inequality.** Compound inequalities often occur when two inequalities are joined by the word "and," implying the intersection ($\cap$) of two sets, or the word "or," implying the union ($\cup$) of two sets. In general, we solve a compound inequality by using the properties of inequality to isolate the variable between two numbers.

EXAMPLE 12 Solve and graph $8 < 2x + 4 \leq 16$.

SOLUTION

$8 < 2x + 4 \leq 16$

$8 - 4 < 2x + 4 - 4 \leq 16 - 4$ Subtract 4 from each part.

$4 < 2x \leq 12$ Combine like terms.

$\dfrac{4}{2} < \dfrac{2x}{2} \leq \dfrac{12}{2}$ Divide each part by 2.

$2 < x \leq 6$

The solution set is (2, 6]. The graph is shown in Figure 2.19.

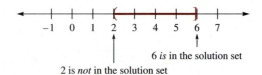

FIGURE 2.19

EXAMPLE 13 Solve and graph $\frac{5}{2} > \frac{-x}{3} + 2 > 1$, where x is an integer.

SOLUTION We begin by multiplying each part by 6, the LCD, to clear the inequality of fractions.

$$\frac{5}{2} > \frac{-x}{3} + 2 > 1$$

$$6\left(\frac{5}{2}\right) > 6\left(\frac{-x}{3} + 2\right) > 6(1) \qquad \text{Multiply } \textit{all parts} \text{ by 6.}$$

$$15 > -2x + 12 > 6 \qquad \text{Distributive property.}$$

$$15 - 12 > -2x + 12 - 12 > 6 - 12 \qquad \text{Subtract 12 from each part.}$$

$$3 > -2x > -6 \qquad \text{Combine like terms.}$$

$$\frac{3}{-2} < \frac{-2x}{-2} < \frac{-6}{-2} \qquad \text{Divide each part by } -2, \text{ changing} > \text{to} <.$$

$$\frac{-3}{2} < x < 3$$

Recall that x can be only an integer. Therefore the solution set is

$$\left\{x \;\middle|\; \frac{-3}{2} < x < 3,\; x \in J\right\} \qquad \text{or} \qquad \{-1, 0, 1, 2\}$$

The graph is shown in Figure 2.20.

FIGURE 2.20

Phrases concerning inequalities occur in many ways in everyday life and their interpretation must be clearly understood. Some of the more common ones are listed in the following table.

Phrase	Interpretation
a exceeds b	$a > b$ or $b < a$
a is no more than b	$a \leq b$
a is at least as much as b	$a \geq b$
a is between c and d	$c < a < d$
a is between c and d, *inclusive*	$c \leq a \leq d$

EXAMPLE 14 The number of male students in an algebra class is 6 more than the number of female students. If there are no more than 44 students in the class, find the maximum number of male and female students.

SOLUTION   We let w = the number of female students, and $w + 6$ = the number of male students. Then the sum of the number of male and female students must be less than or equal to 44:

$$w + (w + 6) \leq 44$$
$$2w + 6 \leq 44 \quad \text{Combine like terms.}$$
$$2w \leq 38 \quad \text{Subtract 6 from each side.}$$
$$w \leq 19 \quad \text{Divide each side by 2.}$$

The number of males is 6 more than the number of females, so the maximum number of females is 19 and the maximum number of males is 25.

EXAMPLE 15   Graph $\{K \mid K > 3\} \cap \{K \mid K \leq 6\}$.

SOLUTION   The solution set for $\{K \mid K > 3\}$ is $(3, \infty)$. The graph is shown in Figure 2.21. The solution set for $\{K \mid K \leq 6\}$ is $(-\infty, 6]$. The graph is shown in Figure 2.22. The intersection of the two sets is $(3, 6]$, shown in Figure 2.23.

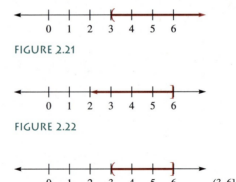

FIGURE 2.21

FIGURE 2.22

FIGURE 2.23

EXAMPLE 16   Solve and graph $\{x \mid x > 3\} \cup \{x \mid x \leq -2\}$.

SOLUTION   Since we are considering the *union* of two sets, any number x that satisfies $x > 3$ or $x \leq -2$ is in the solution set. The graph is shown in Figure 2.24.

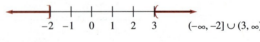

FIGURE 2.24

Do You Remember?

Can you match these?

_____ 1. Equivalent inequalities
_____ 2. Open interval
_____ 3. Subtraction property of inequality
_____ 4. Compound inequality
_____ 5. Division property of inequality
_____ 6. Half-open interval
_____ 7. Symbol for infinity
_____ 8. Addition property of inequality
_____ 9. Multiplication property of inequality
_____ 10. Closed interval

a) $3 < x$ and $x \geq -4$
b) $[a, b]$
c) If $a > b$, then $a + c > b + c$.
d) $(a, b]$
e) ∞
f) If $a < b$, then $a - c < b - c$.
g) If $a < b$, then $a \cdot c < b \cdot c$, if $c > 0$, and $a \cdot c > b \cdot c$, if $c < 0$.
h) (a, b)
i) Inequalities with the same solution set
j) If $a < b$, then $\dfrac{a}{c} < \dfrac{b}{c}$, if $c > 0$, and $\dfrac{a}{c} > \dfrac{b}{c}$, if $c < 0$.

Answers: 1. i 2. h 3. f 4. a 5. j 6. d 7. e 8. c 9. g 10. b

Exercise Set 2.5

Rewrite each of the following sets in interval notation and graph.

1. $\{x \mid x < -7\}$
2. $\{x \mid x < 6\}$
3. $\{x \mid x \geq 0\}$
4. $\{x \mid x \geq -1\}$
5. $\{x \mid x \leq 2\}$
6. $\{x \mid x \leq -5\}$
7. $\{x \mid -9 \leq x \leq 4\}$
8. $\{x \mid 8 \leq x \leq 11\}$
9. $\{x \mid 3 \leq x < 5\}$
10. $\{x \mid -2 < x \leq 4\}$

Solve and graph each of the following inequalities. Write your answers in interval notation. (Recall that J is the set of integers, W is the set of whole numbers, and N is the set of natural numbers.)

11. $x - 3 < 1$
12. $x + 4 > -1$, $x \in J$
13. $x - 5 \leq -1$, $x \in N$
14. $x - 7 \geq -5$
15. $3x \geq 8$
16. $5x \leq -15$
17. $2x - 3 > -4$
18. $6x + 2 < 6$, $x \in N$
19. $7x - 5 \geq 2$, $x \in J$
20. $2x - 7 \leq 3$
21. $\dfrac{3}{5}x \geq -1$
22. $\dfrac{2}{7}x \leq \dfrac{3}{7}$
23. $-6 < 2x \leq 8$
24. $9 > x + 1 \geq 5$, $x \in N$
25. $8 \geq x + 5 \geq 2$, $x \in J$
26. $-2 < 5 - 6x \leq 3$
27. $4 > 2 - x > -1$
28. $\dfrac{3(x + 3)}{5} \geq 5 - x$
29. $\dfrac{2(x - 4)}{7} + 3 \leq x$
30. $\dfrac{6(x + 5)}{7} - 1 \geq \dfrac{-2(3 - x)}{7}$

Solve each of the following inequalities.

31. $-3(2y - 5) + 5y < -4y + 2(7 - y)$
32. $-(5 - 7x) - 3x \geq -8x + 3(1 - x)$, $x \in W$
33. $\dfrac{x}{4} - \dfrac{x}{3} \geq 4 - \dfrac{x - 2}{2}$, $x \in J$
34. $\dfrac{3x}{7} - \dfrac{2x}{7} > 4 + \dfrac{x - 4}{3}$
35. $\dfrac{2}{5}x - \dfrac{1}{3}(x - 4) \leq \dfrac{4}{3}x - \dfrac{1}{5}(x + 2)$, $x \in W$
36. $\dfrac{1}{6}y + \dfrac{1}{2}(3y - 1) \geq \dfrac{1}{3}y + \dfrac{3}{4}(2y - 5)$, $y \in N$

37. $2(6x - 5) - 3(3x + 2) < -(3x - 5) + 2x$

38. $\dfrac{3y - 2}{4} - \dfrac{2y - 1}{3} < \dfrac{y}{12} - \dfrac{1}{4}$

39. $\dfrac{3y + 7}{5} - \dfrac{4y + 1}{3} > \dfrac{y}{6} - \dfrac{1}{5}$

40. $\dfrac{3(2x - 5)}{4} - \dfrac{2(5x - 1)}{3} \leq \dfrac{2x}{3} + \dfrac{1}{12}$

Solve and graph each of the following. Assume all variables represent real numbers.

41. $\{x \mid x > 2\} \cap \{x \mid x < 5\}$
42. $\{y \mid y > -3\} \cap \{y \mid y \leq 5\}$
43. $\{y \mid y \leq 3\} \cup \{y \mid y > 1\}$
44. $\{x \mid x > 3\} \cup \{x \mid x \leq 2\}$
45. $\{x \mid -3 \leq x + 2 < 6\}$
46. $\{y \mid 4 < y + 4 \leq 5\}$

Translate each of the following into an inequality and solve it.

47. The sum of a number and 5 more than the number can be no more than 59.
48. Eight times a number is between -24 and 24.
49. One-third of a number is between -5 and 7.
50. When 2 is added to a number, the result is greater than or equal to 15.
51. One-half of a number subtracted from 5 is at least 11.
52. One-third of a number subtracted from 13 is no more than 12.
53. The difference between twice a number and 6 less than the number is at least 15.
54. The difference between 7 more than a number and one-half of the number is less than 15.
55. Dennis earned 84 and 87 on his first two tests. What is the least he must earn on his third test for his average to be at least 80?
56. Darlene has a grant to help pay her school bill. One stipulation is that her average outside income must be no more than $120 per week. If during the last three weeks she has earned $141, $117, and $123, what is the most she can earn in the fourth week in order to keep her grant?
57. Judy's boutique makes a profit, P, provided her revenue, $R = 25n$, exceeds her cost, $C = 60 + 10n$, where n is the number of units sold. What is the smallest n can be in order to make a profit? (*Note:* $P = R - C$.)
58. Jeff tries to keep his caloric intake at less than 3000 calories per day. If his daily solid food intake contains 2250 calories, what are the most full glasses of juice he can have per day (at 120 calories per glass) and still stay under the desired caloric intake?
59. Nick expects to have an average score on two tests of less than 90 but hopes for at least 80. If his score on the first test is 88, what range of scores will he need to earn on the second test?
60. Tim likes to keep his average golf score between 65 and 75. If his scores on the first two rounds are 72 and 68, what should his range of scores be on the third round?
61. Nell expects to keep an average score on three tests of less than 100 and at least 90. If her scores on the first two tests are 86 and 90, what range of scores does she need to earn on the third test?
62. Marty needs to maintain an average of least 80 in algebra in order to get a B or higher. If he earns 74, 72, 80, and 72 on his first four tests, what is the lowest score he can get on his fifth test? If the maximum possible score on any test is 100, is it even possible for Marty to maintain an average of at least 80?

REVIEW EXERCISES

Solve each of the following equations.

63. $|x - 1| + 5 = 8$
64. $|2x - 3| = |x - 4|$
65. $|3x + 7| + 4 = 11$
66. $|5y - 3| = -(2 - y)$
67. $\left|\dfrac{3 - x}{5}\right| = x - 2$
68. $-|4 - 3x| + 7 = 4$

WRITE IN YOUR OWN WORDS

69. What is an inequality?
70. What is the difference between the symbols "( )" and "[ ]" when they are used in interval notation?
71. What does the expression "equivalent inequalities" mean?
72. Why do we use a parenthesis rather than a bracket next to the infinity symbol (∞) when we write an interval such as $[4, \infty)$?
73. What is the difference between a **closed** interval and an **open** interval?
74. We write $a < x < b$ to indiciate x is between a and b. State what the notation $\{x \mid x > a\} \cap \{x \mid x < b\}$ indicates about x.

2.6 Solving Linear Inequalities Involving Absolute Value

OBJECTIVES

In this section we will learn to
1. solve absolute value inequalities involving $>$ or $\geq$;
2. solve absolute value inequalities involving $<$ or $\leq$; and
3. recognize when an inequality has no solution.

We begin this section with the reminder that the absolute value of a number measures its distance from the origin on a number line. When we state that the absolute value of a number x is 2, that is,

$$|x| = 2$$

we mean that x represents a number that is 2 units from the origin on the number line. There are two numbers that satisfy this condition, 2 and -2. Any number that lies between 2 and -2 would be less than 2 units from the origin and have an absolute value that is less than 2. This leads to the following conclusion:

If $|x| < 2$ then $-2 < x < 2$

For example, -1 lies between -2 and 2 and its absolute value, 1, is less than 2. Similarly, 0.8 lies between -2 and 2 and its absolute value is less than 2.

EXAMPLE 1 Solve and graph $|x| < 9$.

SOLUTION Numbers that are less than 9 units from the origin must lie between -9 and 9. So, since $|x| < 9$,

$$-9 < x < 9$$

The solution set is $\{x \mid -9 < x < 9\}$. In interval notation, this is $(-9, 9)$. The graph is shown in Figure 2.25.

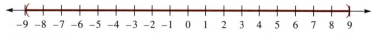

FIGURE 2.25

EXAMPLE 2 Solve and graph the inequality $|a| \leq 3$.

SOLUTION If $|a| \leq 3$, then

$$-3 \leq a \leq 3$$

The solution set is $[-3, 3]$. The graph is shown in Figure 2.26.

FIGURE 2.26

If the absolute value of *x* is *greater* than 2, then *x* is *more than* 2 units from the origin. Numbers that are more than 2 units from the origin are either greater than 2 or less than −2.

EXAMPLE 3 Solve $|x| > 2$ and graph.

SOLUTION If $|x| > 2$, then

$$x < -2 \quad \text{or} \quad x > 2$$

The solution set is $\{x \mid x < -2 \text{ or } x > 2\}$. In interval notation this is written as the union of two intervals: $(-\infty, -2) \cup (2, \infty)$. The graph is shown in Figure 2.27.

FIGURE 2.27

EXAMPLE 4 Solve $|y| \geq 4$ and graph the solution.

SOLUTION The inequality $|y| \geq 4$ means

$$y \leq -4 \quad \text{or} \quad y \geq 4$$

In interval notation, the solution set is $(-\infty, -4] \cup [4, \infty)$. The graph is shown in Figure 2.28.

FIGURE 2.28

The results of the previous examples are summarized in Theorem 2.9.

THEOREM 2.9

For any positive number *p*,

$|x| < p$ is equivalent to $-p < x < p$
$|x| > p$ is equivalent to $x < -p$ or $x > p$

CAUTION

$|x| > p$ cannot be written $p < x < -p$ or $-p > x > p$. Both of these cases imply that a positive number is less than a negative number. It must be written as in Theorem 2.9.

To solve an inequality such as $|a + 7| < 3$, we think of $a + 7$ as the "*x*" in Theorem 2.9. Thus $|a + 7| < 3$ means

$$-3 < a + 7 < 3$$

We solve this by subtracting 7 from each part:

$$-3 - 7 < a + 7 - 7 < 3 - 7$$
$$-10 < a < -7$$

Theorem 2.9 can be generalized as follows.

Theorem 2.10

For any real numbers a, b, and p, where p is positive,

$|ax + b| < p$ means that $-p < ax + b < p$

$|ax + b| > p$ means that $ax + b < -p$ or $ax + b > p$

EXAMPLE 5 Solve $|2y - 3| \leq 5$.

SOLUTION $|2y - 3| \leq 5$ means $-5 \leq 2y - 3 \leq 5$:

$$-5 \leq 2y - 3 \leq 5$$
$$-5 + 3 \leq 2y - 3 + 3 \leq 5 + 3 \qquad \text{Add 3 to each part.}$$
$$-2 \leq 2y \leq 8 \qquad \text{Combine like terms.}$$
$$\frac{-2}{2} \leq \frac{2y}{2} \leq \frac{8}{2} \qquad \text{Divide each side by 2.}$$
$$-1 \leq y \leq 4$$

The solution set is $\{y \mid -1 \leq y \leq 4\}$ or, in interval notation, $[-1, 4]$.

EXAMPLE 6 Solve $|-2x + 3| > 7$.

SOLUTION $|-2x + 3| > 7$ means

$-2x + 3 < -7$	or	$-2x + 3 > 7$	
$-2x < -7 - 3$	or	$-2x > 7 - 3$	Subtract 3 from each side.
$-2x < -10$	or	$-2x > 4$	Combine like terms.
$x > 5$	or	$x < -2$	Divide each side by -2 and reverse the inequality symbol.

The solution set is $(-\infty, -2) \cup (5, \infty)$.

EXAMPLE 7 Solve the inequality $\left|\frac{1}{4}x + 3\right| \leq 2$.

SOLUTION
$$-2 \leq \frac{1}{4}x + 3 \leq 2$$
$$-5 \leq \frac{1}{4}x \leq -1 \qquad \text{Subtract 3 from each part.}$$
$$-20 \leq x \leq -4 \qquad \text{Multiply each part by 4.}$$

The solution set is $\{x \mid -20 \leq x \leq -4\}$, or $[-20, -4]$.

EXAMPLE 8 Solve $|x + 5| \leq 2$, where x is an integer.

SOLUTION   The inequality $|x + 5| \leq 2$ means

$$-2 \leq x + 5 \leq 2$$
$$-7 \leq x \leq -3 \quad \text{Subtract 5 from each part.}$$

Since x must be an integer, the solution set is $\{-7, -6, -5, -4, -3\}$. The solution set can also be written $\{x \mid -7 \leq x \leq -3, x \in J\}$.

EXAMPLE 9   Solve $|7x + 3| + 5 \geq 11$ and graph the solution.

SOLUTION   First we subtract 5 from each side to isolate the absolute value expression:

$$|7x + 3| + 5 \geq 11$$
$$|7x + 3| \geq 11 - 5$$
$$|7x + 3| \geq 6$$
$$7x + 3 \leq -6 \quad \text{or} \quad 7x + 3 \geq 6$$
$$7x \leq -9 \quad \text{or} \quad 7x \geq 3$$
$$x \leq \frac{-9}{7} \quad \text{or} \quad x \geq \frac{3}{7}$$

The graph is shown in Figure 2.29.

FIGURE 2.29

The next two examples illustrate that some inequalities have as a solution set, the set of all real numbers, while others have no solution.

EXAMPLE 10   Solve $|x + 9| > -4$.

SOLUTION   The absolute value of any number is always positive or zero, therefore it is greater than *any* negative number. Thus the solution set is the set of all real numbers, $R = (-\infty, \infty)$.

EXAMPLE 11   Solve $|x + 12| \leq -4$.

SOLUTION   The absolute value of a number is always positive or zero, so for $|x + 12| \leq -4$ to be true, $|x + 12|$ would have to be negative, which is not possible. Therefore the solution set is the empty set, $\emptyset$.

EXAMPLE 12   Translate the following sentence into an absolute value equation and solve it: The absolute value of a number is at most 6.

SOLUTION   "At most 6" means "is less than or equal to 6." So we let x = the number. Then

$$|x| \leq 6 \quad \text{means} \quad -6 \leq x \leq 6$$

The solution set is $\{x \mid -6 \leq x \leq 6\}$, or $[-6, 6]$.

Do You Remember?

Can you match these?

____ 1. $|x| < 4$
____ 2. $|x + 4| < 3$
____ 3. $|x + 2| \geq 4$
____ 4. $|x + 4| < -6$
____ 5. $|x - 2| > 2$

a) All real numbers
b) $x \leq -6$ or $x \geq 2$
c) $-4 < x < 4$
d) The empty set
e) $(-\infty, 0) \cup (4, \infty)$
f) $-7 < x < -1$

Answers: 1. c 2. f 3. b 4. d 5. e

Exercise Set 2.6

Rewrite each of the following inequalities as a compound inequality without absolute value symbols. Do not solve.

1. $|x| < 3$
2. $|y| < 5$
3. $|x| \leq 4$
4. $|y| \leq 0$
5. $|y| > 3$
6. $|x| > 0$
7. $|x - 5| \leq 8$
8. $|3y + 4| \leq 7$
9. $|2x - 1| \geq 4$
10. $|3y - 7| > 3$

Write each of the following inequalities using absolute value symbols. Do not solve.

11. $-2 < x < 2$
12. $-5 \leq x \leq 5$
13. $y < -3$ or $y > 3$
14. $y < -1$ or $y > 1$
15. $-6 \leq x + 1 \leq 6$
16. $-2 < 3y - 2 < 2$
17. $5x - 1 < -3$ or $5x - 1 > 3$
18. $1 - x \leq -6$ or $1 - x \geq 6$
19. $-3 < y < 3$
20. $-6 \leq y \leq 6$
21. $-5 \leq y - 3 \leq 5$
22. $-8 < 2x + 3 < 8$
23. $x + 3 < -5$ or $x + 3 > 5$
24. $2y - 5 < -8$ or $2y - 5 > 8$
25. $6y - 2 < -4$ or $6y - 2 > 4$
26. $3x + 4 < -11$ or $3x + 4 > 11$

Solve and graph each of the following inequalities, if possible. Write your answers in interval notation.

27. $|x| < 2$
28. $|y| < 3$
29. $|x| \geq 5$, $x \in J$
30. $|x| \geq 1$, $x \in J$
31. $|x| \leq 0$
32. $|x| \leq -5$
33. $|x| \leq -3$
34. $|x| \geq -2$
35. $|x| \geq -3$
36. $|x + 2| < 4$
37. $|3y - 2| > 0$, $y \in J$
38. $|2x - 3| \geq 0$, $x \in J$
39. $|2t - 5| \leq 3$
40. $\left|\dfrac{1}{4}t - 1\right| < -5$

41. $\left|\frac{1}{2}x + 1\right| \geq -3$
42. $\left|\frac{1}{2}x + 1\right| \leq \frac{1}{4}$
43. $\left|\frac{1}{4}x - 2\right| < \frac{1}{4}$
44. $|3 - y| \geq -2$
45. $-1 > |6 - x|, x \in W$
46. $1 + |3 - 4x| \leq 3$
47. $6 + |2 - 7x| \geq 7$
48. $0 > |3 - 4x|, x \in W$
49. $-2 \geq |3 - 5x|, x \in W$
50. $6 > |15 - 4y|$
51. $|3x - 8| - 9 \geq -6$
52. $|7y - 5| + 2 \leq 4$
53. $|-3x + 1| + 3 \leq 8$
54. $|-7x - 5| - 1 \geq 3$
55. $|6y + 5| + 4 < 4$
56. $3 + |6y + 5| \geq 2$

Solve each absolute value inequality for x. Assume that a, b, and c are positive real numbers.

57. $|ax + b| \geq c$
58. $|ax + b| \leq c$
59. $\left|\frac{ax}{c} - b\right| > b$
60. $\left|\frac{ax}{c} - b\right| < b$

Translate each of the following sentences into an absolute value inequality and solve it.

61. The absolute value of a natural number is less than 8.
62. The absolute value of a negative integer is at least 1.
63. The absolute value of a number is more than 5.
64. The absolute value of a number is at most 6.
65. The absolute value of a number is less than or equal to 10.
66. The absolute value of a number is greater than or equal to 3.
67. If 5 is subtracted from the absolute value of the sum of a number and 3, the result is at least 0.
68. If −6 is added to the absolute value of the difference of twice a number and 7, the result is no more than negative 2.
69. If $5 is added to twice the price of a pen, the absolute value of the result is no more than $12.
70. If 25 cents is added to the price of a dozen eggs, the absolute value of the result is at least 95 cents.
71. If $6 is subtracted from the price of a shirt, the absolute value of the difference is more than $8.
72. If 10% is deducted from a student's grade, the absolute value of the result is no less than 78%.

REVIEW EXERCISES

Solve each of the following.

73. $|3x - 1| = 4$
74. $|2x + 3| = 9$
75. $|0.6x - 0.11| = 0.05$
76. $|0.8x + 0.1| - 0.4 = 7$
77. $5x + 7 \leq 5$
78. $6x - 9 \geq 4$
79. $11x - 5 > 2$
80. $4x + 1 < 1$
81. $2x + 3 \geq 8$
82. $7x - 6 \leq 1$

WRITE IN YOUR OWN WORDS

83. Why is $|ax + b| < c$ equivalent to the expression $-c < ax + b < c$?
84. Why is multiplying or dividing both sides of an inequality by a negative number correct if and only if the inequality symbol is reversed?
85. In Chapter 1, we defined the equation $|x| = 8$ to be those numbers 8 units from the origin. What do the inequalities $|x| < 8$ and $|x| > 8$ mean?
86. If $|x| > 5$ is described as those numbers more than 5 units from the origin, describe $|x + 1| > 5$.

Show that each of the following is true.

87. $|a + b| \geq |a| + |b|$, if $a = -3$ and $b = -7$.
88. $|a + b| < |a| + |b|$, if $a = -10$ and $b = 41$.

2.7 More Applications

HISTORICAL COMMENT

Applications, often called "verbal" or "word problems," have appeared in mathematical works as far back as mathematical history can be traced. Many types have survived until the present and appear in this text. Compare Example 10 of this section to the problem of a 1788 work that follows.

If during ebb tide, a wherry should set out from Haverhill, to come down the river, and, at the same time, another should set out from Newburyport, to go up the river, allowing the difference to be 18 miles; suppose the current forwards

2.7 · MORE APPLICATIONS   85

one and retards the other $1\frac{1}{2}$ miles per hour; the boats are equally laden, the rowers equally good, and, in the common way of working, in still water, would proceed at the rate of 4 miles per hour; when, in the river will the two boats meet?

OBJECTIVES   In this section we will learn to
1. solve applied problems by using linear equations; and
2. solve applied problems by using linear inequalities.

Applied problems are an essential part of any course in mathematics. As problems become more difficult and varied, they will be easier to understand once we have a formal plan of attack to use in solving them. One such plan follows.

To Solve Applied Problems

Step 1 Read the problem through completely as many times as necessary, until you understand what the problem is about.

Step 2 Reread the problem and determine what it is that you are asked to find.

Step 3 Represent what you are asked to find by a variable or variables. Draw a picture or construct a chart if necessary.

Step 4 Establish the relationship between the variable and other parts of the problem. The relationship may be an equation or an inequality.

Step 5 Solve the equation or inequality and check the solution in the words of the problem.

To make it easier at the beginning, applications will be grouped by type in this section. We begin with problems involving numbers.

Number Problems

EXAMPLE 1   An integer is 1 more than a second integer. Their sum is 99. Find the integers.

SOLUTION   **Step 1** **Read completely.**
An integer is 1 more than a second integer. Their sum is 99. Find the integers.

Step 2 **What are you asked to find?**
An integer is 1 more than a second integer. Their sum is 99. *Find the integers.*

Step 3 **Represent the unknown quantities by a variable or variables.**
Let: $x + 1 =$ the first integer
         $x =$ the second integer

Step 4 **Establish the relationship between the variable and other parts of the problem.**

$$x + (x + 1) = 99$$

Step 5 Solve and check in the words of the problem.

$$x + (x + 1) = 99$$
$$2x + 1 = 99$$
$$2x = 98$$
$$x = 49$$
$$x + 1 = 50$$

The two integers are 49 and 50. The second integer is one more than the first and their sum is 99. The solution checks.

EXAMPLE 2 The sum of three consecutive odd integers is 111. Find the integers.

SOLUTION Consecutive odd (or even) integers differ by 2. Let

x = the smallest odd integer
$x + 2$ = the next consecutive odd integer
$x + 4$ = the largest odd integer

Since their sum is 111, we have

$$x + (x + 2) + (x + 4) = 111$$
$$3x + 6 = 111 \quad \text{Combine like terms.}$$
$$3x = 105 \quad \text{Subtract 6 from each side.}$$
$$x = 35 \quad \text{Divide each side by 3.}$$
$$x + 2 = 37$$
$$x + 4 = 39$$

The numbers are 35, 37, and 39. Their sum is 111. The solution checks.

EXAMPLE 3 A student has test scores of 70 and 83. What score must she make on her next test to have an average of at least 80?

SOLUTION **Step 1** **Read completely.**
A student has test scores of 70 and 83. What score must she make on her next test to have an average of at least 80?
Step 2 **What are you asked to find?**
A student has test scores of 70 and 83. *What score must she make on her next test to have an average of at least 80?*
Step 3 **Represent the unknown quantity by a variable.**
Let s = her next test score.
Step 4 **Establish the relationship.**
An average test score is the sum of the scores divided by the number of scores. We want this average to be at least 80.

$$\frac{70 + 83 + s}{3} \geq 80$$

Step 5 **Solve and check.**

$$\frac{70 + 83 + s}{3} \geq 80$$

$70 + 83 + s \geq 3 \cdot 80$ Multiply each side by 3.

$153 + s \geq 240$

$s \geq 87$ Subtract 153 from each side.

We'll check only the smallest value, 87. If 87 produces an average of 80, then any score *above* 87 will produce an average above 80.

CHECK $\frac{70 + 83 + 87}{3} \stackrel{?}{\geq} 80$

$\frac{240}{3} \stackrel{?}{\geq} 80$

$80 \stackrel{\checkmark}{\geq} 80$

We now turn our attention to problems from geometry. Here are some examples.

Geometry Problems

EXAMPLE 4   A rancher has 2000 yards of fencing to enclose a rectangular pasture. The length of the pasture is 500 yards longer than the width. Find the dimensions of the pasture.

SOLUTION   Let

w = the width

$l = w + 500$ = the length

Since the pasture is rectangular, its *perimeter* (the distance around the outside) is given by the formula $P = 2(w + l)$, or by the distributive property,

$P = 2w + 2l$

$2000 = 2w + 2(w + 500)$ $P = 2000; l = w + 500$

$2000 = 2w + 2w + 1000$ Distributive property.

$2000 = 4w + 1000$ Combine like terms.

$1000 = 4w$ Subtract 1000 from each side.

$250 = w$ Divide each side by 4.

The width w is 250 yards, so

$l = w + 500 = 750$ yards

The check is left to the reader.

EXAMPLE 5   The sum of the interior angles of a triangle is 180°. The largest angle is 20° larger than the smallest. The third angle is 10° larger than the smallest. Find the measure (size) of each angle. See Figure 2.30.

SOLUTION   Let

$$d = \text{the number of degrees in the smallest angle}$$
$$d + 10 = \text{the number of degrees in the next angle}$$
$$d + 20 = \text{the number of degrees in the largest angle}$$

$$d + (d + 10) + (d + 20) = 180$$
$$3d + 30 = 180$$
$$3d = 150$$
$$d = 50$$
$$d + 10 = 60$$
$$d + 20 = 70$$

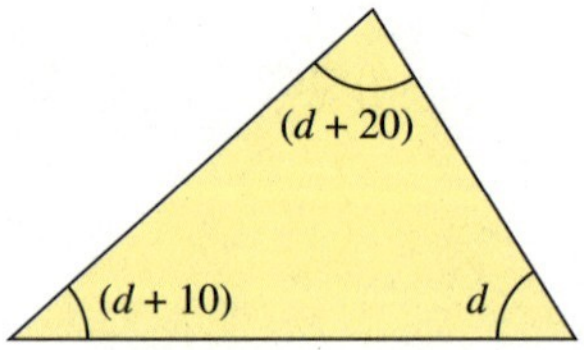

FIGURE 2.30

The measures of the three angles are 50°, 60°, and 70°.

Mixture Problems

The next three examples are **mixture problems,** in which two or more quantities are combined to form a final mixture. *The key to solving a mixture problem is to determine what remains constant throughout the mixing process.*

EXAMPLE 6   The owner of a nut shop receives a new shipment of peanuts and cashews. Rather than sell them separately, she decides to create a mixture to sell for $2.30 a pound. If the peanuts normally sell for $1.50 a pound and the cashews for $3.50 a pound, how many pounds of each should she use if she wants to have 30 pounds of the mixture?

SOLUTION   In this example, the amount of money received by selling the mixture must be the same as the amount that would be received by selling the varieties separately. We begin by letting p represent the number of pounds of peanuts in the mixture. Since the mixture contains both peanuts and cashews, the number of pounds of cashews in the mixture will be the total number of pounds in the mixture minus the number of pounds of peanuts in the mixture. We represent this as $30 - p$. Let's collect the data in a table for ease of reference.

	NUMBER OF POUNDS	PRICE PER POUND	CASH VALUE
Peanuts	p	$1.50	$1.50p$
Cashews	$30 - p$	$3.50	$3.50(30 - p)$
Mixture	30	$2.30	$2.30(30)$

$$\begin{array}{c} \text{CASH VALUE} \\ \text{OF PEANUTS} \end{array} + \begin{array}{c} \text{CASH VALUE} \\ \text{OF CASHEWS} \end{array} = \begin{array}{c} \text{CASH VALUE} \\ \text{OF MIXTURE} \end{array}$$

$$1.50p + 3.50(30 - p) = 2.30(30)$$
$$15p + 35(30 - p) = 23(30)$$

Multiply each side by 10 to eliminate decimals.

$$15p + 1050 - 35p = 690$$
$$-20p = -360$$
$$p = 18$$

The mixture should contain 18 pounds of peanuts and 12 pounds of cashews. The check is left to the reader.

EXAMPLE 7 How many liters of a 20% acid solution must be mixed with 2 liters of a 35% acid solution to make a solution that is 30% acid?

SOLUTION We let x represent the number of liters of the 20% acid solution. When the two solutions are mixed, none of the acid is lost. Only the concentration changes. See Figure 2.31.

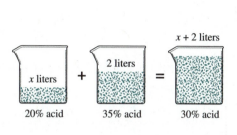

FIGURE 2.31

$$\begin{pmatrix}\text{ACID IN}\\ x \text{ LITERS}\end{pmatrix} + \begin{pmatrix}\text{ACID IN}\\ 2 \text{ LITERS}\end{pmatrix} = \begin{pmatrix}\text{ACID IN THE}\\ \text{MIXTURE}\end{pmatrix}$$

$$0.20x + 0.35(2) = 0.30(x + 2)$$
$$20x + 35(2) = 30(x + 2) \quad \text{Multiply each side by 100.}$$
$$20x + 70 = 30x + 60$$
$$10 = 10x$$
$$1 = x$$

One liter of the 20% acid is required.

CHECK The amount of acid in 1 liter of a 20% solution is $0.20(1) = 0.2$ liters. The amount of acid in 2 liters of a 35% solution is $0.35(2) = 0.7$ liters. The amount of acid in 3 liters of the 30% solution is $0.30(3) = 0.9$ liters. Since $0.2 + 0.7 = 0.9$, the solution checks. No acid was lost in the mixture.

EXAMPLE 8 Ron McGurk had a very successful year as a salesman and received a $20,000 bonus. He invested part of it in a second mortgage paying 12% and the rest in municipal bonds paying 8%. Both of the investments pay simple interest. If the income at the end of one year is $2080, how much did he invest at each rate?

SOLUTION We let x represent the 12% investment. The remainder of the bonus, $20,000 - x$, was invested at 8%. From Exercise Set 2.3, Exercise 1, recall that

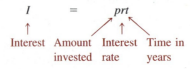

	p	r	t	I
Mortgage	x	12% = 0.12	1	$0.12(x)(1)$
Bonds	$20,000 - x$	8% = 0.08	1	$0.08(20,000 - x)(1)$

The total interest received in one year is $2080.

$$\text{Mortgage Interest} + \text{Bond Interest} = \text{Total Interest}$$
$$0.12x + 0.08(20,000 - x) = 2080$$

We multiply each side by 100 to eliminate decimals:

$$12x + 8(20,000 - x) = 208,000$$
$$12x + 160,000 - 8x = 208,000$$
$$4x + 160,000 = 208,000$$
$$4x = 48,000$$
$$x = 12,000$$
$$20,000 - x = 8000$$

So Ron invested $12,000 at 12% and $8,000 at 8%.

CHECK Interest earned at 12% of $12,000 is $0.12(12,000) = \$1440$
Interest earned at 8% of $8000 is $0.08(8000) = \$640$
The total interest is $1440 + $640 = $2080. The solution checks.

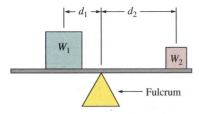

FIGURE 2.32

Fulcrum Problems

In physics, a fulcrum is a point of support on which a lever rests. The seesaw ("teeter-totter") found on playgrounds is an example of a lever resting on a fulcrum. If two weights W_1 and W_2 are placed at distances d_1 and d_2, respectively, from the fulcrum, they will balance each other if $W_1 \cdot d_1 = W_2 \cdot d_2$. See Figure 2.32.

EXAMPLE 9 A 56-pound weight is placed on a balanced beam, 6 feet to the left of the fulcrum. How far to the right must a 40-pound weight be placed to keep the system in balance?

SOLUTION   Let d_1 be the distance of the 40-pound weight from the fulcrum, where $W_1 = 40$, $W_2 = 56$, and $d_2 = 6$. Then

$$W_1 \cdot d_1 = W_2 \cdot d_2$$
$$40d_1 = 56(6)$$
$$40d_1 = 336$$
$$d_1 = 8.4$$

The weight should be placed 8.4 feet from the fulcrum.

Distance-Rate-Time Problems

In Exercise Set 2.3, the formula $d = rt$ (distance = rate · time) was introduced. The next example shows one way that it can be used to solve applied problems.

EXAMPLE 10   Two joggers are headed toward each other on the same road. One jogs at 5 miles per hour, while the other jogs at 8 miles per hour. If they start 6.5 miles apart, how long will it take before they meet?

SOLUTION   We let t stand for the time it takes for them to meet. Since $d = rt$, the first jogger travels $5t$ miles in that time and the second jogger $8t$ miles.

	Rate	Time	Distance
First jogger	5	t	$5t$
Second jogger	8	t	$8t$

When the two joggers meet, the sum of the distances they will have traveled will be 6.5 miles:

$$5t + 8t = 6.5$$
$$13t = 6.5$$
$$t = \frac{1}{2}$$

It will take half an hour for the two joggers to meet.

CHECK   The first jogger covers $0.5(5) = 2.5$ miles in one-half (0.5) hours. The second jogger covers $0.5(8) = 4$ miles in 0.5 hours. Since they are headed toward each other, the distances should add to 6.5 miles. The solution checks.

EXAMPLE 11   Goeff has a boat that can go 8 miles per hour when heading upstream in a river. When the motor is running at the same speed, assuming the rate of the current is constant, he can go 12 miles per hour downstream. If a trip up the river and back takes a total of 12 hours, how far upstream does he go?

SOLUTION We can use the formula $d = rt$ to solve for t. Time equals distance divided by the rate of speed:

$$t = \frac{d}{r}$$

	RATE	TIME	DISTANCE
Upstream	8	$\frac{x}{8}$	x
Downstream	12	$\frac{x}{12}$	x

The total time for the trip is 12 hours.

$\frac{x}{8} + \frac{x}{12} = 12$ Time upstream + time downstream = total time

$3x + 2x = 288$ Multiply each side by 24, the LCD.

$5x = 288$

$x = \frac{288}{5} = 57\frac{3}{5}$ miles

So Geoff goes $57\frac{3}{5}$ miles upstream. The check is left to the reader.

Solution to the Historical Problem at the Beginning of This Section

EXAMPLE 12 If during ebb tide, a wherry should set out from Haverhill, to come down the river, and, at the same time, another should set out from Newburyport, to go up the river, allowing the difference to be 18 miles; suppose the current forwards one and retards the other $1\frac{1}{2}$ miles per hour; the boats are equally laden, the rowers equally good, and, in the common way of working, in still water, would proceed at the rate of 4 miles per hour; when, in the river will the two boats meet?

SOLUTION

DEFINITION OF TERMS
Wherry — A small rowboat used on rivers
Ebb — When the tide returns toward the ocean
Equally laden — Carries the same load
Common way of working — The rate at which they generally row

Since the question asks when the two boats will meet, we let

t = the number of hours that elapse before they meet

The boat coming *down* the river will proceed at $4 + 1\frac{1}{2} = 5\frac{1}{2}$ mph, while the boat moving *up* the river will proceed at $4 - 1\frac{1}{2} = 2\frac{1}{2}$ mph.

	Rate	Time	Distance
Upstream	$2\frac{1}{2}$	t	$2\frac{1}{2}t$
Downstream	$5\frac{1}{2}$	t	$5\frac{1}{2}t$

When the two boats meet, the sum of the distances the two boats have traveled will be equal to the distance between the two towns, which is 18 miles. Thus

$$2\frac{1}{2}t + 5\frac{1}{2}t = 18 \quad \text{The sum of the distances is 18 miles.}$$
$$8t = 18$$
$$t = \frac{18}{8} = 2\frac{1}{4}$$

The two boats will meet $2\frac{1}{4}$ hours after they set out.

CHECK The two boats are approaching each other at $2\frac{1}{2} + 5\frac{1}{2} = 8$ mph. The elapsed time until they meet is $2\frac{1}{4}$ hours. Using the distance formula,

$$d = rt = 8\left(2\frac{1}{4}\right) = 8\left(\frac{9}{4}\right) = 18 \text{ miles}$$

The solution checks.

Exercise Set 2.7

Write an equation or inequality for each of the following sentences, solve it, and check your answer.

Number Problems

1. Find three consecutive integers with a sum of 168.
2. Find two consecutive even integers such that 8 times the smaller one equals 6 times the larger one.
3. One-half of a natural number plus one-third of the next consecutive natural number is 12. Find the first number.
4. Four-thirds of an even integer less one-half of the next consecutive even integer is 9. Find the two integers.
5. The average of three numbers is 21. If the first number is 15 and the second number is 8 more than the first, find the third number.
6. The average of four numbers is 5. The first number is 3, the second number is the square of the first number, and the third number is two less than the second number. Find the fourth number.
7. Find three consecutive integral multiples of 3 with a sum of 45.
8. Find three consecutive integral multiples of 5 with a sum of 60.
9. Lena earned 79, 83, and 80 on her first three tests. What is the least she must earn on her fourth test so that her average is at least 81?
10. Jose earned 85, 89, and 92 on his first three tests. What is the least he must earn on his fourth test so that his average is at least 90?

Geometry Problems

11. The average size of the three angles of a triangle is 60°. If one angle is 40° and the second angle is 20° larger, what is the measure of the third angle?

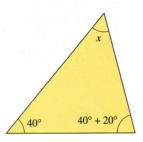

12. The average length of the four sides of a trapezoid is 13.5 in. If the second side is 3 in. longer than the first, the third side is 6 in. longer than the first, and the fourth side is twice as long as the first, find the length of the third side.

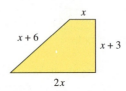

13. The length of a rectangular backyard is 3 m less than twice its width. If the perimeter is 66 m find the dimensions.

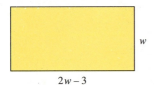

14. Find the dimensions of a rectangle with a perimeter of 112 cm if its width is 10 cm less than three-eights its length.

15. If one side of a triangle is one-third the perimeter P, the second side is 5 ft long, and the third side is one-fourth the perimeter, what is the perimeter?

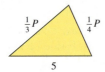

16. A telephone pole is located in a pond. If one-eighth of the height of the pole, H, is in concrete, 3 m in water, and one-half of the height above water, what is the total height of the pole?

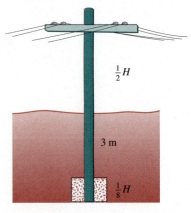

Fulcrum Problems

17. A 30-lb weight (W_1) is on the left side of a balance beam, 3 ft from the fulcrum. What weight (W_2) at a distance of 6 ft from the right side of the fulcrum will balance it? For the weights to balance, $W_1 \cdot d_1 = W_2 \cdot d_2$.

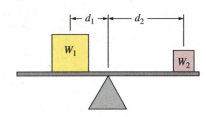

18. A 24-g weight is placed 3 cm farther from the fulcrum on one side of a balance beam than a 36-g weight on the other side. How far is each weight from the fulcrum?

19. Where should the fulcrum of a 9-ft beam be placed so that a 180-lb man will balance a 540-lb weight?

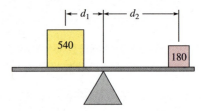

20. Where should the fulcrum of a 3-m beam be placed so that an 18-kg weight balances a 51-kg weight?

Money Problems

21. Eric earned $272, $310, $364, and $284 in four consecutive weeks. What is the most he can earn in the fifth week if his average earnings are to be no more than $276?

22. Mimi earned $27, $33, $26, and $39 in four consecutive days. What is the least she can earn on the fifth day if her average pay is to be no less than $31?

23. One hundred tickets were sold for the school play for a total of $333.00. If the tickets sold for $3.00 and $4.50, how many of each kind were sold?

24. Cindy wishes to realize a return of 11% on her total investments. If she has $10,000 invested at 9%, how much additional money should she invest at 15%?

25. Les invested $25,000 in two lots (pieces of property) last year. He made a profit of 15% on one lot but lost 5% on the second lot. If his profit was equivalent to a return of 8% on the total investment, how much was invested in each lot?

26. Tom deals in antique cars. He bought two cars for a total of $6,000. He sold one of the cars for a profit of 25%, but lost 10% on the second car. If his gross profit was $940, how much was invested in each car?

27. Vonetta treated a client to dinner. The cost of the dinner, including 4% sales tax and a 15% tip, was $76.16. What was the amount of the bill (not including the sales tax and the tip, which was based on the cost of the meal before tax)?

28. Vasquez Clothiers has discounted the price of a topcoat by 40%. If the sale price of the topcoat is $180, what was the original price?

Distance-Rate-Time Problems

29. It takes you three-fifths of an hour to ride your bicycle to school when you ride at 15 mph for part of the way and 10 mph the rest of the way. If the school is 7 mi from your house, for how many minutes did you travel at each rate?

30. Two bicyclists leave Cuesta College at the same time, traveling in opposite directions. One travels at 17 mph and the other at 13 mph. How long does it take before they are 60 mi apart?

31. Two cars leave Atascadero at the same time, traveling in opposite directions. The average rate of one car is 5 mph faster than the other car. If the two cars are 420 mi apart after 4 hr, what is the rate of each car?

32. Two bike riders leave Sacramento traveling south. The first rider leaves at 11:00 A.M. Two hours later the second rider leaves, traveling at an average rate that is 7 mph faster than the first. The second rider catches the first rider at 5:00 P.M. What is the rate of each?

33. Jim walks along a road at 3 mph toward Paso Robles, which is 18 mi away. After 40 min he gets a ride in a car and is in Paso Robles in 15 min. What is the average rate of the car?

34. A boat travels up a river at 4 mph and back down the river at 8 mph. If the round trip takes 3 hr, how far up the river does the boat go?

35. Ida travels east for 2 hr at 55 mph. She then slows down to 40 mph for the next hour and a half. How far does she travel?

36. Shawn walks at x mph for 20 min, and then he walks at $(x + 1)$ mph for 15 min to the park (2 miles from his home). What was his *average* speed during the 20 min?

Mixture Problems

37. How many liters (ℓ) of a 30% salt solution must be added to 40 ℓ of a 12% salt solution to obtain a 20% salt solution?

38. How many cℓ of pure alcohol must be added to 35 cℓ of 20% alcohol solution to obtain a 40% alcohol solution?

39. How many quarts of 40% antifreeze solution must be drained from a 12-qt radiator and replaced with pure antifreeze to obtain a 50% antifreeze solution?

40. How many liters of a 30% chemical solution should be drained from a 60 ℓ tank and replaced with an 80% chemical solution to obtain a 50% chemical solution?

41. How many gallons of pure alcohol must be added to 90 gal of pure gasoline to make "gasohol" (90% gasoline and 10% alcohol)?

42. How many quarts of pure antifreeze must be added to 8 qt of water to make a 50% blend of antifreeze and water (50% antifreeze and 50% water)? Note: This is what is generally kept in a car radiator.

43. A coffee shop blends a $5.00/lb coffee with one priced at $6.50/lb to produce a blend that sells for $6.00/lb. How many pounds of each should be used to obtain 75 lb of the new blend?

44. Dee's Meat Market mixes 600 lb of sausage to sell for $1.65/lb. How many pounds of $1.89/lb hamburger should be mixed with other meat at $1.49/lb to make up the sausage?

2.1 Solving Linear Equations in One Variable

A **linear (first degree) equation in one variable** is one that can be written in the form $ax + b = c$, where a, b, and $c \in R$, with $a \neq 0$. A **solution,** or **root,** to a linear equation is a value that makes it true. To solve an equation means to find the **solution set. Equivalent equations** are equations with the same solution set.

An equation that is true for all permissible replacements of the variable is called an **identity.** One that is true only for specified values of the variable is called a **conditional equation.**

Linear equations are solved by applying a series of six steps:

Step 1 Clear the equation of fractions by using the multiplication property.
Step 2 Remove grouping symbols by using the distributive property.
Step 3 Combine like terms on each side of the equation.
Step 4 Isolate the variable by using the addition and/or subtraction property.
Step 5 Solve for the variable by using the multiplication and/or division property.
Step 6 Check the solution.

2.2 From Words to Symbols: An Introduction to Applied Problems

Applied problems are stated in verbal format. To solve an applied problem, the verbal statement must be translated into a mathematical equation or an inequality.

A **mathematical sentence** is one that involves an equal sign ($=$) or a symbol of inequality, such as "$<$."

2.3 Literal Equations and Formulas

Equations such as $d = rt$, $P = 2(w + l)$, and $A = lw$ are called **literal equations.** Literal equations often occur as well-known formulas that can be solved for one of the variables by using the techniques of this chapter.

2.4 Solving Linear Equations Involving Absolute Value

Equations involving absolute value are solved by writing them in equivalent forms free of absolute value. In other words,

$$|ax + b| = p, \quad p > 0$$

is equivalent to

$$ax + b = p \quad \text{or} \quad ax + b = -p$$

2.5 Solving Linear Inequalities

A **linear inequality in one variable** is one that can be written in the form $ax + b < c$, where a, b, and $c \in R$, with $a \neq 0$. Linear inequalities exhibit properties similar to the properties of equality.

1. If $a < b$, then $a + c < b + c$.
2. If $a < b$, then $a - c < b - c$.
3. If $a < b$, then $a \cdot c < b \cdot c$, if c is positive.
4. If $a < b$, then $a \cdot c > b \cdot c$, if c is negative.
5. If $a < b$, then $\dfrac{a}{c} < \dfrac{b}{c}$ if c is positive.
6. If $a < b$, then $\dfrac{a}{c} > \dfrac{b}{c}$ if c is negative.

2.6 Solving Linear Inequalities Involving Absolute Value

Inequalities involving absolute value are solved by writing them in equivalent forms until the solution set can be determined by inspection:

$|ax + b| < p, \quad p > 0$ is equivalent to $-p < ax + b < p$

$|ax + b| > p, \quad p > 0$ is equivalent to $ax + b < -p$ or $ax + b > p$

If $|x|$ is greater than a negative number, then the solution set is the set of all real numbers. If $|x|$ is less than a negative number, the solution set is the empty set.

COOPERATIVE EXERCISE

The Egglestons are building a new house. The house is already framed and the roof is complete. They simply need to calculate the cost of some of the finish work. Each family member is given a portion of the task, using the following floor plan.

PART 1. BATHROOM TILING

The floor (9 ft by 7 ft, excluding the bathtub area of 2 ft by 7 ft) will be covered by tile. The wall (6 ft high) on the backside and two ends of the bathtub will also be tiled. Each tile costs $0.37 and is 4 in. × 4 in. in size. The cost of the labor to install the tile is $5.25/sq. ft. A tax of $7\frac{1}{4}\%$ is added to the cost of the tile. There is no tax on the labor. Determine the cost of buying and installing the tile.

PART 2. BEDROOM CARPETING

The two bedrooms will have shag carpeting at $17.50/sq. yd. The padding is an additional $4.95/sq. yd. Labor to install the carpet and padding is an additional $2.00/sq. yd. A $7\frac{1}{4}\%$ tax is added to the cost of the pad and carpet. There is no tax on the labor. *Note:* You cannot purchase part of a square yard of carpet or padding. Determine the cost of buying and installing the pad and carpet in the two rooms.

PART 3. LIVING AREA CARPETING

The living room, dining room, and hall floors are to be carpeted with hi/lo carpeting costing $19.95/sq. yd. The padding is an additional $4.95/sq. yd. Labor to install the carpet and padding is an added $2.00/sq. yd. A $7\frac{1}{4}\%$ tax is added to the cost of the padding and carpet. The hall is 3 ft × 31 ft in size. There is no tax on the labor. Determine the cost of buying the carpet and padding and the labor to install it. *Note:* This carpet and padding can be purchased only in whole square yards.

PART 4. KITCHEN/FAMILY ROOM FLOORS

The family room and kitchen floors will be covered with wood that comes in 9 in. × 9 in. squares. Each square costs $3.75. There is an additional cost of $2.35/sq. ft to install the wood. *Note:* The kitchen is 11 ft × 16 ft but there are cabinets on three sides of the room that are 18 in. wide and 3 ft high. The tops of the cabinets will be covered by butcher

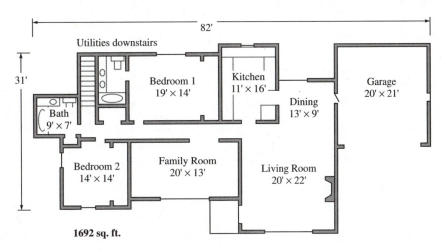

board that can be purchased in 18 in. widths. The butcher board costs $4.90/linear foot, and 41 linear feet will be needed. There is no additional charge for the labor to install the butcher board. Determine the total cost for these two rooms.

PART 5. PAINTING

The wall (8 ft high) of the living room, dining room, hall, and the two bedrooms are to be painted. A gallon of paint will cover approximately 360 sq. ft. In calculating the number of square feet, disregard all windows, doors, closets, and the fireplace. The ceilings need to be painted, too. Paint costs $14.95/gal and will cover in one coat. A painter estimates that it will take her 23 hr to paint this area and she charges $28.00/hr. A $7\frac{1}{4}\%$ tax will be added to the cost of the paint, but there is no tax on the labor. *Note:* This paint can be purchased only in gallon units. Determine the total cost to paint these areas.

PART 6. TOTAL COST

Use the calculations from Parts 1–5 to determine the overall cost to finish this part of the new house.

REVIEW EXERCISES

Complete each of the following.

1. The sum of $6x$ and $5x$ is ____.
2. $8y - 3y$ equals ____.
3. If $a = b$ then $a - c = b - c$ is an example of the ____ property of equality.
4. If $a = b$ then $a + c = b + c$ is an example of the ____ property of equality.
5. The solution set for $3p - 2 - p = p + 5$ is ____.
6. The multiplication property of equality, in words, is ____.
7. The division property of equality, in words, is ____.
8. The solution set for $3(a + 2) - 4(a - 5) = 7$ is ____.
9. The solution set for $\dfrac{h}{3} - \dfrac{h}{6} = 1$ is ____.
10. The solution set for $-\{-5 - [2 - (3x - 1)] + 2\} = -[-4 - 2(2 - x)] + 3$ is ____.
11. The solution set for $r - 4 - 2r = 2(r - 1) - 3r - 2$, is ____.
12. An equation that is true for any permissible values of the variable is called an ____.
13. An equation that is true only for specified values of the variable is called a ____ equation.
14. The solution set for $|y| = p$ is ____.
15. The solution set for $8x - 3(1 - x) \geq 3x + (5 - 7x)$, $x \in W$, is ____.
16. The number $\dfrac{2}{7}$ in decimal form is ____.
17. 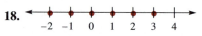 is the graph of $\{x \mid x \text{ ____}\}$.
18. 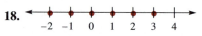 is the graph of $\{x \mid x \text{ ____}\}$.
19. $x + (3 + y) = (x + 3) + y$ is an example of the ____ property.
20. x and $-x$ are additive ____ of each other.
21. $\dfrac{a}{3}$ and $\dfrac{3}{a}$ are multiplicative ____ of each other.
22. The multiplicative identity element is ____.
23. The multiplicative property of equality states that if a, b, and $c \in R$ and $a = b$, then ____.
24. By the definition of subtraction, $a - b =$ ____.
25. The solution set for $3x - 5 = 8$ is ____.
26. The solution set for $|x| = 4$ is ____.
27. The solution set, in interval notation, for $|x - 3| \leq 3$ is ____.
28. The solution set for $|7x - 3| + 5 \leq 2$ is ____.
29. The sign of the product of three negative numbers is ____.
30. Simplified, $\dfrac{3(x - 5) - 2(x + 3)}{15 + (-8)}$ is ____.
31. The translation of ''twice the difference of a number and 6'' is ____.
32. The restrictions on the variable in $\dfrac{1}{x - 2} + \dfrac{3}{x} = \dfrac{5}{x + 4}$ are ____.

Solve and check each of the following. Note any restrictions on the variables. Graph all inequalities.

33. $x - 5 = 7$
34. $6y - 3 = 11$
35. $|3x + 4| + 5 = 6$
36. $y + 3 \leq 4$, $y \in W$

37. $|m| = 6$
38. $|6a + 1| - 3 = 2$
39. $|2y - 1| = |y + 4|$
40. $|k| < 3$
41. $\dfrac{1}{x} + 3 = \dfrac{5}{x}$
42. $\dfrac{3}{4}x \geq -2$
43. $\dfrac{2x + 7}{3} - \dfrac{3x - 1}{2} \geq 1,\ x \in \mathbf{J}$
44. $\dfrac{3y + 2}{3} + \dfrac{y - 1}{2} \geq 2$
45. $|3y - 5| + 6 \leq 7$
46. $|3x - 1| \geq -2,\ x \in \mathbf{N}$

Solve for the specified variable in each exercise.

47. $y = mx + b$ for x
48. $I = prt$ for t
49. $a = p(1 + rt)$ for r
50. $\dfrac{1}{f} = \dfrac{1}{a}$ for a
51. $3x - 8y = 2$ for y
52. $4x + 5y = 7$ for x
53. Rewrite $\{x \mid -5 \leq x \leq 5\}$ using absolute value notation.
54. Rewrite $\{y \mid y \leq -5 \text{ or } y \geq 5\}$ using absolute value notation.

Translate each of the following into a mathematical sentence and solve.

55. The sum of three consecutive even numbers is 114. Find the numbers.
56. Thirty coins, some nickels and the rest quarters, are worth $4.50. How many nickels and quarters are there?
57. A 40-lb mixture of candy has a value of $153.00. Some of the candy sells for $3.75/lb and the rest sells at $4.00/lb. How many pounds of each are in the mixture?
58. A billboard for advertising is to be made in the form of a parallelogram, as shown in the accompanying diagram. If the perimeter is 65 ft, find the dimensions of the sides.

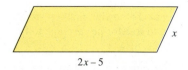

59. An isosceles triangle has dimensions as shown in the figure. If the perimeter is 11 m, find the dimensions of the sides.

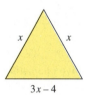

60. A train traveling at 68 km/hr travels 340 km. How long does it take?
61. Don and Jeanette leave San Luis Obispo at the same time, traveling in opposite directions. Jeanette travels at 50 km/hr and Don travels at 65 km/hr. How far apart are they after 5 hr?
62. Chris leaves Mustang at 5:00 P.M., while R.J. leaves from the same place at 7:00 P.M. traveling along the same road. If Chris averages 50 mph and R.J. averages 60 mph, how long will it take for R.J. to catch Chris?
63. Vonetta heads away from Amarillo at 45 mph for 3 hr before she realizes that she has left her wallet in the motel room. She returns to the motel at 55 mph. How long does the return trip take?
64. Two trains travel toward each other along parallel tracks. They start at the same time from towns 255 km apart. One train travels at 100 km/hr and the other at 70 km/hr. How far has the faster train gone when they pass each other?

Chapter Test

1. Simplify: $16x + 2(3x - 4) - 6(x - 1)$.
2. Solve the equation $|x| = 25$.
3. Solve the inequality $|9x - 5| > 4$.
4. Solve $5x - 7 = 2$.
5. Solve $|3x - 1| - 1 = 4$.
6. Solve $A = p(1 + rt)$ for p.
7. Donna flies a small plane to the town of Enid. On the way there she flies at 90 mph with the wind. She then flies back home at 60 mph against the wind. If the round trip takes 7.5 hr, how far away is Enid?
8. Solve: $\dfrac{x + 1}{3} - \dfrac{x + 4}{2} = \dfrac{x - 1}{2}$.
9. Solve: $-2(2x - 3) \leq 7$.
10. The perimeter of a rectangle is 72 in. If the length is twice the width, find the dimensions.

11. If Marcy earned 88 on her first test, what must she earn on her second test for her average to be at least 91?
12. Solve: $|7x + 2| \leq 1$.
13. Solve: The sum of two consecutive integers is 31.
14. Find the value of $x - y + z$ when $x = -4$, $y = -7$, and $z = 3$.
15. The area of a right triangle is 42 sq. m. Find the length of the base if the height is 7 m.
16. Solve: $\frac{1}{4}\left(4x - \frac{1}{3}\right) = \frac{1}{2}x + 5$.
17. Solve: $4(-2x + 1) = 2(x - 4) - 3(x - 5)$.
18. Find two consecutive odd natural numbers with a sum of 132.
19. The perimeter of a rectangular field is 68 km. If the width is 8 km less than the length, find the dimensions.

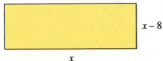

20. The cost of buying and installing linoleum in the Wong's rectangular kitchen is $336.00. The kitchen is 9 ft wide and 12 ft long. What is their cost per sq. yd for buying and installing linoleum?
21. A service station attendant wants to mix 8 qt of a 40% solution of antifreeze with a 70% solution of antifreeze to make a 50% solution of antifreeze. How many quarts of the 70% solution should be used?
22. Brian makes and sells wooden wagons. He took $7000 of his income last year and invested it, part at 7% and the rest at 9%. He earned $570 in interest after one year from the investments. How much did he invest at each rate?
23. Mrs. Wood invests $10,000 at 8%. How much additional should she invest at 11% in order to realize a return of 10% on her total investments?
24. On an 84-mi trip in the country, Jessica and Justine travel at a steady speed for the first hour. Their average speed for the second hour of the trip is 12 mph faster than the first hour. Find their average speed for the first hour. (They cover the entire 84 mi in 2 hours.)
25. In a charity run, Terry averages 5 mph. James leaves 10 min later and averages 6 mph. How long will it take for James to catch up with Terry?

CHAPTER 3

LINEAR EQUATIONS AND FUNCTIONS IN TWO VARIABLES

3.1 The Rectangular Coordinate System; Graphing Linear Equations in Two Variables
3.2 The Slope of a Line; Distance and Midpoint Formulas
3.3 Point–Slope, Slope–Intercept, and the General Form of the Equation of a Line
3.4 Relations, Functions, and Function Notation
3.5 The Algebra of Functions; Inverse Functions

*A*PPLICATION

A tractor manufacturer finds the daily cost function for producing n tractors to be $C(n) = 180 + 272n$. The total revenue from the sale of n tractors is $R(n) = 302n$.

a. Find the profit function $P(n)$, where $P(n) = R(n) - C(n)$.
b. Find the value(s) of n when the profit is zero.
c. For what values of n does the company make money?
d. For what values of n does the company lose money?
e. If this were your company, how many tractors would you try to produce and sell?

See Exercise Set 3.4 number 72.

I don't know how I appear to the world; but to myself I feel like a boy playing on the seashore, and diverting myself now and then by finding another pebble or a prettier shell than is originally found, while a great ocean of truths lies undiscovered all around me.

ISSAC NEWTON (1642)

3.1 The Rectangular Coordinate System; Graphing Linear Equations in Two Variables

HISTORICAL COMMENT René Descartes (1596–1650) was a French mathematician, philosopher, and physician who developed the rectangular coordinate system we use in this section. It is often called the Cartesian coordinate system in honor of him. There are several stories of how the idea first came to him. One is that he was watching a fly crawl across the ceiling near a corner of a room and decided to develop a system that would allow the location of the fly to be known at any time in terms of its distance from the walls. Two other stories have him lying in bed. In one the idea came to him in a dream; in the other it was a cold day and, since the town was under siege, he remained in bed and thought about mathematics.

OBJECTIVES In this section we will learn to
1. locate points on a rectangular coordinate system; and
2. graph a line when its equation is known.

In Chapter 2, we developed methods for solving linear equations and inequalities in one variable. We then used the methods to solve applied problems. This chapter introduces linear equations in two variables and shows how we can use them to solve additional types of applied problems.

Consider the equation

$$y = 2x + 4$$

The value of y is 4 more than twice the value of x. When the value of one variable is known, we can find the other by *substitution*.

EXAMPLE 1 Suppose $y = 2x + 4$. Find the value of y when $x = 5$.

SOLUTION
$y = 2x + 4$
$= 2 \cdot 5 + 4$ Substitute 5 for x.
$= 10 + 4 = 14$

A solution to an equation in *two* variables consists of values for each variable that together make the equation true. One solution to the equation in Example 1 is $x = 5$, $y = 14$. This solution is written as an ordered pair: (5, 14). An **ordered**

pair consists of two numbers written in a specified order within parentheses. In an ordered pair, the *x*-value is always given first and the *y*-value second. *A linear equation in two variables has infinitely many ordered pairs that are solutions.* For each new value of *x* or *y*, a new ordered pair results. It will be shown in Section 3.3 that any equation that can be put in the form

$$ax + by = c, \quad \text{with } a \text{ and } b \text{ not both zero}$$

will have a graph that is a straight line. Equations with graphs that are straight lines are called **linear equations.**

> **DEFINITION**
>
> **Linear Equation in Two Variables**
>
> A **linear equation in two variables,** *x* and *y*, is one that can be put in the form
>
> $$ax + by = c, \quad \text{with } a \text{ and } b \text{ not both zero}$$
>
> where *a*, *b*, and *c* are real numbers.

EXAMPLE 2 Find solutions for $y = 3x + 5$ when $x = 2$, $x = 4$, and $x = -2$.

SOLUTION

$y = 3x + 5$		$y = 3x + 5$		$y = 3x + 5$	
$= 3 \cdot 2 + 5$	$x = 2$	$= 3 \cdot 4 + 5$	$x = 4$	$= 3(-2) + 5$	$x = -2$
$= 6 + 5$		$= 12 + 5$		$= -6 + 5$	
$= 11$		$= 17$		$= -1$	

The solutions are $(2, 11)$, $(4, 17)$, and $(-2, -1)$.

The relationship between the *x*- and *y*-values in an equation can be visualized by plotting the ordered pairs that satisfy the equation on the **rectangular (Cartesian) coordinate system** that was mentioned in the historical comment at the beginning of this section. Descartes' system involves the construction of two perpendicular number lines, one horizontal and one vertical, that intersect at the zero point of both lines. The point of intersection of these lines is called the **origin.** The horizontal line is called the ***x*-axis** and the vertical line is called the ***y*-axis.** Above the *x*-axis or to the right of the *y*-axis is the *positive* direction, while below the *x*-axis or to the left of the *y*-axis is the *negative* direction. The two axes divide the plane into four **quadrants** that are numbered in a counterclockwise direction as shown in Figure 3.1 on the next page. The axes are not considered part of any quadrant.

Every point in the plane can be described in terms of an ordered pair (x, y). The value for *x* indicates the distance to the right or left of the *y*-axis, while the value for *y* indicates the distance above or below the *x*-axis. The *x*- and *y*-distances are known as the **coordinates** of the point. The coordinates of the origin are $(0, 0)$. In the ordered pair $(3, 4)$, the number 3 is the *x*-coordinate and 4 is the *y*-coordinate.

$$(3, 4)$$
x-coordinate ↑ ↑ *y*-coordinate

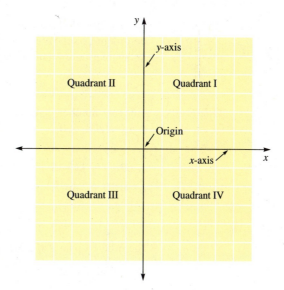

FIGURE 3.1

To locate the point (3, 4), we move 3 units to the right of the origin and then move 4 units up. This point has *x*-coordinate 3 and *y*-coordinate 4. The point (3, 4) and several other points are shown in Figure 3.2.

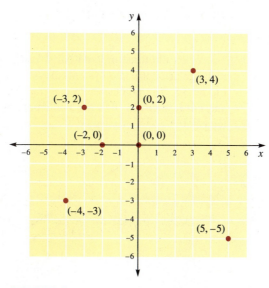

In Quadrant I, both the *x*- and *y*-coordinates are positive.

In Quadrant II, the *x*-coordinate is negative and the *y*-coordinate is positive.

In Quadrant III, both the *x*- and *y*-coordinates are negative.

In Quadrant IV, the *x*-coordinate is positive and the *y*-coordinate is negative.

FIGURE 3.2

Various names are used to describe the *x*- and *y*-axes. The *x*-axis is often called the **horizontal axis** or **axis of abscissas.** The *y*-axis is called the **vertical axis** or **axis of ordinates.** Using the latter terminology suggests that the coordinates of the point (*x, y*) are referred to as the **abscissa** and the **ordinate,** respectively.

$$(x, y)$$
abscissa ↑   ↑ ordinate

We opened this section by observing that solutions to equations in two variables are ordered pairs. We now show how these ordered pairs relate to each other on a rectangular coordinate system.

EXAMPLE 3 Find three ordered pairs that are solutions of the equation $2x + 3y = 6$.

SOLUTION Any real number can be chosen for either x or y. Two of the easier choices are $x = 0$ and $y = 0$.

$2x + 3y = 6$　　　　　　　　　　　　　$2x + 3y = 6$
$2 \cdot 0 + 3y = 6$　　Substitute 0 for x.　$2x + 3 \cdot 0 = 6$　　Substitute 0 for y.
$0 + 3y = 6$　　　　　　　　　　　　　$2x + 0 = 6$
$3y = 6$　　　　　　　　　　　　　　　$2x = 6$
$y = 2$　　　　　　　　　　　　　　　　$x = 3$

The ordered pair is (0, 2).　　　　　　　The ordered pair is (3, 0).

As a third choice, we let $x = -3$.

$2x + 3y = 6$
$2(-3) + 3y = 6$　　Substitute -3 for x.
$-6 + 3y = 6$
$3y = 12$
$y = 4$

The ordered pair is $(-3, 4)$.

NOTE All y-coordinates on the x-axis are zero and all x-coordinates on the y-axis are zero.

When the three ordered pairs found in Example 3 are graphed (plotted) on a rectangular coordinate system, they all lie on the same straight line, as shown in Figure 3.3. The points where the graph crosses the axes are called the **x-intercept** and **y-intercept**. To find the x-intercept, we let $y = 0$. To find the y-intercept, we let $x = 0$.

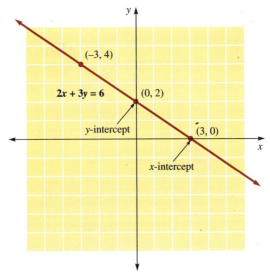

FIGURE 3.3

Only two points are needed to sketch the graph of a straight line. Obtaining a third point provides a check on the accuracy of the other two.

EXAMPLE 4 Graph $-2x + y = 4$ using the x- and y-intercepts together with a check point.

SOLUTION We find the x-intercept by letting $y = 0$:

$-2x + y = 4$
$-2x + 0 = 4$ Substitute 0 for y.
$x = -2$ x-intercept

The ordered pair is $(-2, 0)$. We find the y-intercept by letting $x = 0$:

$-2x + y = 4$
$-2 \cdot 0 + y = 4$ Substitute 0 for x.
$0 + y = 4$
$y = 4$ y-intercept

The ordered pair is $(0, 4)$. As a check point, we let $x = 1$:

$-2x + y = 4$
$-2 \cdot 1 + y = 4$ Substitute 1 for x.
$-2 + y = 4$
$y = 6$

The ordered pair is $(1, 6)$. The graph is shown in Figure 3.4.

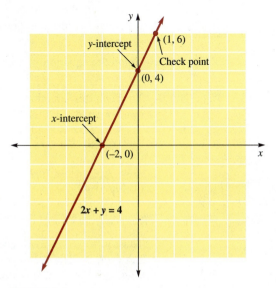

FIGURE 3.4

EXAMPLE 5 Graph the lines $x = 2$ and $y = 1$ on the same coordinate system.

SOLUTION The equation $x = 2$ can be thought of as $x + 0 \cdot y = 2$. The value of x will be 2 for any value of y. Three points are $(2, -2)$, $(2, 0)$, and $(2, 4)$. Similarly, $y = 1$ can be thought of as $0 \cdot x + y = 1$. The points $(-2, 1)$, $(0, 1)$, and $(2, 1)$ all lie on the line $y = 1$. The graphs are shown in Figure 3.5. Notice that the two lines intersect at the point $(2, 1)$.

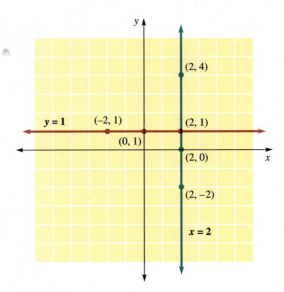

FIGURE 3.5

The graphs of Figure 3.5 illustrate that the graph of a line where x is a constant is a **vertical line**. The graph of a line where y is a constant is a **horizontal line**.

EXAMPLE 6 Graph the equation $y = |x + 2|$.

SOLUTION To graph this equation, we consider the definition of absolute value together with three ordered pairs that are solutions to each part of the definition.

$$|x + 2| = x + 2 \quad \text{if} \quad x + 2 \geq 0, \text{ which means } x \geq -2$$

Thus $y = x + 2$ whenever $x \geq -2$. Three ordered pairs are $(-2, 0)$, $(2, 4)$, and $(4, 6)$. Also,

$$|x + 2| = -(x + 2) \quad \text{if} \quad x + 2 < 0, \text{ which means } x < -2.$$

Thus, $y = -(x + 2)$ whenever $x < -2$. Three ordered pairs are $(-8, 6)$, $(-6, 4)$, and $(-4, 2)$. The graph is shown in Figure 3.6. Since $y = |x + 2|$, we know that y will always be positive or zero. In other words, the graph will not extend below the x-axis.

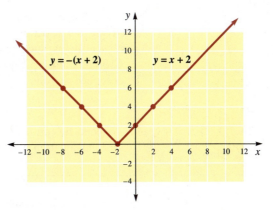

FIGURE 3.6

EXAMPLE 7   Each day a company manufactures n articles that sell for $3.00 each. The costs C associated with the manufacturing process are $25.00 daily. If P represents the daily profit and n represents the number of units sold each day, find the equation relating P to n and sketch its graph. Recall that

   profit = revenue − cost

SOLUTION   Since n items sell for $3.00 each, the revenue is $3n$. The cost is $25.00, so

$$P = 3n - 25$$

The graph can be sketched by arbitrarily letting P be the vertical axis and n the horizontal axis. The coordinates of the points used to construct the graph in Figure 3.7 are shown in the accompanying table of values. No negative values of n were used to construct the graph since a negative number of items cannot be sold.

n	p
0	−25
5	−10
10	5
15	20

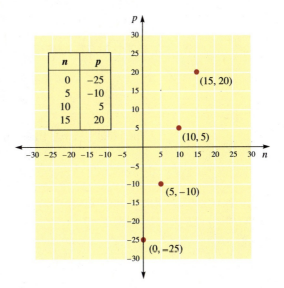

FIGURE 3.7

In Example 7, note that although a straight line could pass through the points satisfying the equation, none was drawn. Only n-values that represent the sold number of items, such as $n = 1, 2, 3, \ldots, 25, \ldots,$ apply here.

The relationship between the solutions of a linear equation and the graph can be described as follows.

> **DEFINITION**
> **The Graph of a Line**
>
> A point **lies on a line** if and only if its coordinates **satisfy** (are solutions to) the equation of the line.

In other words, the graph is a picture of the solutions to an equation.

EXAMPLE 8 Which of the following points lie on the line $2x - 3y = 6$?

(a) $(3, 0)$ (b) $(5, 1)$ (c) $\left(\dfrac{1}{2}, -\dfrac{5}{3}\right)$

SOLUTION (a)
$2x - 3y = 6$
$2 \cdot 3 - 3 \cdot 0 \stackrel{?}{=} 6$ Substitute 3 for x and 0 for y.
$6 \stackrel{\checkmark}{=} 6$

The point $(3, 0)$ lies on the line.

(b)
$2x - 3y = 6$
$2 \cdot 5 - 3 \cdot 1 \stackrel{?}{=} 6$ Substitute 5 for x and 1 for y.
$7 \neq 6$

The point $(5, 1)$ does *not* lie on the line.

(c)
$2x - 3y = 6$
$2\left(\dfrac{1}{2}\right) - 3\left(-\dfrac{5}{3}\right) \stackrel{?}{=} 6$ Substitute $\dfrac{1}{2}$ for x and $-\dfrac{5}{3}$ for y.
$1 + 5 \stackrel{?}{=} 6$
$6 = 6$

The point $\left(\dfrac{1}{2}, -\dfrac{5}{3}\right)$ lies on the line.

Do You Remember?

Can you match these?

___ 1. René Descartes
___ 2. Example of a vertical line
___ 3. Linear equation
___ 4. Abscissa
___ 5. Ordinate
___ 6. y-intercept
___ 7. x-intercept
___ 8. Example of a horizontal line

a) Second element of an ordered pair
b) $ax + by = c$
c) First element of an ordered pair
d) $(0, y)$
e) $x = 6$
f) $y = 3$
g) Developed the rectangular coordinate system
h) $(x, 0)$

Answers: 1. g 2. e 3. b 4. c 5. a 6. d 7. h 8. f

Exercise Set 3.1

1. Plot the points represented by the ordered pairs $A(0, 0)$, $B(5, 1)$, $C(-2, -3)$, $D(4, -3)$, $E(-6, 0)$, and $F(-1, 4)$.
2. Find the coordinates of the points on the given graph.

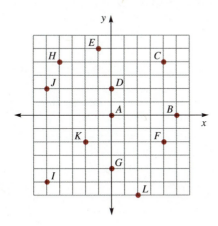

State the quadrant in which each point is located.

3. $(2, 3)$
4. $(4, 1)$
5. $(-1, 2)$
6. $(-3, -2)$
7. $(-8, 3)$
8. $(3, -1)$
9. $(-3, -1)$
10. $(-2, 8)$
11. $(x, y); x < 0, y < 0$
12. $(x, y); x > 0, y < 0$
13. $(-x, y); x < 0, y > 0$
14. $(-x, -y); x < 0, y < 0$

Graph each of the following equations by finding three ordered pairs.

15. $y = x + 1$
16. $y = x + 2$
17. $x - y = 5$
18. $\frac{1}{4}x - \frac{1}{2}y = 1$
19. $-3x + 4y = 12$
20. $-\frac{1}{2}x + y = 1$
21. $2x - 5y = -10$
22. $5x - 3y = -15$
23. $y = 6$
24. $y = -3$
25. $\frac{1}{4}x = -1$
26. $x = 7$
27. $y = x$
28. $y = -x$
29. $\frac{1}{2}x - y = 0$
30. $3x + y = 0$

Which of the given points lie on the line (satisfy the equation of the line)?

31. $7x + y = 8$
   (a) $(1, 1)$
   (b) $(2, -6)$
   (c) $(0, 8)$
   (d) $(-1, -1)$

32. $m - 3n = 6$
   (a) $(0, 2)$
   (b) $(0, -2)$
   (c) $(6, 0)$
   (d) $\left(\frac{3}{2}, -\frac{9}{2}\right)$

33. $\frac{3}{4}x - y = \frac{3}{2}$
   (a) $(2, 0)$
   (b) $\left(0, \frac{3}{2}\right)$
   (c) $\left(1, -\frac{3}{4}\right)$
   (d) $\left(4, \frac{3}{2}\right)$

34. $\frac{1}{3}a - b = -\frac{4}{3}$
   (a) $(1, 1)$
   (b) $(2, 2)$
   (c) $\left(0, \frac{4}{3}\right)$
   (d) $\left(-5, -\frac{1}{3}\right)$

Graph each equation by using the x- and y-intercepts and one check point.

35. $x + 2y = 6$
36. $5x - y = 10$
37. $3x + y = 9$
38. $x - \frac{1}{2}y = 2$
39. $-x + 6y = 3$
40. $\frac{1}{2}y = x$
41. $\frac{1}{4}x - y = 0$
42. $x + 5y = 0$
43. $3x - \frac{1}{2}y = 0$
44. $3x = -y$
45. $4x + 3y = 0$
46. $2x + 7y = 14$
47. $8x + y = 8$
48. $x - 9y = 3$
49. $-x + y = -4$
50. $\frac{1}{3}x - \frac{1}{5}y = 1$
51. $\frac{2}{7}x + \frac{8}{4}y = 2$

Translate each statement into an equation, then find three ordered pairs satisfying each equation and graph the line.

52. The *y*-value equals the *x*-value.
53. The *x*-value is three times the *y*-value.
54. Twice the *y*-value subtracted from three times the *x*-value is −6.
55. If 5 is subtracted from five times the *y*-value, the result is twice the *x*-value.
56. The *y*-value is 2 more than one-half the *x*-value.
57. The *y*-value divided by 2 is 1 more than the *x*-value.

Graph each equation, calculating the ordered pairs in each case for x = −3, −2, −1, 0, 1, 2, 3.

58. $y = |x|$
59. $y = -|x|$
60. $y = |x - 2|$
61. $y = |x + 3|$

Translate each statement into an equation and answer parts (a)–(d).

62. Carl's Transfer Company bases the total charge to the customer on the following: $80 initial fee, $0.50 per mile of traveling distance, and $45 per hour of loading and unloading time.
   (a) Find the cost *C* of a trip involving *n* miles of traveling and 6 hours of loading and unloading time.
   (b) If *C* = $400 and the loading and unloading time is 6 hours, find the number of miles, *n*.
   (c) Find the total cost *C* of a 100-mile trip with a loading and unloading time of 3 hours.
   (d) Construct a graph to illustrate the cost of traveling 10 miles, 20 miles, 40 miles, 60 miles, and 80 miles, if the loading and unloading time remains constant at 2 hours. Use the graph to estimate the cost of traveling 30 miles.

63. Berna's Flowers has a constant cost of $270 per week for rent and materials. Berna can make and sell *n* bouquets of flowers per week at $9 per bouquet. Assume her profit *P* is equal to her revenue less her costs.
   (a) What is her profit if she makes and sells 40 bouquets in one week?
   (b) How many bouquets must she sell to break even (*P* = 0)?
   (c) How many bouquets does she have to sell to earn a profit of $90 in one week?
   (d) How much money would she lose if she sold only 20 bouquets in one week?
   (e) Construct a graph to illustrate the profit or loss for making and selling 20, 30, 40 and 60 bouquets. Use the graph to estimate her profit for making and selling 50 bouquets.

Write in Your Own Words

64. Why is $ax + by = c$ called a linear equation?
65. Describe the Cartesian coordinate system.
66. What is the least number of points needed to graph a straight line? Why does the text suggest finding a *check point*?

67. Given the ordered pair (p, q), the numbers p and q are called the *coordinates* of the point. Why is the word "coordinates" used? (A dictionary may help.)

68. By definition, "A point lies on a line if and only if its coordinates satisfy (are solutions to) the equation of the line." Describe what is meant by "satisfy or are solutions to the equation of the line."

3.2 The Slope of a Line; Distance and Midpoint Formulas

OBJECTIVES

In this section we will learn to
1. find the slope of a line;
2. find the distance between two points; and
3. find the midpoint of a line segment.

In the preceding section, we sketched the graphs of many lines. In some cases the graph of the line rose when moving from left to right. In other cases the graph fell when moving from left to right. In yet others the line was horizontal or vertical. The number that measures how fast a line rises or falls is called its **slope.** The slope is found in the following manner.

Suppose that (x_1, y_1) and (x_2, y_2) are any two distinct points on a line, as shown in Figure 3.8. The small 1 and 2 are called "subscripts" and are used to differentiate between the two points. When we move from the first point, (x_1, y_1), to the second point, (x_2, y_2), the x-value changes from x_1 to x_2 and the y-value changes from y_1 to y_2. The amounts of change are $x_2 - x_1$ and $y_2 - y_1$.

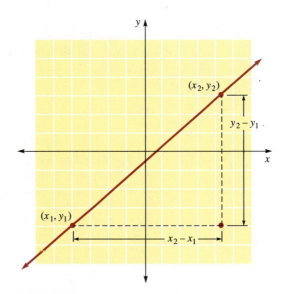

FIGURE 3.8

DEFINITION
Slope

The ratio of the change in y to the change in x between any two points on a line is called the **slope** of a line. The letter m is generally used to designate slope.

$$m = \text{slope} = \frac{\text{change in } y}{\text{change in } x}$$

Thus,

$$m = \frac{y_2 - y_1}{x_2 - x_1}$$

where $x_2 \neq x_1$ since division by zero is undefined.

EXAMPLE 1 Find the slope of the line through the points (2, 5) and (7, 9).

SOLUTION To find the slope, we let $(x_1, y_1) = (2, 5)$ and $(x_2, y_2) = (7, 9)$. Then

$$m = \frac{y_2 - y_1}{x_2 - x_1} = \frac{9 - 5}{7 - 2} \quad \text{Substitute } y_2 = 9, y_1 = 5, x_2 = 7, x_1 = 2.$$

$$= \frac{4}{5}$$

The choice of (x_1, y_1) and (x_2, y_2) was arbitrary. If we had let $(x_1, y_1) = (7, 9)$ and $(x_2, y_2) = (2, 5)$, then

$$m = \frac{y_2 - y_1}{x_2 - x_1} = \frac{5 - 9}{2 - 7} = \frac{-4}{-5} = \frac{4}{5}$$

The two results are the same. Just *be consistent in the order you choose.* If you use $y_2 - y_1$ in the numerator, you must use $x_2 - x_1$ in denominator.

EXAMPLE 2 Find the slope of the line $x + y = 5$ and show the slope on its graph.

SOLUTION To find the slope, we must first determine two points that lie on the line. If we let $y = 2$, then $x + 2 = 5$ and $x = 3$. One point is (3, 2). Similarly, when $x = 0$, then $y = 5$, so a second point is (0, 5). We use these two points to obtain the slope.

$$m = \frac{y_2 - y_1}{x_2 - x_1} = \frac{2 - 5}{3 - 0} \quad (x_1, y_1) = (0, 5), \quad (x_2, y_2) = (3, 2)$$

$$= \frac{-3}{3} = -1$$

The graph is shown in Figure 3.9.

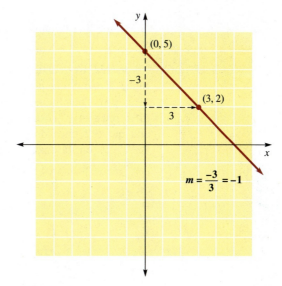

FIGURE 3.9

If two other points on the line had been used, say, (4, 1) and (1, 4), to calculate the slope in Example 2, the result would have been the same:

$$m = \frac{y_2 - y_1}{x_2 - x_1} = \frac{4 - 1}{1 - 4} = \frac{3}{-3} = -1$$

Any two points on a line can be used to determine the slope of the line. In Figure 3.10, the slope between points A and B is the same as the slope between points D and E.

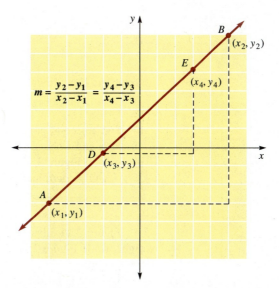

FIGURE 3.10

EXAMPLE 3   Graph the line through $(-2, -3)$ with slope $\dfrac{2}{5}$.

SOLUTION   To do this, we locate a second point by letting y change by 2 and x change by 5. Since 2 and 5 are positive, both changes will be in the positive direction from $(-2, -3)$. Thus the new point is $(3, -1)$; see Figure 3.11. The slope $\tfrac{2}{5}$ can also be interpreted as $\tfrac{-2}{-5}$. To sketch the line using the slope $\tfrac{-2}{-5}$, we start at the point $(-2, -3)$ and move 2 units down and 5 units to the left. Doing this gives a third point, $(-7, -5)$, on the line, as shown in Figure 3.11.

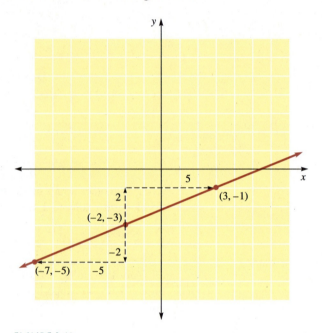

FIGURE 3.11

EXAMPLE 4   Graph the line through $(-2, 3)$ with slope $\dfrac{-4}{3}$.

SOLUTION   We find a second point on the line by letting y change by 4 units in the negative direction and x change by 3 units in the positive direction from $(-2, 3)$. The graph is shown in Figure 3.12 on the next page. The second point has x-coordinate $-2 + 3 = 1$ and y-coordinate $3 + (-4) = -1$.

Figure 3.11 illustrates that a line with **positive slope ($\tfrac{2}{5}$) rises from left to right,** while Figure 3.12 illustrates that a line with **negative slope ($-\tfrac{4}{3}$) falls from left to right.**

If a line is horizontal, the y-coordinates of any two distinct points on the line are equal. For example, the points $(3, 5)$ and $(6, 5)$ lie on the same horizontal line.

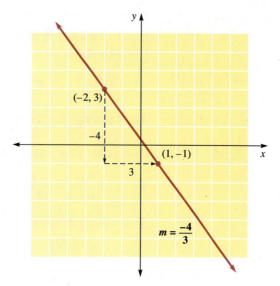

FIGURE 3.12

Therefore

$$m = \frac{y_2 - y_1}{x_2 - x_1} = \frac{5 - 5}{6 - 3} = \frac{0}{3} = 0$$

A horizontal line has zero slope.

On a vertical line, the x-coordinates of any two distinct points are equal. For example, the points (3, 8) and (3, 10) lie on the same vertical line, so

$$m = \frac{y_2 - y_1}{x_2 - x_1} = \frac{8 - 10}{3 - 3} = \frac{-2}{0}$$

Since division by zero is not defined, **a vertical line is said to have undefined slope.**

EXAMPLE 5   Find the slope of each line:

(a) $x = 2$ (b) $y = 1$

SOLUTION   (a) Since the value of x is always 2, the line is vertical. The slope is undefined.
(b) Since the value of y is always 1, the line is horizontal. The slope of the line is zero.

The graphs of the lines $x = 2$ and $y = 1$ are shown in Figure 3.5 on page 108.

> In general, if c represents a constant, the line $x = c$ is a **vertical line with undefined slope** and the line $y = c$ is a **horizontal line with zero slope.**

118 CHAPTER 3 · LINEAR EQUATIONS AND FUNCTIONS IN TWO VARIABLES

> **DEFINITION**
> **Parallel Lines**
>
> Two lines, L_1 and L_2, are **parallel lines**, written $L_1 \parallel L_2$, if and only if their slopes m_1 and m_2 are equal.

EXAMPLE 6   Show that L_1: $x + 2y = 6$ and L_2: $2x + 4y = 8$ are parallel.

SOLUTION   Two points satisfying L_1 are $(0, 3)$ and $(6, 0)$. Thus,

$$m_1 = \frac{0 - 3}{6 - 0} = \frac{-1}{2}$$

Two points satisfying L_2 are $(0, 2)$ and $(4, 0)$. Thus,

$$m_2 = \frac{0 - 2}{4 - 0} = \frac{-1}{2}$$

Since $m_1 = m_2$, we see that $L_1 \parallel L_2$. The graphs of L_1 and L_2 are shown in Figure 3.13.

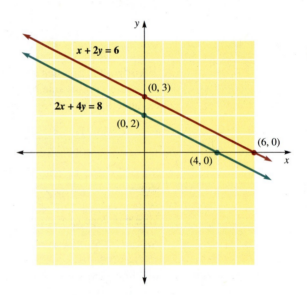

FIGURE 3.13

> **DEFINITION**
> **Perpendicular Lines**
>
> Two intersecting lines, L_1 and L_2, neither of which is vertical, are **perpendicular lines**, written $L_1 \perp L_2$, if and only if their slopes are negative reciprocals of each other:
>
> $$m_1 \cdot m_2 = -1 \quad \text{or} \quad m_1 = -\frac{1}{m_2}$$

EXAMPLE 7   Show that the line L_1 that goes through $(5, -2)$ and $(-2, 3)$ is perpendicular to the line L_2 that goes through $(-2, -1)$ and $(3, 6)$.

SOLUTION   We let m_1 be the slope of L_1 and m_2 the slope of L_2. Then

$$m_1 = \frac{y_2 - y_1}{x_2 - x_1} = \frac{3 - (-2)}{-2 - 5} = \frac{5}{-7} = -\frac{5}{7}$$

$$m_2 = \frac{y_2 - y_1}{x_2 - x_1} = \frac{6 - (-1)}{3 - (-2)} = \frac{7}{5}$$

Since the slopes are negative reciprocals of each other, the lines are perpendicular. Their graphs are shown in Figure 3.14.

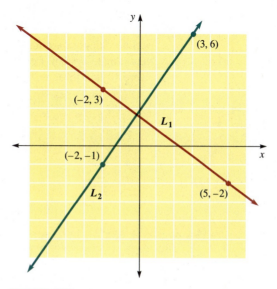

FIGURE 3.14

Three points are **collinear** if they lie on the same line. This means that the slope between any two of the points must be the same.

EXAMPLE 8   Use slopes to determine whether the points $A(-1, 2)$, $B(0, 4)$, and $C(2, 8)$ are collinear.

SOLUTION   To determine if the points are collinear, we must show that the slopes between any two ordered pairs are the same.

$$m_{AB} = \frac{4 - 2}{0 - (-1)} = \frac{2}{1} = 2 \qquad m_{AB} = \text{slope from } A \text{ to } B$$

$$m_{BC} = \frac{8 - 4}{2 - 0} = \frac{4}{2} = 2 \qquad m_{BC} = \text{slope from } B \text{ to } C$$

$$m_{AC} = \frac{8 - 2}{2 - (-1)} = \frac{6}{3} = 2 \qquad m_{AC} = \text{slope from } A \text{ to } C$$

The three points A, B, and C are collinear.

As we continue to work with the rectangular coordinate system, we will sometimes need to find the distance between two points on a line or the midpoint of a line segment joining the two points. Before finding the distance between two points, we will review the **Pythagorean theorem.**

DEFINITION	In a right triangle, the square of the length of the longest side (the hypotenuse) is equal to the sum of the squares of the lengths of the other two sides (the legs):
The Pythagorean Theorem	$c^2 = a^2 + b^2$

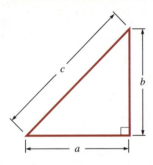

FIGURE 3.15

Figure 3.15 shows a right triangle with legs a and b and hypotenuse c.

We now return to the problem of finding the **distance between two points** on a line. Suppose we let $P_1(x_1, y_1)$ and $P_2(x_2, y_2)$ be any two points on a rectangular coordinate system. Let P_3 be the point where the horizontal line through P_1 and a vertical line through P_2 intersect to form a right triangle; see Figure 3.16.

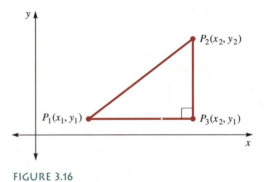

FIGURE 3.16

Since the line through P_1 and P_3 is horizontal, the y-coordinate of both points will be the same, y_1. Since the line through P_2 and P_3 is vertical, the x-coordinate of both points will be the same, x_2. The horizontal distance from P_1 to P_3 is the absolute value of the difference of their x-coordinates, $|x_2 - x_1|$. Similarly, the vertical distance from P_2 to P_3 is the absolute value of the difference of their y-coordinates, $|y_2 - y_1|$. See Figure 3.17.

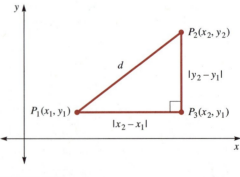

FIGURE 3.17

We use the absolute value because distance (the length of a line segment) is nonnegative. The expressions $|x_2 - x_1|$ and $|x_1 - x_2|$ represent the same length. If we let d represent the distance from P_1 to P_2, then by the Pythagorean theorem,

$$d^2 = |x_2 - x_1|^2 + |y_2 - y_1|^2 = (x_2 - x_1)^2 + (y_2 - y_1)^2$$

The absolute value symbols are not necessary here since the square of any real number is nonnegative. The distance formula is derived by taking the square root of each side. Since distance is nonnegative, we use the positive square root.

THEOREM 3.1

Formula for Distance Between Points

The distance between two points $P_1(x_1, y_1)$ and $P_2(x_2, y_2)$ in a plane is given by

$$d = \sqrt{(x_2 - x_1)^2 + (y_2 - y_1)^2}$$

EXAMPLE 9   Find the distance between the points $(3, 4)$ and $(-2, 6)$.

SOLUTION   We let $(x_1, y_1) = (-2, 6)$ and $(x_2, y_2) = (3, 4)$. Substituting into the distance formula yields

$$d = \sqrt{(x_2 - x_1)^2 + (y_2 - y_1)^2}$$
$$d = \sqrt{[3 - (-2)]^2 + (4 - 6)^2}$$
$$= \sqrt{5^2 + (-2)^2}$$
$$= \sqrt{29}$$

EXAMPLE 10   Find the distance between the points $(3, 2)$ and $(-2, -10)$.

SOLUTION   We let $(x_1, y_1) = (3, 2)$ and $(x_2, y_2) = (-2, -10)$. Then

$$d = \sqrt{(x_2 - x_1)^2 + (y_2 - y_1)^2}$$
$$= \sqrt{(-2 - 3)^2 + (-10 - 2)^2}$$
$$= \sqrt{(-5)^2 + (-12)^2}$$
$$= \sqrt{25 + 144}$$
$$= \sqrt{169}$$
$$= 13$$

The **midpoint** P_m of a line segment joining P_1 and P_2 is the point such that the distance from P_1 to P_m is equal to the distance from P_2 to P_m. The midpoint of a line segment can be found when its endpoints are known. Suppose that $P_1(x_1, y_1)$ and $P_2(x_2, y_2)$ are two points in a plane and $P_m(x_m, y_m)$ is the midpoint of the line joining them, as in Figure 3.18 on page 122.

From geometry we know that triangle P_1P_mS is congruent (same size and shape) to triangle P_mP_2T. Thus, the length of segment P_1S is half as long as the length of segment P_1P_3:

$$P_1S = \frac{P_1P_3}{2} = \frac{x_2 - x_1}{2}$$

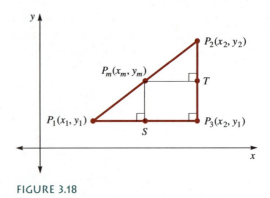

FIGURE 3.18

which is positive because $x_2 > x_1$. The x-coordinate of the midpoint, x_m, can be found by observing that the x-coordinate at the point S is also x_m. The distance $P_1 S$ is $x_m - x_1$. We now have two expressions for the distance $P_1 S$, which must be equal:

$$x_m - x_1 = \frac{x_2 - x_1}{2}$$

$$x_m = x_1 + \frac{x_2 - x_1}{2} \qquad \text{Add } x_1 \text{ to each side.}$$

$$x_m = \frac{2x_1}{2} + \frac{x_2 - x_1}{2} = \frac{x_1 + x_2}{2}$$

In a similar manner, the reader should verify that the y-coordinate of the midpoint, y_m, is

$$y_m = \frac{y_1 + y_2}{2}$$

The results show that the coordinates of the midpoint of a line segment can be found by averaging the coordinates of the end points.

> **DEFINITION**
>
> **The Midpoint of a Line Segment**
>
> The **midpoint of a line segment** joining point (x_1, y_1) to point (x_2, y_2) is given by
>
> $$(x_m, y_m) = \left(\frac{x_1 + x_2}{2}, \frac{y_1 + y_2}{2} \right)$$

EXAMPLE 11 Find the midpoint of the line segment joining $(2, 6)$ to $(-4, 5)$.

SOLUTION

$$x_m = \frac{x_1 + x_2}{2} = \frac{2 + (-4)}{2} = -1$$

$$y_m = \frac{y_1 + y_2}{2} = \frac{6 + 5}{2} = \frac{11}{2}$$

The midpoint is $\left(-1, \frac{11}{2} \right)$.

EXAMPLE 12 Find the midpoint of the line segment joining $\left(\frac{4}{3}, \frac{3}{4}\right)$ to $\left(\frac{2}{3}, \frac{1}{4}\right)$.

SOLUTION
$$x_m = \frac{\frac{4}{3} + \frac{2}{3}}{2} \qquad y_m = \frac{\frac{3}{4} + \frac{1}{4}}{2}$$

$$= \frac{\frac{6}{3}}{2} \qquad\qquad = \frac{\frac{4}{4}}{2}$$

$$= \frac{2}{2} \qquad\qquad = \frac{1}{2}$$

$$= 1$$

The midpoint is $\left(1, \frac{1}{2}\right)$.

EXAMPLE 13 Show that the points $A(3, 4)$, $B(5, 7)$, $C(9, 2)$, and $D(7, -1)$ are the vertices of a parallelogram. See Figure 3.19.

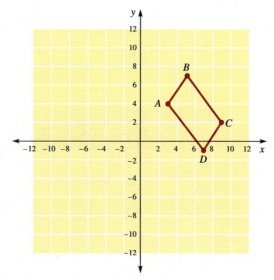

FIGURE 3.19

SOLUTION One way to show that a figure is a parallelogram is to show that the opposite sides are parallel.

$$m_{AB} = \frac{7 - 4}{5 - 3} = \frac{3}{2} \qquad m_{CD} = \frac{-1 - 2}{7 - 9} = \frac{-3}{-2} = \frac{3}{2}$$

$$m_{AD} = \frac{-1 - 4}{7 - 3} = \frac{-5}{4} \qquad m_{BC} = \frac{2 - 7}{9 - 5} = \frac{-5}{4}$$

Since the slopes of the opposite sides are equal, the opposite sides are parallel. The figure is a parallelogram.

EXAMPLE 14 A line through the points (2, 7) and (4, −4) is parallel to the line through (x, 4) and (6, 5). Find x.

SOLUTION The slope of the line through (2, 7) and (4, −4) is

$$\frac{-4-7}{4-2} = \frac{-11}{2}$$

The slope of the line through (x, 4) and (6, 5) is

$$\frac{5-4}{6-x} = \frac{1}{6-x}$$

Since the two lines are parallel, their slopes are equal:

$$\frac{1}{6-x} = \frac{-11}{2}$$

$$2(6-x)\left(\frac{1}{6-x}\right) = 2(6-x)\left(\frac{-11}{2}\right) \qquad \text{Multiply each side by } 2(6-x), \text{ the LCD.}$$

$$2 = -66 + 11x$$

$$68 = 11x \qquad \text{Add 66 to each side.}$$

$$x = \frac{68}{11} = 6\frac{2}{11}$$

Do You Remember?

Can you match these?

_____ 1. Slope formula
_____ 2. Parallel symbol
_____ 3. Perpendicular symbol
_____ 4. Midpoint formula
_____ 5. Distance formula
_____ 6. Zero slope
_____ 7. Undefined slope
_____ 8. Collinear points
_____ 9. Positive slope
_____ 10. Negative slope

a) Horizontal line
b) $\sqrt{(x_2-x_1)^2+(y_2-y_1)^2}$
c) A line that rises from left to right
d) Vertical line
e) $\dfrac{y_2-y_1}{x_2-x_1}$
f) Points that lie on the same line
g) $\left(\dfrac{x_1+x_2}{2},\dfrac{y_1+y_2}{2}\right)$
h) ⊥
i) A line that falls from left to right
j) ∥

Answers: 1. e 2. j 3. h 4. g 5. b 6. a 7. d 8. f 9. c 10. i

Exercise Set 3.2

Find the slope of the line joining each given pair of points. Some pairs may have an undefined slope.

1. $(3, 4), (6, 5)$
2. $(1, 5), (3, 8)$
3. $(4, 1), (3, 2)$
4. $(8, 5), (6, 7)$
5. $(0, 0), (4, -1)$
6. $(3, -2), (0, 0)$
7. $(3, 9), (4, 9)$
8. $(-5, 2), (3, 2)$
9. $(4, -7), (4, 8)$
10. $(-6, -1), (-6, -2)$
11. $(-5, -3), (-1, 0)$
12. $(-7, -7), (-4, -1)$
13. $\left(\frac{1}{2}, \frac{4}{3}\right), \left(\frac{3}{2}, \frac{1}{3}\right)$
14. $\left(-\frac{1}{5}, \frac{3}{8}\right), \left(\frac{1}{4}, -\frac{1}{2}\right)$
15. $(a, b), (c, d)$
16. $(r, s), (t, v)$
17. $(a_1, b_1), (a_2, b_2)$
18. $(x_2, y_2), (x_1, y_1)$

Graph each equation. Find the slope and state whether the graph of the equation is horizontal or vertical, or whether it rises or falls.

19. $y = 6$
20. $y = -8$
21. $x = -4$
22. $x = 7$
23. $y = x + 2$
24. $y = x - 3$
25. $x + y = 1$
26. $2x + y = 5$
27. $y = x$
28. $x + y = 0$

Use slopes to determine whether the three points in each group are collinear.

29. $(3, -5), (10, 7), (5, 1)$
30. $(2, 1), (0, 7), (1, 3)$
31. $(3, 6), (-2, -1), (7, 8)$
32. $(0, 0), (4, -2), (10, 1)$
33. $\left(\frac{1}{2}, 1\right), \left(\frac{1}{4}, 2\right), (1, -1)$
34. $\left(\frac{5}{8}, \frac{1}{4}\right), \left(\frac{3}{8}, \frac{1}{2}\right), \left(\frac{7}{8}, -1\right)$

Find the slope of each line in Exercises 35–43.

35.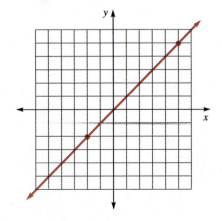

36.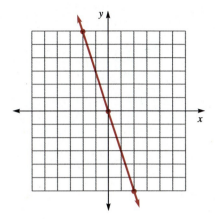

37.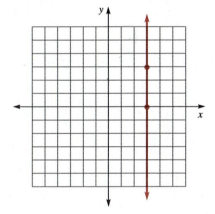

38.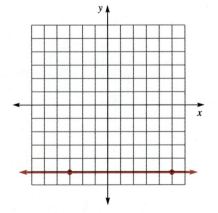

39.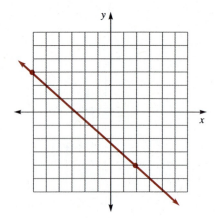

40.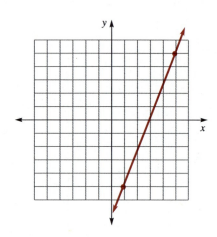

41.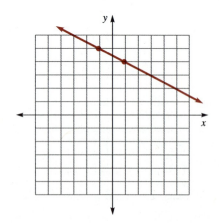

42.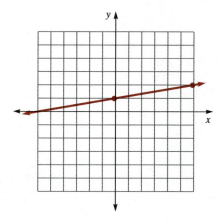

43.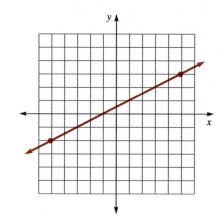

Find the distance between the given points.

44. $(-4, -5)$ and $(2, 3)$ **45.** $(1, 7)$ and $(4, 3)$

46. $(5, 3)$ and $(-4, 3)$ **47.** $(-5, 3)$ and $(1, 11)$

48. $(5, 9)$ and $(0, -3)$ **49.** $(3, 4)$ and $(-2, -8)$

50. (a, b) and (c, d) **51.** (a_1, b_1) and (a_2, b_2)

Find the coordinates of the midpoint of the line segment between the given points.

52. $(3, -5)$ and $(-2, 3)$ **53.** $(1, 6)$ and $(2, 8)$

54. $(-6, -5)$ and $(-4, 3)$ **55.** $(-6, -2)$ and $(-2, 2)$

56. (a, b) and (c, d) **57.** (m, n) and (s, t)

58. $\left(\dfrac{1}{2}, \dfrac{1}{4}\right)$ and $\left(\dfrac{1}{4}, \dfrac{1}{2}\right)$

59. $\left(\dfrac{1}{3}, \dfrac{1}{2}\right)$ and $\left(-\dfrac{1}{2}, -\dfrac{1}{3}\right)$

Determine whether the lines L_1 and L_2 are parallel, perpendicular, or neither in Exercises 60–67 and graph each pair of lines.

60. L_1 goes through (1, 4) and (3, 6);
  L_2 goes through (−3, 1) and (2, 6)
61. L_1 goes through (−7, 3) and (1, 11);
  L_2 goes through (−2, −5) and (3, 0)
62. L_1: $x + 2y = 7$;   **63.** L_1: $5x − 2y = 0$
  L_2: $2x − y = 5$ L_2: $2x − 3 = −5y$
64. L_1 goes through (3, −5) and (−2, 4);
  L_2 goes through (0, 6) and (−3, 1)
65. L_1 goes through $\left(\frac{1}{2}, \frac{1}{4}\right)$ and (−2, −3);
  L_2 goes through $\left(-\frac{1}{3}, 2\right)$ and $\left(4, -\frac{1}{5}\right)$
66. L_1: $3x − y = −9$ **67.** L_1: $8x − 4y = −1$
  L_2: $3y = −x + 2$ L_2: $2x + 4y = 5$

Graph each of the figures in Exercises 68–79 and use the slope to carry out the indicated instructions.

68. The four points $A(8, 4)$, $B(6, 1)$, $C(3, 4)$, and $D(1, 1)$ are vertices of a parallelogram. Show that the pairs of opposite sides are parallel.
69. The points $A(−3, 5)$, $B(0, 8)$, $C(3, 6)$, and $D(−3, 0)$ are the vertices of a trapezoid. Show that one pair of opposite sides is parallel.
70. The three points $A(−1, 2)$, $B(−3, 5)$, and $C(0, 7)$ are the vertices of a right triangle. Show that two sides are perpendicular.
71. The points $A(5, 8)$, $B(−3, 2)$, $C(0, −2)$, and $D(8, 4)$ are the vertices of a rectangle. Show that any two adjacent sides are perpendicular.
72. The line through the points $(x, 3)$ and $(−2, 5)$ is parallel to the line through the points $(8, −4)$ and $(−1, 2)$. Find x.
73. The line through the points $(2, y)$ and $(4, 9)$ is perpendicular to the line through the points $(6, 7)$ and $(8, 0)$. Find y.
74. A line L has an x-intercept of 3 and a y-intercept of 4. Find its slope. Find the slope of any line perpendicular to L.
75. A line L goes through the points $(2, −3)$ and $(5, 1)$. Find the slope of any line parallel to L. Find the slope of any line perpendicular to L.
76. A line L goes through the points $(−2, 3)$ and $(7, 1)$. Find the slope of any line parallel to L. Find the slope of any line perpendicular to L.
77. A line L has an x-intercept of $−5$ and a y-intercept of 7. Find the slope of L. Find the slope of any line parallel to L. Find the slope of any line perpendicular to L.
78. Graph the lines $y = 3x − 2$ and $y = 3x + 1$ on the same coordinate system. What can you say about the slopes of these lines? What can you say about the lines?
79. Graph the lines $2x + 3y = 6$ and $3x − 2y = 6$ on the same coordinate system. What can you say about the slopes of these lines? What can you say about the lines?

DESCRIBE IN YOUR OWN WORDS

80. Parallel lines   **81.** Slope of a line
82. Perpendicular lines **83.** Congruent triangles
84. Collinear points   **85.** Undefined slope
86. Midpoint of a line segment

3.3 Point–Slope, Slope–Intercept, and the General Form of the Equation of a Line

HISTORICAL COMMENT Pierre de Fermat (1601–1665) preceded Descartes in determining the equation of a straight line. He failed to publish his work, however, and consequently Descartes received much of the recognition for the development of it.

OBJECTIVES In this section we will learn to
1. write the equation of a line given its slope and a given point on the line; and
2. write the equation of a line given its slope and its y-intercept.

In Section 3.2 we learned how to draw the graph of a line by finding two or more points on it. In this section we will learn how to reverse that process.

Suppose that we want to find the equation of the line that has slope 2 and passes through (3, 4). See Figure 3.20.

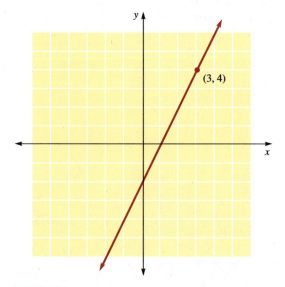

FIGURE 3.20

We know that given any two points on a line, the slope of the line is always the same, and that the slope between any two points (x_1, y_1) and (x_2, y_2) can be found by using the formula

$$m = \frac{y_2 - y_1}{x_2 - x_1}$$

If we let (x, y) be any point on the line other than (3, 4), the slope between (x, y) and (3, 4) must be 2:

$$\frac{y - 4}{x - 3} = 2$$

When the equation is cleared of fractions by multiplying each side by $x - 3$, we have

$$y - 4 = 2(x - 3) = 2x - 6$$

which can be rewritten

$$2x - y = 2$$

If an equation is written in the form $ax + by = c$, it is said to be in **general form.** When the procedure we just illustrated is completed using a point (x_1, y_1) and the slope m, the result is a formula for finding the equation of a line with slope m through any point (x_1, y_1):

$$\frac{y - y_1}{x - x_1} = m$$

$y - y_1 = m(x - x_1)$ Multiply each side by $x - x_1$.

This form is known as the **point–slope form** for the equation of a line.

DEFINITION
Point–Slope Form

The **point–slope form** for the equation of a line is

$$y - y_1 = m(x - x_1)$$

EXAMPLE 1 Find the equation of the line passing through (1, 2) and with slope $= \frac{1}{2}$. Write the answer in general form.

SOLUTION The equation of a line in point–slope form is

$$y - y_1 = m(x - x_1)$$

If we let the point $(x_1, y_1) = (1, 2)$ and $m = \frac{1}{2}$, we have

$$y - 2 = \frac{1}{2}(x - 1)$$

$2y - 4 = x - 1$ Multiply each side by 2.

$x - 2y = -3$ General form for the equation of a line

EXAMPLE 2 Write the equation of the line through (3, 2) and (−5, 4).

SOLUTION First we compute the slope of the line, letting $(x_1, y_1) = (3, 2)$ and $(x_2, y_2) = (-5, 4)$. Then

$$m = \frac{y_2 - y_1}{x_2 - x_1} = \frac{4 - 2}{-5 - 3} = \frac{2}{-8} = \frac{-1}{4}$$

The equation is

$y - 2 = \dfrac{-1}{4}(x - 3)$ $(x_1, y_1) = (3, 2)$

$4y - 8 = -x + 3$ Multiply each side by 4.

$x + 4y = 11$ General form for the equation of a line

It should be noted that (−5, 4) could have been used instead of (3, 2) to find the equation of the line. The equation

$$y - 4 = -\frac{1}{4}(x + 5)$$

simplifies to the same general form.

When the known point on a line is the *y*-intercept, a special form for the equation of the line, called the **slope–intercept form,** results. For example, if the line passes through $(0, b)$, the *y*-intercept, with slope *m*, then

$y - y_1 = m(x - x_1)$

$y - b = m(x - 0)$ $(x_1, y_1) = (0, b)$

$y - b = mx$

$y = mx + b$ Add *b* to each side.

DEFINITION	The **slope–intercept form** for the equation of a line is
Slope–Intercept Form	$y = mx + b$
	slope ↗ ↖ y-intercept
	where m is the slope and b is the y-intercept.

EXAMPLE 3 Find the equation of the line with slope $\frac{2}{3}$ and y-intercept 2. Sketch its graph.

SOLUTION We want an equation of the form

$$y = mx + b$$

$$y = \frac{2}{3}x + 2 \qquad m = \frac{2}{3}, b = 2$$

The graph is shown in Figure 3.21

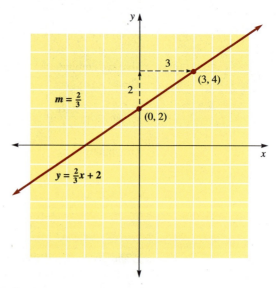

FIGURE 3.21

EXAMPLE 4 Find the slope and y-intercept of the line $-3x - 4y - 6 = 0$.

SOLUTION To find the slope and y-intercept, we can write the equation in the slope–intercept form, $y = mx + b$, by solving the equation for y.

$$-3x - 4y - 6 = 0$$

$$-4y = 3x + 6 \qquad \text{Add } 3x \text{ and } 6 \text{ to each side.}$$

$$y = \frac{3}{-4}x + \frac{6}{-4} \qquad \text{Divide each side by } -4.$$

$$y = -\frac{3}{4}x - \frac{3}{2}$$

The slope is $-\frac{3}{4}$ and the y-intercept is $-\frac{3}{2}$.

EXAMPLE 5 Find the equation of the line that goes through (3, 2) and is perpendicular to the line $4x + 2y = 3$.

SOLUTION The slope–intercept form for $4x + 2y = 3$ is

$$y = -2x + \frac{3}{2}$$

Recall that two lines are perpendicular if their slopes are negative reciprocals. The slope of the given line is -2, which means that the slope of a perpendicular line will be $\frac{1}{2}$ (see Section 3.2). Therefore the equation of the line in point–slope form is

$$y - 2 = \frac{1}{2}(x - 3)$$

In slope–intercept form, the equation of the desired line is

$$y = \frac{1}{2}x + \frac{1}{2}$$

EXAMPLE 6 Find the equation of the line that goes through (6, 2) and has undefined slope.

SOLUTION Recall from Section 3.2 that it is a vertical line that has undefined slope. The point–slope form and the slope–intercept form cannot be used to find its equation. A line with undefined slope is vertical, which means that the x-coordinate of any point on the line is 6. Therefore the equation is

$$x = 6$$

EXAMPLE 7 Find the general form of the equation of the line that has x-intercept 7 and is parallel to the line containing $(\frac{1}{2}, 2)$ and $(5, -\frac{1}{3})$.

SOLUTION The slope of the line is

$$m = \frac{y_2 - y_1}{x_2 - x_1} = \frac{-\frac{1}{3} - 2}{5 - \frac{1}{2}} = \frac{-\frac{7}{3}}{\frac{9}{2}} = \frac{-14}{27}$$

The x-intercept is 7, which means the point (7, 0). Using the point–slope form, we have

$$y - 0 = \frac{-14}{27}(x - 7) \quad \text{Point–slope form}$$
$$27y = -14x + 98 \quad \text{Multiply each side by 27.}$$
$$14x + 27y = 98 \quad \text{General form}$$

EXAMPLE 8 Find the equation of the line that is graphed in Figure 3.22.

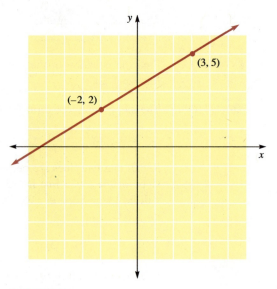

FIGURE 3.22

SOLUTION The line passes through $(-2, 2)$ and $(3, 5)$, so its slope is

$$m = \frac{2-5}{-2-3} = \frac{-3}{-5} = \frac{3}{5}$$

The equation can now be written using the point–slope form together with either of the points, $(-2, 2)$ or $(3, 5)$. We choose $(3, 5)$.

$(y - y_1) = m(x - x_1)$ Point–slope form

$(y - 5) = \frac{3}{5}(x - 3)$ Substitute $\frac{3}{5}$ for m and $(3, 5)$ for (x_1, y_1).

$3x - 5y = -16$

A Summary of Linear Equations

General form	$ax + by = c$
Point–slope form	$y - y_1 = m(x - x_1)$ for point (x_1, y_1), slope m
Slope–intercept form	$y = mx + b$ for slope m, y-intercept b
Horizontal line	$y = c$, with c constant, slope zero
Vertical line	$x = c$, with c constant, slope undefined

Exercise Set 3.3

Write the equation, in general form, of the line that passes through the given point with the given slope.

1. $(0, 2)$; $m = 1$
2. $(0, 3)$; $m = 2$
3. $(4, -1)$; $m = -3$
4. $(3, -2)$; $m = -1$
5. $(-1, -2)$; $m = \frac{1}{2}$
6. $(-3, 0)$; $m = \frac{2}{3}$
7. $(-5, 4)$; $m = -\frac{2}{5}$
8. $(3, -3)$; $m = -\frac{1}{3}$
9. $(4, 4)$; $m = 0$
10. $(3, 1)$; $m = 0$
11. $(2, 5)$; m undefined
12. $(4, -2)$; m undefined
13. $(-3, 4)$; parallel to $3x - y = 1$
14. $(1, 0)$; parallel to $x - y = 2$
15. $(4, 0)$; perpendicular to $y = -x$
16. $(0, 3)$; perpendicular to $y = x$
17. $(6, 2)$; parallel to $y = 5$
18. $(1, -1)$; parallel to $y = -2$
19. $(3, 2)$; perpendicular to $x = -\frac{1}{2}$
20. $(2, -5)$; perpendicular to $x = \frac{3}{4}$

Write the equation, in general form, of the line passing through the given pair of points.

21. $(1, 6), (5, -2)$
22. $(6, 1), (3, -4)$
23. $(2, -3), (-4, 2)$
24. $(-1, 7), (-2, 1)$
25. $(3, 4), (5, 4)$
26. $(-2, 3), (6, 3)$
27. $(-5, 2), (-5, 0)$
28. $(3, 4), (3, -1)$
29. $\left(5, \frac{1}{2}\right), \left(3, \frac{1}{3}\right)$
30. $\left(\frac{1}{2}, 2\right), \left(-\frac{1}{4}, 3\right)$
31. $\left(\frac{3}{5}, \frac{2}{3}\right), \left(-\frac{1}{2}, -\frac{1}{4}\right)$
32. $\left(-\frac{1}{6}, \frac{3}{5}\right), \left(\frac{1}{2}, -\frac{1}{3}\right)$

Write the equation, in slope–intercept form, of the line with the given slope and y-intercept.

33. $m = 2$, $b = 5$
34. $m = 1$, $b = -3$
35. $m = -\frac{1}{3}$, $b = \frac{1}{2}$
36. $m = -\frac{1}{4}$, $b = -4$
37. $m = 0$, $b = -5$
38. $m = 0$, $b = 3$

Write the equation, in general form, of the line that satisfies the given conditions.

39. x-intercept 3; y-intercept does not exist
40. y-intercept -2; x-intercept does not exist
41. x-intercept -4; y-intercept 1
42. x-intercept 2; y-intercept -3
43. x-intercept 0; y-intercept 0; parallel to $2x - y = 4$
44. x-intercept 1; parallel to $y = -x$
45. x-intercept -5; perpendicular to $y = -2x + 7$
46. x-intercept -1; perpendicular to $3x + y = -4$
47. y-intercept 2; parallel to the line containing $(3, 1)$ and $(-2, 4)$
48. y-intercept -1; parallel to the line containing $(0, 2)$ and $(4, -1)$
49. Through $(0, 3)$; perpendicular to $3x - 4y = 1$
50. Through $(2, -1)$; perpendicular to $5x + 3y = 7$
51. x-intercept 3, slope $\frac{1}{2}$
52. x-intercept -2, slope $\frac{1}{3}$
53. y-intercept -1, slope $= -\frac{2}{5}$
54. y-intercept 4, slope $= \frac{1}{2}$
55. Through the origin; undefined slope
56. Through the origin; zero slope

Write the equation of the line in slope–intercept form for Exercises 57–64.

57.

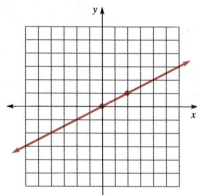

58.

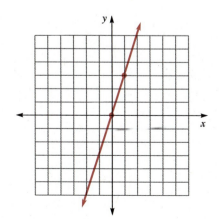

59.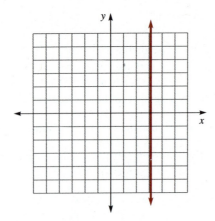

62.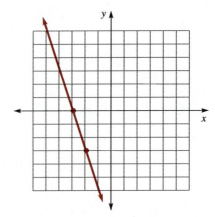

60.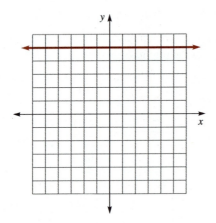

63.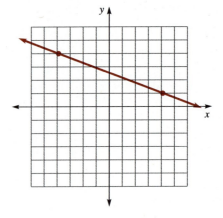

61.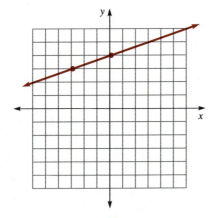

64.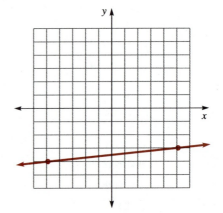

Write the equations, in general form, of the lines described.

65. Two perpendicular lines form the sides of a triangle ABC with vertices $A(1, 2)$, $B(3, 1)$, and $C(3, 6)$.

66. Two lines form the shortest sides of a triangle ABC with vertices $A(3, -1)$, $B(5, 7)$, and $C(-3, 7)$.

67. Two parallel lines form the sides of a trapezoid $ABCD$ with vertices $A(-4, 1)$, $B(3, 0)$, $C(4, 1)$ and $D(-2, 3)$.

68. Two parallel lines form the longest sides of a parallelogram $ABCD$ with vertices $A(-2, -1)$, $B(4, 1)$, $C(4, 5)$, and $D(-2, 3)$.

Write the general equation of the line described.

69. Graph the two lines $x + y = 6$ and $x - y = 2$ on the same coordinate system. Locate the point (ordered pair) where the two lines intersect (cross). Write the equation of the line through this point of intersection and the point $(0, 0)$.

70. Graph the two lines $x + 2y = 4$ and $x - y = 3$ on the same coordinate system. Locate the point (ordered pair) where the two lines intersect (cross). Write the equation of the line through this point of intersection and the point $(1, 1)$.

71. Graph the two lines $x + y = -5$ and $x - 2y = 4$ on the same coordinate system. Locate the point where the two lines intersect. Write the equation of the line through this point of intersection and parallel to the line $x + 3y = 7$.

72. Graph the two lines $2x - y = 2$ and $x + 3y = 8$ on the same coordinate system. Locate the point where the two lines intersect. Write the equation of the line through this point of intersection and perpendicular to the line $y = x$.

Write In Your Own Words

73. Explain the approach you would use to write the equations of the lines forming the sides of a rectangle if you know the vertices. Think of ways to save time and work.

74. If your weekly salary is calculated by the formula $I = 7.25h + 24$, where I = salary in dollars and h = hours worked, explain how you would calculate your salary if you worked a certain number of hours. Suppose you have to work n hours a week to keep your job, but you can't work more than p hours a week and maintain your grades. Write equations indicating your smallest and largest salaries, using these conditions.

75. In writing the equation of a line, we sometimes use the slope–intercept form and sometimes the point–slope form of the equation of a line. Is it possible that we could eliminate one of these forms? If so, which one? If it is possible to eliminate one of the forms, why, in your opinion, do we keep both forms?

76. Explain the approach you would use to write the equations of the lines forming the sides of a parallelogram. Think of ways to save time and work.

77. Suppose you are asked to graph either the linear equation $ax + by = c$ or $y = mx + b$. Which do you think is easier to graph and why? Explain how you would graph $ax + by = c$.

3.4 Relations, Functions, and Function Notation

HISTORICAL COMMENT Leonard Euler (1707–1783) was a Swiss mathematician who was the first to define a function and to use the concept, although others may have used notation similar to his at an earlier date. Euler, without a doubt, was one of the most prolific mathematicians in history. He was blind during the last seventeen years of his life, but even blindness did not slow his ability to do mathematical calculations, nor did it decrease his productivity.

OBJECTIVES In this section we will learn to
1. identify relations and functions;
2. use function notation;
3. find the domain of a relation; and
4. identify linear functions.

Suppose that a family averages 50 miles per hour on a trip between two cities. In 1 hour they will have traveled 50 miles and in 3 hours, 150 miles. The relationship between time and distance can be expressed using ordered pairs, as follows:

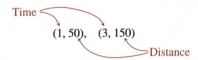

In a similar manner, the ordered pairs (2, 100), (4, 200), and (6, 300) also show how the time and distance are related to each other. The concept just illustrated leads us to define a **relation** as follows.

DEFINITION	
Relation	A **relation** is a set of ordered pairs.

For example, the set $A = \{(1, 6), (4, -4), (5, 7), (9, 2)\}$ is a relation that contains four ordered pairs. The set of ordered pairs described by $B = \{(x, y) \mid y = 2x + 3\}$, read, "The set of ordered pairs x, y such that $y = 2x + 3$," is also a relation. Some of the ordered pairs in B can be found by assigning values to x and using the equation $y = 2x + 3$ to calculate the corresponding values for y:

When $x = -4$, $\quad y = 2(-4) + 3 = -8 + 3 = -5$
When $x = -3$, $\quad y = 2(-3) + 3 = -6 + 3 = -3$
When $x = \dfrac{1}{2}$, $\quad y = 2\left(\dfrac{1}{2}\right) + 3 = 1 + 3 = 4$
When $x = 1$, $\quad y = 2(1) + 3 = 5$

Thus, $(-4, -5)$, $(-3, -3)$, $(\frac{1}{2}, 4)$, and $(1, 5)$ all belong to B.

The set of all first elements in the ordered pairs of a relation is called its **domain.** The set of all second elements in the ordered pairs is called its **range.** In the relation $\{(-3, 2), (1, 2), (5, -6), (9, 11)\}$, the domain is $\{-3, 1, 5, 9\}$ and the range is $\{-6, 2, 11\}$.

EXAMPLE 1 Determine the domain and the range of $C = \{(-2, 7), (4, -2), (3, -2)\}$.

SOLUTION The domain is $\{-2, 3, 4\}$. The range is $\{-2, 7\}$.

The **graph of a relation** is the graph of its ordered pairs. The graph of $y = 2x + 4$ is the straight line shown in Figure 3.23 at the top of the next page. The domain and range of a relation can often be determined when its graph is known.

EXAMPLE 2 Find the domain and range of the relations that are graphed in Figures 3.24(a) and (b).

SOLUTION (a) x can assume any value from -4 to 5. The domain is $[-4, 5]$.
y can assume any value from -2 to 2. The range is $[-2, 2]$.

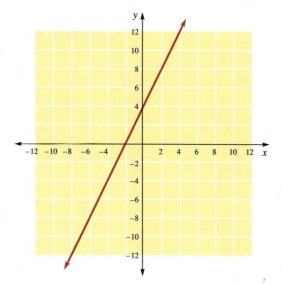

FIGURE 3.23

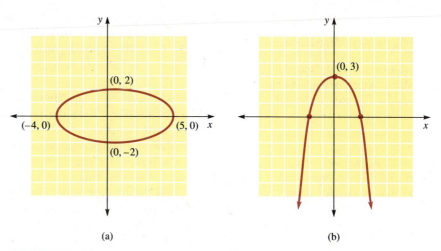

FIGURE 3.24

(b) x can assume any value from $-\infty$ to ∞. The domain is $(-\infty, \infty)$.
y can assume any value from $-\infty$ to 3. The range is $(-\infty, 3]$.

A **function** is a special type of relation in which certain restrictions are placed on the ordered pairs.

DEFINITION **Function**	A **function** is a correspondence or rule that pairs each element in the domain with *exactly one* element in the range.

Another way to describe a function is to say that no first element in an ordered pair can correspond to more than one second element.

The relations $A = \{(1, 2), (3, 4), (5, 6)\}$ and $B = \{(1, 2), (3, 4), (5, 4)\}$ are functions because each element in the domain corresponds to exactly one element in the range. The lines in Figure 3.25(a) illustrate the correspondence for set A, while the lines in Figure 3.25(b) illustrate the correspondence for set B.

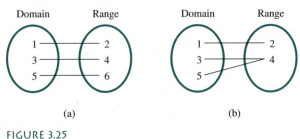

FIGURE 3.25

The relation $C = \{(1, 2), (2, 3), (2, 7)\}$ is *not* a function because the element 2 in the domain corresponds to two elements in the range, 3 and 7, as shown in Figure 3.26.

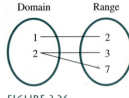

FIGURE 3.26

Functions are often assigned names, such as f or g. To say that "y is a function of x" means that each value of x in the domain corresponds to exactly one value of y in the range. To emphasize that *y is a function of x,* notation such as

$$y = f(x)$$

is used. Such notation is called **function notation.** The notation $f(x)$ is read "f of x" or "a function of x." Similarly, the notation $g(x)$ is read "g of x" or "a function of x." In function notation $y = 2x + 4$ is written $f(x) = 2x + 4$, where $f(x)$ indicates that the value of y depends on the value that is assigned to x and that only one value of y will result for each value of x.

CAUTION

The notation $f(x)$ is reserved to indicate that a function is being considered. It does *not* mean f times x, which is written fx.

In general, $f(x)$ represents the y-coordinate of a point corresponding to any given x-coordinate.

A function is often thought of a machine in which raw material is fed into one end and a finished product comes out of the other. (See Figure 3.27.) In a function, the machine is the equation that relates the domain to the range. The input is elements in the domain and the finished product is corresponding elements in the range.

To find the value of $f(x) = 2x + 4$ when $x = 1$, we substitute 1 for each x:

$$f(1) = 2(1) + 4 = 6.$$

Notice that $f(1) = 6$ represents the y-coordinate when $x = 1$. If $f(x) = 2x + 4$, then

$f(2) = 2(2) + 4 = 4 + 4 = 8$	Replace every x with 2.
$f(3) = 2(3) + 4 = 6 + 4 = 10$	Replace every x with 3.
$f(4) = 2(4) + 4 = 8 + 4 = 12$	Replace every x with 4.
$f(a) = 2 \cdot a + 4 = 2a + 4$	Replace every x with a.

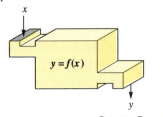

Input = Domain
$y = f(x)$
Output = Range

FIGURE 3.27

EXAMPLE 3 Find the value of the given function.

(a) $f(x) = 7x - 11$; find $f(-2)$.
(b) $g(x) = -4x - 9$; find $g(7)$.
(c) $h(x) = -5x + 11$; find $h(0)$.

SOLUTION (a) $f(-2) = 7(-2) - 11 = -14 - 11 = -25$ Replace every x by -2.
(b) $g(7) = -4(7) - 9 = -28 - 9 = -37$ Replace every x by 7.
(c) $h(0) = -5(0) + 11 = 0 + 11 = 11$ Replace every x by 0.

All of the functions in Example 3 are called **linear functions.** Their graphs are straight lines.

DEFINITION

Linear Function

A **linear function** is one that can be written in the form

$$f(x) = mx + b$$

where m and b represent real numbers.

The only linear equation that is not a function is $x = k$, a vertical line, since there is only one value of x but an infinite number of values for y.

EXAMPLE 4 Write $2x + 3y = 5$ using function notation. Identify its slope and y-intercept.

SOLUTION We will solve the equation for y and then replace y by $f(x)$.

$3y = -2x + 5$ Subtract $2x$ from each side.

$y = -\frac{2}{3}x + \frac{5}{3}$ Divide each side by 3.

$f(x) = -\frac{2}{3}x + \frac{5}{3}$ Replace y by $f(x)$.

The slope is $-\frac{2}{3}$ and the y-intercept is $\frac{5}{3}$.

We will study many other functions that are not linear. The notation, however, does not change.

EXAMPLE 5 Find the value of the given function.

(a) $f(x) = x^2 - 3x - 4$; find $f(2)$.   (b) $h(x) = x^3 - 1$; find $h(4)$.

SOLUTION (a) $f(2) = 2^2 - 3(2) - 4 = 4 - 6 - 4 = -6$ Replace every x with 2.
(b) $h(4) = 4^3 - 1 = 64 - 1 = 63$ Replace every x with 4.

EXAMPLE 6 Given $g(x) = x + 2$, find $g(a)$, $g(b)$, $g(a + b)$, and $g(x + b)$.

SOLUTION

$g(a) = a + 2$ Replace every x with a.

$g(b) = b + 2$ Replace every x with b.

$$g(a+b) = a+b+2 \qquad \text{Replace every } x \text{ with } a+b.$$
$$g(x+b) = x+b+2 \qquad \text{Replace every } x \text{ with } x+b.$$

EXAMPLE 7 The area and circumference of a circle both depend upon the length of its radius. Use function notation to indicate this when $r = 3$ feet.

SOLUTION

AREA		CIRCUMFERENCE
$A(r) = \pi r^2$	and	$C(r) = 2\pi r$
$A(3) = \pi(3^2) = 9\pi$		$C(3) = 2\pi(3) = 6\pi$

When the radius is 3 feet, the area is 9π square feet and the circumference is 6π feet.

EXAMPLE 8 A company that sells parts for repairing alternators has determined that the price in dollars per unit for selling x units is $P(x) = 50 - (0.3)x$, where $1 \leq x \leq 20$. Find the price per unit when 4 units are sold and the price per unit when 20 units are sold.

SOLUTION When 4 units are sold, the price per unit is

$$P(4) = 50 - (0.3)(4) = \$48.80$$

When 20 units are sold, the price per unit is

$$P(20) = 50 - (0.3)(20) = \$44.00$$

EXAMPLE 9 Is the equation $y = \dfrac{2x}{x-3}$ a function? What is its domain?

SOLUTION To determine whether $y = \dfrac{2x}{x-3}$ is a function, we need to answer the question, "Can any value of x yield more than one value of y?" The answer is no. For example, if $x = 1$,

$$y = \frac{2(1)}{1-3} = \frac{2}{-2} = -1 \qquad \text{Replace each } x \text{ by 1.}$$

Whenever $x = 1$, we get $y = -1$. We will always get a unique value for y whenever we substitute a value for x in $y = \dfrac{2x}{x-3}$. There can be only one value of y for each value of x. The equation describes a function.

The domain of a function is the set of all first elements in its ordered pairs. In $y = \dfrac{2x}{x-3}$, if $x = 3$ the denominator will be zero, and division by zero is not defined. The domain is therefore all real numbers except 3.

$$(-\infty, 3) \cup (3, \infty)$$

EXAMPLE 10   Is $y = \dfrac{x-2}{3x+4}$ a function? What is its domain?

SOLUTION   The denominator will be zero if $x = -\frac{4}{3}$, so $-\frac{4}{3}$ must be excluded from the domain. The domain is $(-\infty, -\frac{4}{3}) \cup (-\frac{4}{3}, \infty)$. Each time a particular value of x is used in the formula, a unique value of y will result. The equation describes a function.

When the graph of a relation is known, we can determine whether the relation is a function by looking at the graph. *If a vertical line intersects the graph of the relation in more than one point, the relation is not a function.* This is called the **vertical line test.** In Figure 3.28(a), the x-coordinate is the same for both points where the vertical line crosses the graph. This means there are two values of y for this value of x (two ordered pairs in which the first element corresponds to two different second elements). Therefore the relation is not a function. In Figure 3.28(b), a vertical line will not intersect the graph in more than one point anywhere on the graph. There is one value of y for each value of x, so the graph is that of a function.

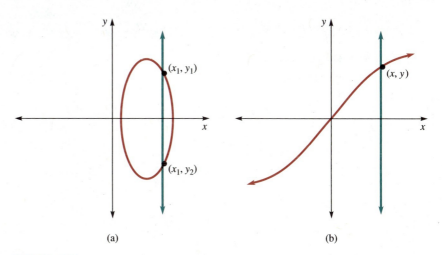

FIGURE 3.28

ℰxploring 𝒢raphics

Graphics calculators and computers can be used to draw the graph of a function. The keystrokes necessary to graph a function will vary depending on the make and model of the calculator you use. It will be necessary to consult the users' manual for this information. Brief instructions for operating the TI-81 and TI-82 are contained in the Appendix. Many different computer programs have been written that support graphing. Figure 3.29 shows the graph of $f(x) = 2x - 2$ drawn by a Macintosh computer.

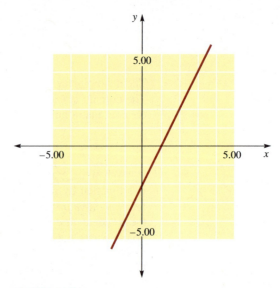

FIGURE 3.29

Do You Remember?

Can you match these?
More than one answer may apply.

_____ 1. A relation
_____ 2. An example of a function
_____ 3. Function notation
_____ 4. The domain of a relation
_____ 5. The range of a relation

a) The set of first elements in a set of ordered pairs
b) A set of ordered pairs
c) The set of second elements in a set of ordered pairs
d) {(1, 2), (3, 4), (5, 6), (7, 8)}
e) $f(x)$
f) {(1, 2), (3, 4), (5, 6), (3, 2)}

Answers: 1. b, f 2. d 3. e 4. a 5. c

Exercise Set 3.4

Determine whether each given relation is a function. Find the domain and range of each.

1. {(1, 3), (2, 7), (3, 8), (−1, 2)}
2. {(−3, −7), (−2, 5), (0, 1), (2, 5)}
3. {(−4, 16), (0, −8), (2, −8), (6, 16)}
4. {(−5, −12), (−2, 3), (0, 3), (−2, 3), (3, −12)}
5. {(3, 7), (2, −5), (3, −7), (6, −5)}
6. {(0, 6), (6, 0), (0, −6), (−6, 0)}
7. {(2, 1), (4, 3), (3, 4), (2, 0)}
8. {(1, 0), (−1, 8), (0, 1), (−1, 0)}

Find the domain and range of each of the following functions.

9. $y = x$
10. $y = -x$
11. $y = 2x - 3$
12. $y = -2x + 7$
13. $y = |x|$
14. $y = -|x|$
15. $y = 2 - x$
16. $y = 3 - x$
17. $y = -7$
18. $y = 2$
19. $y = |x + 2|$
20. $y = |x + 3|$
21. $y = -|x - 4|$
22. $y = -|x - 1|$

Determine whether each given relation is a function and find the domain.

23. $y = \dfrac{1}{x}$
24. $y = \dfrac{5}{x}$
25. $y = \dfrac{1}{x - 1}$
26. $y = \dfrac{2}{x + 3}$
27. $y = \dfrac{x - 4}{x + 3}$
28. $y = \dfrac{2x - 1}{x - 4}$
29. $x = 4$
30. $x = -3$
31. $y = \dfrac{2x - 1}{x - 5}$
32. $y = \dfrac{4x - 7}{x + 8}$

Find the domain and range of each given relation.

33.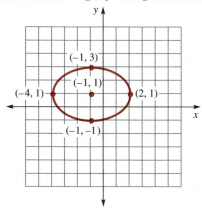

34.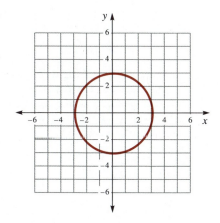

35.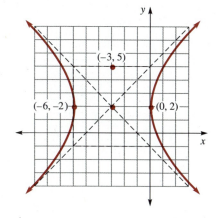

36.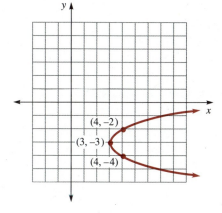

37.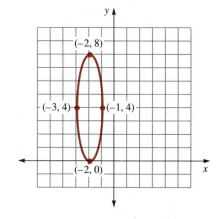

38.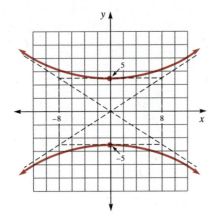

41.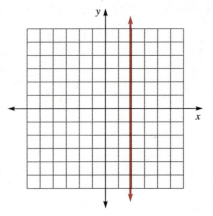

39.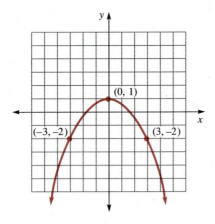

42.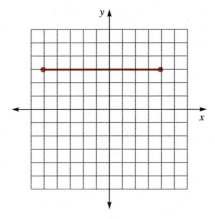

40.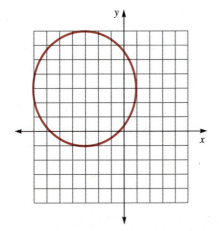

43.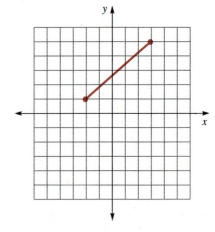

44.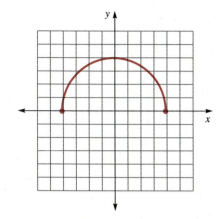

62.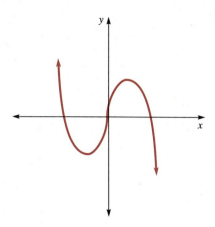

Find the value of each function as indicated.
45. $f(x) = x - 3$; $f(1), f(-1), f(3)$
46. $g(x) = 3x - 4$; $g(1), g(0), g\left(\dfrac{1}{2}\right)$
47. $h(x) = x^2 - 1$; $h(0), h(-1), h(1)$
48. $F(x) = x^2 - 4$; $F(0), F(2), F(-2)$
49. $g(x) = |16 - x|$; $g(4), g(-4), g(0)$
50. $f(x) = |5 - x|$; $f(0), f(2), f(-2)$
51. $f(x) = 3x^2 + x - 8$; $f(0), f(2), f(-1)$
52. $F(x) = 6x^2 - 25x + 14$; $F\left(\dfrac{2}{3}\right), F\left(\dfrac{7}{2}\right), F(0)$
53. $f(x) = ax^2 + bx + c$; $f\left(\dfrac{-b}{2a}\right)$ and $f\left(\dfrac{b}{2a}\right)$
54. $g(x) = 7x - 3$; $g(b), g(x + b), g(x - c), g(c^2 - 1)$

63.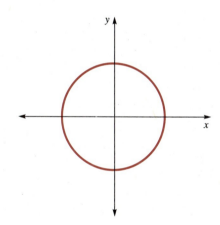

Graph each of the following linear functions.
55. $f(x) = x$ **56.** $f(x) = x - 5$
57. $f(x) = 2x + 7$ **58.** $F(x) = 3x - 2$
59. $f(x) = \dfrac{1}{2}x + 1$ **60.** $f(x) = \dfrac{1}{4}x - 2$

Use the vertical line test to determine whether each of the following graphs represents a function.

61.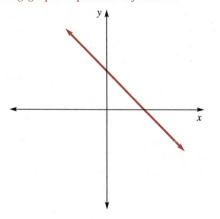

64.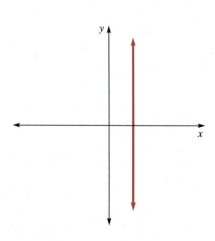

65.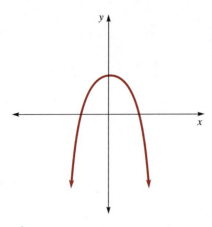

66.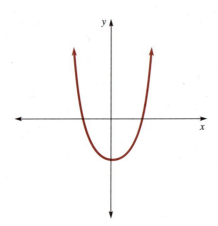

67.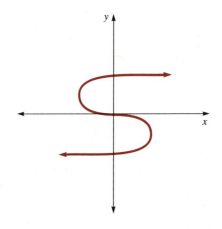

68.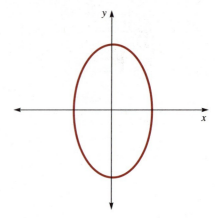

Each of the following is a function. Solve as indicated.

69. The value $V(t)$ in dollars of factory equipment depreciates with time t in years according to the formula
$$V(t) = 15{,}000 - 1000t$$

(a) What is the value of the equipment after 15 years?

(b) What is the value of the equipment when new $(t = 0)$?

(c) At what time t will the equipment have lost one-half of its original value?

70. The cost $C(x)$ in dollars of producing x pens is given by
$$C(x) = \frac{1}{3}x + 212$$

(a) What is the cost of producing 30 pens?

(b) How many pens can be produced for a cost of $272.00?

(c) How many pens should be produced so that the cost for each pen is no more than 74 cents?

71. A company's daily cost function is $C(n) = 100 + 5n$ and its revenue function is $R(n) = 55n$, where n is the number of units produced and sold, and cost and revenue are in dollars.

(a) Recall that $P(n) = R(n) - C(n)$. Find the profit function $P(n)$.

(b) Find the production value, n, such that the profit is zero.

(c) For what production values does the company make money?

(d) For what production values does the company lose money?

(e) If this were your company, how many units would you choose to produce and sell daily?

72. A tractor manufacturer finds the daily cost function for producing n tractors to be $C(n) = 180 + 272n$.

The total revenue from the sale of n tractors is $R(n) = 302n$.

(a) Find the profit function $P(n)$, where $P(n) = R(n) - C(n)$.
(b) Find the production values n that make the profit zero.
(c) For what values of n does the company make money?
(d) For what values of n does the company lose money?
(e) If this were your company, how many tractors would you try to produce and sell each day?

REVIEW EXERCISES

73. Find the equation of the line through the points $(1, 0)$ and $(0, 2)$.
74. Find the equation of the line through $(0, 4)$ and with slope $m = \frac{1}{2}$.
75. Find the coordinates of the midpoint of the line segment joining $(4, 3)$ and $(-1, 5)$.
76. Find the slope of the line segment joining $(4, 3)$ and $(-1, 5)$.
77. Find the distance from $(-3, -5)$ to $(3, 3)$.
78. Is the line through $(-2, 5)$ and $(0, 1)$ parallel or perpendicular to the line through $(4, -1)$ and $(6, 0)$?

3.5 The Algebra of Functions; Inverse Functions

OBJECTIVES In this section we will learn to
1. add, subtract, multiply, and divide functions;
2. find the composition of functions; and
3. find the inverse of a function

Functions are often added, subtracted, multiplied, or divided for a particular value of a variable. For example suppose that $f(x) = x^2 + 5$ and $g(x) = 2x - 3$ and suppose that we wish to find $f(3) + g(3)$. In Section 3.4 we learned how to find the value of a function.

$$f(x) = x^2 + 5 \qquad \text{and} \qquad g(x) = 2x - 3$$
$$f(3) = 3^2 + 5 \qquad\qquad\qquad g(3) = 2(3) - 3$$
$$= 9 + 5 \qquad\qquad\qquad\quad = 6 - 3$$
$$= 14 \qquad\qquad\qquad\qquad = 3$$

Thus, $f(3) + g(3) = 14 + 3 = 17$.

If we add the two function *before* finding $f(3)$ and $g(3)$, the result will be the same. The notation for adding two functions $f(x)$ and $g(x)$ is $(f + g)(x)$.

$$f(x) = x^2 + 5 \qquad \text{and} \qquad g(x) = 2x - 3$$
$$(f + g)(x) = x^2 + 5 + 2x - 3$$
$$= x^2 + 2x + 2$$

When this is evaluated for $x = 3$, we have

$$(f + g)(3) = 3^2 + 2(3) + 2$$
$$= 9 + 6 + 2$$
$$= 17$$

We have illustrated the notation used in the addition of functions. Similar notation indicates subtraction, multiplication, and division as shown in the following box entitled **The Algebra of Functions.**

The Algebra of Functions

If $f(x)$ and $g(x)$ represent functions and if x is in the domain of both $f(x)$ and $g(x)$, then

1. **Addition:** $\quad f(x) + g(x) = (f + g)(x)$
2. **Subtraction:** $\quad f(x) - g(x) = (f - g)(x)$
3. **Multiplication:** $[f(x)] \cdot [g(x)] = (f \cdot g)(x)$
4. **Division:** $\quad \dfrac{f(x)}{g(x)} = \left(\dfrac{f}{g}\right)(x), \quad g(x) \neq 0$

EXAMPLE 1 $f(x) = x + 5$ and $g(x) = 3x - 2$. Find $(f - g)(5)$.

SOLUTION $f(5) = 5 + 5 = 10$ and $g(5) = 3(5) - 2 = 13$, so

$$(f - g)(5) = 10 - 13 = -3$$

EXAMPLE 2 $f(x) = x^3 - 2x$ and $g(x) = x^2 - 1$. Find $(f \cdot g)(2)$.

SOLUTION $f(2) = 2^3 - 2(2) = 8 - 4 = 4$ and $g(2) = 2^2 - 1 = 4 - 1 = 3$, so

$$(f \cdot g)(2) = 4 \cdot 3 = 12$$

EXAMPLE 3 $f(x) = x^3 - x^2$ and $h(x) = 3x - 2$. Find $\left(\dfrac{f}{h}\right)(4)$.

SOLUTION $\quad f(4) = 4^3 - 4^2 = 64 - 16 = 48$
$\quad h(4) = 3(4) - 2 = 12 - 2 = 10$

So $\quad \left(\dfrac{f}{h}\right)(4) = \dfrac{48}{10} = \dfrac{24}{5} = 4\dfrac{4}{5}$

EXAMPLE 4 $g(x) = 2x^2 + x$ and $h(x) = x + 9$. Find $(g + h)\left(\dfrac{2}{3}\right)$.

SOLUTION $g\left(\frac{2}{3}\right) = 2\left(\frac{2}{3}\right)^2 + \frac{2}{3}$ and $h\left(\frac{2}{3}\right) = \frac{2}{3} + 9$

$= \frac{8}{9} + \frac{2}{3} = \frac{8}{9} + \frac{6}{9}$ $= \frac{29}{3}$

$= \frac{14}{9}$

So $(g + h)\left(\frac{2}{3}\right) = \frac{14}{9} + \frac{29}{3} = \frac{14}{9} + \frac{87}{9} = \frac{101}{9} = 11\frac{2}{9}$

The domain of the sum, difference, product, or quotient of functions is the *intersection* of the domains of the individual functions. In other words, it cannot include any values that are excluded from the domains of any of the individual functions.

EXAMPLE 6 If $f(x) = x + 5$ and $g(x) = x + 9$, find the domain of $(f + g)(x)$.

SOLUTION There are no values of x that must be excluded, so the domain of $f(x)$ is $(-\infty, \infty)$. In the same manner, there are no values of x that must be excluded from the domain of $g(x)$; its domain is also $(-\infty, \infty)$. The domain of $(f + g)(x)$ is $(-\infty, \infty)$, or R.

EXAMPLE 7 $f(x) = 4x - 7$ and $g(x) = \frac{1}{x - 4}$. Find the domain of $(f \cdot g)(x)$.

SOLUTION No values of x need be excluded from $f(x)$. The domain of $g(x)$ cannot contain any x for which the denominator is zero, so $x \neq 4$. The domain of $(f \cdot g)(x)$ must also exclude 4, so the domain of $(f \cdot g)(x)$ is $(-\infty, 4) \cup (4, \infty)$.

Functions are often indicated by the use of a single letter, such as f, g, or h. In Examples 8 and 9 following, the functions f and g are each defined by a set of ordered pairs.

EXAMPLE 8 Suppose $f = \{(1, 2), (3, 2), (5, 4), (2, 7)\}$ and $g = \{(1, 9), (2, 4), (3, 6), (-4, 1)\}$. Find the domain of $f - g$.

SOLUTION The domain of $f - g$ is the intersection of the domains of f and g. If we let D_f and D_g represent the domains of f and g, respectively, then $D_f = \{1, 2, 3, 5\}$ and $D_g = \{1, 2, 3, -4\}$. Thus $D_f \cap D_g = \{1, 2, 3\}$, so the domain of $f - g$ is $\{1, 2, 3\}$.

EXAMPLE 9 Let $f = \{(9, -1), (4, -9), (3, 1) (7, 2)\}$ and $g = \{(4, 6), (7, -1), (3, 5), (-4, 1)\}$.

Find: **(a)** $(f + g)(4)$ **(b)** $\left(\frac{f}{g}\right)(7)$ **(c)** $(f + g)(9)$

SOLUTION (a) The ordered pairs are of the form $(x, y) = (x, f(x))$ and $(x, g(x))$. Examining the sets, we see that $f(4) = -9$ and $g(4) = 6$, so $(f + g)(4) = -9 + 6 = -3$.

(b) $f(7) = 2$ and $g(7) = -1$, so $\left(\dfrac{f}{g}\right)(7) = \dfrac{2}{-1} = -2$.

(c) $(f + g)(9) = f(9) + g(9)$. Since 9 is not in the domain of g, we cannot find $g(9)$. The sum of the functions does not exist.

Functions can be related in yet another way, as follows: Suppose a student gets a job at a bookstore that pays $6.00 an hour. She knows that her gross monthly paycheck depends on the number of hours that she works. Her total earnings are a function of the number of hours she works. In function notation, we can write

$g(h) = 6h,$ where h is the number of hours worked

For example, if she works 14 hours, her pay is $g(14) = 6 \cdot 14 = \$84$. She also knows that each day she works a 7-hour shift. This means that the total hours worked is dependent on the number of days worked:

$h(d) = 7d,$ where d is the number of days worked

For example, if she works 3 days, her total hours worked is $h(3) = 7 \cdot 3 = 21$ hours. She also knows that the more days she works, the more money she will make. So this problem involves two functions: pay in terms of hours and hours in terms of days. How much pay will she receive for 21 days of work?

Hours Function	Pay Function
$h(d) = 7d$	$g(h) = 6h$
$h(21) = 7(21) = 147$ hours	$g(147) = 6(147) = \$882$

The process of analyzing interrelated functions such as this is called the **composition of functions.** We can write the composition of the pay function and the hours function as

$g[h(x)] = g(147) = 6(147) = \$882$

DEFINITION

The Composition of Functions

Let $f(x)$ and $g(x)$ be two functions. The expression $f[g(x)]$ is called the **composition of f with g** and $g[f(x)]$ is called the **composition of g with f.** Alternate notation is

$(f \circ g)(x) = f[g(x)]$ and $(g \circ f)(x) = g[f(x)]$

EXAMPLE 10 Let $f(x) = 3x - 1$ and $g(x) = 2x + 4$. Find $f[g(2)]$.

SOLUTION We begin by finding $g(2)$:

$g(x) = 2x + 4$
$g(2) = 2(2) + 4 = 8$ Replace every x in $g(x)$ by 2.

Now

$$f[g(2)] = f(8) \quad \text{Replace } g(2) \text{ by its value, 8.}$$
$$= 3(8) - 1 \quad \text{Replace every } x \text{ in } f(x) \text{ by 8.}$$
$$= 23$$

The operations in Example 10 can be likened to two "function machines" being connected to each other. The first machine is $g(x) = 2x + 4$ and the second machine is $f(x) = 3x - 1$. See Figure 3.30.

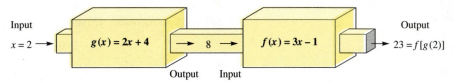

FIGURE 3.30

EXAMPLE 11 Let $f(x) = x^2 - 1$ and $g(x) = x - 6$. Find:

(a) $(f \circ g)(3) = f[g(3)]$ **(b)** $(g \circ f)(3) = g[f(3)]$

SOLUTION **(a)**
$$g(3) = 3 - 6 = -3 \quad \text{Replace every } x \text{ in } g(x) \text{ by 3.}$$
$$f[g(3)] = f(-3) \quad \text{Replace } g(3) \text{ by its value, } -3.$$
$$= (-3)^2 - 1 \quad \text{Replace every } x \text{ in } f(x) \text{ by } -3.$$
$$= 9 - 1$$
$$= 8$$

(b)
$$f(3) = 3^2 - 1 = 8 \quad \text{Replace every } x \text{ in } f(x) \text{ by 3.}$$
$$g[f(3)] = g(8) \quad \text{Replace } f(3) \text{ by its value, 8.}$$
$$= 8 - 6 \quad \text{Replace every } x \text{ in } g(x) \text{ by 8.}$$
$$= 2$$

Notice that $(f \circ g)(3) \neq (g \circ f)(3)$.

The **inverse of a relation** A is the relation obtained when its domain and range are interchanged. For example, if $A = \{(2, 3), (4, -1), (-5, 7)\}$, then its inverse is $\{(3, 2), (-1, 4), (7, -5)\}$. Notice that we have interchanged the first and second coordinates of the ordered pairs of A to find its inverse.

The equation $y = 3x - 2$ is a relation consisting of infinitely many ordered pairs. To find its inverse, we interchange x and y, and solve the resulting equation for y:

$$y = 3x - 2$$
$$x = 3y - 2 \quad \text{Interchange } x \text{ and } y.$$
$$x + 2 = 3y \quad \text{Add 2 to each side.}$$
$$\frac{x + 2}{3} = y \quad \text{Divide each side by 3.}$$

The inverse of $y = 3x - 2$ is $y = \frac{x+2}{3}$.

Notice that the equation $y = 3x - 2$ is a linear function that can be written $f(x) = 3x - 2$. Its inverse, $y = \frac{x+2}{3}$, is also a function. It is designated by $f^{-1}(x)$. Thus, $f^{-1}(x) = \frac{x+2}{3}$.

If the inverse of a function is again a function, it "undoes" the work that the function has done. For example, the output of the function $f(x) = 3x - 2$ is changed back to its input by using its inverse, $f^{-1}(x) = \frac{x+2}{3}$. This too can be pictured as a series of two machines (Figure 3.31):

> **C**AUTION
>
> The notation $f^{-1}(x)$ designates the inverse of $f(x)$. It does *not* mean $\frac{1}{f(x)}$.

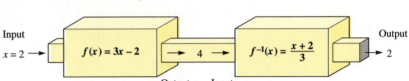

FIGURE 3.31

The input is the same as the output. We can say this in function notation:

$$(f^{-1} \circ f)(x) = x$$

When a linear function has nonzero slope, its inverse it can be found by following these steps.

To Find the Inverse of a Linear Function with Nonzero Slope

Step 1 Replace $f(x)$ by y.
Step 2 Interchange x and y.
Step 3 Solve the equation obtained in Step 2 for y.
Step 4 Replace y by $f^{-1}(x)$.

EXAMPLE 12 Find the inverse of $f(x) = -3x - 2$.

SOLUTION

$f(x) = -3x - 2$
$y = -3x - 2$ **Step 1** Replace $f(x)$ by y.
$x = -3y - 2$ **Step 2** Interchange x and y.
$3y = -x - 2$ **Step 3** Solve the equation for y.
$y = -\frac{1}{3}x - \frac{2}{3}$
$f^{-1}(x) = -\frac{1}{3}x - \frac{2}{3}$ **Step 4** Replace y by $f^{-1}(x)$.

When a function and its inverse are graphed on the same coordinate system, and if we think of $y = x$ as a mirror, the graph of the inverse is a reflection of the graph of the function in the line $y = x$.

EXAMPLE 13 Graph $y = -3x - 2$, its inverse, $y = -\frac{1}{3}x - \frac{2}{3}$, and $y = x$ on the same coordinate system.

SOLUTION The graph of the inverse is the reflection of the graph of the function in the line $y = x$. Their graphs are shown in Figure 3.32.

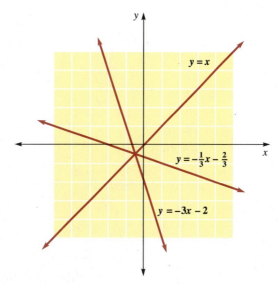

FIGURE 3.32

The algebra of functions and their inverses will be developed more fully in succeeding chapters as we gain additional skills.

Do You Remember?

Can you match these?

___ 1. The domain of the inverse of $\{(1, 2), (3, 4), (5, 6)\}$
___ 2. Alternate notation for $f(x) + g(x)$
___ 3. Alternate notation for $f[g(x)]$
___ 4. The inverse of $\{(-1, 2), (4, 6), (2, -3)\}$
___ 5. If $f(x) = x^2 + 2$ and $g(x) = x - 1$, what is $f[g(3)]$?

a) 6
b) 10
c) $\{1, 4, 5\}$
d) $\{2, 4, 6\}$
e) $(f + g)(x)$
f) $\{(2, -1), (6, 4), (-3, 2)\}$
g) $(f \circ g)(x)$

Answers: 1. d 2. e 3. g 4. f 5. a

Exercise Set 3.5

In Exercises 1–10, given $f(x) = x^2 - 3x + 2$ and $g(x) = x - 2$, find:

1. $(f + g)(1)$
2. $(g + f)(2)$
3. $(g - f)(0)$
4. $(f - g)(0)$
5. $(f \cdot g)(-1)$
6. $(g \cdot f)(-2)$
7. $\left(\dfrac{f}{g}\right)(0)$
8. $\left(\dfrac{g}{f}\right)(3)$
9. $f[g(1)]$
10. $g[f(2)]$

In Exercises 11–20, given $f(x) = x + 3$ and $g(x) = x - 2$, find:

11. $f[g(2)]$
12. $f[g(-1)]$
13. $g[f(-5)]$
14. $g[f(3)]$
15. $f[g(a + 2)]$
16. $f[g(a - 3)]$
17. $g[f(x)]$
18. $f[g(x)]$
19. $f[g(x + 1)]$
20. $g[f(x + 2)]$

For each of the following relations, (a) form the inverse and (b) determine the domain and range of the inverse.

21. $f = \{(0, 1), (1, 2), (2, 3), (3, 4)\}$
22. $g = \{(-4, 2), (-2, 0), (0, 2), (2, 4)\}$
23. $h = \{(-2, 4), (-1, 1), (0, 0), (1, 1), (2, 4)\}$
24. $f = \{(-2, -4), (-1, -1), (0, 0), (1, -1), (2, -4)\}$
25. $h = \{(1, 1), (4, 2), (9, 3), (16, 4)\}$
26. $g = \{(0, 0), (1, 1), (8, 2), (27, 3)\}$
27. $f = \{(-3, -5), (-1, 1), (1, 7), (2, 13)\}$
28. $h = \{(0, 1), (1, -4), (2, -9), (3, -14)\}$

Find the inverse of each function in Exercises 29–36. Sketch the graph of each function, its inverse, and the line $y = x$ on the same coordinate system.

29. $f(x) = 2x$
30. $f(x) = x$
31. $f(x) = 2x - 1$
32. $f(x) = 2x + 3$
33. $f(x) = -2x + 5$
34. $f(x) = -x + 1$
35. $f(x) = -\dfrac{4}{3}x + \dfrac{1}{2}$
36. $f(x) = -\dfrac{1}{2}x + \dfrac{1}{3}$

37. If $f(x) = x + 3$ and $g(x) = x + 1$, find (a) $(f - g)(x)$ and (b) the domain of $(f - g)(x)$.

38. If $f(x) = 2x + 3$ and $g(x) = \dfrac{1}{x + 2}$, find the domain of $(f \cdot g)(x)$.

39. If $f(x) = 5x - 1$ and $g(x) = \dfrac{1}{x}$, find the domain of $\left(\dfrac{f}{g}\right)(x)$.

40. Find the domain of $f + g$, if $f = \{(0, 1), (1, 2), (2, 3)\}$ and $g = \{(-1, 2), (0, 3), (4, 6)\}$.

In Exercises 41–46, if $h = \{(2, 3), (3, 7), (4, 9)\}$ and $k = \{(0, -2), (2, -1), (3, 0), (9, 3)\}$, find:

41. $(h + k)(2)$
42. $(h - k)(3)$
43. $(h \cdot k)(3)$
44. $\left(\dfrac{h}{k}\right)(2)$
45. $k[h(2)]$
46. $(k \circ h)(4)$

Review Exercises

In each of Exercises 47–60, use the given information to write the slope–intercept form of the equation of the line.

47. Goes through $(2, 4)$; $m = \dfrac{2}{3}$
48. Goes through $(0, 5)$; $m = -2$
49. Goes through $(1, 3)$ and $(2, -4)$
50. Goes through $\left(\dfrac{1}{2}, 1\right)$ and $\left(\dfrac{1}{4}, -1\right)$
51. Parallel to $x + 2y = 4$ and goes through $(0, 0)$
52. Perpendicular to $2x + y = -3$ and goes through $(1, 2)$
53. Perpendicular to $x = 4$ and goes through $(2, 5)$
54. Parallel to $y = 2$ and goes through $(5, -3)$
55. Goes through the origin and is parallel to the line through $(0, 2)$ and $(6, 2)$
56. Goes through the origin and is perpendicular to the line through $(-1, 0)$ and $(-1, 4)$
57. Parallel to $2x - 3y = 1$ and goes through the midpoint of the line segment from $(0, 0)$ to $(4, 2)$
58. Perpendicular to $5x - y = 0$ and goes through the midpoint of the line segment from $(2, 6)$ to $(4, 8)$
59. Goes through the x-intercept of the line $x + 2y = 5$ and has a slope of $-\dfrac{1}{2}$
60. Goes through the y-intercept of the line $4x - y = 3$ and has a slope of -2

In Exercises 61 and 62, find the distance between the two points.

61. $(0, 0)$ and $(3, 4)$
62. $(-1, -2)$ and $(5, 6)$

Write In Your Own Words

63. Does the constant function $f(x) = 2$ have an inverse that is a function? Explain.
64. Does the relation $f(x) = 6$ have an inverse that is a function? Explain.

65. Discuss how the graphs of $f(x)$ and $f^{-1}(x)$ are related. Include any information about how they are related to each other and the graph of $y = x$.

66. The inverse of a function may or may not be a function. Explain why.

Summary

3.1
The Rectangular Coordinate System; Graphing Linear Equations in Two Variables

The **rectangular (Cartesian) coordinate system** is formed by two perpendicular number lines (called **axes**) that intersect at the zero point of both lines (called the **origin**). The two axes divide the plane into four **quadrants** that are numbered in a counterclockwise direction.

Equations of the form $ax + by = c$, with a and b not both zero, are known as **linear equations in two variables.** Their graphs are straight lines.

The **solution** to an equation in two variables consists of the values for each variable that together make the equation true. The solution to an equation in two variables is called an **ordered pair.** The first element in an ordered pair is called the **abscissa** and the second element is called the **ordinate.**

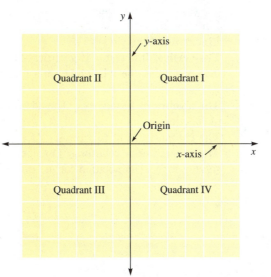

In Quadrant I, both the x- and y-coordinates are positive.

In Quadrant II, the x-coordinate is negative and the y-coordinate is positive.

In Quadrant III, both the x- and y-coordinates are negative.

In Quadrant IV, the x-coordinate is positive and the y-coordinate is negative.

3.2
The Slope of a Line; Distance and Midpoint Formulas

The **slope** of a line between any two points (x_1, y_1) and (x_2, y_2) is given by

$$m = \frac{y_2 - y_1}{x_2 - x_1} \qquad x_2 \neq x_1$$

Parallel lines have the same slope. **Perpendicular lines** have slopes that are negative reciprocals of each other. The slope of a vertical line is always **undefined.** The slope of a horizontal line is always **zero.**

Given two points (x_1, y_1) and (x_2, y_2) in the plane, the distance d between them can be found by using the **distance formula:**

$$d = \sqrt{(x_2 - x_1)^2 + (y_2 - y_1)^2}$$

If the two points lie on the same horizontal or vertical line, the distance between them is simply $|x_2 - x_1|$ or $|y_2 - y_1|$, respectively.

The **midpoint of a line segment** is found by averaging the coordinates of its endpoints. The midpoint of the line segment joining (x_1, y_1) and (x_2, y_2) is given by the formula

$$(x_m, y_m) = \left(\frac{x_1 + x_2}{2}, \frac{y_1 + y_2}{2}\right)$$

3.3 Point–Slope, Slope–Intercept, and the General Form of the Equation of a Line

Equations of lines may be written in many different ways.

General form. $ax + by = c$
Point–slope form. $y - y_1 = m(x - x_1)$ for point (x_1, y_1), slope m
Slope–intercept form. $y = mx + b$ for slope m, y-intercept b
Horizontal line. $y = c$, with c constant, slope zero
Vertical line. $x = c$, with c constant, slope undefined

3.4 Relations, Functions and Function Notation

Relation. A set of ordered pairs.
Domain of a relation. The set of first elements in the ordered pairs.
Range of a relation. The set of second elements in the ordered pairs.
Function. A correspondence that pairs each element in the domain with exactly one element in the range.
Function notation. $y = f(x)$
Linear function. $f(x) = mx + b$
The vertical line test. If a vertical line intersects the graph of a relation in more than one point, the relation is not a function.

3.5 The Algebra of Functions; Inverse Functions

The algebra of functions. The addition, subtraction, multiplication, and division of functions.
The composition of functions. $(f \circ g)(x) = f[g(x)]$
The inverse of a relation. Obtained by interchanging the x- and y-values and solving for y. The notation for the inverse of a function $f(x)$ is $f^{-1}(x)$.

*C*OOPERATIVE *E*XERCISE

Sandra is graduating from college and is looking for a good job. She sends her resume out to 15 top companies. Ten of these companies invite her to come in for an interview and three of these make her a job offer. The terms of the offers are listed as follows.

PART 1

THE HI-TECH COMPANY offers a $12,000 base salary plus an 8% commission on all gross sales. A $7,000 bonus is paid if she reaches a quota of $150,000 in *gross* sales. Her quota increases by 5% each year. Sandra projects that she will be able to sell $130,000 her first year, $160,000 her second year, and $190,000 her third year. These are gross sales. Assuming her projections are correct, do the following:

A. Write linear equations representing her salary (S) in terms of her sales (x) in dollars for sales less than $150,000 and for sales greater than or equal to $150,000 for the first year. Now calculate her salary for the first year using her projected sales.

B. Do the same for the second year. Note the changes.

C. Do the same for the third year. Note the changes.

D. Plot the three ordered pairs (x, S) for each of the three years on a Cartesian coordinate system. Draw a smooth line through these points.

E. Assuming her sales continue along this pattern, estimate her salary for the fourth year from the graph.

PART 2
THE COMPUTER CHIP COMPANY offers a $20,000 base salary plus an 8% commission on net sales. Net sales are equal to gross sales less 25% for various overhead expenses. If she reaches a quota of $100,000 in *net* sales, she receives a bonus of $5,000. Her quota will remain at $100,000 for her first two years on the job and increases by 6% each year thereafter. Sandra projects that she will be able to sell $130,000 her first year, $160,000 her second year, and $190,000 her third year. All sales are gross sales. Assuming her projections are correct, do the following:

A. Write linear equations representing her salary (S) in terms of her net sales (x) in dollars for net sales less than $100,000 and for net sales greater than or equal to $100,000. Now calculate her salary for the first year using her projected sales.

B. Do the same for the second year. Note the changes.

C. Do the same for the third year. Note the changes.

D. Plot the three ordered pairs (x, S) for each of the three years on a Cartesian coordinate system. Draw a smooth line through these points.

E. Assuming her sales continue along this pattern, estimate her salary for the fourth year from the graph.

PART 3
THE SOFTWARE DESIGNER offers a $30,000 base salary plus a 2% commission on gross sales. No performance bonuses are paid, but she is guaranteed a 0.5% increase in commission the year after she reaches a quota of $120,000 in *net* sales (gross sales less 25% for various costs). This quota increases by 5% each year (starting with the third year). Sandra projects that she will be able to sell $130,000 her first year, $160,000 her second year, and $190,000 her third year. Assuming her projections are correct do the following:

A. Write linear equations representing her salary (S) in terms of her sales (x) in dollars for net sales less than $120,000 and for net sales greater than or equal to $120,000. Now calculate her salary for the first year using her projected sales.

B. Do the same for the second year. Note the changes.

C. Do the same for the third year. Note the changes.

D. Plot the three ordered pairs (x, S) for each of the three years on a Cartesian coordinate system. Draw a smooth line through these points.

E. Assuming her sales continue along this pattern, estimate her salary for the fourth year from the graph.

PART 4
Based upon the information from Parts 1, 2, and 3, which job offer should Sandra accept, and *why*? Write a paragraph explaining your choice.

REVIEW EXERCISES

Complete the following

1. The slope of a horizontal line is _____.
2. The formula for the slope m is _____.
3. Two lines with the same slope are said to be _____.
4. The slope of a vertical line is _____.
5. If two lines have slopes that are negative reciprocals, the lines are _____.
6. The slope of the line through the two points (2, 4) and (−3, 5) is _____.
7. The slope of the line $x - 2y - 3$ is _____.
8. The line through (−2, 3) and (3, 2) is _____ to the line through (0, 6) and (−10, 8).
9. Find a point A(__, __) so that points A, B, and C are collinear for B(0, 0) and C(2, 2). Answers may vary.
10. The line through $(a, -2)$ and (2, 4) is parallel to the line $x - y = 4$. The value of a is given by _____.
11. The distance between two points (x_1, y_1) and (x_2, y_2) on a line is given by the formula _____.
12. The formula for the midpoint of the line segment joining (x_1, y_1) to (x_2, y_2) is _____.
13. The slope–intercept form of the equation of a line is _____.
14. The point–slope form of the equation of a line is _____.
15. If $5x - y = 10$ is the equation of a line, then
    (a) (0, __) is a point on the line.
    (b) The slope of the line is _____.

(c) The x-intercept of the line is _____.
(d) The y-intercept of the line is _____.
(e) (___, 12) is a point on the line.

Find the x- and y-intercepts, a check point, and sketch the graph in Exercises 16–19.

16. $y = 3x$
17. $2x - 5y = 10$
18. $2(x - 5) + 3 = 3(y + 2) - 1$
19. $3x = 4$

Write the general form of the equation of the line satisfying the given conditions.

20. Has an *x*-intercept of 2 and a *y*-intercept of 3
21. Goes through (1, 3) and has a slope of 2
22. Goes through (2, 5) and (7, 1)
23. Goes through (0, 2) and is parallel to $x + y = 3$
24. Goes through (5, 1) and is perpendicular to $2x - y = 1$
25. Goes through the origin and has zero slope
26. Goes through (4, 0) and has undefined slope
27. $m = 3$ and $b = -1$
28. Goes through (0, 0) and is parallel to $x + y = 6$
29. Goes through $\left(\dfrac{1}{2}, \dfrac{1}{4}\right)$ and is parallel to $2x + 3y = 5$
30. Goes through (0, 0) and is perpendicular to $x - y = 2$
31. Is parallel to $y = 3$ and goes through (5, −1)
32. Is perpendicular to $x = 4$ and goes through $\left(-3, \dfrac{1}{2}\right)$
33. Is parallel to $x = -7$ and goes through the origin

Use each of the following graphs to write the equation of the line in slope-intercept form.

34.

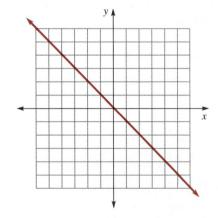

35.

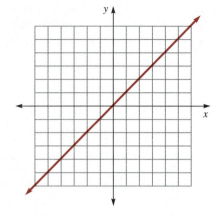

36.

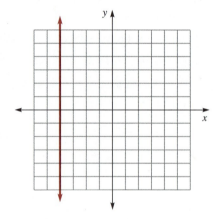

37.

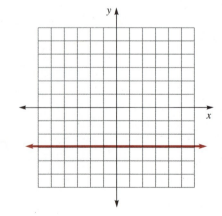

38.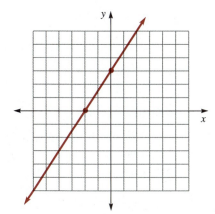

39.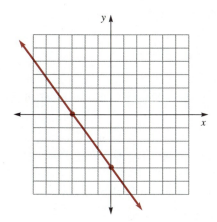

40.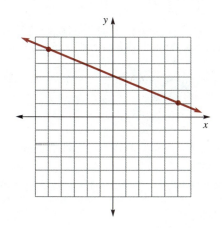

41.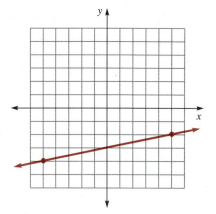

42.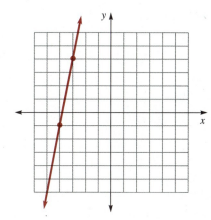

Find (a) the coordinates of the midpoint and (b) the distance between each pair of points in Exercises 43–48.

43. (0, 0) and (3, 5) **44.** (1, 2) and (3, 4)

45. (2, 7) and (5, 4) **46.** (−6, −9) and (−5, 0)

47. $\left(\frac{3}{4}, \frac{5}{2}\right)$ and $\left(\frac{1}{3}, \frac{3}{5}\right)$

48. $\left(-\frac{1}{2}, -\frac{2}{3}\right)$ and $\left(-\frac{2}{7}, -\frac{1}{5}\right)$

Translate each statement into a linear equation. Find the x- and y-intercepts and a check point, then graph each line.

49. The *y*-value is the negative of the *x*-value.

50. The *y*-value is 4 times the *x*-value.

51. The *y*-value is 2 more than the *x*-value.

52. Twice the *y*-value is 5 less than the *x*-value.

53. One-half of the *y*-value is 7 more than one-half of the *x*-value.

54. The *y*-value divided by 5 is one less than twice the *x*-value.

55. Four times the y-value divided by 7 equals the x-value added to 5.

56. The sum of the y-value and 4, divided by 3, is equal to 1 more than the x-value.

Determine whether the pairs of lines in Exercises 57–60 are parallel, perpendicular, or neither.

57. L_1: $x + y = 4$
    L_2: $x - y = 2$.

58. L_1: through (2, 5) and (3, 4)
    L_2: through (4, 7) and (3, 8)

59. L_1: $6x - 2y = 7$
    L_2: $3y = x - 5$

60. L_1: through (1, 1) and $\left(\dfrac{1}{2}, \dfrac{1}{4}\right)$
    L_2: through (2, 1) and $\left(\dfrac{1}{4}, \dfrac{1}{2}\right)$

Find x or y as indicated in Exercises 61 and 62.

61. The line through the points (x, 3) and (5, 7) is parallel to the line through the points (5, 3) and (8, 6).

62. The line through the points (3, y) and (2, 3) is perpendicular to the line through the points (6, 5) and (2, 7).

63. If $f(x) = 5x$, find $f(0), f(-2), f(3)$, and $\dfrac{f(x+h) - f(x)}{h}$.

64. Is $A = \{(0, 1), (0, 2), (1, 4), (2, 5)\}$ a function? Why or why not? Explain.

65. Find the domain and range of $x = 4$. Is $x = 4$ a function?

66. Given $f(x) = x + 5$ and $g(x) = x - 7$, find $f[g(3)]$, $g[f(0)]$, $(f + g)(1)$, and $(f \cdot g)(-2)$.

67. Given $f(x) = 4x - 9$, find $f^{-1}(x)$.

68. Graph $f(x) = 2x + 3$.

69. Find the domain of $f(x) = \dfrac{3x}{2x - 3}$.

Chapter Test

Use the given points P(3, 1) and Q(9, 9) to answer Exercises 1–5.

1. Find the slope of the line from P to Q.
2. Find the distance from P to Q.
3. Find the coordinates of the midpoint of the line segment from P to Q.
4. Find the equation of the line through P and Q.
5. Graph the line using x- and y-intercepts and a check point.
6. Find the equation of the line from the accompanying graph.

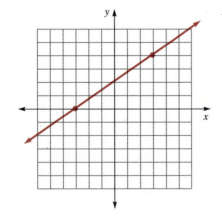

7. Find the equation of the line with slope $\dfrac{1}{2}$ and y-intercept 5.
8. Find the equation of the line through (0, 4) and (6, 4).
9. Find the equation of the line that has x-intercept 6 and goes through the point (6, 5).
10. Find the equation in slope–intercept form of the line through $\left(\dfrac{1}{4}, \dfrac{1}{3}\right)$ and $\left(\dfrac{2}{5}, \dfrac{1}{7}\right)$.
11. Find the equation in slope–intercept form of the line through (2, 1) and perpendicular to $2x + y = 5$.
12. Find the slope and y-intercept of the line $2x + 3y = 9$ and graph it.
13. Find the equation in slope–intercept form of the line that goes through (4, 8) and is parallel to the line $3x - y = 4$.
14. Are the three points $A(3, 5)$, $B(-2, 6)$, and $C(8, 4)$ collinear? Explain.
15. Are the lines L_1: $4x - 3y = 7$ and L_2: $3x + 4y = 7$ parallel, perpendicular, or neither?
16. Find the equation of the line with x-intercept 5 and y-intercept -3.
17. The y-value of a linear equation is 3 less than 4 times the x-value. Find the equation and graph it using 3 points.

18. Three times the sum of the *x*- and *y*-values of a linear equation is 15. Find the equation and determine whether it has a positive or negative slope.

19. Find the equation of a line through (5, 1) and perpendicular to the line through (2, −3) and (−3, 4).

20. Find the equation of a line with a slope of one-half, if it forms one side of a triangle with vertices $A(-5, -2)$, $B(2, 1)$, and $C(-3, -1)$.

21. Graph the line through the point (−2, −3) with slope $m = \dfrac{3}{4}$.

22. Graph $y = -\dfrac{1}{2}x + 4$.

23. Graph $5x + 4y = 2$.

24. Given $f(x) = 3x - 2$, find $f^{-1}(x)$.

25. Find the domain and range of $f(x) = -x + 2$.

26. Find the domain of $f(x) = \dfrac{-7}{4x - 1}$.

27. Given $f(x) = 2 - x$ and $g(x) = x - 3$, find $f[g(1)]$, $(f + g)(2)$, and $(f - g)(-1)$.

28. Given $g(x) = x^2 - 3x + 2$, find $g(0)$, $g(1)$, and $g(2)$.

29. Is $M = \{(-2, 1), (-1, 1), (0, 1), (2, 1)\}$ a function? Explain.

Cumulative Review for Chapters 1–3

Write each of the following sets by listing its elements.

1. $\{x | x \in N, x < 5\}$
2. $\{y | y \in W, y \leq 3\}$
3. $\{p | p \in J, -3 \leq p \leq 2\}$
4. $\{d | d$ is a day of the week starting with the letter M$\}$

Given $A = \{$natural numbers$\}$, $B = \{1, 2, 3\}$, and $C = \{2, 4, 6\}$, find:

5. $A \cap B$
6. $B \cap C$
7. $A \cup C$
8. $B \cup C$

Indicate whether each of the following is true *or* false.

9. $4 \leq 9$
10. $-(-4) > 0$
11. $|-8| < |-1|$
12. The additive inverse of *x* is −*x*.
13. An example of the associative property of addition is $7 + 2 = 2 + 7$.
14. An example of the associative property of multiplication is $7(9p) = (7 \cdot 9)p$.
15. The symbol "≥" means "at least."
16. An example of the distributive property of multiplication is $5(a + b) = 5a + 5b$.
17. $0.725 \in \{$real numbers$\}$
18. An example of the additive inverse property is $-(3x + 2) + (3x + 2) = 0$.
19. Subtraction is commutative.

Simplify each of the following quantities if possible. Some may be undefined.

20. $-3(7 - 2^3) \cdot 6 \div 2$
21. $-2(-3)(-1)^2$
22. $\left(\dfrac{3}{2}\right)\left(\dfrac{-4}{3}\right)(-2)$
23. $\dfrac{-6 - (3 - 5)}{-4}$
24. $\dfrac{-(-6) - 6}{7 - (-3)}$
25. $\dfrac{-15 + 8}{-3 - (-3)}$

Graph the line in Exercises 26–29.

26. It goes through the point (−2, 5) and has a slope of $-\dfrac{1}{3}$.

27. $y = \dfrac{1}{2}x + \dfrac{1}{2}$

28. $2x + 3y = 3$

29. $4y = -3x - 8$

Complete each of the following.

30. The sum of 3 and −11 is ____.
31. The difference of −2 and −5 is ____.
32. The quotient of $\dfrac{2}{3}$ and $\dfrac{3}{8}$ is ____.

Solve each equation.

33. $p - 3 = 4$
34. $m + 7 = 5$
35. $6a + 2 = -5$
36. $3b + 7 = b - 2$
37. $\dfrac{x}{3} = 7$
38. $\dfrac{2y}{3} = 10$
39. $\dfrac{4r}{3} + \dfrac{1}{2} = \dfrac{r}{4} - 5$
40. $m + (m - 1) - 3(m - 2) = 0$
41. $-\{-2 - [-3 + x(4 - 7) + 5x] + 8x\} = 0$
42. $\dfrac{3p + 4}{2} = 4$

43. $-\dfrac{1}{2}(a + 1) - 2a = \dfrac{1}{4}(2a - 1)$

Solve each of the following.

44. The sum of two consecutive integers is 27. Find the integers.

45. Four times a number is -56. Find the number.

46. The perimeter of a square is $64x$ feet. Find the length of a side.

47. A side of an equilateral triangle is $5d$ inches. Find the perimeter.

48. One number is 7 more than another number. If their sum is 57, find the numbers.

49. One number is 5 more than 3 times another number. If the sum of the numbers is 5, find the numbers.

50. The perimeter of a rectangle is 39 meters. If the length is 3 more than the width, find the dimensions of the rectangle.

51. The area of a circle is 25π cm². Find the length of the radius r.

52. Find two consecutive multiples of 4 that add to 52.

Solve for the specified variable.

53. $A = \dfrac{1}{2}bh$, for h

54. $y = mx + b$, for x

55. $\dfrac{1}{R} = \dfrac{1}{R_1} + \dfrac{1}{R_2}$, for R_1

56. $A = p(1 + rt)$, for r

57. $ax + by = c$, for y

58. $C = \dfrac{5}{9}(F - 32)$, for F

Solve each of the following.

59. $|d| = 5$

60. $|2a - 3| + 2 = 1$

61. $|4m + 5| = 0$

62. $|2m + 3| = |m - 3|$

63. $|x - 2| = 2x$

64. $\left|\dfrac{5y + 4}{3}\right| = \dfrac{1}{2}$

65. $2y - 1 \leq 3$

66. $\dfrac{2(x - 3)}{5} \geq 2 - x$

67. $\{x | x > 2\} \cap \{x | x < 6\}$

68. $\{y | y > 4\} \cup \{y | y \leq 3\}$

69. One-third of a number is at least 7.

70. Six times a number is between -8 and 12.

71. Ramon wants to keep a test average of at least 80. If he earns 84 and 74 on his first two tests, what is the least he must earn on his third test?

72. $|m| < 2$

73. $|2a - 7| \geq 2$

74. $|x| \leq 0$

75. $\left|\dfrac{1}{2}y + 1\right| > -1$

76. $|4p - 3| + 4 \geq 7$

77. $2 - |a + 4| \leq 1$

Given the two points in Exercises 78–83, find (a) the coordinates of the midpoint (b) the slope of the line segment (if it exists), and (c) the equation of the line through the points, in general form.

78. $(0, 2)$ and $(3, 5)$

79. $(-2, 4)$ and $(3, -1)$

80. $(0, 4)$ and $(2, 4)$

81. $\left(\dfrac{1}{2}, \dfrac{1}{4}\right)$ and $\left(\dfrac{1}{4}, \dfrac{1}{2}\right)$

82. $\left(3, \dfrac{1}{2}\right)$ and $(3, 4)$

83. $(6, -5)$ and $(-2, 3)$

Write the equation of the line, in slope–intercept form, using the given figure in each of Exercises 84–87.

84.

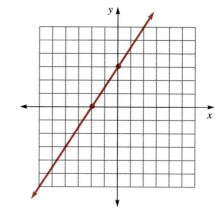

85.

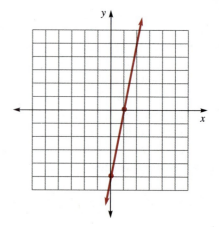

86.

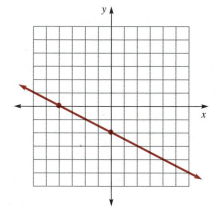

87.

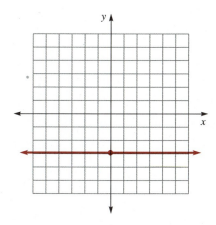

Carry out each of the following.

88. Write the equation of the line that is parallel to $y = 3x - 5$ and that goes through the point $(0, 0)$.

89. Write the equation of the line that is perpendicular to $2x + 3y = 1$ and that goes through the point $(1, 1)$.

90. Write the equation of the line that is parallel to $y = -2$ and that goes through the midpoint of the line segment from $(-3, -1)$ to $(5, 3)$.

91. Is $A = \{(0, 1), (1, 2), (2, 3), (3, 4), (4, 4)\}$ a function?

92. Find the domain and range of $y = 7$.

93. Find the domain of $f(x) = \dfrac{x}{x - 2}$.

94. Given $f(x) = 3x - 7$, find $f(0)$, $f(-2)$, $f(a)$, and $f(a + 2)$.

95. Define "function."

96. Is $B = \{(x, y)/y = 3 - x\}$ a function? Find the domain and range.

Use the vertical line test to determine whether the graph in each of Exercises 97–100 is a function.

97.

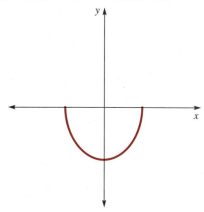

98.

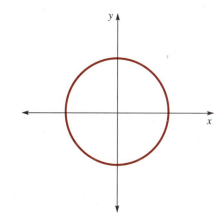

99.

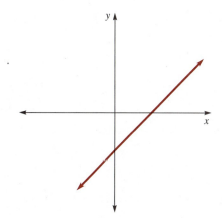

100.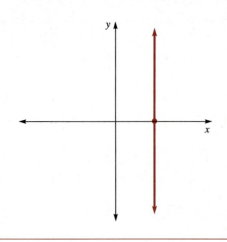

101. Given $g(x) = 3x^2 - 5x$, find $g(0)$, $g(1)$, and $g(a)$.

102. Given $f(x) = 2x - 1$ and $g(x) = 3x - 4$, find $f[g(2)]$, $g[f(0)]$, $(f + g)(3)$, $(f - g)(0)$, $(f \cdot g)(-1)$, and $\left(\dfrac{f}{g}\right)(4)$.

CHAPTER 4
SYSTEMS OF LINEAR EQUATIONS AND INEQUALITIES

4.1 Solving Systems of Linear Equations by Graphing, Elimination, and Substitution
4.2 Solving Systems of Linear Equations in Three or More Variables
4.3 Solving Systems of Linear Equations Using Matrices
4.4 Determinants and Cramer's Rule
4.5 Graphing Linear Inequalities in Two Variables
4.6 Linear Programming
4.7 More Applications

Application

Antonio is preparing for a wilderness backpacking trip on which he plans to eat a combination of cereal and nuts. The recommended daily requirement is at least 3000 calories and 2.4 ounces of protein. Each ounce of cereal supplies 120 calories and 0.04 ounces of protein. Each ounce of nuts provides 60 calories and 0.12 ounces of protein. If the cereal costs 20¢/ounce and each ounce of nuts costs 30¢, what combination should he plan to eat in order to keep the cost to a minimum?

See Section 4.6, Example 4

EMMY NOETHER

Emmy Noether, a German mathematician, made significant contributions to many areas of mathematics. Her father, also a mathematician, supported her early interests in mathematics. At that time, it was rare for a woman to receive such encouragement. When the famous mathematician David Hilbert was fighting to get her a post at Göttingen University, he said to the opposing faculty members: "Meine Herren, I do not see that the sex of the candidate is an argument against her admission as a Privatdozent *[a lecturer paid only by student fees]. After all, the Senate is not a bathhouse." Hilbert eventually got his way and Emmy Noether, more than anyone else in her generation at Göttingen, was to have an enormous influence on the course of mathematics. She became the center of the most fertile circle of research there. In 1933 Noether had to leave Germany to escape the Nazis, and she joined the faculty of Bryn Mawr College in Pennsylvania. There she reached the peak of her professional powers, and along with other mathematicians of her time, she went on to lay the foundations of modern algebra. She died in 1935 following an operation when she was just 53.*

4.1 Solving Systems of Linear Equations by Graphing, Elimination, and Substitution

HISTORICAL COMMENT   Systems of equations were known in Egypt, Greece, India, and China as early as 275 A.D. Various methods were used to indicate such systems, and it was the Hindus who first wrote their equations one below the other, much as we will do in this section. They used the names of colors to indicate the various unknowns. The *elimination method,* which we will discuss in this section, was developed in 1559 when various letters were used to represent different unknowns. The first person to use letters to stand for numbers was François Viéte (1540–1603): He used consonants for known quantities and vowels for unknown quantities.

OBJECTIVES   In this section we will learn to
1. solve systems of equations by graphing, elimination, and substitution;
2. recognize independent, inconsistent, and dependent systems of equations; and
3. use systems of linear equations to solve applied problems.

A **system of linear equations** involves two or more linear equations. In this section we will examine several types of linear systems and concentrate our efforts on finding solutions to these systems.

| **DEFINITION** **Solution to a System of Linear Equations** | A **solution to a system of linear equations** is a set of values for the variables that is a solution to *every* equation in the system. |

Suppose that you had purchased two items for a total of $10 but forgot the price of each. You did, however, remember that the one item cost $2 more than the other. How could you determine the cost of each item?

The problem can be solved by trial and error, but a more systematic approach is to write a system of equations describing what you remember. To do this, we let x represent the cost of the more expensive item and y the less expensive item. Then

$x + y = 10$ The total cost is $10.

$x - y = 2$ The difference between the cost of the two items is $2.

To determine which two numbers are the solution to this system of equations, we draw the graphs of both equations on the same coordinate system, as shown in Figure 4.1. To show that the solution to both equations is the intersection point of the two lines, (6, 4), we need to show that (6, 4) satisfies both equations. Recall that the ordered pair (6, 4) means $x = 6$ and $y = 4$.

$x + y = 10$ $\qquad\qquad\qquad$ $x - y = 2$
$6 + 4 \stackrel{?}{=} 10$ $\quad \leftarrow$ Substitute 6 for x and 4 for y. $\rightarrow \quad$ $6 - 4 \stackrel{?}{=} 2$
$10 \stackrel{\checkmark}{=} 10$ $\qquad\qquad\qquad\qquad\qquad$ $2 \stackrel{\checkmark}{=} 2$

The higher-priced item costs $6.00 and the lower-priced item costs $4.00. There are many other points that satisfy the equations separately, but only one point satisfies both equations. That point is (6, 4) and it is the only (the *unique*) solution to the system.

The system of equations

$x + y = 10$
$x - y = 2$

is an example of a system of linear equations in two variables. An ordered pair (x, y) that is a solution to both equations in the system is a solution to the system itself.

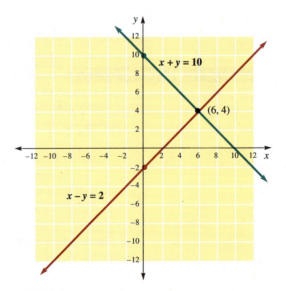

FIGURE 4.1

The next four examples illustrate how we can use the **graphing method** to solve some systems of equations.

The Graphing Method

EXAMPLE 1 Solve the following system by graphing:

$$2x + y = 6$$
$$3x - 2y = 2$$

SOLUTION Recall that two easy points to use in graphing a line are the *x*- and *y*-intercepts. The intercepts and a check point for each of the equations are

$2x + y = 6$		$3x - 2y = 2$
(0, 6)	← *y*-intercepts →	(0, −1)
(3, 0)	← *x*-intercepts →	$\left(\frac{2}{3}, 0\right)$
(2, 2)	← Check points →	(4, 5)

When the two lines are graphed on the same coordinate system, the solution *appears* to be the point (2, 2). See Figure 4.2. To determine whether (2, 2) is the true solution, we must check its coordinates in each equation.

CHECK $2x + y = 6$ ← Substitute 2 for → $3x - 2y = 2$
$2 \cdot 2 + 2 \stackrel{?}{=} 6$ *x* and 2 for *y*. $3 \cdot 2 - 2 \cdot 2 \stackrel{?}{=} 2$
$6 \stackrel{\checkmark}{=} 6$ $2 \stackrel{\checkmark}{=} 2$

Since the ordered pair (2, 2) satisfies both equations, the solution to the system is $\{(2, 2)\}$.

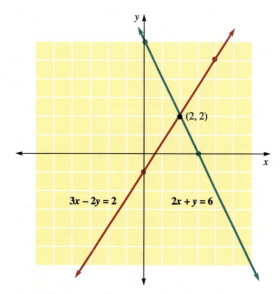

FIGURE 4.2

> Solutions to a system of two linear equations are the coordinates of the points of intersection of their graphs.

We can use set notation to indicate that we want to find the intersection of the graphs of two linear equations. Recall that set-builder notation is read

$$\underbrace{\{(x, y)}_{\text{The set of all ordered pairs}} \underbrace{|}_{\substack{\uparrow \\ \text{such} \\ \text{that}}} \underbrace{2x + 3y = 8\}}_{\text{they lie on the given line}}$$

The next example illustrates this.

EXAMPLE 2. Find the intersection (solution) for the system

$$2x + 3y = 8$$
$$2x - 3y = -4$$

by graphing

$$\{(x, y) \mid 2x + 3y = 8\} \cap \{(x, y) \mid 2x - 3y = -4\}$$

SOLUTION First we determine three points that lie on each line:

$2x + 3y = 8$	$2x - 3y = -4$
$(-2, 4)$	$\left(0, \dfrac{4}{3}\right)$
$(4, 0)$	$(-2, 0)$
$\left(0, \dfrac{8}{3}\right)$	$(4, 4)$

The graphs are shown in Figure 4.3.

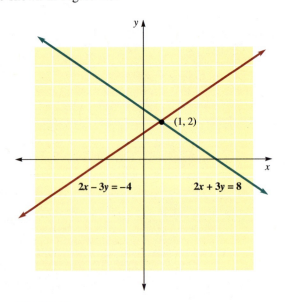

FIGURE 4.3

The solution *appears* to be the ordered pair (1, 2). We verify this observation by substitution.

CHECK $2x + 3y = 8$ $2x - 3y = -4$
           $2 \cdot 1 + 3 \cdot 2 \stackrel{?}{=} 8$ $2 \cdot 1 - 3 \cdot 2 \stackrel{?}{=} -4$
           $8 \stackrel{\checkmark}{=} 8$ $-4 \stackrel{\checkmark}{=} -4$

The solution set is {(1, 2)}.

A linear system consisting of two equations in two variables may not always have a unique (single) solution. If the graphs of the equations are parallel lines, the lines will not intersect. When this happens, there is *no* point that is a solution to both equations. At the other extreme, the graphs of the two equations may coincide (one line lies on top of the other). *When two lines coincide, any point on either line is a solution to the system.*

The Number of Solutions to a System of Two Linear Equations

1. **One solution:**   The two lines intersect in exactly one point. See Figure 4.4(a). A system of equations with a unique solution is said to be **consistent** and **independent.**
2. **No solution:**   The two lines do not intersect. See Figure 4.4(b). A system of equations involving parallel lines is said to be **inconsistent** and **independent.**
3. **Infinitely many solutions:**   The two lines coincide. See Figure 4.4(c). A system in which the lines coincide is said to be **consistent** and **dependent.**

Although systems that have a single solution are both consistent and independent, for the sake of brevity we will call them **independent.** Similarly, systems with no solutions (inconsistent and independent) will be called **inconsistent,** and systems

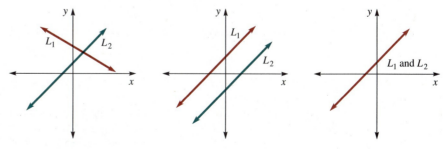

(a) A unique solution: consistent and **independent**

(b) No solution: **inconsistent** and independent

(c) Infinitely many solutions: consistent and **dependent**

FIGURE 4.4

with an infinite number of solutions (consistent and dependent) will be called **dependent.**

EXAMPLE 3   Can a unique solution to the following system be found by graphing?

$$x + y = -2$$
$$x + y = 3$$

SOLUTION   By writing each equation in the slope–intercept form, we can determine whether the lines are parallel or whether they coincide.

$$x + y = -2 \qquad x + y = 3$$
$$y = -x - 2 \qquad y = -x + 3$$

The slope of both lines is -1, but their y-intercepts are different. The lines are parallel, so the system is inconsistent. A unique solution does not exist. Their graphs are shown in Figure 4.5.

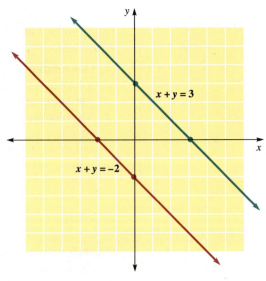

FIGURE 4.5

EXAMPLE 4   Solve the following system of equations by graphing.

$$2x - y = 2$$
$$4x - 2y = 4$$

SOLUTION   The three ordered pairs $(0, -2)$, $(1, 0)$, and $(2, 2)$ are solutions to both equations. Since the same three ordered pairs satisfy each equation, the lines will coincide. The system is dependent. (See Figure 4.6.) The solution set can be expressed using either one of the equations, which means it can be written $\{(x, y) \mid 2x - y = 2\}$ or $\{(x, y) \mid 4x - 2y = 4\}$.

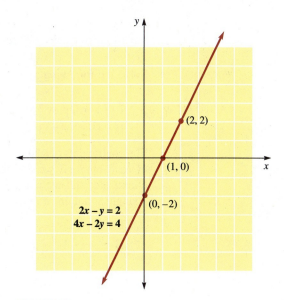

FIGURE 4.6

Notice that when the graphs of two lines coincide, one equation is a multiple of the other. In Example 4, multiplying each side of the first equation, $2x - y = 2$, by 2 yields the second equation, $4x - 2y = 4$. That's why any point that lies on either line is a solution to both equations. An infinite number of solutions exists.

Solving a system of linear equations by graphing has the drawback that if the solution does not involve integers, it may be difficult to read the solution from the graph. We now consider two other methods for solving such systems. They are known as the **elimination method** and the **substitution method.**

The Elimination Method

The elimination method is based on the *addition property of equality* from Section 2.1. This states that we can add the same number to both sides of an equation and obtain an equivalent equation:

If $x = y$ then $x + a = y + a$

Now suppose we know that $a = b$. Then a and b represent the same number, and we can add a to one side and b to the other and still obtain an equivalent equation:

$$\begin{array}{r} x = y \\ \underline{a = b} \\ x + a = y + b \end{array}$$

Recall that *equivalent equations* have the same solution set.

To use the elimination method, two equations must have a pair of numerical coefficients that are additive inverses of each other. (Recall that in the term $3x$, the number 3 is the *numerical coefficient,* or simply the *coefficient,* of x.) Then, by the preceding discussion, we can add the two equations and obtain an equation with one of the variables eliminated.

EXAMPLE 5 Solve the following system of equations by the elimination method.

SOLUTION

I $x + 2y = 7$
II $x - 2y = 3$

The coefficients of y in the two equations are 2 and -2, which are additive inverses of each other. If we add the left and right sides of the two equations, y will be eliminated, and we will have reduced the system of equations in two variables to a single equation in one variable (x), which we know how to solve.

$$
\begin{array}{ll}
\text{I} & x + 2y = 7 \\
\text{II} & \underline{x - 2y = 3} \\
& 2x = 10 \quad \text{Add I to II.} \\
& x = 5
\end{array}
$$

Once we know the value of x or y, the value of the other variable can be found by substituting the known value into either of the original equations. To determine y, we substitute $x = 5$ into either of the original equations.

I $x + 2y = 7$ II $x - 2y = 3$
   $5 + 2y = 7$ ← Substitute 5 for x. → $5 - 2y = 3$
   $2y = 2$ $-2y = -2$
   $y = 1$ $y = 1$

The solution set is $\{(5, 1)\}$. The ordered pair $(5, 1)$ is the intersection point of the two graphs of the system.

The *multiplication property of equality* states that if both sides of an equation are multiplied by any nonzero constant, an equivalent equation results. We use this fact to solve a system of equations in which neither of the variables has coefficients that are additive inverses of each other.

EXAMPLE 6 Solve the following system of equations, beginning with the elimination method.

I $2x + 3y = 11$
II $x - 5y = -14.$

SOLUTION The coefficients of neither x (2 and 1) nor y (3 and -5) are additive inverses of each other. If both sides of equation **II** are multiplied by -2, however, the coefficients of x will then be additive inverses:

$$
\begin{array}{lllll}
\text{I} & 2x + 3y = 11 & \rightarrow & 2x + 3y = 11 & \\
\text{II} & x - 5y = -14 & \rightarrow & \underline{-2x + 10y = 28} & \text{Multiply each side of II by } -2. \\
& & & 13y = 39 & \text{Add the left and right sides of} \\
& & & y = 3 & \text{I and II.}
\end{array}
$$

CAUTION

Always check the solution in the original system. It is easy to make an error when multiplying to get coefficients that are additive inverses. A new *incorrect* system is solved correctly from this point on, the solution would check in this *new* system. However, it would not be a solution to the original system.

We substitute $y = 3$ into either equation **I** or **II** to find x. Let's use equation **I**.

$$2x + 3y = 11$$
$$2x + 3 \cdot 3 = 11$$
$$2x + 9 = 11$$
$$x = 1$$

CHECK
$$2x + 3y = 11 \qquad x - 5y = -14$$
$$2 \cdot 1 + 3 \cdot 3 \stackrel{?}{=} 11 \qquad 1 - 5 \cdot 3 \stackrel{?}{=} -14$$
$$2 + 9 \stackrel{?}{=} 11 \qquad 1 - 15 \stackrel{?}{=} -14$$
$$11 \stackrel{\checkmark}{=} 11 \qquad -14 \stackrel{\checkmark}{=} -14$$

The solution set is $\{(1, 3)\}$.

EXAMPLE 7 Solve the following system of equations beginning with the method of elimination.

I $\quad 2x + 3y = 5$
II $\quad 3x + 8y = 5$

SOLUTION We multiply equation **I** by 3 and equation **II** by -2, so that the coefficients of x are additive inverses:

I $\quad 2x + 3y = 5$	$\rightarrow$	$6x + 9y = 15$	Multiply each side of **I** by 3.
II $\quad 3x + 8y = 5$	$\rightarrow$	$-6x - 16y = -10$	Multiply each side of **II** by -2.
		$-7y = 5$	Add the left and right sides **I** and **II**.
		$y = -\dfrac{5}{7}$	

We choose to solve for x by substituting $y = -\dfrac{5}{7}$ into $2x + 3y = 5$.

$$2x + 3\left(-\frac{5}{7}\right) = 5$$

$$2x - \frac{15}{7} = 5$$

$$7 \cdot 2x - 7 \cdot \frac{15}{7} = 7 \cdot 5 \qquad \text{Multiply each side by 7, the LCD.}$$

$$14x - 15 = 35$$

$$14x = 50$$

$$x = \frac{50}{14} = \frac{25}{7}$$

CHECK
$$2x + 3y = 5 \qquad\qquad\qquad 3x + 8y = 5$$
$$2 \cdot \frac{25}{7} + 3\left(-\frac{5}{7}\right) \stackrel{?}{=} 5 \qquad x = \frac{25}{7}, y = -\frac{5}{7} \qquad 3 \cdot \frac{25}{7} + 8\left(-\frac{5}{7}\right) \stackrel{?}{=} 5$$
$$5 \stackrel{\checkmark}{=} 5 \qquad\qquad\qquad\qquad\qquad\qquad 5 \stackrel{\checkmark}{=} 5$$

The solution set is $\left\{\left(\dfrac{25}{7}, -\dfrac{5}{7}\right)\right\}$.

Recall that *when a system of equations is inconsistent, the lines are parallel and no solution exists.* Trying to solve an inconsistent system by the elimination method results in an outcome like the following.

The lines represented by $x - y = 5$ and $-x + y = 3$ both have a slope of 1 but different y-intercepts. The lines do not intersect.

$$x - y = 5$$
$$\underline{-x + y = 3}$$
$$0 = 8 \quad \text{Add the two equations.}$$

Trying to solve an inconsistent system by elimination will always yield a false statement.

Notice that $0 = 8$ is a false statement. The solution set is $\varnothing$.

When *a system of equations is dependent, the two lines coincide.* The system $x + y = 4$ and $2x + 2y = 8$ is dependent because the second equation is obtained from the first one by multiplying each side by 2. When we try to solve such a system by the elimination method, the following result is obtained:

$$\mathbf{I} \quad x + y = 4 \;\rightarrow\; -2x - 2y = -8 \quad \text{Multiply each side of } \mathbf{I} \text{ by } -2.$$
$$\mathbf{II} \quad 2x + 2y = 8 \;\rightarrow\; \underline{2x + 2y = 8} \quad \text{Add } \mathbf{I} \text{ and } \mathbf{II}.$$
$$0 = 0$$

Trying to solve a dependent system by elimination always yields $0 = 0$.

The solution set is $\{(x, y) \mid x + y = 4\}$.

The Substitution Method

A third method for solving systems of equations, in addition to graphing and elimination, is *the substitution method,* as shown in the next two examples.

EXAMPLE 8 Solve the following system of equations by the substitution method.

$$\mathbf{I} \quad 2x + 3y = 11$$
$$\mathbf{II} \quad x - 5y = -14$$

SOLUTION If the system has a *unique* solution, the ordered pair that is the solution will satisfy both equations. Thus, an alternative to the elimination method is first to solve one of the equations for either variable and then to substitute that result into the other equation.

Note that solving for y in equations $\mathbf{I}$ and $\mathbf{II}$ gives

$$y = \dfrac{11 - 2x}{3} \quad \text{and} \quad y = \dfrac{14 + x}{5}$$

Since the coefficient of x in equation $\mathbf{II}$ is 1, we will solve for x in this equation to avoid working with fractions:

$$x - 5y = -14$$
$$x = -14 + 5y \quad \text{Add } 5y \text{ to each side.}$$

When this value of x is substituted into equation $\mathbf{I}$, we have

$$2x + 3y = 11$$
$$2(-14 + 5y) + 3y = 11$$

$$-28 + 10y + 3y = 11 \qquad \text{Distributive property}$$
$$-28 + 13y = 11 \qquad \text{Combine like terms.}$$
$$13y = 39 \qquad \text{Add 28 to each side.}$$
$$y = 3 \qquad \text{Divide each side by 13.}$$

Substituting $y = 3$ into $x - 5y = -14$, we have

$$x - 5 \cdot 3 = -14$$
$$x - 15 = -14$$
$$x = 1$$

CHECK This system was checked in Example 6. The solution set is $\{(1, 3)\}$.

EXAMPLE 9 Solve the following system of equations by substitution.

I $\quad 2x + 3y = 5$
II $\quad 3x + 6y = 5$

SOLUTION Since no coefficient is 1, we will solve equation **I** for y.

$$2x + 3y = 5$$
$$3y = 5 - 2x \qquad \text{Subtract } 2x \text{ from each side.}$$
$$y = \frac{5 - 2x}{3} \qquad \text{Divide each side by 3.}$$

This value is now substituted for y in equation **II**:

$$3x + 6y = 5$$
$$3x + 6\left(\frac{5 - 2x}{3}\right) = 5$$
$$3x + 10 - 4x = 5$$
$$-x + 10 = 5$$
$$x = 5$$

To obtain y, the value of x is now substituted into the most convenient form of y:

$$y = \frac{5 - 2x}{3}$$
$$= \frac{5 - 2 \cdot 5}{3}$$
$$= \frac{5 - 10}{3}$$
$$= -\frac{5}{3}$$

The solution set is $\left\{\left(5, -\frac{5}{3}\right)\right\}$, which must be checked in the original equations.

If we had solved the first equation for x instead of y, we would have gotten

$$2x + 3y = 5$$

$$x = \frac{5 - 3y}{2}$$

Substituting this value for x into the second equation yields

$$3\left(\frac{5 - 3y}{2}\right) + 6y = 5$$

$$2 \cdot 3\left(\frac{5 - 3y}{2}\right) + 2 \cdot 6y = 2 \cdot 5 \quad \text{Multiply each side by 2.}$$

$$3(5 - 3y) + 12y = 10$$

$$15 - 9y + 12y = 10$$

$$3y = -5$$

$$y = -\frac{5}{3}$$

Substituting $y = -\dfrac{5}{3}$ into $x = \dfrac{5 - 3y}{2}$ yields $x = 5$.

Again the solution set is $\left\{\left(5, -\dfrac{5}{3}\right)\right\}$.

Once you have mastered all three methods—graphing, elimination, and substitution—for solving systems of equations, you can choose the one that is easiest in each case. Example 9 could have been solved much more easily using the elimination method.

EXAMPLE 10 Determine the values of a and b so that the line of the equation $ax + by = 5$ passes through $(1, 1)$ and $(-2, 3)$.

SOLUTION If a line passes through a point, the coordinates of the point must satisfy the equation of the line. This means that when the coordinates of the point are substituted for x and y in the equation of the line, a true statement results.

$a(1) + b(1) = 5$ $(1, 1)$ satisfies the equation $ax + by = 5$.
I $a + b = 5$

$a(-2) + b(3) = 5$ $(-2, 3)$ satisfies the equation $ax + by = 5$.
II $-2a + 3b = 5$

The system of equations

I $a + b = 5$
II $-2a + 3b = 5$

can be solved by elimination or substitution. The solution is $a = 2$ and $b = 3$. In other words, these are the values of a and b that allow the line $ax + by = 5$ to pass through the points $(1, 1)$ and $(-2, 3)$. Thus the equation is $2x + 3y = 5$.

CHECK $\quad 2x + 3y = 5 \qquad\qquad 2x + 3y = 5$
$\quad\quad\quad\; 2(1) + 3(1) \stackrel{?}{=} 5 \quad 2(-2) + 3(3) \stackrel{?}{=} 5$
$\quad\quad\quad\quad\quad\quad\quad 5 \stackrel{\checkmark}{=} 5 \quad\quad\quad\quad\quad\quad 5 \stackrel{\checkmark}{=} 5$

EXAMPLE 11 Solve the system of equations

$$\frac{3}{x} - \frac{4}{y} = -6$$

$$\frac{2}{x} + \frac{3}{y} = 13$$

by the elimination method.

SOLUTION We will not attempt to solve for x and y directly; it is easier to make a substitution. We let

$$a = \frac{1}{x} \quad \text{and} \quad b = \frac{1}{y}$$

The equations then become

I	$3a - 4b = -6 \;\rightarrow$	$9a - 12b = -18$	Multiply each side of **I** by 3.
II	$2a + 3b = 13 \;\rightarrow$	$\underline{8a + 12b = 52}$	Multiply each side of **II** by 4.
		$17a \quad\quad\; = 34$	Add the equations.
		$a \quad\quad\quad = 2$	

Substituting $a = 2$ into equation **I** yields $b = 3$. To find $\frac{1}{x}$ and $\frac{1}{y}$, we substitute

$\frac{1}{x}$ for a $\quad$ and $\quad$ $\frac{1}{y}$ for b

$\frac{1}{x} = 2 \qquad\qquad \frac{1}{y} = 3$

$x = \frac{1}{2} \qquad\qquad y = \frac{1}{3}$

CHECK $\quad \dfrac{3}{x} - \dfrac{4}{y} = -6 \quad$ and $\quad \dfrac{2}{x} + \dfrac{3}{y} = 13$

$\quad\quad\quad\;\; \dfrac{3}{\frac{1}{2}} - \dfrac{4}{\frac{1}{3}} \stackrel{?}{=} -6 \qquad\quad \dfrac{2}{\frac{1}{2}} + \dfrac{3}{\frac{1}{3}} \stackrel{?}{=} 13$

$\quad\quad\quad\;\; 6 - 12 \stackrel{?}{=} -6 \qquad\quad\;\; 4 + 9 \stackrel{?}{=} 13$

$\quad\quad\quad\;\; -6 \stackrel{\checkmark}{=} -6 \qquad\quad\quad\;\; 13 \stackrel{\checkmark}{=} 13$

The solution set is $\left\{\left(\frac{1}{2}, \frac{1}{3}\right)\right\}$.

EXAMPLE 12 Two different types of cattle feed contain protein in the amount of 45% and 20%, respectively. How many pounds of each type must a feed lot manager mix together to obtain 10,000 pounds of a mixture that is 25% protein?

SOLUTION We let x = the number of pounds of feed containing 45% protein
y = the number of pounds of feed containing 20% protein

Since 10,000 pounds of the mixture are desired, $x + y = 10{,}000$. We can find a second equation by considering that the number of pounds of protein in the mixture is equal to the sum of the number of pounds of protein that are in each of the two feeds from which the mixture is made.

$$\underset{\downarrow}{45\% \text{ protein}} \quad \underset{\downarrow}{20\% \text{ protein}} \quad \underset{\downarrow}{25\% \text{ protein}}$$
$$0.45x \ + \ 0.20y \ = \ 0.25(10{,}000)$$

When each side of this equation is multiplied by 100 to eliminate decimals, we have

$$45x + 20y = 250{,}000$$

We now have a system of two equations to solve:

$$\begin{array}{llll}
\mathbf{I} & x + y = 10{,}000 & \rightarrow & -20x - 20y = -200{,}000 \\
\mathbf{II} & 45x + 20y = 250{,}000 & \rightarrow & \underline{45x + 20y = 250{,}000} \\
& & & 25x \hphantom{+ 20y} = 50{,}000 \\
& & & x \hphantom{+ 20y} = 2000
\end{array}$$

Multiply each side of **I** by -20.

Since the mixture contains 10,000 pounds and x is 2000, we get $y = 8000$. So the mixture must contain 2000 pounds of the feed with 45% protein and 8000 pounds of the feed with 20% protein.

$\mathcal{E}$XPLORING $\mathcal{G}$RAPHICS

It is also possible to find solutions to systems of equations in two variables by using a computer or a graphics calculator. In Figure 4.7, the system $2x + y = 4$ and $x - y = -1$ have been graphed on the same set of axes on a computer. The intersection of these two lines appears to be (1, 2), although it may be somewhat difficult to read that from the print-out. Most graphics calculators and computer graphing programs, however, have a way of moving the cursor (indicator) to the intersection and displaying the approximate coordinates of its position. If so, instructions will be found in the user's manual.

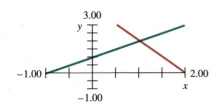

FIGURE 4.7
Graphics solution

To Solve a Linear System of Equations in Two Variables

BY GRAPHING
Step 1 Sketch the graphs of the two equations on the same coordinate system.
Step 2 The solution, if it exists, is (a, b) where a and b are the coordinates of the point where the two lines intersect.

BY THE ELIMINATION METHOD
Step 1 If neither of the variables has coefficients that are additive inverses of each other, use the multiplication property of

Continued

> **BY THE ELIMINATION METHOD**—*Continued*
> equality to generate equivalent equations that satisfy this condition.
>
> **Step 2** Add the two equations to eliminate one variable, then solve for the remaining variable.
>
> **Step 3** Substitute the result of Step 2 into either of the original equations and solve for the remaining variable.
>
> **BY THE SUBSTITUTION METHOD**
>
> **Step 1** Solve for either variable, say x, in either equation. If possible, solve for a variable that has coefficient 1 or -1.
>
> **Step 2** Substitute the value of x from Step 1 in the remaining equation and solve for y.
>
> **Step 3** Substitute the value for y in Step 2 into either of the original equations or their equivalent and solve for x.
>
> **Remember always to check your results in the *original* equations.**

Do You Remember?

Can you match these?

_____ 1. A system of linear equations
_____ 2. A solution to a system of linear equations
_____ 3. The intersection of the graphs of two consistent linear equations
_____ 4. An independent system of equations
_____ 5. An inconsistent system of linear equations
_____ 6. A dependent system of equations
_____ 7. The elimination method
_____ 8. Additive inverses
_____ 9. The graphing method
_____ 10. A numerical coefficient of a variable

a) A system of equations with a unique solution
b) The set of values that are solutions to every equation in the system
c) Two or more linear equations in two or more variables
d) Two numbers that add to zero
e) A method of solving a system of linear equations visually
f) The point where the two lines meet
g) The number in front of a variable
h) A system of equations with graphs that coincide
i) Solving a system by addition or subtraction
j) A system of equations with graphs that do not intersect

Answers: 1. c 2. b 3. f 4. a 5. j 6. h 7. i 8. d 9. e 10. g

Exercise Set 4.1

Solve each system of equations by graphing.

1. $x + y = 10$
   $x - y = 2$

2. $x + y = 2$
   $x - y = 4$

3. $\{(x, y) \mid x + 2y = 3\} \cap \{(x, y) \mid x - 3y = 3\}$

4. $\{(x, y) \mid 2x - y = 8\} \cap \{(x, y) \mid x - 2y = -2\}$

5. $4x - y = 9$
   $x + y = 1$

6. $x = 1$
   $y = -2$

Solve each system of equations by the elimination method. Some may be dependent or inconsistent.

7. $x + y = 1$
   $x - y = 3$

8. $-x + y = 5$
   $x + y = 3$

9. $\{(x, y) \mid x - 2y = 3\} \cap \{(x, y) \mid x + 2y = 2\}$

10. $\{(x, y) \mid 3x - 4y = 6\} \cap \{(x, y) \mid x + 4y = 2\}$

11. $-4x + 3y = 9$
    $4x + 2y = 1$

12. $3x + y = 6$
    $-3x + y = 6$

13. $-2x + 5y = 4$
    $3x + 2y = 13$

14. $2x - y = 3$
    $-6x + 3y = -9$

15. $x - y = -2$
    $-4x + 4y = 2$

16. $3s + t = 15$
    $2s - 5t = 10$

17. $4m - 3n = 12$
    $3m - 4n = 2$

18. $5x - 3y = 1$
    $-3x + 7y = 2$

19. $-x - 2y = 0$
    $5x - y = 27$

20. $2a - 3b = 4$
    $-4a + 6b = 2$

21. $3p + 7q = 17$
    $4p - 2q = 0$

22. $10x + 4y = 10$
    $-4x + 6y = -42$

23. $5x + 2y = 7$
    $3x + 5y = 8$

24. $15x - 7y = 2$
    $4x - 2y = 2$

Solve each system of equations by the substitution method. Some may be dependent or inconsistent.

25. $x - y = 1$
    $x + y = 5$

26. $x + 2y = 8$
    $3x - 4y = 4$

27. $x + 3y = 7$
    $2x + 6y = 14$

28. $2a - b = 5$
    $-4a + 2b = 5$

29. $3m - 6n = 0$
    $2m + 5n = 9$

30. $6x - 5y = 11$
    $5x - 10y = 0$

31. $5x + 7y = 22$
    $x = 5 - 2y$

32. $3x + y = 7$
    $2x = 4y + 14$

33. $\{(x, y) \mid 12x + 4y = 28\} \cap \{(x, y) \mid -3x = y + 9\}$

34. $\{(s, t) \mid 3s + 2t = 10\} \cap \{(s, t) \mid 6s = 9t + 8\}$

35. $3p - q = 18$
    $\frac{3}{2}p - \frac{5}{2}q = 1$

36. $2x - y = 6$
    $\frac{1}{2}x - \frac{3}{2}y = 2$

Solve each system of equations.

37. $\frac{1}{x} + \frac{1}{y} = 2$
    $\frac{1}{x} - \frac{1}{y} = 3$

38. $\frac{1}{x} + \frac{1}{y} = \frac{5}{6}$
    $\frac{1}{x} + \frac{2}{y} = \frac{7}{6}$

39. $\frac{5}{x} - \frac{2}{y} = 4$
    $\frac{3}{x} + \frac{3}{y} = 1$

40. $\frac{7}{x} - \frac{1}{y} = 6$
    $\frac{3}{x} + \frac{5}{y} = 8$

41. $\frac{7}{x} - \frac{3}{y} = 8$
    $\frac{5}{x} + \frac{4}{y} = 14$

42. $\frac{6}{x} - \frac{5}{y} = \frac{1}{2}$
    $\frac{11}{x} + \frac{2}{y} = \frac{13}{2}$

43. $\frac{x-3}{6} + \frac{y+1}{8} = \frac{25}{8}$
    $\frac{x-2}{8} - \frac{y+1}{4} = \frac{3}{4}$

44. $\frac{x+3}{5} - \frac{y-1}{2} = \frac{9}{5}$
    $\frac{x-2}{3} + \frac{y-2}{5} = \frac{2}{15}$

45. $0.02x + 0.05y = 0.41$
    $x + 0.1y = 3.7$

46. $0.6x + 0.5y = 1.3$
    $x + 0.1y = 0.7$

Set up a system of two equations in two unknowns and solve.

47. The sum of two numbers is 74 and their difference is 10. Find the numbers.

48. The sum of two numbers is 69. If one number is twice the other, find the numbers.

49. The width of a rectangle is 5 cm less than its length. The perimeter of the rectangle is 40 cm. Find its length and width.

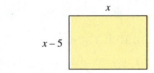

50. The perimeter of the isosceles triangle in the figure is 43.5 in. The length of the side x is 3 in. less than the length of the side y. Find the lengths x and y.

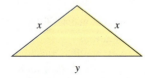

51. Three hamburgers and two orders of french fries cost $6.30. Two hamburgers and three orders of french fries cost $5.70. What is the cost of a hamburger?

52. Mistie bought two pencils and one package of paper for $2.00. Later she bought one pencil and two packages of paper for $2.95. What is the cost of a package of paper?

53. Vonetta invests $10,000. She invests part at 6% per annum and the rest at 7%. How much does she invest at each rate if the total interest earned in one year is $635?

54. For a ball game, 325 tickets were sold for a total of $1140. Front-row-seat tickets cost $4.50 each, and the rest of the tickets cost $3.00. How many of each type of ticket were sold?

55. Two different mixtures contain acid in the amount of 5% and 20%, respectively. How many liters (ℓ) of each type must be mixed to obtain 10 ℓ of a 12.5% acid mixture?

56. Two types of silver ore contain different amounts of silver. When 10 lb of ore A are processed with 20 lb of ore B, then 6 lb of silver are obtained. When 20 lb of ore A are processed with 10 lb of ore B, then 9 lb of silver are obtained. Find the percent of silver in each of the two ores.

57. Thordis is planning a large party. She is making a party mix by adding peanuts that cost $2.50/lb to a cereal mix that costs $1.00/lb. How much of each should she use to get 10 lb of mix at a cost of $1.50/lb?

58. A popular fruit drink is made by mixing pure fruit juice extract and carbonated water. If one mixture with 90% carbonated water is added to another mixture with 70% carbonated water to get 20 ℓ of a mixture with 80% carbonated water, how much of each mixture should be used?

Determine a and b so that each line passes through the given points.

59. $ax + by = 8$; (2, 2) and (6, 2)
60. $ax + by = -10$; (2, 2) and (−4, 1)
61. $ax + by - 6 = 0$; (2, 3) and (1, −3)
62. $ax + by - 15 = 0$; (5, 1) and (10, −1)
63. $ax + by = 6$; x-intercept 2, y-intercept, 3
64. $ax + by = 12$; x-intercept 3, y-intercept, −4

EXPLAIN IN YOUR OWN WORDS

65. What are the differences between consistent, inconsistent, and dependent systems of two equations in two unknowns?
66. What do you find when you solve a system of linear equations?
67. Why is graphing not always a good method for solving a system of two linear equations?
68. Which method for solving systems of equations do you like best? Why?

REVIEW EXERCISES

Simplify and graph each equation.

69. $3(y - 1) - 2(x + 2) = 1 + y + 2x$
70. $2(y + 3) - 6(x + 1) = 8 - 2y$
71. $-x = 12(2y - 4) - 19(y - 2)$
72. $3(x + 5) + 4(y - 2) = 2(x + 3) + 3(y - 1)$
73. $5x + 7y - 8 = 4x + 6(y - 1)$
74. $\frac{1}{2}(x - 4) + \frac{1}{4}(y - 6) = \frac{1}{2}(3x - 2) + \frac{1}{4}(5y - 1)$

Find the equation of the line that satisfies the following conditions.

75. Has a slope of 3 and a y-intercept of 5
76. Has a slope of −1 and a y-intercept of 0
77. Passes through (3, 1) and (5, −4)
78. Passes through (4, 2) and (7, −1)
79. Has an x-intercept of 4 and a y-intercept of −1
80. Has an x-intercept of −2 and a y-intercept of 0

4.2 Solving Systems of Linear Equations in Three or More Variables

HISTORICAL COMMENT Leonardo Fibonacci (1170–1250) was one of the first mathematicians to establish a method for solving three equations in three variables, and his methods of solution were not restricted to solving *linear* equations. At that time in history, many mathematicians did not consider a negative number to be a solution to an equation. Fibonacci, however, allowed that negative numbers could occur if, for example, they showed that a merchant sustained a loss in a business transaction.

OBJECTIVES In this section we will learn to solve systems of linear equations in three or more variables.

An equation such as $3x + 4y + z = 19$ is a **linear equation in three variables.** A **solution** to such an equation consists of values for x, y, and z that simultaneously satisfy it. One such solution to $3x + 4y + z = 19$ is $x = 1$, $y = 2$, and $z = 8$:

$$3 \cdot 1 + 4 \cdot 2 + 8 \stackrel{?}{=} 19$$
$$3 + 8 + 8 \stackrel{?}{=} 19$$
$$19 \stackrel{\checkmark}{=} 19$$

Solutions to linear equations in three variables are called **ordered triples.** The ordered triple $(1, 2, 8)$ in the order (x, y, z) is one solution to $3x + 4y + z = 19$, as we just saw. A linear equation in three variables, however, has infinitely many solutions. The ordered triples $(3, -1, 14)$ and $(2, -3, 25)$ are also solutions of the original equation, since

$$3 \cdot 3 + 4(-1) + 14 = 19$$

and

$$3 \cdot 2 + 4(-3) + 25 = 19$$

The ordered triple $(1, 4, 6)$ is not a solution, however, since

$$3 \cdot 1 + 4 \cdot 4 + 6 \neq 19$$

A system of equations such as

I $x + y + z = 6$
II $2x + y - z = 1$
III $3x + 2y + z = 11$

is called a **system of three linear equations in three variables.** A **solution to the system** is an ordered triple (x, y, z) that is a solution to each equation in the system. The **solution set** to the preceding system is $\{(3, -1, 4)\}$, since

$$3 + (-1) + 4 = 6$$
$$2 \cdot 3 + (-1) - 4 = 1$$
$$3 \cdot 3 + 2(-1) + 4 = 11$$

To solve a system of linear equations in three variables, either the elimination method or the substitution method from the Section 4.1 can be used. The elimination method is generally considered to be easier, so our efforts will be directed toward its use. Although we are given three equations, the method of elimination involves solving only two equations at any one time. The first thing we do is reduce the system to one with two equations in the same two variables. The new system in two variables can then be solved by one of the methods introduced in Section 4.1. The next example illustrates this process.

EXAMPLE 1   Use the method of elimination to solve the system of equations just discussed:

I $x + y + z = 6$
II $2x + y - z = 1$
III $3x + 2y + z = 11$

SOLUTION   To solve this system, we begin by eliminating one of the variables. Let's choose z, since it can be easily eliminated by addition. Starting with equations **I** and **II**, we have

I $x + y + z = 6$
II $\underline{2x + y - z = 1}$
IV $3x + 2y = 7$ **IV** is the sum of **I** and **II**.

Equation **IV** is in terms of the variables x and y. Now we must find a second equation in x and y. This can be accomplished by eliminating z from equations **II** and **III**.

II $2x + y - z = 1$
III $\underline{3x + 2y + z = 11}$
V $5x + 3y = 12$ **V** is the sum of **II** and **III**.

The two equations **IV** and **V** form a linear system in two variables, as we desired. We solve the new system using the elimination method as follows:

IV $3x + 2y = 7$ → $-9x - 6y = -21$ Multiply each side of **IV** by -3.
V $5x + 3y = 12$ → $\underline{10x + 6y = 24}$ Multiply each side of **V** by 2.
                        $x = 3$ Add.

We can substitute 3 for x in either of equations **IV** or **V** to find y. Choosing equation **IV** yields

$3x + 2y = 7$
$3 \cdot 3 + 2y = 7$ $x = 3$
$9 + 2y = 7$
$2y = -2$
$y = -1$

Finally, we can substitute $x = 3$ and $y = -1$ back into any of the original equations, **I, II,** or **III**. Choosing equation **I** yields

$x + y + z = 6$
$3 + (-1) + z = 6$ $x = 3$ and $y = -1$
$2 + z = 6$
$z = 4$

The solution set is $\{(3, -1, 4)\}$.

When a linear equation in two variables is graphed on a plane, its graph is a straight line. When a linear system in two variables is graphed, its solution (if it exists) is the intersection of the two lines. The graph of a linear system in *three* variables is a **plane.** The solution to a linear system in three variables is the intersection of the graphs of the three planes, if it exists. We will not attempt to graph a specific system, but the graphs in Figure 4.8 illustrate some of the possibilities.

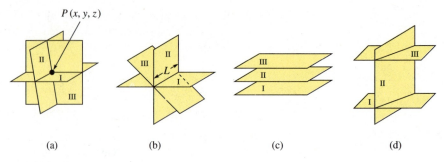

FIGURE 4.8

In Figure 4.8(a), the planes intersect in a single point. There is exactly one solution and the system is said to be **consistent**. In Figure 4.8(b), the planes intersect along the line L. There are infinitely many solutions, since every point on the line L lies on all three planes. We call this system **dependent**. In Figure 4.8(c), the planes are all parallel—no solution exists. The system is called **inconsistent**. In Figure 4.8(d), there are two distinct lines of intersection but no points that are on all three planes. This system is also inconsistent. No solution exists. If the three equations are *equivalent* (each one is a multiple of the other), the planes they represent will coincide and any point that lies on any one of the planes will be a solution. Again, infinitely many solutions will exist and the system will be dependent.

Two Consistent Systems

EXAMPLE 2 Solve the following system:

$$\begin{align} \mathbf{I}\quad & x + y \phantom{{}+z} = 4 \\ \mathbf{II}\quad & \phantom{x+{}} y + z = 6 \\ \mathbf{III}\quad & x \phantom{{}+y} - 2z = -2 \end{align}$$

SOLUTION If we eliminate y from equations **I** and **II**, the resulting equation will have the same variables as equation **III**.

$$\begin{array}{llll} \mathbf{I} & x + y = 4 & \rightarrow & x + y \phantom{{}-z} = 4 \\ \mathbf{II} & y + z = 6 & \rightarrow & \underline{-y - z = -6} \quad \text{Multiply each side of \textbf{II} by } -1. \\ & & \mathbf{IV} & x \phantom{{}+y} - z = -2 \quad \text{Add.} \end{array}$$

Now we eliminate z from equations **III** and **IV**.

$$\begin{array}{llll} \mathbf{III} & x - 2z = -2 & \rightarrow & x - 2z = -2 \\ \mathbf{IV} & x - z = -2 & \rightarrow & \underline{-2x + 2z = 4} \quad \text{Multiply each side of \textbf{IV} by } -2. \\ & & & -x \phantom{{}+2z} = 2 \quad \text{Add.} \\ & & & x \phantom{{}+2z} = -2 \end{array}$$

To find the values of y and z, we substitute the x-value into equations **I** and **III**.

$$\begin{array}{ll} \mathbf{I} \quad x + y = 4 & \mathbf{III} \quad x - 2z = -2 \\ \phantom{\mathbf{I} \quad} -2 + y = 4 & \phantom{\mathbf{III} \quad} -2 - 2z = -2 \\ \phantom{\mathbf{I} \quad x + {}} y = 6 & \phantom{\mathbf{III} \quad x - 2{}} z = 0 \end{array}$$

CHECK   I $x + y = 4$ II $y + z = 6$
             $-2 + 6 \stackrel{?}{=} 4$ $x = -2, y = 6$ $6 + 0 \stackrel{?}{=} 6$ $y = 6, z = 0$
             $4 \stackrel{\checkmark}{=} 4$ $6 \stackrel{\checkmark}{=} 6$

III $x - 2z = -2$
     $-2 - 2 \cdot 0 \stackrel{?}{=} -2$ $x = -2, z = 0$
     $-2 \stackrel{\checkmark}{=} -2$

The solution set is $\{(-2, 6, 0)\}$.

EXAMPLE 3   Solve the following system.

     I $2x + y + z = 7$
     II $4x + 2y - z = 8$
     III $6x - 3y + 4z = -1$

SOLUTION   First we eliminate z from equations **I** and **II,** because the coefficients of z are additive inverses.

     I $2x + y + z = 7$
     II $\underline{4x + 2y - z = 8}$
     IV $6x + 3y \quad\quad = 15$ Add.

Now we eliminate z from equations **II** and **III** by multiplying each side of equation **II** by 4 and adding the result to equation **III.** (Note that we must eliminate the *same variable* from the second pair of equations that we eliminated from the first pair.)

   II $4x + 2y - z = 8$ $\rightarrow$ $16x + 8y - 4z = 32$ Multiply each side of
   III $6x - 3y + 4z = -1$ $\rightarrow$ $\underline{6x - 3y + 4z = -1}$ **II** by 4.
                                     V $22x + 5y \quad\quad = 31$ Add.

To eliminate y from equations **IV** and **V,** we multiply each side of equation **IV** by 5 and each side of **V** by -3.

   IV $6x + 3y = 15$ $\rightarrow$ $30x + 15y = 75$ Multiply each side of **IV** by 5.
   V $22x + 5y = 31$ $\rightarrow$ $\underline{-66x - 15y = -93}$ Multiply each side of **V** by -3.
                              $-36x \quad\quad\quad = -18$ Add.
                                   $x \quad\quad\quad\quad = \dfrac{1}{2}$

Substituting $\dfrac{1}{2}$ for x in equation **IV** yields

        $6x + 3y = 15$
     $6\left(\dfrac{1}{2}\right) + 3y = 15$
        $3 + 3y = 15$
           $3y = 12$
            $y = 4$

Substituting $\frac{1}{2}$ for x and 4 for y in equation **I** yields

$$2x + y + z = 7$$
$$2\left(\frac{1}{2}\right) + 4 + z = 7$$
$$5 + z = 7$$
$$z = 2$$

The solution set is $\left\{\left(\frac{1}{2}, 4, 2\right)\right\}$. The check is left to the reader.

An Inconsistent System

EXAMPLE 4 Solve the following system.

$$\begin{array}{ll} \textbf{I} & x + y - 2z = 4 \\ \textbf{II} & 2x + y = 3 \\ \textbf{III} & 5x + 3y - 2z = 6 \end{array}$$

SOLUTION Since equation **II** involves only x and y, we will eliminate z from equations **I** and **III** to obtain a second equation in x and y.

$$\begin{array}{lllll} \textbf{I} & x + y - 2z = 4 & \rightarrow & -x - y + 2z = -4 & \text{Multiply each side of} \\ \textbf{III} & 5x + 3y - 2z = 6 & \rightarrow & \underline{5x + 3y - 2z = 6} & \text{\textbf{I} by } -1. \\ & & \textbf{IV} & 4x + 2y = 2 & \text{Add.} \end{array}$$

If we eliminate x from equations **II** and **IV**, an impossible statement results.

$$\begin{array}{llll} \textbf{II} & 2x + y = 3 & \rightarrow & -4x - 2y = -6 \quad \text{Multiply each side of \textbf{II} by } -2. \\ \textbf{IV} & 4x + 2y = 2 & \rightarrow & \underline{4x + 2y = 2} \\ & & & 0 = -4 \quad \text{Impossible.} \end{array}$$

The system is inconsistent. The solution set is $\varnothing$.

A Dependent System

EXAMPLE 5 Solve the following system.

$$\begin{array}{ll} \textbf{I} & x + y + z = 3 \\ \textbf{II} & x - y + z = 1 \\ \textbf{III} & 3x + y + 3z = 7 \end{array}$$

SOLUTION We begin by eliminating y first from equations **I** and **II** and then from equations **II** and **III**, as follows:

$$\begin{array}{lll} \textbf{I} & x + y + z = 3 & \\ \textbf{II} & \underline{x - y + z = 1} & \\ \textbf{IV} & 2x + 2z = 4 & \text{Add \textbf{I} to \textbf{II}.} \end{array} \qquad \begin{array}{lll} \textbf{II} & x - y + z = 1 \\ \textbf{III} & \underline{3x + y + 3z = 7} \\ \textbf{V} & 4x + 4z = 8 & \text{Add \textbf{II} to \textbf{III}.} \end{array}$$

Eliminating x from equations **IV** and **V** results in a statement that is true for any value of x:

$$\begin{array}{rlrl}
\textbf{IV} & \quad 2x + 2z = 4 & \rightarrow & \quad -4x - 4z = -8 \qquad \text{Multiply each side of } \mathbf{V} \text{ by } -2. \\
\textbf{V} & \quad 4x + 4z = 8 & \rightarrow & \quad \underline{4x + 4z = 8} \\
& & & \qquad\qquad\qquad 0 = 0 \qquad \text{True.}
\end{array}$$

The system is dependent. There are infinitely many solutions.

EXAMPLE 6 A collection of nickels, dimes, and quarters has a face value of $2.00. The value of the nickels is 90¢ less than that of the dimes and the quarters put together. How many of each are there if the collection contains 21 coins?

SOLUTION We let $n =$ the number of nickels
$d =$ the number of dimes
$q =$ the number of quarters

The total number of coins is 21. This means that

$$n + d + q = 21$$

The value of the coins is $2.00, or 200 cents, so

$$5n + 10d + 25q = 200$$

We now have two equations in three variables. A third equation can be obtained from the fact that the value of the nickels is 90¢ less than the value of the dimes and the quarters together.

$$\underbrace{5n}_{\text{The value of the nickels}} \underbrace{=}_{\text{is}} \underbrace{10d + 25q - 90}_{\text{90 less than the value of the dimes and quarters together}}$$

To solve the system by the elimination method, we write each equation in **standard form** (all terms involving variables stand alone on the left side of the equal sign).

$$\begin{array}{rl}
\textbf{I} & n + d + q = 21 \\
\textbf{II} & 5n + 10d + 25q = 200 \\
\textbf{III} & 5n - 10d - 25q = -90
\end{array}$$

We begin by eliminating d from equations **I** and **III** and also from equations **II** and **III**:

$$\begin{array}{rlrl}
\textbf{I} & \quad n + d + q = 21 & \rightarrow & \quad 10n + 10d + 10q = 210 \qquad \text{Multiply each} \\
\textbf{III} & \quad 5n - 10d - 25q = -90 & \rightarrow & \quad \underline{5n - 10d - 25q = -90} \qquad \text{side of } \mathbf{I} \text{ by 10.} \\
& & \textbf{IV} & \quad 15n - 15q = 120 \qquad \text{Add.}
\end{array}$$

Also,

$$\begin{array}{rl}
\textbf{II} & 5n + 10d + 25q = 200 \\
\textbf{III} & \underline{5n - 10d - 25q = -90} \\
\textbf{V} & 10n = 110 \qquad \text{Add.} \\
& n = 11
\end{array}$$

Substituting 11 for n in equation **IV** yields

$$15n - 15q = 120$$
$$15(11) - 15q = 120$$
$$165 - 15q = 120$$
$$-15q = -45$$
$$q = 3$$

Finally, substituting $n = 11$ and $q = 3$ in equation **I** yields

$$n + d + q = 21$$
$$11 + d + 3 = 21$$
$$d = 7$$

The solution is the ordered triple (11, 7, 3) in the order (n, d, q). In other words, there are 11 nickels, 7 dimes, and 3 quarters in the collection of coins.

CHECK The number of coins in the collection is 21.

$$11 + 7 + 3 \stackrel{\checkmark}{=} 21$$

The value of the collection is $2.00, or 200¢.

$$11(5) + 7(10) + 3(25) = 55 + 70 + 75 \stackrel{\checkmark}{=} 200$$

The value of the nickels is 90¢ less than the value of the dimes and the quarters put together.

$$11(5) \stackrel{?}{=} 7(10) + 3(25) - 90$$
$$55 \stackrel{?}{=} 70 + 75 - 90$$
$$55 \stackrel{\checkmark}{=} 55$$

The solution checks

EXAMPLE 7   When Aglaja, David, and Julian work together, they can complete a job in 5 hours. If Aglaja works on the job for 2 hours and then David works for 5 hours, they can complete half of the job. If Julian works for 8 hours and then David works for 3 hours, they can complete three-fifths of the job. How long would it take each one of them to complete the job working alone?

SOLUTION   Work problems are typically solved by determining what **fractional part of a job** a person can do in **one unit of time.** For example, if a woman can do an entire job in 3 hours, she can do one-third of the job in 1 hour working at a constant rate. If she can do a job in n hours, then she can do one-nth of the job in 1 hour. In 2 hours she can do

$$2\left(\frac{1}{n}\right) = \frac{2}{n}$$

parts of the job, and so forth.

$$\text{quantity of work done} = \frac{\text{time worked}}{\text{time required to do the entire job alone}}$$

We let x = the number of hours for Aglaja to do the job alone
y = the number of hours for David to do the job alone
z = the number of hours for Julian to do the job alone

Based on the preceding discussion,

Aglaja can do $\frac{1}{x}$ of the job in 1 hour.

David can do $\frac{1}{y}$ of the job in 1 hour.

Julian can do $\frac{1}{z}$ of the job in 1 hour.

Aglaja + David + Julian can do $\frac{1}{5}$ of the job in 1 hour
$$\frac{1}{x} + \frac{1}{y} + \frac{1}{z} = \frac{1}{5}$$

Aglaja (2 hrs) + David (5 hrs) can do $\frac{1}{2}$ of the job
$$\frac{2}{x} + \frac{5}{y} = \frac{1}{2}$$

David (3 hrs) + Julian (8 hrs) can do $\frac{3}{5}$ of the job
$$\frac{3}{y} + \frac{8}{z} = \frac{3}{5}$$

To make the solution to the system easier to work with, we let

$$a = \frac{1}{x} \quad b = \frac{1}{y} \quad c = \frac{1}{z}$$

This will yield three alternate equations free of fractions.

	EQUATIONS		ALTERNATES	
I	$a + b + c = \frac{1}{5}$	$\rightarrow$	$5a + 5b + 5c = 1$	⎫
II	$2a + 5b = \frac{1}{2}$	$\rightarrow$	$4a + 10b = 1$	⎬ Clear of fractions.
III	$3b + 8c = \frac{3}{5}$	$\rightarrow$	$15b + 40c = 3$	⎭

To solve this system, we first eliminate a from the alternate forms of equations **I** and **II**.

I	$5a + 5b + 5c = 1$	$\rightarrow$	$-20a - 20b - 20c = -4$	Multiply each side of **I** by -4
II	$4a + 10b = 1$	$\rightarrow$	$\underline{20a + 50b = 5}$	Multiply each side of **II** by 5.
		IV	$30b - 20c = 1$	Add.

Now we eliminate b from equations **III** and **IV**.

IV	$30b - 20c = 1$	$\rightarrow$	$-30b + 20c = -1$	Multiply each side of **IV** by -1.
III	$15b + 40c = 3$	$\rightarrow$	$\underline{30b + 80c = 6}$	Multiply each side of **III** by 2.
			$100c = 5$	Add.

$$c = \frac{1}{20}$$

Substituting this value into equation **III** gives us $b = \frac{1}{15}$. When these two values are substituted into equation **I**, we find that $a = \frac{1}{12}$. Finally, since

$$a = \frac{1}{x} \qquad b = \frac{1}{y} \qquad c = \frac{1}{z}$$

we have $x = 12$, $y = 15$, and $z = 20$. Thus Aglaja can do the job alone in 12 hours, David in 15 hours, and Julian in 20 hours.

Although a linear equation in four or more variables does not represent a physical quantity such as a line or a plane, systems involving such equations can be solved by similar techniques. To solve a system of four equations in four variables, we must first reduce it to a system of three equations in three variables. The next example shows one way it can be done.

EXAMPLE 8 Use the elimination method to solve the following system of equations.

I	$x + y + w + z = 4$
II	$x + 2y - w + z = 7$
III	$2x - y + w - z = -3$
IV	$x - y - w + 2z = 7$

SOLUTION We begin by eliminating w from the system, using equations **I** and **II**:

I	$x + y + w + z = 4$
II	$\underline{x + 2y - w + z = 7}$
V	$2x + 3y + 2z = 11$ Add.

Next we eliminate w from equations **I** and **III**:

I	$x + y + w + z = 4$	$\rightarrow$	$x + y + w + z = 4$	
III	$2x - y + w - z = -3$	$\rightarrow$	$\underline{-2x + y - w + z = 3}$	Multiply each side of **III** by -1.
		VI	$-x + 2y + 2z = 7$	Add.

Finally, we eliminate w from equations **I** and **IV**:

I	$x + y + w + z = 4$
IV	$\underline{x - y - w + 2z = 7}$
VII	$2x + 3z = 11$

The system

$$\begin{aligned} \mathbf{V} \quad & 2x + 3y + 2z = 11 \\ \mathbf{VI} \quad & -x + 2y + 2z = 7 \\ \mathbf{VII} \quad & 2x + 3z = 11 \end{aligned}$$

can now be solved by the methods for systems of three equations in three variables. The reader is encouraged to solve this system to verify that $x = 1$, $y = 1$, and $z = 3$. When these values are substituted into any of equations **I**, **II**, **III**, or **IV**, we find that $w = -1$. The solution set, given in alphabetical order, is the **ordered quadruple** $\{(1, 1, -1, 3)\}$.

To Solve a Linear System of Equations in Three or Four Variables

Step 1 If necessary, rewrite the equations in standard form.
Step 2 Use the elimination method to eliminate any single variable from two equations.
Step 3 Eliminate the same variable from other combinations of equations different from those in Step 2.
Step 4 Solve the system of equations from Step 3 by either elimination or substitution.
Step 5 Substitute the result of Step 4 into any of the original equations or their equivalent to find the remaining variable.

---

*E*XERCISE *S*ET 4.2

Solve each system of equations and check.

1. $\begin{aligned} x - 4y - 4z &= -13 \\ 2x + 9y + z &= 2 \\ -x + 3y + 2z &= 4 \end{aligned}$

2. $\begin{aligned} x + y + 3z &= 10 \\ 2x + z &= 7 \\ x + y - z &= -2 \end{aligned}$

3. $\begin{aligned} 2a - b + c &= -1 \\ a - 4b - 3c &= -4 \\ -a - b + 2c &= -5 \end{aligned}$

4. $\begin{aligned} a + b - c &= 6 \\ a - 3b - c &= 10 \\ a + b + c &= 12 \end{aligned}$

5. $\begin{aligned} 5x + 4y - 6z &= -5 \\ -2x + 3y - 5z &= -11 \\ 4x - 7y + 8z &= 14 \end{aligned}$

6. $\begin{aligned} 2x + 3y + z &= 1 \\ -3x + y - 2z &= -5 \\ 5x + y + 4z &= 8 \end{aligned}$

7. $\begin{aligned} y + z &= 7 \\ x + z &= 6 \\ x + y &= 5 \end{aligned}$

8. $\begin{aligned} y - z &= 1 \\ x - z &= 5 \\ x - y &= 3 \end{aligned}$

9. $\begin{aligned} x + z &= 2 \\ x + y &= 1 \\ y - z &= 3 \end{aligned}$

10. $\begin{aligned} 3a + b + 6c &= 1 \\ -4a - 3b - 5c &= -5 \\ 2a + 3b + 3c &= 9 \end{aligned}$

11. $\begin{aligned} 9x - 5z &= 18 \\ 3x + 7y + 4z &= -45 \\ 5x - 9y - 6z &= 39 \end{aligned}$

12. $\begin{aligned} x + 5y - 3z &= 5 \\ 2x + 10y + 6z &= 18 \\ 4x + 3y - 7z &= 0 \end{aligned}$

13. $\begin{aligned} -2x - 9y &= -8 \\ x + 8z &= 3 \\ -15y + 17z &= 7 \end{aligned}$

14. $\begin{aligned} -5b - c &= -1 \\ 2a + 3c &= 4 \\ -3a + 20b &= 1 \end{aligned}$

15. $\begin{aligned} x + z &= 5 \\ x + y - z &= -1 \\ 2x + 2y &= 4 \end{aligned}$

16. $\begin{aligned} \tfrac{3}{2}x - 3y + 2z &= -2 \\ -\tfrac{5}{8}x + \tfrac{12}{9}y + \tfrac{14}{6}z &= \tfrac{7}{3} \\ 3x - \tfrac{8}{3}y - z &= 1 \end{aligned}$

17. $\frac{1}{2}x + y + \frac{1}{2}z = -\frac{1}{2}$

$x + y - 2z = -4$

$\frac{1}{4}x - \frac{1}{4}y + \frac{1}{2}z = 1$

18. $r - \frac{7}{4}s - 2t = \frac{11}{2}$

$r + \frac{6}{5}s + \frac{2}{10}t = -\frac{132}{35}$

$2r - \frac{5}{2}s + t = \frac{55}{7}$

19. $0.3r + 0.5s - 0.3t = 0.5$
$0.60r + 0.80s - 0.4t = 1.0$
$r + 0.5s - 0.5t = 1.0$

20. $0.05a - 0.01b + 0.04c = 0.16$
$-0.05a + 0.2b - 0.6c = -1.8$
$0.2a - 0.5b - 0.3c = -1.2$

21. $\frac{1}{x} + \frac{4}{y} + \frac{3}{z} = 8$

$\frac{3}{x} - \frac{1}{y} + \frac{1}{z} = 3$

$\frac{2}{x} - \frac{3}{y} + \frac{1}{z} = 0$

22. $\frac{2}{x} - \frac{1}{y} + \frac{4}{z} = -2$

$\frac{3}{x} - \frac{2}{y} - \frac{1}{z} = 2$

$\frac{1}{x} + \frac{1}{y} + \frac{3}{z} = 4$

23. $\frac{2}{x} + \frac{3}{y} - \frac{6}{z} = 11$

$\frac{3}{x} - \frac{4}{y} + \frac{18}{z} = -19$

$\frac{6}{x} - \frac{9}{y} - \frac{2}{z} = -1$

24. $\frac{1}{x} + \frac{1}{3y} + \frac{2}{3z} = \frac{5}{3}$

$\frac{2}{5x} - \frac{3}{5y} + \frac{2}{5z} = -1$

$\frac{1}{3x} + \frac{1}{y} + \frac{1}{3z} = \frac{11}{3}$

25. $x + y - z + 5w = 5$
$6x - 2y + z + w = 0$
$x - y - z - 2w = -4$
$x + 2y + 2z - w = 3$

26. $x - y + z - w = -1$
$2x + 2y + 2z + w = 3$
$x - 2y - 2z - 2w = -9$
$3x - 3y - z + w = 1$

Write a system of three equations in three variables for each of the following applied problems and solve.

27. A collection of nickels, dimes, and quarters is worth $5.55. There are 9 more dimes in the collection than the number of nickels and quarters combined. The number of nickels is 1 less than three times the number of quarters. Find the number of nickels, dimes, and quarters in the collection.

28. The sum of the measures of the angles of a triangle is always 180°. The measure of the first angle is three times the measure of the second angle. The third angle measures 20° more than the sum of the measures of the first two angles. What is the measure of each angle?

29. A company invests $25,000 for one year. Part of it is invested at 5% annual interest, another part at 6% annual interest, and the rest at 8%. The total interest income for the year is $1600. The income from the 8% investment is the same as the sum of the incomes from the other two investments. Find the amount invested at each rate.

30. Adam invested $21,900 for one year. He invested part of it at 9% annual interest, another part at 10% annual interest, and the rest at 14%. The total income from the interest for the year is $2409. The income from the 14% investment is $164 more than the income from the 10% investment. Find the amount invested at each rate.

31. Joe is considering the purchase of a duplicating machine. The salesman supplies the following information on three available models: If all three machines are operating, a particular job can be completed in 50 minutes. If machine A operates for 20 minutes and machine B for 50 minutes, one-half of the job gets finished. If machine B operates for 30 minutes and machine C for 80 minutes, three-fifths of the job gets completed. Which is the fastest machine, and how long would it take for this machine alone to finish the whole job?

32. Carlos, Sam, and Bennie are doing a certain computing job. The job can be completed if Carlos and Sam work together for 2 hours and Bennie works alone for 1 hour, or if Carlos and Sam work together for 1 hour and Bennie works alone for 2½ hours, or if Carlos works alone for 3 hours, Sam works alone for 1 hour, and Bennie works alone for half an hour. Determine the time it takes for each man working alone to complete the job.

WRITE IN YOUR OWN WORDS

33. What are you solving for when you solve a system of three equations in three variables?

34. If the graphs of two equations in three variables intersect, how do you describe the intersection?

35. What is the graph of an equation in three variables?

36. What is the graph of an equation in two variables?

REVIEW EXERCISES

37. Given $f(x) = 3x - 1$, find $f(0), f(1), f(-2)$, and $f(b)$.

38. Is $A = \{(-2, 0), (0, 2), (2, 4), (3, 6)\}$ a function?

39. Given $f(x) = x - 3$ and $g(x) = 2x + 5$, find $f[g(0)]$, $(f + g)(2)$, and $(f \cdot g)(-1)$.

40. Given $f(x) = 5x - 8$, find $f^{-1}(x)$.

41. Is $g(x) = \dfrac{x - 5}{3x + 4}$ a function? Find its domain.

42. Given $f(x) = 7x - 1$ and $g(x) = 4x + 3$, does $f[g(0)]$ equal $g[f(0)]$?

4.3 Solving Systems of Linear Equations Using Matrices

OBJECTIVES In this section we will learn
1. the definition of a matrix;
2. how to use row operations to change the rows of a matrix; and
3. how to apply matrix operations to solve systems of equations.

In Sections 4.1 and 4.2 we solved systems of linear equations by using either graphing, elimination, or substitution. In this section we will introduce the concept of a **matrix**, which can be used as an additional tool for solving such systems. It is important to note that solutions by matrices has both advantages and disadvantages when compared to the methods that we have already developed. If you pursue a career where mathematics is involved, the particular problem that you are solving will largely dictate which method is the most efficient.

DEFINITION

Matrix

A **matrix**, usually designated by a capital letter, is a rectangular array of numbers enclosed by parentheses or brackets. The numbers in a matrix are called its **elements** or **entries**. We say that a matrix has two **dimensions:** (1) the number of rows and (2) the number of columns.

For example,

$$A = \begin{bmatrix} 1 & 5 & 6 \\ 2 & 1 & 2 \end{bmatrix}$$

is a matrix with two rows, three columns, and six elements or entries.

Column 1 Column 2 Column 3

$$A = \begin{bmatrix} 1 & 5 & 6 \\ 2 & 1 & 2 \end{bmatrix} \begin{matrix} \leftarrow \text{Row 1} \\ \leftarrow \text{Row 2} \end{matrix}$$

The dimensions of a matrix are written in the form $m \times n$, which is read "m by n." Matrix A has dimensions

$$m \times n = 2 \times 3$$

Number of rows Number of columns

We say it is a two-by-three matrix. (Note that we say the row dimension first.) Other examples of matrices and their dimensions are

$$B = \begin{bmatrix} 1 & 2 \\ 3 & 4 \\ 5 & 6 \end{bmatrix} \quad \text{and} \quad C = \begin{bmatrix} 1 & 2 & 3 \\ 2 & 3 & 1 \\ 2 & 1 & 3 \end{bmatrix}$$

Here B is a 3×2 matrix and C is a 3×3 matrix. If a matrix has the same number of rows as columns, we call it a **square matrix.** Matrix C is a square matrix.

There are many operations that can be performed on matrices. We will limit our discussion to those operations that are particularly useful in solving systems of equations.

The system of equations

I $\quad 2x + y + z = 7$
II $\quad 4x + 2y - z = 8$
III $\quad 6x - 3y + 4z = -1$

can be written in matrix form by considering only the coefficients of the variables and the constants, 7, 8, and -1:

$$\begin{bmatrix} 2 & 1 & 1 & 7 \\ 4 & 2 & -1 & 8 \\ 6 & -3 & 4 & -1 \end{bmatrix}$$

The matrix corresponding to the first three columns is called the **coefficient matrix.** When the constants, 7, 8, and -1, are included with the coefficient matrix as a fourth column, the result is called the **augmented matrix.**

COEFFICIENT MATRIX $\quad$ AUGMENTED MATRIX

$$\begin{bmatrix} 2 & 1 & 1 \\ 4 & 2 & -1 \\ 6 & -3 & 4 \end{bmatrix} \quad \begin{bmatrix} 2 & 1 & 1 & | & 7 \\ 4 & 2 & -1 & | & 8 \\ 6 & -3 & 4 & | & -1 \end{bmatrix}$$

We solved the preceding system of equations by the elimination method in Example 3 of Section 4.2. It would be a good idea to review the solution at this point. The solution to the system involved multiplying one equation by a constant and adding it to a different equation to eliminate a variable. The matrix operations that achieve the same purpose are called **row operations.** They are outlined in Theorem 4.1.

THEOREM 4.1

Row Transformation Theorem

If a matrix represents a system of linear equations, then each of the following operations determines a system of equivalent linear equations.

1. Any two rows may be interchanged. (Interchanging rows corresponds to writing the equations in a different order.)
2. All of the elements in any row may be multiplied by the same nonzero constant. (This corresponds to multiplying each side of an equation by the same number.)
3. The product of any constant and any row may be added to any other row. (The changes occur in the row to which the product is added.)

EXAMPLE 1 Given the matrix

$$B = \begin{bmatrix} 1 & 2 & 1 \\ 3 & 4 & -2 \end{bmatrix}$$

where the first row is denoted R_1 and the second row R_2, carry out each of the following operations on the matrix:

(a) interchange R_1 and R_2 **(b)** multiply R_1 by -2 **(c)** add $(-3)R_1$ to R_2

SOLUTION **(a)** Interchanging R_1 and R_2 yields

$$\begin{bmatrix} 3 & 4 & -2 \\ 1 & 2 & 1 \end{bmatrix}$$

(b) Multiply R_1 by -2 means that every element in R_1 is multiplied by -2. The result is the matrix

$$\begin{bmatrix} -2 & -4 & -2 \\ 3 & 4 & -2 \end{bmatrix}$$

(c) To add $(-3)R_1$ to R_2, we must first find $(-3)R_1$. By Theorem 4.1, $(-3)R_1 = [-3 \quad -6 \quad -3]$. When $(-3)R_1 = [-3 \quad -6 \quad -3]$ is added to $R_2 = [3 \quad 4 \quad -2]$, the corresponding matrix is

$$\begin{bmatrix} 1 & 2 & 1 \\ 3+(-3) & 4+(-6) & -2+(-3) \end{bmatrix} = \begin{bmatrix} 1 & 2 & 1 \\ 0 & -2 & -5 \end{bmatrix}$$

Notice that R_1 remains unchanged under these operations. The "receiving" row, R_2, is the one that is changed.

EXAMPLE 2 Write the augmented matrix for the system of equations

$$3x + 5y = 9$$
$$2x - y = -3$$

SOLUTION The augmented matrix consists of the coefficient matrix and the constants.

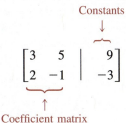

EXAMPLE 3 Write the augmented matrix for the system

$$3x + 2y + z = 5$$
$$x \phantom{{}+2y} + z = 2$$
$$\phantom{x +{}} y - z = 4$$

SOLUTION If a variable does not appear in an equation, the coefficient of that variable is zero. In the second equation, the coefficient of y is zero. In the third equation, the coefficient of x is zero. The matrix is

$$\begin{bmatrix} 3 & 2 & 1 & | & 5 \\ 1 & 0 & 1 & | & 2 \\ 0 & 1 & -1 & | & 4 \end{bmatrix}$$

EXAMPLE 4 Write a system of equations in the variables a, b, and c that corresponds to the augmented matrix

$$\begin{bmatrix} 2 & 1 & 3 & | & 2 \\ 1 & 0 & 4 & | & -1 \end{bmatrix}$$

SOLUTION The matrix represents a system of two equations in three variables. If we call the variables a, b, and c, the equations are

$$2a + b + 3c = 2$$
$$a + 4c = -1$$

We now turn our attention to the use of matrices as a tool for solving systems of equations. To illustrate how this is done, consider the equations

I $x + y = 6$
II $2x + 3y = 16$

To solve this system by the elimination method (Section 4.1), equation **I** can be multiplied by -2 and the result added to equation **II** to obtain a new equation.

I $x + y = 6$ $\rightarrow$ $-2x - 2y = -12$ Multiply each side of **I** by -2.
II $2x + 3y = 16$ $\rightarrow$ $\underline{2x + 3y = 16}$
                                      III $y = 4$ Add: **III** = **II** + (-2)**I**.

The system of equations **I** and **III** is equivalent to the original system. In equation **III**, $y = 4$. When we substitute $y = 4$ into equation **I**, we get $x + 4 = 6$, so x has a value of 2. The solution set is $\{(2, 4)\}$.

We can use these operations to solve the system when it is written in augmented matrix form:

$$\begin{bmatrix} 1 & 1 & | & 6 \\ 2 & 3 & | & 16 \end{bmatrix} \begin{matrix} R_1 \\ R_2 \end{matrix}$$

We begin by multiplying row 1 by -2 and adding the result to row 2:

$$\begin{bmatrix} 1 & 1 & | & 6 \\ 0 & 1 & | & 4 \end{bmatrix} \begin{matrix} R_1 \\ R_2 + (-2)R_1 \end{matrix}$$

The second row represents the equation $0x + y = 4$ or simply $y = 4$, and the first row represents the equation $x + y = 6$.

Solving by Back Substitution

Now let's consider an augmented matrix that represents a system of linear equations in the variables x, y, and z:

$$\begin{bmatrix} 1 & 2 & 3 & | & 1 \\ 0 & 1 & 3 & | & 2 \\ 0 & 0 & 1 & | & 5 \end{bmatrix} \begin{matrix} R_1 \\ R_2 \\ R_3 \end{matrix}$$

When this matrix is translated into a system of equations, the solution set can be found by substituting known values back into the previous equations.

I $x + 2y + 3z = 1$
II $y + 3z = 2$
III $z = 5$

Substituting $z = 5$ into equation **II** yields

$$y + 3z = 2$$
$$y + 3 \cdot 5 = 2$$
$$y + 15 = 2$$
$$y = 2 - 15 = -13$$

Substituting $z = 5$ and $y = -13$ into equation **I** yields

$$x + 2y + 3z = 1$$
$$x + 2(-13) + 3 \cdot 5 = 1$$
$$x - 26 + 15 = 1$$
$$x - 11 = 1$$
$$x = 12$$

This method is called **back substitution.** The solution set is $\{(12, -13, 5)\}$.

EXAMPLE 5   Solve the following system of equations using matrices and row operations.

$$3x - 2y + z = 2$$
$$2x + y + 3z = 6$$
$$x + y - z = 1$$

SOLUTION   First we write the system in augmented matrix form.

$$\begin{bmatrix} 3 & -2 & 1 & | & 2 \\ 2 & 1 & 3 & | & 6 \\ 1 & 1 & -1 & | & 1 \end{bmatrix}$$

We interchange R_1 and R_3 so that the first entry in the matrix is 1:

$$\begin{bmatrix} 1 & 1 & -1 & | & 1 \\ 2 & 1 & 3 & | & 6 \\ 3 & -2 & 1 & | & 2 \end{bmatrix} \begin{matrix} R_1 \\ R_2 \\ R_3 \end{matrix}$$

When using row operations, we will use symbolism such as $R_2 + R_1 \to R_2$ to mean that the sum of rows 1 and 2 becomes the new row 2. The arrow indicates where the result of the row operation is to be placed. We now use row operations to get the first elements in R_2 and R_3 both to be zero.

$$\begin{bmatrix} 1 & 1 & -1 & | & 1 \\ 0 & -1 & 5 & | & 4 \\ 0 & -5 & 4 & | & -1 \end{bmatrix} \begin{matrix} R_1 \\ R_2 + (-2)R_1 \to R_2 \\ R_3 + (-3)R_1 \to R_3 \end{matrix}$$

To get a 1 as the second element in R_2, we multiply R_2 by -1:

$$\begin{bmatrix} 1 & 1 & -1 & | & 1 \\ 0 & 1 & -5 & | & -4 \\ 0 & -5 & 4 & | & -1 \end{bmatrix}$$

Now we add $5R_2$ to R_3 to get a zero as the second element in R_3:

$$\begin{bmatrix} 1 & 1 & -1 & | & 1 \\ 0 & 1 & -5 & | & -4 \\ 0 & 0 & -21 & | & -21 \end{bmatrix} \begin{matrix} R_1 \\ R_2 \\ R_3 + 5R_2 \to R_3 \end{matrix}$$

Finally we divide R_3 by -21 to get a 1 in the third position in R_3.

$$\begin{bmatrix} 1 & 1 & -1 & | & 1 \\ 0 & 1 & -5 & | & -4 \\ 0 & 0 & 1 & | & 1 \end{bmatrix}$$

The third line of the matrix shows that $z = 1$. Now we can use back substitution to find x and y. When $z = 1$ is substituted into the second equation, which from the matrix is $y - 5z = -4$, we find that $y = 1$ also. Finally, when we substitute the values $z = 1$ and $y = 1$ into the equation represented by the first row of the matrix, $x + y - z = 1$, we find that x is also 1. The solution set is $\{(1, 1, 1)\}$.

Gauss–Jordan Elimination Method

If a coefficient matrix can be written in **reduced form,**

$$\begin{bmatrix} 1 & 0 & 0 \\ 0 & 1 & 0 \\ 0 & 0 & 1 \end{bmatrix}$$

by using row operations, then we can use inspection on the augmented form of the solution to solve the system of equations it represents. For example, if the variables of the system are x, y, and z, then augmented in the matrix

$$\begin{matrix} x & y & z \\ \searrow & \downarrow & \swarrow \end{matrix}$$

$$\begin{bmatrix} 1 & 0 & 0 & | & 2 \\ 0 & 1 & 0 & | & 3 \\ 0 & 0 & 1 & | & 2 \end{bmatrix}$$

the first line indicates that $x = 2$, the second line that $y = 3$, and the third that $z = 2$.

EXAMPLE 6 Write the following system of equations,

$$3x - 2y + z = 2$$
$$2x + y + 3z = 6$$
$$x + y - z = 1$$

as a matrix in reduced form.

SOLUTION This is the same system of equations that was solved in Example 5 by using matrices and back substitution. We begin by using the final matrix of that example. The three entries marked by * must be reduced to zero with the use of row operations.

$$\begin{bmatrix} 1 & 1* & -1* & | & 1 \\ 0 & 1 & -5* & | & -4 \\ 0 & 0 & 1 & | & 1 \end{bmatrix} \begin{matrix} R_1 \\ R_2 \\ R_3 \end{matrix}$$

We reduce these entries to zero as indicated:

$$\begin{bmatrix} 1 & 1 & 0 & | & 2 \\ 0 & 1 & 0 & | & 1 \\ 0 & 0 & 1 & | & 1 \end{bmatrix} \begin{matrix} R_1 + R_3 \to R_1 \\ R_2 + 5R_3 \to R_2 \\ R_3 \end{matrix}$$

$$\begin{bmatrix} 1 & 0 & 0 & | & 1 \\ 0 & 1 & 0 & | & 1 \\ 0 & 0 & 1 & | & 1 \end{bmatrix} \begin{matrix} R_1 + (-1)R_2 \to R_1 \\ R_2 \\ R_3 \end{matrix}$$

The solution set is {(1, 1, 1)}.

This process of solving a system of linear equations by changing the coefficient matrix to reduced form is known as the **Gauss–Jordan elimination method.** The steps for using this process are outlined for convenience.

The Gauss–Jordan Elimination Method

Step 1 Write the system of equations in augmented matrix form, so that the first entry in row 1, column 1 is 1 (interchange rows if necessary).

Step 2 If a 1 can't be obtained as the first entry by interchanging rows, divide row 1 by its first entry. (The first entry in row 1 can't be zero.)

Step 3 Multiply row 1 by a suitable constant (or constants) and add to the remaining row(s) so that column 1 has a 1 in row 1 and zeros in all rows below row 1.

Step 4 Use suitable row operations to obtain a 1 in row 2, column 2, and all zeros below it.

Step 5 Continue the procedure to obtain ones on the **main diagonal** (through row 1, column 1, row 2, column 2, etc.). All entries below this diagonal must be zero.

Step 6 Reverse the procedure in Steps 3 through 5 to obtain a zero for each entry above the diagonal.

EXAMPLE 7 Solve the following system by the Gauss–Jordan elimination method.

$$2x + y - z = 5$$
$$2x + 2y + z = 5$$
$$3x + 2y - z = 8$$

SOLUTION The augmented matrix is

$$\begin{bmatrix} 2 & 1 & -1 & | & 5 \\ 2 & 2 & 1 & | & 5 \\ 3 & 2 & -1 & | & 8 \end{bmatrix}$$

None of the entries in the first column is 1. Do we have to divide the first row by 2 and work with fractions? The answer is no! To get a 1 as the first entry, we can interchange column 1 and column 2. This avoids the division and the necessity to begin work with fractions. *However, we must remember that the order of the variables is now y, x, z.*

$$\begin{bmatrix} 1 & 2 & -1 & | & 5 \\ 2 & 2 & 1 & | & 5 \\ 2 & 3 & -1 & | & 8 \end{bmatrix} \rightarrow \begin{bmatrix} 1 & 2 & -1 & | & 5 \\ 0 & -2 & 3 & | & -5 \\ 0 & -1 & 1 & | & -2 \end{bmatrix} \begin{matrix} R_1 \\ R_2 + (-2)R_1 \rightarrow R_2 \\ R_3 + (-2)R_1 \rightarrow R_3 \end{matrix}$$

$$\rightarrow \begin{bmatrix} 1 & 2 & -1 & | & 5 \\ 0 & 1 & -1 & | & 2 \\ 0 & -2 & 3 & | & -5 \end{bmatrix} \quad \text{Multiply } R_3 \text{ by } -1 \text{ and interchange with } R_2.$$

$$\rightarrow \begin{bmatrix} 1 & 2 & -1 & | & 5 \\ 0 & 1 & -1 & | & 2 \\ 0 & 0 & 1 & | & -1 \end{bmatrix} \begin{matrix} R_1 \\ R_2 \\ R_3 + 2R_2 \rightarrow R_3 \end{matrix}$$

$$\rightarrow \begin{bmatrix} 1 & 2 & 0 & | & 4 \\ 0 & 1 & 0 & | & 1 \\ 0 & 0 & 1 & | & -1 \end{bmatrix} \begin{matrix} R_1 + R_3 \rightarrow R_1 \\ R_2 + R_3 \rightarrow R_2 \\ R_3 \end{matrix}$$

$$\rightarrow \begin{bmatrix} 1 & 0 & 0 & | & 2 \\ 0 & 1 & 0 & | & 1 \\ 0 & 0 & 1 & | & -1 \end{bmatrix} \begin{matrix} R_1 + (-2)R_2 \rightarrow R_1 \\ R_2 \\ R_3 \end{matrix}$$

This last matrix is now in reduced form. But remember that the entries are in the order y, x, z. Thus we have $y = 2$, $x = 1$, and $z = -1$.

CHECK

$$2x + y - z = 5 \qquad 2x + 2y + z = 5$$
$$2(1) + 2 - (-1) \stackrel{?}{=} 5 \qquad 2(1) + 2(2) + (-1) \stackrel{?}{=} 5$$
$$5 \stackrel{\checkmark}{=} 5 \qquad 5 \stackrel{\checkmark}{=} 5$$

$$3x + 2y - z = 8$$
$$3(1) + 2(2) - (-1) \stackrel{?}{=} 8$$
$$8 \stackrel{\checkmark}{=} 8$$

The solution set is $\{(1, 2, -1)\}$.

EXAMPLE 8 Solve systems (a) and (b) by Gauss–Jordan elimination.

(a) $x + y = 2$ (b) $x + y = 4$
$\,x + y = 4$ $\,2x + 2y = 8$

SOLUTION (a) When an attempt is made to solve this system by matrix methods, we get

$$\begin{bmatrix} 1 & 1 & | & 2 \\ 1 & 1 & | & 4 \end{bmatrix} \to \begin{bmatrix} 1 & 1 & | & 2 \\ 0 & 0 & | & 2 \end{bmatrix} \begin{array}{l} R_1 \\ R_2 + (-1)R_1 \to R_2 \end{array}$$

The last row implies that $0y = 2$, which is impossible for any y. Therefore the lines are parallel and the system is inconsistent. The solution set is $\emptyset$.

$$\begin{bmatrix} 1 & 1 & | & 4 \\ 2 & 2 & | & 8 \end{bmatrix} \to \begin{bmatrix} 1 & 1 & | & 4 \\ 0 & 0 & | & 0 \end{bmatrix} \begin{array}{l} R_1 \\ R_2 + (-2)R_1 \to R_2 \end{array}$$

The last row implies that $0y = 0$, which is true for any value of y. Therefore the lines coincide and the system is dependent. In this case the solution set is $\{(x, y) \,|\, x + y = 4\}$.

Fortunately, matrix methods are adaptable to both the computer and the programmable calculator, since they aren't always the easiest methods to use when the calculations must be done by hand.

Exercise Set 4.3

Perform the indicated row operations on the matrix

$$\begin{bmatrix} 1 & -3 & 2 \\ 4 & -6 & -8 \end{bmatrix}$$

and write the resulting matrix.

1. Interchange R_1 and R_2.
2. Multiply R_1 by -4.
3. Add $(-2)R_2$ to R_1.
4. Add $(-2)R_1$ to R_2.
5. Add $4R_1$ to R_2.
6. Add $(-4)R_1$ to R_2.

Perform the indicated row operations on the matrix

$$\begin{bmatrix} 1 & -1 & 3 & 0 \\ 2 & 1 & 4 & 1 \\ -1 & 2 & 1 & -3 \end{bmatrix}$$

and write the resulting matrix.

7. Interchange R_1 and R_2.
8. Interchange R_2 and R_3.
9. Multiply R_1 by 2.
10. Multiply R_1 by -3.
11. Add R_1 to R_3.
12. Add $(-2)R_2$ to R_3.
13. Add $(-2)R_1$ to R_2.
14. Add $2R_1$ to R_3.

Write the augmented matrix for each of the following systems of equations.

15. $2x - y = 0$
$\,-x + y = 1$

16. $a + b = 5$
$\,2a - b = 4$

17. $x + y + z = 3$
$\,2x + y - z = 2$
$\,x + y - 2z = 5$

18. $2x - 3y + z = 0$
$\,x + y - z = 1$
$\,x - y + z = 2$

Write a system of equations using x, y, and z (as needed) that has the given augmented matrix.

19. $\begin{bmatrix} 1 & 2 & | & 5 \\ 3 & 1 & | & 4 \end{bmatrix}$

20. $\begin{bmatrix} -2 & 1 & | & -1 \\ 4 & 3 & | & 0 \end{bmatrix}$

21. $\begin{bmatrix} 1 & 0 & 1 & | & 2 \\ 1 & -1 & 1 & | & 3 \\ 0 & 0 & 1 & | & 4 \end{bmatrix}$

22. $\begin{bmatrix} 2 & 3 & -1 & | & 4 \\ 1 & 1 & 2 & | & 2 \\ 3 & 1 & -1 & | & -1 \end{bmatrix}$

Solve each of the following systems of equations, if possible, using matrices.

23. $3x + 4y = 1$
$x - 2y = 7$

24. $2x - 3y = 6$
$6x + 8y = 1$

25. $2a + 6b = -3$
$a + 3b = 2$

26. $-2m + 6n = 6$
$3m - 9n = -9$

Solve each of the following systems of equations using Gauss–Jordan elimination.

27. $x + 3y = 1$
$3x - 2y = 14$

28. $x - 3y = -5$
$3x + y = -5$

29. $2x - 3y = -2$
$4x + 6y = 8$

30. $3x - y = -5$
$x + 3y = 5$

Solve each of the following systems of equations using matrices and back substitution.

31. $2x + 8y + 6z = 20$
$4x + 2y - 2z = -2$
$3x - y + z = 11$

32. $2x + 2y - z = 1$
$x + y - 2z = -1$
$-x + 2y + z = -6$

33. $x - y + 4z = 20$
$2x + 2y - z = -3$
$3x - 2y + 3z = 29$

34. $2x + 4y - z = 0$
$x - 2y - 2z = 2$
$5x + 8y - 3z = 2$

35. $x - 2y = 8$
$x + y = -1$
$2x - y = 7$

36. $2x - y = 4$
$x + 4y = 2$
$-x - 3y = -2$

Solve each of the following systems of equations, if possible, using Gauss–Jordan elimination.

37. $2x - y = 0$
$3x + 2y = 7$

38. $3x - 4y - z = 1$
$2x - 3y + z = 1$
$x - 2y + 3z = 2$

39. $x + y - z + 2w = 0$
$2x + y - w = -2$
$3x + 2z = -3$
$-x + 2y + 3w = 1$

40. $2y + 3z = 4$
$4x + y + 8z + 15w = -14$
$x - y + 2z = 9$
$-x - 2y - 3z - 6w = 10$

Solve Exercises 41–44, rounding answers to the nearest hundredth.

41. $2.56x - 7.34y = 4$
$1.48x + 2.01y = -3$

42. $\sqrt{3}x - \sqrt{7}y = -\sqrt{2}$
$\sqrt{2}x + \sqrt{5}y = \sqrt{3}$

43. $\pi x + \pi y = 1$
$\sqrt{2}x - \sqrt{3}y = -1$

44. $1.45x + 3.17y = 7$
$2.01x - 1.02y = 2$

In exercises 45–48, solve for the ordered pair (x, y) in terms of the nonzero constants a and b.

45. $2ax + y = b$
$3ax - y = 2b$

46. $3ax + by = 5$
$2ax - by = 1$

47. $ax + by = 6$
$bx - ay = 2$

48. $4ax - 3by = 4$
$2ax + 5by = 11$

In Exercises 49 and 50, find the values of a and b that make the ordered pair (x, y) = (2, 1) a solution to each system of equations.

49. $ax + by = 4$
$2ax + by = 5$

50. $5ax - by = 0$
$2ax + 3by = 15$

In Exercises 51 and 52, find the values of a, b, and c that make the ordered triple (x, y, z) = (2, 1, 1) a solution to each system of equations.

51. $ax + by + cz = 4$
$ax - by - cz = -2$
$2ax + 3by + 4cz = 12$

52. $ax - 2by + 3cz = 20$
$3ax - by + cz = 10$
$2ax + 3by - cz = -4$

For Extra Thought

53. A right circular cylinder encloses a sphere. The sphere is the same height as the cylinder. Show that the volume of the sphere is two-thirds the volume of the cylinder.

54. Use the information from Exercise 53 to show that the surface area of the sphere is equal to the *lateral* surface area of the cylinder.

4.4 Determinants and Cramer's Rule

HISTORICAL COMMENT Carl Friedrich Gauss (1777–1855) was a German mathematician who first entered the university to study ancient languages. At the age of seventeen, however, he became interested in mathematics while trying to develop geometric methods for constructing certain polygons. He was so successful in his endeavors that he gave up the study of languages and turned to mathematics. The Gauss–Jordan method of using matrices to solve systems of equations bears his name, as do many other processes in mathematics, physics, and astronomy. The determinants that we will use to solve systems of equations in this section were used as early as 1683 in Japan by Seki Kōwa and seem to have originated at about the same time in Europe. It was Gauss, however, who coined the name "determinant."

OBJECTIVES In this section we will learn to
1. expand determinants to find their value; and
2. use determinants to solve systems of linear equations.

In Section 4.3, we used matrices to solve systems of linear equations. We will now show how *determinants* can do the same thing.

Associated with each **square matrix** is a real number called its **determinant.** (Recall that a matrix is *square* if it has the same number of rows as columns.) The determinant of the two-by-two (2 × 2) matrix

$$\begin{bmatrix} a_1 & b_1 \\ a_2 & b_2 \end{bmatrix} \quad \text{is written} \quad \begin{vmatrix} a_1 & b_1 \\ a_2 & b_2 \end{vmatrix}$$

The determinant of the three-by-three (3 × 3) matrix

$$\begin{bmatrix} a_1 & b_1 & c_1 \\ a_2 & b_2 & c_2 \\ a_3 & b_3 & c_3 \end{bmatrix} \quad \text{is written} \quad \begin{vmatrix} a_1 & b_1 & c_1 \\ a_2 & b_2 & c_2 \\ a_3 & b_3 & c_3 \end{vmatrix}$$

Note that a matrix is denoted by using brackets, while a determinant is denoted by using vertical lines.

Determinants can be used to solve a system of equations whenever the number of equations is equal to the number of variables.

The Value of a 2 × 2 Determinant

If a_1, b_1, a_2, and b_2 represent real numbers, then

$$\begin{vmatrix} a_1 & b_1 \\ a_2 & b_2 \end{vmatrix} = a_1 b_2 - b_1 a_2$$

EXAMPLE 1 Find the value of each of the following 2 × 2 determinants:

(a) $\begin{vmatrix} 1 & 3 \\ 2 & 5 \end{vmatrix}$ (b) $\begin{vmatrix} 3 & 1 \\ 3 & -2 \end{vmatrix}$ (c) $\begin{vmatrix} -4 & 5 \\ -2 & \frac{1}{2} \end{vmatrix}$

SOLUTION

(a) $\begin{vmatrix} 1 & 3 \\ 2 & 5 \end{vmatrix} = 1 \cdot 5 - 3 \cdot 2 = 5 - 6 = -1$ $a_1 = 1, b_1 = 3, a_2 = 2, b_2 = 5$

(b) $\begin{vmatrix} 3 & 1 \\ 3 & -2 \end{vmatrix} = 3(-2) - 1 \cdot 3 = -6 - 3 = -9$ $a_1 = 3, b_1 = 1, a_2 = 3, b_2 = -2$

(c) $\begin{vmatrix} -4 & 5 \\ -2 & \frac{1}{2} \end{vmatrix} = -4\left(\frac{1}{2}\right) - 5(-2) = -2 + 10 = 8$

EXAMPLE 2 Find x in each 2 × 2 determinant.

(a) $\begin{vmatrix} 2 & x \\ 3 & 4 \end{vmatrix} = 20$ (b) $\begin{vmatrix} x & 3 \\ -2 & 4 \end{vmatrix} = 15$

SOLUTION

(a) $\begin{vmatrix} 2 & x \\ 3 & 4 \end{vmatrix} = 20$

$2 \cdot 4 - 3 \cdot x = 20$

$8 - 3x = 20$

$-3x = 12$

$x = -4$

(b) $\begin{vmatrix} x & 3 \\ -2 & 4 \end{vmatrix} = 15$

$4 \cdot x - 3(-2) = 15$

$4x + 6 = 15$

$4x = 9$

$x = \dfrac{9}{4}$

The Value of a 3 × 3 Determinant

If $a_1, b_1, c_1, a_2, b_2, c_2, a_3, b_3$, and c_3 represent real numbers, then

$$\begin{vmatrix} a_1 & b_1 & c_1 \\ a_2 & b_2 & c_2 \\ a_3 & b_3 & c_3 \end{vmatrix} = a_1 b_2 c_3 - a_1 b_3 c_2 - a_2 b_1 c_3 + a_3 b_1 c_2 + a_2 b_3 c_1 - a_3 b_2 c_1$$

It's difficult to memorize this complicated definition. Fortunately that's not necessary. In practice, the value of a 3 × 3 determinant is found by using something called **minors.** When we use minors, we find the value of a 3 × 3 determinant by using three 2 × 2 determinants, a much simpler task.

There is a *minor* associated with every entry within a determinant.

DEFINITION
Minor

The *minor* of any element in a 3 × 3 determinant is the 2 × 2 determinant that remains after deleting the row and column containing that element.

The following example shows how to delete rows and columns and how to find the minors of several elements. Let's start with the determinant

$$\begin{vmatrix} a_1 & b_1 & c_1 \\ a_2 & b_2 & c_2 \\ a_3 & b_3 & c_3 \end{vmatrix}$$

(a) To find the minor of a_1, we delete row 1 and column 1:

$$\begin{vmatrix} a_1 & b_1 & c_1 \\ a_2 & b_2 & c_2 \\ a_3 & b_3 & c_3 \end{vmatrix}$$

The minor of a_1 is then the 2 × 2 determinant

$$\begin{vmatrix} b_2 & c_2 \\ b_3 & c_3 \end{vmatrix} = b_2c_3 - c_2b_3$$

(b) The minor of b_3 in the determinant $\begin{vmatrix} a_1 & b_1 & c_1 \\ a_2 & b_2 & c_2 \\ a_3 & b_3 & c_3 \end{vmatrix}$ is $\begin{vmatrix} a_1 & c_1 \\ a_2 & c_2 \end{vmatrix} = a_1c_2 - c_1a_2.$

(c) The minor of c_2 in the determinant $\begin{vmatrix} a_1 & b_1 & c_1 \\ a_2 & b_2 & c_2 \\ a_3 & b_3 & c_3 \end{vmatrix}$ is $\begin{vmatrix} a_1 & b_1 \\ a_3 & b_3 \end{vmatrix} = a_1b_3 - b_1a_3.$

We can evaluate a 3 × 3 determinant using minors. To do this, we must determine the correct sign to use with each minor. The signs associated with each entry are shown in the following display:

$$\begin{vmatrix} + & - & + \\ - & + & - \\ + & - & + \end{vmatrix}$$

When we multiply the entries in any one row or column of the determinant by their minors, affix the proper sign, then add these signed products, the result is the same as the expansion given in the definition of the value of a 3 × 3 determinant. Using row 1 with the signs +, −, +, for example,

$$\begin{vmatrix} a_1 & b_1 & c_1 \\ a_2 & b_2 & c_2 \\ a_3 & b_3 & c_3 \end{vmatrix} = a_1 \cdot \begin{vmatrix} b_2 & c_2 \\ b_3 & c_3 \end{vmatrix} - b_1 \cdot \begin{vmatrix} a_2 & c_2 \\ a_3 & c_3 \end{vmatrix} + c_1 \cdot \begin{vmatrix} a_2 & b_2 \\ a_3 & b_3 \end{vmatrix}$$

$$= a_1(b_2c_3 - b_3c_2) - b_1(a_2c_3 - a_3c_2) + c_1(a_2b_3 - a_3b_2)$$

$$= a_1b_2c_3 - a_1b_3c_2 - a_2b_1c_3 + a_3b_1c_2 + a_2b_3c_1 - a_3b_2c_1$$

Expansion about column 2 uses the signs $-, +, -$, found in the preceding sign display. Except for the order of the terms, the result is the same as when we used row 1.

$$\begin{vmatrix} a_1 & b_1 & c_1 \\ a_2 & b_2 & c_2 \\ a_3 & b_3 & c_3 \end{vmatrix} = -b_1 \cdot \begin{vmatrix} a_2 & c_2 \\ a_3 & c_3 \end{vmatrix} + b_2 \cdot \begin{vmatrix} a_1 & c_1 \\ a_3 & c_3 \end{vmatrix} - b_3 \cdot \begin{vmatrix} a_1 & c_1 \\ a_2 & c_2 \end{vmatrix}$$

$$= -b_1(a_2 c_3 - c_2 a_3) + b_2(a_1 c_3 - c_1 a_3) - b_3(a_1 c_2 - c_1 a_2)$$
$$= -a_2 b_1 c_3 + a_3 b_1 c_2 + a_1 b_2 c_3 - a_3 b_2 c_1 - a_1 b_3 c_2 + a_2 b_3 c_1$$

The process we use to find the value of a determinant by using minors is called **expansion by minors.** We say that we **expand about a row** or **column.**

EXAMPLE 3 Expand the determinant by minors:

$$\begin{vmatrix} 1 & 2 & -3 \\ 2 & 1 & -1 \\ 3 & 4 & 3 \end{vmatrix}$$

(a) about row 1   **(b)** about row 2   **(c)** about column 3

SOLUTION

(a) $\begin{vmatrix} 1 & 2 & -3 \\ 2 & 1 & -1 \\ 3 & 4 & 3 \end{vmatrix} = 1 \cdot \begin{vmatrix} 1 & -1 \\ 4 & 3 \end{vmatrix} - 2 \cdot \begin{vmatrix} 2 & -1 \\ 3 & 3 \end{vmatrix} + (-3) \cdot \begin{vmatrix} 2 & 1 \\ 3 & 4 \end{vmatrix}$

$$= 1[3 - (-4)] - 2[6 - (-3)] - 3(8 - 3)$$
$$= 1(7) - 2(9) - 3(5)$$
$$= 7 - 18 - 15 = -26$$

(b) $\begin{vmatrix} 1 & 2 & -3 \\ 2 & 1 & -1 \\ 3 & 4 & 3 \end{vmatrix} = -2 \cdot \begin{vmatrix} 2 & -3 \\ 4 & 3 \end{vmatrix} + 1 \cdot \begin{vmatrix} 1 & -3 \\ 3 & 3 \end{vmatrix} - (-1) \cdot \begin{vmatrix} 1 & 2 \\ 3 & 4 \end{vmatrix}$

$$= -2[6 - (-12)] + 1[3 - (-9)] + 1[4 - 6]$$
$$= -2(18) + 1(12) + (-2)$$
$$= -36 + 12 - 2 = -26$$

(c) $\begin{vmatrix} 1 & 2 & -3 \\ 2 & 1 & -1 \\ 3 & 4 & 3 \end{vmatrix} = -3 \cdot \begin{vmatrix} 2 & 1 \\ 3 & 4 \end{vmatrix} - (-1) \cdot \begin{vmatrix} 1 & 2 \\ 3 & 4 \end{vmatrix} + 3 \cdot \begin{vmatrix} 1 & 2 \\ 2 & 1 \end{vmatrix}$

$$= -3(8 - 3) + 1(4 - 6) + 3(1 - 4)$$
$$= -3(5) + 1(-2) + 3(-3)$$
$$= -15 - 2 - 9 = -26$$

EXAMPLE 4 Find x in the equation

$$\begin{vmatrix} 1 & 0 & 4 \\ 2 & 4 & x \\ 5 & 6 & 7 \end{vmatrix} = -16$$

SOLUTION   If we expand about a row or column containing a zero, there will be one less minor to evaluate. Therefore let's choose row 1:

$$1 \cdot \begin{vmatrix} 4 & x \\ 6 & 7 \end{vmatrix} - 0 \cdot \begin{vmatrix} 2 & x \\ 5 & 7 \end{vmatrix} + 4 \cdot \begin{vmatrix} 2 & 4 \\ 5 & 6 \end{vmatrix} = -16$$

$$1(28 - 6x) - 0 + 4(12 - 20) = -16$$

$$28 - 6x - 32 = -16$$

$$-6x = -12$$

$$x = 2$$

CHECK $\begin{vmatrix} 1 & 0 & 4 \\ 2 & 4 & 2 \\ 5 & 6 & 7 \end{vmatrix} \stackrel{?}{=} -16$

Again using row 1 to evaluate the determinant, we have

$$1 \cdot \begin{vmatrix} 4 & 2 \\ 6 & 7 \end{vmatrix} - 0 \cdot \begin{vmatrix} 2 & 2 \\ 5 & 7 \end{vmatrix} + 4 \cdot \begin{vmatrix} 2 & 4 \\ 5 & 6 \end{vmatrix} \stackrel{?}{=} -16$$

$$1(28 - 12) - 0 + 4(12 - 20) \stackrel{?}{=} -16$$

$$16 - 32 \stackrel{?}{=} -16$$

$$-16 \stackrel{\checkmark}{=} -16$$

In the historical comment at the beginning of this section, we stated that we can use determinants to solve systems of linear equations. The technique that we use is called *Cramer's rule*. Before stating the rule as a formal theorem, we will solve the system of equations

I $a_1 x + b_1 y = c_1$
II $a_2 x + b_2 y = c_2$

by the elimination method. Cramer's rule will then summarize the results of the process.

To eliminate x, we multiply equation **I** by $-a_2$ and equation **II** by a_1 so that the coefficients of x are additive inverses of each other:

I $a_1 x + b_1 y = c_1 \quad \rightarrow \quad -a_1 a_2 x - a_2 b_1 y = -a_2 c_1$ Multiply both sides of **I** by $-a_2$.

II $a_2 x + b_2 y = c_2 \quad \rightarrow \quad \underline{a_1 a_2 x + a_1 b_2 y = a_1 c_2}$ Multiply both sides of **II** by a_1.

III $a_1 b_2 y - a_2 b_1 y = a_1 c_2 - a_2 c_1$ Add.

Applying the distributive property to the left side of equation **III** gives us

$$(a_1 b_2 - a_2 b_1) y = a_1 c_2 - a_2 c_1$$

If $a_1 b_2 - a_2 b_1 \neq 0$, we can divide each side of the equation by this quantity:

$$y = \frac{a_1 c_2 - a_2 c_1}{a_1 b_2 - a_2 b_1}, \quad a_1 b_2 - a_2 b_1 \neq 0$$

To eliminate y, we multiply equation **I** by b_2 and **II** by $-b_1$:

I	$a_1x + b_1y = c_1$	$\rightarrow$	$a_1b_2x + b_1b_2y = b_2c_1$	Multiply each side of **I** by b_2.
II	$a_2x + b_2y = c_2$	$\rightarrow$	$-a_2b_1x - b_1b_2y = -b_1c_2$	Multiply each side of **II** by $-b_1$.

$$a_1b_2x - a_2b_1x = b_2c_1 - b_1c_2 \quad \text{Add.}$$
$$(a_1b_2 - a_2b_1)x = b_2c_1 - b_1c_2 \quad \text{Distributive property.}$$
$$x = \frac{b_2c_1 - b_1c_2}{a_1b_2 - a_2b_1}, \quad a_1b_2 - a_2b_1 \neq 0$$

Notice that the solutions for both x and y have the same denominator. In fact, both x and y can be expressed in terms of determinants. The solution (x, y) can be written in the form

$$x = \frac{b_2c_1 - b_1c_2}{a_1b_2 - a_2b_1} = \frac{\begin{vmatrix} c_1 & b_1 \\ c_2 & b_2 \end{vmatrix}}{\begin{vmatrix} a_1 & b_1 \\ a_2 & b_2 \end{vmatrix}} = \frac{D_x}{D}$$

$$y = \frac{a_1c_2 - a_2c_1}{a_1b_2 - a_2b_1} = \frac{\begin{vmatrix} a_1 & c_1 \\ a_2 & c_2 \end{vmatrix}}{\begin{vmatrix} a_1 & b_1 \\ a_2 & b_2 \end{vmatrix}} = \frac{D_y}{D}$$

where D is the determinant formed by the coefficients of the variables of the two equations, D_x is the determinant formed by replacing the coefficients of x in D by the constants, and D_y is the determinant formed by replacing the coefficients of y in D by the constants.

We now state **Cramer's rule** for a system of linear equations in two variables.

THEOREM 4.2(A)

Cramer's Rule

The solution to the system of equations

$$a_1x + b_1y = c_1$$
$$a_2x + b_2y = c_2$$

is

$$x = \frac{D_x}{D} \quad \text{and} \quad y = \frac{D_y}{D}, \quad D \neq 0^*$$

where

$$D_x = \begin{vmatrix} c_1 & b_1 \\ c_2 & b_2 \end{vmatrix} \quad D_y = \begin{vmatrix} a_1 & c_1 \\ a_2 & c_2 \end{vmatrix} \quad D = \begin{vmatrix} a_1 & b_1 \\ a_2 & b_2 \end{vmatrix}$$

*If $D = 0$ but $D_x \neq 0$ and $D_y \neq 0$, the system is **inconsistent** and the solution set is $\emptyset$. If $D = 0$, $D_x = 0$, and $D_y = 0$, the system is **dependent** and there are infinitely many solutions.

EXAMPLE 5 Use Cramer's rule to solve the following system of equations:

$$3x + 4y = 11$$
$$2x + 7y = 16$$

SOLUTION

$$D = \begin{vmatrix} 3 & 4 \\ 2 & 7 \end{vmatrix} = 21 - 8 = 13$$

$$D_x = \begin{vmatrix} 11 & 4 \\ 16 & 7 \end{vmatrix} = 77 - 64 = 13 \qquad D_y = \begin{vmatrix} 3 & 11 \\ 2 & 16 \end{vmatrix} = 48 - 22 = 26$$

$$x = \frac{D_x}{D} = \frac{13}{13} = 1 \qquad\qquad y = \frac{D_y}{D} = \frac{26}{13} = 2$$

CHECK

$$3x + 4y = 11 \qquad\qquad 2x + 7y = 16$$
$$3(1) + 4(2) \stackrel{?}{=} 11 \qquad 2(1) + 7(2) \stackrel{?}{=} 16$$
$$3 + 8 \stackrel{?}{=} 11 \qquad\qquad 2 + 14 \stackrel{?}{=} 16$$
$$11 \stackrel{\checkmark}{=} 11 \qquad\qquad 16 \stackrel{\checkmark}{=} 16$$

The solution set is $\{(1, 2)\}$.

Notice in Example 5 that we solved for D first. If D had been equal to zero, there would have been no need to go on, since division by zero is undefined. A system in which D is zero is either dependent or inconsistent. If D_x and D_y are also both zero, the system is dependent; otherwise it is inconsistent.

EXAMPLE 6 Use Cramer's rule to solve the system

$$x + 3y = -2$$
$$2x + y = 7$$

SOLUTION

$$D = \begin{vmatrix} 1 & 3 \\ 2 & 1 \end{vmatrix} = 1 - 6 = -5$$

$$D_x = \begin{vmatrix} -2 & 3 \\ 7 & 1 \end{vmatrix} = -2 - 21 = -23$$

$$D_y = \begin{vmatrix} 1 & -2 \\ 2 & 7 \end{vmatrix} = 7 - (-4) = 11$$

$$x = \frac{D_x}{D} = \frac{-23}{-5} = \frac{23}{5}, \qquad y = \frac{D_y}{D} = \frac{11}{-5} = -\frac{11}{5}$$

CHECK

$$x + 3y = -2 \qquad\qquad 2x + y = 7$$
$$\frac{23}{5} + 3\left(-\frac{11}{5}\right) \stackrel{?}{=} -2 \qquad 2\left(\frac{23}{5}\right) + \left(-\frac{11}{5}\right) \stackrel{?}{=} 7$$
$$\frac{23}{5} - \frac{33}{5} \stackrel{?}{=} -2 \qquad\qquad \frac{46}{5} - \frac{11}{5} \stackrel{?}{=} 7$$

$$-\frac{10}{5} \stackrel{?}{=} -2 \qquad\qquad \frac{35}{5} \stackrel{?}{=} 7$$
$$-2 \stackrel{\checkmark}{=} -2 \qquad\qquad 7 \stackrel{\checkmark}{=} 7$$

The solution set is $\left\{\left(\dfrac{23}{5}, -\dfrac{11}{5}\right)\right\}$.

It can be shown that Cramer's rule holds for a linear system of equations in three or more variables. The determinants are computed in a manner similar to Theorem 4.2(A).

THEOREM 4.2(B)

Cramer's Rule

The solution to the system of equations

$$a_1x + b_1y + c_1z = k_1$$
$$a_2x + b_2y + c_2z = k_2$$
$$a_3k + b_3y + c_3z = k_3$$

is

$$x = \frac{D_x}{D}, \quad y = \frac{D_y}{D}, \quad \text{and} \quad z = \frac{D_z}{D}, \qquad D \neq 0$$

where

$$D_x = \begin{vmatrix} k_1 & b_1 & c_1 \\ k_2 & b_2 & c_2 \\ k_3 & b_3 & c_3 \end{vmatrix} \qquad D_y = \begin{vmatrix} a_1 & k_1 & c_1 \\ a_2 & k_2 & c_2 \\ a_3 & k_3 & c_3 \end{vmatrix} \qquad D_z = \begin{vmatrix} a_1 & b_1 & k_1 \\ a_2 & b_2 & k_2 \\ a_3 & b_3 & k_3 \end{vmatrix}$$

and

$$D = \begin{vmatrix} a_1 & b_1 & c_1 \\ a_2 & b_2 & c_2 \\ a_3 & b_3 & c_3 \end{vmatrix}$$

A unique solution exists only if $D \neq 0$.

EXAMPLE 7 Use Cramer's rule to solve the system

$$x + y + z = 2$$
$$3x - y + z = 6$$
$$2x + 3z = 8$$

SOLUTION Notice that the coefficient of y in the third equation is zero because y is missing.

$$D = \begin{vmatrix} 1 & 1 & 1 \\ 3 & -1 & 1 \\ 2 & 0 & 3 \end{vmatrix} \qquad D_x = \begin{vmatrix} 2 & 1 & 1 \\ 6 & -1 & 1 \\ 8 & 0 & 3 \end{vmatrix}$$

$$D_y = \begin{vmatrix} 1 & 2 & 1 \\ 3 & 6 & 1 \\ 2 & 8 & 3 \end{vmatrix} \qquad D_z = \begin{vmatrix} 1 & 1 & 2 \\ 3 & -1 & 6 \\ 2 & 0 & 8 \end{vmatrix}$$

We use the minors of row 3 to find D:

$$D = 2 \cdot \begin{vmatrix} 1 & 1 \\ -1 & 1 \end{vmatrix} - 0 \cdot \begin{vmatrix} 1 & 1 \\ 3 & 1 \end{vmatrix} + 3 \cdot \begin{vmatrix} 1 & 1 \\ 3 & -1 \end{vmatrix}$$

$$= 2[1-(-1)] - 0 + 3(-1-3)$$

$$= 4 - 12 = -8 \neq 0$$

The reader should now verify that $D_x = -8$, $D_y = 8$, and $D_z = -16$. Thus,

$$x = \frac{D_x}{D} = \frac{-8}{-8} = 1, \qquad y = \frac{D_y}{D} = \frac{8}{-8} = -1, \qquad z = \frac{D_z}{D} = \frac{-16}{-8} = 2$$

The solution set is $\{(1, -1, 2)\}$. The check is left to the reader.

EXAMPLE 8 Use Cramer's rule to determine whether a unique solution exists for the system

$$r - s + 3t = 3$$
$$3r - 5s + 2t = -1$$
$$2r - s + 6t = 8$$

SOLUTION A unique solution will exist only if $D \neq 0$. We treat r, s, and t as x, y, and z in Theorem 4.2(B). Let's expand about row 1.

$$D = \begin{vmatrix} 1 & -1 & 3 \\ 3 & -5 & 2 \\ 2 & -1 & 6 \end{vmatrix} = 1 \cdot \begin{vmatrix} -5 & 2 \\ -1 & 6 \end{vmatrix} - (-1) \cdot \begin{vmatrix} 3 & 2 \\ 2 & 6 \end{vmatrix} + 3 \cdot \begin{vmatrix} 3 & -5 \\ 2 & -1 \end{vmatrix}$$

$$= 1(-30 + 2) + 1(18 - 4) + 3(-3 + 10)$$

$$= -28 + 14 + 21 = 7$$

Since $D \neq 0$, a unique solution exists.

EXAMPLE 9 Solve the following system of equations using Cramer's rule.

$$\frac{1}{x} + \frac{1}{y} - \frac{1}{z} = 2$$

$$\frac{1}{x} - \frac{1}{y} + \frac{1}{z} = -1$$

$$\frac{-2}{x} + \frac{2}{y} \phantom{+ \frac{1}{z}} = 0$$

SOLUTION We can use Cramer's rule to solve the system for $\frac{1}{x}, \frac{1}{y}$, and $\frac{1}{z}$:

$$D = \begin{vmatrix} 1 & 1 & -1 \\ 1 & -1 & 1 \\ -2 & 2 & 0 \end{vmatrix} = -4 \qquad D_{\frac{1}{x}} = \begin{vmatrix} 2 & 1 & -1 \\ -1 & -1 & 1 \\ 0 & 2 & 0 \end{vmatrix} = -2$$

This means that $\frac{1}{x} = \frac{-2}{-4} = \frac{1}{2}$, so $x = 2$. When $D_{\frac{1}{y}}$ and $D_{\frac{1}{z}}$ are found in a similar manner, we get $y = 2$ and $z = -1$. The solution set is $\{(2, 2, -1)\}$.

Some calculators can evaluate determinants. Many TI and HP models have this feature, as do several other popular brands. If your calculator does, consult the owner's manual to see how it is used and then try it on some of the determinants we found in this section.

Exercise Set 4.4

Evaluate each determinant.

1. $\begin{vmatrix} 1 & 2 \\ 1 & 3 \end{vmatrix}$

2. $\begin{vmatrix} 2 & 0 \\ 3 & 1 \end{vmatrix}$

3. $\begin{vmatrix} 5 & \frac{1}{2} \\ 4 & \frac{2}{5} \end{vmatrix}$

4. $\begin{vmatrix} \frac{3}{4} & -1 \\ \frac{2}{3} & -4 \end{vmatrix}$

5. $\begin{vmatrix} 0 & 7 & -5 \\ -2 & 6 & 8 \\ 3 & 4 & 1 \end{vmatrix}$

6. $\begin{vmatrix} 1 & 0 & 0 \\ 0 & 1 & 3 \\ 0 & 4 & 6 \end{vmatrix}$

7. $\begin{vmatrix} 2 & 4 & -1 \\ 3 & -1 & 2 \\ 5 & -1 & 0 \end{vmatrix}$

8. $\begin{vmatrix} 8 & -1 & 0 \\ 2 & 3 & 0 \\ 0 & 0 & -2 \end{vmatrix}$

9. $\begin{vmatrix} -4 & 2 & 5 \\ -1 & 3 & -2 \\ 1 & -3 & -1 \end{vmatrix}$

10. $\begin{vmatrix} -2 & -3 & -4 \\ 6 & 1 & 5 \\ 2 & 1 & -1 \end{vmatrix}$

Solve for the variable in each of the following equations.

11. $\begin{vmatrix} r & 1 & 0 \\ 0 & 2 & -8 \\ 3 & 0 & 1 \end{vmatrix} = -16$

12. $\begin{vmatrix} a & 0 & 1 \\ 3 & 2 & 0 \\ 2 & -1 & 3 \end{vmatrix} = -1$

13. $\begin{vmatrix} x & -2 & 3 \\ x & 1 & 2 \\ 4 & -8 & 12 \end{vmatrix} = 0$

14. $\begin{vmatrix} y & y & 2 \\ 1 & 3 & 3 \\ 3 & 1 & -2 \end{vmatrix} = 0$

Solve each system of equations using Cramer's rule.

15. $3x - 4y = 0$
    $-6x + 7y = -3$

16. $2x + y = 0$
    $x + 3y = 4$

17. $3a - b = 3$
    $a + 2b = 5$

18. $a + 3b = -14$
    $a - 2b = 8$

19. $\frac{1}{3}x - \frac{1}{2}y = 0$
    $\frac{1}{2}x + \frac{1}{4}y = 4$

20. $\frac{2}{3}x + 3y = -5$
    $x + \frac{1}{3}y = 4$

21. $-x + 3y - 2z = 5$
    $2x + y + 3z = 3$
    $-x - y - z = -2$

22. $4x + 8y - z = -6$
    $x + 7y - 3z = -8$
    $3x - 3y + 2z = 0$

23. $r + s = 2$
    $2s - 3t = -1$
    $2r - t = 1$

24. $r + 4t = 3$
    $2r + 5s - 5t = -5$
    $s + 3t = 9$

25. $5x + y - 3z = 1$
    $3x + y + z = 5$
    $-2x - 2y + z = 4$

26. $5x - y + z = 6$
    $y + z = 5$
    $2x + 4y - 3z = 1$

27. $2x - 4y + 4z = 1$
    $3x - 3y - 3z = 4$
    $x - 2y + 2z = 3$

28. $x - y + 6z = 21$
    $2x + 2y + z = 1$
    $3x + 2y - z = -4$

29. $\frac{1}{2}x - \frac{1}{3}y - \frac{1}{4}z = 0$
    $\frac{12}{3}x + 5y - \frac{6}{2}z = -1$
    $2x - \frac{2}{3}y + z = 2$

30. $\frac{1}{8}x \quad - \frac{1}{2}z = -\frac{1}{8}$
    $\frac{1}{4}x + \frac{3}{12}y \quad = \frac{1}{6}$
    $x \quad + \frac{4}{3}z = 5$

Consider each of the following as systems in $\frac{1}{x}$ and $\frac{1}{y}$; solve using Cramer's rule.

31. $\frac{3}{x} + \frac{6}{y} = \frac{3}{4}$
    $\frac{4}{3x} + \frac{3}{2y} = \frac{5}{8}$

32. $\frac{5}{x} - \frac{2}{y} = \frac{1}{3}$
    $\frac{8}{x} - \frac{4}{y} = \frac{1}{5}$

33. $\frac{2}{x} + \frac{5}{y} = \frac{3}{4}$
    $\frac{1}{3x} - \frac{1}{2y} = \frac{1}{5}$

34. $\frac{3}{4x} - \frac{1}{2y} = \frac{1}{2}$
    $\frac{1}{5x} + \frac{1}{6y} = \frac{3}{10}$

Show that each statement is true.

35. $\begin{vmatrix} x & 0 \\ y & 0 \end{vmatrix} = 0$

36. $\begin{vmatrix} x & 3 \\ x & 1 \end{vmatrix} = -2x$

37. $\begin{vmatrix} a & b \\ kc & kd \end{vmatrix} = k\begin{vmatrix} a & b \\ c & d \end{vmatrix}$

38. $\begin{vmatrix} a & b \\ c & d \end{vmatrix} = -\begin{vmatrix} c & d \\ a & b \end{vmatrix}$

39. $\begin{vmatrix} a & b & c \\ d & e & f \\ g & h & i \end{vmatrix} = \begin{vmatrix} a+kd & b+ke & c+kf \\ d & e & f \\ g & h & i \end{vmatrix}$

40. $\begin{vmatrix} a & b & c \\ 3a & 3b & 3c \\ d & e & f \end{vmatrix} = 0$

41. Show that

$\begin{vmatrix} x & y & 1 \\ 1 & 2 & -1 \\ 3 & 4 & 1 \end{vmatrix} = 0$

is the equation of the line containing the points (1, 1) and (−3, −5).

42. Show that

$\begin{vmatrix} x & y & 0 \\ 3 & 5 & -2 \\ 2 & 0 & 1 \end{vmatrix} = 35$

is the equation of the line containing the points (0, −5) and (7, 0).

Write each exercise as a system of two or three equations and solve it using Cramer's rule.

43. Starr swims to maintain her weight. Doing the crawl stroke for 1 min burns 10 calories, while doing the breast stroke for 1 min burns 8 calories. If she burns 140 calories by swimming for 15 min, how much time does she spend on each type of stroke?

44. Kim jogs and walks in order to stay physically fit. If walking burns 7 calories/min while jogging burns 10 calories/min, and he burns 164 calories in 20 min, how many minutes does he spend jogging?

45. Sherry, Jason, and Lisa go out to lunch together. By pooling all of their money, they come up with $12. Twice Sherry's contribution less twice Jason's is equal to the amount of Lisa's contribution. Lisa's lunch cost one-half of her contribution. Jason's lunch cost one and a half times his, and Sherry's lunch cost three-fifths of hers. The total cost of their lunches was $7.95. How much money did each one originally have?

46. Helen often goes to the post office to buy three different types of stamps (A, B, and C) for mailing parcels. One time she bought 30 of stamp A, 3 of B, and 2 of C at a cost of $5.41. Another time she bought 40 of A, 12 of B, and 3 of C at a cost of $8.64. Today she bought 15 of A, 5 of B, and 4 of C at a cost of $3.90. What is the cost of each type of stamp?

REVIEW EXERCISES

Solve by any method.

47. $5x + 2y = -2$
    $3x + 4y = -1$

48. $\frac{1}{2}x - \frac{2}{3}y = \frac{3}{4}$
    $\frac{1}{3}x + 2y = \frac{5}{6}$

49. $x + 2z = 5$
    $-x - 2y + 3z = 8$
    $-3x + 4y + 6z = 30$

50. $-3x - 2y + 7z = 4$
    $2x + 5y - z = 3$
    $-x + 3y + 6z = 0$

4.5 Graphing Linear Inequalities in Two Variables

OBJECTIVES

In this section we will learn to
1. display the solution set of a linear inequality; and
2. display the solution set of a system of linear inequalities.

In Section 4.1 we decided that graphing is not the best method for solving a system of linear equations. *Linear inequalities* are quite a different matter. Graphing is the *only* method for displaying the solution set. In this section, we will consider graphs of linear inequalities such as

$$y > x + 1, \quad x \leq 2, \quad y > 5, \quad \text{and} \quad 2x + y \leq 4$$

as well as *systems* of linear inequalities, such as

$$2x + y > 4$$
$$y < x + 1$$

A **linear inequality in two variables,** like a linear equation, has solutions that are ordered pairs. For example, (3, 7) is a solution to $y > x + 2$ because a true statement results when x is replaced by 3 and y by 7:

$$y > x + 2$$
$$7 \stackrel{?}{>} 3 + 2 \quad x = 3, y = 7$$
$$7 \stackrel{\checkmark}{>} 5 \quad \text{True}$$

There are infinitely many ordered pairs other than (3, 7) that satisfy $y > x + 2$.

The graph of the solution set of a linear inequality in two variables is a region in the plane called a **half-plane,** which consists of all points on one side of a line called the **boundary line.** This line is found by replacing the inequality symbol by an equal sign and graphing the resulting line. The half-plane of solutions may or may not include the boundary line. When the inequality symbol is $<$ or $>$, the boundary line is *not* included and is shown as a **broken line.** When the inequality is $\leq$ or $\geq$, the boundary line *is* included and is shown as a **solid line.**

$$y > x + 2 \quad \text{Boundary not included}$$

$$y \geq x + 2 \quad \text{Boundary included}$$

The solution set of an inequality is indicated by shading. The solution set of $y > x + 2$ is shown in Figure 4.9. The solution set to $y \geq x + 2$ is shown in Figure 4.10.

In each half-plane, one of the following must be true:

1. All points are solutions to the inequality.
2. No points are solutions to the inequality.

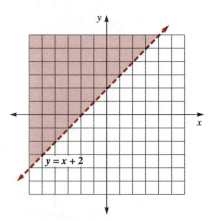

FIGURE 4.9
$\{(x, y) \mid y > x + 2\}$

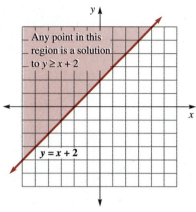

FIGURE 4.10
$\{(x, y) \mid y \geq x + 2\}$

To decide which half-plane contains the solution set, we select a *test point*. A test point is any point in the plane that **does not** lie on the boundary line. If the test point satisfies the inequality (makes it true), the solution set is the entire half-plane on the same side of the line as the test point. If the test point does not satisfy the inequality (makes it false), the solution set is the half-plane on the opposite side of the line.

EXAMPLE 1   Use a graph to illustrate the solution set of $2x + 3y \leq 6$.

SOLUTION   We begin by drawing the graph of the line $2x + 3y = 6$; see Figure 4.11. Three ordered pairs that satisfy this equation are $(0, 2)$, $(3, 0)$, and $(6, -2)$. We draw a *solid* line, because the solution set to $2x + 3y \leq 6$ includes the boundary line $2x + 3y = 6$. To determine which half-plane is the graph of the solution set, we choose $(0, 0)$ as the test point, since it does not lie on the boundary line.

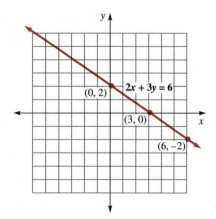

FIGURE 4.11

$$2x + 3y \leq 6$$
$$2 \cdot 0 + 3 \cdot 0 \stackrel{?}{\leq} 6 \quad \text{Test point } x = 0, y = 0$$
$$0 \stackrel{\checkmark}{\leq} 6 \quad \text{True}$$

The solution set is the half-plane containing (0, 0), since $0 \leq 6$. (See Figure 4.12.)

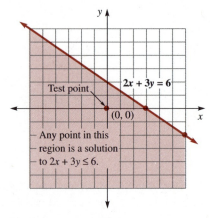

FIGURE 4.12
$\{(x, y) \mid 2x + 3y \leq 6\}$

It is common practice to indicate the solution set of a linear inequality with a graph. We will use this practice throughout the remainder of this text.

EXAMPLE 2   Graph the inequality $y > 2x$.

SOLUTION   Since $y > 2x$ does not include the line $y = 2x$, the boundary of the half-plane will be a broken line. The line $y = 2x$ has a slope of 2 and a y-intercept of 0 (it passes

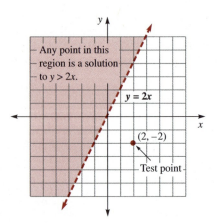

FIGURE 4.13
$\{(x, y) \mid y > 2x\}$

through the origin). Therefore we cannot use the origin as a test point. Let's try $(2, -2)$:

$$y > 2x$$
$$-2 \stackrel{?}{>} 2 \cdot 2 \quad \text{Test point } x = 2, y = -2$$
$$-2 \stackrel{?}{>} 4 \quad \text{False}$$

The test has resulted in a false statement, so the solution set is the half-plane on the opposite side of the line from $(2, -2)$, as shown in Figure 4.13.

EXAMPLE 3 Graph the inequality $x > -1$.

SOLUTION The inequality symbol is $>$. Therefore the boundary line $x = -1$ is not a part of the graph and so must be shown as a broken line. Any point that lies to the right of $x = -1$ has an x-coordinate that is greater than -1. The graph is shown in Figure 4.14.

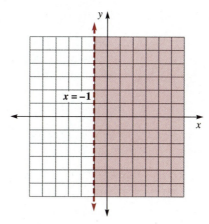

FIGURE 4.14
$\{(x, y) \mid x > -1\}$

As we saw in Example 3, inequalities of the form $x > c$ or $x < c$ have graphs that are half-planes to the right or left of the vertical line $x = c$. Graphs of $y > c$ or $y < c$ are half-planes that are above or below the horizontal line $y = c$.

EXAMPLE 4 Graph the inequality $y \geq |x + 2|$.

SOLUTION Recall that the absolute value of a number is always positive or zero. (See Example 6, Section 3.1, to review the graph of $y = |x + 2|$.) Since the inequality symbol is $\geq$, the boundary is solid, or included. The graph is shown in Figure 4.15. We use the origin as a test point:

$$y \geq |x + 2|$$
$$0 \stackrel{?}{\geq} |0 + 2| \quad \text{Test point } x = 0, y = 0$$
$$0 \stackrel{?}{\geq} |2|$$
$$0 \stackrel{?}{\geq} 2 \quad \text{False}$$

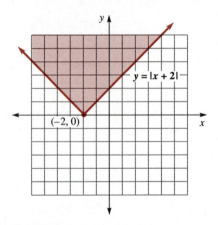

FIGURE 4.15

Systems of Linear Inequalities

We solve systems of linear inequalities by using the same graphing procedures that we used to solve a single linear inequality. The solution set is the *intersection* of the graphs of the half-planes.

EXAMPLE 5   Graph the system of inequalities $y \leq x + 1$ and $2y + x \geq 4$.

SOLUTION   First we graph both boundary lines on the same set of axes (see Figure 4.16). Neither line passes through the origin, so we can use $(0, 0)$ as a test point for both half-planes.

$$y \leq x + 1 \qquad\qquad 2y + x \geq 4$$
$$0 \stackrel{?}{\leq} 0 + 1 \qquad\qquad 2 \cdot 0 + 0 \geq 4$$
$$0 \stackrel{\checkmark}{\leq} 1 \quad \text{True} \qquad 0 \geq 4 \quad \text{False}$$

Now we shade the origin side of $y \leq x + 1$ and the side *opposite* the origin of $2y + x \geq 4$. The region with double shading is the solution (see Figure 4.17).

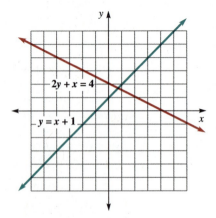

FIGURE 4.16

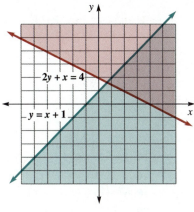

FIGURE 4.17
$\{(x, y) \mid y \leq x + 1\} \cap \{(x, y) \mid 2y + x \geq 4\}$

EXAMPLE 6   Graph the system of inequalities $2y - 3x \leq 6$, $x - 3y \leq 3$, and $x + y \leq 5$.

SOLUTION   We sketch each line associated with the inequality and shade the region that forms its solution set. The region where the shading from each of the inequalities intersects is the solution set to the system, as shown in Figure 4.18.

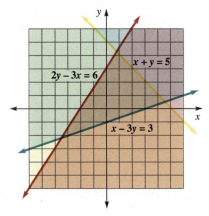

FIGURE 4.18
$\{(x, y) \mid 2y - 3x \leq 6\} \cap \{(x, y) \mid x - 3y \leq 3\} \cap \{(x, y) \mid x + y \leq 5\}$

Inequalities are often used to solve problems related to business. The goal of business is providing a quality product for the consumer while returning a fair profit for the manufacturer. There are often limitations on a plant's capacity to produce certain items, and these limitations affect manufacturing decisions. If the limitations can be expressed in terms of a system of inequalities and the objective is to maximize profits, the problem falls into a class of problems solved by a method called *linear programming*. The next section will address this type of problem.

Exercise Set 4.5

Solve each inequality by graphing.

1. $x > 5$
2. $x < -2$
3. $y \geq 1$
4. $y \leq -3$
5. $y < 3x$
6. $y > x$
7. $y \geq -2x$
8. $y \leq -x$
9. $y \leq x + 1$
10. $y > x - 2$
11. $x + y < 1$
12. $x + y < 3$
13. $3x - y \geq 2$
14. $2x - 3y \leq 1$
15. $-x + 2y > 1$
16. $-2x + 3y < 2$
17. $y \leq |x|$
18. $y \geq |x|$
19. $y \geq |x + 1|$
20. $y \leq |x - 1|$

Solve each of the following systems of inequalities by graphing.

21. $y \geq 2$
    $x \leq 4$
22. $y \geq -3$
    $x \geq 2$
23. $x + y > 3$
    $x - y > 2$
24. $y - x > 0$
    $y - x \leq 3$
25. $x + y < 2$
    $2x + 2y \geq 0$
26. $3x - y > 0$
    $x < y$
27. $x > -y$
    $x - 2y < 0$
28. $x \geq 0$
    $y \leq 0$
29. $x < 0$
    $y > 0$
30. $x + y < 5$
    $x - y \geq 5$
31. $3x + y \geq 4$
    $2x - y < 1$
32. $4x - y > 3$
    $4x - 2y \leq 6$
33. $3x - 5y < 2$
    $x + 2y < 0$
34. $5x + 2y < 10$
    $-10x - 4y < 4$
35. $4x - 3y < 8$
    $8x - 6y \leq -9$
36. $x + y \leq 3$
    $x - y \leq 1$
    $x \geq 0$
    $y \geq 0$
37. $x - y < 0$
    $x + y \geq 1$
    $x < 4$
38. $x + y < 2$
    $y - 3x \leq 3$
    $x - y < 0$
39. $x + 3y \leq 4$
    $-2x - 6y \leq 3$
    $0 \leq x \leq 3$
40. $x + y \leq 4$
    $-3x - 3y \leq 3$
    $0 \leq y \leq 3$
41. $x - y \geq 0$
    $x - y \leq 3$
    $-2 \leq y \leq 0$
42. $x - y \leq 2$
    $3x - 3y \geq 0$
    $-2 \leq x \leq 2$

Solve each of the following systems of inequalities by graphing.

43. $\{(x, y) \mid x - y > 3\} \cap \{(x, y) \mid 2x + y \leq 8\}$
44. $\{(x, y) \mid y - x < 6\} \cap \{(x, y) \mid x + y < -2\}$
45. $\{(x, y) \mid 3x - y > 3\} \cap \{(x, y) \mid 6x - 2y > 9\}$
46. $\{(x, y) \mid 2x + y < 3\} \cap \{(x, y) \mid 4x - 2y > 11\}$

Determine whether each ordered pair is in the solution set of the accompanying system of inequalities.

47. $(1, 1)$; $x - y < 2$
    $x + y > 1$
48. $(3, 1)$; $y - 3x < 3$
    $6x - 2y < 12$
49. $(2, 6)$; $y > x + 3$
    $y > -x + 3$
    $x > 1$
    $y > 5$
50. $(1, 1)$; $x < y + 2$
    $x > y - 1$
    $x > 0$
    $y \geq 0$
51. $(1, 1)$; $y - x < 0$
    $y - x \geq 3$
    $-2 \leq x \leq 2$
    $-2 \leq y \leq 2$
52. $(3, 0)$; $x + y \leq 4$
    $x - y > 0$
    $y \geq 0$
    $-6 \leq x \leq 6$

Write an inequality representing the shaded region in each of Exercises 53–58.

53.

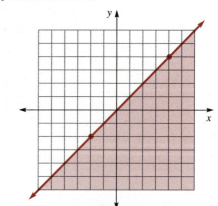

54.

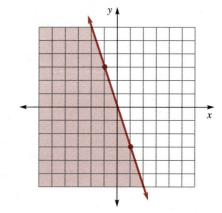

55.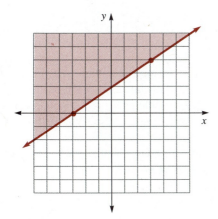

56.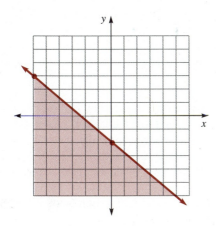

57.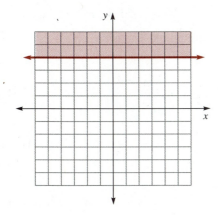

58.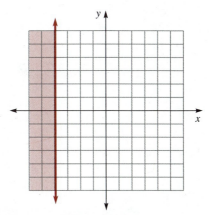

Applications

59. The perimeter of a rectangle of length y and width x cannot be less than 8 ft. Write a linear inequality for this and sketch its graph. Find three ordered pairs (x, y) from the graph that satisfy this condition.

60. The perimeter of an isosceles triangle with dimensions x (the length of the equal sides) and y cannot exceed 6 in. Write a linear inequality for this and sketch its graph. Find three ordered pairs (x, y) from the graph that satisfy this condition.

61. A classroom has 480 sq. ft of floor space. The space required for each chair is 10 sq. ft and, for each table, 16 sq. ft. Write a system of linear inequalities for this if x is the number of chairs and y is the number of tables. Sketch the graph.

62. An office has 1000 sq. ft of floor space. Each chair uses 10 sq. ft of floor space and each table, 15 sq. ft. Write a system linear inequalities for this if x is the number of chairs and y is the number of tables. Sketch the graph.

Review Exercises

Solve each of the following systems of equations by the method indicated.

63. Graphing
$2x - y = 5$
$3x + 2y = 11$

64. Elimination
$3x - 2y = 5$
$x + y = 7$

65. Substitution
$4a - b = 4$
$2a + 5b = 8$

66. Cramer's rule
$6x - \frac{1}{2}y = 4$
$\frac{1}{4}x + y = 2$

67. Matrices
$x + y - z = 1$
$2x - y + z = 2$
$x + 2y + 3z = 6$

68. Any method
$x + y = 5$
$2x + z = 3$
$y - z = 2$

4.6 Linear Programming

HISTORICAL COMMENT During World War II, much attention was given to the allocation of resources to the war effort. Mathematicians were put to work developing methods of solving allocation problems using something called "linear programs." In this section we will examine geometric methods of solving problems of this type. In the middle of the 1940s, however, George Dantzig developed an algebraic method of solution that is used to the present time. It has been estimated that over 20% of scientific computer use today is devoted to solving such problems.

OBJECTIVE In this section we will learn to use the graphs of linear inequalities as an aid to solving applied problems.

In business and economics, it is often desirable to design a manufacturing process resulting in a business that operates at a minimum cost while producing a quality product. At the same time, however, the profit obtained from the sale of the product should be maximized. One method of determining this process is called **linear programming.** To see how linear programming is applied, let's consider the next example.

EXAMPLE 1 A small company has decided to manufacture skateboards for use in competition. Two models will be produced. It is estimated that the custom model will require 2 hours of labor for shaping and 4 hours of labor for final detail work. The standard model will require 2 hours for shaping and 2 hours for detail work. The company has 2 people who can do shaping and 3 who can do detail work, and each can work a maximum of 40 hours/week.

It is estimated that the standard model will return a profit of $20 each, while the custom model will return a profit of $30 each.

(a) Write a system of inequalities describing the company's possible output and graph the region that represents the possible production schedule.
(b) Determine a production schedule that will return the maximum possible profit.

SOLUTION Inequalities that describe the limitations of the problem are called the **constraints.**

(a) We let x = the number of custom models made
 y = the number of standard models made

	CUSTOM MODEL	STANDARD MODEL
Shaping time hours	2	2
Detailing time hours	4	2

Two people are assigned to shaping. Each works a maximum of 40 hours per week, so a maximum of 80 work hours are available for this task. How should the time be used? The x custom models and the y standard models require 2 hours each for shaping. Thus, $2x$ represents the shaping time needed for the custom models and $2y$ the time needed for the standard models. The sum of the times must be less than or equal to 80 hours, so

$$2x + 2y \leq 80$$

Three persons are assigned to detailing at 40 hours each, which gives the company a maximum of $3 \cdot 40 = 120$ hours for detailing. The x custom models require 4 hours each for a total of $4x$ hours. At 2 hours each, the y standard models require a total of $2y$ hours. Thus,

$$4x + 2y \leq 120$$

It is also true that $x \geq 0$ and $y \geq 0$, or the company would not be in business. The region in the plane that indicates the options for production must satisfy each of the following inequalities:

$$\left. \begin{array}{r} 2x + 2y \leq 80 \\ 4x + 2y \leq 120 \\ x \geq 0 \\ y \geq 0 \end{array} \right\} \text{Constraints}$$

Since $x \geq 0$ is the region to the right of and including the y-axis, and $y \geq 0$ is the region above and including the x-axis, the graph is limited to the first quadrant and the positive x- and y-axes, as shown in Figure 4.19. Any point (x, y) in the shaded region, including the boundary, represents a possible production schedule for x custom models and y standard models.

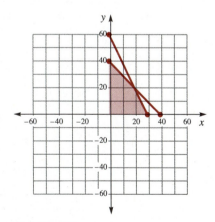

FIGURE 4.19

(b) We must first write an equation that describes the profit for the production schedule. The x custom models each yields a profit of $30 for a total of $30x$. The y standard models each yields a profit of $20 for a total of $20y$. The equation for determining the profit for both models is

$$P = 30x + 20y$$

This equation defines what is called the **objective function.** How does the profit change for the various production schedules? The following chart shows some of the possible production schedules together with the profits they would yield.

Point	Number of Custom Models	Number of Standard Models	Objective Function $P = 30x + 20y$	Profit
(0, 0)	0	0	$30 \cdot 0 + 20 \cdot 0$	$0
(10, 10)	10	10	$30 \cdot 10 + 20 \cdot 10$	$500
(10, 20)	10	20	$30 \cdot 10 + 20 \cdot 20$	$700
(0, 40)	0	40	$30 \cdot 0 + 20 \cdot 40$	$800
(20, 10)	20	10	$30 \cdot 20 + 20 \cdot 10$	$800
(20, 20)	20	20	$30 \cdot 20 + 20 \cdot 20$	$1000
(30, 0)	30	0	$30 \cdot 30 + 20 \cdot 0$	$900

Among those points at which the profit was computed, the maximum profit is $1000/week, when 20 of each model are manufactured.

Although we shall not do so, it can be shown that the maximum and minimum values of the objective function always occur at a **vertex** (the intersection of the boundary lines) of the region that represents the solution set. In Example 1 there are four vertices: (0, 0), (0, 40), (20, 20), and (30, 0); see Figure 4.20.

EXAMPLE 2 Find the maximum value for the function $S(x, y) = 3x + 7y$ over the region shown in Figure 4.21.

SOLUTION If we let $S(0, 7)$ represent the value of S at the vertex $(0, 7)$, then the maximum will occur at a vertex. The region has four vertices: $(0, 7)$, $(9, 13)$, $(15, 3)$, and $(8, 0)$.

$$S(0, 7) = 3 \cdot 0 + 7 \cdot 7 = 49$$
$$S(9, 13) = 3 \cdot 9 + 7 \cdot 13 = 118$$

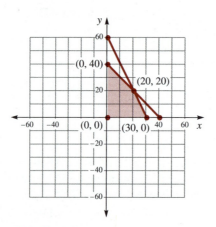

FIGURE 4.20

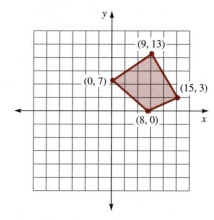

FIGURE 4.21

$$S(15, 3) = 3 \cdot 15 + 7 \cdot 3 = 66$$
$$S(8, 0) = 3 \cdot 8 + 7 \cdot 0 = 24$$

The maximum value of S occurs at $(9, 13)$ and is 118.

EXAMPLE 3   Maximize the objective function $P(x, y) = 4x + y$ subject to the constraints

$$x \geq 0$$
$$y \geq 0$$
$$2x + y \leq 5$$
$$2x + 3y \leq 12$$

SOLUTION   We first sketch the graph of the region satisfying all four constraints and then find the vertices of the region bounding the solution set. To find the vertices, we find the solution to each pair of equations associated with the inequalities, using any of the methods introduced earlier in this chapter. Each solution corresponds to the intersection point of the graphs of the two equations; which is a vertex. The vertices are $(0, 0)$, $(\frac{5}{2}, 0)$, $(\frac{3}{4}, \frac{7}{2})$, and $(0, 4)$. The region is shown in Figure 4.22.

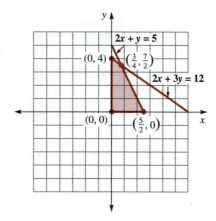

FIGURE 4.22

The objective function is $P(x, y) = 4x + y$:

$$P(0, 0) = 4 \cdot 0 + 0 = 0$$
$$P(0, 4) = 4 \cdot 0 + 4 = 4$$
$$P\left(\frac{3}{4}, \frac{7}{2}\right) = 4\left(\frac{3}{4}\right) + \frac{7}{2} = 3 + \frac{7}{2} = \frac{13}{2} = 6\frac{1}{2}$$
$$P\left(\frac{5}{2}, 0\right) = 4\left(\frac{5}{2}\right) + 0 = 10 + 0 = 10$$

The maximum value of P is 10 at $\left(\frac{5}{2}, 0\right)$.

EXAMPLE 4   Antonio is preparing for a wilderness backpacking trip on which he plans to eat a combination of cereals and nuts. The recommended daily requirement is at least 3000 calories and 2.4 ounces of protein. Each ounce of cereal supplies 120 calories and 0.04 ounces of protein. Each ounce of nuts provides 60 calories and 0.12 ounces of protein. If the cereal costs 20¢/ounce and each ounce of nuts costs 30¢, what combination should he plan to eat in order to keep the cost to a minimum?

SOLUTION   We let x = the number of ounces of cereal in his daily diet
y = the number of ounces of nuts in his daily diet

	CEREAL	NUTS		TOTAL
Calories/oz	120	60	Calories	$120x + 60y$
Protein/oz	0.04	0.12	Protein	$0.04x + 0.12y$

Since his caloric intake must be at least 3000 calories and his protein intake at least 2.4 ounces, two constraints are

$120x + 60y \geq 3000$

$0.04x + 0.12y \geq 2.4$

The other constraints are $x \geq 0$ and $y \geq 0$. The graph of the region satisfying the constraints is shown in Figure 4.23.

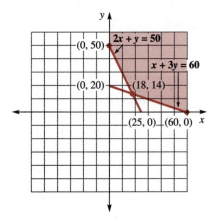

FIGURE 4.23

One vertex, (18, 14), is found by solving the system of equations

$120x + 60y = 3000$ → $2x + y = 50$ Divide each side by 60.
$0.04x + 0.12y = 2.4$ → $x + 3y = 60$ Divide each side by 0.04.

that forms two of the boundaries. Since Antonio's cost is the function we want to minimize, the objective function is based on the cost per ounce for each of the foods. The cereal costs 20¢/oz and the nuts cost 30¢/oz. The total cost in cents is therefore

$C(x, y) = 20x + 30y$

When the cost is computed at each vertex, (0, 50), (18, 14), and (60, 0), of the graph of the solution set, the minimum cost is found to be at (18, 14):

$$C(18, 14) = 20 \cdot 18 + 30 \cdot 14 = 780¢ \quad \text{or} \quad \$7.80$$

So Antonio should plan to eat 18 ounces of cereal and 14 ounces of nuts a day on his backpacking trip.

To Solve a Linear Programming Problem By Graphing

Step 1 Determine the constraints as specified by the conditions of the problem.

Step 2 Determine the objective function (the expression to be maximized or minimized).

Step 3 Graph the region that illustrates the constraints and determine the coordinates of each vertex on its boundary.

Step 4 Substitute the coordinates of each vertex from Step 3 into the objective function to determine its maximum or minimum value.

As we stated in the historical comment in the beginning of this section, there are algebraic methods for solving linear programming problems. One process is called the *simplex method*, which is typically introduced in a course called finite mathematics, for business majors.

Exercise Set 4.6

Solve each of the following systems by graphing. Find the coordinates of each vertex.

1. $5x + y \geq 42$
   $2x + y \geq 30$
   $x \geq 0$
   $y \geq 0$

2. $3x + y \geq 18$
   $x + 2y \geq 16$
   $x \geq 0$
   $y \geq 0$

3. $2x + y \leq 20$
   $2x + 5y \geq 36$
   $10x + y \geq 36$

4. $x - y \leq 2$
   $3x + 4y \leq 48$
   $x \geq 4$

Find the maximum values of the given function over the regions with the given vertices in the appropriate figures that follow.

5. $P = x + 2y$
6. $C = 3x + y$
7. $C = 5x - y$
8. $P = 2x + 5y$

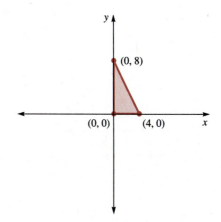

FIGURE TO ACCOMPANY EXERCISE 5

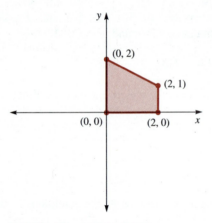

FIGURE TO ACCOMPANY EXERCISE 6

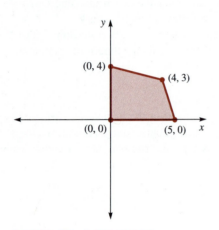

FIGURE TO ACCOMPANY EXERCISE 7

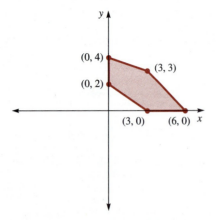

FIGURE TO ACCOMPANY EXERCISE 8

9. Maximize $P = 2x + 3y$ subject to the constraints

$$x + y \leq 5$$
$$3x + y \leq 6$$
$$x \geq 0$$
$$y \geq 0$$

10. Maximize $P = 2x + 3y$ subject to the constraints

$$x + 2y \leq 5$$
$$2x + 3y \geq 6$$
$$x \geq 0$$
$$y \geq 0$$

11. Dawson's Candy Company makes two types of candy bars, each weighing 4 oz. Bar A has 3 oz of caramel and 1 oz of chocolate, while bar B has 2 oz of caramel and 2 oz of chocolate. Bar A sells for 30¢ and B sells for 54¢. Dawson's has 144 lb of caramel in stock and 90 lb of chocolate. How many bars of each type should the company make in order to maximize the revenue?

12. A certain food is 90% carbohydrates and 10% protein by weight and costs 50¢/lb. Another food is 60% carbohydrates and 40% protein by weight and costs $1.00/lb. What diet of these two foods provides at least 2 lb of carbohydrates and 1 lb of protein at a minimum cost? What is the cost per pound?

13. Donna's company is packing and selling food packets that (a) weigh at least 16 oz; (b) have at least 32 g of protein; (c) have at least 12 mg of iron; and (d) have no more than 42 g of fat. Two food types are available to mix in the packets. Type A contains 3 mg of iron, 1 g of protein, and 2 g of fat per ounce. Type B contains 1 mg of iron, 5 g of protein, and 3 g of fat per ounce. The profit is 3¢/oz for type A and 5¢/oz for type B. How many ounces of each type of food should be in each packet to maximize profit? What is the profit per packet?

14. Brian's Boats builds and sells one-person and three-person inflatable boats. Each one-person boat requires 0.9 hr of labor to cut and 0.8 hr to assemble. Each three-person boat requires 1.8 hr to cut and 1.2 hr to assemble. The maximum numbers of hours available for cutting and assembling each month are 864 and 672, respectively. If all boats sell at a profit of $12.00 for each one-person boat and $16.00 for each three-person boat, how many of each should the company build and sell to maximize profit?

15. Kelsey wants to take a daily dose of at least 16 units of vitamin A, at least 5 units of vitamin D, and at least 20 units of vitamin C. Tablet M costs 10¢/tablet and contains 8 units of A, 1 unit of D, and 2 units of C; tablet P costs 20¢/tablet and contains 2 units of A, 1 unit of B, and 7 units of C. How many of each tablet

should she take a day in order to minimize her cost? What is her cost per day?

16. Gino is taking a test that includes true–false and completion-type questions. True–false questions take 3 min to answer and are worth 10 points each. Completion questions take 6 min and are worth 15 points each. Gino is allowed 60 min to take the test and may answer no more than 16 questions. Assuming all of his answers are correct, how many of each type should he answer to obtain the highest possible score? What is that score?

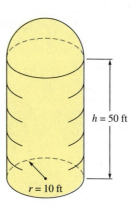

FIGURE TO ACCOMPANY EXERCISE 17 AND 18

For Extra Thought

A silo is formed by placing a hemisphere on top of a right circular cylinder (see the figure). A silo is used to store grain in a farming community.

17. Find the volume of the silo shown in the figure. Use $\pi = 3.1416$ and round your answer to the nearest whole number.

18. Find the total surface area of the silo. (The total surface area is made up of the lateral surface area of the cylinder, the area of the circular base, and the surface area of the hemisphere on the top.) Use $\pi = 3.1416$ and round your answer to the nearest whole number.

4.7 More Applications

OBJECTIVES

In this section we will learn how to use systems of linear equations and inequalities to solve applied problems.

Many applied problems are easier to solve when two or more variables are used instead of one. All that is necessary is to be able to write two or more equations or inequalities describing the conditions of the problem. The five steps for solving applied problems are still useful to us.

To Solve Applied Problems

Step 1 Read the problem through completely as many times as necessary until you understand what the problem is about.

Step 2 Reread the problem and determine what it is that you are asked to find.

Step 3 Represent what you are asked to find by a variable or variables. Draw a picture or construct a chart if necessary.

Step 4 Establish the relationship between the variable and other parts of the problem. The relationship may be an equation or an inequality.

Step 5 Solve the equation or inequality and check the solution in the words of the problem.

EXAMPLE 1   The sum of two integers is 6. Twice the smaller integer is 24 less than the larger one. Find the integers.

SOLUTION   We let x represent the smaller integer and y the larger one. Since their sum is 6, one equation is

I $x + y = 6$

Twice the smaller integer is $2x$. When this product is subtracted from the larger integer, y, the difference is 24. Therefore a second equation is

II $y - 2x = 24$

We can solve the system by any of the methods we learned in this chapter. Let's use substitution and solve equation **I** for y and substitute the result in equation **II**.

$$y = 6 - x$$
$$6 - x - 2x = 24$$
$$-3x = 18$$
$$x = -6$$

Since $y = 6 - x$ and $x = -6$, we have $y = 6 - (-6) = 12$. The smaller integer is -6 and the larger one is 12.

EXAMPLE 2   Roger, a chemistry lab technician, has two different concentrations of hydrochloric acid in stock. One is a 15% solution and the other is a 40% solution. Mr. Brundage's Chem 1A class is conducting an experiment that requires 200 milliliters (ml) of a 25% solution. If Roger is to make the 25% solution by mixing the two available solutions, how many milliliters of each must he use?

SOLUTION   We will use two variables, which means that we must find two equations. With that in mind, we let

$x =$ the number of ml of the 15% solution
$y =$ the number of ml of the 40% solution

Since Mr. Brundage needs 200 ml of the solution, which Roger will obtain by mixing the 15% solution and the 40% solution, the following equation must be satisfied:

Number of ml of 15% sol'n	mixed with	number of ml of 40% sol'n	makes	200 ml
↓	↓	↓	↓	↓
I x	$+$	y	$=$	200 ml

We obtain a second equation by observing that the total amount of acid present before mixing remains the same after mixing. No acid is lost, but its concentration changes, as shown in Figure 4.24.

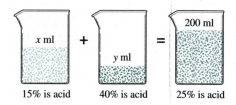

FIGURE 4.24

Figure 4.24 suggests the following equation.

$$\underbrace{\text{Amount of acid in } x \text{ ml}}_{0.15x} \underbrace{\text{plus}}_{+} \underbrace{\text{amount of acid in } y \text{ ml}}_{0.40y} \underbrace{\text{equals}}_{=} \underbrace{\text{amount of acid in mixture}}_{0.25(200)}$$

II $0.15x + 0.40y = 0.25(200)$

If both sides of equation **II** are multiplied by 100, we obtain an equivalent equation free of decimals:

II $0.15x + 0.40y = 0.25(200)$ →   **III** $15x + 40y = 5000$

Solving equations **I** and **III** simultaneously by the elimination method yields

I $x + y = 200$ →	$-15x - 15y = -3000$	Multiply each side of **I** by -15.
III $15x + 40y = 5000$ →	$\underline{15x + 40y = 5000}$	
	$25y = 2000$	Add.
	$y = 80$ ml	

Since the mixture contains 200 ml, we conclude that $x = 120$ ml. Therefore Roger will use 120 ml of the 15% solution and 80 ml of the 40% solution.

EXAMPLE 3   A manufacturing company produces items that require both casting and finishing. The maximum number of hours available for casting each week is 48 and the maximum number of hours available for finishing is 32. They produce two items, A and B, that require both casting and finishing. The times required on each machine for each item are listed in the following chart.

	ITEM A (hours)	ITEM B (hours)	MAXIMUM HOURS AVAILABLE
Casting	6	4	48
Finishing	5	2	32

If the machines work at full capacity, how many items of each type can the company produce in a single week?

SOLUTION   We let a = the number of items of type A that it produces weekly
b = the number of items of type B that it produces weekly.

The total time available for casting is 48 hours. Item A requires 6 hours for each of the a units, so a total of $6a$ hours. Item B requires 4 hours for each of the b units, and so a total of $4b$ hours. Therefore one equation is

THE SUM OF THE CASTING TIMES IS 48 HOURS.

I $\quad 6a + 4b = 48$

Similarly, there are 32 hours available for finishing. Five hours are required for each of the a units of item A, a total of $5a$ hours. Two hours are required for each of the b units of item B, a total of $2b$ hours. A second equation is

THE SUM OF THE FINISHING TIMES IS 32 HOURS.

II $\quad 5a + 2b = 32$

The equations can be solved by the elimination method:

$$
\begin{array}{lllll}
\textbf{I} & 6a + 4b = 48 & \rightarrow & 6a + 4b = 48 & \\
\textbf{II} & 5a + 2b = 32 & \rightarrow & -10a - 4b = -64 & \text{Multiply each side of \textbf{II} by } -2. \\
& & & \overline{-4a \quad\quad = -16} & \text{Add.} \\
& & & a \quad\quad = 4 \text{ items} &
\end{array}
$$

Substituting $a = 4$ into equation **I** yields

$6 \cdot 4 + 4b = 48$

$b = 6$ items

The company can produce 4 units of item A and 6 units of item B.

EXAMPLE 4 Suppose that the demand for the items in Example 3 does not always require maximum use of the machines. Graph the area that shows all possible combinations of output for items A and B.

SOLUTION The equations in Example 3 now become inequalities:

$6a + 4b \leq 48$

$5a + 2b \leq 32$

$\left.\begin{array}{l} a \geq 0 \\ b \geq 0 \end{array}\right\}$ A negative number of items cannot be produced.

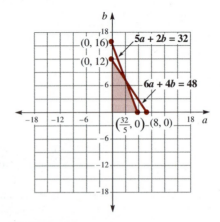

FIGURE 4.25

We graph the inequalities in Figure 4.25. Any ordered pair of integers in the shaded region represents a possible production schedule.

EXAMPLE 5 An airplane can fly 360 miles into the wind in 3 hours. When the plane reverses direction, and if the wind conditions remain the same, the return trip to the airport takes 2 hours. Find the speed of the wind and the speed of the plane in still air.

SOLUTION One of the best ways to organize information in motion problems of this type is to summarize the information in a chart. We let

w = the speed (rate) of the wind

p = the speed of the plane in still air

When an airplane flies *against* the wind, its still-air speed is *decreased* by an amount equal to the speed of the wind, w, so the plane's speed in miles per hour is $p - w$. When an airplane flies *with* the wind, its still-air speed is *increased* by the speed of the wind, so its speed in miles per hour is $p + w$. We know that the flying time against the wind is 3 hours, and that the flying time with the wind is only 2 hours. The distance for both trips is 360 miles.

DIRECTION	DISTANCE (d)	RATE (r)	TIME (t)
Against the Wind	360	$p - w$	3
With the Wind	360	$p + w$	2

We can obtain two equations using the distance formula:

RATE · TIME = DISTANCE

$r \cdot t = d$

The equations are

AGAINST THE WIND

I $\quad 3(p - w) = 360 \quad \rightarrow \quad p - w = 120 \quad$ Divide each side of **I** by 3.

WITH THE WIND

II $\quad 2(p + w) = 360 \quad \rightarrow \quad \underline{p + w = 180} \quad$ Divide each side of **II** by 2.

$\qquad\qquad\qquad\qquad\qquad\qquad\quad 2p = 300 \quad$ Add.

$\qquad\qquad\qquad\qquad\qquad\qquad\quad p = 150$ mph

Using $p = 150$ and substitution, we find that $w = 30$ mph. Thus the speed of the plane in still air is 150 mph and the speed of the wind is 30 mph.

EXAMPLE 6 Last year Mrs. Laguna invested her life savings of $100,000 in three different ways: mutual funds, government bonds, and the money market. For the year, the mutual funds returned 7% on her investment, the bonds returned 8%, and the money market only 4%, due to falling interest rates. She has as much in the money market as in the other two investments put together. Her total interest income from the investments for this last year was $5800. How much did she invest at each rate?

SOLUTION   Recall that

$$\text{INTEREST} = \text{PRINCIPAL} \cdot \text{RATE} \cdot \text{TIME}$$
$$I = p \cdot r \cdot t$$

Since we're talking about one year, $t = 1$. We let

x = the amount invested in mutual funds
y = the amount invested in bonds
z = the amount invested in the money market

Type of Investment	Amount Invested (p)	Rate of Return (r)	Income ($I = prt$)
Mutual Funds	x	7%	$0.07x(1) = 0.07x$
Government Bonds	y	8%	$0.08y(1) = 0.08y$
Money Market	z	4%	$0.04z(1) = 0.04z$

THE TOTAL AMOUNT INVESTED IS $100,000.

I $x + y + z = 100{,}000$

MONEY MARKET IS EQUAL TO THE SUM OF THE OTHER TWO INVESTMENTS.

II $z = x + y$

LAST YEAR'S INCOME FROM INTEREST WAS $5800.

III $0.07x + 0.08y + 0.04z = 5800$

We rewrite equations **I**, **II**, and **III** in a convenient form for solving by the elimination method:

$x + y + z = 100{,}000$
$-x - y + z = 0$ Subtract x and y from each side of **II**.
$7x + 8y + 4z = 580{,}000$ Multiply each side of **III** by 100.

The reader should verify that the solution is $x = \$20{,}000$, $y = \$30{,}000$, and $z = \$50{,}000$.

In economics, a manufacturer is free to set any price p (in dollars) for each unit of the product that is produced. If the price is too high, not enough people will want to buy the product. This means that the **demand** (the quantity q, or the number of units) for the product is low and that the **supply** is high. The opposite will occur if the price is too low. The demand will be high, because people are always looking for a bargain. In general, as the price increases, the demand decreases; conversely, as the price decreases, the demand increases. The ideal situation is to have the demand for the items *equal* to the number of units that the manufacturer supplies. The price that achieves this ideal is called the **equilibrium price**.

EXAMPLE 7   The supply and demand equations for a certain product are given by

I Supply: $q = 2p + 5$ **II** Demand: $q = -p + 14$

where p is the price in dollars for each unit and q is the quantity (the number in hundreds) of units to be sold.

(a) Find the equilibrium price and the number of units that will sell at this price.
(b) Sketch both the supply and the demand equations on the same coordinate system.

SOLUTION   (a) We can find the equilibrium price by solving the supply and the demand equations simultaneously. To do this, we substitute $q = 2p + 5$ from equation **I** into equation **II**:

$$q = -p + 14$$
$$2p + 5 = -p + 14$$
$$3p = 9$$
$$p = 3$$

So if $p = 3$, then $q = 11$ units. In other words, at a price of $3.00, the manufacturer can expect to sell 1100 units.

(b) The graph is shown in Figure 4.26, where the q-axis is in hundreds of units and the p-axis is in dollars.

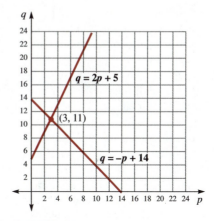

FIGURE 4.26

We can make some observations based on the results of Example 7.

> When both the supply and the demand equations are graphed on the same coordinate system:
>
> 1. The two lines representing supply and demand intersect at the point that represents the equilibrium price.
> 2. The line representing demand falls, while the line representing supply rises.

Exercise Set 4.7

Write a system of equations or inequalities for each of the following applied problems and solve it.

Number Problems

1. The sum of two numbers is −7. When the second number is subtracted from the first one, the difference is 10. Find the two numbers.
2. The difference of two integers is 36. When twice the smaller integer is added to three times the larger one, the resulting sum is −117. Find the two integers.
3. The sum of two consecutive odd integers is less than 37, and their difference is less than 3. Find the largest consecutive odd integers that satisfy these conditions.
4. The sum of two consecutive even integers is at least 46. When the larger one is subtracted from twice the smaller one, the resulting difference is at least 20. Find the two smallest consecutive even integers that satisfy these conditions.

Distance-Rate-Time Problems

5. A boat can travel 60 mi downstream in 3 hr or 30 mi upstream in 2 hr. What is the rate of the current and the speed of the boat in still water?
6. In 2 hr an airplane can travel 600 mi *with* the wind or 550 mi *against* the wind. Find the rate of the wind and the speed of the airplane in still air.
7. Two cars leave a gas station and travel in opposite directions. The difference in the speeds of the two cars is 10 km/hr. At what speed is each car traveling if the two cars are 200 km apart after 2 hr?
8. When Ron drives his truck, he averages 10 km/hr faster in level country than he does in the mountains. After traveling 4 hr in level country and 2 hr in the mountains, Ron has covered 295 km. What was his speed in level country? What was his speed in the mountains?

Mixture Problems

9. Marie bought 3 small colas and 2 large colas for a total of $3.60. Had she bought 2 small colas and 3 large colas, the cost would have been $3.90. What is the cost of each size of cola?
10. Cuesta Theater sold $410 worth of tickets one night. Adult tickets cost $2.00 each and children's tickets cost $1.00 each. If 310 tickets were sold, how many of each did the theater sell?
11. Jan invested $3000. She invested part at 6% per annum and the rest at 9%. Her annual total income from both of these investments is equivalent to 8% of her total investment. How much did she invest at 9%?
12. Iva invested her money in two parts at annual rates of 8% and 9% and received total interest earnings of $260 per year from these investments. If she were to reverse the amount invested at each rate, her yearly income would be $250. How much did she invest at each rate?
13. Mr. Asire needs 2 ℓ of a 65% acid solution. He has a 40% solution and an 80% solution on hand with which to make the mixture. How much of each should he use?
14. How many gallons of gasoline and gasahol (90% gasoline, 10% alcohol) does it take to make a 20-gal mixture that is 7% alcohol and 93% gasoline?
15. Peanuts worth $1.29/lb are mixed with pecans worth $1.99/lb to make a 10-lb mixture worth $1.71/lb. How many pounds of each are used?
16. The main floor of the Topliff Office Building contains 20,000 square feet and is divided into two office complexes. The rent in one office complex is $1.00/sq. ft, and in the other it is $1.20/sq. ft. If the total rent collected for the main floor is $21,000, how many square feet are in each office complex?
17. A 24-in. bookshelf holds 12 dictionaries and 6 encyclopedias. A 15-in. bookshelf holds 4 dictionaries and 15 encyclopedias. How thick is each dictionary? How thick is each encyclopedia?
18. Four cartons of note pads and 6 cartons of paper together weigh 98 lb. Three cartons of paper and 6 cartons of note pads weigh 93 lb. How much does a carton of note pads weigh?
19. Anne sells insurance for two different companies. She receives a 15% commission on all of the premiums from her sales for the first company. She receives a salary of $800/month and a 5% commission from her sales for the second company. The total premiums from all of Anne's sales for both companies was $7000 one month. Her income that month was $1400. What were the premiums worth for each company?

Inequality Problems

Write a linear inequality or a system of linear inequalities to represent each of the following applied problems and graph the region that represents the solution set.

20. The perimeter of a rectangle of length *l* and width *w* cannot be more than 100 ft.

21. A classroom has at most 1200 sq. ft of floor space. The space required for each chair is 10 sq. ft and the space needed for each table is 15 sq. ft.

22. The perimeter of a rectangle of length l and width w is at least 60 ft. The width is no more than one-half of the length.

23. At most, Scott can spend M hours studying math and P hours studying physics each day for a total of 6 hr/day. He wants to spend at least 1 hour more each day studying math than he does studying physics.

Supply-and-Demand Problems

24. The supply and demand equations for a computer manufacturer in San Luis Obispo are

    Supply: $q = 2p - 12$
    Demand: $q = -3p + 28$

    where p is the price (in hundreds of dollars) of one computer and q is the quantity (the number in tens) of computers to be sold.

    (a) Find the equilibrium price and the number of computers that will sell at that price.
    (b) Sketch both the supply and the demand equations on the same coordinate system.

25. The Fiesta Shirt Co. makes handmade shirts. The supply and demand equations for its shirts are

    Supply: $q = 4p - 3$
    Demand: $q = -3p + 11$

    where p is the price (in tens of dollars) per shirt and q is the quantity (the number in hundreds) of shirts to be made and sold.

    (a) Find the equilibrium price and the number of shirts that will be made and sold at this price.
    (b) Sketch both the supply and the demand equations on the same coordinate system.

26. The weekly supply and demand equations for packages of computer disks manufactured by Software Supplies, Inc., are

    Supply: $-5p + q = -71$
    Demand: $2p + q = 34$

    where p is the price (in dollars) per package and q is the quantity (the number in thousands) of packages to be supplied and sold.

    (a) Find the equilibrium price and the number of packages that will sell at that price.
    (b) Sketch both the supply and the demand equations on the same coordinate system.

27. The supply and demand equations of a certain brand of ball-point pen are

    Supply: $q = 3p - 8$
    Demand: $q = -5p + 24$

    where p is the price per pen in dollars and q is the quantity sold.

    (a) Find the equilibrium price and the number that will sell at that price.
    (b) When the pens are priced at $4.00 apiece, will the supply exceed the demand? If so, by how much?
    (c) When the pens are priced at $2.50 apiece, will the demand exceed the supply? If so, by how much?

28. For the Practical Paper Co., the supply and demand equations of a ream of paper are

    Supply: $4p - q = 10$
    Demand: $2p + q = 20$

    where p is the price in dollars of a ream and q is the quantity (in hundreds) of reams supplied and sold.

    (a) Find the equilibrium price and the number of reams that will sell at that price.
    (b) If the paper is $7.00/ream, does the supply exceed the demand? If so, by how much?
    (c) If the paper is priced at $4.50/ream, does the supply exceed the demand? If so, by how much?

Write a system of three equations in three variables for each of the following applied problems and solve it.

29. Several years ago, Lester invested $15,000 in three parts. One investment pays 6% annual interest, a second pays 5% annual interest, and Lester invested the remaining amount in a business. Two years ago he was charged 3% of his business investment to cover some losses, and his total income that year was only $400. Last year the business paid 9% in dividends, and his income from the three investments was $1000. How much of the $15,000 does he have in each investment?

30. Nina has $25,000 invested—part at 5% per annum, another part at 6% per annum, and the rest at 8%. The total yearly income from the three investments is $1600. The income from the 8% investment is the same as the sum of the incomes from the other two. How much has Nina invested at each rate?

Linear Programming

31. Wood's Publishing Company publishes a certain title in both paperback and hardback. The paperback book yields a profit of $400/day, while the hardback yields $700/day. The company uses only one printing crew,

which works no more than a 5-day week. Equipment limitations require that the paperback be produced no more than 3 days a week, whereas hardback production is limited to no more than 4 days a week.

(a) Determine the possible production schedules by graphing the system of inequalities.

(b) Find the coordinates of each vertex.

(c) Calculate the profit at each vertex.

(d) If you owned the company, how would you assign your printing crew?

32. Trett's Machine Shop makes both drum and disc brakes. Both types require a lathe and a grinder in production. Drum brakes require 2 hr on the lathe and 4 hr on the grinder. Disc brakes require 4 hr on the lathe and 2 hr on the grinder. The shop makes a profit of $14 on each set of drum brakes and $18 on each set of disc brakes. There is only one lathe and one grinder, and both machines can be used a maximum of 16 hr/day.

(a) Determine the production schedules by graphing the system of inequalities.

(b) Find the coordinates of each vertex.

(c) Calculate the profit at each vertex.

(d) If you owned the company, how many sets of each type of brakes would you produce?

SUMMARY

4.1

Solving Systems of Linear Equations by Graphing, Elimination, and Substitution

A **system of linear equations** involves two or more linear equations. A **solution** to a system of equations is the ordered pair, ordered triple, etc., with entries that are solutions to every equation in the system. If the graphs of a linear system *intersect in a single point,* the system is said to be **consistent and independent;** if they are *parallel,* the system is **inconsistent;** and if they *coincide,* the system is **consistent and dependent.**

Three ways of solving systems of linear equations are by **graphing,** the **elimination method,** and the **substitution method.**

4.2

Solving Systems of Linear Equations in Three or More Variables

The graph of a linear equation in three variables is a **plane.**

A **system of linear equations in three variables** may have none, one, or an infinite number of solutions.

4.3

Solving Systems of Linear Equations Using Matrices

A **matrix** is a rectangular array of numbers enclosed by brackets. A matrix has two **dimensions:** the number of rows and the number of columns. When a system of equations is written in matrix form, the matrix consists of the coefficients of the variables augmented by the constants. When a particular variable does not appear in an equation, its coefficient is zero. Zero coefficients must appear in the augmented matrix.

The **row transformation theorem** specifies what operations can be performed on a matrix representing a system of linear equations:

1. Any two rows may be interchanged.
2. All of the elements in any row may be multiplied by the same nonzero constant.
3. The product of any constant and any row may be added to any other row.

There are two different methods for solving systems of equations by matrices. In the first method, **row operations** introduce zeros into all elements below the diagonal that passes from the upper left corner to the lower right corner of the coefficient matrix. The solution to the system can then be found by **back substitution.** The second method, called the **Gauss–Jordan elimination method,** transforms all entries both above and below the main diagonal to zero and all the elements along the diagonal to 1. Such a matrix is said to be in **reduced form.** The solution can be read directly from the matrix.

4.4 Determinants and Cramer's Rule

A **square matrix** is one that has the same number of rows as columns. Associated with each square matrix is a real number called its **determinant.** The determinant of a matrix is written within vertical bars and, in the 2×2 case, is evaluated by the formula

$$\begin{vmatrix} a & c \\ b & d \end{vmatrix} = ad - bc$$

A 3×3 determinant is evaluated in terms of the **minors** associated with the elements of a chosen row or column.

When determinants are used to solve systems of linear equations, the technique is known as **Cramer's rule.**

4.5 Graphing Linear Inequalities in Two Variables

The graph of the solution set of a **linear inequality in two variables** is a **half-plane** on one side of the **boundary line** obtained by replacing the inequality symbol with an equal sign. The boundary line is part of the solution set if the original inequality symbol is $\leq$ or $\geq$. The solution set of a **system of linear inequalities** has as its graph that portion of the plane that is the intersection of all the half-planes of the system.

4.6 Linear Programming

Inequalities are often used to solve problems related to business. The goal of a business is to provide a quality product for the consumer and at the same time to return a fair profit for the manufacturer. There are often limitations on a plant's capacity to produce certain items, and these limitations affect manufacturing decisions. If the limitations, called **constraints,** can be expressed in terms of a system of linear inequalities, and the objective is to maximize profits or minimize costs (generating an **objective function**), then the problem falls into a class of problems solved by a method called **linear programming.**

*C*OOPERATIVE *E*XERCISE

You have just graduated from college and you're looking for a good job. You have sent your resume out to 15 top companies. Ten of the companies invite you for an interview and three of these make you a job offer. The companies and their job offers are as follows.

A. The Hi-Tech Company offers a $12,000 base salary plus an 8% commission on gross sales. They pay a $7,000 bonus if you reach your quota of $125,000 in gross sales. Your quota increases by 5% each year.

B. North Publishing Company offers a $20,000 base salary plus a 6% commission on all net sales. (Net sales are equal to gross sales less 20% for various overhead expenses.) If you reach your quota of $100,000 in net sales, you will receive a bonus of $5,000. Your quota will remain at $100,000 for the first two years and will increase by 6% each year thereafter.

C. The Software Design Company offers a $30,000 base salary plus a 2% commission on net sales. No performance bonuses are paid, but you are guaranteed a 0.5% increase in commission each year that you reach your quota in net sales (gross sales less 20%). The first year your quota is $100,000 and then it increases by 5% each year.

You project that you will be able to sell $130,000 your first year, $160,000 your second year, and $190,000 your third year. All sales are gross sales. You want to choose the best of the three jobs. *Assume that your sales projections are correct for the following exercises.*

GROUP 1
What would your total salary be for the first three years from each of the three companies? If you were to make your choice of a job based upon this information, which company would you choose? Write a paragraph explaining your choice.

GROUP 2
In looking over the job offers, you decide that you would really like to live in the vicinity of North Publishing Company or the Software Design Company.

A. Write a linear equation representing your salary (S) in terms of sales (x) in dollars for each of the two companies.

B. Graph these linear equations on the same Cartesian coordinate system. Use the graph to estimate the point of intersection of the two linear equations.

C. Use the graph to estimate your fourth-year salary at each of the two companies. (Assume that the projected sales pattern continues in the same manner.)

D. Solve the system of two equations in two unknowns for the point of intersection. What is your salary at this point? What are the sales at this point?

E. Which of these two jobs would you choose based upon this information? Write a paragraph explaining your choice.

GROUP 3
In looking over the job offers, you find that the Hi-Tech Company and North Publishing Company have the best reputation for treatment of their employees.

A. Write a linear equation representing your salary (S) in terms of sales (x) in dollars for each of the two companies.

B. Graph these linear equations on the same Cartesian coordinate system. Use the graph to estimate the point of intersection of the two linear equations. What is your salary at this point? What are the sales at this point?

C. Use the graph to estimate your fifth-year salary at each of the two companies. (Assume that the projected sales pattern continues in the same manner.)

D. Solve the system of two equations in two unknowns for the point of intersection. What is your salary at this point? What are the sales at this point? During which year does this occur?

E. Which of the two jobs would you choose based upon this information? Why? Write a paragraph explaining your choice.

GROUP 4
A. Write a linear equation representing your salary (S) in terms of sales (x) in dollars for each of the three companies.

B. Graph these three linear equations on the same Cartesian coordinate system.

C. Calculate your fifth-year salary at each of the three companies. (Assume that the projected sales pattern continues in the same manner.)

D. Based upon this information, which of the three jobs would you choose? If you were looking for a short-term job (less than five years), which of the jobs would you choose? If you were looking for a long-term job (more than five years), which of the jobs would you choose? Write a paragraph explaining your choices.

In Exercises 1–4, complete the following sentences.

1. The solution to the system of equations $3x - y = 5$ and $2x + 3y = 6$ is _____.

2. When the graphs of a system of equations coincide, the system is said to be _____.

3. When the graphs of a system of equations do not intersect, the system is said to be _____.

4. When the graphs of a system of equations intersect in a unique point, the system is said to be _____.

5. Solve the system of equations by the method of graphing:
$4x + 2y = 6$ and $3x + y = 2$.

6. Solve the system of equations by substitution:
$2x + y = 4$ and $4x - y = 1$.

7. Solve the system of equations by the elimination method:
   $4a + 3b = 11$ and $a - 2b = 0$.

8. Solve the system of equations by substitution:
   $4m + 2n = 3$ and $m - n = 5$.

9. Solve the system of inequalities by graphing:
   $2x + y \leq 6$ and $6x + 4y \geq 8$.

10. Solve the system of equations by the elimination method:
    $6s + t = 10$ and $10s + 2t = 20$.

11. Solve the system of inequalities by graphing:
    $2x + y \leq 1$ and $x - y > 1$.

12. Solve the system of inequalities by graphing:
    $y < x$, $y > 0$, and $x \leq 2$.

13. Solve the system
    $$2x - 4y - z = 10$$
    $$x - 2y + z = 2$$
    $$2x - y + z = 3$$

14. Solve the system $5x + 7y = 8$ and $2x - 3y = 4$.

15. Solve by graphing:
    $\{(x, y) \mid y \leq x + 3\} \cap \{(x, y) \mid y \geq -x - 2\}$
    $\cap \{(x, y) \mid x \leq 3\}$
    $\cap \{(x, y) \mid y \geq -1\}$

16. Solve the system
    $$\frac{5}{x} - \frac{2}{y} = \frac{1}{2}$$
    $$\frac{3}{x} + \frac{3}{y} = \frac{1}{4}$$

17. Write the augmented matrix for the system:
    $3x - 2y = 1$ and $2x - 5y = -3$.

18. Write a system of equations for the augmented matrix
    $\begin{bmatrix} 3 & 1 & | & 4 \\ 2 & 3 & | & 5 \end{bmatrix}$

19. Solve the system $2x - y = 4$ and $x + y = 1$ using matrices and back substitution.

20. Solve the system $6x + y = 5$ and $x - 4y = 5$ using Cramer's rule.

21. Evaluate the determinant
    $\begin{vmatrix} 4 & -1 & 2 \\ 2 & 1 & -6 \\ -3 & 0 & 5 \end{vmatrix}$

22. Use Gauss–Jordan elimination to solve the system
    $$x + y - 2z = 1$$
    $$3x - y + z = 2$$
    $$x + y + z = 4$$

23. Solve the augmented matrix $\begin{bmatrix} 2 & 1 & | & 2 \\ 1 & 1 & | & 1 \end{bmatrix}$.

24. Evaluate the determinant
    $\begin{vmatrix} 2 & 0 & 1 \\ 3 & 1 & 0 \\ 1 & 2 & -1 \end{vmatrix}$

25. Solve the following system using matrices:
    $$2x + z = 7$$
    $$ y + 2z = 1$$
    $$3x + y + z = 9$$

26. Solve the following system using Cramer's rule:
    $$2x + y - z = 3$$
    $$x - 2y + 2z = 1$$
    $$x - y + 2z = 3$$

27. Solve for x: $\begin{vmatrix} x & 3 \\ -1 & 2 \end{vmatrix} = 5$.

28. Evaluate the determinant
    $\begin{vmatrix} 2 & 3 & 1 \\ 0 & 0 & 0 \\ -1 & 2 & 0 \end{vmatrix}$

Use row operations to determine whether the second matrix can be obtained from the first one in Exercises 29–32.

29. $\begin{bmatrix} 1 & 1 \\ 2 & 1 \end{bmatrix} \to \begin{bmatrix} 1 & 1 \\ 5 & 4 \end{bmatrix}$

30. $\begin{bmatrix} 1 & 2 & 3 \\ 4 & 5 & 6 \\ 7 & 8 & 9 \end{bmatrix} \to \begin{bmatrix} 1 & 2 & 3 \\ 16 & 5 & 6 \\ 25 & 8 & 9 \end{bmatrix}$

31. $\begin{bmatrix} 2 & -1 & 1 & -3 \\ 1 & 3 & -4 & 2 \\ 1 & 0 & -2 & 1 \\ 3 & -1 & 5 & 2 \end{bmatrix} \to \begin{bmatrix} 2 & -1 & 1 & -3 \\ 7 & 0 & -1 & -7 \\ 1 & 0 & -2 & 1 \\ 3 & -1 & 5 & 2 \end{bmatrix}$

32. $\begin{bmatrix} 1 & 0 & 3 \\ 2 & 1 & 7 \\ 1 & 0 & 2 \end{bmatrix} \rightarrow \begin{bmatrix} 1 & 0 & 3 \\ 2 & 1 & 7 \\ 3 & 1 & 9 \end{bmatrix}$

33. Solve using Gauss–Jordan elimination:
$$2x + 3y = -4$$
$$-3x + y = 17$$

34. Show that
$$\begin{vmatrix} 1 & a & a^2 \\ 1 & b & b^2 \\ 1 & c & c^2 \end{vmatrix} = (a - b)(c - a)(b - c)$$

Write a system of equations and solve it.

35. Lonnie's Farms sold 25 bushels of oats and 30 bushels of rye to the local co-op for $257.50; later they sold 50 bushels of oats and 10 bushels of rye (at the same unit price) to a local feed store for $315.00. What was the price per bushel of each kind of grain?

36. Three hot dogs and a shake cost $3.50. Two hot dogs and two shakes cost $3.40. What is the cost of each item?

37. Two shirts and three pairs of socks cost $24.40. One shirt and five pairs of socks cost $18.50. What is the cost of each?

38. Two different kinds of milk contain 6% and 2% butterfat, respectively. How many gallons of each type must be mixed together to obtain 10 gallons of milk with 4% butterfat?

39. The sum of the speeds of two airplanes is 470 mph. The difference of their speeds is 120 mph. Find their respective speeds.

40. A collection of dimes and quarters has a value of $2.85. The total value of the dimes is 65¢ less than the total value of the quarters. Find the number of dimes and quarters in the collection.

41. James has $3000 to invest. He invests part at 5% annual interest, another part at 6% annual interest, and the rest at 8%. The total interest earned for one year is $192. If the interest earned on the 8% investment is twice that earned on the 5% investment, how much did he invest at each rate?

42. A boat can travel 36 mi upstream in 3 hr and 32 mi downstream in 2 hr. What is the speed of the current?

43. An airplane can travel 525 mi in 3 hr with the wind and 310 mi in 2 hr against the wind. What is the speed of the airplane in still air?

44. Darlene averages 35 mph faster in the desert than in the city. After traveling 5 hr in the desert and $\frac{1}{2}$ hr in the city, she has covered 315 miles. What was her average speed in the desert?

45. A research company requires a survey using 600 telephone contacts and 400 house contacts. Company A can do 30 telephone contacts and 10 house contacts in one hour. Company B can do 20 telephone contacts and 20 house contacts in one hour. How many hours should the research company schedule for each of Companies A and B in order to obtain exactly the number of contacts needed?

46. A college wants to hire 10 new full-time staff members for the coming year, and it has exactly $165,000 for salaries. Beginning instructors can be hired for $15,000 and experienced instructors for $18,000. How many of each can be hired in order to use the exact sum of money allocated for salaries?

Write a system of inequalities for each of the following applied problems, and solve it.

47. Jeff's Electronics produces two models of VCRs. Each unit of model A requires 3 hr of work on assembly line 1 and 3 hr of work on assembly line 2. Model B requires 10 hr of work on assembly line 1 and 2 hr of work on assembly line 2. In any given week, assembly line 1 has at most 150 work hours and assembly line 2 has at most 54 work hours to be used for VCR production. How should production be allotted to maximize the total profit if model A nets a profit of $40/unit and model B nets a profit of $60/unit? Assume that all units of both types are sold.

48. Art's Shop produces two models of a product, x and y. Two raw materials, m and n, are used in the production. At least 18 lb of m and 12 lb of n must be used daily. Also, at most 34 hr of labor can be utilized. Two pounds of m are needed for each unit of x and 1 lb of m for each unit of y. One pound of n is needed for each unit of both x and y. It takes 3 hr of labor to manufacture each unit of x and 2 hr of labor to manufacture each unit of y. If the profit is $5 for each unit of x and $3 for each unit of y, how many units of each model should be produced and sold to maximize the profit?

Chapter Test

Solve Exercises 1–5 by the indicated method.

1. Graphing: $x + 2y = 8$
   $3x - 5y = -9$

2. Elimination: $8x + y = 10$
   $5x - 3y = -1$

3. Substitution: $2x + y = 7$
   $3x - 2y = 0$

4. Graphing: $x - y < 1$
   $2x + y \geq 4$

5. Elimination: $x - y + z = -1$
   $x + y + z = 5$
   $2x - y + z = 2$

Write an appropriate equation and solve each applied problem.

6. Nick invests $3000, part at 8% per annum and the rest at 6%. If, after one year, his income from the two investments is $200, how much has Nick invested at each rate?

7. The sum of two integers is 16. When the second integer is subtracted from the first one, the result is 34. Find the integers.

Solve Exercises 8 and 9 by the Gauss–Jordan elimination method.

8. $4x + 3y = -2$
   $3x + 2y = -1$

9. $2x + y + 2z = 5$
   $3x - 2y + 2z = 3$
   $x - y + z = 1$

Solve for x in Exercises 10 and 11.

10. $\begin{vmatrix} 3 & x \\ 2 & -1 \end{vmatrix} = 2$

11. $\begin{vmatrix} 2 & 1 & -1 \\ 3 & 4 & x \\ 1 & 2 & -3 \end{vmatrix} = 4$

12. Show that
$$\begin{vmatrix} 3 & 5 \\ 7 & 1 \end{vmatrix} = -\begin{vmatrix} 7 & 1 \\ 3 & 5 \end{vmatrix}$$

Solve Exercises 13–16 using Cramer's rule.

13. $3x + y = 3$
    $6x - 2y = -2$

14. $4x + 2y + 3z = 6$
    $x + 4y + z = 3$
    $2x - 6y - z = 2$

15. $\dfrac{2}{x} - \dfrac{2}{y} = -1$

    $\dfrac{4}{x} + \dfrac{3}{y} = 5$

16. $x + y + z = 1$
    $2x - y - 4z = 0$
    $4x + y + 3z = 2$

Write an appropriate equation and solve each applied problem.

17. Mr. Martinez wants 25 ℓ of a 25% acid solution. He has both 20% and 30% solutions in the stockroom. How many liters of each should he use?

18. In 8 hr a boat can travel up a river and back to the starting point, which is 60 mi altogether. What are the speed of the current and the speed of the boat in still water if it takes 6 hr just to go up the river (*against the current*)?

19. The length of a rectangle is 3 in. more than twice its width. If the perimeter is 36 in., find the dimensions.

20. Ingrid bought 2 lb of cashews and 3 lb of peanuts for a family reunion, at a cost of $14.50. Later that day she went back to the same store and bought another pound of cashews and 2 more pounds of peanuts at a total cost of $8.50. Assuming the prices per pound stayed the same, what was the cost of a pound of cashews? A pound of peanuts?

21. Ricardo's Transport Service delivered a 5-lb package for a fee of $4.90, and an 11-lb package for a fee of $9.10. If there is one price for the first pound and a second price for each additional pound, find these two prices.

22. When Hilda, Angelina, and Erika all work on an assembly line together, they produce 222 items per day. On the day that Erika works in another department, Hilda and Angelina produce 159 items. When Hilda does not work with them, Angelina and Erika produce 147 items a day. Their boss offers to give a bonus to the worker who produces the most items per day. Which worker should receive the bonus and how many items would this worker produce in a day if she were working by herself?

CHAPTER 5
LAWS OF EXPONENTS, POLYNOMIALS, AND POLYNOMIAL FUNCTIONS

5.1 Exponents and Scientific Notation
5.2 Adding and Subtracting Polynomials
5.3 Multiplying Polynomials
5.4 Dividing Polynomials
5.5 Synthetic Division

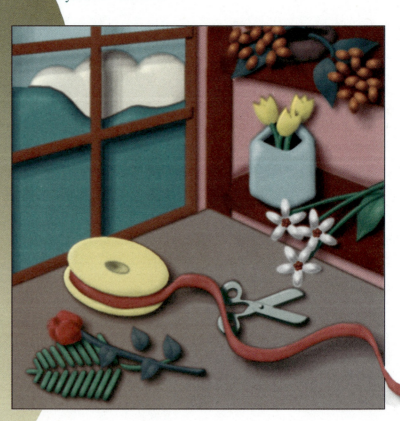

Application

Mitsie makes and sells silk flower arrangements. Her cost per week is $C(x) = 20 + 14x$, where x is the number of flower arrangements she makes and sells. Her revenue per week is $R(x) = 20x + x^2$. Find the polynomial function $P(x)$ that represents her profit per week if $P(x) = R(x) - C(x)$. Find her cost, revenue, and profit to make and sell 10 arrangements per week.

See Section 5.2, Exercise 75.

I have so many mathematical ideas that may be of some use in future times if others more penetrating than I will go deeply into them some day and join the beauty of their minds to the labor of mine.

GOTTFRIED WILHELM LEIBNIZ, 1646–1716

5.1 Exponents and Scientific Notation

HISTORICAL COMMENT René Descartes (1596–1650), although most well known for his development of the Cartesian coordinate system, also contributed significantly to the notation that we use in modern mathematics, including the notation that we use today for writing exponents. Before his time there were numerous schemes for indicating a power of a number. In 1484, Chuquet used 3^0, 3^1, and 3^2 to mean 3, $3x$, and $3x^2$, in 1572 Bombelli used 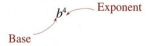 to mean the same thing. Descartes used x, xx, x^3, and x^4. The use of the repeated letter xx for the second power of a variable continued for many years. Gauss (1777–1855) was a supporter of the notation xx, saying that it took no more space than x^2 did.

OBJECTIVES In this section we will learn to
1. use exponents in multiplication and division;
2. interpret and use exponents that are integers; and
3. use exponents to write expressions in scientific notation.

In Chapter 1 we learned that the expression b^4 means $b \cdot b \cdot b \cdot b$. The letter b is called the **base,** and 4 is the **exponent,** or **power.**

$$b^4 \quad \text{Exponent}$$
Base

When we write $b \cdot b \cdot b \cdot b$ as b^4, we say it is in **exponential form.**

Recall from arithmetic that if a number is written as the product of two or more numbers, we say it is **factored.** *When an exponent is a natural number, it indicates the number of times the base is used as a factor.* As such, it provides a convenient way to indicate a product of repeated factors.

Exponent is 4: $\quad b^4 = b \cdot b \cdot b \cdot b \qquad$ b is used as a **factor** 4 times.

Exponent is n: $\quad b^n = b \cdot b \cdots b \qquad$ b is used as a **factor** n times, where n is a natural number.

When the exponent is 1, it is usually not written, since $b^1 = b$.

The expression b^4 is read, "b to the fourth power." The expression b^n is read, "b to the nth power." The expressions b^2 and b^3 can be read, "b to the second power" and "b to the third power," respectively, but they are also known by shorter names:

b^2 is called **"b squared,"** and b^3 is called **"b cubed."**

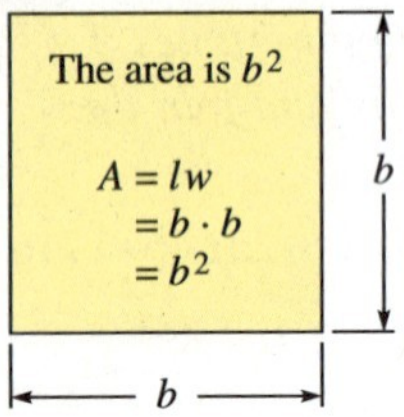

The area is b^2

$A = lw$
$= b \cdot b$
$= b^2$

FIGURE 5.1

The expression b^2 and b^3 can be illustrated geometrically, as shown in Figures 5.1 and 5.2. Figure 5.1 illustrates that the area of a square of side b is $b \cdot b$, or b^2. Figure 5.2 illustrates that the volume of a cube of side b is $b \cdot b \cdot b$, or b^3.

EXAMPLE 1 Evaluate each of the following:

(a) $5^2 = 5 \cdot 5 = 25$
(b) $4^3 = 4 \cdot 4 \cdot 4 = 64$
(c) $\left(\dfrac{1}{2}\right)^4 = \dfrac{1}{2} \cdot \dfrac{1}{2} \cdot \dfrac{1}{2} \cdot \dfrac{1}{2} = \dfrac{1}{16}$
(d) $5^1 = 5$
(e) $2 \cdot 3^2 = 2 \cdot (3 \cdot 3) = 2 \cdot 9 = 18$ The exponent applies only to 3, as indicated by the order of operations.
(f) $2^3 \cdot 3^2 = (2 \cdot 2 \cdot 2)(3 \cdot 3) = 8 \cdot 9 = 72$

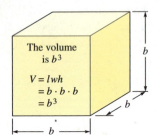

FIGURE 5.2

EXAMPLE 2 Identify the base and exponent in each of the following quantities.

(a) 7^2 The base is 7 and the exponent is 2.
(b) p The base is p and the exponent is 1.
(c) $(3x)^2$ The base is $3x$ and the exponent is 2.
(d) -2^3 The base is 2 and the exponent is 3. Think of -2^3 as $-1 \cdot 2^3$. The exponent does not apply to the factor -1.
(e) $(-2)^4$ The parentheses indicates that the exponent applies to the negative sign as well as the 2. The base is -2 and the exponent is 4.
(f) $(x + 7)^6$ The base is $x + 7$ and the exponent is 6.
(g) b^n The base is b and the exponent is n.

EXAMPLE 3 Write each of the following products in exponential form.

(a) $x \cdot x \cdot x \cdot x = x^4$
(b) $x \cdot x + 3 \cdot y \cdot y = x^2 + 3y^2$
(c) $3t \cdot 3t \cdot 3t = (3t)^3$
(d) $-2x \cdot 2x \cdot 2x \cdot 2x = -(2x)^4$
(e) $(a + b + c)(a + b + c) = (a + b + c)^2$
(f) $(-3x)(-3x)(-3x)(-3x) = (-3x)^4$

NOTE Throughout this book we will assume that the base and the exponent cannot be zero at the same time. We will discuss this restriction further on page 250.

Several definitions and theorems govern the way that we can use exponents to simplify expressions. For example, consider the product of b^2 and b^3:

$$b^2 \cdot b^3 = (b \cdot b)(b \cdot b \cdot b) = b \cdot b \cdot b \cdot b \cdot b = b^5$$

This observation leads us to the **product rule,** which is given as Theorem 5.1.

THEOREM 5.1

Product Rule for Exponents

If b is any real number, then

$$b^m \cdot b^n = b^{m+n}, \qquad m, n \text{ are integers}$$

CAUTION

The product rule can be used only when the bases are the same.

$$a^2 \cdot b^3 = a^2 b^3$$

The bases are not the same; the exponents cannot be added.

$$a^3 \cdot a^4 = a^7$$

The bases are the same; add the exponents.

We will prove the theorem for the case where m and n are natural numbers.

PROOF By the definition of a natural-number exponent,

$$b^m = \underbrace{(b \cdot b \cdots b)}_{m \text{ factors}} \qquad b^n = \underbrace{(b \cdot b \cdots b)}_{n \text{ factors}}$$

Therefore

$$b^m \cdot b^n = \underbrace{(b \cdot b \cdots b)}_{m \text{ factors}}\underbrace{(b \cdot b \cdots b)}_{n \text{ factors}}$$

$$= \underbrace{(b \cdot b \cdots b)}_{m + n \text{ factors}} \qquad \text{Associative property}$$

$$= b^{m+n} \qquad \text{Definition of an exponent}$$

EXAMPLE 4

Use the product rule for exponents to simplify each of the following products.

(a) $x^7 \cdot x^9 = x^{7+9} = x^{16}$
(b) $a^2 \cdot a^3 \cdot a^5 = a^{2+3+5} = a^{10}$
(c) $(x + y)^3(x + y)^4 = (x + y)^{3+4} = (x + y)^7$ The base is $(x + y)$.
(d) $-2y^2 \cdot 3y^3 = (-2 \cdot 3)(y^2 \cdot y^3)$ Associative and commutative properties
$\qquad = -6y^{2+3} = -6y^5$
(e) $10^3 \cdot 10^5 = 10^{3+5} = 10^8$

Notice that even when the base is numerical, we find the product by adding the exponents and retaining the same base.

CAUTION

Unless grouping symbols indicate otherwise, the exponent applies only to the quantity immediately to its left.

Right
$$2x^3 = 2 \cdot x \cdot x \cdot x$$
$$-5^2 = -(5 \cdot 5) = -25$$
$$-3 \cdot 4^2 = -3 \cdot 4 \cdot 4 = -48$$
$$(-3 \cdot 4)^2 = (-3 \cdot 4)(-3 \cdot 4)$$
$$= 144$$

Wrong
$$2x^3 = 2x \cdot 2x \cdot 2x$$
$$-5^2 = (-5)(-5)$$
$$-3 \cdot 4^2 = (-3 \cdot 4)(-3 \cdot 4)$$

We can simplify exponential expressions involving quotients much as we do those involving products. For example, if $b \neq 0$,

$$\frac{b^5}{b^2} = \frac{b \cdot b \cdot b \cdot b \cdot b}{b \cdot b} \qquad \text{Definition of an exponent}$$

$$= \frac{\overset{1}{\cancel{b}} \cdot \overset{1}{\cancel{b}} \cdot b \cdot b \cdot b}{\cancel{b} \cdot \cancel{b}} \qquad \text{Divide out common factors.}$$

$$= 1 \cdot 1 \cdot b \cdot b \cdot b$$

$$= 1 \cdot b^3 = b^3$$

Thus $\frac{b^5}{b^2} = b^3$.

We could easily have achieved this result by subtracting the exponent in the denominator from the exponent in the numerator:

$$\frac{b^5}{b^2} = b^{5-2} = b^3$$

Notice that the exponent in the numerator was greater than the exponent in the denominator. What if the exponent in the denominator is greater than the one in the numerator? Can we still divide exponential numbers with like bases by subtracting exponents? Consider the following quotient:

$$\frac{b^2}{b^4} = \frac{b \cdot b}{b \cdot b \cdot b \cdot b} \qquad \text{Definition of an exponent}$$

$$= \frac{\overset{1}{\cancel{b}} \cdot \overset{1}{\cancel{b}}}{\cancel{b} \cdot \cancel{b} \cdot b \cdot b} \qquad \text{Divide out common factors.}$$

$$= \frac{1}{b^2}$$

If we complete the division by subtracting the exponents, we have

$$\frac{b^2}{b^4} = b^{2-4} = b^{-2}$$

Now if

$$\frac{b^2}{b^4} = \frac{1}{b^2} \qquad \text{and} \qquad \frac{b^2}{b^4} = b^{-2}$$

then by the transitive property of Section 1.2,

$$\frac{1}{b^2} = b^{-2}$$

A generalization of this last observation leads to the following definition.

DEFINITION

Negative Exponent

If b represents any number except zero and n is an integer, then

$$b^{-n} = \frac{1}{b^n}$$

Now let's consider the case where the exponents in the numerator and the denominator are the same. Remember that any nonzero quantity divided by itself is 1, so

$$\frac{b^n}{b^n} = 1, \qquad b \neq 0$$

When the division is carried out by subtracting exponents, we get

$$\frac{b^n}{b^n} = b^{n-n} = b^0$$

Now if

$$\frac{b^n}{b^n} = 1 \qquad \text{and} \qquad \frac{b^n}{b^n} = b^0$$

then by the transitive property, $b^0 = 1$. Since b can represent *any nonzero algebraic expression*, we have the following definition.

***N**OTE* $\frac{0}{0} = 0^{1-1} = 0^0$ is not defined, since it indicates division by zero.

DEFINITION
Zero Exponent

If b represents any real number except zero, then $b^0 = 1$.

With the last two definitions in mind, we now state the **quotient rule for exponents** as Theorem 5.2.

THEOREM 5.2
Quotient Rule for Exponents

If b represents any real number except zero, then

$$\frac{b^m}{b^n} = b^{m-n}, \qquad m, n \in J$$

EXAMPLE 5 Evaluate each of the following powers.

(a) $10^3 = 10 \cdot 10 \cdot 10 = 1000$
(b) $10^2 = 10 \cdot 10 = 100$
(c) $10^1 = 10$
(d) $10^0 = 1$
(e) $10^{-1} = \dfrac{1}{10^1} = \dfrac{1}{10}$
(f) $10^{-2} = \dfrac{1}{10^2} = \dfrac{1}{100}$
(g) $10^{-3} = \dfrac{1}{10^3} = \dfrac{1}{1000}$

In the remainder of this section we will assume that no denominator is equal to zero.

EXAMPLE 6 Simplify each expression.

(a) $\dfrac{x^7}{x^2} = x^{7-2} = x^5$

(b) $\dfrac{14x^8}{-7x^6} = \dfrac{14}{-7} \cdot \dfrac{x^8}{x^6} = -2 \cdot x^{8-6} = -2x^2$

(c) $\dfrac{(x+y)^{12}}{(x+y)^{-8}} = (x+y)^{12-(-8)} = (x+y)^{12+8} = (x+y)^{20}$

(d) $(x+9)^0 = 1$, provided $x \neq -9$. (If $x = -9$, then $(x+9)^0 = 0^0$.)

(e) $\dfrac{y^{-3}}{y^{-6}} = y^{-3-(-6)} = y^{-3+6} = y^3$

EXAMPLE 7 Simplify each of the following quantities.

(a) $-2x^{-2} = -2 \cdot x^{-2} = -2 \cdot \dfrac{1}{x^2} = \dfrac{-2}{x^2}$

(b) $(2a)^{-1} = \dfrac{1}{(2a)^1} = \dfrac{1}{2a}$

(c) $2a^{-1} = 2 \cdot a^{-1} = 2 \cdot \dfrac{1}{a} = \dfrac{2}{a}$

(d) $\dfrac{(x+y)^4}{(x+y)^6} = (x+y)^{4-6} = (x+y)^{-2} = \dfrac{1}{(x+y)^2}$

> **CAUTION**
> An algebraic expression is considered to be in **simplest form** if it does not contain negative or zero exponents.

Work with negative exponents can frequently be made easier by observing the following facts.

$\dfrac{a^{-n}}{b^{-m}} = \dfrac{\dfrac{1}{a^n}}{\dfrac{1}{b^m}}$ Definition of a negative exponent

$= \dfrac{1}{a^n} \cdot \dfrac{b^m}{1}$ Definition of division of fractions

$= \dfrac{b^m}{a^n}$ Definition of multiplication of fractions

Thus $\dfrac{a^{-n}}{b^{-m}} = \dfrac{b^m}{a^n}$.

EXAMPLE 8 Simpify the following quotients.

(a) $\dfrac{3x^{-2}}{4^{-1}y^{-3}} = \dfrac{3 \cdot 4y^3}{x^2} = \dfrac{12y^3}{x^2}$

(b) $\dfrac{2a^{-6}y^4}{5ay^{-6}} = \dfrac{2y^4(y^6)}{5a(a^6)} = \dfrac{2y^{10}}{5a^7}$

We must consider some additional rules for exponents at this point; we introduce them in the form of examples.

EXAMPLE 9 Simplify the following powers.

(a) $(b^3)^2 = b^3 \cdot b^3 = b^{3+3} = b^6$

(b) $(a \cdot b)^2 = (a \cdot b)(a \cdot b) = (a \cdot a)(b \cdot b) = a^2b^2$

(c) $\left(\dfrac{a}{b}\right)^2 = \left(\dfrac{a}{b}\right)\left(\dfrac{a}{b}\right) = \dfrac{a^2}{b^2}$

(d) $\left(\dfrac{a}{b}\right)^{-1} = \dfrac{a^{-1}}{b^{-1}} = \dfrac{b}{a}$

THEOREM 5.3
Power Rules for Exponents

If a and b represent real numbers and m and n are integers,

(a) $(b^m)^n = b^{mn}$

(b) $(a \cdot b)^m = a^m \cdot b^m$

(c) $\left(\dfrac{a}{b}\right)^m = \dfrac{a^m}{b^m}, \quad b \neq 0$

(d) $\left(\dfrac{a}{b}\right)^{-m} = \dfrac{a^{-m}}{b^{-m}} = \dfrac{b^m}{a^m}, \quad a \neq 0, b \neq 0$

We will prove parts (a) and (b) for n a natural number.

PROOF

(a) $(b^m)^n = \overbrace{b^m \cdot b^m \cdots b^m}^{n \text{ factors of } b^m}$ Definition of the exponent n

$ = b^{m+m+\cdots+m}$ There are n exponents of m.

$ = b^{m \cdot n}$ Theorem 5.1 for n factors b^m

(b) $(a \cdot b)^m = \overbrace{(a \cdot b)(a \cdot b) \cdots (a \cdot b)}^{m \text{ factors of } (a \cdot b)}$ Definition of the exponent m

$ = \overbrace{(a \cdot a \cdots a)}^{m \text{ factors of } a}\overbrace{(b \cdot b \cdots b)}^{m \text{ factors of } b}$ Associative and commutative properties

$ = a^m \cdot b^m$ Definition of an exponent

CAUTION
$(a + b)^n \neq a^n + b^n,$ unless $n = 1$

Parts (a), (b), and (c) of Theorem 5.3 are often interpreted as follows:

(a) To raise a power to a power, multiply the exponents.
$$(b^m)^n = b^{mn}$$

(b) The mth power of a product is the product of the mth powers.
$$(a \cdot b)^m = a^m \cdot b^m$$

(c) The mth power of a quotient is the quotient of the mth powers.
$$\left(\dfrac{a}{b}\right)^m = \dfrac{a^m}{b^m}$$

EXAMPLE 10 Simplify each of the following powers.

(a) $(3^2)^2 = 3^{2 \cdot 2} = 3^4 = 81$ $\qquad (b^m)^n = b^{mn}$

(b) $(a^2)^4 = a^{2 \cdot 4} = a^8$ $\qquad (b^m)^n = b^{mn}$

(c) $(3^{-2})^2 = 3^{(-2) \cdot 2} = 3^{-4} = \dfrac{1}{3^4} = \dfrac{1}{81}$ $\qquad (b^m)^n = b^{mn}$ and $b^{-n} = \dfrac{1}{b^n}$

(d) $\left(\dfrac{a}{b}\right)^2 = \dfrac{a^2}{b^2} \qquad \left(\dfrac{a}{b}\right)^m = \dfrac{a^m}{b^m}$

(e) $(-3x^2)^{-1} = \dfrac{1}{-3x^2} = -\dfrac{1}{3x^2} \qquad b^{-n} = \dfrac{1}{b^n}$

(f) $\begin{aligned}(x^2y^3)^2 &= (x^2)^2(y^3)^2 \\ &= x^{2\cdot 2}y^{3\cdot 2} \\ &= x^4y^6\end{aligned} \qquad \begin{aligned}(a\cdot b)^m &= a^m\cdot b^m \\ (b^m)^n &= b^{mn}\end{aligned}$

(g) $\begin{aligned}(-3x^2y^3)^2 &= (-3)^2(x^2)^2(y^3)^2 \\ &= 9x^{2\cdot 2}y^{3\cdot 2} \\ &= 9x^4y^6\end{aligned} \qquad \begin{aligned}(a\cdot b)^m &= a^m\cdot b^m \\ (b^m)^n &= b^{mn}\end{aligned}$

(h) $\left(\dfrac{-a^2}{b^3}\right)^2 = \left(\dfrac{-1\cdot a^2}{b^3}\right)^2 = \dfrac{(-1)^2(a^2)^2}{(b^3)^2} = \dfrac{a^4}{b^6}$

(i) $(x^3y^4)^n = x^{3n}y^{4n}$

(j) $\dfrac{-a^{-2}}{b^{-3}} = \dfrac{-1\cdot a^{-2}}{b^{-3}} = \dfrac{\dfrac{-1}{a^2}}{\dfrac{1}{b^3}} = \dfrac{-1}{a^2}\cdot\dfrac{b^3}{1} = \dfrac{-b^3}{a^2} = -\dfrac{b^3}{a^2}$

Any or all of the rules for simplifying exponential expressions can occur in any problem. The order in which they are applied makes little difference as long as we apply them correctly and follow the order of algebraic operations. The fields of astronomy and chemistry make frequent use of exponents. For example, astronomy deals with very large numbers while chemistry often has to describe very small quantities. To do this, they use a special format called **scientific notation** for writing such numbers.

> **DEFINITION**
>
> **Scientific Notation**
>
> A number is in **scientific notation** when it is in the form
>
> $a \times 10^n$
>
> where $1 \leq a < 10$ and n is an integer.

For example, the distance from the earth to the sun is approximately 93,000,000 miles. In scientific notation we write this as

$9.3 \times 10^7, \qquad 1 \leq 9.3 < 10$

Recall that multiplying a number by 10 moves the decimal point one place to the *right*, so multiplying by 10^7 moves it seven places to the right, since 10^7 is the product of seven 10s.

EXAMPLE 11 Write each of the following numbers in scientific notation.

(a) $927,600,000 = 9.276 \times 10^8$
   $\underbrace{}_{\text{8 places}}$

(b) $26,000 = 2.6 \times 10^4$
   $\underbrace{}_{\text{4 places}}$

(c) $53 \times 10^4 = 5.3 \times 10^1 \times 10^4 = 5.3 \times 10^5$

(d) $675 \times 10^5 = 6.75 \times 10^2 \times 10^5 = 6.5 \times 10^7$

Similarly, we can write very small numbers in scientific notation by using negative exponents. For example, the mass of a hydrogen atom is approximately

0.000000000000000000000000167 gram

$\underbrace{}_{24 \text{ places}}$

In scientific notation, this is written

1.67×10^{-24}

Recall that dividing a number by 10, is the same as multiplying it by 10^{-1}, and this moves the decimal point one place to the *left*. Since

$10^{-24} = \dfrac{1}{10^{24}}$

multiplying by 10^{-24} is the same as dividing by 10 twenty-four times. The decimal point is shifted 24 places to the left.

EXAMPLE 12   Write the following numbers in scientific notation.

(a) $0.000097 = 9.7 \times 10^{-5}$
     5 places

(b) $0.000854 = 8.54 \times 10^{-4}$
     4 places

(c) $0.937 \times 10^{-6} = 9.37 \times 10^{-1} \times 10^{-6} = 9.37 \times 10^{-7}$

(d) $0.0026 \times 10^{5} = 2.6 \times 10^{-3} \times 10^{5} = 2.6 \times 10^{2}$

(e) $53 \times 10^{-2} = 5.3 \times 10^{1} \times 10^{-2} = 5.3 \times 10^{-1}$

> **To Change a Number from Scientific Notation to Decimal Notation**
>
> Move the decimal point the number of places indicated by the power of 10. If the exponent is **positive,** move the decimal point to the **right.** If the exponent is **negative,** move the decimal point to the **left.**

EXAMPLE 13   Write each of the following numbers in decimal notation.

(a) $5.78 \times 10^{6} = 5,780,000$
    $\phantom{5.78 \times 10^{6} = 5,78}$ 6 places

(b) $3.294 \times 10^{9} = 3,294,000,000$
    $\phantom{3.294 \times 10^{9} = 3,294,}$ 9 places

(c) $3.25 \times 10^{-5} = 0.0000325$
    $\phantom{3.25 \times 10^{-5} = 0.000}$ 5 places

(d) $8.238 \times 10^{-7} = 0.0000008238$
    $\phantom{8.238 \times 10^{-7} = 0.00000}$ 7 places

EXAMPLE 14 Simplify $\dfrac{4.5 \times 10^7}{1.5 \times 10^3}$ using Theorem 5.2.

SOLUTION First we divide 4.5 by 1.5 and then apply the rule for division of exponential numbers.

$$\dfrac{4.5 \times 10^7}{1.5 \times 10^3} = 3 \times 10^{7-3}$$
$$= 3 \times 10^4 \quad \text{Scientific notation}$$
$$= 30{,}000 \quad \text{Decimal notation}$$

EXAMPLE 15 Use scientific notation to evaluate the following quantity without the use of a calculator:

$$\dfrac{(360{,}000)(0.00093)}{(0.00009)(310{,}000)}$$

SOLUTION We write each number in scientific notation first.

$$\dfrac{(360{,}000)(0.00093)}{(0.00009)(310{,}000)} = \dfrac{(3.6 \times 10^5)(9.3 \times 10^{-4})}{(9 \times 10^{-5})(3.1 \times 10^5)}$$
$$= \dfrac{(3.6)(9.3) \times 10^{5-4}}{(9)(3.1) \times 10^{-5+5}}$$
$$= \dfrac{(0.4)(3) \times 10^1}{1 \times 10^0}$$
$$= 1.2 \times 10^1 \quad \text{Scientific notation}$$
$$= 12 \quad \text{Decimal notation}$$

Do You Remember?

Can you match these?

___ 1. $a^m \cdot a^n$
___ 2. $(a^m)^n$
___ 3. $\dfrac{a^m}{a^n}$
___ 4. $(a + b)^0$
___ 5. $\left(\dfrac{a}{b}\right)^m$
___ 6. $(a \cdot b)^m$
___ 7. $-2a^{-2}$
___ 8. $(-2a)^2$
___ 9. a^{-n}
___ 10. 0.00034
___ 11. $34{,}000$

a) 3.4×10^{-4}
b) 3.4×10^4
c) $\dfrac{-2}{a^2}$
d) $\dfrac{a^m}{b^m}$
e) $\dfrac{1}{a^n}$
f) $4a^2$
g) a^{m-n}
h) a^{m+n}
i) a^{mn}
j) 1
k) $a^m \cdot b^m$

Answers: 1. h 2. i 3. g 4. j 5. d 6. k 7. c 8. f 9. e 10. a 11. b

Exercise Set 5.1

Evaluate each of the following powers and simplify where possible.

1. 2^3
2. 3^2
3. 10^{-1}
4. 10^{-3}
5. 10^0
6. $(6 \cdot 3)^0$
7. -2^5
8. -10^2
9. $(-10)^2$
10. $(-2)^5$
11. -5^{-1}
12. -6^{-2}
13. $(-3)^{-2}$
14. $(-2)^{-4}$
15. $(0.2)^2$
16. $(0.3)^2$
17. $\left(\dfrac{2}{3}\right)^2$
18. $\left(\dfrac{3^2}{4}\right)^2$
19. $\left(\dfrac{3^2}{5}\right)^{-2}$
20. $\left(\dfrac{-2}{3}\right)^{-3}$
21. $\dfrac{4^5}{4^3}$
22. $\dfrac{10^{-2}}{10^4}$
23. $\dfrac{10^{-3}}{10^{-4}}$
24. $\dfrac{2^{-5}}{2^{-5}}$
25. $(5^2 - 4^2)^{-2} \cdot 3^4$
26. $(6^2 - 5^2)^{-1} \cdot 11^2$
27. $\left(\dfrac{1}{3}\right)^{-2} \cdot 5^{-3} \cdot 10^0$
28. $\left(\dfrac{1}{3}\right)^{-3} \cdot 3^{-2} \cdot 3$
29. $\left(\dfrac{6^{-8} \cdot 5^{-3} \cdot 2^{-4}}{6^{-8} \cdot 5^{-4} \cdot 2^{-5}}\right)^{-1}$
30. $\left(\dfrac{10^{-7} \cdot 10^5 \cdot 10^{-3}}{10^{-5} \cdot 10^4 \cdot 10^{-4}}\right)^0$

Write each of the following numbers in scientific notation.

31. 18,000
32. 3600
33. 0.08
34. 0.00006
35. 43,000,000
36. 1,643,000
37. 0.0000000081
38. 0.0000014

Write each of the following numbers in decimal notation.

39. 8×10^3
40. 6×10^4
41. 1.1×10^{-2}
42. 5.8×10^{-3}
43. 0.21×10^2
44. 0.15×10^4
45. 0.317×10^{-1}
46. 0.018×10^{-2}
47. 7.8×10^3
48. 3.5×10^4
49. 1.92×10^{-4}
50. 2.183×10^{-3}

Simplify each of the following products or quotients. Leave your answers in decimal notation.

51. $(3 \times 10^3)(4 \times 10^{-2})$
52. $(5 \times 10^6)(6 \times 10^{-5})$
53. $(1.1 \times 10^1)(7.0 \times 10^1)$
54. $(1.3 \times 10^4)(5.0 \times 10^{-2})$
55. $(10^8)(10^{-6})(10^3)$
56. $(10^5)(10^{-13})(10^7)$
57. $\dfrac{6.3 \times 10^{-6}}{9.0 \times 10^{-5}}$
58. $\dfrac{8.4 \times 10^{-8}}{1.2 \times 10^{-9}}$
59. $\dfrac{1.56 \times 10^{-17}}{1.30 \times 10^{-16}}$
60. $\dfrac{1.80 \times 10^7}{1.50 \times 10^6}$
61. $\dfrac{75{,}000{,}000}{1{,}500{,}000}$

Simplify each of the following expressions. Your answers should have only positive exponents. Assume that no variable in the denominator is equal to zero.

62. $x^2 \cdot x^3$
63. $a^4 \cdot a^5$
64. $m^7 \cdot m^{-5}$
65. $t^3 \cdot t^{-2}$
66. $(d^2)^3$
67. $(p^{-2})^{-4}$
68. $(b^{-4})^2$
69. $(c^{-5})^3$
70. $\dfrac{y^5}{y^2}$
71. $\dfrac{s^9}{s^5}$
72. $\left(\dfrac{x^2}{x^7}\right)^{-1}$
73. $\left(\dfrac{c^5}{c^9}\right)^3$
74. $(ab)^4$
75. $s^6 \cdot t^{-5} \cdot s^{-4} \cdot t^6$
76. $(2p^4)(3r^{-5})(2p^{-3})(r^{11})$
77. $\dfrac{(3x^{-5})(2y^{-7})(2x^{10})^2(y^{13})}{9x^6 y^7}$
78. $\dfrac{(9a^3)^2(-3a)^{-2}(6a)^{-3}}{(-3a)^3(12a)^{-1}}$
79. $\dfrac{(-2x^3)^3(-3x^2)^2}{-9x^{15}}$
80. $\dfrac{-x^2}{(-x)^2}$
81. $\dfrac{-y^3}{(-y)^3}$
82. $\dfrac{(-2a^2)^3}{(2^2 a)^4}$
83. $\dfrac{(-p^2)^2}{(-p^3)^2}$
84. $\dfrac{(18m^3)^0}{(3m^2)^{-2}}$
85. $\dfrac{8a^{-3}b^{-1}}{6a^2 b^{-4}}$
86. $\dfrac{9s^{-4}t^3}{12s^{-1}t^{-1}}$
87. $\dfrac{2m^6 n^{-2}}{(-2)^4 m^{-3} n^{-4}}$
88. $\left(\dfrac{6ab^{-2}}{3a^{-1}b^2}\right)^{-3}$
89. $\left(\dfrac{2p^{-3}q^{-2}}{(-2)^2 pq^{-1}}\right)^{-2}$
90. $\left[\left(\dfrac{a^{-2}b^3 c}{a^{-3}b^{-2}c^2}\right)^2\right]^{-1}$

Simplify each of the following quantities. Assume that no denominator is equal to zero.

91. $x^n \cdot x^m$
92. $y^n \cdot y^{2n}$
93. $a^{2n+1} \cdot a^{3n-2}$
94. $(c^{n+1})^n$

95. $\dfrac{p^{m+1}}{p^{m-1}}$

96. $\dfrac{u^{n+3}}{u^{2n-1}}$

97. $\dfrac{t^m}{t^n}$

98. $(r^{3n+2})^n$

99. $\dfrac{(a+b)^3}{(a+b)^2}$

100. $\dfrac{(x-y)^5}{(x-y)^3}$

101. $\dfrac{(m-n-p)^{-2}}{(m-n-p)^{-2}}$

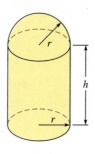

FIGURE TO ACCOMPANY EXERCISE 107

Solve each of the following applied problems.

102. The mass of the earth is 6×10^{24} kg. If 1 kg is 1.1×10^{-3} tons, what is the mass of the earth in tons?

103. The width of a particular asteroid belt is 2.8×10^8 km. The space probe Pioneer 10 passed through this belt at a speed of 1.4×10^5 km/hr. How long did it take Pioneer 10 to pass through the asteroid belt?

104. The nematode sea worm is the most plentiful form of sea life. It is estimated that there are 4.0×10^{25} sea worms in the oceans. There are about 3.16×10^9 cubic mi of ocean and about 1.10×10^{13} gal of water per cubic mile. Assuming that the sea worms are uniformly distributed in all of the earth's oceans, about how many are there per gallon?

105. A light year is the distance light travels in a year's time. One light year is about 9.6×10^{12} km. If a star is 220 light years away, how far is it in kilometers?

106. The amount A of money in a bank after n years is given by the formula $A = p(1 + r)^n$, where p is the original amount invested (the principal) and r is the annual rate of interest. To the nearest dollar, how much money would you have after 10 years if you deposited $1000 at 9% annual interest?

107. The volume of the silo shown in the accompanying figure is given by the formula $V = \pi r^2 h + \tfrac{2}{3}\pi r^3$, where r is the radius of the base and the hemispherical top, and h is the height of the cylinder. The dimensions are $r = 1.5 \times 10^1$ ft and $h = 3.2 \times 10^1$ ft. Use $\pi = 3.1416$ to find the volume V.

108. The volume of the square-based tower shown in the accompanying figure is given by the formula $V = w^2 h + \tfrac{1}{3}w^3$, where w is the length of a side of the base and h is the height of a side of the tower. The dimensions are $w = 1.2 \times 10^1$ m and $h = 6.0 \times 10^1$ m. Find the volume V.

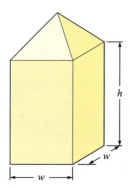

109. Show that

$$\left(\dfrac{x}{y}\right)^{-m} = \left(\dfrac{y}{x}\right)^m,$$ where m is a natural number

110. Show that

$$\dfrac{1}{a^{-m}} = a^m,$$ where m is a natural number

111. Show by numerical example that $(a+b)^2 \ne a^2 + b^2$.

112. Show by numerical example that $(a+b)^3 \ne a^3 + b^3$.

5.2 Adding and Subtracting Polynomials

OBJECTIVES

In this section we will learn to
1. classify polynomials; and
2. add and subtract polynomials.

In the earlier chapters we often referred to "algebraic expressions." A *polynomial* is a special type of algebraic expression. Before defining a polynomial, however, we first need to consider the concept of an *algebraic term*. An **algebraic term** is a number, a variable, or the product of a number and one or more variables, for example,

$$5, \quad 5y, \quad 3x^2, \quad \frac{1}{4}x^5y, \quad 9x^3yz, \quad x^2y, \quad \sqrt{2}x^4.$$

Recall that in $5y$, the number 5 is called the *numerical coefficient* of y. If a term does not have a written numerical coefficient, it is understood to be 1. Thus, $x^2y = 1x^2y$. If a single term or the sum or difference of several terms meets the following conditions, the result is what we call a **polynomial.**

> **DEFINITION**
> **Polynomial**
>
> A **polynomial** in simplified form is a single algebraic term or the sum or difference of algebraic terms satisfying the following conditions:
>
> 1. All exponents must be whole numbers.
> 2. No term may contain a variable in the denominator.
> 3. No variable may appear under a radical sign.

Thus, $13x$, $3x + 5$, $x^3 - 7x + \frac{1}{2}$, and $x^6 + 3x^2 - \frac{1}{2}x + 7$ are all polynomials, since they satisfy the conditions of the definition. The following algebraic expressions are *not* polynomials:

$x^2 + 2x^{-1}$ The exponent -1 is not a whole number.
$\sqrt{x} + 3xy + x^2$ A variable appears under the radical sign.
$6xy^2 + 4y - \dfrac{6}{x^2}$ A variable appears in the denominator.

Polynomials may contain a single term or any number of terms. We have specific names for polynomials with one, two, or three terms.

A **monomial** is a polynomial with one term.

A **binomial** is a polynomial with two terms.

A **trinomial** is a polynomial with three terms.

EXAMPLE 1   Classify each of the following polynomials.

(a) $6x^2$, $3y$, and 7 are monomials.
(b) $-2x + \dfrac{1}{2}$, and $\dfrac{1}{5}x^2y + y^3$ are binomials.
(c) $-6x^3 + 9x - 3$ is a trinomial.
(d) $4x^2 + 9xy + 3x^2$ at first appears to be a trinomial. However, when it is simplified, it becomes $7x^2 + 9xy$, which is a binomial.

CAUTION

Before a polynomial can be classified as a monomial, a binomial, or a trinomial, it must first be simplified.

Polynomials can also be classified by their *degree*. When only one variable is present in a polynomial, the **degree of the polynomial** is the largest exponent that appears on that variable among all the terms. If the polynomial is written in *descending powers* of the variable, we say it is in **standard form.** This means that the term with the largest exponent appears first, the term with the next largest exponent appears next, and so on.

EXAMPLE 2 Write each of the following polynomials in standard form and determine its degree.

(a) $x^5 + 2x^3 + 5$ is a fifth-degree trinomial that is already in standard form. The first term, x^5, is a fifth-degree term; the second term, $2x^3$, is a third-degree term, and the third term, 5, is a *zero-degree term*. (Recall that $5 = 5 \cdot 1 = 5x^0$.)

(b) $\frac{2}{5}x^7 - \frac{1}{8}x^6 + 5x^2 - 8$ is a seventh-degree polynomial in standard form.

(c) 3 is a zero-degree monomial, since $3 = 3x^0$.

(d) $-x^2 + 7x + x^9 + 5$, when written in standard form, is $x^9 - x^2 + 7x + 5$, so it is a ninth-degree polynomial.

(e) $-x + 8$ is a first-degree binomial in standard form.

When a term of a polynomial contains more than one variable, the degree of that term is the sum of the exponents on the variables. Thus, $3x^2y^5z$ is an eighth-degree term, since the sum of the exponents on the variables is $2 + 5 + 1 = 8$.

Polynomials are added and subtracted by combining *like* or *similar terms*. Like or similar terms differ only by their numerical coefficients. Except for order, *their variable parts must be identical*.

EXAMPLE 3 Determine whether the terms are like or unlike.

(a) $3x^2$, $\frac{1}{2}x^2$, and $9x^2$ are like terms, since they all involve only x^2.

(b) $-5xy$, $4yx$, and $-\frac{3}{7}xy$ are like terms, since $xy = yx$ by the commutative property of multiplication.

(c) $4x^2y$ and $5xy^2$ are unlike terms, since x^2y and xy^2 are not identical.

(d) $13x$ and $13y$ are unlike terms, since x is not identical to y.

If we assume that the coefficients of the variables in the polynomial represent real numbers, we can use the properties of real numbers stated in Chapter 1 to add or subtract them. For example,

$$2x^2 + 5x^2 = (2 + 5)x^2 = 7x^2 \qquad \text{Distributive property}$$

EXAMPLE 4 Add $2x^2 + 7x + 9$ to $4x^2 + 5x + 7$.

SOLUTION
$(2x^2 + 7x + 9) + (4x^2 + 5x + 7)$
$= 2x^2 + 7x + 9 + 4x^2 + 5x + 7$ Remove grouping symbols.
$= 2x^2 + 4x^2 + 7x + 5x + 9 + 7$ Commutative property
$= (2x^2 + 4x^2) + (7x + 5x) + (9 + 7)$ Associative property
$= (2 + 4)x^2 + (7 + 5)x + (9 + 7)$ Distributive property
$= 6x^2 + 12x + 16$

EXAMPLE 5 Add $-x^3 + 3x^2$ to $5x^3 + x^2 - x - 7$.

SOLUTION
$$(-x^3 + 3x^2) + (5x^3 + x^2 - x - 7)$$
$$= -x^3 + 3x^2 + 5x^3 + x^2 - x - 7 \quad \text{Remove grouping symbols.}$$
$$= (-x^3 + 5x^3) + (3x^2 + x^2) - x - 7 \quad \text{Commutative and associative properties}$$
$$= (-1 + 5)x^3 + (3 + 1)x^2 - x - 7 \quad \text{Distributive property}$$
$$= 4x^3 + 4x^2 - x - 7$$

EXAMPLE 6 Simplify $(x^2 + xy) + (3x^2 + 2xy + 5y^2) + y^3$.

SOLUTION
$$(x^2 + xy) + (3x^2 + 2xy + 5y^2) + y^3$$
$$= (x^2 + 3x^2) + (xy + 2xy) + 5y^2 + y^3$$
$$= 4x^2 + 3xy + 5y^2 + y^3$$

> **𝒞AUTION**
>
> Avoid the temptation to add terms such as $2x^3$ and $3x^2$ to get $5x^5$. Notice that while x^3 and x^2 do have the same *base,* they have *different exponents.* **They are not like terms.**

Polynomials can be subtracted as well as added. To discover one method for doing this, recall that

$$-(a + b) = -1 \cdot (a + b) = (-1) \cdot a + (-1) \cdot b$$
$$= -a + (-b) = -a - b$$

This shows that when we remove a grouping symbol that is preceded by a negative sign, all signs within the grouping symbol are changed. When removing a grouping symbol that is preceded by a positive sign (or no sign at all), the signs within the grouping symbol remain unchanged.

$$a + (b - c) = a + b - c$$

EXAMPLE 7 Remove the grouping symbols and change the signs accordingly.

SOLUTION
(a) $-(7x^2 + 5x - 3) = -7x^2 - 5x + 3$
(b) $-(-4y^2 + 6xy - 9y) = 4y^2 - 6xy + 9y$

Extending this concept leads to a rule for subtracting polynomials. We carry out subtraction in a manner similar to that used with real numbers.

> **To Subtract One Polynomial from Another**
>
> Change the sign of every term in the polynomial being subtracted and follow the rules for addition.

EXAMPLE 8 Subtract $4x^3 - 3x^2 + 2x$ from $6x^3 + 4x^2 + x - 9$.

SOLUTION Writing this with grouping symbols, we have

$$(6x^3 + 4x^2 + x - 9) - (4x^3 - 3x^2 + 2x)$$

We change the sign of every term in the polynomial being subtracted and then add.

$(6x^3 + 4x^2 + x - 9) - (4x^3 - 3x^2 + 2x)$
$= (6x^3 + 4x^2 + x - 9) + (-4x^3 + 3x^2 - 2x)$
$= [6x^3 + (-4x^3)] + (4x^2 + 3x^2) + [x + (-2x)] + (-9)$ Commutative and associative properties
$= 2x^3 + 7x^2 - x - 9$

EXAMPLE 9   Simplify $(3x^3 + 2x - 4) - (-5x^3 + x^2 - 6x + 9)$.

SOLUTION $(3x^3 + 2x - 4) - (-5x^3 + x^2 - 6x + 9)$
$= (3x^3 + 2x - 4) + [5x^3 + (-x^2) + 6x + (-9)]$ Change signs.
$= (3x^3 + 5x^3) + (-x^2) + (2x + 6x) + [-4 + (-9)]$
$= 8x^3 + (-x^2) + 8x + (-13)$ Combine like terms.
$= 8x^3 - x^2 + 8x - 13$

In practice, many of the steps shown in Examples 8 and 9 can be carried out mentally. Compare the following example with Example 9 to see which steps can usually be eliminated.

EXAMPLE 10   Simplify $(3x^3 + 2x - 4) - (-5x^3 + x^2 - 6x + 9)$.

SOLUTION
$(3x^3 + 2x - 4) - (-5x^3 + x^2 - 6x + 9)$
$= 3x^3 + 2x - 4 + 5x^3 - x^2 + 6x - 9$ Change signs.
$= 8x^3 - x^2 + 8x - 13$ Combine like terms.

EXAMPLE 11   Simplify $-\{x^2 + x - [2x^2 + 2x - (3x^2 + 3x)]\}$.

SOLUTION   We begin by removing the *innermost grouping symbols* first. We must be careful to change all signs when a symbol is preceded by a negative sign.

$-\{x^2 + x - [2x^2 + 2x - (3x^2 + 3x)]\}$
$= -\{x^2 + x - [2x^2 + 2x - 3x^2 - 3x]\}$ Remove parentheses and change signs.
$= -\{x^2 + x - [-x^2 - x]\}$ Combine like terms.
$= -\{x^2 + x + x^2 + x\}$ Remove brackets and change signs.
$= -\{2x^2 + 2x\}$ Combine like terms.
$= -2x^2 - 2x$ Remove braces and change signs.

We can use function notation to represent polynomials in a single variable with real coefficients.

EXAMPLE 12   If $p(x) = x^4 - 3x^2 - 2x - 5$ and $g(x) = 2x^5 - 2x^4 + 6x - 5$, find $(p + g)(x)$.

SOLUTION   Recall that $(p + g)(x) = p(x) + g(x)$. Therefore
$(p + g)(x) = (x^4 - 3x^2 - 2x - 5) + (2x^5 - 2x^4 + 6x - 5)$
$= 2x^5 - x^4 - 3x^2 + 4x - 10$

EXAMPLE 13 Let $f(x) = x^{2m} - 4x^m - x$ and $g(x) = 2x^{4m} - 3x^{2m} - 7x^m - 9$, where m represents a natural number. Find $f(x) - g(x)$.

SOLUTION
$$f(x) - g(x) = (x^{2m} - 4x^m - x) - (2x^{4m} - 3x^{2m} - 7x^m - 9)$$
$$= x^{2m} - 4x^m - x - 2x^{4m} + 3x^{2m} + 7x^m + 9$$
$$= -2x^{4m} + 4x^{2m} + 3x^m - x + 9$$

EXAMPLE 14 A potter makes and sells x vases at a cost of C dollars, expressed in function notation as $C(x) = 7x + 50$. Her profit function, in dollars, is $P(x) = x^2 + 11x - 5$.

(a) Find a polynomial to express her revenue, $R(x)$, if the revenue is equal to the cost plus the profit.

(b) Find the revenue in dollars when $x = 25$ vases.

SOLUTION (a) The cost, profit, and revenue are all dependent on x, the number of items produced and sold. In other words, they are all functions of x. In function notation,

$$R(x) = C(x) + P(x)$$
$$= (7x + 50) + (x^2 + 11x - 5) \quad C(x) = 7x + 50, \text{ and}$$
$$= x^2 + 18x + 45 \quad\quad\quad\quad\quad\quad P(x) = x^2 + 11x - 5$$

The revenue function is $R(x) = x^2 + 18x + 45$.

(b) When $x = 25$,
$$R(x) = x^2 + 18x + 45$$
$$R(25) = (25)^2 + 18(25) + 45$$
$$= 625 + 450 + 45$$
$$= \$1120$$

Her revenue is $1120.

Do You Remember?

Can you match these?

More than one answer may apply, and some answers may be used more than once.

_____ 1. Trinomial

_____ 2. Numerical coefficient of $4xy$

_____ 3. Like terms

_____ 4. Seventh-degree polynomial

_____ 5. Binomial

_____ 6. Zero-degree polynomial

_____ 7. Fifth-degree polynomial

_____ 8. Not a polynomial

a) $5x^2 + \dfrac{3}{x} + 5$

b) $5x^3 + 2$

c) 4

d) $4x^2y + 9x^2y^2z + 2z^4 + 2$

e) $2x^7 + 6x^2 + 4$

f) $4x^5 - 3x - 9$

g) $4x^2$ and $3x^3$

h) $7xy^3$ and $-3xy^3$

Answers: 1. e, f 2. c 3. h 4. e 5. b 6. c 7. d, f 8. a

EXERCISE SET 5.2

Determine whether each of the following is a polynomial.

1. $x + 2$
2. $-3y + 4$
3. $\dfrac{1}{x}$
4. $\dfrac{3}{m} + 2$
5. $t^6 - 5t + 1$
6. $m^7 + 3m^2 - 5$
7. $a^{-3} + 5$
8. $c^{-4} + c^2 + 1$
9. $\dfrac{2s^3}{3} + \dfrac{4s}{3} - 2$
10. $\dfrac{-5r^2}{4} + \dfrac{7r}{3} - \dfrac{1}{5}$
11. $3^x + 5$
12. $8\sqrt{x} - 11$

For each of the following polynomials: (a) write it in standard form and simplify; (b) determine its degree; (c) determine the coefficient of the highest-degree term; and (d) state if it is a monomial, a binomial, or a trinomial.

13. $5p^2 + p$
14. $t^2 - 7t$
15. $19r + 4r$
16. $-16m^3 + 7m^3$
17. $5q^0 - 9q^3 - q^5$
18. $b^0 - 8b^4 - 11b^5$
19. $x^2 - 5x + 2x^3 - 4x + 4x^2$
20. $3a - a^2 + 4a - a^3 + 2a^2 + 4a^3$

Perform the indicated operations on the given polynomials.

21. $6a + a$
22. $8x + 3x$
23. $9y - 6y$
24. $p^2 + 4p - 2p^2 + 6p$
25. $3m - 7m^2 - 6m + m^2$
26. $8m^2n + 3m^2n$
27. $9ab^2 + 3ab^2$
28. $\dfrac{1}{3}s - \dfrac{1}{2}s$
29. $\dfrac{1}{4}t - \dfrac{1}{2}t$
30. $(3a + 4) + (2a - 5)$
31. $(6x^2 + x) + (2x^2 - x)$
32. $(y^2 - 1) + (2y^2 + 1)$
33. $(7b + 4) + (-8b - 4)$
34. $(3m - 2) - (m + 7)$
35. $(2t - 4) - (5t - 7)$
36. $(5y + 1) - (y^2 + 3y - 5)$
37. $b - (b^2 - 2b + 1)$
38. $(5x^2 + 2x - 5) + (x^2 - 3x + 4) - (3x^2 - 7x + 1)$
39. $(3a + 5) - (a^2 + a + 1) + (a^2 + a + 1)$
40. $(m + n - 1) - (2m + n - 7) - (5m - 7n + 3)$
41. $(7k - 5) - (8k + 5) + (3k + 4) - (6k + 11)$
42. $(9x - 3) - (2x + 7) - (8x + 9) + (x - 11)$
43. $(x^2y - 2xy^2 - 3) - (xy^2 + 3x^2y + 1)$
44. $(5s^2t - 9st^2 + 4st) - (8st^2 + s^2t + 6st)$

45. Subtract $8r^2 - 7r + 3$ from $2r^2 + 9r - 2$.
46. Subtract $5s^2 - 6st + 4t^2 + 7t - 2$ from $5s^2 - 8st + 6s - 9t + 4$.
47. Subtract $x^3y^2 - 5xy^2 + 6x^2y - 11xy$ from $xy^2 - 8x^2y + x^3y^2 + 7xy$.
48. Subtract $9a^2b - 13ab^3 + 7ab$ from $16a^3b - 18ab^2 + 8ab$.

Simplify each polynomial by removing all grouping symbols and performing the indicated operations.

49. $x + 3y - [3x - (4x + y)]$
50. $5a + (2b - a) - [(2b - 4) - a]$
51. $8 - \{2x - [3y + (2y - 5x)]\}$
52. $4a + 3 - [6b - (3 + 5a - 2)]$
53. $5m - [m - (2m + 1)]$
54. $8 - [(1 + 8p) - (5 - 2p)]$
55. $7(3s - 2t) - [s - t - (s - t)]$
56. $-(k + 3) - 4[3k - 2(4 + 5k) - 8]$
57. $3[6r - (r + s)] + 3[-(2r + 4s) + 5(r - s)]$
58. $11x^2 - [6 - (4x - x^2)] - 2[3 - 3(4x - x^2)]$
59. $-3(-x^2 + 2xy - 3y^2) - (-6x^2 + 4xy + 2y^2) + [3 - (5xy + y^2)]$
60. $7(-a^2 - 5ab + 2b^2) - (-3a^2 - 7ab + 6b^2) - 2[1 - 4(ab + b^2)]$

Add or subtract the following polynomial functions as indicated. Assume all exponents are natural numbers.

61. Given $p(x) = 2x^2 + 5x - 8$ and $g(x) = 3x^2 - 4x + 9$, find $(p + g)(x)$.
62. Given $p(x) = x^2 + 7x + 5$ and $g(x) = 7x^2 - x + 2$, find $(p + g)(x)$.
63. Given $p(x) = 4x^{3m} + x^{2m} + 9x + 4$ and $g(x) = x^{2m} - 9$, find $(p + g)(x)$.
64. Given $p(x) = 3x^{3a} - x^{2a} + 7x^a + 6x - 11$ and $g(x) = x^{3a} - x^{2a} + 7x^a + 6x - 6$, find $(p - g)(x)$.
65. Given $p(x) = 11x^{3a} - 19x$ and $g(x) = 8x^{3a} + 7x^a + 4$, find $(p - g)(x)$.
66. Given $p(x) = 0.3x^{3m} + 0.6x^{2m} - 8.1x^m$ and $g(x) = 0.7x^{3m} - 0.4x^{2m} + 5.3x - 7$, find $p(x) + g(x)$.
67. Given the expressions $f(x) = 3x^{5b} - 7x^{2b} - 4$ and $g(x) = 2x^{5b} - 6x^{4b} - 3x^{3b} + 6x$, find $f(x) - g(x)$.
68. Given $f(x) = 13x^{6b} - 5x^{4b} + 2x - 15$ and $g(x) = 8x^{5b} + 2x^{3b} + 6x$, find $f(x) - g(x)$.

Use a calculator, if needed, to perform the indicated operations.

69. $(-6.85x^2 + 3.241x - 9.173)$
    $- (-0.79x^2 + 2.099x + 8.676)$

70. $(0.001y^3 - 17.842y^2 + 1.111y)$
    $- (-1.899y^3 - 16.001y^2 - 0.018y)$

71. Evaluate $0.02p^2 + 0.3p - 0.5$ for $p = 0.2$.

72. Evaluate $2a^3 - 7a^2 + 8a - 10$ for $a = 3.1$.

Solve each of the following applied problems.

73. A company's cost in dollars for producing and marketing a certain product is given by $C(x) = 50 + 30x$, where x is the number of units produced and marketed. Its revenue in dollars from this product is given by the function $R(x) = 90x$. Find the polynomial representing profit $P(x) = R(x) - C(x)$. If the company produces and markets 50 units/day, what is its daily profit?

74. Togs, Inc., manufactures and sells x pairs of custom-fitted tennis shoes for a profit, in dollars, of $P(x) = x^2 + 31x - 200$. The total revenue, in dollars, for x pairs is $R(x) = x^2 + 49x + 600$. If the cost to manufacture and sell x pairs is represented by $C(x) = R(x) - P(x)$, find the polynomial $C(x)$. Find the cost and profit for manufacturing and selling 100 pairs.

75. Mitsie makes and sells silk flower arrangements. Her cost per week is $C(x) = 20 + 14x$, where x is the number of arrangements she makes and sells. Her revenue is $R(x) = 20x + x^2$. Find the polynomial function that represents her profit per week. Find the cost, revenue, and profit for making and selling 10 arrangements.

76. Lewis makes and sells inline skates. His profit is given by $P(x) = x^2 + 23x - 200$. The total revenue for x pairs of skates is given by $R(x) = x^2 + 32x + 400$, in dollars. Find the cost function and his cost and profit for making and selling 10 pairs of skates.

REVIEW EXERCISES

77. Solve by substitution:
    $5x - y = 2$
    $3x + 3y = 5$

78. Solve by the elimination method:
    $x - 3y = 8$
    $3x - 2y = 5$

79. Graph the region that is the solution to
    $\{(x, y) \mid y - x \leq 3\} \cap \{(x, y) \mid x + y < 5\}$

80. Graph the region that is the solution to
    $\{(x, y) \mid x - y > -6\} \cap \{(x, y) \mid -x - y > 2\}$

Simplify each of the following.

81. $\dfrac{(3x^{-4}y^{-3})^{-1}}{(-9)^{-1}(x^3y^{-2})}$

82. $\dfrac{[-(10)^4(10^{-2})^2(3^{-3})]^2}{[(-10)^2(-10)^3(9^2)]^0}$

83. $\dfrac{(x^{5n-4}y^{2n+3})^{-2}}{(x^{5n-5}y^{2n-5})^{-2}}$

84. $\dfrac{(x^{-2n}y^{3n})^{-2}(x^{4n}y^{-2n})^{-1}}{x^{-n}y^{-2n}}$

5.3 Multiplying Polynomials

HISTORICAL COMMENT Gottfried Wilhelm Leibnez introduced the dot (·) as a symbol for multiplication in place of the cross (×), arguing that × was too easily confused with the unknown quantity x. He experimented with many other symbols, such as ⌢, the comma itself, the semicolon, and the asterisk (∗). Of these, only ×, ·, and ∗ have survived to be used today.

OBJECTIVES In this section we will learn to
1. multiply polynomials by various means; and
2. square a binomial.

Consider the product of $3x^2y$ and $2xy^5$. To find it, we apply the commutative and associative properties together with the rules for exponents.

$$3x^2y \cdot 2xy^5 = 3 \cdot 2 \cdot x^2 \cdot x \cdot y \cdot y^5 \quad \text{Commutative property}$$
$$= (3 \cdot 2)(x^2 \cdot x)(y \cdot y^5) \quad \text{Associative property}$$
$$= 6x^3y^6 \quad b^m \cdot b^n = b^{m+n}$$

EXAMPLE 1 Multiply.

(a) $(2rs^2)(8r^7s^9) = (2 \cdot 8)(r \cdot r^7)(s^2 \cdot s^9) = 16r^8s^{11}$
(b) $(-4x^2y^5)(7x^7y^8) = [(-4) \cdot 7](x^2 \cdot x^7)(y^5 \cdot y^8) = -28x^9y^{13}$
(c) $(-4mn)(-3m^2n)(2m^2n^6) = [(-4)(-3)(2)](m \cdot m^2 \cdot m^2)(n \cdot n \cdot n^6)$
$= 24m^5n^8$

The product of two polynomials, other than two monomials, is found by using the distributive property:

$$a(b + c + d + e + \cdots + z)$$
$$= a \cdot b + a \cdot c + a \cdot d + a \cdot e + \cdots + a \cdot z$$

EXAMPLE 2 Find the product of $2x^2$ and $x^3 + 3x^2 + 6$.

SOLUTION $2x^2(x^3 + 3x^2 + 6) = 2x^2 \cdot x^3 + 2x^2 \cdot 3x^2 + 2x^2 \cdot 6$
$= 2x^5 + 6x^4 + 12x^2$

EXAMPLE 3 Find the product of $-3x^5$ and $x^2 - 2xy - 9$.

SOLUTION In order to understand how we find the signs of the product, remember that subtraction is defined as addition of the additive inverse.

$$-3x^5(x^2 - 2xy - 9) = -3x^5[x^2 + (-2xy) + (-9)] \quad \text{Definition of subtraction}$$
$$= -3x^5(x^2) + (-3x^5)(-2xy) \quad \text{Distributive property}$$
$$+ (-3x^5)(-9)$$
$$= -3x^7 + 6x^6y + 27x^5$$

The first two steps in finding the product in Example 3 are usually completed mentally. With this approach, we can separate all the products by plus signs.

EXAMPLE 4 Find the product of $-2c^3$ and $c^2 + 5c - 8$ without rewriting the subtraction as addition.

SOLUTION $-2c^3(c^2 + 5c - 8) = -2c^5 + (-10c^4) + 16c^3$

The product of $-2c^3$ and c^2 The product of $-2c^3$ and $5c$ The product of $-2c^3$ and -8

We can then rewrite the last step in the product as $-2c^5 - 10c^4 + 16c^3$.

The product of two binomials is found by using the distributive property twice. To see how we do this, consider the product of $2x + 5$ and $x^2 + 6$. Remember that $a(b + c) = a \cdot b + a \cdot c$:

$$\overset{a}{(2x} + \overset{b}{5})\overset{c}{(x^2 + 6)} = \overset{a}{(2x + 5)} \cdot \overset{b}{x^2} + \overset{a}{(2x + 5)} \cdot \overset{c}{6}$$

Distribute $2x + 5$ over x^2 and 6.

$$= 2x \cdot x^2 + 5 \cdot x^2 + 2x \cdot 6 + 5 \cdot 6$$
$$= 2x^3 + 5x^2 + 12x + 30$$

Distribute x^2 and 6 over $2x + 5$.

Notice in the last line that the product, except for order, is the result of multiplying each term of the second polynomial by each term of the first polynomial. This observation leads us to the following rule.

> **To Find the Product of Two Polynomials**
>
> Multiply each term in the second polynomial by each term in the first polynomial and combine like terms.

EXAMPLE 5 Multiply $2x + 3$ times $x^2 - 6x - 9$.
We multiply each term in $x^2 - 6x - 9$ by each term in $2x + 3$:

$$(2x + 3)(x^2 - 6x - 9) = 2x(x^2) + 2x(-6x) + 2x(-9) + 3(x^2) + 3(-6x) + 3(-9)$$
$$= 2x^3 - 12x^2 - 18x + 3x^2 - 18x - 27$$
$$= 2x^3 - 9x^2 - 36x - 27 \qquad \text{Combine like terms.}$$

One way to find the product of two binomials is to observe a certain pattern. This pattern, called **FOIL,** is illustrated in the following diagram.

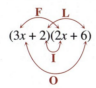

F is the product of the *first* terms.
O is the product of the *outer* terms.
I is the product of the *inner* terms.
L is the product of the *last* terms.

$$\textbf{FOIL} = \textbf{F} + \textbf{O} + \textbf{I} + \textbf{L} = 6x^2 + 18x + 4x + 12 = 6x^2 + 22x + 12$$

EXAMPLE 6 Find each of the following products by the FOIL method.

$$\overset{\textbf{F}}{\downarrow} + \overset{\textbf{O}}{\downarrow} + \overset{\textbf{I}}{\downarrow} + \overset{\textbf{L}}{\downarrow}$$

(a) $(x + 2)(x + 3) = x^2 + 3x + 2x + 6 = x^2 + 5x + 6$
(b) $(2x - 1)(x + 6) = 2x^2 + 12x - x - 6 = 2x^2 + 11x - 6$
(c) $(3x - 2)(x - 4) = 3x^2 - 12x - 2x + 8 = 3x^2 - 14x + 8$

Two types of products occur so frequently that the pattern of multiplication should be memorized, symbolically as well as in words. The first of these is the product of the sum and difference of the same two quantities. Consider

$$\underset{\text{Sum}}{(a+b)} \cdot \underset{\text{Difference}}{(a-b)}$$

When we find the product using the FOIL method, we get

$$(a+b)(a-b) = \overset{F}{a^2} \overset{O}{- ab} \overset{I}{+ ab} \overset{L}{- b^2}$$
$$= a^2 - b^2 \quad \text{Combine like terms.}$$

We call this product **the difference of the squares of the two quantities** or **the difference of two squares,** for short. This observation leads to the following rule.

> The product of the sum and the difference of the same two quantities is the difference of their squares:
>
> $(a+b)(a-b) = a^2 - b^2$

EXAMPLE 7 Multiply the given pairs of binomials.

(a) $(x+5)(x-5) = x^2 - 5^2 = x^2 - 25$

    The product of the sum and difference is the difference of their squares.

(b) $(2a+3)(2a-3) = (2a)^2 - 3^2 = 4a^2 - 9$
(c) $(y^2+7)(y^2-7) = (y^2)^2 - 7^2 = y^4 - 49$
(d) $(y^n - x^t)(y^n + x^t) = (y^n)^2 - (x^t)^2 = y^{2n} - x^{2t}$

When two binomials are identical, such as $a+b$ and $a+b$, or $a-b$ and $a-b$, another definite pattern emerges.

$$(a+b)(a+b) = (a+b)^2 = \overset{F}{a^2} \overset{O}{+ ab} \overset{I}{+ ab} \overset{L}{+ b^2} = a^2 + 2ab + b^2$$
$$(a-b)(a-b) = [a+(-b)]^2 = a^2 - ab - ab + b^2 = a^2 - 2ab + b^2$$

First term   Last term   The square of the first term   Twice the product of the terms   The square of the last term

We summarize the rule that describes these products as follows.

AUTION

Be careful not to make the following common error!

RIGHT
$(a + b)^2 = a^2 + 2ab + b^2$

WRONG
$(a + b)^2 = a^2 + b^2$

In general, $(a + b)^n \neq a^n + b^n$ unless $n = 1$, assuming $a \neq 0$ and $b \neq 0$.

> The **square of a binomial** is equal to the square of the first term plus twice the product of the first and the last terms plus the square of the last term.
>
> $(a + b)^2 = a^2 + 2ab + b^2$
>
> $(a - b)^2 = a^2 - 2ab + b^2$

To emphasize this caution, consider $(2 + 3)^2$:

$(2 + 3)^2 \neq 2^2 + 3^2$

$5^2 \neq 4 + 9$

$25 \neq 13$

EXAMPLE 8 Find the following products.

(a) $(x + 3)^2 = x^2 + 2(3 \cdot x) + 9 = x^2 + 6x + 9$

First term — Last term — Square of the first term — Twice the product of the terms — Square of the last term

(b) $(2x - 1)^2 = (2x)^2 + 2(2x)(-1) + (-1)^2 = 4x^2 - 4x + 1$

(c) $(4x - 3)^2 = (4x)^2 + 2(4x)(-3) + (-3^2) = 16x^2 - 24x + 9$

EXAMPLE 9 Find $(x^n + y^{n+1})^2$.

SOLUTION

$(x^n + y^{n+1})^2 = (x^n)^2 + 2(x^n \cdot y^{n+1}) + (y^{n+1})^2$

$= x^{2n} + 2x^n y^{n+1} + y^{2(n+1)}$ $(b^m)^n = b^{mn}$

$= x^{2n} + 2x^n y^{n+1} + y^{2n+2}$ $(b^m)^n = b^{mn}$

EXAMPLE 10 Multiply $[(x + y) + 6][(x + y) - 6]$.

SOLUTION

$[(x + y) + 6][(x + y) - 6] = (x + y)^2 - 6^2$ $(a - b)(a + b) = a^2 - b^2$

$= x^2 + 2xy + y^2 - 36$ $(a + b)^2 = a^2 + 2ab + b^2$

Recall from Chapter 3 that the **product of two functions,** $f(x)$ and $g(x)$, is indicated by $(f \cdot g)(x)$. If $f(x)$ and $g(x)$ are both polynomials, then $(f \cdot g)(x) = f(x) \cdot g(x)$ will also be a polynomial.

EXAMPLE 11 If $f(x) = x - 2$ and $g(x) = x + 3$, find $(f \cdot g)(x)$ and $(f \cdot g)(5)$.

SOLUTION To find $(f \cdot g)(x)$, we multiply $f(x)$ by $g(x)$:

$(f \cdot g)(x) = f(x) \cdot g(x) = (x - 2)(x + 3)$

$= x^2 + x - 6$

To find $(f \cdot g)(5)$, we replace each x in $(f \cdot g)(x)$ with 5:

$(f \cdot g)(x) = x^2 + x - 6$
$(f \cdot g)(5) = 5^2 + 5 - 6$ Replace each x with 5.
$ = 25 + 5 - 6$
$ = 24$

EXAMPLE 12 Let $f(x) = x^2 + x - 2$ and $g(x) = x - 4$. Show that $(f \cdot g)(7) = f(7) \cdot g(7)$.

SOLUTION

$(f \cdot g)(x) = f(x) \cdot g(x) = (x^2 + x - 2)(x - 4)$
$ = x^3 - 4x^2 + x^2 - 4x - 2x + 8$
$ = x^3 - 3x^2 - 6x + 8$

Therefore

$(f \cdot g)(7) = 7^3 - 3(7^2) - 6(7) + 8$ Replace each x in $(f \cdot g)(x)$ with 7.
$ = 343 - 147 - 42 + 8$
$ = 162$

Also, replacing each x in $f(x)$ and $g(x)$ with 7, we get

$f(7) = 7^2 + 7 - 2 = 54$
$g(7) = 7 - 4 = 3$
$f(7) \cdot g(7) = 54 \cdot 3 = 162$

Thus $(f \cdot g)(7) = f(7) \cdot g(7)$.

Exploring Graphics

A scientific calculator or graphics calculator can be very useful for evaluating polynomials for certain values of x. For example, to evaluate $7^3 - 3(7^2) - 6(7) + 8$, which is $f(7)$ from Example 12, we follow these keystrokes:

TI 81 GRAPHICS CALCULATOR

7 [^] 3 [−] 3 [×] 7 [^] 2 [−] 6 [×] 7 [+] 8 [Enter]

The result is 162.

SCIENTIFIC CALCULATOR

The result is 162.

We can expand the rules for multiplying polynomials to more complex products. The next three examples illustrate this.

EXAMPLE 13 Multiply $(x + 1)(x + 2)(x - 2)$.

SOLUTION Since the last two factors are the sum and difference of the same two quantities, we find their product first to make it easier to complete the intermediate steps:

$$(x + 1)(x + 2)(x - 2) = (x + 1)(x^2 - 4) \qquad (x + 2)(x - 2) = x^2 - 4$$
$$= x^3 + x^2 - 4x - 4$$

The same product results when we multiply $x + 1$ and $x + 2$ first.

$$(x + 1)(x + 2)(x - 2) = (x^2 + 3x + 2)(x - 2)$$
$$= x^3 - 2x^2 + 3x^2 - 6x + 2x - 4 \qquad \text{Combine like terms.}$$
$$= x^3 + x^2 - 4x - 4$$

EXAMPLE 14 Multiply $(x - 2)(x + 4)(x - 5)$.

SOLUTION
$$(x - 2)(x + 4)(x - 5) = (x^2 + 2x - 8)(x - 5) \qquad (x - 2)(x + 4) = x^2 + 2x - 8$$

Now we multiply each term in the second polynomial by each term in the first polynomial.

$$(x^2 + 2x - 8)(x - 5) = x^3 - 5x^2 + 2x^2 - 10x - 8x + 40$$
$$= x^3 - 3x^2 - 18x + 40 \qquad \text{Combine like terms.}$$

EXAMPLE 15 Multiply $[(2x + 3) + y][(2x + 3) - y]$.

SOLUTION If we let $2x + 3 = a$ and $y = b$, the multiplication takes the form $(a + b)(a - b)$.

$$[(2x + 3) + y][(2x + 3) - y] = (2x + 3)^2 - y^2 \qquad (a + b)(a - b) = a^2 - b^2$$
$$= (4x^2 + 12x + 9) - y^2 \qquad (a + b)^2 = a^2 + 2ab + b^2$$
$$= 4x^2 + 12x + 9 - y^2$$

EXAMPLE 16 Multiply $(3x^2 + 2y + z)(x^2 + 3y - 2z)$.

SOLUTION We multiply each term in the second polynomial by each term in the first polynomial:

$$(3x^2 + 2y + z)(x^2 + 3y - 2z)$$
$$= 3x^4 + 9x^2y - 6x^2z + 2x^2y + 6y^2 - 4yz + x^2z + 3yz - 2z^2$$
$$= 3x^4 + 11x^2y - 5x^2z + 6y^2 - yz - 2z^2 \qquad \text{Combine like terms.}$$

EXAMPLE 17 Multiply and simplify:

$$-x(3x - 4)^2 + 2x(6 - 3x)^2 - (5x^3 - 8)$$
$$= -x(9x^2 - 24x + 16) + 2x(36 - 36x + 9x^2) - (5x^3 - 8)$$
$$\qquad\qquad\qquad\qquad\qquad\qquad\qquad (a - b)^2 = a^2 - 2ab + b^2$$

$$= -9x^3 + 24x^2 - 16x + 72x - 72x^2 + 18x^3 - 5x^3 + 8 \quad \text{Distributive property}$$

$$= 4x^3 - 48x^2 + 56x + 8$$

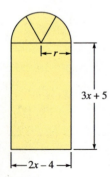

FIGURE 5.3

EXAMPLE 18 A Norman window has the form of a rectangle surmounted by a semicircle, as shown in Figure 5.3.

(a) Find the area of the window in terms of x.
(b) Find the area if $x = 18$ inches.

SOLUTION (a) The bottom portion of the window is a rectangle, so its area is

$$A_R = l \cdot w$$

Since the area of a circle is πr^2 and the top portion is a semicircle, its area is

$$A_S = \frac{1}{2} \pi r^2$$

Consulting Figure 5.3, the area of the rectangle is

$$A_R = (2x - 4)(3x + 5) = 6x^2 - 2x - 20$$

The radius of the semicircle is one-half the base of the rectangle, or $\frac{1}{2}(2x - 4) = x - 2$. The area of the semicircle is therefore

$$A_S = \frac{1}{2} \pi (x - 2)^2 = \frac{1}{2} \pi (x^2 - 4x + 4)$$

The area of the window, A_W, is the sum of the two areas A_R and A_S:

$$A_W = (6x^2 - 2x - 20) + \frac{1}{2} \pi (x^2 - 4x + 4)$$

$$= 6x^2 - 2x - 20 + \frac{\pi x^2}{2} - 2\pi x + 2\pi$$

Using the distributive property, this becomes

$$A_W = \left(6 + \frac{\pi}{2}\right) x^2 - (2 + 2\pi) x + (2\pi - 20)$$

(b) To find the area of the window when $x = 18$ inches, we substitute 18 for x and 3.14 for π:

$$A_W = \left(6 + \frac{3.14}{2}\right)(18)^2 - [2 + 2(3.14)](18) + [2(3.14) - 20]$$

$$\approx 2289.92 \text{ sq. inches}$$

Exercise Set 5.3

Multiply each of the following polynomials and simplify where possible.

1. $(2x^2)(3x^3)$
2. $(-4a^3)(2a^4)$
3. $(-7m)(-8m^5)$
4. $(-x^2y^3)^2$
5. $(-p^3q^4)^3$
6. $3^2x(-xy)^3$
7. $(-t^3)^2(t)$
8. $(-2^2xy^2)^2(3x^2y)^2$
9. $-7^2(a^2)^3(-a)^3$
10. $3x(x^2 + 2)$
11. $2y^2(y - 3)$
12. $-6a(3a^2 + 7a - 9)$
13. $15t^2(-t^2 - 2t - 3)$
14. $6m(m + 7)$
15. $-3s(s^3 - 9)$
16. $-2xy(x^2 + 3 - y^2)$
17. $a^2b(a^2 + 7ab - b^2)$
18. $5u^3v(u^2 + uv - v^3)$
19. $x^3(3x^2 + 7x - 9) - 6x(x^4 - 2x^3 + 8x^2)$
20. $(4k)^2(k^3 + k^2 - k + 1) - (k)^2(3k^3 + 2k^2 - 5k + 6)$
21. $a^n(a^{2n} + 3a^n - 4)$
22. $t^{2n}(3t^{2n} - 4t^n - 6)$
23. $(u + v)(u^2 + uv + v^2)$
24. $(2m + n)(m^2 - mn + n^2)$
25. $(3x^2 - y^2)(2x^3 + x^2y + y^3)$
26. $(2s^2 + t^2)(5s^2 + 6st - t^2)$
27. $(x^n - y^n)(x^{2n} + x^ny^n + 2y^{2n})$
28. $(x^{2n} + y^{2n})(x^n + xy + y^n)$

Find each of the following special types of products.

29. $(x + 1)^2$
30. $(a + 3)^2$
31. $(2p - 3)(2p - 3)$
32. $(3m - 4)(3m - 4)$
33. $(r + s)(r - s)$
34. $(p^2 + q^2)(p^2 - q^2)$
35. $(3k + 5)(3k - 5)$
36. $(9t - 4)(9t + 4)$
37. $(2h^2 - 9)(2h^2 - 9)$
38. $(3u^2 + 5)(3u^2 + 5)$
39. $(4y^2 - 3)(4y^2 + 3)$
40. $(5m^4 - n^2)(5m^4 + n^2)$
41. $(x + 2)^3 = (x + 2)^2(x + 2)$
42. $(x - 1)^3 = (x - 1)^2(x - 1)$
43. $(x - 2)^3$
44. $(x + 3)^3$
45. $[(a + b) - 4][(a + b) + 4]$
46. $[(m - n) - 3][(m - n) + 3]$
47. $[(5p + q) - 5]^2$
48. $[(3x - y) + 2]^2$

In Exercises 49–54, work as indicated and simplify if possible.

49. If $f(x) = x + 3$ and $g(x) = x - 1$, find $(f \cdot g)(x)$ and $(f \cdot g)(-2)$.
50. If $f(a) = a + 5$ and $g(a) = a - 2$, find $(f \cdot g)(a)$ and $(f \cdot g)(4)$.
51. If $f(k) = k + 4$ and $g(k) = k + 7$, find $(f \cdot g)(k)$.
52. If $f(m) = 3m + 2$ and $g(m) = 5m - 6$, find $(f \cdot g)(m)$.
53. Given $f(t) = 6t + 7$ and $g(t) = t + 1$, show that $(f \cdot g)(2) = f(2) \cdot g(2)$.
54. If $f(m) = 9m + 4$ and $g(m) = m + 8$, show that
$$(f \cdot g)\left(-\frac{1}{2}\right) = f\left(-\frac{1}{2}\right) \cdot g\left(-\frac{1}{2}\right)$$

Multiply each of the following polynomials and simplify.

55. $(7y - 5)(4y + 8)$
56. $(9a + 4b)(2a - 7b)$
57. $(7r - 5s)(8r + 6s)$
58. $(x^2 - y)(x^2 + 2y)$
59. $(7x^2 - y^2)(3x^2 + 5y^2)$
60. $(m^3 - n^2)(2m^3 + 11n^2)$
61. $(6m^2 + 7n)(11m^2 - 9n)$
62. $(2x + 3y)(x^2 - 5x + 6)$
63. $(5p - 3)(2p^2 - 7p + 2)$
64. $(13u - 5)(3u^2 - u + 7)$
65. $(8a - 5b)(3a^2 - 7ab + 2b^2)$
66. $(6s - 7t)(s^2 - 13st + t^2)$
67. $(q^2 - r^2)(q^3 - 3q^2r + 3qr^2 + r^3)$
68. $(m^2 + n^2)(m^3 + 3m^2n - 3mn^2 + n^3)$
69. $(a^n + b^n)(a^{2n} - a^nb^n + b^{2n})$
70. $(x^m - y^m)(x^{2m} + x^my^m + y^{2m})$
71. $(a^2 - 3a + 5)(2a^2 + a - 2)$
72. $(x^2 - 2xy + y^2)(x^2 + 2xy + y^2)$
73. $(3x^2 + 4xy + y^2)(2x^2 - 7xy - 5y^2)$
74. $(8p^2 - 5pq + 2q^2)(3p^2 + pq - 8q^2)$
75. $(x - 3)^2 - (x + 3)^2$
76. $(5x - 2)^2 - (5x + 2)^2$
77. $(3x^n - y^n)^2 - (3x^n + y^n)^2$
78. $(r^n - 2s^n)^2 - (r^n + 2s^n)^2$
79. $(2u + v)(u - v)(2u + 3v)$
80. $(3a - b)(a + b)(3a + 4b)$
81. $(p^{2m+5} - q^{m+3})(p^m - 2p^{m+1}q^{m+1} + q^3)$
82. $(t^{5a+4} + w^{3a+2})(t^{2a+4} + 4tw - 8w^{2a-1})$

For Extra Thought

A flower garden is divided into five areas, A, B, C, D, and E, with the dimensions given in the accompanying diagram. Find each of the areas in terms of x and simplify. A small box in a corner indicates a right angle (90°). Area E is one-fourth of a circle.

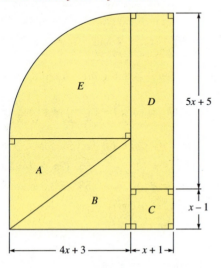

83. $A + B$

84. $D + C$

85. $A + D + E$

86. $A + E - C$

87. $2A + D + C + E$

88. $A + B + C + D$

🖩 *Use a calculator and round to two decimal places.*

89. If $x = 1.6$ yd, find the area in Exercise 83.

90. If $x = 2.89$ m, find the area in Exercise 84.

REVIEW EXERCISES

91. $[(x)^{-4}]^2$

92. $x^5 \cdot y^7 \cdot x^{-3} \cdot y^{-6}$

93. $\left(\dfrac{x^5}{x^3}\right)^{-2}$

94. $\dfrac{-y^2}{(-y)^2}$

95. $\dfrac{-3y^2}{(-3y)^2}$

96. $(6a^2 - 7) - (5a + 3)$

97. $\dfrac{1}{4}a - \dfrac{1}{2}a$

98. $7 - (2x - 3) - [-(x + 4) + (3x - 2)]$

99. $b - [b^2 - (4 - 3b)]$

100. Solve the system of inequalities by graphing:
$2x - y \leq 3$ and $x + y \leq 4$.

5.4 Dividing Polynomials

OBJECTIVES   In this section we will learn to
1. divide a polynomial by a monomial;
2. divide a polynomial by a binomial; and
3. divide a polynomial by a trinomial.

Section 5.3 focussed on how to find the product of two or more polynomials. This section will be devoted to division of polynomials. A monomial can be divided by a monomial by using the rules for exponents that we developed in Section 5.1. For example, $6x^2y^3 \div 3xy^2$ is written

$$\dfrac{6x^2y^3}{3xy^2} = \dfrac{6}{3}x^{2-1}y^{3-2} = 2xy$$

Division of a polynomial by a monomial is carried out by dividing each term of the polynomial by the monomial. For example, to divide $3x^4 - 18x^2$ by $3x^2$, we think of the two terms of the polynomial as $3x^4$ and $-18x^2$. We then write the division

$$\dfrac{3x^4 + (-18x^2)}{3x^2} = \dfrac{3x^4}{3x^2} + \dfrac{-18x^2}{3x^2}$$

$$= x^2 + (-6)$$
$$= x^2 - 6$$

Recall that if two fractions have a common denominator, $\frac{a}{c}$ and $\frac{b}{c}$, their sum is

$$\frac{a}{c} + \frac{b}{c} = \frac{a+b}{c}$$

Applying the symmetric property, we have

$$\frac{a+b}{c} = \frac{a}{c} + \frac{b}{c}$$

This observation provides the basis for dividing a polynomial by a monomial.

To Divide a Polynomial by a Monomial
Divide each term of the polynomial by the monomial.

EXAMPLE 1 Divide $6x^5 + 9x^2$ by $3x$.

SOLUTION
$$\frac{6x^5 + 9x^2}{3x} = \frac{6x^5}{3x} + \frac{9x^2}{3x} \quad \text{Divide each term by } 3x.$$
$$= 2x^{5-1} + 3x^{2-1} \quad \text{Divide the coefficients and subtract the exponents.}$$
$$= 2x^4 + 3x$$

EXAMPLE 2 Divide $4x^2y^5 + 18x^5y^7 - 7x^2y$ by $2xy^3$.

SOLUTION
$$\frac{4x^2y^5 + 18x^5y^7 - 7x^2y}{2xy^3} = \frac{4x^2y^5}{2xy^3} + \frac{18x^5y^7}{2xy^3} - \frac{7x^2y}{2xy^3} \quad \text{Divide each term by } 2xy^3.$$
$$= 2xy^2 + 9x^4y^4 - \frac{7xy^{-2}}{2} \quad \text{Divide the coefficients and subtract the exponents.}$$
$$= 2xy^2 + 9x^4y^4 - \frac{7x}{2y^2} \quad b^{-n} = \frac{1}{b^n}$$

Notice that the final result is written so as to be free of negative exponents.

EXAMPLE 3 Divide: $\dfrac{-8x^3y - 5xy^4 + x^9y^2z - xy^2}{-xy^2}$.

SOLUTION
$$\frac{-8x^3y - 5xy^4 + x^9y^2z - xy^2}{-xy^2} = \frac{-8x^3y}{-xy^2} + \frac{-5xy^4}{-xy^2} + \frac{x^9y^2z}{-xy^2} + \frac{-xy^2}{-xy^2} \quad \text{Divide each term by } -xy^2.$$
$$= 8x^2y^{-1} + 5y^2 + (-x^8z) + 1 \quad \text{Since } \frac{-xy^2}{-xy^2} = 1.$$
$$= \frac{8x^2}{y} + 5y^2 - x^8z + 1$$

EXAMPLE 4   Divide $x^{3n+4} + x^{n+5}$ by x^n.

SOLUTION
$$\frac{x^{3n+4} + x^{n+5}}{x^n} = \frac{x^{3n+4}}{x^n} + \frac{x^{n+5}}{x^n}$$ Divide each term by x^n.
$$= x^{3n+4-n} + x^{n+5-n}$$ Subtract exponents.
$$= x^{2n+4} + x^5$$

Division by a Binomial

Division of a polynomial by a binomial is similar to the long division process in arithmetic. We show the similarities side by side for comparison.

DIVIDE 552 BY 24.

Step 1 Write the problem in standard form.

$$24\overline{)552}$$

Step 2 24 divides into 55 a total of *two* times.

$$2 \cdot 24 = 48$$

$$\begin{array}{r} 2 \\ 24\overline{)552} \\ 48 \end{array}$$

Step 3 Subtract and bring down the next number.

$$\begin{array}{r} 2 \\ 24\overline{)552} \\ 48 \\ \hline 72 \end{array}$$

Step 4 24 divides into 72 *three* times.

$$3 \cdot 24 = 72$$

Subtract.

$$\begin{array}{r} 23 \\ 24\overline{)552} \\ 48 \\ \hline 72 \\ 72 \\ \hline 0 \end{array}$$

The remainder is 0. Since there are no more numbers to bring down, the division is complete.

DIVIDE $x^2 + 7x + 12$ BY $x + 4$.

Step 1 Write the problem in standard form.

$$x + 4\overline{)x^2 + 7x + 12}$$

Step 2 x divides into x^2 a total of x times.

$$x(x + 4) = x^2 + 4x$$

$$\begin{array}{r} x \phantom{{}+x^2+7x+2} \\ x + 4\overline{)x^2 + 7x + 2} \\ x^2 + 4x \end{array}$$

Step 3 Subtract by changing the signs on $x^2 + 4x$ and adding. Bring down the next term.

$$\begin{array}{r} x \phantom{{}+x^2+7x+12} \\ x + 4\overline{)x^2 + 7x + 12} \\ x^2 + 4x \phantom{{}+12} \\ \hline 3x + 12 \end{array}$$

Step 4 x divides into $3x$ *three* times.

$$3(x + 4) = 3x + 12$$

Subtract by changing the signs on $3x + 12$ and adding.

$$\begin{array}{r} x + 3 \phantom{{}+12} \\ x + 4\overline{)x^2 + 7x + 12} \\ x^2 + 4x \phantom{{}+12} \\ \hline 3x + 12 \\ 3x + 12 \\ \hline 0 \end{array}$$

The remainder is 0. Since there are no more numbers to bring down, the division is complete.

Step 5 Check by multiplying:

$24 \cdot 23 = 552$

Step 5 Check by multiplying:

$(x + 4)(x + 3) = x^2 + 7x + 12$

It is interesting to note that when x is replaced by 20, the second problem reduces to the first one, showing that the first problem is a special case of the second one. Observe what happens to each polynomial when $x = 20$:

$$x + 3 = 20 + 3 = 23$$
$$x + 4 = 20 + 4 = 24$$
$$x^2 + 7x + 12 = 20^2 + 7 \cdot 20 + 12 = 400 + 140 + 12 = 552$$

EXAMPLE 5 Divide $x^2 - 6x - 7$ by $x + 1$.

SOLUTION

$$\begin{array}{r} x \\ x+1 \overline{\smash{)}x^2 - 6x - 7} \\ \underline{x^2 + x } \end{array}$$ ← x divides into x^2 a total of x times.

← $x(x + 1) = x^2 + x$

$$\begin{array}{r} x \\ x+1 \overline{\smash{)}x^2 - 6x - 7} \\ \underline{x^2 + x } \downarrow \\ -7x - 7 \end{array}$$

Subtract and then bring down the next term. Remember: to subtract, change the sign on $x^2 + x$ and add.

$$\begin{array}{r} x - 7 \\ x+1 \overline{\smash{)}x^2 - 6x - 7} \\ \underline{x^2 + x } \\ -7x - 7 \\ \underline{-7x - 7} \\ 0 \end{array}$$

← x divides into $-7x$ a total of -7 times.

← $-7(x + 1) = -7x - 7$
← The remainder is zero.

CHECK $(x + 1)(x - 7) \stackrel{\checkmark}{=} x^2 - 6x - 7$

EXAMPLE 6 Divide $x^2 + 2x^3 + 6x + 1$ by $x + 2$.

SOLUTION

First we arrange the terms of the dividend in descending powers of the variable:

$$x^2 + 2x^3 + 6x + 1 = 2x^3 + x^2 + 6x + 1$$

x divides into $2x^3$ a total of $2x^2$ times.
x divides into $-3x^2$ a total of $-3x$ times.
x divides into $12x$ a total of 12 times.

$$\begin{array}{r} 2x^2 - 3x + 12 \\ x+2 \overline{\smash{)}2x^3 + x^2 + 6x + 1} \\ \underline{2x^3 + 4x^2 } \\ -3x^2 + 6x \\ \underline{-3x^2 - 6x } \\ 12x + 1 \\ \underline{12x + 24} \\ -23 \end{array}$$

← The remainder is -23.

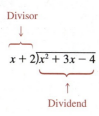

CAUTION

Before carrying out a long division problem, arrange both the dividend and the divisor in descending powers of the variable.

We write the remainder as the numerator of the fraction $\frac{-23}{x+2}$ to indicate that it has not been divided by $x + 2$. Thus

$$\frac{x^2 + 2x^3 + 6x + 1}{x + 2} = 2x^2 - 3x + 12 + \frac{-23}{x + 2}$$

CHECK
$$(x + 2)(2x^2 - 3x + 12) + (-23) \stackrel{?}{=} x^2 + 2x^3 + 6x + 1$$
$$2x^3 - 3x^2 + 12x + 4x^2 - 6x + 24 - 23 \stackrel{?}{=} x^2 + 2x^3 + 6x + 1$$
$$2x^3 + x^2 + 6x + 1 \stackrel{\checkmark}{=} x^2 + 2x^3 + 6x + 1$$

How do we know when a long division problem is complete? The answer is given in the following generalization.

> Whenever the degree of the remainder is less than the degree of the divisor, the division is complete.

EXAMPLE 7 Divide $4x^4 - x$ by $2x^2 + 3$.

SOLUTION Since there are no third-, second-, or zero-degree terms in $4x^4 - x$, we consider the coefficients of these terms to be zero:

$$4x^4 - x = 4x^4 + 0x^3 + 0x^2 - x + 0$$

In the division display, we use place holders with coefficients of zero or an open space.

$2x^2$ divides into $4x^4$ a total of $2x^2$ times.
$2x^2$ divides into $-6x^2$ a total of -3 times.

$$\begin{array}{r} 2x^2 - 3 \\ 2x^2 + 3 \overline{\smash{)}4x^4 + 0x^3 + 0x^2 - x + 0} \\ \underline{4x^4 + 6x^2 } \quad \leftarrow \text{Change the signs and add.} \\ -6x^2 - x \\ \underline{-6x^2 - 9} \\ -x + 9 \quad \leftarrow \text{The degree is 1.} \end{array}$$

Since the degree of the remainder, $-x + 9$, is 1, which is less than the degree of the divisor, $2x^2 + 3$, which is 2, the division is complete. We write the result

$$2x^2 - 3 + \frac{-x + 9}{2x^2 + 3}$$

CHECK Divisor · Quotient + Remainder = Dividend
$$(2x^2 + 3) \cdot (2x^2 - 3) + (-x + 9) \stackrel{?}{=} 4x^4 - x$$
$$4x^4 - 9 - x + 9 \stackrel{?}{=} 4x^4 - x$$
$$4x^4 - x \stackrel{\checkmark}{=} 4x^4 - x$$

EXAMPLE 8 Divide $3x^3 + 2x^2 - x + 6$ by $2x - 1$.

SOLUTION

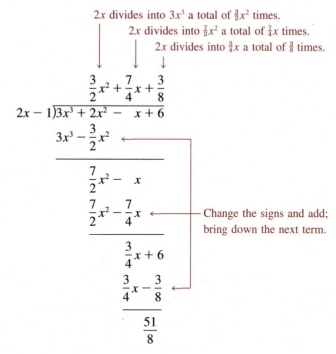

We write the result

$$\frac{3}{2}x^2 + \frac{7}{4}x + \frac{3}{8} + \frac{\frac{51}{8}}{2x - 1}$$

Division by a Trinomial

EXAMPLE 9 Divide $2x^5 + 5x^3 + x + 7$ by $x^2 - 6x - 1$.

SOLUTION The terms involving x^4 and x^2 are missing, so zero coefficients are placed in the dividend to accommodate them.

$$\begin{array}{r}
2x^3 + 12x^2 + 79x + 486 \\
x^2 - 6x - 1 \overline{\smash{)}2x^5 + 0x^4 + 5x^3 + 0x^2 + x + 7} \\
\underline{2x^5 - 12x^4 - 2x^3 } \\
12x^4 + 7x^3 \\
\underline{12x^4 - 72x^3 - 12x^2 } \\
79x^3 + 12x^2 + x \\
\underline{79x^3 - 474x^2 - 79x } \\
486x^2 + 80x + 7 \\
\underline{486x^2 - 2916x - 486} \\
2996x + 493
\end{array}$$

The result is

$$\frac{2x^5 + 5x^3 + x + 7}{x^2 - 6x - 1} = 2x^3 + 12x^2 + 79x + 486 + \frac{2996x + 493}{x^2 - 6x - 1}$$

EXAMPLE 10 Suppose we were to illustrate a division problem involving polynomials as follows:

$$D(x)\overline{)P(x)}^{Q(x)}$$

If $D(x) = x + 7$ and $Q(x) = x - 5$, and the remainder is zero, find $P(x)$.

SOLUTION Since the dividend $P(x)$ must be the product of the quotient and the divisor,

$$P(x) = Q(x) \cdot D(x) = (x - 5)(x + 7) = x^2 + 2x - 35$$

Notice that since there is no remainder, both $x + 7$ and $x - 5$ are factors of the dividend, $x^2 + 2x - 35$.

EXAMPLE 11 Given the dividend $P(y) = 9y^2 + 9y - 3$ and the divisor $D(y) = 3y - 4$, find the quotient $Q(y)$ and the remainder $R(y)$ such that $P(y) = D(y) \cdot Q(y) + R(y)$.

SOLUTION

$$\begin{array}{r} 3y + 7 \\ 3y - 4 \overline{)9y^2 + 9y - 3} \\ \underline{9y^2 - 12y} \\ 21y - 3 \\ \underline{21y - 28} \\ 25 \end{array}$$

Therefore $Q(y) = 3y + 7$ and $R(y) = 25$.

CHECK
$$P(y) = D(y) \cdot Q(y) + R(y)$$
$$9y^2 + 9y - 3 \stackrel{?}{=} (3y - 4) \cdot (3y + 7) + 25$$
$$9y^2 + 9y - 3 \stackrel{?}{=} 9y^2 + 21y - 12y - 28 + 25$$
$$9y^2 + 9y - 3 \stackrel{\checkmark}{=} 9y^2 + 9y - 3$$

EXAMPLE 12 If $f(x) = x^3 + 3x^2 + 5x + 3$ and $g(x) = x + 1$, find $\left(\dfrac{f}{g}\right)(x)$ and show that

$$\left(\frac{f}{g}\right)(2) = \frac{f(2)}{g(2)}$$

SOLUTION To find $\left(\dfrac{f}{g}\right)(x)$, we divide $f(x)$ by $g(x)$:

$$\begin{array}{r}
x^2 + 2x + 3 \\
x+1 \overline{\smash{\big)}\, x^3 + 3x^2 + 5x + 3} \\
\underline{x^3 + x^2 } \\
2x^2 + 5x \\
\underline{2x^2 + 2x } \\
3x + 3 \\
\underline{3x + 3} \\
0
\end{array}$$

Therefore $\left(\dfrac{f}{g}\right)(x) = x^2 + 2x + 3$ and so $\left(\dfrac{f}{g}\right)(2) = 2^2 + 2 \cdot 2 + 3 = 11$.

On the other hand,

$$\dfrac{f(x)}{g(x)} = \dfrac{x^3 + 3x^2 + 5x + 3}{x+1} \quad \text{and} \quad \dfrac{f(2)}{g(2)} = \dfrac{2^3 + 3(2^2) + 5(2) + 3}{2+1}$$

$$= \dfrac{8 + 12 + 10 + 3}{3}$$

$$= \dfrac{33}{3} = 11$$

Thus, $\left(\dfrac{f}{g}\right)(2) = \dfrac{f(2)}{g(2)} = 11$.

EXAMPLE 13 Divide $x^{2n} - 3x^n y^n + 2y^{2n}$ by $x^n - y^n$.

x^n divides into x^{2n} a total of x^n times.
x^n divides into $-2x^n y^n$ a total of $-2y^n$ times.

$$\begin{array}{r}
x^n \phantom{{}^{2n}} - 2y^n \phantom{{}^{2n}} \\
x^n - y^n \overline{\smash{\big)}\, x^{2n} - 3x^n y^n + 2y^{2n}} \\
\underline{x^{2n} - x^n y^n \phantom{+ 2y^{2n}}} \quad \leftarrow \; x^n \cdot x^n = x^{n+n} = x^{2n} \\
-2x^n y^n + 2y^{2n} \\
\underline{-2x^n y^n + 2y^{2n}} \quad \leftarrow \; y^n \cdot y^n = y^{n+n} = y^{2n} \\
0
\end{array}$$

Exercise Set 5.4

Perform each of the following divisions.

1. $\dfrac{3a + 6}{3}$
2. $\dfrac{4m + 12}{4}$
5. $\dfrac{4t^2 - 6t + 2}{2}$
6. $\dfrac{12u^2 + 15u - 3}{3}$
3. $\dfrac{15t - 20}{-5}$
4. $\dfrac{36u - 27}{-9}$
7. $\dfrac{64x^3 + 32x^2 - 8x}{8x}$
8. $\dfrac{72y^3 - 54y^2 + 63y}{9y}$

9. $\dfrac{8x^2 + 24x - 32}{-8x}$

10. $\dfrac{55r^2 - 66r + 33}{-11r}$

11. $\dfrac{18a^3 - 22a^2 + 3}{-2a^2}$

12. $\dfrac{39b^3 - 91b^2 - 20}{-13b^2}$

13. $\dfrac{42a^3b^3 + 28a^2b^2}{7a^2b}$

14. $\dfrac{60r^2s^4 - 54rs^2}{6r^2s}$

15. $\dfrac{a^{6m} - a^{4m}}{a^{2m}}$

16. $\dfrac{16t^{9n} - 12t^{6n}}{-4t^{3n}}$

17. $\dfrac{56t^{7n} - 36t^{3n}}{-4t^{-5n}}$

18. $\dfrac{81v^{-2n} + 99v^{5n}}{9v^{-3n}}$

19. $\dfrac{-7a^{11}b^9c^8 + 4a^8b^{10}c^{15}}{a^4b^5c^3}$

20. $\dfrac{15x^{10}y^6z^4 - 3x^8y^{12}z^7}{2x^4y^4z^3}$

21. $\dfrac{-90b^{10}c^8d^5 + 30b^7c^8d^9 + 15b^2cd^4}{-66b^2cd^4}$

22. $\dfrac{-44r^{10}s^8t^{12} - 48r^7s^9t^6 + 28r^3s^8t^5}{-44r^2s^2t^3}$

23. $\dfrac{x^2 + 9x + 8}{x + 8}$

24. $\dfrac{y^2 - 5y + 6}{y - 3}$

25. $(a^2 + 13a + 30) \div (a + 10)$

26. $(m^2 - 3m - 18) \div (m + 3)$

27. $(t^2 - 2t - 15) \div (t - 5)$

28. $(s^2 + s - 56) \div (s - 7)$

29. $\dfrac{x^3 + 2x^2 - 21x + 18}{x - 3}$

30. $\dfrac{m^3 + 5m^2 + 6m + 8}{m + 4}$

31. $(5v^2 - 13v + 6) \div (5v - 3)$

32. $(8k^3 - 6k^2 - 5k + 3) \div (4k + 3)$

33. $(35a^3 + 53a^2 - 19a - 1) \div (7a - 2)$

34. $(-12y^3 - 2y^2 + 21y + 5) \div (2y + 3)$

35. If $f(x) = 8x^3 - 1$ and $g(x) = 2x - 1$, find

$\left(\dfrac{f}{g}\right)(x)$ and $\left(\dfrac{f}{g}\right)(-1)$

36. If $f(y) = 125y^3 - 1$ and $g(y) = 5y + 1$, find

$\left(\dfrac{f}{g}\right)(y)$ and $\left(\dfrac{f}{g}\right)(1)$

37. $(15s^4 + 8s^2 + 3s^3 - 1) \div (5s^2 + 1)$

38. $(-5k + 10k^5 - 5k^2 - 4k^3) \div (2k^2 - 3)$

39. $(2x^4 - 5x^3 + 5x^2 + 11x - 10) \div (2x^2 + x - 2)$

40. $(10x^3 - 41x^2 + 31x - 6) \div (2x^2 - 7x + 2)$

41. $(24x^3 - 71x^2y + 51xy^2 - 10y^3) \div (3x^2 - 7xy + y^2)$

42. $(6x^4 - 13x^3y - 41x^2y^2 - 27xy^3 - 5y^4) \div (3x^2 + 4xy + y^2)$

43. $(x^{2n} - y^{2n}) \div (x^n - y^n)$

44. $(x^{2n} + 2x^ny^n + y^{2n}) \div (x^n + y^n)$

45. $(x^{2n} - 2x^ny^n + y^{2n}) \div (x^n - y^n)$

46. $(x^{3n} - y^{3n}) \div (x^n - y^n)$

47. $(x^{3n} + y^{3n}) \div (x^n + y^n)$

48. $(x^{2n} + x^ny^n - 6y^{2n}) \div (x^n + 3y^n)$

49. Given polynomials $P(x) = x^2 + 7x + 13$ and $D(x) = x + 3$, find polynomials $Q(x)$ and $R(x)$ such that $P(x) = D(x) \cdot Q(x) + R(x)$. See Examples 10 and 11.

50. Given polynomials $P(y) = 2y^3 + 5y^2 - y - 6$ and $D(y) = y + 2$, find polynomials $Q(y)$ and $R(y)$ such that $P(y) = D(y) \cdot Q(y) + R(y)$. See the example.

51. Given polynomials $P(x) = 5x^2 + 7x - 9$ and $D(x) = 4x - 1$, find polynomials $Q(x)$ and $R(x)$ such that $P(x) = D(x) \cdot Q(x) + R(x)$.

52. Given polynomials $P(t) = t^3 - t^2 - 1$ and $D(t) = 5t + 3$, find polynomials $Q(t)$ and $R(t)$ such that $P(t) = D(t) \cdot Q(t) + R(t)$.

For Extra Thought

In the accompanying figure, the area of the rectangle ABCD is $15x^2 + 12x - 3$ and the area of the trapezoid EBCD is $10x^2 + 8x - 2$. Use the formulas of a triangle, a rectangle, and a trapezoid to complete Exercises 53–56.

53. Find the length of side BC in terms of x.

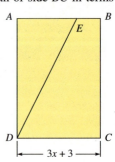

54. Find the length of side AE in terms of x.

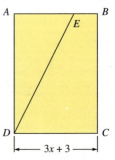

55. Find the area of triangle AED in terms of x.

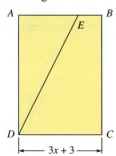

56. Find the square of the length of side ED in terms of x.

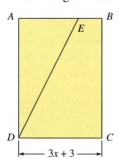

Use the given figures to find the indicated lengths.

57. If a box has a volume of $V(x) = 6x^3 - x^2 - 9x + 4$, find the height, $h(x)$, in terms of x if the length is $l(x) = 3x + 4$ and the width is $w(x) = x - 1$.

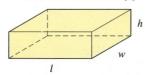

58. A box has volume $V(x) = 2x^3 + 9x^2 - 27$. Find the length, $l(x)$, in terms of x if the height is $h(x) = x + 3$ and the width is $w(x) = 2x - 3$.

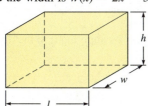

59. Given a right triangle with an area of $A(x) = x^2 + 2x - 63$, find the height, $a(x)$, in terms of x if the base is $b(x) = 2x + 18$.

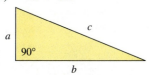

60. Given a trapezoid with an area of $A(x) = x^2 + 5x + 4$, find the height, $h(x)$, if the larger base is $B(x) = x + 5$ and the smaller base is $b(x) = x + 3$.

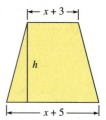

REVIEW EXERCISES

Perform the following indicated operations.

61. Find $(f - g)(x)$, if $f(x) = 6x^3 + 7x^2 - 5x + 2$ and $g(x) = x^3 - 8x^2 - 4x + 3$.

62. Find $(f - g)(a)$, if $f(a) = -2(a^2 - 7a + 5)$ and $g(a) = 3(2a^2 + 8a - 6)$.

63. $(3m + 2n)^2$

64. $(7s - 5t)^2$

65. $(11a - 6b)(11a + 6b)$

66. $(5r^2 + 2s^2)(5r^2 - 2s^2)$

67. If $f(y) = y^2 - 5y + 7$ and $g(y) = y^2 - 5y - 7$, find $(f \cdot g)(y)$.

68. If $f(k) = k^2 + 8k - 3$ and $g(k) = k^2 + 8k + 3$, find $(f \cdot g)(k)$.

69. $(3x + 4y)(7x - 9y)$

70. $(7m + 11n)(5m + 6n)$

71. $(2x - y)^3$

72. $(3x + 2y)^3$

73. Graph $3x - 5y = 15$ using the x- and y-intercepts and one other point.

74. Graph $y \leq 3x - 4$.

75. Graph the system of inequalities:
$y - x \geq 3$ and $2y + x \leq 5$.

76. Solve the system of equations by the method of graphing:
$5x - 2y = 5$ and $x + y = 1$.

77. Solve the system of equations by the method of elimination:
$3x \quad y = 4$ and $2x + 3y - 5$.

78. Solve the system of equations by the method of substitution:
$3x + 5y = 8$ and $4x - y = 2$.

79. Solve the system of equations $4x + 7y = 15$ and $3x - 2y = 4$ using Cramer's rule.

Write in Your Own Words

80. What are the steps involved in doing a long division problem?
81. What do the words "divisor" and "dividend" mean?
82. What do the words "quotient" and "remainder" mean?
83. How do you know when to end the division process?

5.5 Synthetic Division

OBJECTIVES In this section we will learn how to divide polynomials by using synthetic division.

In certain cases, the long division process introduced in Section 5.4 can be carried out more efficiently by a process called **synthetic division.** To see how it works, consider the quotient when $2x^3 + 8x^2 + 11x + 7$ is divided by $x + 2$.

$$
\begin{array}{r}
2x^2 + 4x + 3 \\
x+2 \overline{\smash{)}\,2x^3 + 8x^2 + 11x + 7} \\
\underline{2x^3 + 4x^2} \\
4x^2 + 11x \\
\underline{4x^2 + 8x} \\
3x + 7 \\
\underline{3x + 6} \\
1
\end{array}
$$

Divisor → $x + 2$; ← Quotient ($2x^2 + 4x + 3$); ← Dividend; ← Remainder (1)

Observe that

1. The divisor is linear (first-degree).
2. The coefficient of the lead term in the divisor, $x + 2$, is 1.
3. The coefficient of the first terms in the dividend and the quotient are the same.
4. The degree of the quotient is one less than the degree of the dividend.

Keeping these four observations in mind, we can omit all of the variables in the preceding division and still carry out the division correctly! However, the following caution must be observed.

We outline the synthetic division process by applying it to the division problem we just did.

> **CAUTION**
>
> In order to use synthetic division, the divisor must be of the form
>
> $x + a$ or $x - a$
>
> Note that both the **coefficient** of x and the **power** on x are **1.**

$$
\begin{array}{r}
2 + 4 + 3 \\
1+2 \overline{\smash{)}\,2 + 8 + 11 + 7} \\
2 + 4 \\
\hline
4 + \boxed{11} \\
\boxed{4} + 8 \\
3 + \boxed{7} \\
\boxed{3} + 6 \\
1
\end{array}
$$

The circled items repeat the coefficients in the dividend and may be omitted.

The boxed items repeat the terms immediately above and may be omitted.

These omissions yield the following streamlining:

$$\begin{array}{r}2 + 4 + 3 \\ 1 + 2 \overline{) 2 + 8 + 11 + 7} \\ 4 \\ 4 \\ + 8 \\ 3 \\ + 6 \\ 1 \end{array}$$

All the remaining numbers and their signs can be shifted upward.

We now have a division display that looks like this:

$$\begin{array}{r}2 + 4 + 3 \\ 1 + 2 \overline{) 2 + 8 + 11 + 7} \\ 4 + 8 + 6 \\ \hline 2 + 4 + 3 + \boxed{1} \end{array}$$

← If the lead term in the quotient, 2, is moved to the bottom row, then the first three terms of the bottom row duplicate the coefficients in the quotient, and the fourth term in the last row, 1, is the remainder.

This yields

$$\begin{array}{r}2 + 4 + 3 \\ 1 + 2 \overline{) 2 + 8 + 11 + 7} \\ 4 + 8 + 6 \\ \hline 2 + 4 + 3 + 1 \end{array}$$

← Since the row above the division symbol is now repeated in the last row, we can omit it.

← Notice that the bottom row is obtained by subtracting the second row from the first one.

Omitting the row above the division symbol simplifies our diagram still further:

$$\begin{array}{r}1 + 2 \overline{) 2 + 8 + 11 + 7} \\ 4 + 8 + 6 \\ \hline 2 + 4 + 3 + 1 \end{array}$$

Finally, since the variable has been omitted, we can drop the 1 in the divisor and change the 2 to −2. By making this change in sign, every sign in the product will be changed, and each term in the second row will automatically be replaced by its additive inverse. Now we should add the two rows rather than subtract them. The division symbol $\overline{)}$ is no longer needed and is replaced by $\underline{|}$ to separate the divisor from the dividend.

$$\begin{array}{r} \text{Coefficients of the dividend} \\ -2 \underline{| 2 + 8 + 11 + 7} \\ -4 - 8 - 6 \\ \hline 2 + 4 + 3 + 1 \end{array}$$

← The bottom row is the sum of the second row and the first one.

↑ Coefficients of the quotient ↑ Remainder

We can read the result directly from the bottom row:

$$2x^2 + 4x + 3 + \frac{1}{x+2}$$

It is common to omit all plus signs in synthetic division. Thus

$$\begin{array}{r|rrr} -2 & 2 + 8 + 11 + 7 \\ & -4 - 8 - 6 \\ \hline & 2 + 4 + 3 + 1 \end{array}$$

will be written

$$\begin{array}{r|rrr} -2 & 2 \quad 8 \quad 11 \quad 7 \\ & -4 \; -8 \; -6 \\ \hline & 2 \quad 4 \quad 3 \quad 1 \end{array}$$

EXAMPLE 1 Use synthetic division to divide $3x^3 + x^2 - 10$ by $x - 2$.

SOLUTION We think of $3x^3 + x^2 - 10$ as $3x^3 + x^2 + 0x - 10$.

When using synthetic division where the dividend is missing terms, **coefficients of zero must be supplied** for each missing term so that like terms will be written in the same column.

Step 1 We write down the coefficients of the dividend, including a zero as the coefficient of any missing term.

$$3 \quad 1 \quad 0 \quad -10$$

Step 2 Change the sign of the constant term in the divisor and place it to the left of the coefficients in the dividend, separated by $\rfloor$.

$$2 \rfloor 3 \quad 1 \quad 0 \quad -10$$

Step 3 We bring down the lead coefficient, 3, from the dividend.

$$\begin{array}{r|rrr} 2 & 3 \quad 1 \quad 0 \quad -10 \\ & \downarrow \\ \hline & 3 \end{array}$$

Step 4 We multiply 3 by 2 and write the product, 6, under the 1. We add these to get 7.

$$\begin{array}{r|rrr} 2 & 3 \quad 1 \quad 0 \quad -10 \\ & \quad 6 \\ \hline & 3 \quad 7 \end{array}$$

Step 5 We multiply 7 by 2 and write the product, 14, under the 0. We add these to get 14.

$$\begin{array}{r|rrr} 2 & 3 \quad 1 \quad 0 \quad -10 \\ & \quad 6 \quad 14 \\ \hline & 3 \quad 7 \quad 14 \end{array}$$

Step 6 We multiply 14 by 2 and write the product, 28, under the -10. We add these to get 18.

$$\begin{array}{r|rrr} 2 & 3 \quad 1 \quad 0 \quad -10 \\ & \quad 6 \quad 14 \quad 28 \\ \hline & 3 \quad 7 \quad 14 \quad 18 \end{array} \leftarrow \text{The remainder is eighteen.}$$

We started with a third-degree dividend, so the quotient will be a second-degree polynomial. The result is

$$3x^2 + 7x + 14 + \frac{18}{x-2}$$

EXAMPLE 2 Divide $x^4 - 12x^2 + 3x + 18$ by $x - 3$.

$$\underline{3\,|}\ \begin{array}{ccccc} 1 & 0 & -12 & 3 & 18 \\ & 3 & 9 & -9 & -18 \\ \hline 1 & 3 & -3 & -6 & 0 \end{array}\ \begin{array}{l}\leftarrow \text{Fourth-degree} \\ \\ \leftarrow \text{The remainder is zero.}\end{array}$$

Third-degree →

The quotient is $x^3 + 3x^2 - 3x - 6$.

EXAMPLE 3 Divide $x^5 - 1$ by $x - 1$.

SOLUTION We begin by writing

$$x^5 - 1 = x^5 + 0x^4 + 0x^3 + 0x^2 + 0x - 1$$

Then

$$\underline{1\,|}\ \begin{array}{cccccc} 1 & 0 & 0 & 0 & 0 & -1 \\ & 1 & 1 & 1 & 1 & 1 \\ \hline 1 & 1 & 1 & 1 & 1 & 0 \end{array}$$

The quotient is $x^4 + x^3 + x^2 + x + 1$.

EXAMPLE 4 Divide $x^3 + \frac{3}{2}x^2 - 2x + 1$ by $x + \frac{1}{2}$.

SOLUTION

$$\underline{-\tfrac{1}{2}\,|}\ \begin{array}{cccc} 1 & \tfrac{3}{2} & -2 & 1 \\ & -\tfrac{1}{2} & -\tfrac{1}{2} & \tfrac{5}{4} \\ \hline 1 & 1 & -\tfrac{5}{2} & \tfrac{9}{4} \end{array}$$

The result is $x^2 + x - \frac{5}{2} + \dfrac{\frac{9}{4}}{x + \frac{1}{2}}$.

EXAMPLE 5 Divide $x^3 + x^4 - 1 + x^2 + x$ by $x - 4$.

SOLUTION We first rearrange the dividend so that it is in descending powers of x:

$$x^3 + x^4 - 1 + x^2 + x = x^4 + x^3 + x^2 + x - 1$$

$$\underline{4\,|\,}\begin{array}{rrrrr} 1 & 1 & 1 & 1 & -1 \\ & 4 & 20 & 84 & 340 \\ \hline 1 & 5 & 21 & 85 & 339 \end{array}$$

The quotient is $x^3 + 5x^2 + 21x + 85 + \dfrac{339}{x - 4}$.

Caution
When using synthetic division, the dividend must be arranged in descending powers of the variable and the divisor must be linear, with a lead coefficient of 1.

Exercise Set 5.5

Use synthetic division to find the quotient and remainder for each of the following.

1. $(x^2 - 9x + 20) \div (x - 4)$
2. $(y^2 - 11y + 30) \div (y - 6)$
3. $(a^3 + 4a^2 - 7a + 5) \div (a - 2)$
4. $(p^3 - p^2 - 6p + 5) \div (p + 2)$
5. $(3t^3 - t^2 + 8) \div (t + 1)$
6. $(6k^4 - k^2 + 4) \div (k - 3)$
7. $(x^5 - 32) \div (x - 2)$
8. $(y^5 + 32) \div (y + 2)$
9. $(2y^4 - 3y^3 + y^2 - 3y - 6) \div (y - 2)$
10. $(2s^4 + 5s^3 + 6s^2 + 3s - 10) \div (s + 2)$
11. $(r^4 - r^3 + 3r^2 - 7r - 6) \div (r - 2)$
12. $(2m^4 - 3m^3 - m^2 - m - 2) \div (m - 2)$
13. $(a^6 - 3) \div (a - 1)$
14. $(x^6 + 2) \div (x + 1)$
15. $(2t^4 - 7t^3 - t^2 + 7) \div (t - 5)$
16. $(d^4 + 2d^3 - 15d^2 - 32d - 12) \div (d - 4)$
17. $(5y^4 - 18y^2 - 6y + 3) \div (y + 2)$
18. $(2k^3 + 5k^2 - 8k + 6) \div (k + 2)$
19. $(2m^4 - 3m^2 + 5m - 7) \div (m + 3)$
20. $(2x^4 - 10x^2 - 23x + 6) \div (x - 3)$

Use synthetic division to find the quotient and remainder. Rewrite the dividend in standard form, if needed.

21. $\dfrac{4x^3 - 17x + 11x^2 + 12}{x + 4}$

22. $\dfrac{6x^4 + 123x + 25x^2 - 10}{x + 5}$

23. $\dfrac{5x^3 - 12x^2 - 17x + 26}{x - 3}$

24. $\dfrac{2x^5 - 5x^4 - 18 - 7x^2}{x - 3}$

25. $\dfrac{-3x^4 + 18x^3 - 32x^2 + 26x - 6}{x - 3}$

26. $\dfrac{2x^4 - 129x + 4}{x - 4}$

27. $\dfrac{-2x^4 + 36x - 51x^2 - 22x^3}{x + 7}$

28. $\dfrac{-25x^4 + 17x^2 + 8 + 5x^5 - 3x^3 - 11x}{x - 5}$

29. $\dfrac{7x^2 + 18 + 3x^3}{x + 3}$

30. $\dfrac{38x^2 - 29x + 56 - 5x^3}{x - 7}$

Use synthetic division to find the quotient and remainder.

31. $\left(x^2 - \dfrac{1}{6}x - \dfrac{1}{6}\right) \div \left(x - \dfrac{1}{2}\right)$

32. $\left(x^3 - \dfrac{11}{3}x^2 + \dfrac{14}{3}x - 2\right) \div \left(x - \dfrac{1}{3}\right)$

33. $\left(x^5 + 2x^4 + \dfrac{1}{2}x^3 + x^2 + \dfrac{5}{4}x + \dfrac{5}{2}\right) \div (x + 2)$

34. $\left(\dfrac{1}{2}x^4 - \dfrac{1}{5}\right) \div (x - 1)$

35. $\left(7x^3 + \frac{1}{2}x - 5\right) \div \left(x - \frac{3}{2}\right)$

36. $\left(\frac{5}{3}x^2 - 8\right) \div \left(x + \frac{1}{4}\right)$

Use long division in Exercises 37–40 to find $\left(\frac{f}{g}\right)(x)$.

37. $f(x) = 2x^2 + 13x + 15$; $g(x) = 4x + 6$
38. $f(x) = 5x^2 + 74x - 20$; $g(x) = 25x - 5$
39. $f(x) = 6x^2 + 43x - 34$; $g(x) = 6x - 5$
40. $f(x) = x^2 - 5x + 1$; $g(x) = 3x - 1$

REVIEW EXERCISES

41. $-xy(x^2 + 5x - 7) + 3xy(-5x^2 - 2x + 3) - 4xy(x^2 - x - 1)$
42. $3ac(a^2 + 8a - 9) - 5ac(a^2 - 2a + 4) - ac(-2a^2 + 5a - 7)$
43. $-3[-5 - 2x(x - 7)] + 5[-3x - 7(x^2 - 9)]$
44. $-8t[t - 5(2t - 9)] - [-2 + t(8t - 1)]$
45. $[(3a + 2b) - c][(3a + 2b) + c]$
46. $[(r - 5s) + 3t][(r - 5s) - 3t]$
47. $(3x^2 - 2y)^2$
48. $(7a - 3k^2)^2$
49. $(x^{m+5} - y^{2m-1})(x^{m-1} + 4y^{3m-2})$
50. $(5x^{-m} + 7y^m)(3x^{2m+3} - 5y^{3-m})$
51. $(3x^2 - 2y^2x)^2$
52. $(7a^2bc + 8ab^2c^2)^2$
53. Graph:
$\{(x, y) \mid y < x\} \cap \{(x, y) \mid x < 4\} \cap \{(x, y) \mid y > 1\}$.
54. Graph:
$\{(x, y) \mid x + y \geq 1\} \cap \{(x, y) \mid y \geq x - 4\} \cap \{(x, y) \mid y < 3\}$

FOR EXTRA THOUGHT

55. The formula for the volume of a right circular cylinder is $V = \pi r^2 h$. If $V(x) = x^3 - 2x^2 - 13x - 10$ and

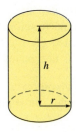

$h(x) = x - 5$, find r^2 in terms of x and π. See the accompanying figure.

56. The formula for the volume of a right circular cone is $V = \frac{1}{3}\pi r^2 h$. If $V(x) = \frac{1}{3}\pi r^2 h$ and $h(x) = x + 3$, find r^2 in terms of x and π. See the accompanying figure.

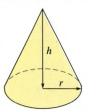

57. The formula for the volume of a sphere is $V = \frac{4}{3}\pi r^3$. If $V(x) = x^3 + 3x^2 + 3x + 1$ and $r^3(x) = x + 1$, find π in terms of x. See the given figure.

58. The formula for the volume of a pyramid is $V = \frac{1}{3}Bh$, where B is the area of the base and h is the height. If $V(x) = x^3 - x^2 - 7x - 20$ and $h(x) = x - 4$, find $B(x)$. See the given figure.

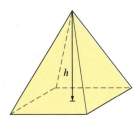

WRITE IN YOUR OWN WORDS

59. What is the purpose or value of using synthetic division?
60. Is it possible to use synthetic division on all division problems? If not, explain when it is or is not possible and why.
61. Why are the variables removed when doing synthetic division?
62. In doing long division, we may or may not use zeros as placeholders in the dividend, whereas in doing synthetic division, we *must* use zeros as placeholders in the dividend. Explain why.

Summary

5.1 Exponents and Scientific Notation

Complex expression can often be simplified by using exponents. An exponential expression consists of two parts, the **base** and the **exponent.**

Base $\rightarrow a^n \leftarrow$ Exponent

If a and b are real numbers and m and n are integers, the following rules hold. If any base is zero, the exponent cannot be zero.

Product rule: $b^m \cdot b^n = b^{m+n}$

Quotient rule: $\dfrac{b^m}{b^n} = b^{m-n}, \quad b \neq 0$

Power rules: $(b^m)^n = b^{m \cdot n}$
$(a \cdot b)^n = a^n \cdot b^n$
$\left(\dfrac{a}{b}\right)^n = \dfrac{a^n}{b^n}, \quad b \neq 0$

Negative exponent: $b^{-n} = \dfrac{1}{b^n}, \quad b \neq 0$

$\left(\dfrac{a}{b}\right)^{-m} = \dfrac{b^m}{a^m}$

Zero exponent: $b^0 = 1, \quad b \neq 0$

Numbers can be written in exponential form by the use of **scientific notation.** A number is written in scientific notation when it is in the form $a \times 10^n$, where $1 \leq a < 10$ and n is an integer.

5.2 Adding and Subtracting Polynomials

An **algebraic term** is a number, a variable, or the product of a number and one or more variables. A **polynomial** in simplified form is a single algebraic term or the sum or difference of algebraic terms satisfying the following conditions:

1. All exponents must be whole numbers.
2. No term may contain a variable in the denominator.
3. No variable may appear under a radical sign.

A polynomial having one term is called a **monomial;** a polynomial having two terms is a **binomial;** and a polynomial having three terms is a **trinomial.**

A polynomial is in **standard form** when it is written in descending powers of the variable.

The **degree** of a polynomial in one variable is the largest exponent that appears on the variable.

To **add** or **subtract** two polynomials, add or subtract the coefficients of *like terms* (same base and exponent in the variable parts). The variable parts remain unchanged. All polynomials with real coefficients represent **functions.** When adding two polynomials $f(x)$ and $g(x)$, their sum is designated $(f + g)(x)$. When subtracting two polynomials $f(x)$ and $g(x)$, their difference is designated $(f - g)(x)$.

5.3 Multiplying Polynomials

To **multiply** two polynomials, use the distributive property together with the rules for exponents. When the polynomials are binomials, three procedures can be used to carry out the multiplication.

1. **FOIL method:** Add the products of the first terms (**F**), the outer terms (**O**), the inner terms (**I**), and the last terms (**L**).
2. **A sum times a difference:** $(a - b)(a + b) = a^2 - b^2$
3. **The square of a binomial:** $(a + b)^2 = a^2 + 2ab + b^2$
   $(a - b)^2 = a^2 - 2ab + b^2$

The notation $(f \cdot g)(x)$ indicates that two polynomials designated $f(x)$ and $g(x)$ are to be multiplied.

5.4 Dividing Polynomials

To divide a polynomial by a monomial, divide each term of the polynomial by the monomial.

The process used to carry out **division of a polynomial by a polynomial** is essentially the same as the long division process used in arithmetic. The division of one polynomial by another is considered to be complete when the degree of the remainder is less than the degree of the divisor.

The notation $\left(\dfrac{f}{g}\right)(x)$ indicates that two polynomials designated $f(x)$ and $g(x)$ are to be divided.

5.5 Synthetic Division

If a polynomial in x is to be divided by a linear divisor that is of the form $x - a$ or $x + a$, **synthetic division** can be used. Synthetic division is a shorthand form of division that is dependent only on the coefficients of the dividend and the constant term in the divisor.

COOPERATIVE EXERCISE

The Acme Toaster Oven Company has decided to start making microwave ovens. The Research Department has compiled data for three different possible sites for the microwave factory (see table). The Research Department has also noted that to be competitive, Acme should sell the microwave ovens for $90: at that price the company could sell all of the microwave ovens that it makes.

The Research Department's data include fixed costs (rent, advertising, etc.), variable costs (materials, labor, etc.), and an estimate of the maximum number of units that can be built at the site.

DATA CATEGORY	SITE A	SITE B	SITE C
Fixed costs	$25,000	$31,500	$40,000
Variable costs (per oven)	$40	$45	$50
Maximum number made	650	850	900

1. Write a linear equation representing revenue (R) in terms of the number x of microwaves sold. *Note:* This will be the same for all three sites.
2. For each site, write a linear equation representing cost (C) in terms of the number x of microwaves made.
3. For each site, graph the revenue equation from Number 1 and the cost equation from Number 2.
4. The break-even point is where cost equals revenue (the company breaks even). Label this point on each graph and solve each system of equations to find the coordinates of the break-even points.
5. Consider each graph and its break-even point. How many ovens does the company have to make in order to break even for each site?
6. If the company makes more ovens than the number needed to break even, will they make money or lose money? Explain.
7. For each site, there is a maximum number of ovens

that can be made there. If that maximum number is made, what will the cost and revenue be for each site?

8. The profit equation is

$$\text{profit} = \text{revenue} - \text{cost}$$

Find the possible profit at each site.

9. The *percent return* equation is

$$\frac{\text{profit}}{\text{cost}} \times 100 = \text{percent return}$$

Calculate the percent return for each site.

10. Write a paragraph recommending a site and give the reasons for your recommendation.

Review Exercises

Simplify each of the following expressions.

1. $7^2 \cdot 14^{-1}$
2. $8^{-1} \cdot 16^1$
3. $-5^2 \cdot (-3)^{-2}$
4. $(5 \times 10^1)(7 \times 10^{-1})$
5. $(0.3)^{-2}(10)^2$
6. $(4^2 - 3^{-2})^2$
7. $x^4 \cdot x^5$
8. $y^5 \cdot y^{-6}$
9. $x^m \cdot x^n$
10. $\dfrac{-p^2}{(-p)^2}$
11. $\dfrac{6x^5y^2}{3x^3y^3}$
12. $(x^{n+1})^2$
13. $15x - x$
14. $\left(\dfrac{1}{2}y - \dfrac{1}{4}y\right)^3$
15. $5x - [-2x - (3x - 4)]$
16. $(x^2 - 5x - 1) - (x^2 - 4x + 3)$
17. $(3x^{-2}y^3z^{-4})^{-2}$
18. $\dfrac{(6x^2y^3)^{-1}}{(3x^{-2}y^{-1})^{-3}}$
19. $\left(\dfrac{2x^{-4}y^{-3}}{(-2)^2x^2y^{-4}}\right)^2$

Find each of the following products and simplify if possible.

20. $3x^3(5x)$
21. $3^2x^2(x - 1)$
22. $(a - 2)^2$
23. $(2x - 3)(3x - 7)$
24. $(p - 1)^3$
25. $(7x + 2y)(7x - 2y)$
26. $(2x^2 - 11y^2)(2x^2 + 3y^2)$
27. $(3x - 1)^2 - (2x + 1)^2$
28. $[(2x - 1) + 3][(2x - 1) - 3]$
29. $(2p^2 - 3)(p^3 + 2p + 5)$

Perform each of the following divisions.

30. $(4a + 8) \div (2a)$
31. $(6t^2 - 8t + 2) \div (2)$
32. $(15x - 45) \div (-15)$
33. $(x^2 - 7x + 5) \div (x + 2)$
34. $(m^3 - 27) \div (m - 3)$

35. $(64a^3 - 1) \div (4a - 1)$
36. $(x^4 - 5x^3 + 2x^2 - 7x + 9) \div (x^2 + 2x + 3)$

Use synthetic division to find the quotient and remainder in Exercises 37–40.

37. $(p^4 + 2p^3 + 11p^2 + 17p - 10) \div (p + 2)$
38. $(3x^4 - 7x^2 + 9) \div (x - 3)$
39. $\left(\dfrac{1}{2}x^3 - 2x^2 + \dfrac{1}{4}x - 7\right) \div \left(x - \dfrac{1}{2}\right)$
40. $(17x^2 + 33x + 16) \div (x + 1)$

Work each of the following geometry problems as indicated.

41. Given a parallelogram with short side $a(x) = x + 3$, long side $b(x) = x + 5$, and altitude $h(x) = x + 1$:
    (a) find the perimeter $P(x)$;
    (b) find the area $A(x)$.

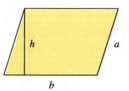

42. Given a triangle with sides $a(x) = x + 3$, $b(x) = x + 5$, and $c(x) = x + 3$ and altitude $h(x) = x$:
    (a) find the area $A(x)$;
    (b) find the perimeter $P(x)$.

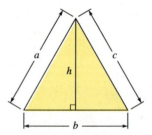

43. Given a rectangle of width $w(x) = x + 5$ and a semicircle of radius $r(x) = x + 3$:

(a) find the perimeter $P(x)$ of the accompanying figure;

(b) find the area $A(x)$ of the figure;

(c) find the area $A(x)$ of just the semicircle.

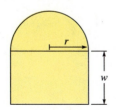

44. Given the rectangle pictured in the accompanying figure with two circles of radius $r(x) = x$ inscribed side by side:

(a) find the area $A_c(x)$ of *one* circle;

(b) find the area $A_r(x)$ of the shaded region.

(c) find the perimeter $P_c(x)$ of one circle.

(d) find the perimeter $P_r(x)$ of the rectangle.

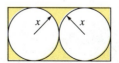

Chapter Test

Multiply or divide as indicated and simplify if possible.

1. $(15x)(2x^2)$
2. $(35x^3 - 49x^2 + 14x) \div (-7x)$
3. $(3x - 2)^2$
4. $(7x - 3)(7x + 3)$
5. Use synthetic division: $(x^2 + 10x - 7) \div (x - 3)$
6. Use long division: $(5x^2 - 15x + 6) \div (5x - 1)$
7. $(5m + 8)(4m - 7)$
8. $(2y^2 - 7)(3y^2 + 1)$
9. $a^m(a^{m+1} + a^m + 2)$
10. $(m - n)(m^2 + mn + n^2)$

For each of the following polynomials: (a) write it in standard form and simplify it; (b) state the degree; and (c) state whether it is a monomial, a binomial, or a trinomial.

11. $3 - x^2$
12. $x^3 - x + 7x^2 + 2x - 5x^2$
13. $15x - 3(5x)$
14. $y^0 - 11y^2 + 15y^3$

Simplify each of the following expressions.

15. $(5p - 1) - [3(2p + 5)]$
16. $(x^2 + 1) - (1 - x) + (3 - 2x^2)$
17. $4y - 3 - [6x - (3 + 5y - x)]$
18. $x + x^5 - 5(x + 1)$
19. $\left(\dfrac{10xy^2}{3x^2y}\right)^{-1}$
20. $\dfrac{(-2r^2)^2}{-4r^2}$
21. $\dfrac{6.3 \times 10^{-7}}{4.2 \times 10^{-8}}$
22. $\dfrac{(22{,}000{,}000)(3{,}000{,}000)}{110{,}000}$

23. Divide $7x^4 - 8x^3 + 4x^2 - 3$ by $x^2 - 4$.

24. $\left(x^3 - \dfrac{1}{2}x^2 + \dfrac{1}{3}x - 5\right) \div \left(x - \dfrac{1}{2}\right)$

25. Simplify:

$$-xy(x^2 - 3y + 2) + xy(-7y + x^2 - 7) - 2y(x^3 - xy + 4)$$

26. Expand and simplify: $(2x - 1)^3$.

27. Expand and simplify: $[(3x - 1) + 2][(3x - 1) - 2]$.

28. $(x^3 - 3x^2 + 5x - 3) \div (x^2 - 2x + 3)$

29. The box in the accompanying figure has volume $V(x) = x^3 + 6x^2 + 12x + 8$. Find the height $h(x)$ if the length is $l(x) = x + 2$ and the width is $w(x) = x + 2$.

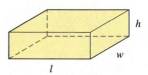

30. Find the area $A(x)$ of the trapezoid in the figure if it has altitude $h(x) = x - 1$, larger base $B(x) = 2x + 3$, and smaller base $b(x) = x + 4$.

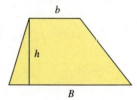

CHAPTER 6
FACTORING POLYNOMIALS; FACTORING BY GROUPING

6.1 Monomial Factors; Factoring by Grouping
6.2 Factoring Type I Trinomials
6.3 Factoring Type II Trinomials; A Test for Factorability
6.4 Factoring Special Types of Polynomials; More Factoring by Grouping
6.5 General Methods of Factoring
6.6 Solving Nonlinear Equations by Factoring
6.7 More Applications; Solving Applied Problems by Factoring

APPLICATION

A square garden is surrounded by a brick walk 2 feet wide. The combined area of the garden and the walk is 196 square feet. Find the length of one side of the garden.

See Section 6.7, Example 1.

Leonard Euler was a great mathematician known for his ingenious algorithms. An algorithm *is a systematic method for solving a problem. Euler would not be satisfied with the solution to a problem until he had reduced it to one of his algorithms. It has been said that he could make discoveries in three days that were forever beyond the reach of other mathematicians. Besides being a mathematical genius, Euler was also an ordinary family man with thirteen children.*

LEONHARD EULER (1707–1783)

6.1 Monomial Factors; Factoring by Grouping

OBJECTIVES   In this section we will learn how to
1. factor polynomials by removing the greatest common factor; and
2. factor basic expressions by grouping.

In Section 5.3 we discussed methods for finding the product of two polynomials. Factoring is the reverse operation of finding products. We begin with a polynomial and we rewrite it as the product of polynomials.

The simplest type of factoring associated with polynomials involves the **greatest common factor (GCF).** To understand the concept of the GCF, recall from arithmetic that every integer can be expressed as the product of *prime numbers*.

DEFINITION

Greatest Common Factor of Integers

The **greatest common factor, abbreviated GCF, of two or more integers** is the greatest integer that is a factor of each integer.

For example, 6 is the greatest common factor of 30 and 42:

$$30 = 2 \cdot 3 \cdot 5 = 6 \cdot 5$$

while

$$42 = 2 \cdot 3 \cdot 7 = 6 \cdot 7$$

EXAMPLE 1   Find the greatest common factor of $30x^4$, $42x^6$, and $12x^2$.

SOLUTION
$$30x^4 = 2 \cdot 3 \cdot 5 \cdot x^2 \cdot x^2$$
$$42x^6 = 2 \cdot 3 \cdot 7 \cdot x^2 \cdot x^4$$
$$12x^2 = 2 \cdot 2 \cdot 3 \cdot x^2$$

The greatest common factor is $2 \cdot 3 \cdot x^2 = 6x^2$.

Finding the greatest common factor of a polynomial is based on the distributive property:

$$a(b + c) = ab + ac$$

When we interchange the left- and right-hand sides by use of the symmetric property, we have

$$ab + ac = a(b + c)$$

Here a is "common" to both the terms ab and ac and so is a *common factor* of $ab + ac$. Applying this reasoning to the expression $6x + 21$, the *greatest common factor* of both $6x$ and 21 is 3. Thus $6x + 21$ can be written

$$6x + 21 = 3 \cdot 2x + 3 \cdot 7$$
$$= 3(2x + 7) \quad \text{Distributive property}$$

EXAMPLE 2 Factor $8x + 20$ by removing the GCF.

SOLUTION
$8x + 20 = 4 \cdot 2x + 4 \cdot 5$ 4 is the GCF.
$= 4(2x + 5)$ Distributive property

Notice that 2 is also a common factor of $8x + 20$. It is not, however, the *greatest* common factor.

EXAMPLE 3 Factor each of the following polynomials by removing the GCF.

(a) $12y + 9 = 3(4y + 3)$ The GCF is 3.
(b) $7x^2 + 14y^3 = 7(x^2 + 2y^3)$ The GCF is 7.
(c) $8z + 8 = 8(z + 1)$ The GCF is 8.
(d) $6x + 5 = 1(6x + 5)$ The GCF is 1.

NOTE Whenever an entire term is removed as a factor, it must be replaced by a 1.

The greatest common factor can involve variables, as was the case in Example 1, where we looked at $30x^4$, $42x^6$, and $12x^2$. If these were terms of the polynomial $30x^4 + 42x^6 + 12x^2$, then the factorization would be

$$30x^4 + 42x^6 + 12x^2 = 6x^2(5x^2 + 7x^4 + 2) \quad 6x^2 \text{ is the GCF.}$$

In general, we have the following definition.

DEFINITION

Greatest Common Factor of Algebraic Expressions

The **greatest common factor of two or more algebraic expressions** includes any common variable with an exponent that is equal to the smallest exponent on that variable in addition to the greatest common numerical factor.

EXAMPLE 4 Factor each polynomial by removing the GCF.

(a) $4x^3 + 2x^2 + 8x$
(b) $6a^4 + 3a^3 + 12a^2$
(c) $7x^4y + 3x^2y - 15x^3y^3$

SOLUTION (a) The greatest common numerical factor is 2, and the smallest exponent on x is 1. Therefore the greatest common factor is $2x$.

$$4x^3 + 2x^2 + 8x = 2x \cdot 2x^2 + 2x \cdot x + 2x \cdot 4$$
$$= 2x(2x^2 + x + 4) \qquad \text{Distributive property}$$

(b) The greatest common numerical factor is 3, while the smallest exponent on a is 2, so the greatest common factor is $3a^2$.

$$6a^4 + 3a^3 + 12a^2 = 3a^2 \cdot 2a^2 + 3a^2 \cdot a + 3a^2 \cdot 4$$
$$= 3a^2(2a^2 + a + 4)$$

(c) The greatest common numerical factor is 1, the smallest exponent on x is 2, and the smallest exponent on y is 1. Therefore the greatest common factor is $1x^2y = x^2y$.

$$7x^4y + 3x^2y - 15x^3y^3 = (x^2y)(7x^2) + (x^2y)(3) + (x^2y)(-15xy^2)$$
$$= x^2y(7x^2 + 3 - 15xy^2)$$

When the leading coefficient of a polynomial is negative, it is common practice to seek a common factor that is negative.

EXAMPLE 5 Factor and check the result: $-2x^4 + 6x^2 - 4x$.

SOLUTION The greatest common numerical factor is 2 and the smallest exponent on x is 1. Because the leading coefficient is negative, we use $-2x$ as the common factor:

$$-2x^4 + 6x^2 - 4x = -2x(x^3) + (-2x)(-3x) + (-2x)(2)$$
$$= -2x[x^3 + (-3x) + 2]$$
$$= -2x(x^3 - 3x + 2)$$

To check this factorization, we must find the product of the factors. Such a check, however, does not guarantee that the *largest* common factor was removed—only that a common factor was removed.

CHECK $-2x(x^3 - 3x + 2) \stackrel{?}{=} -2x^{3+1} + 6x^{1+1} - 4x$
$\stackrel{\checkmark}{=} -2x^4 + 6x^2 - 4x$

EXAMPLE 6 Factor out the indicated factor:

$$6x^2y + \frac{1}{3}x^2 + \frac{2}{9}x^4 = \frac{1}{9}x^2(\qquad)$$

SOLUTION To remove a factor of $\frac{1}{9}$, each numerical coefficient will have to be thought of in terms of ninths.

$$6x^2y + \frac{1}{3}x^2 + \frac{2}{9}x^4 = \frac{54}{9}x^2y + \frac{3}{9}x^2 + \frac{2}{9}x^4$$
$$= \frac{1}{9}x^2(54y + 3 + 2x^2)$$

EXAMPLE 7 Factor: $5x^{2m-2}y^{m+1} + 10x^{2m}y^{m+3} + 25x^{2m-2}$.

SOLUTION The greatest common numerical factor is 5 and the smallest exponent on the variable x is $2m - 2$, but there is no common factor of y. Therefore the greatest common factor is $5x^{2m-2}$.

$$5x^{2m-2}y^{m+1} + 10x^{2m}y^{m+3} + 25x^{2m-2}$$
$$= 5x^{2m-2}(y^{m+1}) + 5x^{2m-2}(2x^2y^{m+3}) + 5x^{2m-2}(5) \quad\quad 10x^{2m} = 5x^{2m-2}2x^2$$
$$= 5x^{2m-2}(y^{m+1} + 2x^2y^{m+3} + 5)$$

EXAMPLE 8 Factor: $6x^{m+1} + 3x^{m-1}$.

SOLUTION The greatest common numerical factor is 3. The exponent $m + 1$ is 2 greater than the exponent $m - 1$. Thus, the smallest exponent on x is $m - 1$. The greatest common factor is $3x^{m-1}$.

$$6x^{m+1} + 3x^{m-1} = 3x^{m-1}(2x^2) + 3x^{m-1}(1) \quad\quad x^{m+1} = x^{m-1}(x^2)$$
$$= 3x^{m-1}(2x^2 + 1)$$

EXAMPLE 9 Factor by removing the greatest common binomial factor: $3(x + 1)^2 - (x + 1)$.

SOLUTION
$$3(x + 1)^2 - (x + 1) = (x + 1)[3(x + 1) - 1] \quad\quad \text{GCF} = x + 1$$
$$= (x + 1)(3x + 3 - 1) \quad\quad \text{Distributive property}$$
$$= (x + 1)(3x + 2)$$

EXAMPLE 10 Factor: $(x + 5)^4(2x - 1)^5 - 3(x + 5)^3(2x - 1)^6$.

SOLUTION The greatest common factor is $(x + 5)^3(2x - 1)^5$.

$$(x + 5)^4(2x - 1)^5 - 3(x + 5)^3(2x - 1)^6 = (x + 5)^3(2x - 1)^5[(x + 5) - 3(2x - 1)]$$
$$= (x + 5)^3(2x - 1)^5[x + 5 - 6x + 3]$$
$$= (x + 5)^3(2x - 1)^5[-5x + 8]$$

EXAMPLE 11 Factor: $(m + n)^5(x - y)^3z^3 - (m + n)^2(x - y)^4z^5 - (m + n)(x - y)^8z^7$.

SOLUTION The greatest common factor is $(m + n)(x - y)^3z^3$, so

$$(m + n)^5(x - y)^3z^3 - (m + n)^2(x - y)^4z^5 - (m + n)(x - y)^8z^7$$
$$= (m + n)(x - y)^3z^3[(m + n)^4 - (m + n)(x - y)z^2 - (x - y)^5z^4]$$

We now consider a special type of factoring called **factoring by grouping**. This name originates from the idea that how the terms of a polynomial are grouped may help us see how to factor it. For example, $ax + by + bx + ay$ can be factored by grouping the first and third terms and the second and fourth terms as follows:

$$ax + by + bx + ay = (ax + bx) + (ay + by) \quad \text{Commutative and associative properties}$$
$$= x(a + b) + y(a + b) \quad \text{Distributive property}$$

We now have a common factor of $a + b$.

$$x(a + b) + y(a + b) = (a + b)(x + y) \quad \text{Distributive property}$$

We could also have factored the expression by grouping it in terms of a and b:

$$ax + by + bx + ay = (ax + ay) + (bx + by)$$
$$= a(x + y) + b(x + y) = (x + y)(a + b)$$

EXAMPLE 12 Factor: $x^2 - 4x + 2x - 8$.

SOLUTION Since x is a factor of the first two terms and 2 is a factor of the last two terms, we remove these common factors:

$$x^2 - 4x + 2x - 8 = (x^2 - 4x) + (2x - 8) \quad \text{Associative property}$$
$$= x(x - 4) + 2(x - 4) \quad \text{Distributive property}$$

Because $x - 4$ is now a common factor, we remove it.

$$x(x - 4) + 2(x - 4) = (x - 4)(x + 2) \quad \text{Distributive property}$$

EXAMPLE 13 Factor: $x^2 + 2x + xy + 2y$.

SOLUTION Since x is a factor of the first two terms and y is a factor of the last two terms, we remove these common factors.

$$x^2 + 2x + xy + 2y = (x^2 + 2x) + (xy + 2y) \quad \text{Associative property}$$
$$= x(x + 2) + y(x + 2) \quad \text{Distributive property}$$

Now $x + 2$ is a common factor, so we remove it.

$$x(x + 2) + y(x + 2) = (x + 2)(x + y) \quad \text{Distributive property}$$

EXAMPLE 14 Factor: $m^2 - 5m - 3m + 15$.

SOLUTION Notice that m is a factor of the first two terms while -3 is a factor of the last two terms.

When removing a common negative factor, change all of the signs in the other factor.

$$m^2 - 5m - 3m + 15 = m(m - 5) - 3(m - 5)$$
$$= (m - 5)(m - 3) \quad m - 5 \text{ is a common factor.}$$

Notice in Example 14 that 3 was also a factor of the last two terms, but if we factor it out, we will not find a common factor:

$$m^2 - 5m - 3m + 15 = m(m - 5) + 3(-m + 5)$$

If we use 3 as a factor of the last two terms, the resulting expression cannot be factored further, since $m - 5$ is not the same as $-m + 5$. Using -3 is better.

EXAMPLE 15 Factor: $m^2 - mn - m + n$.

SOLUTION
$$m^2 - mn - m + n = (m^2 - mn) + (-m + n)$$
$$= m(m - n) - 1(m - n)$$
$$= (m - n)(m - 1)$$

EXAMPLE 16 Factor: $15x^2 - 20x + 6x - 8$.

SOLUTION
$$15x^2 - 20x + 6x - 8 = 5x(3x - 4) + 2(3x - 4)$$
$$= (3x - 4)(5x + 2)$$

We can also factor this expression by arranging the terms in a different order.

$$15x^2 - 20x + 6x - 8 = 15x^2 + 6x - 20x - 8 \quad \text{Commutative property}$$
$$= 3x(5x + 2) - 4(5x + 2)$$
$$= (5x + 2)(3x - 4)$$

EXAMPLE 17 Factor: $(x - 4)^2 + 4 - x$.

SOLUTION First, $x - 4$ is a factor of $(x - 4)^2$. Next, since $4 - x$ is $= -1(x - 4)$, we can create a common factor of $x - 4$ by factoring -1 from the last two terms. Thus,

$$(x - 4)^2 + 4 - x = (x - 4)(x - 4) - 1(x - 4)$$
$$= (x - 4)[(x - 4) - 1]$$
$$= (x - 4)(x - 4 - 1)$$
$$= (x - 4)(x - 5)$$

Exercise Set 6.1

Remove the indicated factor.

1. $3x + 6 = 3(\quad)$
2. $5a + 10 = 5(\quad)$
3. $14m - 21 = 7(\quad)$
4. $15p - 20 = 5(\quad)$
5. $7s^2 + s = s(\quad)$
6. $13r^2 - r = r(\quad)$
7. $-y^2 + 3y = -y(\quad)$
8. $-c^2 - 11c = -c(\quad)$
9. $-5x^2 - 10x + 15 = -5(\quad)$
10. $-16a^2 + 24a - 8 = -8(\quad)$
11. $-18t^3 + 24t^2 - 12t = -6t(\quad)$
12. $-28m^3 - 49m^2 + 42m = -7m(\quad)$
13. $-\frac{1}{5}a^2 + a - \frac{1}{10} = -\frac{1}{10}(\quad)$
14. $-\frac{1}{2}c^2 + 2c - \frac{1}{4} = -\frac{1}{4}(\quad)$

15. $\frac{3}{8}p^3 + 2p^2 - \frac{5}{4}p = \frac{1}{8}p(\quad)$

16. $\frac{5}{7}t^3 - 3t^2 - \frac{1}{5}t = \frac{1}{35}t(\quad)$

17. $-\frac{4}{9}s^4 + \frac{2}{3}s^3 - \frac{1}{9}s^2 = -\frac{1}{9}s^2(\quad)$

18. $-\frac{2}{5}y^4 - \frac{1}{10}y^3 + 2y^2 = -\frac{1}{10}y^2(\quad)$

19. $-\frac{2}{7}r^5 + \frac{3}{4}r^4 - \frac{1}{2}r^3 = -\frac{1}{28}r^3(\quad)$

20. $-\frac{2}{11}x^5 + \frac{1}{2}x^4 - \frac{1}{3}x^3 = -\frac{1}{66}x^3(\quad)$

Factor by removing the greatest monomial factor. If the expression cannot be factored, write prime.

21. $2x + 10$
22. $3x + 21$
23. $11y - 55$
24. $13y - 39$
25. $15a^2 - 50a$
26. $12m^2 - 30m$
27. $8 - 16t$
28. $7 - 14s$
29. $3a + 9b - 6$
30. $7x + 14y - 21$
31. $x^2y - xy^2$
32. $r^2s - 2rs^2$
33. $12m^2p - 6mp^2 + 18mp$
34. $30x^2y - 15xy^2 + 45xy$
35. $at + bt - ct$
36. $ry - ay + my$
37. $-2x^2 + 10x - 5$
38. $-3x^2 + 15x - 2$
39. $15r^5s^3 - 24r^2s^4$
40. $18a^6b^4 - 30a^3b^2$
41. $5a + 7a^2 - 9a^3$
42. $7s^5 - 18s^3 + 24s$
43. $28x^4 - 56x^3 - 49x^2$
44. $65t^5 - 39t^3 - 91t^2$
45. $8r^5s^3 - 6r^4s^5 - 20r^4s^3$
46. $34x^3y^5 - 51x^2y^4 - 102x^2y^6$
47. $48xyz - 64x^2yz^2 - 80xy^2z$
48. $44x^2y^2z - 77xy^2z^2 - 99xy^2z$
49. $63a^3c - 81a^2d + 72acd$
50. $88ab^2c - 56abc^2 + 80c$

Factor by removing the greatest common binomial factor.

51. $x(x + 1) + 2(x + 1)$
52. $y(y - 3) - 2(y - 3)$
53. $a(2a + 1) - (2a + 1)$
54. $2t(3t - 4) - (3t - 4)$
55. $m(a + c) + n(a + c)$
56. $r(e + f) - t(e + f)$
57. $x(x - y) - y(x - y)$
58. $2x(2x - 3y) - 3y(2x - 3y)$
59. $t(t - 1)^2 - (t - 1)$
60. $a(a^2 + 1) - (a^2 + 1)$
61. $(m - 5)4 - (m - 5)m$
62. $(2y - 3)5 - (2y - 3)3y$
63. $\left(\frac{1}{2} - a\right)\frac{1}{3} - \left(a - \frac{1}{2}\right)a$
64. $\left(\frac{1}{4} - m\right)\frac{1}{5} - \left(m - \frac{1}{4}\right)m$
65. $(x - 1)^2 + (1 - x)$
66. $(y + 2)^2 - (y + 2)$
67. $(2a + 3)^2 + 2(2a + 3)$
68. $(3c - 1)^2 - 5(1 - 3c)$
69. $(a - b)^4(a^2 + b)^3 - 3(a - b)^3(a^2 + b)^2$
70. $(x - y)^5(x^2 - 2y)^3 - 2(x - y)^4(x^2 - 2y)^2$

Factor by removing the greatest monomial factor. Assume m is a natural number.

71. $a^{m+1} + a^m$
72. $y^{m+2} + y^m$
73. $x^{2m+1} - x^{2m}$
74. $t^{3m+1} - t^{3m}$
75. $y^m + y$
76. $s^m + s^{m-1}$
77. $r^{2m+3} - r^{2m+2}$
78. $x^{5m-4} - x^{5m-5}$

Factor by grouping.

79. $cx + cy + dx + dy$
80. $5x + 5y + sx + sy$
81. $2a^2 + 2ab - 4a - 4b$
82. $2m^2 + 2mn - 6m - 6n$
83. $mr + 2nr + 2ms + 4ns$
84. $4ac - 6bc + 4ad - 6bd$
85. $ax - bx - ay + by$
86. $2ax - 4bx - 3ay + 6by$
87. $4abxy - 6abz - 10cxy + 15cz$
88. $15acmn - 25acx - 6mnz + 10xz$
89. $x^2 + 2x - 5x - 10$
90. $x^2 + 4x - 2x - 8$
91. $5x^2 + 15x - x - 3$
92. $4y^2 + 6y - 6y - 9$
93. $28y^2 - 21y - 4y + 3$
94. $8p^2 - 20p - 4p + 10$
95. $8x^2 + 4x + 2x + 1$
96. $6p^2 + 18p + 2p + 6$
97. $30x^2 - 75x + 12x - 30$
98. $42y^2 + 6y - 28y - 4$

REVIEW EXERCISES

Carry out the indicated operations.

99. $(x + y)(x - y)$
100. $(4x^2 + 6x + 11) \div (2x - 3)$
101. $(18x^2y + 6xy^2 - 24xy) \div (-6xy)$
102. $(a^2 - 7a - 30) \div (a - 10)$
103. $(3m + 2n)^2$
104. $(x^m - y^n)(x^m + y^n)$
105. $(a - b)(a^2 + ab + b^2)$
106. $(r + s)(r^2 - rs + s^2)$
107. $(x^5 - 32) \div (x - 2)$
108. $(-54a^3b^2c - 30a^2b^3c^2 + 18ab^4c) \div (-6ab^2c)$
109. $(a^m + a^n)^2$
110. $(x^3 + 3x^2 + 3x + 1) \div (x + 1)$
111. Find a polynomial that can be factored as $-2x(3x + 2)(x - 5)$.
112. Find a polynomial that can be factored as $-(5x - 1)(4x + 7)$.
113. Find a polynomial that can be factored as $2(x - 3)^3$.
114. Find a polynomial that can be factored as $2(x + 2)^3$.

WRITE IN YOUR OWN WORDS

115. Is $(3x + 6)(2x + 1)$ the completely factored form of $6x^2 + 15x + 6$? Why or why not?
116. Which is (are) the completely factored form(s) of $10 - 3x - x^2$?
    (a) $(2 - x)(x + 5)$; **(b)** $-(x + 5)(x - 2)$; or both **(a)** and **(b)**?
117. Is $(x - 3)(x - y)^2$ the completely factored form of $(x - 4)(x - y)^2 + (x - y)^2$? Explain.
118. Is $(p - 3)(m + n)^2$ the completely factored form of $(m + n)^2(p + 6) - 3(m + n)^2$? Explain.

6.2 Factoring Type I Trinomials

OBJECTIVES In this section we will learn to factor trinomials with leading coefficient 1.

The expressions

$$x^2 + 6x + 5, \quad y^3 - 2y^2 - y, \quad \text{and} \quad m^4 - 3m^2 - 4$$

are all trinomials, in which the **leading coefficient,** the coefficient of the highest-degree term, equal to 1. For example,

$$x^2 + 6x + 5 = 1x^2 + 6x + 5$$

A trinomial is often the product of two binomials. Therefore, if you try to factor $x^2 + 6x + 5$, it is reasonable to expect that the factors may each be a binomial. How do we go about finding them? To factor $x^2 + 6x + 5$, we want to find two numbers m and n such that

$$(x + m)(x + n) = x^2 + 6x + 5$$

However,

$$(x + m)(x + n) = x^2 + mx + nx + mn \quad \text{FOIL method}$$
$$= x^2 + (m + n)x + mn \quad \text{Distributive property}$$

Thus

$$x^2 + (m + n)x + mn = x^2 + 6x + 5$$

This shows that $m \cdot n = 5$ and $m + n = 6$. So we are seeking two numbers m and n that have a product $m \cdot n = 5$ and a sum $m + n = 6$. The numbers are 1 and 5. We let $m = 1$ and $n = 5$:

$$x^2 + 6x + 5 = (x + 1)(x + 5)$$

These results can be quickly checked by means of the FOIL method to make sure the sum of the inner and outer products is $6x$ and that the product of the last terms is 5.

CHECK $(x + 1)(x + 5) \stackrel{?}{=} x^2 + 6x + 5$

EXAMPLE 1 Factor: $x^2 + 8x + 12$.

SOLUTION To factor $x^2 + 8x + 12$, note that the leading coefficient is 1 and then look for two numbers with a product of 12 and a sum of 8. What possibilities exist?

Product	Sum
$12 \cdot 1 = 12$	$12 + 1 = 13$
$6 \cdot 2 = 12$	$6 + 2 = 8$
$4 \cdot 3 = 12$	$4 + 3 = 7$

Only 6 and 2 satisfy both conditions. Thus,

$$x^2 + 8x + 12 = (x + 2)(x + 6)$$

EXAMPLE 2 Factor: $a^2 - 7a - 18$.

SOLUTION If we think of $a^2 - 7a - 18$ as $a^2 - 7a + (-18)$, then the last term is negative. This means that the signs of the two numbers we're looking for will be different. We need to find two numbers with a product of -18 and a sum of -7. The possibilities are as follows:

Product	Sum
$18(-1) = -18$	$18 + (-1) = 17$
$9(-2) = -18$	$9 + (-2) = 7$
$6(-3) = -18$	$6 + (-3) = 3$
$3(-6) = -18$	$3 + (-6) = -3$
$2(-9) = -18$	$2 + (-9) = -7$
$1(-18) = -18$	$1 + (-18) = -17$

The numbers are -9 and 2.

$$a^2 - 7a - 18 = (a - 9)(a + 2)$$

EXAMPLE 3 Factor: $3y^3 + 12y^2 - 15y$.

SOLUTION The coefficient of the first term is not 1. However, $3y$ is a common factor that we can remove first.

$$3y^3 + 12y^2 - 15y = 3y(y^2 + 4y - 5)$$

Now we can factor $y^2 + 4y - 5$ by paying special attention to the sign before the 5, like we did in Example 2.

$$3y^3 + 12y^2 - 15y = 3y(y^2 + 4y - 5)$$
$$= 3y(y + 5)(y - 1)$$

*C*AUTION

When factoring polynomials, always remove the greatest common factor first.

EXAMPLE 4 Factor: $y^4 + 15y^2 + 54$.

SOLUTION The polynomial $y^4 + 15y^2 + 54$ fits the pattern we established in Examples 1 and 2. To see this, we substitute m for y^2. If $m = y^2$ then $m^2 = y^4$, which means we substitute m^2 for y^4.

$$y^4 + 15y^2 + 54 = m^2 + 15m + 54$$
$$= (m + 6)(m + 9)$$

Now we substitute y^2 back for m, and we have

$$y^4 + 15y^2 + 54 = (y^2 + 6)(y^2 + 9)$$

Neither $y^2 + 6$ nor $y^2 + 9$ can be factored further with our present knowledge.

EXAMPLE 5 Factor: $x^2(x + 2) - 8x(x + 2) - 20(x + 2)$.

SOLUTION We remove the common factor $x + 2$ first:

$$x^2(x + 2) - 8x(x + 2) - 20(x + 2) = (x + 2)(x^2 - 8x - 20)$$
$$= (x + 2)(x + 2)(x - 10)$$
$$= (x + 2)^2(x - 10) \quad (x + 2)(x + 2) = (x + 2)^2$$

EXAMPLE 6 Factor: $-2a^2 - 4a + 70$.

SOLUTION Since -2 is a common factor, we remove it first.

$$-2a^2 - 4a + 70 = -2(a^2 + 2a - 35)$$
$$= -2(a + 7)(a - 5)$$

EXAMPLE 7 Factor: $15 - 2x - x^2$.

SOLUTION There are two ways to factor this expression. If we factor it in the form in which it is given, then we get

$$15 - 2x - x^2 = (5 + x)(3 - x)$$

> **CAUTION**
>
> When factoring polynomial expressions involving fractions, do not eliminate fractions by multiplying through by the LCD. Fractions are eliminated from an *equation* to be solved by multiplying each side by the LCD. Right now the focus is on factoring, not equation-solving.

We can also change the order of the terms by using the commutative property for addition and removing a common factor of -1.

$$15 - 2x - x^2 = -x^2 - 2x + 15$$
$$= -1(x^2 + 2x - 15)$$
$$= -1(x + 5)(x - 3)$$

-1 is a common factor that must appear in the final factorization.

Thus,

$$(5 + x)(3 - x) = -1(x + 5)(x - 3) \quad \text{Notice that } -1(x - 3) = 3 - x.$$

Either $(5 + x)(3 - x)$ or $-1(x + 5)(x - 3)$ is an acceptable result.

EXAMPLE 8 Factor: $x^2 + \frac{5}{3}x + \frac{4}{9}$.

SOLUTION We need to find two numbers with a product of $\frac{4}{9}$ and a sum of $\frac{5}{3}$. The numbers are $\frac{1}{3}$ and $\frac{4}{3}$.

$$x^2 + \frac{5}{3}x + \frac{4}{9} = \left(x + \frac{1}{3}\right)\left(x + \frac{4}{3}\right) \qquad \frac{1}{3} \cdot \frac{4}{3} = \frac{4}{9} \text{ and } \frac{1}{3} + \frac{4}{3} = \frac{5}{3}$$

EXAMPLE 9 Factor: $x^{2n} + 3x^n - 4$.

SOLUTION We think of x^{2n} as $(x^n)^2$. Then

$$x^{2n} + 3x^n - 4 = (x^n + 4)(x^n - 1)$$

We could also have used substitution as an aid in determining the factors. If we substitute y for x^n then $x^{2n} = y^2$. Thus

$$x^{2n} + 3x^n - 4 = y^2 + 3y - 4 \qquad \text{Substitute } y \text{ for } x^n \text{ and } y^2 \text{ for } x^{2n}.$$
$$= (y + 4)(y - 1)$$
$$= (x^n + 4)(x^n - 1) \qquad \text{Substitute } x^n \text{ for } y.$$

EXAMPLE 10 Factor: $x^{m+2} + 8x^{m+1} + 7x^m$.

SOLUTION Since x^m is a common factor, we remove it first.

$$x^{m+2} + 8x^{m+1} + 7x^m = x^m(x^2 + 8x + 7) \qquad x^m \cdot x^2 = x^{m+2}; \quad x^m \cdot x = x^{m+1}$$
$$= x^m(x + 7)(x + 1)$$

Not all polynomials can be factored by the means we've used so far. Some polynomials are **prime polynomials**.

> **DEFINITION**
> **Prime Polynomial**
>
> A polynomial that is not factorable is said to be **prime.**

EXAMPLE 11 Factor: $x^2 - 6x + 7$.

SOLUTION To factor this polynomial, we have to find two numbers with a product of 7 and a sum of -6. Since the sign before the last term is *positive,* both of the desired numbers have to have the same sign, and since the middle term is *negative,* the last terms must both be negative. Thus the only possibilities are -7 and -1. But -7 and -1 don't add to -6, so the polynomial is not factorable.

Exercise Set 6.2

Factor each of the following trinomials.

1. $x^2 + 3x + 2$
2. $y^2 + 4y + 3$
3. $m^2 + 7m + 6$
4. $a^2 + 9a + 8$
5. $y^2 - 2y + 1$
6. $x^2 - 4x + 3$
7. $p^2 - 5p + 6$
8. $r^2 - 4r + 4$
9. $x^2 - x - 6$
10. $y^2 - y - 12$
11. $t^2 + 2t - 3$
12. $b^2 + 3b - 28$
13. $y^2 - 8y - 20$
14. $x^2 - 5x - 24$
15. $m^2 + m - 30$
16. $a^2 + 2a - 48$
17. $r^2 + 6ar - 16a^2$
18. $m^2 + mn - 72n^2$
19. $a^2 - 7ab - 60b^2$
20. $p^2 - 8pq - 48q^2$

Factor each of the following trinomials, removing any common factors first. If the trinomial is not factorable, write prime.

21. $2x^2 + 20x + 48$
22. $2y^2 - 12y + 10$
23. $a^2 + 4a + 2$
24. $m^2 + 3m + 1$
25. $5p^2 - 45p + 40$
26. $6r^2 - 30r - 84$
27. $10 - 3s - s^2$
28. $15 - 3x - x^2$
29. $x^3 - 4x^2 - 32x$
30. $y^3 + 9y^2 - 36y$
31. $4a^3 - 16a^2b - 20ab^2$
32. $2x^3 - 2x^2y - 84xy^2$
33. $r^2 + \frac{8}{15}r + \frac{1}{15}$
34. $s^2 + \frac{1}{20}s - \frac{1}{20}$
35. $k^2 - \frac{1}{6}k - \frac{1}{6}$
36. $m^2 - \frac{1}{12}m - \frac{1}{12}$
37. $y^2 - 7y - 1$
38. $a^2 - 5a - 2$
39. $a^2b^2 - 4ab + 4$
40. $x^2y^2 - 6xy + 9$
41. $-3m^2 - 3m + 36$
42. $-5z^2 - 25z + 70$
43. $-7a^3 - \frac{7}{2}a^2 + \frac{7}{16}a$
44. $-2r^2 - 2r + \frac{1}{2}$
45. $-x^2 + 8x - 12$
46. $-y^2 - 5y + 14$
47. $3st^3 + 33s^2t^2 + 54s^3t$
48. $4a^3b - 4a^2b^2 - 48ab^3$
49. $5p^3q + 10p^2q^2 + 5pq^3$
50. $2x^3y - 12x^2y^2 + 18xy^3$

Factor each polynomial, using substitution if needed. (See Example 4.)

51. $x^4 + 6x^2 + 8$
52. $y^4 + 7y^2 + 12$
53. $a^4 + 8a^2 + 16$
54. $b^4 + 10b^2 + 25$
55. $2m^4 + 18m^2 + 40$
56. $3s^4 + 12s^2 + 12$
57. $-5t^4 - 5t^2 + 150$
58. $-2c^4 + 16c^2 + 130$
59. $-k^3 + \frac{11}{30}k^2 - \frac{1}{30}k$
60. $-z^3 + \frac{10}{21}z^2 - \frac{1}{21}z$

Factor each of the following trinomials. If there is a common factor, remove it first.

61. $x^2(x + 3) - x(x + 3) - 2(x + 3)$
62. $y^2(y - 2) + 3y(y - 2) + 2(y - 2)$
63. $a^2(a - 3) - a(a - 3) - 6(a - 3)$
64. $z^2(z + 5) + 10z(z + 5) + 25(z + 5)$
65. $t^2(t + 7) + 14t(t + 7) + 49(t + 7)$
66. $s^2(s - 4) - 7s(s - 4) + 12(s - 4)$
67. $k^2\left(k - \frac{1}{3}\right) - \frac{2}{3}k\left(k - \frac{1}{3}\right) + \frac{1}{9}\left(k - \frac{1}{3}\right)$
68. $c^2\left(c - \frac{1}{5}\right) - c\left(c - \frac{1}{5}\right) + \frac{12}{49}\left(c - \frac{1}{5}\right)$

Factor each of the following trinomials. If there is a common factor, remove it first.

69. $x^{2m} + 8x^m + 12$
70. $x^{2m} + 9x^m + 20$
71. $y^{2m} - 0.3y^m + 0.02$
72. $a^{2m} - 0.9a^m + 0.2$
73. $c^{m+2} + 11c^{m+1} + 30c^m$
74. $y^{m+2} - 7y^{m+1} + 12y^m$
75. $s^{2m+2} - s^{m+2} - 2s^2$
76. $t^{2m+2} - 5t^{m+2} - 24t^2$
77. $k^{2m+3} - k^{m+3} + 0.21k^3$
78. $z^{2m+3} - 1.3z^{m+3} + 0.42z^3$

Use a calculator to remove the indicated common factor.

79. $0.585y^2 - 0.0715y;\ 0.65y$
80. $0.018x^2 + 0.0204x;\ 0.12x$
81. $-0.1023t^2 - 0.837t;\ -0.93t$
82. $-0.1881a^2 + 0.1767a;\ -0.57a$

REVIEW EXERCISES

Factor each of the following polynomials.

83. $6x^2 - 3x$
84. $4ab^2 - 8ab + 2ab^2$
85. $x(x + 3) + x^2(x + 3)$
86. $(x - 2)(x^2 + 2x) - (x - 2)(3x + x^2)$

Perform each of the following divisions.

87. $(15x^2 + 20x - 10) \div 5$
88. $(x^3 - x^2 + 2) \div (x^2 - 1)$
89. $(x^3 + x^2 - 2x - 1) \div (x - 2)$
90. $(x^5 - 32) \div (x - 2)$

Simplify each of the following expressions.

91. $(x^5)^3$
92. $y^{3n} \div y$
93. $x^{5n} \cdot x^6 \cdot x^n \cdot x^{-4}$
94. $(-2x^{-2})^3$
95. $[-(-3x)^2]^2$

ᴵor ℰxtra 𝒯hought

96. The formula for the area of a circle is $A = \pi r^2$, where r is the radius. If a circle has an area of $A(x) = 4\pi x^2 + 12\pi x + 9\pi$, explain how to find $r(x)$, the radius. Find $r(x)$, and find $r(3)$.

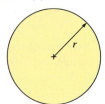

97. The formula for the area of a square is $A = s^2$, where s is a side. Given a square with an area of $A(x) = 25x^2 - 20x + 4$, explain how to find the length of $s(x)$, a side. Find $s(x)$, then find $s(2)$.

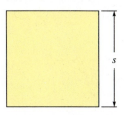

98. The area of a trapezoid is given by the formula $A = \frac{1}{2}h(B + b)$, where h is the height, B is the larger base, and b is the smaller base. If a trapezoid has an area of $A(x) = \frac{1}{2}(5x^2 + 15x)$, with a height of $h(x) = x + 3$ and a large base of $B(x) = 3x - 1$, explain how to find $b(x)$, the small base. Find $b(\frac{1}{2})$, $B(\frac{1}{2})$, and $h(\frac{1}{2})$.

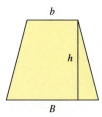

99. The formula for the volume of a right circular cone is $V = \frac{1}{3}\pi r^2 h$, where r is the radius of the base and h is the height of the cone. If a right circular cone has a volume of $V(x) = \frac{1}{3}(5\pi x^2 - \frac{5}{2}\pi x + 5\pi)$, explain how to find the radius, $r(x)$. Find $r(4)$.

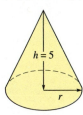

6.3 Factoring Type II Trinomials; A Test for Factorability

OBJECTIVES In this section we will learn to
1. factor trinomials with leading coefficient not 1;
2. determine if a trinomial is factorable; and
3. use substitutions to simplify factoring of those that are factorable.

Section 6.2 showed us ways to factor trinomials in which the coefficient of the highest-degree term was 1. This section will be devoted to factoring trinomials in which the leading coefficient is not 1. For example, to factor $3x^2 + 7x + 2$, we must find two binomials of the form $ax + b$ and $cx + d$ such that

$$3x^2 + 7x + 2 = (ax + b)(cx + d)$$

The factors of the first term, $3x^2$, are $3x$ and x. The last term, 2, has factors of 1 and 2 as well as -1 and -2. So there are four possible combinations to consider:

$(3x + 2)(x + 1)$
$(3x + 1)(x + 2)$
$(3x - 1)(x - 2)$
$(3x - 2)(x - 1)$

Only one of these can be correct. We must use the combination that yields a middle term of $7x$.

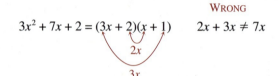

Now let's exchange the 2 and the 1 to see how the middle term is affected.

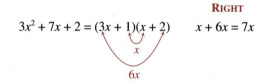

CHECK $(3x + 1)(x + 2) = 3x^2 + 6x + x + 2 = 3x^2 + 7x + 2$ FOIL method

The combinations

$(3x - 1)(x - 2) = 3x^2 - 7x + 2$ and $(3x - 2)(x - 1) = 3x^2 - 5x + 2$

yield the wrong middle terms.

The next example shows that if there is more than one way to factor the first and last terms of a trinomial, the number of possible ways to arrange the numbers within the factors increases.

EXAMPLE 1 Factor: $6x^2 + 11x + 4$.

SOLUTION The first term, $6x^2$, has two possible choices for factors:

$3x \cdot 2x$ or $6x \cdot x$

The last term, 4, also has two possible choices:

$2 \cdot 2$ or $4 \cdot 1$

All signs of the terms in the trinomial are positive, so only positive signs can occur in the factors. The middle term will be $11x$ when we find the correct arrangement.

Possible Factors	Middle Term	Conclusion
$(6x + 1)(x + 4)$	$24x + x = 25x$	Wrong
$(3x + 4)(2x + 1)$	$3x + 8x = 11x$	**Right**

Thus,

$$6x^2 + 11x + 4 = (3x + 4)(2x + 1)$$

Notice that we did not try the other two combinations in Example 1:

$$(6x + 4)(x + 1) \quad \text{and} \quad (3x + 2)(2x + 2)$$

Why? Since 2 is a factor of both of these expressions,

$$(6x + 4)(x + 1) = 2(3x + 2)(x + 1)$$

and

$$(2x + 2)(3x + 2) = 2(x + 1)(3x + 2)$$

Thus 2 would also have to be a factor of the original expression, $6x^2 + 11x + 4$. But $6x^2 + 11x + 4$ has no common factors, so these last two combinations are impossible.

*C*AUTION

In general, if a polynomial has no common factors, then none of its factors will have a common factor.

EXAMPLE 2 Factor: $2y^2 + y - 15$.

SOLUTION
$$2y^2 = 2y \cdot y$$
$$15 = 15 \cdot 1 \quad \text{or} \quad 5 \cdot 3$$

If we think of $2y^2 + y - 15$ as $2y^2 + y + (-15)$, then the last term is negative. This means that the signs of the second terms in the two factors will be different. Let's try various combinations of the factors until we find the one that produces the correct middle term in the trinomial. Remember that the middle term is found by adding the inner and outer products from the two factors.

POSSIBLE FACTORS	MIDDLE TERM	CONCLUSION
$(2y + 1)(y - 15)$	$-30y + y = -29y$	Wrong
$(2y - 1)(y + 15)$	$30y - y = 29y$	Wrong
$(2y - 15)(y + 1)$	$2y - 15y = -13y$	Wrong
$(2y + 15)(y - 1)$	$-2y + 15y = 13y$	Wrong
$(2y - 3)(y + 5)$	$10y - 3y = 7y$	Wrong
$(2y + 3)(y - 5)$	$-10y + 3y = -7y$	Wrong
$(2y + 5)(y - 3)$	$-6y + 5y = -y$	Wrong
$(2y - 5)(y + 3)$	$6y - 5y = y$	**Right**

Therefore

$$2y^2 + y - 15 = (2y - 5)(y + 3)$$

EXAMPLE 3 Factor: $20x^2 - 47x + 21$.

SOLUTION
$$20x^2 = 20x \cdot x \quad \text{or} \quad 10x \cdot 2x \quad \text{or} \quad 5x \cdot 4x$$
$$21 = 21 \cdot 1 \quad \text{or} \quad 7 \cdot 3$$

The sign before the last term in $20x^2 - 47x + 21$ is positive, so the signs before the last terms in the two factors must be the same, either both positive or both negative. The sign in front of the middle term is negative, so the signs in the two factors must both be negative. Numerous trials yielded the following factorization:

$$20x^2 - 47x + 21 = (5x - 3)(4x - 7)$$

> **To Determine the Signs of the Binomial Factors of a Trinomial**
>
> Where the trinomial is written in descending powers of the variable:
>
> 1. If all signs of the trinomial are positive, all signs in the binomial factors will be positive.
> 2. If the leading term of the trinomial is positive and the last term is negative, the signs between the terms of the binomial factors will differ.
> 3. If the leading term of the trinomial is positive, the middle term negative, and the last term positive, the signs between the terms of the binomial factors will both be negative.
> 4. If the leading term of the trinomial is negative, factor out a negative common factor and follow Step 1, 2, or 3. The negative common factor may be -1.

EXAMPLE 4 Factor: $-14x^2 - 34x - 12$.

SOLUTION The leading coefficient is negative, so we factor out a common factor of -2.

$$-14x^2 - 34x - 12 = -2(7x^2 + 17x + 6)$$
$$= -2(7x + 3)(x + 2)$$

EXAMPLE 5 Factor: $3x^2 + 2xy - 8y^2$.

SOLUTION $3x^2 + 2xy - 8y^2 = (3x - 4y)(x + 2y)$ $-4xy + 6xy = 2xy$

$-4xy$

$6xy$

EXAMPLE 6 Factor: $4x^{2m} + 7x^m + 3$.

SOLUTION The polynomial is second-degree in terms of x^m.

$$4x^{2m} + 7x^m + 3 = 4(x^m)^2 + 7x^m + 3$$

All terms in the trinomial are positive, so all terms in the binomials will also be positive.

Possible Factors	Middle Term		Conclusion
$(2x^m + 3)(2x^m + 1)$	$2x^m + 6x^m =$	$8x^m$	Wrong
$(4x^m + 1)(x^m + 3)$	$12x^m + x^m =$	$13x^m$	Wrong
$(4x^m + 3)(x^m + 1)$	$4x^m + 3x^m =$	$7x^m$	**Right**

Thus, $4x^{2m} + 7x^m + 3 = (4x^m + 3)(x^m + 1)$.

EXAMPLE 7 Factor: $\frac{1}{9}x^2 - \frac{2}{15}x + \frac{1}{25}$.

SOLUTION Since

$$\frac{1}{9}x^2 = \left(\frac{1}{3}x\right)^2 \quad \text{and} \quad \frac{1}{25} = \left(\frac{1}{5}\right)^2$$

the expression may be the square of a binomial. Recall that $(a - b)^2 = a^2 - 2ab + b^2$. The middle term *must* be $-2ab$. Let's see:

$$-2ab = -2\left(\frac{1}{3}x\right)\left(\frac{1}{5}\right) = -\frac{2}{15}x$$

Thus,

$$\frac{1}{9}x^2 - \frac{2}{15}x + \frac{1}{25} = \left(\frac{1}{3}x - \frac{1}{5}\right)\left(\frac{1}{3}x - \frac{1}{5}\right) = \left(\frac{1}{3}x - \frac{1}{5}\right)^2$$

A polynomial like the one in Example 7 is called a **perfect-square trinomial** because it is the square of a binomial. We also could have factored it by trial and error, as we did in the previous examples. We will say more about factoring perfect-square trinomials in the next section.

EXAMPLE 8 Factor $2r^2 + r + 1$ using integral coefficients (integers).

SOLUTION The only possibility to try is $(2r + 1)(r + 1)$. However,

$$(2r + 1)(r + 1) = 2r^2 + 3r + 1$$

which is not correct. Therefore the expression is *prime* (not factorable).

Example 8 shows us that some polynomials cannot be factored using just integers. When a large number of trials are necessary, such a search can prove to be very time-consuming. Fortunately, there is a very simple test for factorability. It is called the ***ac* test**, since it depends on the coefficients of the trinomial

$$ax^2 + bx + c$$

The *ac* Test for Factorability

The trinomial $ax^2 + bx + c$ can be factored as the product of two binomials if the following conditions are met:

1. The product ac can be factored into two numbers m and n; that is, $m \cdot n = ac$.
2. $m + n = b$.

This test also provides a means to factor the trinomial after it is determined to be factorable. If the numbers m and n exist, we can substitute their sum for b:

$$ax^2 + bx + c = ax^2 + (m + n)x + c$$

Then, using the distributive property, we can write

$$ax^2 + (m + n)x + c = ax^2 + mx + nx + c$$

We will factor trinomials using this information in the next three examples.

EXAMPLE 9 Factor $12x^2 + 11x + 2$ using the *ac* test.

SOLUTION $a = 12, \quad c = 2, \quad a \cdot c = 24; \quad b = 11$

We need to find two numbers m and n such that $m \cdot n = 24$ and $m + n = 11$. Since all terms in the trinomial are positive, we will try only positive numbers.

$m \cdot n = 24$	$m + n = 11$	Conclusion
$24 \cdot 1 = 24$	$24 + 1 = 25$	Wrong
$12 \cdot 2 = 24$	$12 + 2 = 14$	Wrong
$8 \cdot 3 = 24$	$8 + 3 = 11$	**Right**
$6 \cdot 4 = 24$	$6 + 4 = 10$	Wrong

The third combination, 8 and 3, satisfies the requirements of the test. We substitute $8 + 3$ for 11 and break up the middle term.

$$12x^2 + 11x + 2 = 12x^2 + (8 + 3)x + 2 \qquad b = 11 = 8 + 3$$
$$= 12x^2 + 8x + 3x + 2 \qquad \text{Distributive property}$$

Now we regroup the terms as follows:

$$12x^2 + 8x + 3x + 2 = (12x^2 + 8x) + (3x + 2) \qquad \text{Associative property}$$
$$= 4x(3x + 2) + 1(3x + 2) \qquad \text{Remove common factors of } 4x \text{ and } 1.$$

Here $3x + 2$ is a common factor in the two terms. Thus,

$$12x^2 + 11x + 2 = (3x + 2)(4x + 1)$$

EXAMPLE 10   Factor: $6x^2 - 7x - 5$.

SOLUTION   Here $a = 6$, $c = -5$, and $b = -7$.

$ac = mn = -30$	$m + n = -7$	CONCLUSION
$30(-1) = -30$	$30 + (-1) = 29$	Wrong
$15(-2) = -30$	$15 + (-2) = 13$	Wrong
$10(-3) = -30$	$10 + (-3) = 7$	Wrong

Notice that in the last trial, only the sign of b is wrong. If we interchange the signs of m and n, we have

$$(-10)(3) = -30 \quad \text{and} \quad -10 + 3 = -7$$

Then

$$\begin{aligned} 6x^2 - 7x - 5 &= 6x^2 + (-10 + 3)x - 5 & &\text{Substitute } (-10 + 3) \text{ for } -7. \\ &= 6x^2 - 10x + 3x - 5 & &\text{Distributive property} \\ &= (6x^2 - 10x) + (3x - 5) & &\text{Associative property} \\ &= 2x(3x - 5) + 1(3x - 5) & &\text{Remove common factors of } 2x \text{ and } 1. \\ &= (3x - 5)(2x + 1) & &\text{Remove common factor } 3x - 5. \end{aligned}$$

EXAMPLE 11   Factor: $4x^2 - 7x + 9$.

The signs of the trinomial indicate that *if* it is factorable, the signs between the terms of the binomial factors will both have to be negative.

$ac = mn = 36$	$m + n = -7$	CONCLUSION
$(-36)(-1) = 36$	$-36 + (-1) = -37$	Wrong
$(-18)(-2) = 36$	$-18 + (-2) = -20$	Wrong
$(-12)(-3) = 36$	$-12 + (-3) = -15$	Wrong
$(-9)(-4) = 36$	$-9 + (-4) = -13$	Wrong
$(-6)(-6) = 36$	$-6 + (-6) = -12$	Wrong

Since none of the integer factors of 36 produces a sum of -7, the trinomial is prime and can't be factored.

EXAMPLE 12   Factor: $6(x - y)^2 - 9(x - y)(x + y) - 15(x + y)^2$.

SOLUTION   We can use a substitution that will make the expression simpler to view. If we let $a = x - y$ and $b = x + y$, then

$$6(x - y)^2 - 9(x - y)(x + y) - 15(x + y)^2$$

becomes

$$6a^2 - 9ab - 15b^2$$

We factor out the common factor 3. Then

$$6a^2 - 9ab - 15b^2 = 3(2a^2 - 3ab - 5b^2) = 3(a + b)(2a - 5b)$$

Now we substitute back $x - y = a$ and $x + y = b$:

$$3(a + b)(2a - 5b) = 3(x - y + x + y)[2(x - y) - 5(x + y)]$$
$$= 3(2x)[2x - 2y - 5x - 5y]$$
$$= 6x(-3x - 7y)$$
$$= -6x(3x + 7y) \quad \text{Remove a common factor of } -1.$$

Exercise Set 6.3

Factor each polynomial using integral coefficients.

1. $2a^2 + 3a + 1$
2. $3p^2 + 4p + 1$
3. $3x^2 + 7x + 2$
4. $4y^2 + 13y + 3$
5. $2s^2 + 7s - 4$
6. $2t^2 + 5t - 6$
7. $12r^2 + 10r - 8$
8. $10x^2 - 5x - 5$
9. $8m^2 - 6m - 5$
10. $4c^2 - 12c - 7$
11. $8x^2 + 26x - 34$
12. $24y^2 + 92y - 16$
13. $4a^2 + 20a + 25$
14. $10d^2 + 13d + 4$
15. $8y^5 + 20y^4 - 100y^3$
16. $6s^4 - 24s^3t + 24s^2t^2$
17. $9c^2 - 24cd - 20d^2$
18. $12x^6 - 25x^5y + 12x^4y^2$
19. $12 + k - 6k^2$
20. $3 + 10b - 8b^2$
21. $2 - 5t - 12t^2$
22. $3 - 16r - 12r^2$
23. $6a^2b^2 - 17ab - 10$
24. $9m^3n + 9m^2n^2 - 4mn^3$
25. $-2k^2x^2 - 14kax^2 - 24a^2x^2$
26. $-3ax^2 - 12axy - 12ay^2$
27. $15y^2 + 49y + 40$
28. $-30t^2 - 17t + 35$
29. $-32x^2 - 4xy + 6y^2$
30. $-27a^2 + 72ab + 60b^2$
31. $a^2b^2 + 5abcd + 6c^2d^2$
32. $x^2y^2 - 5xyuv - 6u^2v^2$

*Use the ac test to determine whether each polynomial is factorable using the set of **integers**. Do not actually factor.*

33. $3x^2 + 4x + 2$
34. $5y^2 + y + 1$
35. $2a^2 - 7a + 9$
36. $6m^2 - 11m + 3$
37. $8p^2 - 6p - 5$
38. $4c^2 + 8c - 5$
39. $6c^2 + 13cd - 8d^2$
40. $4x^2 + 8x - 21$
41. $24 - 26x - 8x^2$
42. $8 + 12x - 36x^2$

*Factor each polynomial completely. If it is not factorable using the set of **rational** numbers, write prime.*

43. $2x^4 - 16x^2 + 32$
44. $4a^4 + 7a^2 - 2$
45. $3a^4 - 30a^2b^2 + 27b^4$
46. $8x^4 - 74x^2y^2 + 18y^4$
47. $c^6 + 7c^3d^3 - 8d^6$
48. $27m^6 + 26m^3n^3 - n^6$
49. $9x^4 - 85x^2y^2 + 36y^4$
50. $4s^4 - 45s^2t^2 + 81t^4$
51. $98c^4 - 100c^2d^2 + 2d^4$
52. $16x^6 - 130x^3y^3 + 16y^6$
53. $\dfrac{4}{9}y^2 + \dfrac{5}{9}y + \dfrac{1}{6}$
54. $\dfrac{3}{14}a^2 + \dfrac{5}{42}ab - \dfrac{2}{9}b^2$
55. $\dfrac{3}{10}m^2 + \dfrac{1}{60}mn - \dfrac{1}{12}n^2$
56. $\dfrac{2}{15}p^2 + \dfrac{41}{225}pq - \dfrac{1}{15}q^2$
57. $3x^{2m} + 10x^m y^n - 8y^{2n}$
58. $9a^{2m} - 6a^m b^n - 8b^{2n}$
59. $12c^{2m} + 23c^m d^n - 9d^{2n}$
60. $4r^{2m} + 21r^m s^n - 18s^{2n}$

Factor by any method.

61. $(x + 3)^2 + (x + 3) - 2$
62. $(a - 4)^2 - 4(a - 4) + 3$
63. $24(m + n)^2 + 38(m + n) + 15$
64. $24(t - 2)^2 - 82(t - 2) + 48$
65. $6x(13b + 4a) + 4y(4a + 13b)$
66. $-11z(3b + 2a) + 9y(2a + 3b)$
67. $2x(2x - 1) + y(1 - 2x)$
68. $y(5y + 1) + x(-1 - 5y)$
69. $(a - b)^2 - b(b - a)$
70. $(x - 2z)^2 - 2z(2z - x)$
71. $72x(y - x) + (x - y)$
72. $27b(a - 2b) + (2b - a)$
73. $(x - 2)^3 + 5(x - 2)^2 + 6(x - 2)$
74. $(a + 1)^3 - 7(a + 1)^2 + 6(a + 1)$
75. $6(x - y)^2 - 11(x - y)(a + b) - 35(a + b)^2$

76. $6(2x - 3)^2 - 11(2x - 3)(x + 1) - 10(x + 1)^2$
77. $(5y + 1)^2 - 4(5y + 1)(y - 4) - 5(y - 4)^2$
78. $6(a - 1)^2 + 3(a - 1)(a + 1) - 18(a + 1)^2$

REVIEW EXERCISES

Factor each of the following trinomials, if possible. If not factorable, write prime.

79. $x^2 - 2x + 1$ **80.** $x^2 - 3x + 2$
81. $x^2 + 11x + 28$ **82.** $x^2 - x - 30$
83. $y^2 - 4y + 3$ **84.** $y^2 - 6y + 5$
85. $x^2 - 9x + 14$ **86.** $x^2 - 4x - 21$
87. $y^2 + 4yz + 4z^2$ **88.** $x^2 + 14xy + 28y^2$
89. $x^2 + 6xy - 27y^2$ **90.** $y^2 - 5y - 10$
91. $x^2 + 30xy + 29y^2$ **92.** $y^2 - 7y + 4$
93. $b^2 + 11b + 18$ **94.** $b^2 - 5b - 24$

WRITE IN YOUR OWN WORDS

95. Explain each step as you factor $-3x^4 + 12x^2 + 15$.
96. Explain each step as you factor $-\pi r^2 + 7\pi r + 8\pi$.
97. Explain each step as you factor:
$2(3y - 5)^3(4x - 1)^2 - 4(3y - 5)^2(4x - 1)$.
98. Explain each step as you factor:
$-7(2x - 1)^4(y + 3) - 14(2x - 1)^3(y + 3)^2$.
99. Explain each step as you attempt to factor:
$ax^2 + bx + c$. (*Note:* Remember the *ac* test.)
100. Explain each step as you attempt to factor:
$dx^2 + ex + f$. (*Note:* Remember the *ac* test.)

6.4 Factoring Special Types of Polynomials; More Factoring by Grouping

OBJECTIVES In this section we will learn to
1. factor polynomials of the forms $a^2 - b^2$, $a^3 - b^3$, and $a^3 + b^3$; and
2. factor polynomials by grouping terms.

One of the easiest types of factoring situations to recognize involves the **difference of two squares.** Recall that

$$(a - b)(a + b) = a^2 - b^2$$

Applying the symmetric property of equality to this expression gives us

$$a^2 - b^2 = (a - b)(a + b)$$

EXAMPLE 1 Factor: (a) $x^2 - 25$; (b) $y^2 - 16$.

SOLUTION (a) $x^2 - 25 = x^2 - 5^2 = (x - 5)(x + 5)$
(b) $y^2 - 16 = y^2 - 4^2 = (y - 4)(y + 4)$

EXAMPLE 2 Factor: $9x^2 - 16y^2$.

SOLUTION $9x^2 - 16y^2 = (3x)^2 - (4y)^2 = (3x - 4y)(3x + 4y)$

EXAMPLE 3 Factor: $m^4 - n^6$.

SOLUTION This expression is the difference of two squares, since m^4 is the square of m^2 and n^6 is the square of n^3. That is, $m^4 = (m^2)^2$ and $n^6 = (n^3)^2$. Therefore

$$m^4 - n^6 = (m^2)^2 - (n^3)^2 = (m^2 - n^3)(m^2 + n^3)$$

EXAMPLE 4 Factor: $x^2y^4 - z^2$.

SOLUTION
$$\begin{aligned} x^2y^4 - z^2 &= (xy^2)^2 - z^2 & (b^m)^n = b^{mn} \\ &= (xy^2 - z)(xy^2 + z) \end{aligned}$$

EXAMPLE 5 Factor: $s^{2n} - t^{4n}$.

SOLUTION
$$\begin{aligned} s^{2n} - t^{4n} &= (s^n)^2 - (t^{2n})^2 & (b^m)^n = b^{mn} \\ &= (s^n - t^{2n})(s^n + t^{2n}) \end{aligned}$$

The difference-of-squares method of factoring often can be used more than once in the same problem. For example,

$$\begin{aligned} x^4 - y^4 &= (x^2 - y^2)(x^2 + y^2) \\ &= (x - y)(x + y)(x^2 + y^2) \end{aligned}$$

Can the *sum* of two squares, such as $x^2 + 4 = x^2 + 2^2$, be factored? The only possibilities are $(x + 2)(x + 2)$, $(x - 2)(x - 2)$, and $(x + 1)(x + 4)$. When we use the FOIL method to compute these products, we get

$$(x + 2)(x + 2) = x^2 + 4x + 4$$
$$(x - 2)(x - 2) = x^2 - 4x + 4$$
$$(x + 1)(x + 4) = x^2 + 5x + 4$$

Since none of these products is equal to $x^2 + 4$, we conclude it cannot be factored.

> **CAUTION**
> The **sum of two squares,** such as $x^2 + y^2$, cannot be factored as the product of two binomials using only real numbers.

EXAMPLE 6 Factor: $x^8 - y^8$.

SOLUTION
$$\begin{aligned} x^8 - y^8 &= (x^4)^2 - (y^4)^2 \\ &= (x^4 - y^4)(x^4 + y^4) \\ &= (x^2 - y^2)(x^2 + y^2)(x^4 + y^4) & (x^4 - y^4) = (x^2 - y^2)(x^2 + y^2) \\ &= (x - y)(x + y)(x^2 + y^2)(x^4 + y^4) & (x^2 - y^2) = (x - y)(x + y) \end{aligned}$$

EXAMPLE 7 Factor: $(s + t)^2 - (x + y)^2$.

SOLUTION
$$a^2 - b^2 = (a - b) \cdot (a + b)$$
$$(s + t)^2 - (x + y)^2 = [(s + t) - (x + y)] \cdot [(s + t) + (x + y)]$$
$$= (s + t - x - y)(s + t + x + y)$$

Two other types of factoring involving binomials are **the sum** and **the difference of two cubes.** The rules for factoring them are as follows:

$$a^3 + b^3 = (a + b)(a^2 - ab + b^2)$$
$$a^3 - b^3 = (a - b)(a^2 + ab + b^2)$$

These factorizations are easy to verify by multiplication:

$$(a + b)(a^2 - ab + b^2) = a^3 - a^2b + ab^2 + a^2b - ab^2 + b^3$$
$$= a^3 + b^3$$
$$(a - b)(a^2 + ab + b^2) = a^3 + a^2b + ab^2 - a^2b - ab^2 - b^3$$
$$= a^3 - b^3$$

Notice that even though the sum of two squares is not factorable, **the sum of two cubes is factorable.**

Factoring Binomials of the Form $x^n + y^n$

In general, the binomial $x^n + y^n$ is factorable if n is odd and is not factorable if n is even.

EXAMPLE 8 Factor: **(a)** $x^3 - 125$; **(b)** $8x^3 + 27$.

SOLUTION **(a)** $x^3 - 125 = x^3 - 5^3$ $\qquad a = x, b = 5$
$$= (x - 5)(x^2 + 5x + 25)$$
$$(a - b)(a^2 + ab + b^2)$$

(b) $8x^3 + 27 = (2x)^3 + 3^3$ $\qquad a = 2x, b = 3$
$$= (2x + 3) \cdot [(2x)^2 - (2x)(3) + 3^2]$$
$$(a + b) \cdot (a^2 - ab + b^2)$$
$$= (2x + 3)(4x^2 - 6x + 9)$$

EXAMPLE 9 Factor: $x^6 - y^6$.

SOLUTION There are two different approaches that we can use to factor this expression. We can express $x^6 - y^6$ as the difference of two squares:

Method 1 $x^6 - y^6 = (x^3)^2 - (y^3)^2$ $\qquad (b^m)^n = b^{mn}$
$$= (x^3 - y^3)(x^3 + y^3)$$
$$= (x - y)(x^2 + xy + y^2)(x + y)(x^2 - xy + y^2)$$

We can also write it as the difference of two cubes.

Method 2 $x^6 - y^6 = (x^2)^3 - (y^2)^3$
$$= (x^2 - y^2)(x^4 + x^2y^2 + y^4)$$
$$= (x - y)(x + y)(x^4 + x^2y^2 + y^4)$$

Methods 1 and 2 are both legitimate factorizations of $x^6 - y^6$ and as such must be equal:

$$(x - y)(x + y)(x^2 + xy + y^2)(x^2 - xy + y^2) = (x - y)(x + y)(x^4 + x^2y^2 + y^4)$$

This means that

$$(x^2 - xy + y^2)(x^2 + xy + y^2) = (x^4 + x^2y^2 + y^4)$$

This can be verified by multiplication, as follows:

$$(x^2 - xy + y^2)(x^2 + xy + y^2)$$
$$= x^4 + x^3y + x^2y^2 - x^3y - x^2y^2 - xy^3 + x^2y^2 + xy^3 + y^4$$
$$= x^4 + x^2y^2 + y^4 \quad \text{Combine like terms.}$$

Since the result of Method 1 is more completely factored than the one in Method 2, it is preferable.

> In general, if a polynomial can be initially factored into even or odd powers, begin with the even powers.

In this text we will not cover methods of factoring expressions that are of the form $x^4 + x^2y^2 + y^4$. However, another type of factoring that is easy to recognize involves the **square of a binomial.** Recall from Section 5.3 that

$$a^2 + 2ab + b^2 = (a + b)^2$$

and

$$a^2 - 2ab + b^2 = (a - b)^2$$

Such expressions are called **perfect-square trinomials** because they are trinomials that factor into squares of binomials. The key to recognizing perfect-square trinomials is that the first and last terms, a^2 and b^2, are perfect squares:

$$a^2 \pm 2ab + b^2 = (a \pm b)^2$$

The square of a The square of b

and the middle term, $\pm 2ab$, is twice the product of the numbers a and b.

EXAMPLE 10   Factor: (a) $x^2 + 6x + 9$; (b) $x^2 - 8x + 16$.

SOLUTION   (a) $x^2 + 6x + 9 = (x + 3)^2$ $6x = 2 \cdot x \cdot 3$
(b) $x^2 - 8x + 16 = (x - 4)^2$ $8x = 2 \cdot x \cdot 4$

EXAMPLE 11   Factor: $4x^2 - 20xy + 25y^2$.

SOLUTION $4x^2 - 20xy + 25y^2 = (2x)^2 - 20xy + (5y)^2$ $20xy = 2 \cdot 2x \cdot 5y$
$= (2x - 5y)^2$

EXAMPLE 12 Factor: $2x^5 + 36x^4 + 162x^3$.

SOLUTION We remove common factors first:
$$2x^5 + 36x^4 + 162x^3 = 2x^3(x^2 + 18x + 81)$$
$$= 2x^3(x + 9)^2$$

*C*AUTION

One of the factors of $a^3 \pm b^3$ is often confused with the square of a binomial:

$a^3 + b^3 = (a + b)(a^2 - ab + b^2)$

Coefficient is 1.

$a^3 - b^3 = (a - b)(a^2 + ab + b^2)$

$(a + b)^2 = a^2 + 2ab + b^2$

Coefficient is 2.

$(a - b)^2 = a^2 - 2ab + b^2$

Summary of Special Types of Factoring

Difference of two squares	$a^2 - b^2 = (a - b)(a + b)$
Sum of two cubes	$a^3 + b^3 = (a + b)(a^2 - ab + b^2)$
Difference of two cubes	$a^3 - b^3 = (a - b)(a^2 + ab + b^2)$
Perfect-square trinomials	$a^2 + 2ab + b^2 = (a + b)^2$
	$a^2 - 2ab + b^2 = (a - b)^2$

In Section 6.1 we introduced a method of factoring called *factoring by grouping*, which we then used to factor an expression that we had shown to be factorable by the *ac* test for factorability. We will now consider more complex expressions that can be factored by grouping.

EXAMPLE 13 Factor: $x^2 + 2xy + y^2 - z^2$.

SOLUTION Note that the first three terms, $x^2 + 2xy + y^2$, form a perfect-square trinomial. When three of the terms of a polynomial form a perfect-square trinomial, those terms are grouped together in the factoring process.

$x^2 + 2xy + y^2 - z^2 = (x^2 + 2xy + y^2) - z^2$ Associative property
$\qquad\qquad\qquad\quad = (x + y)^2 - z^2$ $x^2 + 2xy + y^2$ is a perfect square.

This is now the difference of two squares, $(x + y)^2$ and z^2.

$(x + y)^2 - z^2 = (x + y - z)(x + y + z)$

EXAMPLE 14 Factor: $a^2x^2 - a^2y^2 - b^2x^2 + b^2y^2$.

SOLUTION We factor a^2 from the first two terms and $-b^2$ from the next two terms.

$a^2x^2 - a^2y^2 - b^2x^2 + b^2y^2 = a^2(x^2 - y^2) - b^2(x^2 - y^2)$
$\qquad\qquad\qquad\qquad\qquad\quad = (x^2 - y^2)(a^2 - b^2)$ $x^2 - y^2$ is a common factor.
$\qquad\qquad\qquad\qquad\qquad\quad = (x - y)(x + y)(a - b)(a + b)$ Difference of squares

EXAMPLE 15 Factor: $a^3 - a - b^3 + b$.

SOLUTION We rewrite the expression as $a^3 - b^3 - a + b$. The first two terms then form the difference of two cubes.

$a^3 - b^3 - a + b = (a - b)(a^2 + ab + b^2) - 1(a - b)$
$\qquad\qquad\qquad\; = (a - b)(a^2 + ab + b^2 - 1)$

The polynomial $a^2 + ab + b^2 - 1$ does not factor, so the factorization is complete.

EXAMPLE 16 Factor: $ax^2 + 4ax + 4a - bx^2 - 4bx - 4b$.

SOLUTION
$$ax^2 + 4ax + 4a - bx^2 - 4bx - 4b = (ax^2 + 4ax + 4a) + (-bx^2 - 4bx - 4b)$$
$$= a(x^2 + 4x + 4) - b(x^2 + 4x + 4)$$
$$= (x^2 + 4x + 4)(a - b)$$
$$= (x + 2)^2(a - b)$$

EXAMPLE 17 Factor: $x^2 - x + y - y^2$.

SOLUTION
$$x^2 - x + y - y^2 = (x^2 - y^2) - (x - y)$$
$$= (x - y)(x + y) - (x - y)$$
$$= (x - y)(x + y - 1)$$

To Factor Polynomials

1. Always remove common factors first.
2. **a.** If the polynomial contains only two terms, consider the patterns established for $a^2 - b^2$, $a^3 + b^3$, and $a^3 - b^3$.
   b. If the polynomial contains three terms, decide if it is a perfect-square trinomial. If not, factor it by the methods discussed in Sections 6.2 and 6.3.
   c. If the polynomial contains four or more terms, consider factoring by grouping. If three of the terms form a perfect-square trinomial, group them together.
3. Examine each resulting factor to make sure that it is factored completely.

Do You Remember?

Can you match these?

_____ 1. $a^3 - b^3$
_____ 2. $a^2 + b^2$
_____ 3. $a^2 + 2ab + b^2$
_____ 4. $a^3 + b^3$
_____ 5. $a^2 - 2ab + b^2$
_____ 6. A factor of $a^3 + b^3$
_____ 7. A factor of $a^3 - b^3$

a) $a^2 - ab + b^2$
b) $(a + b)(a^2 - ab + b^2)$
c) $(a - b)^2$
d) $a^2 + ab + b^2$
e) $(a + b)^2$
f) $(a - b)(a^2 + ab + b^2)$
g) Not factorable

Answers: 1. f 2. g 3. e 4. b 5. c 6. a 7. d

Exercise Set 6.4

Factor each difference of squares completely and simplify.

1. $x^2 - 1$
2. $y^2 - 4$
3. $2a^2 - 50$
4. $3m^2 - 27$
5. $x^2y^2 - 16z^2$
6. $r^2s^2 - 49t^2$
7. $121p^2 - 144q^2$
8. $169m^2 - 100n^2$
9. $a^4 - b^4$
10. $18x^4 - 2y^4$
11. $3r^{2m} - 3s^{2m}$
12. $x^{4m} - 4y^{2m}$
13. $\frac{1}{16}a^2b^3 - \frac{1}{9}b^5$
14. $\frac{1}{25}x^3y - \frac{1}{4}xy^3$
15. $0.64a^3b - 0.49ab^3$
16. $0.25m^4n - 0.81n^3$
17. $(a + b)^2 - (a - b)^2$
18. $(x - y)^2 - (x + y)^2$
19. $(2m - 3n)^2 - (2m + 3n)^2$
20. $(x^2 - 1)^2 - (x^2 + 1)^2$
21. $a^8 - b^8$
22. $r^8 - s^8$

Factor each sum or difference of cubes.

23. $x^3 - y^3$
24. $m^3 - n^3$
25. $p^3 + q^3$
26. $r^3 + s^3$
27. $27a^3 - 8b^3$
28. $64x^3 + 125y^3$
29. $x^3y^3 + w^3z^3$
30. $p^3q^3 + r^3s^3$
31. $a^{3m} - b^{3m}$
32. $c^{3m} + d^{3m}$
33. $x^3 + (x - 1)^3$
34. $8a^3 + (2a + 1)^3$
35. $(y + 2)^3 - y^3$
36. $(m + 1)^3 - m^3$
37. $0.008x^3 + 0.001y^3$
38. $0.027x^3 - 0.125y^3$
39. $16x^4y - 54xy^4$
40. $24x^4y^2 + 3xy^5$
41. $(a + b)^3 - (a - b)^3$
42. $(c - d)^3 + (c + d)^3$
43. $\frac{1}{27}s^3 - \frac{1}{8}t^3$
44. $\frac{1}{64}p^3 - \frac{1}{125}q^3$

Factor each perfect-square trinomial completely.

45. $a^2 + 4a + 4$
46. $m^2 + 8m + 16$
47. $x^2 - 2x + 1$
48. $y^2 - 10y + 25$
49. $4x^2 + 12xy + 9y^2$
50. $25r^2 + 60rs + 36s^2$
51. $3p^3 - 6p^2q + 3pq^2$
52. $5m^3 - 20m^2n + 20mn^2$
53. $c^{2m} + 14c^m d^n + 49d^{2n}$
54. $w^{2m} + 16w^m z^n + 64z^{2n}$
55. $y^4 - 18y^2 + 81$
56. $x^4 - 8x^2 + 16$
57. $(a + 2)^2 + 2(a + 2) + 1$
58. $(b - 1)^2 - 6(b - 1) + 9$
59. $3m^5 - 30m^3 + 75m$
60. $2p^5 - 12p^3 + 18p$
61. $(x - y)^2 - 4(x - y)(a - b) + 4(a - b)^2$
62. $(r + s)^2 + 22(r + s)(r - s) + 121(r - s)^2$
63. $\frac{1}{16}w^2 + \frac{1}{10}wz + \frac{1}{25}z^2$
64. $\frac{1}{49}a^2 + \frac{2}{63}ab + \frac{1}{81}b^2$

Factor each of the following polynomials by grouping.

65. $a^2 + 2a + 1 - b^2$
66. $x^2 + 4x + 4 - y^2$
67. $y^2 - 9x^2 + 6y + 9$
68. $m^2 - 4n^2 + 2m + 1$
69. $s^2 - 4t^2 - 16t - 16$
70. $25 - y^2 - 4x^2 - 4xy$
71. $6ab - 3bc - 14a + 7c$
72. $8rs + 3t - 8rt - 3s$
73. $8m^3 + 27n^3 + 2m + 3n$
74. $5y - x + 125y^3 - x^3$
75. $128x^3 - 2y^3 + 2y - 8x$
76. $24a^3 - 3b^3 + 3b - 6a$
77. $3x^2 + 18x + 27 - 3y^2 - 12y - 12$
78. $2p^2 - 16p + 32 - 2q^2 + 4q - 2$
79. $ax + bx - cx - ay - by + cy$
80. $dx - ex - fx - dy + ey + fy$
81. $3abc - 6abd + 10c^2d - 5c^3$
82. $ax^2 - 4ay^2 - x^2 + 4y^2$
83. $5s^2x^2 - 5s^2y^2 - 5t^2x^2 + 5t^2y^2$
84. $7cz^2 - 7c - 7dz^2 + 7d$
85. $3ab^2 + c^2x - 3ac^2 - b^2x$
86. $4a^2x^2 - a^2y^2 - 4b^2x^2 + b^2y^2$
87. $144a^2y^2 - 36a^2z^2 - 64b^2y^2 + 16b^2z^2$
88. $m^2u^2 - 4p^2u^2 - 4m^2v^2 + 16p^2v^2$
89. $x^3 - x - xy^2 - 2xy$
90. $y^2 - 4z^2 + x^2 + 2xy$
91. $9x^4 - 9y^2 - 4 + 12y$
92. $16x^2 - y^4 - 4 + 4y^2$
93. $x^3 - x - y^3 + y$
94. $x^3 + 5y - 125y^3 - x$
95. $x^3 + 3x^2 - 9x - 27$
96. $4x^3 + 2 - x - 8x^2$
97. $x^4 + x^3 + x + 1$
98. $x^4 - 54 - 2x^3 + 27x$
99. $64x^3 - 4x + y^3 - y$
100. $8x^3 - 6x + y^3 - 3y$
101. $8x^3 + 2x - y^3 - y$
102. $27x^3 + y^2 - 9x^2 + y^3$

For Extra Thought

103. An object is thrown upward according to the formula $h(t) = -16t^2 + 48t - 32$. Find $h(1)$, $h(2)$, and $h(\frac{3}{2})$. Explain what each of these values tells us about the position of the object.

104. The revenue R from selling n pens is given by the formula $R(n) = 0.7n^2 + n$. The cost C of producing n pens is given by the $C(n) = 0.5n^2 + n + 5$. Find the profit function $P(n) = R(n) - C(n)$. Find $P(5)$. Explain what the answer means.

6.5 General Methods of Factoring

OBJECTIVES In this section we will learn to apply the rules for factoring various polynomials.

The polynomials in this section will encompass a random selection from the types we have discussed in the first four sections of this chapter. Many will require the use of more than one technique to factor them completely. You will find the factoring much easier if you follow the suggestions for factoring that were given at the end of Section 6.4. They are repeated here for convenience.

To Factor Polynomials

1. Always remove common factors first.
2. a. If the polynomial contains only two terms, consider the patterns established for $a^2 - b^2$, $a^3 + b^3$, and $a^3 - b^3$.
   b. If the polynomial contains three terms, decide if it is a perfect-square trinomial. If not, factor it by the methods discussed in Sections 6.2 and 6.3.
   c. If the polynomial contains four or more terms, consider factoring by grouping. If three of the terms form a perfect-square trinomial, group them together.
3. Examine each resulting factor to make sure that it is factored completely.

EXAMPLE 1 Factor: $7x^2 - 7x - 42$.

SOLUTION The trinomial $7x^2 - 7x - 42$ has a common factor of 7.

$$7x^2 - 7x - 42 = 7(x^2 - x - 6) = 7(x - 3)(x + 2)$$

EXAMPLE 2 Factor: $(x + 4)^2 - (x + 4)$.

SOLUTION The polynomial $(x + 4)^2 - (x + 4)$ has a common factor of $(x + 4)$.

$$(x + 4)^2 - (x + 4) = (x + 4)[(x + 4) - 1] = (x + 4)(x + 3)$$

A first impulse might be to immediately multiply out $(x + 4)^2 - (x + 4)$ before factoring. There is nothing wrong with doing this other than it requires extra work. Let's see what happens in Example 3.

EXAMPLE 3 Factor: $(x + 4)^2 - (x + 4)$.

SOLUTION
$$(x + 4)^2 - (x + 4) = x^2 + 8x + 16 - x - 4$$
$$= x^2 + 7x + 12$$
$$= (x + 4)(x + 3)$$

EXAMPLE 4 Factor: $7.5x^3 - 0.3x$.

SOLUTION
$$7.5x^3 - 0.3x = 0.3x(25x^2 - 1)$$
$$= 0.3x(5x - 1)(5x + 1) \qquad a^2 - b^2 = (a - b)(a + b)$$

EXAMPLE 5 Factor: $2y^3 - 16$.

SOLUTION
$$2y^3 - 16 = 2(y^3 - 8) \qquad \text{Remove common factors first.}$$
$$= 2(y^3 - 2^3)$$
$$= 2(y - 2)(y^2 + 2y + 4) \qquad a^3 - b^3 = (a - b)(a^2 + ab + b^2)$$

EXAMPLE 6 Factor: $s^4 - 3s^2 + 2$.

SOLUTION
$$s^4 - 3s^2 + 2 = (s^2)^2 - 3s^2 + 2$$
$$= (s^2 - 1)(s^2 - 2) \qquad \text{Trial and error}$$
$$= (s - 1)(s + 1)(s^2 - 2) \qquad (s^2 - 1) = (s - 1)(s + 1)$$

EXAMPLE 7 Factor: $15 + 2y - y^2$.

SOLUTION We can factor this polynomial in either of two ways. The first requires removing a common factor of -1. The second involves straightforward trial-and-error factorization.

Method 1 $15 + 2y - y^2 = -1(y^2 - 2y - 15) = -1(y - 5)(y + 3)$
Method 2 $15 + 2y - y^2 = (5 - y)(3 + y)$

EXAMPLE 8 Factor: $a^2x^2 - b^2x^2 - a^2y^2 + b^2y^2$.

SOLUTION This polynomial has four terms. Let's try factoring by grouping.
$$a^2x^2 - b^2x^2 - a^2y^2 + b^2y^2 = (a^2x^2 - b^2x^2) - (a^2y^2 - b^2y^2)$$
$$= x^2(a^2 - b^2) - y^2(a^2 - b^2)$$

$$= (a^2 - b^2)(x^2 - y^2)$$
$$= (a - b)(a + b)(x - y)(x + y)$$

EXAMPLE 9 Factor: $9x^2 - 30xy + 25y^2$.

The first and last terms, $9x^2$ and $25y^2$, are perfect squares. It will be a perfect-square trinomial if the middle term is correct. Since $30xy = 2 \cdot 3x \cdot 5y$ it is correct. Thus

$$9x^2 - 30xy + 25y^2 = (3x - 5y)^2$$

EXAMPLE 10 Factor: $x^2 - \frac{5}{9}x + \frac{2}{27}$.

SOLUTION The last term is positive, so we need two numbers of the same sign with a product of $\frac{2}{27}$ and a sum of $-\frac{5}{9}$.

$$x^2 - \frac{15}{27}x + \frac{2}{27} = \left(x - \frac{1}{3}\right)\left(x - \frac{2}{9}\right)$$

EXAMPLE 11 Factor: $(x - 1)^4 + 8(x - 1)$.

SOLUTION
$(x - 1)^4 + 8(x - 1)$
$= (x - 1)[(x - 1)^3 + 8]$ $x - 1$ is a common factor.
$= (x - 1)[(x - 1)^3 + 2^3]$
$= (x - 1)[(x - 1) + 2][(x - 1)^2 - 2(x - 1) + 4]$ $a^3 + b^3 = (a + b)(a^2 - ab + b^2)$
$= (x - 1)(x + 1)(x^2 - 2x + 1 - 2x + 2 + 4)$ $[(x - 1) + 2] = (x + 1)$
$= (x - 1)(x + 1)(x^2 - 4x + 7)$

Exercise Set 6.5

Factor each of the following polynomials. Use any method

1. $18x + 27$
2. $25x - 20$
3. $2.4 - 1.6x$
4. $3.6 - 1.8y$
5. $15x^2y + 35xy^2$
6. $21xy^2 + 49x^2y$
7. $1.2x^3 - 0.3x$
8. $5.0y^3 - 0.2y$
9. $8x^2 + 4x - 12$
10. $9x^2 + 3x - 12$
11. $3(x + 5) + y(x + 5)$
12. $2(y - 1) - x(y - 1)$
13. $(x + 6)^2 + (x + 6)$
14. $(3x - 1)^2 + (3x - 1)$
15. $(3x + 2)^2 + 2(3x + 2)$
16. $(5x - 1)^2 + 3(5x - 1)$
17. $0.09x^2 - 0.16$
18. $0.25y^2 - 0.36$
19. $12y^3 - 75y$
20. $18p^3 - 12p$
21. $\frac{1}{2}a^2 + a + \frac{1}{2}$
22. $\frac{1}{4}t^2 - t + 1$
23. $9x^2 - 12x + 4$
24. $25y^2 - 20y + 4$
25. $2x^2(2x - 1) + x(1 - 2x)$
26. $3y^2(3y - 1) + y(1 - 3y)$
27. $a(a - 4)^3 - 4(a - 4)^2$
28. $x(x - 1)^3 - (x - 1)^2$
29. $ax + by - bx - ay$

30. $mx - my - nx + ny$
31. $3x^3 - 24$
32. $2y^3 - 54$
33. $1 + 0.008x^3$
34. $1 + 0.125t^3$
35. $x^4 - 1$
36. $m^4 - 16$
37. $3y^3 + 192$
38. $2p^3q - 128q$
39. $54a^3 + 16b^3$
40. $81y^3 + 24x^3$
41. $x^4 + x^2 - 2$
42. $x^4 - 7x^2 - 18$
43. $0.1y^2 + 1.8 + 0.9y$
44. $0.01t^2 + 0.25 + 0.10t$
45. $-ax^4 + 6ax^3 + 91ax^2$
46. $-x^2y^2 + 21xy^3 + 72y^4$
47. $-3x^3y + 33x^2y - 72xy$
48. $-b^4x^2 + 19b^3x - 70b^2$
49. $0.2y^2 - 0.1y - 0.6$
50. $0.03y^2 - 0.05y - 0.02$
51. $6x^2 + 11x - 7$
52. $4y^2 - 9y - 9$
53. $5x^2 + 25x - 15$
54. $8y^2 - 56y + 16$
55. $3 + x - 10x^2$
56. $6 + 7x - 10x^2$
57. $12y - 74xy + 12x^2y$
58. $12x - 76x^2 + 24x^3$
59. $x^2 + \frac{4}{3}x - \frac{5}{9}$
60. $\frac{1}{2}y^2 - \frac{9}{4}y + 1$
61. $\frac{1}{3}p^2 + \frac{1}{6}p - \frac{1}{2}$
62. $\frac{3}{4}m^2 - \frac{1}{2}m - \frac{2}{3}$
63. $\frac{2}{9}a^2 - \frac{5}{6}a + \frac{1}{2}$
64. $2y^2 - \frac{2}{3}y - \frac{1}{6}$
65. $x^3 + x^2 + 2x + 2$
66. $y^3 - y^2 - 3y + 3$
67. $24ax - 40bx - 18ay + 30by$
68. $14ax + 7by - 14ay - 7bx$
69. $x^2 + 2xy + y^2 - z^2$
70. $9x^2 - y^2 - 25z^2 + 10yz$
71. $x^2 - 25y^2 + 6x + 9$
72. $x^2 - y^2 - z^2 - 2yz$
73. $x^3 + 3x^2 - 9x - 27$
74. $x^3 - 2x^2 + 8 - 4x$
75. $(x + y)^2 + (x + y) - 2$
76. $(x - y)^2 - 4(x - y) + 3$
77. $(2x + y)^2 + 2(2x + y) - 15$
78. $(3x - y)^2 - 13(3x - y) + 12$
79. $3(m - 2n)^2 + 5(m - 2n) - 12$
80. $3(2a - b)^2 + 14(2a - b) - 24$
81. $2x^4 - 162$
82. $5y^4 - 80$
83. $3x^4 + 14x^2 - 5$
84. $6x^2y^2 - 17xy - 10$
85. $x^6 - 64$
86. $y^6 - x^6$
87. $x^3 - 27(y + 1)^3$
88. $(2a - 1)^3 + 8b^3$
89. $4m^3 - 4(p - 4)^3$
90. $6(x - 1)^3 - 6y^3$
91. $x^3 - 2x + x^2 - 8$
92. $y^2 + y^3 - 9x^2 + 27x^3$
93. $x^3 + 1 - (x + 1)$
94. $p^3 + p^2 - (p + 1)$
95. $x^3 - 6x + 6y - y^3$
96. $x^2 - 25y^2 + 125y^3 + x^3$
97. $120ax + 48ay - 90bx - 36by$
98. $60ax + 70ay - 112by - 96bx$
99. $8x^6 + 7x^3 - 1$
100. $27y^6 + 19y^3 - 8$
101. $x(x - 1)^3(2y + 1)^2 - (x - 1)^3(2y + 1)^2$
102. $p(p - 2)^3(3q - 5)^2 - 2(p - 2)^3(3q - 5)^2$
103. $x^2 + 4z^2 - y^2 - 4xz$
104. $x^2 + 9 - 6y + y^2$
105. $9x^2 + 25y^2 - 16 - 30xy$
106. $16y^2 - 16x^2 + 56x - 49$

For Extra Thought

107. The formula for the volume of a cube is $V = s^3$, where s is the length of a side. If the volume is given by the polynomial $V(x) = x^3 - 3x^2 + 3x - 1$, how would you find $s(x)$, the length of a side? Find a formula for $s(x)$. Find $V(1)$ and $s(1)$.

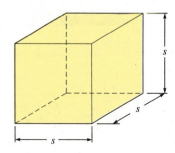

108. The formula for the area of a rectangle is $A = lw$, where l is the length and w is the width. If a rectangle has an area given by $A(x) = x^3 - 64$, how would you find $l(x)$ and $w(x)$? [Assume $w(x)$ is smaller than $l(x)$.] Find $A(5)$, $l(5)$, and $w(5)$.

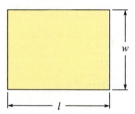

109. The formula for the area of a triangle is $A = \frac{1}{2}bh$, where b is the length of the base and h is the height. If a triangle has an area given by the formula $A(x) = (x + 1)^2 - (x - 1)^2 + 4x$, how would you find b and $h(x)$? (Assume that b is a constant). Find $A(12)$ and $h(12)$.

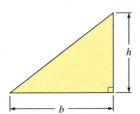

110. The formula for the volume of a right circular cylinder is $V = \pi r^2 h$, where r is the radius of the base and h is the height. If $V(x) = 2\pi x^3 + 6\pi x^2 + 6\pi x + 2\pi$, how would you find $r(x)$ and $h(x)$ when $h(x)$ is twice $r(x)$? Find $V(1)$, $r(1)$, and $h(1)$. Use $\pi = 3.1416$ with $x = 1$. Find $V(x)$, $r(x)$, and $h(x)$ when $x = 1$ cm. Again use $\pi = 3.1416$.

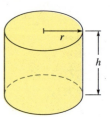

6.6 Solving Nonlinear Equations by Factoring

HISTORICAL COMMENT Many attempts have been made to solve nonlinear equations since the earliest times. There is evidence that some attention was given to special cases as early as 2100 B.C. It is interesting to note that the method of factoring that is commonly used today to solve such equations did not appear until Thomas Harriot's work in the early fifteenth century.

OBJECTIVES In this section we will learn to
1. solve certain nonlinear equations by factoring; and
2. write an equation when its solution set is known.

Chapter 2 introduced methods for solving linear (first-degree) equations. This section will introduce one method of solving equations of degree greater than 1. A **second-degree equation in one variable** is an equation of the form

$$ax^2 + bx + c = 0, \qquad a, b, c \in R, \quad a \neq 0$$

Such an equation is called a **quadratic equation.** Some examples are

$$3x^2 + 5x + 7 = 0, \qquad 4x^2 - 9 = 0, \qquad 5x^2 - 25x = 0$$

If a quadratic equation is factorable, its solution set can be found by applying a theorem called the **zero-factor theorem.**

THEOREM 6.1 **Zero-Factor Theorem**	If a and b represent real numbers and $a \cdot b = 0$, then $a = 0$ or $b = 0$.

PROOF

Assume $a \neq 0$. If $a \neq 0$ then its inverse $\frac{1}{a}$ exists and is nonzero. Thus, the zero-factor theorem is equivalent to

$\frac{1}{a} \cdot a \cdot b = \frac{1}{a} \cdot 0$ Multiply each side of $a \cdot b = 0$ by $\frac{1}{a}$.

$1 \cdot b = 0$ $\frac{1}{a} \cdot a = 1$; $\frac{1}{a} \cdot 0 = 0$ by the multiplication property of zero (Section 1.2)

$b = 0$ Multiplicative identity property

So if one factor is not zero but the product is zero, the other factor must be zero. The theorem is proved.

Although Theorem 6.1 is stated for *real numbers,* it is equally true for *complex numbers,* which we will introduce in Section 8.6.

EXAMPLE 1 Solve: $x^2 + 4x + 3 = 0$.

SOLUTION We first *factor* $x^2 + 4x + 3$:

$x^2 + 4x + 3 = (x + 3)(x + 1) = 0$

If $(x + 3)(x + 1) = 0$, then one or both of the factors must be zero by the zero-factor theorem.

$x + 3 = 0$ or $x + 1 = 0$
$x = -3$ or $x = -1$

To check the solutions, they must be substituted into the original equation.

CHECK

$x^2 + 4x + 3 = 0$ $\qquad\qquad$ $x^2 + 4x + 3 = 0$
$(-3)^2 + 4(-3) + 3 \stackrel{?}{=} 0$ $x = -3$ $\qquad$ $(-1)^2 + 4(-1) + 3 \stackrel{?}{=} 0$ $x = -1$
$9 + (-12) + 3 \stackrel{?}{=} 0$ $\qquad\qquad$ $1 + (-4) + 3 \stackrel{?}{=} 0$
$0 \stackrel{\checkmark}{=} 0$ $\qquad\qquad\qquad\qquad$ $0 \stackrel{\checkmark}{=} 0$

The solution set is $\{-3, -1\}$.

EXAMPLE 2 Solve $2t^2 - 5t = 12$ by factoring.

SOLUTION First we rewrite the equation in **standard form,** $ax^2 + bx + c = 0$.

$2t^2 - 5t - 12 = 0$
$(2t + 3)(t - 4) = 0$ Factor.
$2t + 3 = 0$ or $t - 4 = 0$ Zero-factor theorem
$2t = -3$ or $t = 4$
$t = \dfrac{-3}{2}$ or $t = 4$

CHECK

$$2t^2 - 5t - 12 = 0$$
$$2\left(\frac{-3}{2}\right)^2 - 5\left(\frac{-3}{2}\right) - 12 \stackrel{?}{=} 0 \qquad t = \frac{-3}{2}$$
$$2\left(\frac{9}{4}\right) - 5\left(\frac{-3}{2}\right) - 12 \stackrel{?}{=} 0$$
$$\frac{9}{2} + \frac{15}{2} - 12 \stackrel{?}{=} 0$$
$$0 \stackrel{\checkmark}{=} 0$$

$$2t^2 - 5t - 12 = 0$$
$$2(4)^2 - 5\cdot 4 - 12 \stackrel{?}{=} 0 \qquad t = 4$$
$$2(16) - 20 - 12 \stackrel{?}{=} 0$$
$$32 - 20 - 12 \stackrel{?}{=} 0$$
$$0 \stackrel{\checkmark}{=} 0$$

The solution set is $\left\{-\frac{3}{2}, 4\right\}$.

To Solve a Nonlinear Equation by Factoring

Step 1 Rewrite the equation so that it is in standard form.
Step 2 Factor the polynomial. Remember to remove any common factors first.
Step 3 Set each factor involving variables equal to zero.
Step 4 Solve the equations created in Step 3.
Step 5 Check the solution(s) back in the original equation.

EXAMPLE 3 Solve: $x^2 - 4 = 0$.

SOLUTION
$$x^2 - 4 = 0$$
$$(x - 2)(x + 2) = 0 \qquad \text{Factor.}$$
$$x - 2 = 0 \quad \text{or} \quad x + 2 = 0$$
$$x = 2 \quad \text{or} \quad x = -2$$

CHECK
$$x^2 - 4 = 0$$
$$(2)^2 - 4 \stackrel{?}{=} 0 \qquad x = 2$$
$$4 - 4 \stackrel{?}{=} 0$$
$$0 \stackrel{\checkmark}{=} 0$$

$$x^2 - 4 = 0$$
$$(-2)^2 - 4 \stackrel{?}{=} 0 \qquad x = -2$$
$$4 - 4 \stackrel{?}{=} 0$$
$$0 \stackrel{\checkmark}{=} 0$$

The solution set is $\{-2, 2\}$.

EXAMPLE 4 Solve: $3x^2 - 9x - 30 = 0$.

SOLUTION
$$3x^2 - 9x - 30 = 0$$
$$3(x^2 - 3x - 10) = 0 \qquad \text{Remove common factors first.}$$
$$3(x - 5)(x + 2) = 0$$

Since $3 \neq 0$, we can ignore it. Thus,

$$x - 5 = 0 \quad \text{or} \quad x + 2 = 0$$
$$x = 5 \quad \text{or} \quad x = -2$$

CHECK

$3x^2 - 9x - 30 = 0$		$3x^2 - 9x - 30 = 0$	
$3(5)^2 - 9(5) - 30 \stackrel{?}{=} 0$	$x = 5$	$3(-2)^2 - 9(-2) - 30 \stackrel{?}{=} 0$	$x = -2$
$75 - 45 - 30 \stackrel{?}{=} 0$		$12 + 18 - 30 \stackrel{?}{=} 0$	
$0 \stackrel{\checkmark}{=} 0$		$0 \stackrel{\checkmark}{=} 0$	

The solution set is $\{-2, 5\}$.

EXAMPLE 5 Solve: $4m^2 + 8m = 0$.

SOLUTION
$$4m^2 + 8m = 0$$
$$4m(m + 2) = 0$$
$$4m = 0 \quad \text{or} \quad m + 2 = 0$$
$$m = 0 \quad \text{or} \quad m = -2$$

The solution set is $\{-2, 0\}$. The check is left to the reader.

EXAMPLE 6 Solve: $s^2 - 6s + 9 = 0$.

SOLUTION
$$s^2 - 6s + 9 = 0$$
$$(s - 3)(s - 3) = 0$$
$$s - 3 = 0 \quad \text{or} \quad s - 3 = 0$$
$$s = 3 \quad \text{or} \quad s = 3$$

The equation has two identical solutions.

CHECK
$$s^2 - 6s + 9 = 0$$
$$(3)^2 - 6(3) + 9 \stackrel{?}{=} 0 \quad s = 3$$
$$9 - 18 + 9 \stackrel{?}{=} 0$$
$$0 \stackrel{\checkmark}{=} 0$$

The solution set is $\{3\}$.

When an equation has a double solution, as in Example 6, the equation is said to have a **root (solution) of multiplicity 2.** If a root occurs three times, the multiplicity is three; and so on.

The zero-factor theorem also applies to equations of degree higher than 2.

Generalized Zero-Factor Theorem	If $a_1, a_2, a_3, \ldots, a_n$ represent real numbers and $a_1 \cdot a_2 \cdot a_3 \cdots a_n = 0$, then at least one of $a_1, a_2, a_3, \ldots, a_n$ is zero.

EXAMPLE 7 Solve: $y^3 + y^2 - 12y = 0$.

SOLUTION

$$y^3 + y^2 - 12y = 0$$
$$y(y^2 + y - 12) = 0 \quad \text{Remove any common factors first.}$$
$$y(y + 4)(y - 3) = 0$$

Now we set each factor equal to 0:

$$y = 0 \quad \text{or} \quad y + 4 = 0 \quad \text{or} \quad y - 3 = 0$$
$$y = 0 \quad \text{or} \quad y = -4 \quad \text{or} \quad y = 3$$

The solution set is $\{-4, 0, 3\}$.

The generalized zero-factor theorem implies that an nth-degree polynomial equation has n roots.

EXAMPLE 8 Solve: $y^4 - 13y^2 + 36 = 0$.

SOLUTION Since $y^4 - 13y^2 + 36 = 0$ is a fourth-degree equation, it will have four solutions. To factor it, we let $y^2 = a$. Then $y^4 = a^2$.

$$a^2 - 13a + 36 = 0 \quad \text{Substitute } a \text{ for } y^2 \text{ and } a^2 \text{ for } y^4.$$
$$(a - 4)(a - 9) = 0$$
$$(y^2 - 4)(y^2 - 9) = 0 \quad y^2 = a$$
$$(y - 2)(y + 2)(y - 3)(y + 3) = 0$$
$$y - 2 = 0 \quad \text{or} \quad y + 2 = 0 \quad \text{or} \quad y - 3 = 0 \quad \text{or} \quad y + 3 = 0$$
$$y = 2 \quad \text{or} \quad y = -2 \quad \text{or} \quad y = 3 \quad \text{or} \quad y = -3$$

The solution set is $\{-3, -2, 2, 3\}$.

EXAMPLE 9 Find a second-degree equation that has roots -1 and 2. Assume no multiple roots.

SOLUTION When the solutions to a equation are known, we can find the equation by reversing the steps that we use to solve an equation by factoring. We are given that $x = -1$ or $x = 2$. This means that $x + 1 = 0$ or $x - 2 = 0$. Now we can reconstruct the equation from its factors:

$$(x + 1)(x - 2) = 0$$
$$x^2 - x - 2 = 0$$

EXAMPLE 10 Find an equation with solution set $\{-3, 0, 4\}$. Assume no multiple roots.

SOLUTION Since none of the roots are repeated, we are seeking a third-degree equation. Since $-3, 0,$ and 4 are solutions, we have

$$x = -3 \quad \text{or} \quad x = 0 \quad \text{or} \quad x = 4$$

Therefore $x + 3 = 0$ or $x = 0$ or $x - 4 = 0$. Now, by the multiplication property of zero (Theorem 1.1 in Section 1.2),

$$x(x + 3)(x - 4) = 0$$
$$x(x^2 - x - 12) = 0$$
$$x^3 - x^2 - 12x = 0$$

This is not the only equation with the solution set $\{-3, 0, 4\}$. Multiplying both sides of $x^3 - x^2 - 12x = 0$ by any constant will yield an equivalent equation.

EXAMPLE 11 Find an equation with solution set $\left\{\dfrac{1}{2}, \dfrac{1}{3}, 2\right\}$. Assume no multiple roots.

SOLUTION Given the roots, we know that

$$x = \frac{1}{2} \quad \text{or} \quad x = \frac{1}{3} \quad \text{or} \quad x = 2$$

We first clear each equation of fractions.

$$2x = 1 \quad \text{or} \quad 3x = 1 \quad \text{or} \quad x = 2$$
$$2x - 1 = 0 \quad \text{or} \quad 3x - 1 = 0 \quad \text{or} \quad x - 2 = 0$$
$$(2x - 1)(3x - 1)(x - 2) = 0$$
$$(6x^2 - 5x + 1)(x - 2) = 0$$
$$6x^3 - 17x^2 + 11x - 2 = 0$$

Many applied problems involve quadratic equations, as shown in Examples 12 and 13.

EXAMPLE 12 The product of two consecutive even integers is 288. Find the integers.

SOLUTION We let

$$x = \text{the first even integer}$$
$$x + 2 = \text{the second even integer}$$

Their product is 288, so we write

$$x(x + 2) = 288$$
$$x^2 + 2x = 288$$
$$x^2 + 2x - 288 = 0 \qquad \text{Write in standard form.}$$

Now we factor and solve.

$$(x - 16)(x + 18) = 0$$
$$x - 16 = 0 \quad \text{or} \quad x + 18 = 0$$
$$x = 16 \quad \text{or} \quad x = -18$$

There are two possible solutions. If $x = 16$, then $x + 2 = 18$. If $x = -18$, then $x + 2 = -16$.

CHECK The problem asked us to find two consecutive even integers with a product of 288.

$$(16)(18) \stackrel{\checkmark}{=} 288 \quad \text{and} \quad (-16)(-18) \stackrel{\checkmark}{=} 288$$

The solutions check.

EXAMPLE 13 A rock is tossed vertically upward with a velocity of 32 feet per second from the edge of a building 48 feet high. Neglecting air resistance, the distance s of the rock above the ground at any time $t > 0$ is given by the formula

$$s = -16t^2 + 32t + 48$$

At what time t will the rock strike the ground?

SOLUTION The rock will strike the ground when its distance s above the ground is zero. Thus we want

$$-16t^2 + 32t + 48 = 0$$
$$-16(t^2 - 2t - 3) = 0 \quad \text{Remove common factors first.}$$
$$-16(t - 3)(t + 1) = 0$$

Since $-16 \neq 0$, either $t - 3 = 0$ or $t + 1 = 0$.

$$t - 3 = 0 \quad \text{or} \quad t + 1 = 0$$
$$t = 3 \quad \text{or} \quad t = -1$$

Time cannot be negative, so we reject $t = -1$. Thus the rock will strike the ground 3 seconds after it is tossed upward.

Exercise Set 6.6

Solve each equation by factoring. Reminder: Remove common factors first.

1. $x^2 + 3x = 0$
2. $y^2 + 5y = 0$
3. $a^2 - 3a + 2 = 0$
4. $p^2 - 5p + 4 = 0$
5. $3m^2 - 27 = 0$
6. $7x^2 - 28 = 0$
7. $y^2 - 10y + 21 = 0$
8. $c^2 - c - 72 = 0$
9. $t^2 - 8t - 48 = 0$
10. $s^2 + 4s - 45 = 0$
11. $q^2 - 18q + 81 = 0$
12. $r^2 - 14r + 49 = 0$
13. $4x^2 - 25x - 21 = 0$
14. $2y^2 - 21y + 40 = 0$
15. $10a^2 + a - 24 = 0$
16. $10d^2 - 7d - 6 = 0$

17. $2s^2 - 2s - 12 = 0$
18. $3x^2 - 3x - 60 = 0$
19. $x^3 + 20x^2 + 36x = 0$
20. $a^3 - 13a^2 - 48a = 0$
21. $-96k^3 + 36k^2 + 6k = 0$
22. $-24m^3 - 2m^2 + 15m = 0$
23. $-24p^4 + 10p^3 + 4p^2 = 0$
24. $-9t^4 - 33t^3 + 12t^2 = 0$
25. $(y - 1)(y^2 - 7y + 10) = 0$
26. $(2k + 3)(k^2 + 7k + 12) = 0$
27. $(x^2 - 1)(9x^2 - 24x + 16) = 0$
28. $(p^2 - 9)(21p^2 - 79p - 20) = 0$
29. $\left(x + \frac{1}{4}\right)\left(x - \frac{1}{3}\right)\left(x - \frac{1}{2}\right) = 0$
30. $\left(x - \frac{1}{5}\right)\left(x - \frac{1}{4}\right)\left(x + \frac{3}{7}\right) = 0$
31. $t^2 - \frac{2}{3}t + \frac{1}{9} = 0$
32. $r^2 - \frac{1}{2}r + \frac{1}{16} = 0$
33. $a^2 - \frac{1}{4}a - \frac{1}{8} = 0$
34. $c^2 - \frac{2}{15}c - \frac{1}{15} = 0$
35. $\frac{1}{6}p^2 - \frac{13}{36}p + \frac{1}{6} = 0$
36. $\frac{1}{10}s^2 - \frac{3}{20}s + \frac{1}{18} = 0$

Determine an equation for each given solution set. Assume no multiple roots.

37. $\{2, 3\}$
38. $\{5, 1\}$
39. $\{-3, 0\}$
40. $\{-7, 0\}$
41. $(-1, 0, 2)$
42. $\{-3, 0, 4\}$
43. $\left\{5, -\frac{1}{2}\right\}$
44. $\left\{-6, \frac{3}{4}\right\}$
45. $\left\{-\frac{1}{3}, 0, \frac{2}{7}\right\}$
46. $\left\{-\frac{1}{2}, 0, \frac{4}{5}\right\}$

For each equation: (a) simplify it; (b) rewrite it in standard form, $ax^2 + bx + c = 0$; and (c) solve it by factoring.

47. $4a^2 + 4a = -1$
48. $3c^2 - 10 = 13c$
49. $15y^2 = 13 - 2y$
50. $4t^2 = 21 - 8t$
51. $2x(3x - 5) = x + 10$
52. $y(3y + 4) = 2 + 3y$
53. $(s + 1)^2 = 16$
54. $(k + 2)^2 = 25$
55. $2 + m(m + 7) = m(3 - m)$
56. $s(s - 3) - 3 = -s(2 + s)$
57. $3(y^2 - 1) - 5y = 3 - 2y$
58. $9a + 12 - 3(a + 4) = a^2 - 16$
59. $2(2x + 11)(x - 4) + 2(x - 2)(x + 4) = 32 + 7(x^2 - 16)$
60. $2y(5y + 2) - 3(2y - 1) = 8y + 23$

Write each of the following statements as a quadratic equation and solve.

61. The sum of the squares of two consecutive odd positive integers is 290. Find the integers.

62. The product of two consecutive negative integers is 342. Find the integers.

63. The area of the pictured rectangle is 2100 sq. m. Find its dimensions.

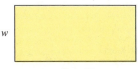

64. The area of the pictured square is numerically 5 more than its perimeter. Find the length of a side.

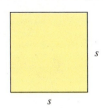

65. The product of two consecutive positive integral multiples of 3 is 108. Find the two integers.

66. The sum of the squares of two consecutive negative integral multiples of 5 is 325. Find the two integers.

67. The area of a triangle is 27 cm². Find the base and height if the height is 3 cm more than twice the base.

68. The height of a triangle is 6 m greater than the base. Find the base and height if the area of the triangle is 80 m².

69. The lengths of the three sides of a right triangle are represented by three consecutive even natural numbers. Find the lengths of the three sides.

70. The lengths of the three sides of a right triangle are represented by three consecutive natural numbers. Find the lengths of the three sides.

6.7 More Applications; Solving Applied Problems by Factoring

HISTORICAL COMMENT   Applied problems have appeared in mathematical works dating back to the early Greeks and Egyptians, not to mention the Arabs and the Hindus. Many of these ancient applied problems no longer have any current meaning, while others continue to appear in works today after being updated to reflect modern conditions and times. Variations of some of the same problems appear in writings everywhere, and it has become very difficult to credit the original authors of many of them. Bhāskara, one of the four great Hindu mathematicians from India (c. 1150), is credited with authoring the following problem, which is an application of the Pythagorean theorem from approximately 500 years earlier.

> A snake's hole is at the foot of a pillar, and a peacock is perched on its summit. Seeing a snake, at a distance of thrice the pillar gliding towards his hole, he pounces obliquely upon him. Say quickly at how many cubits from the snake's hole do they meet, both proceeding an equal distance?

OBJECTIVES   In this section we will learn to
1. use the techniques of multiplying polynomials to solve applied problems; and
2. use the principles of factoring to solve applied problems.

In this section we will continue to use the five-step process introduced in Section 2.7, page 85, for solving applied problems. Since the steps are just as applicable to quadratic equations as to linear equations, it would be a good idea to reread them before proceeding further.

The possible applications of quadratic equations are far too numerous to allow us to provide an example of each type. Study the following examples and then apply the knowledge you gain from those to the variety of problems in the exercise set.

EXAMPLE 1   A square garden is surrounded by a brick walk 2 feet wide. The combined area of the garden and the walk is 196 square feet. Find the length of one side of the garden.

SOLUTION   We let the length of one side of the garden be represented by x. Since the walk is 2 feet wide, the dimensions of the garden plus the walk are $x + 4$ by $x + 4$. (See Figure 6.1.) Because the formula for the area of a square of side s is $A = s^2$, the combined area of the garden and the walk is $(x + 4)^2$, where x is the length of the side of the garden. Thus,

$$(x + 4)^2 = 196 \qquad A = 196;\ s = x + 4$$
$$x^2 + 8x + 16 = 196$$
$$x^2 + 8x - 180 = 0$$
$$(x + 18)(x - 10) = 0$$
$$x + 18 = 0 \quad \text{or} \quad x - 10 = 0$$
$$x = -18 \quad \text{or} \quad x = 10$$

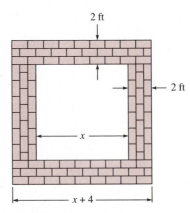

FIGURE 6.1

A measured length cannot be negative, so we reject -18 as a solution. Each side of the garden is 10 feet long.

EXAMPLE 2   The sum of the squares of two positive integers that differ by 2 is 202. Find the two integers.

SOLUTION   We let

x = the first integer

$x + 2$ = the other integer

The sum of their squares is 202.

$$x^2 + (x + 2)^2 = 202$$
$$x^2 + x^2 + 4x + 4 = 202$$
$$2x^2 + 4x - 198 = 0$$
$$2(x^2 + 2x - 99) = 0$$
$$2(x - 9)(x + 11) = 0$$

Since $2 \neq 0$, either

$x - 9 = 0$ or $x + 11 = 0$
$x = 9$ or $x = -11$

The problem asked for positive integers, so we reject -11 as a solution. The two integers are $x = 9$ and $x + 2 = 11$.

As we mentioned earlier, the **Pythagorean theorem** states that in any right triangle, the square of the length of the longest side **(the hypotenuse)** is equal to the sum of the squares of the lengths of the other two sides **(the legs)**. (See Figure 6.2.)

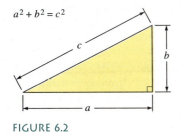

FIGURE 6.2

EXAMPLE 3   The long leg of a right triangle is 7 inches longer than the short leg. The hypotenuse is 8 inches longer than the short leg. Find the length of each leg.

SOLUTION We let

x = the length of the short leg
$x + 7$ = the length of the long leg
$x + 8$ = the length of the hypotenuse

By the Pythagorean theorem,

$x^2 + (x + 7)^2 = (x + 8)^2$ $a = x, b = x + 7, c = x + 8; \ a^2 + b^2 = c^2$
$x^2 + x^2 + 14x + 49 = x^2 + 16x + 64$ Square each binomial.
$2x^2 + 14x + 49 = x^2 + 16x + 64$
$x^2 - 2x - 15 = 0$
$(x - 5)(x + 3) = 0$ Factor.
$x - 5 = 0$ or $x + 3 = 0$
$x = 5$ or $x = -3$

We reject -3 because a length cannot be negative. The sides are $x = 5$ in., $x + 7 = 12$ in., and $x + 8 = 13$ in.

EXAMPLE 4 If we ignore air resistance, the position s of an object projected vertically upward from ground level with a velocity of 96 feet per second is given by the formula $s = -16t^2 + 96t$, where s is the distance above the ground after t seconds. How long will it take for the projectile to go up, come to a stop, and return to the ground?

SOLUTION The projectile is on the ground when its distance above the ground is zero ($s = 0$).

$s = -16t^2 + 96t$
$0 = -16t^2 + 96t$ $s = 0$
$0 = -16t(t - 6)$ Factor.
$-16t = 0$ or $t - 6 = 0$
$t = 0$ or $t = 6$

The time $t = 0$ refers to the projectile's position on the ground, before it is thrown. The projectile will be back on the ground 6 seconds after it is thrown.

EXAMPLE 5 An open box can be made by cutting equally sized squares from the corners of a square piece of sheet metal and folding up the sides. (See Figure 6.3.) What size piece of sheet metal is necessary to make a box 4 inches high and with volume 576 cubic inches?

SOLUTION We let x denote the length of one side of the square piece of sheet metal. If the height is to be 4 inches, then a 4-inch square must be cut from each corner. The side of the box is then $x - (2 \cdot 4) = x - 8$ inches. (See Figure 6.4.) Recall that the formula for the volume of a rectangular box is

volume = length · width · height
$576 = (x - 8) \cdot (x - 8) \cdot 4$

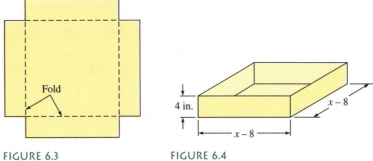

FIGURE 6.3   FIGURE 6.4

We divide each side by 4 to simplify before multiplying:

$$144 = (x - 8)(x - 8)$$
$$144 = x^2 - 16x + 64$$
$$0 = x^2 - 16x - 80$$
$$0 = (x + 4)(x - 20)$$
$$x + 4 = 0 \quad \text{or} \quad x - 20 = 0$$
$$x = -4 \quad \text{or} \quad x = 20$$

We reject -4. The length of one side of the square piece of sheet metal is 20 inches.

Solution to the Historical Problem at the Beginning of this Section

A snake's hole is at the foot of a pillar, and a peacock is perched on its summit. Seeing a snake, at a distance of thrice the pillar gliding towards his hole, he pounces obliquely upon him. Say quickly at how many cubits from the snake's hole do they meet, both proceeding an equal distance?

SOLUTION We let the distance that the snake glides to the point of interception be d cubits, which is also the distance that the peacock flies to the same point. Then we let the height of the pillar be h cubits. The snake is $3h - d$ feet from the foot of the pillar when he is intercepted by the peacock. See Figure 6.5.

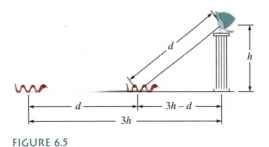

FIGURE 6.5

By the Pythagorean theorem,

$$d^2 = h^2 + (3h - d)^2$$
$$d^2 = h^2 + 9h^2 - 6hd + d^2$$
$$0 = 10h^2 - 6hd$$
$$0 = 2h(5h - 3d)$$

We set each factor equal to zero:

$$2h = 0 \quad \text{or} \quad 5h - 3d = 0$$
$$h = 0 \qquad\qquad 3d = 5h$$
$$d = \frac{5}{3}h$$

We reject $h = 0$ since the height of the pillar cannot be zero. The distance of the snake from its hole is $3h - d$, as shown in Figure 6.5. So

$$3h - d = 3h - \frac{5}{3}h$$
$$= \frac{9h - 5h}{3} = \frac{4}{3}h$$

The peacock catches the snake at $\frac{4}{3}$ the height of the pillar from its base.

Exercise Set 6.7

Number Problems

Write each of the following as a quadratic equation and solve it.

1. The square of a number is $\frac{1}{9}$ less than $\frac{2}{3}$ of the number. Find the number.
2. The square of a positive number less $\frac{4}{3}$ of the number is $\frac{5}{9}$. What is the number?
3. The square of twice an integer is 18 more than 6 times the integer. Find the integer.
4. The square of 3 times an integer is 3 more than 6 times the integer. Find the integer.
5. The square of a negative number added to the number is 132. Find the negative number.
6. Four times the square of a positive number less 21 is equal to 8 times the number. Find the positive number.
7. The sum of the squares of two consecutive odd positive integers is 202. Find the integers.
8. The sum of the squares of two consecutive odd natural numbers is 394. Find the natural numbers.

Geometry Problems

9. The length of a rectangle is 2 in. more than twice the width. The diagonal is 13 in. Find the dimensions of the rectangle.
10. The length of a rectangle is 1 dm more than the width. The diagonal is 5 dm. Find the dimensions of the rectangle.
11. At a point 12 yd from the base of a pine tree, the distance to the top of the tree is 3 more than twice the height. Find the height of the tree.
12. A guy wire is needed to hold a power pole in a vertical position. If the pole is 16 ft high and the guy wire is to be connected to the top of the pole from a point 12 ft from the base of the pole (see figure on page 339), how long must the guy wire be?
13. Two airplanes flying at right angles to each other pass at noon. How far apart are they at 2:00 P.M. if one is traveling at 300 mph and the other at 400 mph?
14. Louise leaves Newark traveling due north at 45 mph. Johan leaves Newark at the same time traveling due

6.7 • MORE APPLICATIONS; SOLVING APPLIED PROBLEMS BY FACTORING   339

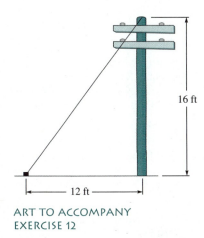

ART TO ACCOMPANY
EXERCISE 12

west at 60 mph. How far apart are they $\frac{2}{3}$ of an hour later?

15. A 200-sq.-ft area needs to be fenced. Find the amount of fence needed if the length is 10 ft more than the width.

16. Myrna's dining room is $\frac{3}{4}$ as wide as it is long. She buys a rug that covers 140 ft² of space. If a 1-ft wide space of floor is left uncovered all around the rug, find the dimensions of the room.

17. Emma has a rectangular back yard with an area of 1000 ft². If the length is 10 ft more than twice the width, find the dimensions.

18. It takes 65 yd² of linoleum to cover the floor in a new building. The length of the rectangular floor is 6 ft less than 3 times the width. Find the dimensions in feet.

19. Find the lengths of the three sides of a right triangle using the Pythagorean theorem when the hypotenuse is $a + 2$ and the legs are a and $a + 1$.

20. A 3 in. × 5 in. picture is going to be framed as shown. (The width of the frame is x.) If the total area of the frame and the picture is 63 in.², find the outside width of frame and picture.

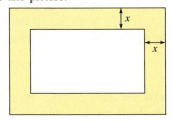

In Exercises 21–25, a square of equal size is cut out of each corner and the sides are folded up to make a box. (See Example 5).

21. A square piece of sheet metal is used to construct a box 4 in. deep, with a volume of 256 in.³ Find the length of one side of the sheet metal.

22. A square piece of sheet metal is used to construct a box 2 cm deep, with a volume of 162 cm³. Find the length of one side of the sheet metal.

23. A rectangular piece of cardboard twice as long as it is wide is used to construct a box 3 in. deep, with a volume of 1080 in.³ What are the dimensions of the cardboard?

24. A rectangular sheet of cardboard 5 times as long as it is wide is used to construct a box 5 cm deep, with a volume of 1625 cm³. What are the dimensions of the cardboard?

25. A rectangular piece of tin is twice as long as it is wide. It is used to construct a box 2 cm deep, with a volume of 480 cm³. Find the dimensions of the tin.

Area/volume Problems

26. A strip of uniform width is to be added to all four sides of a small rectangular garden that is 4 ft × 8 ft, in order to increase the area to 45 ft². Find the width of the strip.

27. A solid cube of metal (see figure) has dimensions as shown. A rectangular solid is removed from the center of the cube. The volume of the remaining metal is given by $V = A^3 - Ab^2$. Factor and solve for b in terms of V and A if the volume $V = 1620$ in.³ and $A = 12$ in².

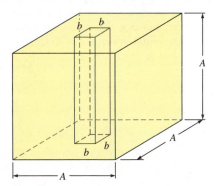

28. A circular steel blade has a hole cut out of the center. (See figure.) The remaining area of the blade is given by $A = \pi R^2 - \pi r^2$. If the area $A = 84\pi$ cm² and $R = 10$ cm, solve for r.

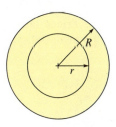

29. The volume of the wall of a hollow cylinder is given by $V = 10\pi R^2 - 10\pi r^2$. (See figure.) If $V = 910\pi$ cm^3 and $r = 3$ cm, solve for R.

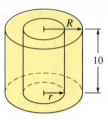

30. A rectangular piece of aluminum is cut into two pieces. (See figure). The area of the upper portion is given by $A = \frac{1}{2}xy + x$. If $A = 30$ in.^{2} and $y = 5$ in., solve for x.

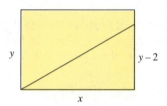

Miscellaneous Problems

31. A ball is thrown upward according to the equation $s(t) = -16t^2 + vt$, where s is the height in feet at time t in seconds, and v is the initial velocity in feet per second. If the initial velocity is 64 ft/sec, how long will it take the ball to reach a height of 64 ft?

32. If the initial velocity is 64 ft/sec, how long will it take the ball in Exercise 31 to reach a height of 48 ft?

33. A marble is dropped from the top of a building 256 ft above the ground. The distance s of the marble from the ground at any time t seconds is given by $s(t) = -16t^2 + 256$. Find the time it takes the marble to hit the ground.

34. A rock is thrown vertically downward from the top of a building 640 ft high. The distance s of the rock from the ground at any time t in seconds is given by $s(t) = -16t^2 - 48t + 640$. Find the time it takes the rock to hit the ground.

35. A ball is thrown vertically downward from a tower. The ball's distance s from the ground is estimated by $s(t) = 25 + 20t - 5t^2$, where s is in feet and t is the time in seconds. At what time will the ball hit the ground?

36. The recommended caloric intake c for a man between ages 20 and 30 is estimated by the equation $c(a) = -5a^2 + 200a + 1200$, where a is the man's age in years. If Paul follows this recommendation and consumes 3180 calories/day, what is his age?

37. The profit P in dollars from producing and selling x items is given by $P(x) = x^2 - 19x + 238$. How many items must be produced and sold to earn a profit of $190?

38. The concentration c of aspirin in the bloodstream t hr after ingestion is estimated by $c(t) = -t^2 + 12t$. Determine the time when the concentration is zero.

Summary

6.1
Monomial Factors; Factoring by Grouping

The simplest type of factoring associated with polynomials involves removing the **greatest common factor,** the **GCF.** Once this factor has been removed, the remaining polynomial can sometimes be factored again, often by regrouping the terms.

6.2–6.3
Factoring Trinomials; A Test for Factorability

Trinomials are often factored by a trial-and-error process. However, using a procedure that is basically the reverse of the FOIL method is helpful. That procedure leads to the **ac test for factorability,** which determines whether a polynomial of the form $ax^2 + bx + c$ can be factored. It reads:

If there are two numbers m and n such that $m \cdot n = ac$ and $m + n = b$, then the trinomial is factorable.

6.4
Factoring Special Types of Polynomials; More Factoring by Grouping

Polynomials are often factored by recognizing known patterns. One easily recognizable pattern is the **difference of two squares,**

$$a^2 - b^2 = (a - b)(a + b)$$

Another easily recognized pattern is a **perfect-square trinomial,** of which there are two:

$$a^2 + 2ab + b^2 = (a + b)^2$$
$$a^2 - 2ab + b^2 = (a - b)^2$$

The **sum** and **difference of two cubes** factor as follows:

$$a^3 + b^3 = (a + b)(a^2 - ab + b^2)$$
$$a^3 - b^3 = (a - b)(a^2 + ab + b^2)$$

Factoring by grouping is often used to factor polynomials of more than three terms.

6.5
General Methods of Factoring

There is a series of steps that can often make factoring polynomials easier.

To Factor Polynomials
1. Always remove common factors first.
2. a. If the polynomial contains only two terms, consider the patterns established for $a^2 - b^2$, $a^3 + b^3$, and $a^3 - b^3$.
   b. If the polynomial contains three terms, decide if it is a perfect-square trinomial. If not, factor it by the methods discussed in Sections 6.2 and 6.3
   c. If the polynomial contains four or more terms, consider factoring by grouping. If three of the terms form a perfect-square trinomial, group them together.
3. Examine each resulting factor to make sure that it is factored completely.

6.6
Solving Nonlinear Equations by Factoring

Many **quadratic equations** are solved by use of the **zero-factor theorem.** It asserts that for real numbers a and b, if $a \cdot b = 0$, then either $a = 0$, or $b = 0$.

To solve a nonlinear equation by factoring:

Step 1 Rewrite the equation so that it is in standard form.
Step 2 Factor the polynomial. Remember to remove any common factors first.
Step 3 Set each factor involving variables equal to zero.
Step 4 Solve the equations created in Step 3.
Step 5 Check the solution(s) back in the original equation.

6.7
More Applications; Solving Applied Problems by Factoring

The **Pythagorean theorem** states that for any right triangle, the square of the length of the hypotenuse, c, is equal to the sum of the squares of the lengths of the other two sides, a and b:

$$a^2 + b^2 = c^2$$

Cooperative Exercise

Business equipment loses value with age and may be depreciated according to I.R.S. rules. Many businesses use what is called a "straight-line depreciation" method to take advantage of this option. To do this, a business must have verification of the cost of the equipment, the date of purchase, its expected *salvage value* (the value at the end of the depreciation period), and the length of the depreciation period.

Wood's Electronics purchases a machine for $13,500 and sells it for a salvage value of $750 after 12 years.

1. Find the equation of the depreciation line.
2. What is the economic meaning of the slope of the line?
3. What is the economic meaning of the y-intercept?
4. How much is the company able to depreciate each year?
5. How much is the company able to depreciate over the 12-year period?
6. If there were a fire and the machine was ruined when it was $3\frac{1}{2}$ years old, what would be the loss to the company?
7. What would be the value of the machine after 9 years?
8. Can the value of the machine ever be zero? If so, approximately how long would this take?
9. What is the economic meaning of the x-intercept?

Review Exercises

Factor each of the following polynomials completely.

1. $5x - 10$
2. $3x^3 - 12x^2 - 3x$
3. $16x^2y + 28xy^2$
4. $(x + 1)^2 + (x + 1)$
5. $16x(2y - 3) - 10x^2(2y - 3)$
6. $4ab + 6ac - 18ad$
7. $91x^2y^3z - 65xy^2z^2 - 39x^2y^2z^3$
8. $y^{3m+1} - y^{2m}$
9. $s^2 - 64$
10. $2a^3 - 16$
11. $6x^2 - 36x + 30$
12. $16m^2 - 24m + 9$
13. $3r^3s^2 - 3r^2s^3 - 36rs^4$
14. $x^4 - 16y^4$
15. $b^2(b - 3) - 6b(b - 3) + 9(b - 3)$
16. $k^{2m} - 5k^m r^m + 4r^{2m}$
17. $3y^2 + \frac{9}{4}y + \frac{3}{8}$
18. $s^{3m} - 8t^{3m}$
19. $25a^2 + 60ab + 36b^2$
20. $x^2 + 3x + xy + 3y$
21. $a^2 + 4ab + 4b^2 - c^2$
22. $3k^2 - 2km + 6k - 4m$
23. $6 - 13x + 2x^2$
24. $-7y^2 + 35y + 42$
25. $9k^2 - m^2 - 6m - 9$
26. $64a^6 - b^6$
27. $6ax^2 - 12ax + 6a + bx^2 - 2bx + b$
28. $\frac{1}{9}t^2 - \frac{1}{6}t + \frac{1}{16}$
29. $27s^3 + (2s - 1)^3$
30. $(a - b)^2 - (a + b)^2$
31. $0.49m^2n^2 - 1.21n^4$
32. $x^8 - y^8$
33. $\frac{1}{81}a^4 - \frac{1}{16}b^4$
34. $(x - 3)^2 - 10(x - 3)(x - 2) - 11(x - 2)^2$
35. $6xy - 4wx + 6yz - 4wz$
36. $10a^3 - 40a^2b + 40ab^2$
37. $s^4 - 9s^2 + 18$
38. $2m^5 - 20m^3 + 50m$
39. $az - bz + cz - at + bt - ct$
40. $a^3 - 8 - (a - 2)$
41. $4xy^3 - 9x^3y + 9x^2y^2$
42. $6k^2 - 32m^2 - 4km$
43. $a^6 - 27b^6 - 26a^3b^3$
44. $18x^{2m} - 4y^{2n} - 21x^m y^n$

Solve each equation by factoring.

45. $x^2 + 3x + 2 = 0$
46. $3y^3 + 2y^2 - 8y = 0$
47. $(k + 4)(k - 4) = 9k - 3(k + 4) + 12$
48. $6a^2 + 13a - 5 = 0$
49. $\frac{1}{7}t^2 - \frac{9}{21}t - \frac{1}{4} = 0$
50. $2b^{m+2} + 14b^{m+1} + 24b^m = 0$

Determine the equations with the indicated solution sets. Assume no multiple roots.

51. $\{-1, 2\}$
52. $\{-3, 0, 1\}$
53. $\left\{-\frac{1}{3}, \frac{1}{2}\right\}$
54. $\left\{-\frac{1}{5}, 0, 7\right\}$

Factor each of the following.

55. $x^2 - 4$
56. $2x^2 - 2$
57. $x^3 - 1$
58. $x^3 + 27$

Write each of the following applied problems as a quadratic equation and solve.

59. The product of two consecutive integers is 72. Find the integers.
60. A 20-foot-tall tree casts a shadow 15 ft long. How far is it from the tip of the shadow to the top of the tree?
61. Megan has a pool 30 ft wide and 50 ft long. She decides to put a sidewalk of uniform width around the pool. If the total area including the sidewalk and pool is 1836 ft^2, what is the width of the sidewalk? See the accompanying figure.
62. Barbara leaves Durango at 3:00 P.M. heading north at an average of 30 mph through the mountains. At 4:00 P.M. Jim leaves Durango heading east at an average of 45 mph. How far apart are they at 5:00 P.M.?

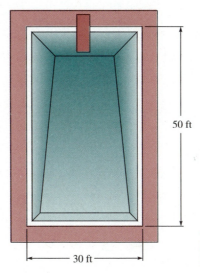

ART TO ACCOMPANY EXERCISE 61

Chapter Test

Factor each of the following polynomials.

1. $3x^2 + 3x$
2. $-5x + 10y - 10xz + 20yz$
3. $x^2 + 3x + 2$
4. $m^2(m - 4) - 2m(m - 4) - 8(m - 4)$
5. $9x^2 + 4y^2 - 12xy - m^2$
6. $15x^5y^3 - 24x^2y^4$
7. $x^2 - 7xy - 60y^2$
8. $my^2 + 4my + 4m - py^2 - 4py - 4p$
9. $4y - 3x + 64y^3 - 27x^3$
10. $2x^2 - 24xy + 72y^2$
11. $2ab + cx^2 - 2ac - bx^2$
12. $8x^2 - 10x - 3$

Solve each of the following by factoring.

13. $2x^2 - 21x + 40 = 0$
14. $3x^2 - 15x + 12 = 0$
15. $2x^4 - 16x^2 + 32 = 0$
16. $(a - 2)^3 + 5(a - 2)^2 + 6(a - 2) = 0$
17. $(x - 5)(x - 3) = 3x(x - 3)$
18. $3y^3 - 5y^2 - 2y = 0$
19. $x^3 + 4x^2 - 9x - 36 = 0$
20. $-3y^2 + 27y = 0$
21. $5y^2 + 28y = 3y^3$
22. $(2y + 3)(y + 2) = 10$
23. $2x^2 = 8$
24. Find an equation with $\{-\frac{1}{2}, 0, 3\}$ as a solution set. Assume no multiple solutions.
25. Find an equation with $\{-1, -2, 2\}$ as a solution set. Assume no multiple solutions.
26. The length of a rectangle is 8 in. If the diagonal of the rectangle is 10 in., find the measure of the width.
27. A square piece of cardboard is used to construct a box 2 cm high. If the volume of the box is 128 cm^3, find the length of a side.
28. Jill's boat goes 80 mi up a river and back in 6 hr. What is the speed of the current and the speed of the boat in still water if it takes 4 hr just to go up the river?

Cumulative Review for Chapters 4–6

Solve each system of equations by the indicated method.

1. $\left.\begin{array}{r} x - y = 4 \\ x + y = 6 \end{array}\right\}$ graphing

2. $\left.\begin{array}{r} x - y = 0 \\ x + y = 4 \end{array}\right\}$ elimination

3. $\left.\begin{array}{r} y = x - 2 \\ 2x + y = 8 \end{array}\right\}$ substitution

4. $\begin{aligned} 6x + 7y &= 0 \\ 8x - 2y &= 0 \end{aligned}\Big\}$ substitution

5. $\begin{aligned} 2x + 3y &= 5 \\ 5x - 2y &= 3 \end{aligned}\Big\}$ elimination

Solve each system of equations.

6. $\begin{aligned} 3x + y - z &= 3 \\ 2x + 3y - 4z &= 1 \\ x - y + 2z &= 2 \end{aligned}$

7. $\begin{aligned} \frac{1}{x} + \frac{2}{y} &= 2 \\ \frac{2}{x} + \frac{4}{y} &= 4 \end{aligned}$

8. Solve the system
$$\begin{aligned} x - y &= -2 \\ 2x - y &= 6 \end{aligned}$$
by the Gauss–Jordan elimination method.

9. Solve the system
$$\begin{aligned} x + y + z &= 1 \\ 2x + 3y - z &= 6 \\ x - 2y + 4z &= -5 \end{aligned}$$
by the Gauss–Jordan elimination method.

10. Solve by Cramer's rule:
$$\begin{aligned} x + 2y &= 9 \\ 2x - y &= 3 \end{aligned}$$

11. Solve by Cramer's rule:
$$\begin{aligned} x + y + z &= 5 \\ x + y - z &= 1 \\ 2x - y + 2z &= 7 \end{aligned}$$

12. Graph: $y \leq x - 3$.

13. Graph the system $y \leq x + 4$ and $x + y > 1$.

Write a system of equations or inequalities for each of the following applied problems and solve it.

14. The sum of two numbers is 24. If the second number is subtracted from twice the first, the result is 23. Find the numbers.

15. The sum of two consecutive even integers is at least 66. If twice the smaller is added to the larger, the result is at least 98. Find the two *smallest* consecutive even integers satisfying these conditions.

16. An airplane can travel 400 mi in 2 hr with the wind and 480 mi in 3 hr against the wind. Find the rate of the wind and the speed of the airplane in still air.

17. Peanuts worth $3.00/lb are going to be mixed with pecans worth $4.20/lb in order to make an 8 lb mixture worth $26.25. How many pounds of each kind of nut are needed?

18. The supply and demand equations for a particular leather belt are given by

   Supply: $q = 7p - 2$
   Demand: $q = -p + 14$

   where p is the price in dollars and q is the quantity sold. Find the equilibrium price and the number sold at that price.

19. Trang has invested $10,000, part at 5% annual interest, another part at 7%, and the rest at 8%. The total yearly interest from the three investments is $620.00. The income from the first two investments less the income from the third one is $460.00. How much did Trang invest at each rate?

20. A test is handed out to an algebra class. Some questions are worth 8 points and take 2 minutes to work. The rest of the questions are worth 12 points and take 6 minutes to work. The class is allowed 60 minutes to take the test, but they are not supposed to answer more than 20 questions. Assuming you know how to work the questions and will get them all correct, how many of each type of question should you answer in order to obtain the most points? What is the most points you can earn?

Carry out the indicated operations. Assume no variable is equal to zero.

21. $\dfrac{x^a}{x^b}$

22. $x^a \cdot x^b$

23. $x^m \cdot x^{2m}$

24. $\dfrac{(-2y^3)^{-1}(3x)^{-2}}{(4y^2)^{-2}(6x^2)^{-1}}$

25. $(6x)^2(2xy)^2(3y)^3(x^2y^2)^{-2}$

Simplify and leave the answer in scientific notation.

26. $\dfrac{48{,}000{,}000}{1.6 \times 10^4}$

27. $\dfrac{9.1 \times 10^5}{13{,}000}$

Simplify each polynomial and (a) write it in standard form; (b) determine its degree; and (c) state whether it is a monomial, a binomial, or a trinomial.

28. $6x^2 - 8 + x^2 - 2x$

29. $\dfrac{1}{2}y + \dfrac{1}{3}y^2 + y - y^2$

Add or subtract, as indicated.

30. $8x - 3y - x$

31. $a^2 + 2ab + 6a^2 - ab$

32. $a - (a^2 - 3a + 4)$

33. $(m^2 + 3m) - (2m + 4) - (5 - m^2)$

34. Find $f(x) - g(x)$ for $f(x) = 3x^2 + 7x - 9$ and $g(x) = x^2 + x + 1$.

35. Find $(f + g)(t)$ given that $f(t) = t^3 - 7$ and $g(t) = t^2 + t + 7$.

36. Given $f(x) = x^2 + 3x + 2$ and $g(x) = x + 1$, find $(f \cdot g)(2)$ and $\left(\dfrac{f}{g}\right)(0)$

Multiply or divide, as indicated.

37. $(x - 4)(x + 4)$
38. $(2x + 1)^2$
39. $(3t + 4)(2t - 3)$
40. $a^n(a^{2n} - 3a^n + 4)$
41. $-2^2(3x^2)^2(-x)^3$
42. $(t^n - w^{n-1})(t^n + w^{n-1})$
43. $\dfrac{36a + 48}{-2a}$
44. $\dfrac{72r^3 - 64r^2 + 48r - 8}{8r}$
45. $(x^2 + 15x + 50) \div (x + 10)$
46. $(27t^3 - 1) \div (3t - 1)$
47. Divide $2a^3 + 8a^2 + 11a - 9$ by $a + 2$ using synthetic division.
48. Use synthetic division to find the quotient and remainder:

$$\left(x^2 - \dfrac{1}{3}x - \dfrac{1}{6}\right) \div \left(x - \dfrac{1}{3}\right)$$

Factor each of the following polynomials.

49. $6n - 3$
50. $45x^2y - 15xy^2 + 30xy$
51. $p(p - 4) - 4(p - 4)$
52. $ax + ay + bx + by$
53. $6by - 4bx + 2ax - 3ay$
54. $(x - y)^2(x + y) - (x + y)^2(x - y)$
55. $y^2(y - 1) - 2y(y - 1) + (y - 1)$
56. $x^2 - 6x + 5$
57. $a^2 + ab - 42b^2$
58. $8 - 2t - t^2$
59. $x^4 - 13x^2 + 36$
60. $18ab^3 - 12a^2b^2 + 2a^3b$
61. $r^2 + \dfrac{7}{12}r + \dfrac{1}{12}$
62. $x^{2n} + 9x^n + 18$
63. $5 - 15a - 10a^2$
64. $60y^2 + 72xy - 27x^2$
65. $\dfrac{1}{15}x^2 - \dfrac{39}{225}xy + \dfrac{2}{15}y^2$
66. $10(x + 1)^2 + 11(x + 1)(2x + 1) - 6(2x + 1)^2$
67. $x^2 - 25$
68. $x^3 - 27$
69. $36a^2 - 60ab + 25b^2$
70. $x^3 + (x - 2)^3$
71. $0.36x^2y - 0.49y^3$
72. $a^4 + 2a^3 + a + 2$
73. $ex + fx - fy - ax - ey + ay$
74. $x^2 + 2x + 1 - 4y^2$

Solve each of the following polynomial equations by factoring.

75. $3x^2 - 12 = 0$
76. $3y^2 - 3y - 36 = 0$
77. $(x^2 - 4)(9x^2 - 24x + 16) = 0$
78. $5y^3 - 45y = 0$
79. $m^2 - 12m + 36 = 0$
80. $25x^2 + 70x + 49 = 0$

Find an equation having the given solution set. Assume no multiple solutions.

81. $\{0, 7\}$
82. $\{2, -3\}$
83. $\{-2, 0, 1\}$

Write each of the following applied problems as a quadratic equation and solve.

84. The sum of two numbers is 19 and their product is 84. Find the numbers.

85. The height of a triangle is 3 m more than the base. The area of the triangle is 20 m². Find the base and height.

86. A square piece of sheet metal is used to construct a box 3 in. deep by cutting a square from each corner and folding up the sides. If the volume of the box is 75 in³, find the length of a side of the sheet metal.

87. A ball is thrown upward from ground level according to the equation $h(t) = -16t^2 + 48t$, where $h(t)$ is its position at time t. How long will it take for the ball to fall back to the ground?

88. A strip of uniform width is to be added to all four sides of a rectangular 16 × 20 ft lawn to increase the area of the lawn to 396 ft². Find the width of the strip.

CHAPTER 7
Rational Expressions, Equations, and Functions

7.1 Reducing and Building Rational Expressions
7.2 Multiplying and Dividing Rational Expressions
7.3 Adding and Subtracting Rational Expressions
7.4 Simplifying Complex Rational Expressions
7.5 Solving Rational Equations
7.6 More Applications

Application

A motorboat can travel 10 mph upstream and 16 mph downstream. If Juan rents a boat for 4 hours, how far can he go upstream before he must head back?

See Section 7.6, Example 3.

When Frederick the Great wrote to Louis Lagrange that "the greatest king in Europe" wanted "the greatest mathematician in Europe" as a part of his court, he may not have been exaggerating in either instance. Lagrange was born of wealthy parents, but by the time he reached his teens the fortune was gone. Lagrange said, "If I had remained rich, I probably would not have cast my lot with mathematics." Some of Lagrange's greatest work occurred between the ages of sixteen and twenty-two. During his lifetime he held numerous posts in prestigious schools, and at one time he served as the president of the French commission that worked out the metric system.

JOSEPH LOUIS LAGRANGE (1736–1813)

7.1 Reducing and Building Rational Expressions

OBJECTIVES In this section we will learn to
1. recognize rational expressions;
2. recognize values that cannot be used as replacements for the variables; and
3. reduce and build rational expressions.

Chapter 1 dealt with the set of real numbers and its subsets. One subset was the set of *rational numbers,* which were defined to be the set of numbers of the form

$$\frac{a}{b}, \quad \text{where } a \text{ and } b \text{ are integers}, \quad b \neq 0$$

In this chapter we consider a particular type of algebraic expression called a **rational expression.** We will show that the theorems we use in working with rational expressions are very much like those that we use with rational numbers.

DEFINITION
Rational Expression

A **rational expression** is the quotient of two polynomials such that the denominator cannot equal zero. Expressed in symbols:

If A and B are polynomials and $B \neq 0$, then $\dfrac{A}{B}$ is a rational expression.

For example,

$$\frac{x+3}{x-2} \quad x \neq 2$$

$$\frac{y^2 + 3y}{2y - 1} \quad y \neq \frac{1}{2}$$

$$\frac{y^2 + 3y + 1}{2y^2 + 7y + 9} \quad \text{and} \quad x^3 + 2 = \frac{x^3 + 2}{1}$$

are all rational expressions.

A rational expression can be evaluated for permissible values of the variable in much the same way that polynomials are evaluated.

EXAMPLE 1 Find the value of $\dfrac{x+3}{x-2}$ when (a) $x = 3$; (b) $x = 2$.

SOLUTION (a) To find the value of the expression when $x = 3$, we replace each x by 3:

$$\dfrac{x+3}{x-2} = \dfrac{3+3}{3-2} \quad \text{Substitute 3 for } x.$$

$$= \dfrac{6}{1} = 6$$

(b) $\dfrac{x+3}{x-2} = \dfrac{2+3}{2-2} \quad \text{Substitute 2 for } x.$

$$= \dfrac{5}{0}$$

𝒞AUTION

The denominator of a rational expression cannot equal zero. To find the excluded values that make the denominator zero:

1. Set the denominator equal to 0.
2. Solve the equation obtained in Step 1.

Since division by zero is undefined, the value of x cannot be 2. In fact, $\dfrac{x+3}{x-2}$ is not a rational expression at $x = 2$.

Any value of a variable that makes the denominator of a rational expression zero must be excluded.

EXAMPLE 2 Evaluate $\dfrac{x^2 + 6x + 3}{x^3 - 2x + 4}$ for (a) $x = -1$; (b) $x = 3$.

SOLUTION (a) $\dfrac{x^2 + 6x + 3}{x^3 - 2x + 4} = \dfrac{(-1)^2 + 6(-1) + 3}{(-1)^3 - 2(-1) + 4} \quad x = -1$

$$= \dfrac{1 - 6 + 3}{-1 + 2 + 4} = \dfrac{-2}{5}$$

(b) $\dfrac{x^2 + 6x + 3}{x^3 - 2x + 4} = \dfrac{3^2 + 6 \cdot 3 + 3}{3^3 - 2 \cdot 3 + 4} \quad x = 3$

$$= \dfrac{9 + 18 + 3}{27 - 6 + 4} = \dfrac{30}{25} = \dfrac{6}{5}$$

EXAMPLE 3 What values of the variables must be excluded in each of the following expressions?

(a) $\dfrac{x+3}{x-2}$ (b) $\dfrac{y-4}{2y-10}$ (c) $\dfrac{a+5}{a^2 + 3a + 2}$ See the caution on this page.

SOLUTION (a) First we set $x - 2 = 0$, which gives us $x = 2$. Since $x - 2$ must not equal zero, $x \neq 2$.
(b) We set $2y - 10 = 0$, giving us $y = 5$. Since $2y - 10$ must not equal zero, $y \neq 5$.

(c) We set $a^2 + 3a + 2 = 0$:

$$(a + 2)(a + 1) = 0 \quad \text{Factor.}$$

$$a + 2 = 0 \quad \text{or} \quad a + 1 = 0$$
$$a = -2 \quad \text{or} \quad a = -1$$

Since $a^2 + 3a + 2$ must not equal zero, $a \neq -2$ and $a \neq -1$.

EXAMPLE 4 Determine the values of the variable that must be excluded in each of the following rational expressions:

(a) $\dfrac{3a - 2}{a^2 - b^2}$ (b) $\dfrac{2y + 5}{y^2 + 4}$

SOLUTION (a) We set $a^2 - b^2 = 0$. Then

$$(a - b)(a + b) = 0 \quad \text{Factor.}$$

$$a - b = 0 \quad \text{or} \quad a + b = 0$$
$$a = b \quad \text{or} \quad a = -b$$

Thus $a \neq b$ and $a \neq -b$. We can write this as $a \neq \pm b$.

(b) We set $y^2 + 4 = 0$. Since $y^2 \geq 0$ for all real numbers y, we conclude that $y^2 + 4 > 0$. No real values of y need be excluded.

*C*AUTION

Values that make only the numerator zero are acceptable, for example,

$$\dfrac{x - 3}{x + 6} = \dfrac{0}{9} = 0$$

when $x = 3$

We can simplify rational expressions by applying the **fundamental principle of rational expressions,** which states that an *equivalent rational expression* is created when:

1. The numerator and denominator of a rational expression are **divided** by the same nonzero polynomial.
2. The numerator and denominator of a rational expression are **multiplied** by the same nonzero polynomial.

THEOREM 7.1

Fundamental Principle of Rational Expressions

If A, B, and C are polynomials and $B \neq 0$, $C \neq 0$, then

1. $\dfrac{A \cdot C}{B \cdot C} = \dfrac{\frac{A \cdot C}{C}}{\frac{B \cdot C}{C}} = \dfrac{A}{B}$ Divide numerator and denominator by C.

2. $\dfrac{A}{B} = \dfrac{A}{B} \cdot \dfrac{C}{C} = \dfrac{A \cdot C}{B \cdot C}$ Multiply numerator and denominator by C.

Note in Theorem 7.1 that multiplying by $\dfrac{C}{C}$ is the same as multiplying by 1. The value of the expression is not changed.

CAUTION

Rational expressions can be simplified only when the numerator and the denominator have a common factor.

$$\frac{xy + x}{x} = \frac{x(y + 1)}{x} = y + 1$$

x is a factor of $xy + x$.

$\dfrac{xy + 1}{x}$ cannot be simplified

x is *not* a factor of $xy + 1$.

EXAMPLE 5 Simplify the following rational expressions.

(a) $\dfrac{3x}{6y} = \dfrac{3 \cdot x}{3 \cdot 2y} = \dfrac{x}{2y}$ Factor the numerator and the denominator, then divide both by the common factor, 3. Set $y \neq 0$.

(b) $\dfrac{x^2 + 2x}{4x^2 + 8x} = \dfrac{x(x + 2)}{4x(x + 2)} = \dfrac{1}{4}$ Factor the numerator and the denominator, then divide both by the common factor, $x(x + 2)$. Set $x \neq 0, -2$.

(c) $\dfrac{x^2 + 7x + 6}{2x^2 + 9x - 18} = \dfrac{(x + 1)(x + 6)}{(2x - 3)(x + 6)}$ Factor the numerator and the denominator, then divide both by the common factor, $x + 6$. Set $x \neq \frac{3}{2}, -6$.

$\phantom{(c)\ \dfrac{x^2 + 7x + 6}{2x^2 + 9x - 18}} = \dfrac{x + 1}{2x - 3}$

As illustrated in Example 5, we can simplify rational expressions by factoring both the numerator and the denominator and then dividing out any common factors that result. Two polynomials with no common factors other than 1 are said to be **relatively prime**.

EXAMPLE 6 Simplify each of the following rational expressions.

(a) $\dfrac{a^2 + 12a - 45}{2a^2 - 3a - 35} = \dfrac{(a + 15)(a - 3)}{(2a + 7)(a - 5)}$ $a \neq 5, -\dfrac{7}{2}$

The numerator and the denominator are relatively prime because they have no common factors. The rational expression is already in lowest terms.

(b) $\dfrac{x^2 - 16}{x^3 - 64} = \dfrac{(x - 4)(x + 4)}{(x - 4)(x^2 + 4x + 16)}$

$\phantom{(b)\ \dfrac{x^2 - 16}{x^3 - 64}} = \dfrac{x + 4}{x^2 + 4x + 16}$ Divide numerator and denominator by $x - 4$. Set $x \neq 4$.

(c) $\dfrac{wy + xy + wz + xz}{(y^2 + z^2)(w - x)(w + x)} = \dfrac{(y + z)(w + x)}{(y^2 + z^2)(w - x)(w + x)}$ Factor by grouping, then divide numerator and denominator by $w + x$. Set $w \neq \pm x$; y and z not both zero.

$\phantom{(c)\ \dfrac{wy + xy + wz + xz}{(y^2 + z^2)(w - x)(w + x)}} = \dfrac{y + z}{(y^2 + z^2)(w - x)}$

Some algebraic expressions have factors in the numerator and denominator that are negatives of each other. These factors can be divided out, leaving a quotient of -1. For example,

$$\dfrac{b - a}{a - b} = \dfrac{-1(a - b)}{a - b} = -1$$

Each term in the numerator must have the opposite sign of its corresponding term in the denominator for this to work.

EXAMPLE 7 Simplify:

$$\frac{r-s}{s^2-r^2} = \frac{r-s}{(s-r)(s+r)}$$
$$= \frac{-1(s-r)}{(s-r)(s+r)} \qquad s \neq \pm r$$
$$= \frac{-1}{s+r}$$

As with addition of rational numbers or fractions, rational expressions also must have the same denominator to be added. When we change a rational expression to an equivalent rational expression by multiplying both the numerator and the denominator by the same nonzero polynomial, as explained in the fundamental principle of rational expressions, the process is called **building a rational expression.**

EXAMPLE 8 Use the fundamental principle to write each rational expression so that it has the indicated numerator or denominator.

(a) $\dfrac{x}{x+3} = \dfrac{?}{2x+6}$ (b) $\dfrac{m+3}{m-4} = \dfrac{m^2+m-6}{?}$

SOLUTION (a) Since $2x + 6 = 2(x + 3)$, we multiply numerator and denominator of the first expression by 2:

$$\frac{x}{x+3} = \frac{x \cdot 2}{(x+3)2} \qquad \text{Multiply by 2. Set } x \neq -3.$$
$$= \frac{2x}{2x+6}$$

(b) Since $m^2 + m - 6 = (m + 3)(m - 2)$, we multiply numerator and denominator of the first expression by $m - 2$:

$$\frac{m+3}{m-4} = \frac{(m+3)(m-2)}{(m-4)(m-2)} \qquad \text{Multiply by } m - 2. \text{ Set } m \neq 2, 4.$$
$$= \frac{m^2+m-6}{m^2-6m+8}$$

EXAMPLE 9 Evaluate $\dfrac{x^2 - 2xy + y^2}{x^2 - y^2}$ for $x = 2, y = 3$.

SOLUTION
$$\frac{x^2 - 2xy + y^2}{x^2 - y^2} = \frac{2^2 - 2 \cdot 2 \cdot 3 + 3^2}{2^2 - 3^2} \qquad x = 2, y = 3$$
$$= \frac{4 - 12 + 9}{4 - 9} = \frac{1}{-5} = \frac{-1}{5}$$

If we simplify before evaluating, then we get

$$\frac{x^2 - 2xy + y^2}{x^2 - y^2} = \frac{(x-y)^2}{(x-y)(x+y)} = \frac{x-y}{x+y}$$
$$= \frac{2-3}{2+3} = \frac{-1}{5}$$

Exercise Set 7.1

Simplify each rational expression. Assume no denominator can be zero.

1. $\dfrac{5x}{15x}$
2. $\dfrac{91y}{13y}$
3. $\dfrac{15xy}{20xyz}$
4. $\dfrac{26ab}{39abc}$
5. $\dfrac{a-b}{b-a}$
6. $\dfrac{s-t}{t-s}$
7. $\dfrac{m^2-n^2}{m-n}$
8. $\dfrac{q^2-r^2}{q+r}$
9. $\dfrac{-17a^2(b-c)}{68a(c-b)}$
10. $\dfrac{-36x^2(y+z)}{54x(y+z)}$
11. $\dfrac{a^2-4a}{a^2-16}$
12. $\dfrac{72a^2-72b^2}{6a^2+12ab+6b^2}$
13. $\dfrac{15r(s-t)(s+t)}{65a^2(s^2-t^2)}$
14. $\dfrac{w^2-3w}{w^2-9}$
15. $\dfrac{x^2y+xy^2}{x^2-y^2}$
16. $\dfrac{p^2-1}{p^2-2p+1}$
17. $\dfrac{(3-2a)(4-a)}{(a-4)(2a-3)}$
18. $\dfrac{(5+b)(2-b)}{(b-2)(b+5)}$
19. $\dfrac{m^2-3m-4}{m^2+3m+2}$
20. $\dfrac{34x^3(x-1)^2(x+5)^3}{17x^3(x-1)^2(x+5)^2}$

Determine which pairs of rational expressions are equal. Assume no denominator can be zero.

21. $\dfrac{x}{3y},\ \dfrac{x^2y^2}{3xy^3}$
22. $\dfrac{ab}{ac},\ \dfrac{a^2b^2}{a^2c^2}$
23. $\dfrac{st-5}{2s-t},\ \dfrac{3s^2t-15s}{6s^2-3st}$
24. $\dfrac{2m+3}{m-1},\ \dfrac{2m^2+5m+3}{m^2+1}$
25. $\dfrac{3p-5}{2p+1},\ \dfrac{3p^2+p-10}{2p^2+5p+2}$
26. $\dfrac{-3-y}{2y+1},\ \dfrac{9-y^2}{2y^2-5y-3}$

Specify any restrictions on the variable(s) so that each rational expression is defined.

27. $\dfrac{3}{x}$
28. $\dfrac{-5}{a-1}$
29. $\dfrac{3b}{2b^2+3b}$
30. $\dfrac{y(y+4)}{y^3+4y^2}$
31. $\dfrac{p^2-3p+2}{p^2-3p-4}$
32. $\dfrac{m^2-m-2}{m^2-m-6}$
33. $\dfrac{(x-2)(y+3)}{(x-3)(y-1)}$
34. $\dfrac{(r-1)(s+4)}{(s-3)(r+2)}$
35. $\dfrac{3ab}{a-b}$
36. $\dfrac{-18vw}{9(v+w)}$
37. $\dfrac{-xy+2x+3y-6}{xy-x-3y+3}$
38. $\dfrac{xy+2y+3x+6}{xy+2y+4x+8}$

Supply the missing numerator or denominator to make the rational expressions equal. Specify any restrictions on the variable.

39. $\dfrac{5a^2}{8b}=\dfrac{?}{48ab^3}$
40. $\dfrac{-10(xy)^2}{3m^2}=\dfrac{?}{-27m^2xy}$
41. $\dfrac{3}{a+3}=\dfrac{?}{a^2+4a+3}$
42. $\dfrac{5}{x-3}=\dfrac{?}{2x^2-x-15}$
43. $\dfrac{-5p}{p^2-q^2}=\dfrac{?}{p^4-q^4}$
44. $\dfrac{2r}{r-s}=\dfrac{?}{r^2-s^2}$
45. $\dfrac{7}{m-4}=\dfrac{7m+28}{?}$
46. $\dfrac{4a}{2a-b}=\dfrac{8a^2+4ab}{?}$
47. $\dfrac{y+1}{y-1}=\dfrac{?}{3xy-3x+2y-2}$
48. $\dfrac{a-8}{3b+4}=\dfrac{?}{6ab+8a-15b-20}$
49. $\dfrac{-6}{-p+q}=\dfrac{6}{?}$
50. $\dfrac{-(a+3)}{7-a}=\dfrac{a+3}{?}$

Evaluate each rational expression for the given value(s). If a rational expression is not defined for the given value(s), write undefined.

51. $\dfrac{5x-4}{3x+7}$ for $x=2$
52. $\dfrac{7y-2}{5y+1}$ for $y=-1$
53. $\dfrac{a^2-3a+4}{a^2+5a+4}$ for $a=-1$
54. $\dfrac{p^2+6p-1}{p^2-3p+2}$ for $p=3$
55. $\dfrac{13s^2t^3}{s^2+7st+t^2}$ for $s=-1, t=0$
56. $\dfrac{m^2-mn+10n^2}{m^2-3mn-10n^2}$ for $m=1, n=-4$

Simplify, if possible. Assume no denominator can be zero.

57. $\dfrac{6x^2 - 6x^2y}{18x^3y^2 - 18x^2y^3}$

58. $\dfrac{15p^3q^2 + 5pq^2}{10p^2q^2 - 20p^2q^3}$

59. $\dfrac{x^3 - y^3}{x^2 + xy + y^2}$

60. $\dfrac{m^3 - n^2}{m - n}$

61. $\dfrac{-3a^2 + 2b + 6a - ab}{3a^2 - b^2 - 2ab}$

62. $\dfrac{st + 2s + rt + 2r}{st + rt + r + s}$

63. $\dfrac{a^2 + 6a + 9}{2a(a + 3) - (a + 3)^2}$

64. $\dfrac{xy^2 - 7xy + 10x}{x(y - 5)^2 - x^2(y - 5)}$

65. $\dfrac{x^2 + y^2}{x^2 - y^2}$

66. $\dfrac{p^3 - q^3}{p^3 + q^3}$

67. $\dfrac{x^3 + y^3}{x^2 - y^2}$

68. $\dfrac{x^3 + xy^3}{x^4 + y^4}$

69. $\dfrac{2a^4 + a^3 - 15a^2}{a^4 + 3a^3}$

70. $\dfrac{3p^4 + 2p^3 - 16p^2}{-p^4 - 7p^3 + 18p^2}$

Write in Your Own Words

71. What does the statement, "Two expressions are relatively prime" mean?

72. Describe what a "rational expression" is.

73. State the fundamental principle of rational expressions in your own words.

74. What does "simplify" mean?

75. In the rational expression $\dfrac{4}{x(2x - 3)}$, what values of x must be excluded and why?

7.2 Multiplying and Dividing Rational Expressions

OBJECTIVES

In this section we will learn to
1. multiply rational expressions; and
2. divide rational expressions.

Section 1.1 showed us that the set of rational numbers is a subset of the set of real numbers, and Section 1.3 stated the rules for the multiplication and division of rational numbers. In general, if $\dfrac{a}{b}$ and $\dfrac{c}{d}$ are rational numbers, then

$$\dfrac{a}{b} \cdot \dfrac{c}{d} = \dfrac{a \cdot c}{b \cdot d} \qquad \text{Multiplication of fractions, } b, d \neq 0$$

and

$$\dfrac{a}{b} \div \dfrac{c}{d} = \dfrac{a}{b} \cdot \dfrac{d}{c} = \dfrac{a \cdot d}{b \cdot c} \qquad \text{Division of fractions, } b, c, d \neq 0$$

The same methods apply to rational expressions.

To Multiply Two Rational Expressions

If $\dfrac{A}{B}$ and $\dfrac{C}{D}$ are rational expressions, then

$$\dfrac{A}{B} \cdot \dfrac{C}{D} = \dfrac{A \cdot C}{B \cdot D}, \quad B, D \neq 0$$

Division by zero is not defined. In each of the following examples, assume that no denominator can be equal to zero.

EXAMPLE 1 Multiply and simplify: $\dfrac{7a}{8b} \cdot \dfrac{3r}{5s}$.

SOLUTION $\dfrac{7a}{8b} \cdot \dfrac{3r}{5s} = \dfrac{21ar}{40bs}$ ← Multiply numerators.
← Multiply denominators.

EXAMPLE 2 Multiply and simplify: $\dfrac{8a^2b^2}{3ab} \cdot \dfrac{6ac}{2ac^2}$.

SOLUTION $\dfrac{8a^2b^2}{3ab} \cdot \dfrac{6ac}{2ac^2} = \dfrac{48a^3b^2c}{6a^2bc^2}$

$= \dfrac{8ab(6a^2bc)}{c(6a^2bc)}$ Factor numerator and denominator; then divide out the common factor, $6a^2bc$.

$= \dfrac{8ab}{c}$

The work in this example can be done another way. If we divide out common factors in either of the numerators with those in either of the denominators before multiplying, the result is the same.

$\dfrac{8a^2b^2}{3ab} \cdot \dfrac{6ac}{2ac^2} = \dfrac{8ab}{\cancel{3}} \cdot \dfrac{\cancel{3}}{c} = \dfrac{8ab}{c}$

EXAMPLE 3 Multiply and simplify.

$\dfrac{x+2}{x^2+4x+3} \cdot \dfrac{x^2+x-6}{2x^2+5x+2} = \dfrac{\cancel{x+2}}{(x+3)(x+1)} \cdot \dfrac{\cancel{(x+3)}(x-2)}{(2x+1)\cancel{(x+2)}}$ Factor first; then divide out common factors.

$= \dfrac{x-2}{(x+1)(2x+1)}$ Multiply.

We can find the product of several rational expressions by applying the procedure for multiplying two rational expressions repeatedly.

EXAMPLE 4 Multiply and simplify.

(a) $\dfrac{x^2-y^2}{x^3-y^3} \cdot \dfrac{x^2+2xy+y^2}{2x+6y} \cdot \dfrac{x^2+xy+y^2}{(x+y)^3}$

$= \dfrac{(x-y)(x+y)}{(x-y)(x^2+xy+y^2)} \cdot \dfrac{(x+y)^2}{2(x+3y)} \cdot \dfrac{x^2+xy+y^2}{(x+y)^3}$ Factor first.

$= \dfrac{\cancel{(x-y)}\cancel{(x+y)}}{\cancel{(x-y)}\cancel{(x^2+xy+y^2)}} \cdot \dfrac{\cancel{(x+y)^2}}{2(x+3y)} \cdot \dfrac{\cancel{x^2+xy+y^2}}{(x+y)\cancel{(x+y)^2}}$ Divide out common factors.

$= \dfrac{1}{2(x+3y)}$ Multiply.

CAUTION

The following caution from Section 7.1 is so important that we repeat it. Rational expressions can be simplified only when the numerator and the denominator have a common factor.

$$\frac{xy + x}{x} = \frac{x(y + 1)}{x} = y + 1$$

x is a factor of $xy + x$.

$$\frac{xy + 1}{x} \text{ cannot be simplified}$$

x is *not* a factor of $xy + 1$.

(b) $\dfrac{ax + bx + ay + by}{x^2 - y^2} \cdot \dfrac{cy - cx + dy - dx}{c^2 + 2cd + d^2} \cdot \dfrac{c + d}{a + b}$

$= \dfrac{(a + b)(x + y)}{(x - y)(x + y)} \cdot \dfrac{(c + d)(y - x)}{(c + d)^2} \cdot \dfrac{c + d}{a + b}$

$= \dfrac{y - x}{x - y} = \dfrac{-1(x - y)}{x - y} = -1$ When two factors differ only by sign, their quotient is -1.

To Divide Two Rational Expressions

If $\dfrac{A}{B}$ and $\dfrac{C}{D}$ are rational expressions, then

$$\frac{A}{B} \div \frac{C}{D} = \frac{A}{B} \cdot \frac{D}{C} = \frac{A \cdot D}{B \cdot C}, \quad B, C, D \neq 0$$

EXAMPLE 5 Divide and simplify: $\dfrac{y + 1}{y} \div \dfrac{y^2 - 2y - 3}{y^2 - 3y}$.

SOLUTION

$\dfrac{y + 1}{y} \div \dfrac{y^2 - 2y - 3}{y^2 - 3y} = \dfrac{y + 1}{y} \cdot \dfrac{y^2 - 3y}{y^2 - 2y - 3}$ $\dfrac{A}{B} \div \dfrac{C}{D} = \dfrac{A}{B} \cdot \dfrac{D}{C}$

$= \dfrac{y + 1}{y} \cdot \dfrac{y(y - 3)}{(y - 3)(y + 1)}$ Factor.

$= 1$ Divide out common factors and multiply.

CAUTION

When all factors can be divided out, **the quotient is 1.**

EXAMPLE 6 Simplify: $\dfrac{y - x}{x - z} \cdot \dfrac{(x + y)^2}{x^2 - y^2} \div \dfrac{x + y}{z - x}$.

SOLUTION

$\dfrac{y - x}{x - z} \cdot \dfrac{(x + y)^2}{x^2 - y^2} \div \dfrac{x + y}{z - x}$

$= \dfrac{y - x}{x - z} \cdot \dfrac{(x + y)^2}{(x - y)(x + y)} \cdot \dfrac{z - x}{x + y}$ Multiply by the reciprocal of $\dfrac{x + y}{z - x}$.

$= \dfrac{y - x}{x - z} \cdot \dfrac{z - x}{x - y}$ Divide out common factors.

$= (-1)(-1) = 1$ $\dfrac{z - x}{x - z} = -1; \dfrac{y - x}{x - y} = -1$

Rational expressions are often written using function notation. For example, if $f(x) = x^2 - 9$ and $g(x) = x + 3$, then the rational expression $\left(\dfrac{f}{g}\right)(x)$ also represents a function, which we will call $r(x)$:

$$r(x) = \left(\frac{f}{g}\right)(x) = \frac{x^2 - 9}{x + 3} = \frac{(x + 3)(x - 3)}{x + 3} = x - 3$$

The domain of $r(x)$ consists of the set of values for which $g(x) \neq 0$. Since $x + 3$ is zero when $x = -3$, we must exclude -3 from the domain. Therefore the domain is $(-\infty, -3) \cup (-3, \infty)$.

EXAMPLE 7 Let $r(y) = \dfrac{y^2 - 9}{y^2 - 49}$ and $s(y) = \dfrac{y + 7}{y + 3}$.

(a) Find $(r \cdot s)(y)$ and simplify the answer.
(b) What values must be excluded from the domain of $(r \cdot s)(y)$?
(c) Find $(r \cdot s)(2)$ and show that it is equal to $r(2) \cdot s(2)$.

SOLUTION (a) $(r \cdot s)(y) = \dfrac{y^2 - 9}{y^2 - 49} \cdot \dfrac{y + 7}{y + 3}$

$= \dfrac{(y - 3)(y + 3)}{(y - 7)(y + 7)} \cdot \dfrac{y + 7}{y + 3}$ Factor first; then divide out common factors and multiply.

$= \dfrac{y - 3}{y - 7}$

(b) Both 7 and -7 are excluded from the domain of $r(y)$ and -3 is excluded from the domain of $s(y)$. Therefore -3, -7, and 7 must be ecluded from the domain of $(r \cdot s)(y)$.

(c) $(r \cdot s)(y) = \dfrac{y - 3}{y - 7}$ $r(y) \cdot s(y) = \dfrac{y^2 - 9}{y^2 - 49} \cdot \dfrac{y + 7}{y + 3}$

$(r \cdot s)(2) = \dfrac{2 - 3}{2 - 7}$ $r(2) \cdot s(2) = \dfrac{2^2 - 9}{2^2 - 49} \cdot \dfrac{2 + 7}{2 + 3}$

$= \dfrac{-1}{-5}$ $= \dfrac{4 - 9}{4 - 49} \cdot \dfrac{9}{5}$

$= \dfrac{1}{5}$ $= \dfrac{-5}{-45} \cdot \dfrac{9}{5} = \dfrac{1}{5}$

EXAMPLE 8 What polynomial will have a quotient of $x^2 + 7x + 10$ when divided by $\dfrac{x + 3}{x + 5}$?

SOLUTION We recall that we check a division problem by multiplying the quotient by the divisor if there is no remainder. So let the unknown polynomial be $P(x)$. The divisor is $D(x) = \dfrac{x + 3}{x + 5}$ and the quotient is $Q(x) = x^2 + 7x + 10$.

Polynomial = Quotient · Divisor

$P(x) = (x^2 + 7x + 10) \cdot \dfrac{x + 3}{x + 5}$

$= (x + 2)(x + 5) \cdot \dfrac{x + 3}{x + 5}$ Factor where possible.

$= (x + 3)(x + 2) = x^2 + 5x + 6$ Simplify then multiply.

The polynomial is $x^2 + 5x + 6$. Therefore,

$(x^2 + 5x + 6) \div \left(\dfrac{x + 3}{x + 5}\right) = x^2 + 7x + 10$

Exercise Set 7.2

Write each product as a single rational expression and simplify it. Assume that no denominator can be zero.

1. $\dfrac{5}{x} \cdot \dfrac{x^2}{10}$
2. $\dfrac{7}{y} \cdot \dfrac{y^2}{21}$
3. $\dfrac{8p^2}{3a} \cdot \dfrac{9a^2}{4p}$
4. $\dfrac{3b^2}{7c} \cdot \dfrac{35c}{12b}$
5. $\dfrac{3xy}{x-y} \cdot \dfrac{(x-y)^2}{6xy}$
6. $\dfrac{a^2-a}{a-1} \cdot \dfrac{a+1}{a}$
7. $\dfrac{m-3}{2p^2} \cdot \dfrac{8p^3}{m-3}$
8. $\dfrac{q-1}{q+2} \cdot \dfrac{q^2-4}{q^2-1}$

Divide and simplify. Assume that no denominator can be zero.

9. $\dfrac{6a}{5b} \div \dfrac{3a}{10b}$
10. $\dfrac{4x}{y} \div \dfrac{12x^2}{y^2}$
11. $3mn \div \dfrac{m}{n}$
12. $15s \div \dfrac{3s}{t}$
13. $\dfrac{p-q}{m-n} \div \dfrac{q-p}{n-m}$
14. $\dfrac{a-b}{c-d} \div \dfrac{b-a}{d-c}$
15. $\dfrac{3y^2-12}{6y} \div (y-2)$
16. $\dfrac{5x^2-45}{7x} \div (x+3)$

Perform the indicated operations and simplify. Assume that no denominator can be zero.

17. $\dfrac{a^2-1}{x+3} \cdot \dfrac{x+3}{a^2+a-2}$
18. $\dfrac{y^2-4}{m-2} \cdot \dfrac{m^2-4}{y^2-4y+4}$
19. $\dfrac{t^2-t}{t-3} \div t$
20. $\dfrac{p^2+p}{p-4} \div (p+1)$
21. $\dfrac{a^2+7a+6}{a+2} \div (a+6)$
22. $\dfrac{x}{y} \cdot \dfrac{a}{b} \div \dfrac{xb}{ya}$
23. $\dfrac{a^2b}{p-3} \cdot \dfrac{(p-3)^2}{ab} \div [a(p-3)]$
24. $\dfrac{(-7y^2z)^3}{25a^2x^3} \div \dfrac{49yz}{(-5ax)^2}$
25. $\dfrac{(-3m^2n)^3}{32x^2y^3} \cdot \dfrac{xy}{-9m} \div \dfrac{m^5n^2}{(-2x)^4}$
26. $\dfrac{a^2-ab}{(b-a)^2} \div \dfrac{a^2}{b^2-a^2} \cdot (b^2+a^2)$
27. $\dfrac{w^3-w^2x}{wy-w} \div \left(\dfrac{w-x}{y-1}\right)^2$
28. $\dfrac{a^3-a^2b}{ac-a} \div \left(\dfrac{a-b}{c-1}\right)^2$
29. $\dfrac{r^3-s^3}{s-r} \cdot \dfrac{(r+s)^2}{r^2+s^2}$
30. $\dfrac{p^3+q^3}{q-p} \div \dfrac{(p+q)^2}{p^2-q^2}$
31. $\dfrac{2a^2+9ab+4b^2}{a^2+2ab+b^2} \cdot \dfrac{3a^2+2ab-b^2}{3a^2+11ab-4b^2}$
32. $\dfrac{4y^2+4y+1}{2y+1} \cdot \dfrac{25(3y^2-19y+20)}{5y^2-5y-100}$
33. $(a+2)(a-1) \cdot \dfrac{3}{a^2+a-2}$
34. $p(p-3)(p+4) \cdot \dfrac{-2}{p^2+p-12}$
35. $\dfrac{mx+my+2x+2y}{6x^2-5xy-4y^2} \div \dfrac{2mx-4x+my-2y}{3mx-6x-4my+8y}$
36. $\dfrac{ax-2a+2xy-4y}{ax+2a-2xy-4y} \div \dfrac{ax+2a+2xy+4y}{ax-2a-2xy+4y}$
37. $\dfrac{2p+1}{3p^2} \cdot \dfrac{5p^2q}{p-7} \div \dfrac{4p+2}{3p-21}$
38. $\dfrac{r^2-s^2}{r^2+rs+s^2} \cdot \dfrac{r^3-s^3}{r^2+s^2} \div \dfrac{r^4-s^4}{r^2-s^2}$
39. $\dfrac{8(x+2)}{4x^2+16x+16} \cdot \dfrac{3(x^2-4)}{-3x^2+192} \div \dfrac{6(x^2-x-6)}{7(x+8)}$
40. $\dfrac{a^2+3a-40}{5a^2+39a-8} \div \dfrac{6a-3}{a^2-25} \div \dfrac{2a^2+16a+30}{8a^2-100}$
41. $\dfrac{k^2-1}{k^3+1} \div \dfrac{k^3-1}{k-1} \cdot (k^2+k+1)$
42. $\dfrac{n^2-1}{n^3+1} \cdot \dfrac{n^3-1}{n-1} \div (n^2+n+1)$
43. $\dfrac{x^3-3x^2+2x-6}{x+5} \div \dfrac{x^2+2}{x^2+7x+10}$
44. $\dfrac{9y^3-18y^2-4y+8}{3y^2-4y-4} \cdot \dfrac{1}{8-12y}$
45. $\dfrac{b}{b-a} \cdot \left(\dfrac{a^2-2ab+b^2}{a^2-b^2} \div \dfrac{1}{a^2+2ab+b^2}\right)$
46. $\dfrac{27x^3-y^3}{xz+yz} \div \left(\dfrac{9x^2-y^2}{z^4} \cdot \dfrac{xz^3-yz^3}{3x^2-2xy-y^2}\right)$

Complete each of the following computations. Assume no denominator can be zero.

47. $\dfrac{6a^2b^2}{4wz^2} \div \left(\underline{\quad}\right) = \dfrac{60a^6b^7}{8w^4z^5}$

48. $\dfrac{3p^3 - 6p^2}{p^2q} \div \left(\dfrac{}{}\right) = \dfrac{-12p}{2 + p}$

49. $\dfrac{x}{x + 1} \cdot \left(\dfrac{}{}\right) = \dfrac{3x^2 + 5x}{x^2 - 1}$

50. $\dfrac{2xy^2}{x + y} \cdot \left(\dfrac{}{}\right) = \dfrac{x + y}{6xy}$

51. $\dfrac{x^2 - 1}{x + 2} \div \left(\dfrac{}{}\right) = x + 1$

52. $(2 - x) \div \left(\dfrac{}{}\right) = \dfrac{-(x + 2)}{x + 3}$

53. $\dfrac{p^2 - p}{p - 3} \cdot \left(\dfrac{}{}\right) = p$

54. $\dfrac{y^2}{2x + 4} \div \dfrac{y}{x + 2} \cdot \left(\dfrac{}{}\right) = \dfrac{x^2}{6y^2}$

For each rational function, find what is indicated.

55. If $f(x) = \dfrac{x + 2}{x - 5}$, **(a)** find $f(3)$ and $f(-2)$, and **(b)** find the domain.

56. For $g(a) = \dfrac{a^2 - 9}{a^2 - 4a + 4}$, **(a)** find $g(1)$ and $g(-1)$, and **(b)** find the domain.

57. If $f(t) = \dfrac{3}{t^2 + t - 2}$ and $g(t) = t + 2$, find $(f \cdot g)(t)$.

58. For $r(s) = \dfrac{s^2 - 4}{s^2 - 5s + 4}$ and $t(s) = \dfrac{s - 4}{s + 2}$, find $(r \cdot t)(s)$.

59. If the quotient is $Q(x) = x^2 + x + 1$ and the divisor is $D(x) = x - 1$, find $P(x)$, where $P(x) = Q(x) \cdot D(x)$.

60. If the quotient is $Q(x) = x^2 + 3x + 2$ and the divisor $D(x) = x - 1$, find $P(x)$, where $P(x) = Q(x) \cdot D(x)$.

61. If $r(t) = \dfrac{t^2 + 4t + 3}{t - 1}$ and $s(t) = \dfrac{t^2 - 1}{t + 3}$, find **(a)** $(r \cdot s)(t)$ and **(b)** $(r \cdot s)(0)$. **(c)** Show that $r(0) \cdot s(0) = (r \cdot s)(0)$.

62. Find the domain of $h(x) = \dfrac{1}{x^3 - x^2 - 2x}$.

63. Find the domain of $g(y) = \dfrac{-5}{7y^2 - 14y}$.

64. If $f(x) = \dfrac{3}{4x^2 - 1}$ and $g(x) = \dfrac{1}{x}$, find $f[g(1)]$.

REVIEW EXERCISES

Solve each equation by factoring.

65. $x^2 + 9x + 8 = 0$
66. $2x^2 - x - 15 = 0$
67. $25x^2 + 10x + 1 = 0$
68. $8x^2 - 7x = 0$
69. $3x^2 - 3x - 18 = 0$
70. $36y^2 - 25 = 0$
71. $5x^3 + 15x^2 + 10x = 0$
72. $x^2(x - 3) - 5x(x - 3) - 6(x - 3) = 0$
73. $15x^2 + 2x - 8 = 0$
74. $30x^2 + 89x + 63 = 0$
75. $\dfrac{1}{2}x^2 + \dfrac{7}{12}x - \dfrac{1}{4} = 0$
76. $x^4 - 4x^2 = 0$
77. $12x^3 + 34x^2 + 10x = 0$

For Extra Thought

78. Mr. Garcia's garden is in the shape of a rectangle with an area of 84 sq. ft. The length of the garden is 5 ft more than the width.
    (a) Find the dimensions of the garden.
    (b) Find the perimeter of the garden.
    (c) Find the cost to fence the garden if fencing costs $1.40 per linear foot.
    (d) Find the cost to fertilize the garden if a sack of fertilizer costs $6.78 and covers 504 sq. ft.

79. Maxine has a flower bed in the shape of a right triangle with an area of 96 sq. ft. See the figure.
    (a) Find the dimensions of the flower bed.
    (b) Find the perimeter of the flower bed.
    (c) Find the cost to fence the flower bed if fencing costs $3.00 per linear foot.
    (d) Find the cost to replant the flower bed if a tray of flowers costs $4.80 and covers 12 sq. ft.

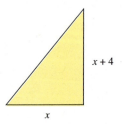

Write In Your Own Words

80. How do you multiply two rational expressions?
81. How do you divide two rational expressions?
82. In the directions "multiply and simplify," what does the word "simplify" mean?
83. How do you divide three rational expressions, such $A \div B \div C$?
84. How do you multiply four rational expressions?

7.3 Adding and Subtracting Rational Expressions

OBJECTIVES In this section we will learn to
1. find the least common denominator of two or more rational expressions; and
2. add and subtract rational expressions.

In Section 1.3 we saw that two fractions with the same denominator are added or subtracted according to the definition

$$\frac{a}{b} + \frac{c}{b} = \frac{a+c}{b}, \qquad \frac{a}{b} - \frac{c}{b} = \frac{a-c}{b}, \qquad b \neq 0$$

We add and subtract rational expressions by a similar method.

To Add Two Rational Expressions

If $\frac{A}{C}$ and $\frac{B}{C}$ are any two rational expressions, then

$$\frac{A}{C} + \frac{B}{C} = \frac{A+B}{C} \qquad \text{and} \qquad \frac{A}{C} - \frac{B}{C} = \frac{A-B}{C}, \qquad C \neq 0$$

EXAMPLE 1 Add: **(a)** $\frac{2x}{3y} + \frac{z}{3y}$, $y \neq 0$; **(b)** $\frac{x^2+3}{x+2} + \frac{x^2-2x}{x+2}$, $x \neq -2$.

SOLUTION **(a)** Since the denominators are the same, we add the numerators:

$$\frac{2x}{3y} + \frac{z}{3y} = \frac{2x+z}{3y}$$

(b) $\dfrac{x^2+3}{x+2} + \dfrac{x^2-2x}{x+2} = \dfrac{(x^2+3)+(x^2-2x)}{x+2}$

$$= \frac{2x^2 - 2x + 3}{x+2}$$

CAUTION
Whenever you add or subtract rational expressions, always check to see if your work can be simplified.

EXAMPLE 2 Add and simplify: $\dfrac{x^2}{x-2} - \dfrac{4}{x-2}$, $x \neq 2$.

SOLUTION $\dfrac{x^2}{x-2} - \dfrac{4}{x-2} = \dfrac{x^2-4}{x-2} = \dfrac{(x-2)(x+2)}{x-2}$

$$= x + 2$$

EXAMPLE 3 Carry out the indicated operations and simplify:

$$\frac{x^2-2}{x-3} - \frac{2x^2-x}{x-3} + \frac{2x^2-10}{x-3}, \qquad x \neq 3$$

CAUTION

When subtracting be particularly careful with the signs.

RIGHT

$$\frac{2x}{x-2} - \frac{x+3}{x-2}$$
$$= \frac{2x - (x+3)}{x-2}$$ Do not eliminate this step.
$$= \frac{2x - x - 3}{x-2}$$ Watch these signs!

WRONG

$$\frac{2x}{x-2} - \frac{x+3}{x-2}$$
$$= \frac{2x - x + 3}{x-2}$$ A sign error occurred here.

SOLUTION

$$\frac{x^2 - 2}{x-3} - \frac{2x^2 - x}{x-3} + \frac{2x^2 - 10}{x-3}$$
$$= \frac{x^2 - 2 - (2x^2 - x) + (2x^2 - 10)}{x-3}$$ Do not eliminate this step with the parentheses.

Watch these signs!

$$= \frac{x^2 - 2 - 2x^2 + x + 2x^2 - 10}{x-3}$$
$$= \frac{x^2 + x - 12}{x-3} = \frac{(x+4)(x-3)}{x-3} = x + 4$$

To add or subtract two different rational expressions, they must have the same denominator. If the denominators are not the same, we must rewrite the expressions so that their denominators are the same. As in work with fractions, the denominator is called the **least common denominator.**

DEFINITION

LCD of Rational Expressions

The **least common denominator (LCD)** of two or more rational expressions is the polynomial of smallest degree that is a multiple of their denominators.

For example, the LCD of $\frac{1}{4}$ and $\frac{1}{6}$ is 12, because 12 is the smallest number that is a multiple of both 4 and 6.

EXAMPLE 4 Find the LCD of $\frac{1}{8}, \frac{1}{12},$ and $\frac{1}{20}.$

SOLUTION We begin by factoring each denominator into its prime factors. Since the LCD is a multiple of each number, it must contain each distinct prime factor the greatest number of times it occurs in any one of the factorizations of 8, 12, and 20.

$$8 = 2 \cdot 2 \cdot 2 \qquad\qquad 8 = 2 \cdot 2 \cdot 2$$
$$12 = 2 \cdot 2 \cdot 3 \qquad\qquad 12 = 2 \cdot 2 \quad\;\; \cdot 3$$
$$20 = 2 \cdot 2 \cdot 5 \qquad\qquad 20 = 2 \cdot 2 \qquad\quad \cdot 5$$
$$\qquad\qquad\qquad\qquad\;\; \text{LCD} = 2 \cdot 2 \cdot 2 \cdot 3 \cdot 5$$

The greatest number of times that 2 appears is three times, in 8. The greatest number of times that 3 appears is once, in 12. And the greatest number of times that 5 appears is once, in 20.

$$\text{LCD} = 2 \cdot 2 \cdot 2 \cdot 3 \cdot 5 = 2^3 \cdot 3 \cdot 5 = 120$$

The LCD of 8, 12, and 20 is 120, since it is the smallest number that is a multiple of each.

EXAMPLE 5 Find the LCD of $\dfrac{1}{x^2 + 2x}$, $\dfrac{1}{x^2 - 4}$, and $\dfrac{1}{x^2 - x - 6}$.

SOLUTION To find the LCD, we must write each denominator in factored form:

$$x^2 + 2x = x(x + 2)$$
$$x^2 - 4 = (x - 2)(x + 2)$$
$$x^2 - x - 6 = (x + 2)(x - 3)$$

The greatest number of times each of x, $x - 2$, $x + 2$, and $x - 3$ occurs in any one denominator is once. Therefore LCD is $x(x - 2)(x + 2)(x - 3)$.

> **To Find the Least Common Denominator of Two or More Denominators**
>
> 1. Factor each denominator completely.
> 2. Include each distinct prime factor the greatest number of times it occurs in any denominator.
> 3. The LCD is the product of the factors in Step 2.

In the remainder of this section, we assume that no denominator is equal to zero, since division by zero is not defined.

EXAMPLE 6 Rewrite $\dfrac{2}{x + 3}$, $\dfrac{4}{x^2 + 5x + 6}$, and $\dfrac{x}{x + 2}$ so that each denominator is the LCD of the three denominators.

SOLUTION To find the LCD, we write each denominator in factored form:

$$x + 3 = x + 3$$
$$x^2 + 5x + 6 = (x + 2)(x + 3)$$
$$x + 2 = x + 2$$

The LCD is $(x + 2)(x + 3)$. Now we rewrite each given rational expression.

$$\frac{2}{x + 3} = \frac{2(x + 2)}{(x + 3)(x + 2)}$$ Multiply by $\frac{x + 2}{x + 2}$ to find an equivalent expression.

$$\frac{4}{x^2 + 5x + 6} = \frac{4}{(x + 2)(x + 3)}$$

$$\frac{x}{x + 2} = \frac{x(x + 3)}{(x + 2)(x + 3)}$$ Multiply by $\frac{x + 3}{x + 3}$ to find an equivalent expression.

EXAMPLE 7 Let $f(x) = \dfrac{3}{x - 2}$, $g(x) = \dfrac{5}{x + 2}$, and $h(x) = \dfrac{3}{2 - x}$. Find:

(a) $(f + g)(x)$ **(b)** $(f - h)(x)$

SOLUTION **(a)** $(f + g)(x) = \dfrac{3}{x - 2} + \dfrac{5}{x + 2}$

$= \dfrac{3(x + 2)}{(x - 2)(x + 2)} + \dfrac{5(x - 2)}{(x - 2)(x + 2)}$ The LCD is $(x - 2)(x + 2)$.

$= \dfrac{3(x + 2) + 5(x - 2)}{(x - 2)(x + 2)}$ Add.

$= \dfrac{3x + 6 + 5x - 10}{(x - 2)(x + 2)}$ Distributive property.

$= \dfrac{8x - 4}{(x - 2)(x + 2)}$ Combine like terms.

$= \dfrac{4(2x - 1)}{(x - 2)(x + 2)}$

(b) $(f - h)(x) = \dfrac{3}{x - 2} - \dfrac{3}{2 - x}$

Notice that $2 - x = -(x - 2)$. Therefore the denominators can be made the same by multiplying the numerator and denominator of either rational expression by -1.

$(f - h)(x) = \dfrac{3}{x - 2} - \dfrac{3(-1)}{(2 - x)(-1)}$ Multiply numerator and denominator by -1.

$= \dfrac{3}{x - 2} - \dfrac{-3}{x - 2}$

$= \dfrac{3 - (-3)}{x - 2}$

$= \dfrac{3 + 3}{x - 2}$

$= \dfrac{6}{x - 2}$

7.3 · ADDING AND SUBTRACTING RATIONAL EXPRESSIONS

EXAMPLE 8 Let $f(a) = \dfrac{a^n}{a^{2n} - 4}$ and $g(a) = 4 - \dfrac{2}{a^n + 2}$, where n is a natural number. Find $(f + g)(a)$.

SOLUTION

$$(f + g)(a) = \frac{a^n}{a^{2n} - 4} + 4 - \frac{2}{a^n + 2}$$

$$= \frac{a^n}{(a^n - 2)(a^n + 2)} + 4 - \frac{2}{a^n + 2} \qquad \text{Factor the denominator.}$$

$$= \frac{a^n}{a^{2n} - 4} + \frac{4(a^{2n} - 4)}{a^{2n} - 4} - \frac{2(a^n - 2)}{a^{2n} - 4} \qquad \begin{array}{l}\text{The LCD is}\\ (a^n - 2)(a^n + 2) = a^{2n} - 4.\end{array}$$

$$= \frac{a^n + 4a^{2n} - 16 - 2a^n + 4}{a^{2n} - 4}$$

$$= \frac{4a^{2n} - a^n - 12}{a^{2n} - 4} \qquad \text{This answer is simplified.}$$

EXAMPLE 9 Carry out the indicated operations and simplify:

$$\left(\frac{s + 2t}{s^2 - 4st + 3t^2} - \frac{s}{s^2 + st - 2t^2}\right) \div \frac{7s + 4t}{s + 2t}$$

SOLUTION We begin by factoring the denominators:

$$s^2 - 4st + 3t^2 = (s - 3t)(s - t)$$
$$s^2 + st - 2t^2 = (s + 2t)(s - t)$$
$$s + 2t = s + 2t$$

The LCD is $(s - 3t)(s - t)(s + 2t)$.

$$\left(\frac{s + 2t}{(s - 3t)(s - t)} - \frac{s}{(s + 2t)(s - t)}\right) \div \frac{7s + 4t}{s + 2t}$$

$$= \left(\frac{(s + 2t)(s + 2t) - s(s - 3t)}{(s - 3t)(s - t)(s + 2t)}\right) \div \frac{7s + 4t}{s + 2t} \qquad \begin{array}{l}\text{Rewrite the}\\ \text{expression}\\ \text{using the LCD.}\end{array}$$

$$= \left(\frac{s^2 + 4st + 4t^2 - s^2 + 3st}{(s - 3t)(s - t)(s + 2t)}\right) \div \frac{7s + 4t}{s + 2t} \qquad \text{Multiply.}$$

$$= \left(\frac{7st + 4t^2}{(s - 3t)(s - t)(s + 2t)}\right) \div \frac{7s + 4t}{s + 2t} \qquad \begin{array}{l}\text{Combine like}\\ \text{terms.}\end{array}$$

$$= \frac{t(7s + 4t)}{(s - 3t)(s - t)(s + 2t)} \cdot \frac{s + 2t}{7s + 4t} \qquad \begin{array}{l}\text{Factor the numerator. To}\\ \text{divide, multiply by the inverse}\\ \text{of the divisor.}\end{array}$$

$$= \frac{t}{(s - 3t)(s - t)} \qquad \text{Simplify.}$$

EXAMPLE 10 Subtract and simplify:

$$\frac{x^n}{x^{2n} - y^{2n}} - \frac{y^n}{x^n - y^n}$$

SOLUTION   First we factor the denominators to find the LCD.

$$x^{2n} - y^{2n} = (x^n - y^n)(x^n + y^n)$$
$$x^n - y^n = x^n - y^n$$

The LCD is $(x^n - y^n)(x^n + y^n)$. Therefore

$$\frac{x^n}{x^{2n} - y^{2n}} - \frac{y^n}{x^n - y^n} = \frac{x^n}{(x^n - y^n)(x^n + y^n)} - \frac{y^n}{x^n - y^n}$$
$$= \frac{x^n}{(x^n - y^n)(x^n + y^n)} - \frac{y^n(x^n + y^n)}{(x^n - y^n)(x^n + y^n)}$$
$$= \frac{x^n - y^n(x^n + y^n)}{(x^n - y^n)(x^n + y^n)}$$
$$= \frac{x^n - x^n y^n - y^{2n}}{(x^n - y^n)(x^n + y^n)}$$

Exercise Set 7.3

Add or subtract as indicated and simplify. Assume no denominator can be zero.

1. $\dfrac{2}{5x} + \dfrac{3}{5x}$
2. $\dfrac{5}{8y} + \dfrac{3}{8y}$
3. $\dfrac{9}{7y} - \dfrac{2}{7y}$
4. $\dfrac{11}{3x} - \dfrac{2}{3x}$
5. $\dfrac{13a}{4b} + \dfrac{7a}{4b}$
6. $\dfrac{15x}{7y} + \dfrac{6x}{7y}$
7. $\dfrac{36ab}{5a} - \dfrac{ab}{5a}$
8. $\dfrac{41xy}{6y} - \dfrac{5xy}{6y}$
9. $\dfrac{x}{x-y} - \dfrac{y}{x-y}$
10. $\dfrac{a^2}{a+b} - \dfrac{b^2}{a+b}$
11. $\dfrac{a}{a^2-b^2} + \dfrac{b}{a^2-b^2}$
12. $\dfrac{a}{x} + \dfrac{b}{x}$
13. $\dfrac{r+s}{r} + \dfrac{r-s}{r}$
14. $\dfrac{x+y}{x} + \dfrac{x-y}{x}$
15. $\dfrac{a+b}{b} - \dfrac{a-b}{b}$
16. $\dfrac{x+y}{y} - \dfrac{x-y}{y}$
17. $\dfrac{x+6}{x-y} - \dfrac{y+6}{x-y}$
18. $\dfrac{r+7}{r-s} - \dfrac{s+7}{r-s}$
19. $\dfrac{x}{x^2-4} + \dfrac{2}{x^2-4}$
20. $\dfrac{p}{p^2-9} + \dfrac{3}{p^2-9}$
21. $\dfrac{r}{r^2-t^2} - \dfrac{t}{r^2-t^2}$
22. $\dfrac{u}{u^2-v^2} - \dfrac{v}{u^2-v^2}$
23. $\dfrac{m}{m^3-1} - \dfrac{1}{m^3-1}$
24. $\dfrac{a^2}{a^3-b^3} + \dfrac{ab+b^2}{a^3-b^3}$

Perform the indicated operations and simplify. Assume no denominator can be zero.

25. $\dfrac{1}{a} + \dfrac{2}{5a} - \dfrac{3}{2a}$
26. $\dfrac{5}{x} + \dfrac{5}{2x} - \dfrac{5}{3x}$
27. $\dfrac{1}{x^2} - \dfrac{2}{3x} + \dfrac{5}{4x}$
28. $\dfrac{7}{p^2} - \dfrac{3}{2p} + \dfrac{1}{p}$
29. $\dfrac{3a-5}{10a} - \dfrac{3a-4}{15a}$
30. $\dfrac{3w-1}{7w} - \dfrac{w-2}{14w}$
31. $\dfrac{m}{m+3} - \dfrac{m-1}{m-4}$
32. $\dfrac{y-3}{y-2} - \dfrac{y}{y-4}$
33. $\dfrac{7a}{4b} + \dfrac{3b}{8a} - \dfrac{12b^2 + 56a^2}{32ab}$
34. $\dfrac{4p}{27q} + \dfrac{2q}{9p} - \dfrac{36p^2 + 54q^2}{54pq}$
35. $\dfrac{2}{m-2} + \dfrac{4}{m+2} + \dfrac{m+6}{4-m^2}$
36. $\dfrac{4}{2a+1} - \dfrac{10a-3}{4a^2-1} - \dfrac{3}{1-2a}$
37. $\dfrac{x^2+1}{x^2-1} - 1$
38. $\dfrac{r^2+9}{r^2-9} - 1$
39. $\dfrac{4k+1}{2k+3} + \dfrac{2}{k}$
40. $\dfrac{2}{1-5v} + \dfrac{3}{4v}$
41. $\dfrac{x+1}{x+2} - \dfrac{x-1}{x-2}$
42. $\dfrac{6}{4-y} - \dfrac{2}{y-4}$
43. $\dfrac{3c}{c^2-c-12} - \dfrac{2c}{c^2-9c+20}$

44. $\dfrac{11-w}{w^2+3w-4} - \dfrac{-22-7w}{w^2+6w+8}$

45. $\dfrac{18s-19}{4s^2+27s-7} - \dfrac{12s-41}{3s^2+17s-28}$

46. $\dfrac{42-22y}{3y^2-13y-10} - \dfrac{21-13y}{2y^2-9y-5}$

47. $ab - \dfrac{a}{b}$

48. $\dfrac{3b}{a} + 2$

49. $m + 1 + \dfrac{1}{m-1}$

50. $3r - 2 + \dfrac{2}{r-1}$

51. $\dfrac{2k+1}{k^2+2k+1} - 1$

52. $2x + \dfrac{xy-y^2}{x+y} - 3y$

Perform the indicated operations and simplify. Assume no denominator can be zero.

53. $\left(\dfrac{a^2-b^2}{u^2-v^2}\right)\left(\dfrac{av-au}{b-a}\right) + \left(\dfrac{a^2-av}{u+v}\right)\left(\dfrac{1}{a}\right)$

54. $\left(\dfrac{6r^2}{r^2-1}\right)\left(\dfrac{r+1}{3}\right) - \dfrac{2r^2}{r-1}$

55. $\left(\dfrac{m^2-5m-6}{3m}\right)\left(\dfrac{-9m^2}{6-m}\right) + \dfrac{10m^2+10m}{5}$

56. $\left(\dfrac{1}{y^2-1} \div \dfrac{1}{y^2+1}\right)\left(\dfrac{y^3+1}{y^4-1}\right) + \dfrac{1}{(y+1)^2(y-1)}$

57. $\left(\dfrac{x^3-1}{x^2+2x+1} \div \dfrac{x^3+1}{x^2-1}\right) \cdot \left(\dfrac{x^2-x+1}{x^2-2x+1}\right)$
$+ \dfrac{-(x^2-x+1)}{x^2+2x+1}$

58. $\left(\dfrac{a^3-64}{a^2-16} \div \dfrac{a^2-4a+16}{a^2-4} \div \dfrac{a^2+4a+16}{a^3+64}\right) + 4 - a^2$

Perform the indicated operations on the given rational functions and simplify. Assume no denominator can be zero.

59. If $f(a) = \dfrac{1}{4a^2} + \dfrac{1}{3a^2+3a}$ and $g(a) = \dfrac{1}{3a^2+6a+3}$, find $(f+g)(a)$.

60. Find $(f+g)(x)$, given
$f(x) = \dfrac{3}{12+7x+x^2}$, $g(x) = \dfrac{2}{x^2+5x+6} + \dfrac{7}{x^2+6x+8}$

61. For $f(k) = \dfrac{2k-5}{3k^2-3k} - \dfrac{k+5}{k^2-1}$ and $g(k) = \dfrac{4}{k-1}$, find $(f+g)(k)$.

62. If $f(w) = \dfrac{4}{2-w} - \dfrac{1}{w-2}$ and $g(w) = 2 - \dfrac{w}{(w-2)^2}$, find $(f+g)(w)$.

63. For $f(x) = \dfrac{x^n}{x^{2n}-1}$ and $g(x) = \dfrac{x^n}{x^{3n}-1}$, find $(f+g)(x)$. Here n is a natural number.

64. For $f(x) = \dfrac{x^n}{x^{3n}-1}$, and $g(x) = \dfrac{1}{x^{2n}-1}$, find $(f-g)(x)$. Here n is a natural number.

REVIEW EXERCISES

Multiply or divide, as indicated.

65. $\dfrac{ax+bx-ay-by}{a+b} \cdot \dfrac{3}{x-y}$

66. $\dfrac{x^3-8}{x^2+2x+4} \cdot \dfrac{1}{x-2}$

67. $\dfrac{x^3-y^3}{x^2+xy+y^2} \div \dfrac{2y-2x}{x^2+xy+y^2}$

68. $\dfrac{x^2-3x+2}{x+2} \cdot \dfrac{3x}{x-2} \div \dfrac{x^2-5x}{2x+4}$

69. $\dfrac{xy^3}{a^2b} \div \dfrac{x^2y^4}{a^4b^3} \cdot \dfrac{x^3y}{ab}$

70. $\dfrac{15x^3}{14y^2} \div \dfrac{20x^2y}{21x} \cdot \dfrac{16y^5}{9x}$

71. $\dfrac{x^6+y^6}{x^4+4x^2y^2+3y^4} \div \dfrac{x^4+3x^2y^2+2y^4}{x^4+5x^2y^2+6y^4}$

72. $\dfrac{x^2+9xy+8y^2}{x^2+7xy-8y^2} \div \dfrac{x^2-y^2}{x^2+5xy-6y^2}$

73. $\dfrac{x^4-5x^2-36}{x^4-14x^2-32} \cdot \dfrac{x^3-8-2x^2+4x}{x^3+1+4x^2+4x}$

74. $\dfrac{4x^4-21x^2+27}{9x^4-49x^2+20} \cdot \dfrac{3x^3+2x^2-15x-10}{2x^3-3x^2-6x+9}$

WRITE IN YOUR OWN WORDS

75. What is an LCD?
76. How do you find the LCD of two rational expressions?
77. How do you add two rational expressions?
78. How do you subtract two rational expressions?

7.4 Simplifying Complex Rational Expressions

OBJECTIVES In this section we will learn to
1. identify complex rational expressions; and
2. simplify complex rational expressions by various means.

A **complex rational expression,** also called a **complex fraction,** is a rational expression in which the numerator, the denominator, or both, contain rational expressions. For example,

$$\frac{\frac{x+1}{3}}{\frac{x+2}{x}}$$

is a complex rational expression: Both the numerator, $\frac{x+1}{3}$, and the denominator, $\frac{x+2}{x}$, are rational expressions.

EXAMPLE 1 Simplify $\dfrac{\frac{3}{4}}{\frac{2}{5}}$.

SOLUTION

$\frac{3}{4}$ is a rational number. $\rightarrow$ $\dfrac{\frac{3}{4}}{\frac{2}{5}}$ $\leftarrow$ The quotient of two rational numbers

$\frac{2}{5}$ is a rational number. $\rightarrow$

There are two ways we can simplify this expression. The first is to view it as a division problem and multiply the numerator by the reciprocal of the denominator:

$$\frac{\frac{3}{4}}{\frac{2}{5}} = \frac{3}{4} \cdot \frac{5}{2} = \frac{15}{8}$$

A second method is to find the LCD of the numerator, $\frac{3}{4}$, and the denominator, $\frac{2}{5}$. The LCD is 20. Now we multiply both the numerator and denominator by this number:

$$\frac{\frac{3}{4}}{\frac{2}{5}} = \frac{\frac{3}{4}}{\frac{2}{5}} \cdot \frac{20}{20} \qquad \text{Fundamental principle of rational expressions;} \frac{20}{20} = 1$$

$$= \frac{\frac{3}{4} \cdot 20}{\frac{2}{5} \cdot 20}$$

$$= \frac{3 \cdot 5}{2 \cdot 4} = \frac{15}{8}$$

EXAMPLE 2 Simplify $\dfrac{\frac{x+1}{3}}{\frac{x+2}{x}}$ in two different ways.

SOLUTION **Method 1** Multiply the numerator by the reciprocal of the denominator.

$$\dfrac{\frac{x+1}{3}}{\frac{x+2}{x}} = \dfrac{x+1}{3} \cdot \dfrac{x}{x+2}$$

$$= \dfrac{(x+1) \cdot x}{3(x+2)} = \dfrac{x^2 + x}{3x + 6}$$

Method 2 Find the LCD of the numerator and the denominator.

The LCD of the numerator, $\dfrac{x+1}{3}$, and the denominator, $\dfrac{x+2}{x}$, is $3x$.

Multiplying both the numerator and the denominator by $3x$ is the same as multiplying by 1. The new expression is equivalent to the original one.

$$\dfrac{\frac{x+1}{3}}{\frac{x+2}{x}} = \dfrac{\left(\frac{x+1}{3}\right) \cdot 3x}{\left(\frac{x+2}{x}\right) \cdot 3x} \quad \textcolor{red}{\text{Multiply by } 3x, \text{ the LCD.}}$$

$$= \dfrac{(x+1) \cdot x}{(x+2) \cdot 3} \quad \textcolor{red}{\text{Simplify before multiplying.}}$$

$$= \dfrac{x^2 + x}{3x + 6}$$

The second method, which depends on the fundamental principle of rational expressions, is sometimes easier than the first. We will use it for the remainder of this section.

EXAMPLE 3 Simplify

$$\dfrac{\frac{1}{x^2} - 2}{3 - \frac{2}{x}}$$

SOLUTION
$$\dfrac{\frac{1}{x^2} - 2}{3 - \frac{2}{x}} = \dfrac{\left(\frac{1}{x^2} - 2\right) \cdot x^2}{\left(3 - \frac{2}{x}\right) \cdot x^2} \quad \textcolor{red}{\text{Multiply by } x^2, \text{ the LCD.}}$$

$$= \dfrac{\frac{1}{x^2} \cdot x^2 - 2 \cdot x^2}{3 \cdot x^2 - \frac{2}{x} \cdot x^2} \quad \textcolor{red}{\text{Distributive property}}$$

$$= \dfrac{1 - 2x^2}{3x^2 - 2x}$$

CHAPTER 7 · RATIONAL EXPRESSIONS, EQUATIONS, AND FUNCTIONS

EXAMPLE 4 Simplify

$$\frac{\dfrac{1}{a} - \dfrac{1}{b}}{\dfrac{1}{a^2} - \dfrac{1}{b^2}}$$

SOLUTION The LCD of all of the fractions in the numerator and the denominator is a^2b^2.

$$\frac{\dfrac{1}{a} - \dfrac{1}{b}}{\dfrac{1}{a^2} - \dfrac{1}{b^2}} = \frac{\left(\dfrac{1}{a} - \dfrac{1}{b}\right) \cdot a^2b^2}{\left(\dfrac{1}{a^2} - \dfrac{1}{b^2}\right) \cdot a^2b^2} \quad \text{Multiply by } a^2b^2, \text{ the LCD.}$$

$$= \frac{\dfrac{1}{a} \cdot a^2b^2 - \dfrac{1}{b} \cdot a^2b^2}{\dfrac{1}{a^2} \cdot a^2b^2 - \dfrac{1}{b^2} \cdot a^2b^2} \quad \text{Distributive property}$$

$$= \frac{ab^2 - a^2b}{b^2 - a^2} \quad \text{Simplify.}$$

$$= \frac{ab(b - a)}{(b - a)(b + a)} \quad \text{Factor.}$$

$$= \frac{ab}{b + a} \quad \text{Simplify.}$$

EXAMPLE 5 Simplify

$$\frac{\dfrac{x - 2}{x - 3} - \dfrac{x - 3}{x - 2}}{\dfrac{2}{x^2 - 5x + 6}}$$

SOLUTION The LCD is $x^2 - 5x + 6 = (x - 2)(x - 3)$.

$$\frac{\dfrac{x - 2}{x - 3} - \dfrac{x - 3}{x - 2}}{\dfrac{2}{x^2 - 5x + 6}} = \frac{\left(\dfrac{x - 2}{x - 3} - \dfrac{x - 3}{x - 2}\right) \cdot (x - 2)(x - 3)}{\left(\dfrac{2}{x^2 - 5x + 6}\right) \cdot (x - 2)(x - 3)} \quad \text{Multiply by the LCD.}$$

$$= \frac{\dfrac{x - 2}{x - 3} \cdot (x - 2)(x - 3) - \dfrac{x - 3}{x - 2} \cdot (x - 2)(x - 3)}{\left(\dfrac{2}{x^2 - 5x + 6}\right) \cdot (x - 2)(x - 3)}$$

$$= \frac{(x - 2)^2 - (x - 3)^2}{2} \quad \text{Simplify before multiplying.}$$

$$= \frac{x^2 - 4x + 4 - (x^2 - 6x + 9)}{2}$$

$$= \frac{x^2 - 4x + 4 - x^2 + 6x - 9}{2}$$

$$= \frac{2x - 5}{2} \qquad \text{Combine like terms.}$$

EXAMPLE 6 Simplify $\dfrac{x^{-1} + x^{-2}y}{x^{-2} + y^{-1}}$.

SOLUTION Recalling that $a^{-n} = \dfrac{1}{a^n}$, we rewrite the expression so that it is free of negative exponents:

$$\frac{x^{-1} + x^{-2}y}{x^{-2} + y^{-1}} = \frac{\dfrac{1}{x} + \dfrac{y}{x^2}}{\dfrac{1}{x^2} + \dfrac{1}{y}}$$

The LCD here is x^2y. The process for simplifying is the same as the one we used in Examples 1 through 5. The reader should verify that the simplified form is

$$\frac{xy + y^2}{y + x^2}$$

If we had multiplied the numerator and the denominator of the original expression in Example 6 by the LCD, x^2y, and applied the rules for exponents, we would have obtained the same result as shown in Example 7.

EXAMPLE 7 Simplify $\dfrac{x^{-1} + x^{-2}y}{x^{-2} + y^{-1}}$.

SOLUTION
$$\frac{x^{-1} + x^{-2}y}{x^{-2} + y^{-1}} = \frac{(x^{-1} + x^{-2}y) \cdot x^2y}{(x^{-2} + y^{-1}) \cdot x^2y} \qquad \text{Multiply by } x^2y, \text{ the LCD.}$$

$$= \frac{x^{-1}(x^2y) + x^{-2}y(x^2y)}{x^{-2}(x^2y) + y^{-1}(x^2y)} \qquad \text{Distributive property}$$

$$= \frac{x^{-1+2}y + x^{-2+2}y^{1+1}}{x^{-2+2}y + x^2y^{-1+1}}$$

$$= \frac{xy + x^0y^2}{x^0y + x^2y^0} = \frac{xy + y^2}{y + x^2}$$

Can you discover a rule for finding the LCD in Example 7 without first eliminating the negative exponents?

EXAMPLE 8 Simplify

$$\frac{a - \dfrac{1}{1 + \dfrac{1}{a}}}{a + \dfrac{1}{a - \dfrac{1}{a}}}$$

SOLUTION A complex rational expression of this type contains complex rational expressions within complex rational expressions. In the numerator,

$$\frac{1}{1 + \dfrac{1}{a}}$$

is a complex fraction, as is

$$\frac{1}{a - \dfrac{1}{a}}$$

in the denominator. We must simplify them first.

$$\frac{a - \dfrac{1}{1 + \dfrac{1}{a}}}{a + \dfrac{1}{a - \dfrac{1}{a}}} = \frac{a - \dfrac{1 \cdot a}{\left(1 + \dfrac{1}{a}\right) \cdot a}}{a + \dfrac{1 \cdot a}{\left(a - \dfrac{1}{a}\right) \cdot a}} \qquad \text{The LCD of each is } a.$$

$$= \frac{a - \dfrac{a}{a + 1}}{a + \dfrac{a}{a^2 - 1}}$$

The LCD of this complex rational expression is $a^2 - 1 = (a - 1)(a + 1)$.

$$\frac{a - \dfrac{a}{a + 1}}{a + \dfrac{a}{a^2 - 1}} = \frac{\left(a - \dfrac{a}{a + 1}\right) \cdot (a - 1)(a + 1)}{\left(a + \dfrac{a}{a^2 - 1}\right) \cdot (a - 1)(a + 1)} \qquad \text{Multiply by the LCD.}$$

$$= \frac{a(a^2 - 1) - a(a - 1)}{a(a^2 - 1) + a}$$

$$= \frac{a^3 - a^2}{a^3} = \frac{a^2(a - 1)}{a^3}$$

$$= \frac{a - 1}{a}$$

Exercise Set 7.4

Simplify each complex fraction. Assume no denominator can be zero.

1. $\dfrac{\frac{1}{2}}{\frac{1}{4}}$

2. $\dfrac{\frac{3}{5}}{\frac{9}{10}}$

3. $\dfrac{\frac{a}{b}}{\frac{a^2}{bc}}$

4. $\dfrac{\frac{xy}{z}}{\frac{x}{yz}}$

5. $\dfrac{\frac{1}{4}+\frac{1}{3}}{\frac{1}{2}-\frac{1}{8}}$

6. $\dfrac{\frac{3}{5}-\frac{4}{9}}{\frac{2}{3}+\frac{1}{5}}$

7. $\dfrac{2+\frac{3}{x}}{x+\frac{3}{2}}$

8. $\dfrac{r-\frac{1}{r}}{\frac{2}{3r}}$

9. $\dfrac{1-\frac{1}{3}}{1-\frac{1}{9}}$

10. $\dfrac{1-\frac{9}{4}}{1-\frac{3}{2}}$

11. $\dfrac{1-\frac{1}{k}}{1-\frac{1}{k^2}}$

12. $\dfrac{1-\frac{m^2}{n^2}}{1-\frac{m}{n}}$

13. $\dfrac{\frac{1}{a}+\frac{1}{b}}{\frac{1}{a^2}-\frac{1}{b^2}}$

14. $\dfrac{\frac{1}{x}-\frac{1}{y}}{\frac{1}{y^2}-\frac{1}{x^2}}$

15. $\dfrac{\frac{1}{a}-\frac{1}{a^2}}{a}$

16. $\dfrac{\frac{1}{s}+\frac{1}{t}}{st}$

17. $\dfrac{9}{\frac{1}{b}-\frac{1}{4b}}$

18. $\dfrac{b^2}{\frac{1}{b}-\frac{1}{b^2}}$

19. $\dfrac{\frac{1}{x}+1}{1-\frac{1}{x^2}}$

20. $\dfrac{\frac{1}{y}-\frac{1}{2}}{\frac{1}{4}-\frac{1}{y^2}}$

21. $\dfrac{v}{1+\frac{v+1}{v+1}}$

22. $\dfrac{p}{1-\frac{p+1}{p-1}}$

23. $\dfrac{\frac{3ab^2}{7cd^2}}{\frac{12ab^2}{28cd^2}}$

24. $\dfrac{\frac{25m^3n^2}{24cd^2}}{\frac{35mn}{18cd^2}}$

25. $\dfrac{\frac{3a}{2b}+\frac{2a}{3b}}{\frac{3a}{2b}+\frac{2a}{3b}}$

26. $\dfrac{\frac{x}{y}+\frac{y}{x}}{\frac{x}{y}-\frac{y}{x}}$

27. $\dfrac{\frac{1}{q}-\frac{1}{q-1}}{\frac{1}{q}-\frac{1}{q-1}}$

28. $\dfrac{\frac{3}{r-1}-\frac{1}{r}}{\frac{1}{r}+\frac{3}{r-1}}$

29. $\dfrac{\frac{1}{x+1}-\frac{1}{x-1}}{\frac{1}{1-x^2}}$

30. $\dfrac{\frac{4}{w-2}-\frac{4}{w+2}}{\frac{4}{4-w^2}}$

31. $\dfrac{1+\dfrac{5}{\frac{1}{x}+2}}{1-\dfrac{1}{\frac{1}{x}+2}}$

32. $\dfrac{1+\dfrac{1}{1-\frac{1}{a}}}{1-\dfrac{3}{1-\frac{1}{a}}}$

33. $\dfrac{4t^2-9}{1-\dfrac{1-t}{t-2}}$

34. $\dfrac{\frac{4}{x^2}-1}{\frac{4}{x^2}-\frac{4}{x}+1}$

35. $\dfrac{6-\frac{9}{b}}{1-\frac{4}{b}+\frac{3}{b^2}}$

36. $\dfrac{\frac{1+y}{1-y}-\frac{1-y}{1+y}}{\frac{1}{1-y}-\frac{1}{1+y}}$

37. $\dfrac{\frac{p-1}{p-2}-\frac{p-2}{p-1}}{\frac{1}{p-1}-\frac{1}{p-2}}$

38. $\dfrac{\frac{z}{z-1}-\frac{z}{z+1}}{\frac{2}{z+1}-\frac{1}{z-1}}$

39. $\dfrac{\frac{1}{m-4}+\frac{1}{m-5}}{\frac{1}{m^2-9m+20}}$

40. $\dfrac{\frac{1}{k^2-7k+12}}{\frac{1}{k-3}+\frac{1}{k-4}}$

41. $\dfrac{s+3-\frac{8}{s-3}}{s+9-\frac{12s}{s-3}}$

42. $\dfrac{\frac{t}{t-1}+1}{\frac{3}{t-2}+2}$

43. $\dfrac{a^2+\frac{1}{a}}{1-\dfrac{a}{1-\frac{a}{a-1}}}$

44. $x-\dfrac{1}{1-\dfrac{1}{1-\frac{1}{1-x}}}$

45. $\dfrac{1+\dfrac{1}{1-\frac{1}{p}}}{1-\dfrac{3}{1-\frac{1}{p}}}$

46. $\dfrac{2r+\dfrac{r-1}{2r+1}}{2r-\dfrac{r+3}{2r+1}}$

47. $\dfrac{w-\dfrac{15}{w+2}}{w-1-\dfrac{10}{w+2}}$

48. $\dfrac{k - 5 + \dfrac{18}{k+4}}{k + \dfrac{3}{k+4}}$

49. $\left(\dfrac{1}{x+y} - \dfrac{1}{x-y}\right) \div \left(\dfrac{4y}{x^2 + 2xy + y^2}\right)$

50. $\left(\dfrac{a-1}{a+2} + \dfrac{2-a}{a+2}\right) \div \left(\dfrac{a}{a-1} - \dfrac{a-1}{a+2}\right)$

Simplify each fraction. Assume no denominator can be zero.

51. $\dfrac{3+4}{3^{-1} + 4^{-1}}$

52. $\dfrac{12 + 6}{12^{-1} + 6^{-1}}$

53. $\dfrac{x^{-1} + y^{-1}}{x^{-1} - y^{-1}}$

54. $\dfrac{a^{-1} + b^{-1}}{(ab)^{-1}}$

55. $(r^{-1} - s^{-1})^{-1} \div \dfrac{1}{(rs)^2}$

56. $\left(\dfrac{m}{n^{-1}} + \dfrac{m^{-1}}{n}\right)^{-1}$

57. $\dfrac{p^{-1}}{p^{-1} + 1}$

58. $\dfrac{2c^{-1} + 6d^{-1}}{4c^{-2} - 4d^{-2}}$

59. $\dfrac{s^{-1} - 2r^{-1}}{r^{-1} + 5s^{-1}}$

60. $\dfrac{2x^{-1} - 3x^{-1}}{x^{-2}}$

61. $\dfrac{6x^{-1} - 3y^{-1}}{12x^{-1} - 6y^{-1}}$

62. $\dfrac{(q^{-2} - t^{-2})^{-1}}{(t^{-1} - q^{-1})^{-1}}$

63. $\left(\dfrac{a^2 - 1}{3} \cdot \dfrac{a}{a^2 + 2a + 1} - \dfrac{4}{a+1}\right) \div \left(a - \dfrac{1}{a+1}\right)$

64. $\left(\dfrac{p^2 - 1}{3} \cdot \dfrac{p}{p^2 + 2p + 1}\right) - \left(\dfrac{4}{p+1} \div p - \dfrac{1}{p+1}\right)$

REVIEW EXERCISES

Solve each of the following equations, systems of equations, or inequalities.

65. $3(x - 3) = 5(x - 4)$

66. $5y - 2(y - 1) \geq 4y + 3$

67. $\begin{cases} x - 2y = 4 \\ x + 3y = 1 \end{cases}$

68. $10a^2 - a - 21 = 0$

69. $|3p - 2| \leq 4$

70. $\dfrac{1}{p} = \dfrac{1}{q} + \dfrac{1}{r}$; for q

For Extra Thought

71. In the given right triangle, the area is $A = \dfrac{x-2}{x^2 - 1}$ and the base is $b = \dfrac{x-2}{x+1}$. Find the height h.

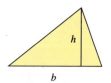

72. If the area of the given parallelogram is $A = \dfrac{x^3 - 8}{x^2 - 4}$ and the height is $h = \dfrac{x^2 + 2x + 4}{x + 2}$, find the base, b.

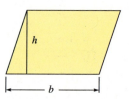

73. If the volume of the given cylinder is $V = \pi\left(\dfrac{2x^2 - 4x + 3}{3x^2 - 7x + 4}\right)$ and the height is $h = \dfrac{2x - 3}{3x - 4}$ find r^2, where r is the radius of the base.

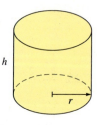

74. The given pyramid has a volume of $V = \dfrac{x^3 - 7x^2 + 12x}{x - 1}$, while the area of the square base is $A = 3x^2 - 18x + 27$. Find the height h.

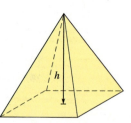

7.5 Solving Rational Equations

OBJECTIVE In this section we will learn to solve equations involving rational expressions.

The methods we use to solve linear equations in one variable are discussed in Section 2.1. Some of the equations we saw there involved rational expressions. We solved them by clearing the equation of fractions by multiplying each side by the least common denominator, the LCD. In this section we will use the same process to solve rational equations with more complicated LCDs. We begin with a review problem.

EXAMPLE 1 Solve: $\dfrac{x}{3} + \dfrac{x}{5} = \dfrac{1}{10}$.

SOLUTION

$30 \cdot \left(\dfrac{x}{3} + \dfrac{x}{5}\right) = 30 \cdot \left(\dfrac{1}{10}\right)$ Multiply each side by 30, the LCD.

$30 \cdot \dfrac{x}{3} + 30 \cdot \dfrac{x}{5} = 30 \cdot \dfrac{1}{10}$ Distributive property

$10x + 6x = 3$

$16x = 3$

$x = \dfrac{3}{16}$

CHECK $\dfrac{x}{3} + \dfrac{x}{5} = \dfrac{1}{10}$

$\dfrac{\frac{3}{16}}{3} + \dfrac{\frac{3}{16}}{5} \stackrel{?}{=} \dfrac{1}{10}$ $x = \dfrac{3}{16}$

$\dfrac{1}{16} + \dfrac{3}{80} \stackrel{?}{=} \dfrac{1}{10}$

$\dfrac{8}{80} \stackrel{?}{=} \dfrac{1}{10}$

$\dfrac{1}{10} \stackrel{\checkmark}{=} \dfrac{1}{10}$

The solution set is $\left\{\dfrac{3}{16}\right\}$.

Recall from Section 7.1 that we should always note any restrictions on the variable before solving an equation.

EXAMPLE 2 Solve: $\dfrac{3}{x-2} = \dfrac{2}{x+2}$, $x \neq 2, -2$.

SOLUTION We multiply each side by $(x-2)(x+2)$, the LCD.

$$(x-2)(x+2)\left(\dfrac{3}{x-2}\right) = (x-2)(x+2)\left(\dfrac{2}{x+2}\right)$$

$$3(x+2) = 2(x-2)$$

$$3x + 6 = 2x - 4$$

$$3x - 2x = -4 - 6 \qquad \text{\color{red}Subtract } 2x \text{ and } 6 \text{ from each side.}$$

$$x = -10$$

CHECK
$$\dfrac{3}{x-2} = \dfrac{2}{x+2}$$

$$\dfrac{3}{-10-2} \stackrel{?}{=} \dfrac{2}{-10+2}$$

$$\dfrac{3}{-12} \stackrel{?}{=} \dfrac{2}{-8}$$

$$\dfrac{-1}{4} \stackrel{\checkmark}{=} \dfrac{-1}{4}$$

The solution set is $\{-10\}$.

EXAMPLE 3 Solve: $\dfrac{x}{x-4} - 6 = \dfrac{4}{x-4}$.

SOLUTION First we note that $x \neq 4$, then we multiply each side by $x - 4$, the LCD.

> **CAUTION**
> When multiplying both sides of an equation by an expression involving variables, be careful to note any restrictions on the variables in the original equation. Eliminate any restricted variables from the solution set.

$$(x-4)\left(\dfrac{x}{x-4} - 6\right) = (x-4)\left(\dfrac{4}{x-4}\right)$$

$$x - 6(x-4) = 4$$

$$x - 6x + 24 = 4$$

$$-5x = -20$$

$$x = 4$$

Since $x \neq 4$, the equation has no solution. The solution set is therefore $\varnothing$.

EXAMPLE 4 Solve: $\dfrac{3}{k-1} - \dfrac{2}{k-2} = \dfrac{4}{k^2 - 3k + 2}$.

SOLUTION   When we factor the denominators on the right side of the equation to find the LCD, we get $k^2 - 3k + 2 = (k - 1)(k - 2)$. The LCD is therefore $(k - 1)(k - 2)$, $k \neq 1, 2$. First we multiply each side by the LCD:

$$(k-1)(k-2)\left(\frac{3}{k-1} - \frac{2}{k-2}\right) = (k-1)(k-2)\left(\frac{4}{k^2 - 3k + 2}\right)$$

$$3(k-2) - 2(k-1) = 4$$

$$3k - 6 - 2k + 2 = 4$$

$$k - 4 = 4$$

$$k = 8$$

CHECK
$$\frac{3}{k-1} - \frac{2}{k-2} = \frac{4}{k^2 - 3k + 2}$$

$$\frac{3}{8-1} - \frac{2}{8-2} \stackrel{?}{=} \frac{4}{8^2 - 3 \cdot 8 + 2}$$

$$\frac{3}{7} - \frac{1}{3} \stackrel{?}{=} \frac{4}{42} = \frac{2}{21}$$

$$\frac{2}{21} \stackrel{\checkmark}{=} \frac{2}{21}$$

The solution set is {8}.

Literal equations often involve rational expressions that are more complicated than those found in Chapter 2.

EXAMPLE 5   Solve $\dfrac{x}{b+c} + \dfrac{y}{b-c} = \dfrac{2}{b+c}$ for b.

SOLUTION   The LCD is $(b - c)(b + c)$, $b \neq c, -c$. First we multiply each side of the equation by the LCD:

$$(b-c)(b+c)\left(\frac{x}{b+c} + \frac{y}{b-c}\right) = (b-c)(b+c)\left(\frac{2}{b+c}\right)$$

$$x(b-c) + y(b+c) = 2(b-c)$$

$$bx - cx + by + cy = 2b - 2c$$

Now we isolate all terms involving b on one side of the equation.

$$bx + by - 2b = cx - cy - 2c$$

$$b(x + y - 2) = cx - cy - 2c \qquad \text{Factor.}$$

$$b = \frac{cx - cy - 2c}{x + y - 2} \qquad \text{Divide by the coefficient of } b.$$

or

$$b = \frac{c(x - y - 2)}{x + y - 2}$$

The solution set is $\left\{\dfrac{c(x - y - 2)}{x + y - 2}\right\}$.

EXAMPLE 6   Solve the equation $\dfrac{1}{x-1} + \dfrac{3}{x+1} = 2$.

SOLUTION   We multiply each side by $(x-1)(x+1)$, the LCD.

$$(x-1)(x+1)\left(\dfrac{1}{x-1} + \dfrac{3}{x+1}\right) = (x-1)(x+1)(2)$$

$x + 1 + 3(x-1) = (x^2 - 1)(2)$	$(x-1)(x+1) = x^2 - 1$
$x + 1 + 3x - 3 = 2x^2 - 2$	Remove grouping symbols.
$4x - 2 = 2x^2 - 2$	Combine like terms.
$0 = 2x^2 - 4x$	
$0 = 2x(x-2)$	Remove common factors.
$2x = 0$ or $x - 2 = 0$	Zero-factor theorem.
$x = 0$ or $x = 2$	

CHECK

$\dfrac{1}{x-1} + \dfrac{3}{x+1} = 2$ $\qquad$ $\dfrac{1}{x-1} + \dfrac{3}{x+1} = 2$

$\dfrac{1}{0-1} + \dfrac{3}{0+1} \stackrel{?}{=} 2$ $\quad x = 0$ $\qquad$ $\dfrac{1}{2-1} + \dfrac{3}{2+1} \stackrel{?}{=} 2$ $\quad x = 2$

$\dfrac{1}{-1} + \dfrac{3}{1} \stackrel{?}{=} 2$ $\qquad\qquad$ $\dfrac{1}{1} + \dfrac{3}{3} \stackrel{?}{=} 2$

$-1 + 3 \stackrel{?}{=} 2$ $\qquad\qquad$ $1 + 1 \stackrel{?}{=} 2$

$2 \stackrel{\checkmark}{=} 2$ $\qquad\qquad$ $2 \stackrel{\checkmark}{=} 2$

The solution set is $\{0, 2\}$.

EXAMPLE 7   Given $y^2 = \dfrac{x+2}{x+3}$, interchange x and y, then solve for y.

SOLUTION

$y^2 = \dfrac{x+2}{x+3}$

$x^2 = \dfrac{y+2}{y+3}$	Interchange x and y.
$x^2(y+3) = y + 2$	Multiply each side by $y+3$.
$x^2 y + 3x^2 = y + 2$	Distributive property.
$x^2 y - y = 2 - 3x^2$	Isolate y.
$y(x^2 - 1) = 2 - 3x^2$	Factor.
$y = \dfrac{-3x^2 + 2}{x^2 - 1}$	Divide by $x^2 - 1$.
$y = \dfrac{3x^2 - 2}{1 - x^2}$	Multiply numerator and denominator by -1.

Either of the last two results is acceptable.

Exercise Set 7.5

Solve each of the following equations. State any restrictions on the variable. Check your answers.

1. $\dfrac{x}{2} - 4 = x$
2. $a + 5 = \dfrac{a}{3}$
3. $\dfrac{3p}{4} = \dfrac{5p}{7} + 2$
4. $\dfrac{5t}{6} = \dfrac{1}{12} + \dfrac{2t}{3}$
5. $\dfrac{m}{2} + \dfrac{4m}{3} = 11$
6. $\dfrac{s}{6} + \dfrac{3s}{5} = 2$
7. $\dfrac{7}{4}x - \dfrac{1}{2}x = 5$
8. $\dfrac{3}{2}y - \dfrac{1}{3}y = 7$
9. $\dfrac{5k - 3}{7} = \dfrac{3k - 1}{5}$
10. $\dfrac{4b - 5}{9} = \dfrac{2b - 1}{5}$
11. $\dfrac{8}{r} - \dfrac{1}{3} = \dfrac{5}{r}$
12. $\dfrac{7}{q} + \dfrac{1}{4} = \dfrac{9}{q}$
13. $\dfrac{1}{6} + \dfrac{5}{6y} = \dfrac{6}{7y}$
14. $\dfrac{3}{4} - \dfrac{3}{4y} = \dfrac{-4}{5y}$
15. $\dfrac{x + 1}{x + 3} = 3$
16. $\dfrac{p - 2}{p - 6} = 5$
17. $\dfrac{2s + 3}{3s + 2} = \dfrac{5}{6}$
18. $\dfrac{2k - 3}{4k - 5} = \dfrac{2}{5}$
19. $\dfrac{3}{m - 2} = \dfrac{3}{m - 4}$
20. $\dfrac{5}{t - 4} = \dfrac{5}{t - 2}$
21. $\dfrac{1}{5} + \dfrac{4}{x} = \dfrac{16}{3} - \dfrac{1}{5x}$
22. $\dfrac{7}{8} - \dfrac{1}{n} = \dfrac{1}{2n} - \dfrac{1}{4}$
23. $\dfrac{x^2 - 8x + 16}{x^2 - 16} = 1$
24. $\dfrac{y^2 + 12y + 36}{y^2 + 2y - 24} = 1$
25. $\dfrac{4x + 1}{2x - 1} = 5 + \dfrac{3}{2x - 1}$
26. $\dfrac{4}{b + 3} - \dfrac{3}{b - 4} = 0$
27. $\dfrac{7x - 5}{3x^2 - 15x} - 2 = \dfrac{2}{x - 5}$
28. $\dfrac{2}{q} - \dfrac{4}{q + 3} = \dfrac{9}{q + 3}$
29. $\dfrac{3x}{x - 3} = 2 + \dfrac{9}{x - 3}$
30. $9 - \dfrac{4}{2m + 5} = \dfrac{-3}{2m + 5}$
31. $\dfrac{-x^2 + 10}{x^2 - 1} + \dfrac{3x}{x - 1} = \dfrac{2x}{x + 1}$
32. $\dfrac{x + 1}{5} - 2 = \dfrac{-4}{x}$
33. $\dfrac{x + 1}{x} - \dfrac{x - 1}{x} = 0$
34. $\dfrac{3 - 5y}{2 + y} = \dfrac{3 + 5y}{2 - y}$
35. $\dfrac{3}{y + 1} - \dfrac{y - 2}{2} = \dfrac{y - 2}{y + 1}$
36. $\dfrac{a + 4}{a + 7} - \dfrac{a}{a + 3} = \dfrac{3}{8}$
37. $\dfrac{2}{a - 2} - \dfrac{1}{a + 1} = \dfrac{1}{a^2 - a - 2}$
38. $\dfrac{5}{2x^2 + x - 3} - \dfrac{2}{2x + 3} = \dfrac{x + 1}{x - 1} - 1$

Solve each literal equation for the specified variable. Assume no denominator can be zero.

39. $p = \dfrac{A}{1 + rt}$; for A
40. $s = \dfrac{a - rl}{1 - r}$; for r
41. $\dfrac{2}{x} + \dfrac{4}{x + a} = \dfrac{-6}{a - x}$; for x
42. $\dfrac{1}{b - x} - \dfrac{1}{x} = \dfrac{-2}{b + x}$; for x
43. $\dfrac{s - vt}{t^2} = -16$; for v
44. $A = p \cdot \left(1 + \dfrac{r}{n}\right)$; for r
45. $\dfrac{1}{p} = \dfrac{1}{f} + \dfrac{1}{F}$; for f
46. $h = \dfrac{v^2}{2g} + \dfrac{p}{c}$; for p
47. $ab - 3c + 3ac = 2c$; for c
48. $k^2 - ky + 1 = m(p - m)$; for p

Interchange x and y and then solve for y. Assume no denominator can be zero.

49. $y^3 = \dfrac{2x - 3}{3x - 1}$
50. $y^2 + 3 = \dfrac{x - 1}{x - 3}$
51. $y^2 - 2y = \dfrac{x - 2}{x + 1}$
52. $y^2 - 2y + 1 = \dfrac{x + 9}{x}$

Solve each of the following equations. State any restrictions on the variable.

53. $\dfrac{a^2 + 9}{a^2 - 9} - \dfrac{3}{a + 3} = \dfrac{-a}{3 - a}$
54. $\dfrac{x}{3x - 2} - \dfrac{8}{2x + 3} = \dfrac{2x^2 - 24x + 18}{6x^2 + 5x - 6}$
55. $\dfrac{r - 2}{r^2 - 10r + 9} - \dfrac{3r - 12}{r^2 - 4r - 45} = \dfrac{1 - 2r}{r^2 + 4r - 5}$
56. $\dfrac{y - 6}{y^2 + 3y - 28} + \dfrac{y + 8}{y^2 + 16y + 63} = \dfrac{2y - 5}{y^2 + 5y - 36}$

57. $\dfrac{k+2}{5k^2-27k-18} + \dfrac{2-k}{4k^2-29k+30}$
    $= \dfrac{-(k+1)}{20k^2-13k-15}$

58. $\dfrac{2-q}{30-29q+4q^2} + \dfrac{q+1}{15+13q-20q^2}$
    $= \dfrac{q+2}{5q^2-27q-18}$

REVIEW EXERCISES

Rewrite each of the following statements as an algebraic expression.

59. The sum of a number and its reciprocal

60. The portion of a job done in 1 hr if it takes 7 hr to do the whole job

61. The distance traveled in 1 hr if you traveled 600 mi in x hr

62. The time it takes to travel 725 km at y km/hr

63. The rate of interest if N dollars earns $126 dollars simple interest in 1 year

64. The part of a tank filled by a pipe in 3 hr if the tank can be filled by the pipe in x hr

WRITE IN YOUR OWN WORDS

65. What is a rational expression?

66. What is an equation containing rational expressions?

67. What method is used to simplify rational expressions?

68. What method is used to solve rational equations?

69. What is a complex fraction?

70. Can you add the following expressions? Why or why not?

$$\dfrac{3x}{x+5} + \dfrac{5}{9x}$$

7.6 More Applications

HISTORICAL COMMENT The number problems, distance problems, and work problems that we will cover in this section date back in history to the Roman legions. The marching stride of the Roman soldier was so standardized, that the amount of time they would take to go from one place to another could be easily figured and problems involving the times easily written down. Other favorites were the time it would take a man and his wife to drink a quantity of beer if they drank alone and if they drank together. The *Greek Anthology* (c. 500) contains the following problem about a fountain.

> *I am a bronze lion; my spouts are my two eyes, my mouth, and the flat of my right foot. My right eye fills a jar in two days, my left in three, and my foot in four. My mouth is capable of filling the jar in 6 hours. How long would it take to fill the jar if all four spouts are working at the same time?*

OBJECTIVE In this section we will learn to apply the techniques for solving rational equations to applied problems.

Applications involving rational expressions are numerous and varied. Many of the applications are, unfortunately, beyond the scope of this text. For this reason we will include only a few special types. As you study the examples and solve the problems in the exercise set, remember the step-by-step process we used in the previous chapters.

Number Problems

EXAMPLE 1 If a number is added to its reciprocal, the sum is $\frac{13}{6}$. Find the number.

SOLUTION We let n represent the number; then its reciprocal is $\frac{1}{n}$.

$$\underbrace{n}_{\text{A number}} + \underbrace{\frac{1}{n}}_{\text{its reciprocal}} = \underbrace{\frac{13}{6}}_{\frac{13}{6}}$$

The LCD here is $6n$.

$$6n\left(n + \frac{1}{n}\right) = 6n \cdot \frac{13}{6}$$
$$6n^2 + 6 = 13n$$
$$6n^2 - 13n + 6 = 0$$
$$(3n - 2)(2n - 3) = 0$$
$$3n - 2 = 0 \quad \text{or} \quad 2n - 3 = 0$$
$$n = \frac{2}{3} \quad \text{or} \quad n = \frac{3}{2}$$

If the number is $\frac{2}{3}$, its reciprocal is $\frac{3}{2}$. If the number is $\frac{3}{2}$, its reciprocal is $\frac{2}{3}$.

CHECK The sum of the number and its reciprocal must be $\frac{13}{6}$.

$$\frac{2}{3} + \frac{3}{2} \stackrel{?}{=} \frac{13}{6}$$
$$\frac{4}{6} + \frac{9}{6} \stackrel{?}{=} \frac{13}{6}$$
$$\frac{13}{6} \stackrel{\checkmark}{=} \frac{13}{6}$$

The solution checks. So there are two solutions: The number is either $\frac{2}{3}$ or $\frac{3}{2}$.

EXAMPLE 2 The denominator of a fraction is 3 more than the numerator. If 1 is subtracted from both the numerator and the denominator, the resulting fraction has a value of $\frac{2}{3}$. Find the original fraction.

SOLUTION We let the numerator of the fraction be x; then the denominator is $x + 3$. The original fraction is therefore $\frac{x}{x+3}$. When 1 is subtracted from both the numerator and the denominator, the resulting fraction is

$$\frac{x-1}{x+3-1} = \frac{x-1}{x+2}$$

Thus

$$\frac{x-1}{x+2} = \frac{2}{3}$$

$3(x - 1) = 2(x + 2)$ The LCD is $3(x + 2)$.

$3x - 3 = 2x + 4$

$x = 7$

So the original fraction is $\dfrac{x}{x+3} = \dfrac{7}{10}$.

CHECK When 1 is subtracted from both the numerator and the denominator, the resulting fraction must have a value of $\frac{2}{3}$.

$$\frac{7-1}{10-1} = \frac{6}{9} \stackrel{\checkmark}{=} \frac{2}{3}$$

Distance-Rate-Time

When setting up an equation to solve a distance-rate-time problem, it's important to remember that the formula $d = r \cdot t$ can be written in two alternate forms:

$$t = \frac{d}{r} \quad \text{and} \quad r = \frac{d}{t}$$

EXAMPLE 3 A motorboat can travel 10 mph upstream and 16 mph downstream. If Juan rents the boat for 4 hours, how far can he go upstream before he must head back?

SOLUTION We let x = the number of miles Juan can travel upstream. Since we don't know the time but we do know the distance and rate, we use $t = \dfrac{d}{r}$.

Direction	Distance (d)	Rate (r)	Time $\left(t = \dfrac{d}{r}\right)$
Upstream	x	10	$\dfrac{x}{10}$
Downstream	x	16	$\dfrac{x}{16}$

Since Juan has rented the boat for 4 hours, the sum of the times upstream and downstream is 4.

$$\frac{x}{10} + \frac{x}{16} = 4$$

$$80\left(\frac{x}{10} + \frac{x}{16}\right) = 80 \cdot 4 \qquad \text{LCD} = 80$$

$$8x + 5x = 320$$
$$13x = 320$$
$$x = \frac{320}{13} = 24\frac{8}{13} \text{ miles}$$

Therefore Juan can travel $24\frac{8}{13}$ miles upstream.

CHECK If Juan travels $\frac{320}{13}$ miles upstream at 10 mph, it will take him

$$\frac{\frac{320}{13}}{10} = \frac{32}{13} \text{ hr}$$

Traveling the same distance downstream at 16 mph will take

$$\frac{\frac{320}{13}}{16} = \frac{20}{13} \text{ hr}$$

The sum of the two times must be 4 hours:

$$\frac{32}{13} + \frac{20}{13} = \frac{52}{13} \stackrel{\checkmark}{=} 4$$

The solution checks.

EXAMPLE 4 An airplane can travel 150 miles *against* the wind *in the same time* it takes it to fly 180 miles *with* the wind. If the wind speed is 15 mph, how fast would the plane be moving in still air?

SOLUTION When the plane moves *against* the wind, its speed is *decreased* by an amount equal to the wind speed. When it flies *with* the wind, its speed is increased by an amount equal to the wind speed. So if we let x be the speed of the plane in still air; then

$x - 15$ = speed against the wind
$x + 15$ = speed with the wind

Direction	Distance (d)	Rate (r)	Time $\left(t = \frac{d}{r}\right)$
Against wind	150	$x - 15$	$\frac{150}{x - 15}$
With wind	180	$x + 15$	$\frac{180}{x + 15}$

Since the flying time with the wind is the same as the time against the wind,

$$\frac{150}{x-15} = \frac{180}{x+15}$$

We multiply each side by $(x - 15)(x + 15)$, the LCD.

$$(x-15)(x+15)\left(\frac{150}{x-15}\right) = (x-15)(x+15)\left(\frac{180}{x+15}\right)$$
$$(x+15) \cdot 150 = (x-15) \cdot 180$$
$$150x + 2250 = 180x - 2700$$
$$4950 = 30x$$
$$165 = x$$

The plane would move at 165 mph in still air. The check is left to the reader.

Work Problems

EXAMPLE 5 Randy and Todd contract to paint 10 identical tract homes. Randy can paint a house in 6 days with power equipment. When Todd joins him without power equipment, they can paint a house in 4 days. How long does it take Todd to paint a house working alone?

SOLUTION We let x be the number of days Todd requires to paint a house working alone. If Randy takes 6 days to paint a house, he can paint $\frac{1}{6}$ of it in one day. If Todd takes x days to paint a house, he can paint $\frac{1}{x}$ of it in one day. Since together they take 4 days to paint a house, they can paint $\frac{1}{4}$ of it in one day.

	TOTAL TIME IN DAYS	AMOUNT IN ONE DAY
Randy	6	$\frac{1}{6}$
Todd	x	$\frac{1}{x}$
Together	4	$\frac{1}{4}$

What they can do together in one day must equal the sum of what they can do individually in one day.

The sum of what they can do individually per day **is** what they can do together per day.

$$\frac{1}{6} + \frac{1}{x} = \frac{1}{4}$$

$$12x\left(\frac{1}{6} + \frac{1}{x}\right) = 12x \cdot \frac{1}{4} \quad \text{The LCD is } 12x.$$

$$2x + 12 = 3x$$

$$12 = x$$

Todd requires 12 days to paint a house working alone.

CHECK Since Todd requires 12 days working alone, he completes $\frac{1}{12}$ of the job in one day and $\frac{4}{12}$ of the job in 4 days. Randy can complete $\frac{1}{6}$ of the job in one day and therefore $\frac{4}{6}$ of the job in 4 days. Since $\frac{4}{12} + \frac{4}{6} = 1$, which represents a complete job, the solution checks.

EXAMPLE 6 Two pipes are used to fill a swimming pool. The larger pipe can fill the pool in 12 hours and the smaller pipe in 16 hours. How long will it take to fill the pool if both pipes are used?

	TOTAL TIME IN HOURS	AMOUNT IN 1 HOUR
Larger Pipe	12	$\frac{1}{12}$
Smaller Pipe	16	$\frac{1}{16}$
Together	x	$\frac{1}{x}$

The sum of what they can do individually per hour **is** what they can do together per hour.

$$\frac{1}{12} + \frac{1}{16} = \frac{1}{x}$$

$$4x + 3x = 48 \quad \text{The LCD is } 48x.$$

$$7x = 48$$

$$x = \frac{48}{7} = 6\frac{6}{7}$$

Together the pipes can fill the pool in $6\frac{6}{7}$ hours.

Ratio-Proportion Problems

The comparison of two numbers using division is called a **ratio**. Ratios are often used to illustrate common statistical data. If a math class has 21 females and 23 males, then the ratio of females to males is 21 to 23, or

$$\frac{21}{23}$$

The ratio of males to females is

$$\frac{23}{21}$$

> **CAUTION**
> When ratios are used, the units should be meaningful. Feet should be compared to feet, inches to inches, etc. In general, when using ratios, the units in the numerator and the denominator must be the same.

A statement that two ratios are equal to each other is called a **proportion.** The next two examples show how proportions can be used to solve applied problems.

EXAMPLE 7 The sum of two angles of a triangle is 105°. The ratio of the smaller angle to the larger one is $\frac{7}{14}$. Find the size of each angle. We let

x = the measure of the smaller angle

$105 - x$ = the measure of the larger angle

The ratio of the angles is $\frac{7}{14}$. Therefore

$$\frac{x}{105-x} = \frac{7}{14}$$

$14x = 7(105 - x)$ The LCD is $14(105 - x)$.

$14x = 735 - 7x$

$21x = 735$

$x = 35$

The smaller angle is 35°, and the larger angle is $105° - 35° = 70°$.

CHECK The sum of the two angles is $35° + 70° = 105°$. The ratio of the smaller one to the larger one is

$$\frac{35}{70} = \frac{1}{2} = \frac{7}{14}$$

The solution checks.

EXAMPLE 8 An investor has $21,060 invested in stocks and bonds in a ratio of 4 to 9. How much does she have invested in each?

SOLUTION We let x denote the amount she has invested in stocks; then $\$21{,}060 - x$ is the amount invested in bonds. Since they're in a ratio of 4 to 9,

$$\frac{x}{21{,}060 - x} = \frac{4}{9}$$

$9x = 4(21{,}060 - x)$ The LCD is $9(21{,}060 - x)$.

$9x = 84{,}240 - 4x$

$$13x = 84{,}240$$
$$x = 6480$$

So the amount in stocks is \$6480, and the amount in bonds is \$21,060 − \$6480 = \$14,580. The check is left to the reader.

Solution to the Historical Problem at the Beginning of This Section

I am a bronze lion; my spouts are my two eyes, my mouth, and the flat of my right foot. My right eye fills a jar in two days, my left in three, and my foot in four. My mouth is capable of filling the jar in 6 hours. How long would it take to fill the jar if all four spouts are working at the same time?

Spout	Time to Fill Jar in Days	Portion of Jar Filled in One Day
Right eye	2	$\frac{1}{2}$
Left eye	3	$\frac{1}{3}$
Foot	4	$\frac{1}{4}$
Mouth	6 hr = $\frac{1}{4}$ day	$\frac{1}{\frac{1}{4}} = 4$
Together	d	$\frac{1}{d}$

The sum of the amounts each part of the lion can fill in one day is equal to the amount they can fill together in one day.

$$\frac{1}{2} + \frac{1}{3} + \frac{1}{4} + 4 = \frac{1}{d}$$

We multiply each side by $12d$, the LCD.

$$6d + 4d + 3d + 48d = 12$$
$$61d = 12$$
$$d = \frac{12}{61} \text{ days}$$

The time in hours is $\frac{12}{61} \cdot 24 = \frac{288}{61} \approx 4.72$ hours.

Exercise Set 7.6

Write each of the following applied problems as an equation and solve it.

Number Problems

1. A number added to 7 times its reciprocal is -8. Find the number.
2. A number added to 20 times its reciprocal is 9. Find the number.
3. A natural number decreased by 27 times its reciprocal is 6. Find the natural number.
4. A natural number decreased by 30 times its reciprocal is 1. Find the natural number.
5. The denominator of a certain fraction is 6 more than the numerator. If the numerator is increased by 7 and the denominator decreased by 3, the value of the new fraction is $\frac{5}{4}$. Find the original fraction.
6. The denominator of a certain fraction is 4 more than the numerator. If 3 is subtracted from the numerator and added to the denominator, the value of the new fraction is $\frac{4}{9}$. Find the original fraction.
7. What number can be subtracted from the numerator and denominator of the fraction $\frac{8}{11}$ so that the value of the new fraction will be $\frac{3}{5}$?
8. If a certain number is subtracted from the numerator of the fraction $\frac{19}{30}$ and twice the number is subtracted from the denominator, the value of the new fraction is $\frac{2}{3}$. What is the number?
9. What number must be subtracted from the numerator and denominator of the fraction $\frac{12}{7}$ to make the value of the new fraction be $\frac{5}{3}$?
10. What number must be added to the numerator and denominator of the fraction $\frac{5}{7}$ to make a new fraction equal to $\frac{5}{3}$?

Distance-Rate-Time Problems

11. Two students leave Portland at the same time, heading south. One travels in a sports car at 65 mph. The second travels on a motorcycle at 55 mph. How long will it be before they are 55 miles apart?
12. Dave and Fran leave home at the same time going in opposite directions. Fran travels at 64 km/hr and Dave travels at 54 km/hr. How far apart will they be after 30 min?
13. The Paytons leave El Paso heading east at 45 mph. Their daughter Joy leaves El Paso $\frac{1}{2}$ hr later, also traveling east, at 60 mph. How long does it take Joy to catch up with her parents? How far does Joy have to go to catch them?
14. An airplane travels 1260 mi in the same time that a car travels 420 mi. If the speed of the car is 120 mph less than the speed of the airplane, find the speed of each.
15. San Antonio and Baton Rouge are 460 mi apart. Penny leaves San Antonio at 8:00 A.M. for Baton Rouge. Bill leaves Baton Rouge for San Antonio at 10:00 A.M., traveling 20 mph faster than Penny. If they meet 210 mi from Baton Rouge, how fast is each traveling?
16. By increasing his usual average speed by 20 km/hr, Tim finds that he can cut his time on a 240 km drive by 2 hr. Find Tim's typical average speed.
17. Janet can row a canoe 2 mph faster than Linda. If Janet rows 5 mi in the same amount of time that Linda rows 4 mi, how fast can Janet row her canoe?
18. Rex can walk 11 mi in the same amount of time that Marvin can walk 8 mi. Rex walks 1 mph faster than Marvin. How fast can Rex walk?
19. Darwin's boat goes 15 mph in still water. If it takes him as much time to travel 6 mi downstream on the Missouri River as it does to travel 4 mi upstream, what is the speed of the current?
20. Madalyn's plane travels at a speed of 350 mph in still air. If she travels 160 mi against the wind in the same time it takes her to travel 190 mi with the wind, what is the speed of the wind?

Work Problems

21. A painter can prepare and paint a room in 10 hr. Her son requires 18 hr to do the same job. How long will it take them to prepare and paint the room working together?
22. Bill can prepare a manual in 6 hr. Joan can prepare it in 4 hr. How long will it take them to prepare the manual working together?
23. Gary can spade his yard in 6 hr, but with his son helping it only takes 4 hr. How long will it take Gary's son working alone to spade the yard?
24. A pool can be filled by a pipe in 10 hr. If a second pipe is also opened, it only takes 4 hr to fill the pool. How long would it take to fill the pool using only the second pipe?
25. It takes Mac $\frac{2}{3}$ as long as it takes Ben to paint a compact car. If together Mac and Ben take 12 hr to paint the car, how long does it take Ben working alone?

26. A hot tub can be filled by the cold-water pipe in 21 min and by the hot-water pipe in 28 min. How long does it take to fill the tub using both pipes?

27. A bathtub can be filled with cold water in 9 min and then drained in 12 min. If the cold water is turned on and the drain left open by mistake, how long will it take to fill the bathtub?

28. A vat can be filled by an inlet pipe in 8 hr. How long will it take to drain the vat if, when both the inlet and drain pipes are opened, the vat is filled in 24 hr?

29. Lewis can clean the house in 12 hr. After he and his wife Minnie have both been cleaning for 1 hr, they are joined by their daughter Rachel and they complete the cleaning in 3 more hours. If it takes Minnie 10 hr to clean the house alone, how long does it take Rachel to do it alone?

30. Bob can clean the kitchen in 45 min. Karen can clean it in 30 min. If both start cleaning and 5 minutes later their daughter Kelsey joins them, it takes the three of them 7 min more to finish. How quickly can Kelsey clean the kitchen alone?

Ratio-Proportion Problems

31. A board 20 ft long is cut into two pieces that have lengths with a ratio of 4 to 6. Find the length of each piece.

32. The ratio of the measure of two supplementary angles is 4 to 5. Find the measure of each angle.

33. The ratio of married to single students at Cimarron College is 4 to 9, and there are 6630 students in the college. How many married students are in the college?

34. The ratio of math majors to engineering students at Benton University is 2 to 7. There is a total of 1827 math and engineering students in the university. How many engineering students are there in Benton University?

35. Uta invests $15,000 in a combination of variable and fixed mutual funds. The ratio of dollars invested in variable funds to the dollars invested in fixed funds is 9 to 6. Find the amount invested in variable funds.

36. Gary invests $28,000 in money market and bond funds. The ratio of dollars invested in money market to the dollars invested in bond funds is 7 to 9. Find the amount invested in bond funds.

For Extra Thought

37. Similar triangles have proportional corresponding sides. Given the similar triangles ABC and DEF, solve for x and find the lengths of the unknown sides.

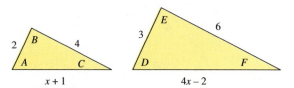

38. Triangles ABC and DEF are similar. Solve for y and find the lengths of the unknown sides. See Exercise 37.

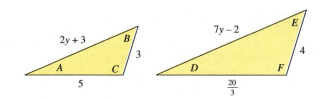

Use the optics formula $\frac{1}{f} = \frac{1}{p} + \frac{1}{q}$ for Exercises 39 and 40.

39. If $p = 2f$ and $q = 40$, find f.

40. If $f = p^2$ and $q = p$, find f in terms of q.

The formula for an electrical circuit with resistors R_1, R_2, and R_3 in parallel is $\frac{1}{R} = \frac{1}{R_1} + \frac{1}{R_2} + \frac{1}{R_3}$

41. If $R_1 = 2$, $R_2 = \frac{2}{3}$, and $R_3 = \frac{3}{4}$ find R.

42. Find R in terms of x if $R_1 = x$, R_2 is twice R_1, and R_3 is three times R_1.

Summary

7.1 Reducing and Building Rational Expressions

A **rational expression** is the quotient of two polynomials such that the denominator cannot be zero. When working with a rational expression, it is important first to note and exclude any values of the variable that make the denominator zero.

A rational expression can have many equivalent forms, most of which can be discovered by using the **fundamental principle of rational expressions,** which states that if A, B, and C are polynomials, $B \neq 0$, $C \neq 0$, then

1. $\dfrac{A \cdot C}{B \cdot C} = \dfrac{A}{B}$ Divide numerator and denominator by C.

2. $\dfrac{A}{B} = \dfrac{A}{B} \cdot \dfrac{C}{C} = \dfrac{A \cdot C}{B \cdot C}$ Multiply numerator and denominator by C.

Some of these equivalent forms are simplifications of the original rational expressions.

If a rational expression has no common factors in the numerator and denominator other than 1, the numerator and denominator are said to be **relatively prime.** When this is the case, the rational expression is said to be in **lowest terms.**

7.2 Multiplying and Dividing Rational Expressions

The **product of two rational expressions** $\dfrac{A}{B}$ and $\dfrac{C}{D}$, $B, C, D \neq 0$, is $\dfrac{A \cdot C}{B \cdot D}$.

The **quotient** of the same two rational expressions is $\dfrac{A \cdot D}{B \cdot C}$.

7.3 Adding and Subtracting Rational Expressions

The **sum of two rational expressions** $\dfrac{A}{C}$ and $\dfrac{B}{C}$, $C \neq 0$, is $\dfrac{A + B}{C}$.

The **difference** of the same two rational expressions is $\dfrac{A - B}{C}$.

If the denominators are different, two rational expressions must first be rewritten in terms of their **LCD (least common denominator)** in order to add or subtract. The LCD is found by factoring each denominator completely and forming the product of each distinct factor to the highest power that it occurs in any denominator.

7.4 Simplifying Complex Rational Expressions

A **complex rational expression** (or **complex fraction**) is one in which the numerator or the denominator (or both) contains rational expressions. Complex rational expressions are simplified by using the fundamental principle of rational expressions. To do this, both the numerator and the denominator of the original complex rational expression are multiplied by the LCD of all the denominators within the expression.

7.5 Solving Rational Equations

To solve an equation involving rational expressions, restrictions on the variable must first be noted. Then the equation is cleared of fractions by multiplying each side of the equation by the LCD. Each proposed solution should be checked against the list of excluded values.

7.6 More Applications

There are numerous applications involving rational expressions. These include number problems, distance-rate-time problems, work problems, and ratio-proportion problems, to name a few.

Cooperative Exercise

PART A
DEFINITION: A number x is said to be a **factor** of y if x divides y without a remainder. A number y is **divisible** by x if x divides y without a remainder.

1. Use a reference, if needed, to write the rules for divisibility by 2, 3, 4, 5, 6, 7, 8, 9, 10, 11, and 12.
2. Determine whether the number 374,832 is divisible by 2, 3, 4, 5, 6, 8, 9, 10, and 12.
3. Determine whether the number 39,837 is divisible by 7.
4. Determine whether the number 6,185,854 is divisible by 11.

PART B
DEFINITION: The **greatest common divisor**, or **GCD** (sometimes referred to as the **greatest common factor**, or **GCF**) of a set of natural numbers is the largest natural number that divides every number in the set without a remainder.

1. Use a reference, if needed, to write the necessary steps for finding the GCD of two or more natural numbers.
2. Find the GCD of 96 and 112.

Another method for finding the GCD is known as the **Euclidean algorithm.** We will illustrate it by finding the GCD of 96 and 112. First we divide 112 by 96. Disregarding, the quotient, we next divide the divisor, 96, by the remainder, 16. We repeat this process until the remainder is zero.

The *divisor* in the last division, the one in which the remainder is zero, is the GCD.

$$\begin{array}{r}1\\96\overline{)112}\\96\\\hline 16\end{array} \qquad \begin{array}{r}6\\16\overline{)96}\\96\\\hline 0\end{array} \qquad \text{The GCD is 16.}$$

3. Find the GCD of 756 and 1764 using the Euclidean algorithm.
4. Find the GCD of 24 and 1684 using the Euclidean algorithm.

PART C
A **palindromic number** is one that reads the same forward and backward, such as 88, 535, or 1221.

1. Check the palindromic number 8778 for divisibility by 11. See Part A.
2. Explain why a four-digit palindromic number must be divisible by 11.
3. Is every five-digit palindromic number divisible by 11? Use an example to support your answer.
4. Is every six-digit palindromic number divisible by 11? Use an example to support your answer.
5. Make a conjecture about whether every palindromic number with an even number of digits is divisible by 11. Explain.

Review Exercises

Complete each of the following.

1. The quotient of two polynomials is called a _____ expression.
2. Two polynomials having no common factors are said to be _____.
3. By the fundamental principle of rational expressions, if A, B, and C are polynomials, $B \neq 0$ and $C \neq 0$, then $\dfrac{A}{B} = $ _____.
4. If $f(x) = x^2 - 4x$ and $g(x) = x^2 - 16$, then $\left(\dfrac{f}{g}\right)(x) = $ _____.
5. If $f(a) = a - 2$ and $g(a) = 4 - a^2$, then $\left(\dfrac{f}{g}\right)(a) = $ _____.
6. Evaluate $\dfrac{r^2 - 2r + 4}{r^2 + 6r + 3}$ for $r = -1$.
7. $\dfrac{x}{x-4} = \dfrac{?}{x^2 - 5x + 4}$
8. $\dfrac{2p}{q-p} = \dfrac{?}{q^2 - p^2}$
9. The rational expression $\dfrac{y^2 - y - 6}{y^2 - y - 12}$ is undefined when $y = $ _____.
10. $\dfrac{3p - 2}{6 - p} = \dfrac{2 - 3p}{?}$

Perform the indicated operation and simplify.

11. $\dfrac{7x}{3y} \cdot \dfrac{15y}{21x}$

12. $\dfrac{a^2 - 9}{b - 3} \cdot \dfrac{b^2 - 9}{a^2 - 6a + 9}$

13. $\left(\dfrac{m^3 - 27}{n + 2} \cdot \dfrac{n^2 - 4}{m^2 + 3m + 9}\right) \div (n + 2)$

14. $\left(\dfrac{2}{3q + 1} - \dfrac{5}{3q - 1}\right) \div \dfrac{1}{9q^2 - 1}$

15. $\dfrac{a - b}{d - c} + \dfrac{a - b}{c - d}$

16. $\left(\dfrac{27x^3y^2}{16x^2z^2} \div \dfrac{3x^2}{8z^2}\right) \cdot \left(\dfrac{yz}{9x}\right)$

17. $3s - \dfrac{rs - s^2}{r + s} - 2r$

18. $1 - \dfrac{-(-1 - 2p)}{p^2 - 2p + 1}$

19. If $f(x) = 5x - 2$ and $g(x) = \dfrac{2 - x}{x - 3}$, find $(f - g)(x)$.

20. Given $f(x) = \dfrac{1}{x^2 - 4}$ and $g(x) = \dfrac{1}{x - 2} - \dfrac{1}{x + 2}$, find $(f + g)(x)$.

Simplify.

21. $\dfrac{x + \dfrac{1}{2}}{1 + \dfrac{1}{x}}$

22. $\dfrac{p - \dfrac{1}{p}}{\dfrac{5}{3p}}$

23. $\dfrac{\dfrac{1}{y} + \dfrac{1}{y - 1}}{\dfrac{1}{y} - \dfrac{1}{y - 1}}$

24. $\dfrac{\dfrac{9}{9 - x^2}}{\dfrac{3}{x - 3} - \dfrac{3}{x + 3}}$

25. $\dfrac{\dfrac{1}{t^2} - 1}{\dfrac{5}{t^2} - \dfrac{5}{t} - 10}$

26. $\dfrac{x^{-1} - y^{-1}}{x^{-1} + y^{-1}}$

27. $\left(\dfrac{1}{a + b} + \dfrac{1}{a - b}\right) \div \left(\dfrac{1}{a + b} - \dfrac{1}{a - b}\right)$

28. $\left(\dfrac{x^2 - y^2}{a^2 - b^2}\right) \div \left(\dfrac{x - y}{a + b}\right)$

Solve each of the following equations. Check your answers.

29. $\dfrac{x}{3} - \dfrac{x}{2} = 5$

30. $\dfrac{1}{x} - \dfrac{3}{x} = 2$

31. $\dfrac{x - 3}{x - 4} = -2$

32. $\dfrac{3b - 2}{2b + 1} = -1$

33. $\dfrac{1}{4} + \dfrac{3}{s} = \dfrac{7}{3} - \dfrac{1}{5s}$

34. $\dfrac{t^2 - 7t + 12}{t^2 - 16} = 1$

35. $\dfrac{x^2 - 3x + 2}{x^2 - 8} = 2$

36. $\dfrac{a^2 + 16a + 3}{a^2 - 5} = 3$

37. $\dfrac{1}{x^2 - 2} - \dfrac{1}{x^2 - 3x + 1} = 0$

38. $\dfrac{m}{m^2 - 9} - \dfrac{1}{m + 2} = 0$

39. $\dfrac{6}{y} - \dfrac{4}{y + 2} = \dfrac{9}{y + 2}$

40. $\dfrac{7}{q} - \dfrac{3}{q - 1} = \dfrac{5}{q - 1}$

Write each of the following applied problems as a rational equation and solve.

41. Three times an integer decreased by 5 times its reciprocal is 2. Find the integer.

42. The denominator of a number is 4 more than the numerator. If both the numerator and denominator are increased by 5, the result is $\tfrac{7}{9}$. Find the number.

43. Mort leaves Kudafuld, ME, at 11:00 A.M., heading south at an average of 56 mph. Helen leaves Kudafuld at 11:30 A.M. on the same day to overtake Mort, traveling at an average of 64 mph. How long does it take for Helen to overtake Mort? How far will she have traveled?

44. Todd runs a one-mile track in 5 min. Rhonda runs the same track in 7 min. If Todd and Rhonda start at opposite ends of the track and run until they meet, how long does it take?

45. Because of the fog, Ray decreases his usual speed by 20 mph for a 240-mi trip. He finds that this increases his usual time by 2 hr. What is his usual speed?

46. A tank can be filled by an inlet pipe in 7 hr. It takes 10 hr to drain the tank. How long will it take to fill the tank if the inlet pipe is opened and the drain pipe is left half open?

Solve for y.

47. $\dfrac{1}{a - y} + \dfrac{4}{y} = \dfrac{3}{a + y}$

48. $ax - 3y + 5xy = 2x$

Chapter Test

Simplify. Assume no denominator can be zero.

1. $\dfrac{16(x^2 - y^2)}{20(x - y)(x + y)}$

2. $\dfrac{x^2 - 4}{x^2 - 4x + 4}$

3. $\dfrac{4(x - 3)}{x^2 - 9} + \dfrac{2}{x + 3}$

4. $\dfrac{x^2 - y^2}{xz - x} \div \dfrac{(x - y)^2}{z - 1}$

5. $\dfrac{5}{x - 2} + \dfrac{4}{x + 2}$

6. $\dfrac{x}{x^2 - 4} + \dfrac{2}{4 - x^2}$

7. $\dfrac{\dfrac{1}{x^2}}{\dfrac{1}{x} - \dfrac{1}{x^2}}$

Solve each of the following equations. State any restrictions on the variables. Check your answers.

8. $\dfrac{1}{x^2 - 5x + 6} + \dfrac{3}{x - 3} = \dfrac{2}{x - 2}$

9. $\dfrac{x}{x + 2} = 1 - \dfrac{3x + 2}{x^2 + 4x + 4}$

10. $\dfrac{9}{x + 3} + \dfrac{4}{x + 3} = \dfrac{2}{x}$

11. $8 + \dfrac{3}{2x + 5} = \dfrac{4}{2x + 5}$

12. $\dfrac{3 + 2a}{a^2 + 6 + 5a} + \dfrac{2 - 5a}{a^2 - 4} = \dfrac{2 - 3a}{a^2 - 6 + a}$

13. $\dfrac{4}{y - 1} - \dfrac{3}{y} = \dfrac{5}{y - 1}$

14. $\dfrac{15x + 15}{5} = \dfrac{9x^2}{x - 6} \cdot \dfrac{x^2 - 5x - 6}{3x^2}$

15. Simplify: $\dfrac{4}{y - 1} + y + 1$.

Solve each of the following applied problems.

16. A natural number decreased by 15 times its reciprocal is equal to 2. Find the natural number.

17. Felix can cut a cord of wood in 4 hr. His daughter can cut a cord of wood in 7 hr. How long will it take them, working together, to cut a cord of wood?

18. Olivia can jog 8 miles in the same time that Luis can jog 6 miles. If Olivia jogs 2 mph faster than Luis, how fast does Olivia jog?

19. Larry can clean the yard working alone in 12 hr. Manny also can clean the yard working alone in 12 hr. It takes Tamara 15 hours working alone to clean the yard. How long will it take them to do the job working together?

20. Gordon invests $22,000 in a first and a second trust deed in a ratio of 8 to 3. How much is in the second trust deed?

21. Shane works one and a half times as fast as his brother Mike. If it takes them 12 hr to do a job together, how long does it take Shane, working alone, to do the same job?

22. Simplify: $\dfrac{\dfrac{2}{x + 2}}{\dfrac{1}{x + 2} + \dfrac{2}{x}}$. State any restrictions on the variable.

CHAPTER 8
Radicals and Rational Exponents

8.1 Rational Exponents
8.2 Radicals and Rational Exponents
8.3 Arithmetic Operations Involving Radicals
8.4 Division by Radicals; Rationalizing
8.5 Solving Equations Involving Radicals
8.6 Complex Numbers
8.7 More Applications

Application

A string stretched with a tension of 50 newtons produces the note C with a frequency of 256 cycles per second (cps). If f and F are two frequencies, and t and T are the corresponding tensions, then frequency and tension are related by the proportion

$$\frac{f}{F} = \frac{\sqrt{t}}{\sqrt{T}}$$

Find the tension T required to produce a frequency of $F = 512$ cps.

See Exercise Set 8.7, number 15.

The symbolism used in mathematics has not always been the same as that used today. It has undergone a period of evolution with standardization gradually taking place. Mathematics is a living science and it is reasonable to expect that further evolution will take place, particularly as computers continue to develop and research continues to explore new areas. Some of the symbolism used by early mathematicians will be introduced in the historical comments in this chapter to increase your appreciation of the conciseness of today's notation.

8.1 Rational Exponents

HISTORICAL COMMENT   Nicole Orseme (1320–1385) was one of a group of early writers whose work bore little fruit in his lifetime or in the next generations because of the Hundred Years' War and the plague that stifled the development of European universities. Orseme was a French writer of that period whose work was far ahead of his time; as a result, his ideas did not find common usage for about 400 years. In his development of exponents, Orseme used $\frac{1}{2}4^p$ in the same way that $4^{1/2}$ is used today.

OBJECTIVES   In this section we will learn to

1. interpret the rational exponent $\dfrac{m}{n}$; and
2. use the rules of exponents with rational numbers.

Chapter 5 examined the rules of exponents as they apply to integers. We began by defining the exponent as a natural number that indicated the number of times the base was to be used as a factor. It soon became necessary to give meaning to zero as an exponent:

$b^0 = 1$

and to negative integers as exponents:

$b^{-n} = \dfrac{1}{b^n}$

In this section we will examine exponential expressions such as

$b^{\frac{m}{n}}$

where the exponent $\frac{m}{n}$ is a rational number in lowest terms, with $n \neq 0$. We will show how the rules of exponents are used when the exponents represent any rational number.

If the rules of exponents are to hold for exponents that are rational numbers, we will have to give meaning to expressions such as

$4^{1/2}$ or $(-27)^{1/3}$

If we assume that the rule $(b^m)^n = b^{m \cdot n}$ is true for any rational numbers m and n;

$(4^{1/2})^2$ and $4^{(1/2) \cdot 2} = 4^{2/2} = 4$

393

This implies that $4^{1/2}$ is a number such that its square is 4. Similarly, we would have

$$[(-27)^{1/3}]^3 = (-27)^{(1/3) \cdot 3} = (-27)^{3/3} = -27$$

Thus $(-27)^{1/3}$ must be a number such that its cube is -27.

Recall from elementary algebra that $\sqrt{4} = 2$ and that $\sqrt[3]{-27} = -3$. Thus $4^{1/2}$ is another way of indicating the positive square root of 4 and $(-27)^{1/3}$ another way of indicating the cube root of -27. In general,

> $b^{1/n}$ indicates an nth root of b if n is a natural number.

There may be more than one nth root of a number:

The number 4 has two square roots, 2 and -2, since $2^2 = 4$ and $(-2)^2 = 4$.
The number 9 has two square roots, 3 and -3, since $3^2 = 9$ and $(-3)^2 = 9$.
The number 16 has two square roots, 4 and -4, since $4^2 = 16$ and $(-4)^2 = 16$.

Some numbers have only one real nth root:

The number -1 has only one cube root, -1, since $(-1)^3 = -1$ but $1^3 = 1$.
The number -8 has only one cube root, -2, since $(-2)^3 = -8$ but $2^3 = 8$.
The number -27 has only one cube root, -3, since $(-3)^3 = -27$ but $3^3 = 27$.

In fact, every nonzero number has exactly n nth roots, although not all of them may be real numbers, as we shall see later in this chapter. In this text we will use what is known as the **principal nth root.**

DEFINITION **Principal nth Root**	When n is even, the **principal nth root** of a positive number is its positive nth root.

For example:

The principal square root of 4 is 2.
The principal square root of 9 is 3.
The principal fourth root of 16 is 2.

The nth root of a negative number may or may not be a real number. For example:

The cube root of -27 is -3, since $(-3)^3 = -27$.
The cube root of -64 is -4, since $(-4)^3 = -64$.
The square root of -4 is not a real number, since $(-2)^2 \neq -4$ and $2^2 \neq -4$.
The square root of -9 is not a real number, since $(-3)^2 \neq -9$ and $3^2 \neq -9$.

To Find the nth Root of b

	b Positive or Zero	b Negative
n even	$b^{1/n}$ is the principal nth root of b or 0	$b^{1/n}$ is not a real number
n odd	$b^{1/n}$ is the positive nth root of b or zero	$b^{1/n}$ is the negative nth root of b

EXAMPLE 1 Find the indicated root of each number.

(a) $4^{1/2} = 2$, since $2^2 = 4$.
(b) $8^{1/3} = 2$, since $2^3 = 8$.
(c) $(-64)^{1/3} = -4$, since $(-4)^3 = -64$.
(d) $(-16)^{1/2} \neq 4$, since $4^2 \neq -16$, and $(-16)^{1/2} \neq -4$, since $(-4)^2 \neq -16$. Thus $(-16)^{1/2}$ is not a real number.

EXAMPLE 2 Find the indicated root of each real number, if possible.

(a) $-4^{1/2} = -2$, since $-4^{1/2} = -1 \cdot 4^{1/2} = -1 \cdot 2 = -2$.
(b) $(-81)^{1/4}$ is not a real number.
(c) $(-32)^{1/5} = -2$, since $(-2)^5 = -32$.

Caution

Be particularly careful to use grouping symbols correctly when writing the sign of a number in a problem involving roots.

$(-16)^{1/2}$ is not a real number.
$-16^{1/2} = -1 \cdot 16^{1/2}$
$= -1 \cdot 4 = -4$

Thus $-16^{1/2}$ *is* a real number. The exponent applies only to the 16 and not to the negative sign.

We now turn our attention to the interpretation of the rational exponent $\dfrac{m}{n}$. If the rules of exponents are to hold, then

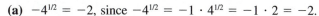

$$b^{m/n} = b^{(1/n) \cdot m} = (b^{1/n})^m = (b^m)^{1/n}, \qquad b \geq 0 \text{ when } n \text{ is even}$$

For example,

$8^{2/3} = (8^{1/3})^2 = 2^2 = 4$ $8^{1/3} = 2$, since $2^3 = 8$.

and

$8^{2/3} = (8^2)^{1/3} = 64^{1/3} = 4$

In general, it is better to treat $b^{m/n}$ as $(b^{1/n})^m$, since it is usually easier to find the root first. For example,

$(32)^{4/5} = (32^{1/5})^4 = 2^4 = 16$ The 5th root of 32 is 2.

while

$(32)^{4/5} = (32^4)^{1/5} = (1{,}048{,}576)^{1/5} = 16$ The 5th root of 1,048,576 is 16.

Notice that it is much more difficult to find the 5th root of 1,048,576 than the 5th root of 32.

EXAMPLE 3 Simplify: (a) $64^{5/6}$ (b) $81^{3/4}$ (c) $(0.125)^{2/3}$

SOLUTION
(a) $64^{5/6} = (64^{1/6})^5 = 2^5 = 32$ $\qquad$ $64^{1/6} = 2$, since $2^6 = 64$.
(b) $81^{3/4} = (81^{1/4})^3 = 3^3 = 27$ $\qquad$ $81^{1/4} = 3$, since $3^4 = 81$.
(c) $(0.125)^{2/3} = [(0.125)^{1/3}]^2 = [0.5]^2 = 0.25$ $\qquad$ $0.125^{1/3} = 0.5$, since $(0.5)^3 = 0.125$.

We now turn our attention to rational exponents in general. The laws of exponents hold as long as we do not apply them to even roots of negative numbers. We review them here.

Rules for Exponents

If m and n are rational numbers, then

$$a^m \cdot a^n = a^{m+n} \qquad\qquad (ab)^m = a^m b^m$$

$$(a^m)^n = a^{m \cdot n} \qquad\qquad \frac{a^m}{a^n} = a^{m-n}, \quad a \neq 0$$

$$a^{-n} = \frac{1}{a^n}, \quad a \neq 0 \qquad\qquad \left(\frac{a}{b}\right)^m = \frac{a^m}{b^m}, \quad b \neq 0$$

$$\left(\frac{a}{b}\right)^{-n} = \left(\frac{b}{a}\right)^n, \quad a \neq 0, b \neq 0 \qquad\qquad a^0 = 1, \quad a \neq 0$$

EXAMPLE 4 Rewrite each of the following powers using the rules for exponents. Simplify the result.

(a) $9^{-1/2} = \dfrac{1}{9^{1/2}} = \dfrac{1}{3}$ $\qquad\qquad$ $b^{-n} = \dfrac{1}{b^n}$

(b) $\left(\dfrac{1}{64}\right)^{-1/6} = (64)^{1/6}$ $\qquad\qquad$ $\left(\dfrac{a}{b}\right)^{-n} = \left(\dfrac{b}{a}\right)^n$

$\qquad\qquad\quad\; = 2$ $\qquad\qquad\qquad\qquad\;\;\;$ $(64)^{1/6} = 2$, since $2^6 = 64$.

(c) $x^{1/2} \cdot x^{1/3} = x^{(1/2)+(1/3)} = x^{(3/6)+(2/6)}$ $\qquad$ $b^m \cdot b^n = b^{m+n}$

$\qquad\quad\;\; = x^{5/6}, \quad x > 0$

(d) $\dfrac{x^{2/3}}{x^{3/5}} = x^{(2/3)-(3/5)} = x^{(10/15)-(9/15)}$ $\qquad$ $\dfrac{b^m}{b^n} = b^{m-n}$

$\quad\;\; = x^{1/15}$

(e) $(xy)^{2/3} = x^{2/3} y^{2/3}$ $\qquad\qquad$ $(ab)^m = a^m b^m$

(f) $\left(\dfrac{a^{2/3}}{b^{1/2}}\right)^{4/5} = \dfrac{a^{(2/3)(4/5)}}{b^{(1/2)(4/5)}}$ $\qquad\qquad$ $\left(\dfrac{a}{b}\right)^m = \dfrac{a^m}{b^m}$

$\qquad\quad\;\; = \dfrac{a^{8/15}}{b^{2/5}}$

EXAMPLE 5 Simplify: $\left(\dfrac{-64x^2}{x^{-1}y^6}\right)^{-2/3}$

SOLUTION

$\left(\dfrac{-64x^2}{x^{-1}y^6}\right)^{-2/3} = \left(\dfrac{-64x^{2-(-1)}}{y^6}\right)^{-2/3}$ $\qquad \dfrac{a^m}{a^n} = a^{m-n}$

$= \left(\dfrac{-64x^{2+1}}{y^6}\right)^{-2/3}$

$= \left(\dfrac{-64x^3}{y^6}\right)^{-2/3}$

$= \left(\dfrac{y^6}{-64x^3}\right)^{2/3}$ $\qquad \left(\dfrac{a}{b}\right)^{-n} = \left(\dfrac{b}{a}\right)^{n}$

$= \dfrac{y^{6(2/3)}}{(-64)^{2/3}x^{3(2/3)}}$ $\qquad \left(\dfrac{a}{b}\right)^{m} = \dfrac{a^m}{b^m};\ (a^m)^n = a^{m \cdot n};\ (ab)^m = a^m b^m$

$= \dfrac{y^4}{[(-64)^{1/3}]^2 x^2}$

$= \dfrac{y^4}{[-4]^2 x^2} = \dfrac{y^4}{16x^2}$

EXAMPLE 6 Simplify each of the following.

(a) $\dfrac{9a^{-2/3}}{3a^{-7/3}} = 3a^{(-2/3)-(-7/3)} = 3a^{(-2/3)+(7/3)} = 3a^{5/3}$

(b) $3x^{2/3}(4x^{7/3}) = 12x^{(2/3)+(7/3)} = 12x^{9/3} = 12x^3$

(c) $x^{1/3}(x^{1/2} + x^{2/3}) = x^{1/3} \cdot x^{1/2} + x^{1/3} \cdot x^{2/3}$
$\qquad = x^{(1/3)+(1/2)} + x^{(1/3)+(2/3)} = x^{5/6} + x$

EXAMPLE 7 Multiply: (a) $(x^{2/3} - y^{2/3})(x^{2/3} + y^{2/3})$ (b) $(x^{1/3} + y^{1/2})^2$

SOLUTION (a) $(x^{2/3} - y^{2/3})(x^{2/3} + y^{2/3}) = (x^{2/3})^2 - (y^{2/3})^2$ $\qquad (a-b)(a+b) = a^2 - b^2$
$\qquad\qquad = x^{4/3} - y^{4/3}$

(b) $(x^{1/3} + y^{1/2})^2 = (x^{1/3})^2 + 2x^{1/3}y^{1/2} + (y^{1/2})^2$ $\qquad (a+b)^2 = a^2 + 2ab + b^2$
$\qquad\qquad = x^{2/3} + 2x^{1/3}y^{1/2} + y$

EXAMPLE 8 Factor: (a) $x^{3/5} + x^{4/5}$ (b) $6x^{-1/2}y + 9xy^{1/4}$

SOLUTION (a) $x^{3/5} + x^{4/5} = x^{3/5}(1 + x^{1/5})$ $\qquad$ The greatest common factor is $x^{3/5}$.

(b) The greatest common numerical factor in $6x^{-1/2}y + 9xy^{1/4}$ is 3. The greatest common factor for each variable has the smallest exponent that is common to each. Since $x = x^{1/2 + 3/2} = x^{-1/2}x^{3/2}$ and $y = y^{1/4 + 3/4} = y^{1/4}y^{3/4}$, for x it is $x^{-1/2}$ and for y it is $y^{1/4}$. The GCF is $3x^{-1/2}y^{1/4}$. Therefore

$6x^{-1/2}y + 9xy^{1/4} = 3x^{-1/2}y^{1/4}(2y^{3/4} + 3x^{3/2})$

EXAMPLE 9 Factor $2a^{2/3} - 4a^{3/5}$.

SOLUTION We write each rational exponent in terms of the LCD of the exponents.

$$2a^{2/3} - 4a^{3/5} = 2a^{10/15} - 4a^{9/15} \qquad \text{The LCM of 3 and 5 is 15.}$$

The greatest common factor is $2a^{9/15} = 2a^{3/5}$. Therefore

$$\begin{aligned} 2a^{2/3} - 4a^{3/5} &= 2a^{10/15} - 4a^{9/15} \\ &= 2a^{9/15}(a^{1/15} - 2) \qquad a^{m+n} = a^m \cdot a^n \\ &= 2a^{3/5}(a^{1/15} - 2) \end{aligned}$$

EXAMPLE 10 Simplify: $\dfrac{x^{2/3}y^{-1/2}z^{1/4}}{x^{1/3}y^{3/2}z^{3/4}}$

SOLUTION $\dfrac{x^{2/3}y^{-1/2}z^{1/4}}{x^{1/3}y^{3/2}z^{3/4}} = x^{(2/3)-(1/3)}y^{(-1/2)-(3/2)}z^{(1/4)-(3/4)} \qquad \dfrac{a^m}{a^n} = a^{m-n}$

$$= x^{1/3}y^{-2}z^{-1/2} = \dfrac{x^{1/3}}{y^2 z^{1/2}}$$

EXAMPLE 11 Simplify: $\left(\dfrac{8x^{2m}}{x^{-4m}}\right)^{-1/3}$

SOLUTION $\left(\dfrac{8x^{2m}}{x^{-4m}}\right)^{-1/3} = (8x^{2m+4m})^{-1/3} = (8x^{6m})^{-1/3}$

$$= \dfrac{1}{(8x^{6m})^{1/3}} = \dfrac{1}{(8)^{1/3}x^{6m(1/3)}} = \dfrac{1}{2x^{2m}}$$

Exercise Set 8.1

Assume all variables in the exercises for this section represent positive real numbers.

Simplify each of the following quantities.

1. $25^{1/2}$
2. $36^{1/2}$
3. $27^{1/3}$
4. $8^{1/3}$
5. $16^{-1/4}$
6. $81^{-1/4}$
7. $4^{3/2}$
8. $9^{3/2}$
9. $(-27)^{1/3}$
10. $(-64)^{1/3}$
11. $27^{2/3}$
12. $125^{2/3}$
13. $(8)^{1/3} \cdot (8)^{2/3}$
14. $(5)^{1/2} \cdot (5)^{3/2}$
15. $(3)^{3/2} \cdot (3)^{-1/2}$
16. $\left(\dfrac{9}{4}\right)^{-1/2}$
17. $\left(-\dfrac{8}{27}\right)^{2/3}$
18. $\left(\dfrac{25}{36}\right)^{3/2}$
19. $\left(\dfrac{1}{32}\right)^{-1/5}$
20. $\left(\dfrac{1}{81}\right)^{-1/4}$
21. $\left(\dfrac{4}{49}\right)^{3/2}$

Simplify each of the following quantities. Express each answer using only positive exponents.

22. $(0.00243)^{2/5}$
23. $(0.0256)^{-1/4}$
24. $(-0.125)^{-1/3}$
25. $(-0.81)^{-1/4}$
26. $y^{3/4} \cdot y^{1/4}$
27. $p^{3/5} \cdot p^{1/5}$
28. $q^{4/3} \cdot q^{-1/3}$
29. $(2t^{1/3})(5t^{2/3})$
30. $(7s^{3/2})(2s^{-1/2})$
31. $(a^4)^{1/2}$
32. $(c^3)^{4/3}$
33. $6m^{2/3} \cdot 3m^{1/6}$

34. $2x^{1/4} \cdot x^{-1/3}$

35. $3k^{-5/2} \cdot k^{1/3}$

36. $9y^{-3/4} \cdot 4y^{1/3}$

37. $\dfrac{a^{3/4}}{a^{1/4}}$

38. $\dfrac{p^{5/8}}{p^{-3/8}}$

39. $\dfrac{6r^{-2/3}}{3r^{-5/3}}$

40. $\dfrac{15s^{-1/4}}{5s^{-5/4}}$

41. $\left[\left(\dfrac{x^{1/2}}{x^{1/4}}\right)^4\right]^0$

42. $\left(\dfrac{m^{3/2}}{m^{-1/2}}\right)^{1/2}$

43. $\left(\dfrac{17y^{8/3}}{34y^{-1/3}}\right)^{1/3}$

44. $\left(\dfrac{18a^{-2/5}}{27a^{3/5}}\right)^{-1}$

45. $(18x^{4/3}y^{7/3}) \div (54x^{1/3}y^{1/3})$

46. $(63a^{9/5}b^{7/5}) \div (7a^{4/5}b^{2/5})$

47. $(16^{2/3}p^{5/7}q^{-1/3}) \div (64^{1/3}p^{-2/7}q^{4/3})$

48. $(-91s^{-1/3}t^{-2/5}) \div (13s^{1/3}t^{2/5})$

49. $x^{1/2}(x^{1/2} + x^{1/4})$

50. $a^{1/3}(a^{2/3} + a^{1/4})$

51. $k^{-1/2}(k^{3/2} - k^{1/2})$

52. $p^{-1/4}(p^{5/4} - p^{1/4})$

53. $3t^{-2/3}(5t^{1/4} - 2t^{3/2})$

54. $7m^{-3/4}(2m^{-1/3} + 3m^{1/2})$

55. $(x^{1/2} - y^{1/2})(x^{1/2} + y^{1/2})$

56. $(a^{1/2} + b^{1/2})(a^{1/2} - b^{1/2})$

57. $(a^{1/2} + b^{1/4})^2$

58. $(y^{1/2} - y^{1/4})^2$

Factor each of the following expressions as indicated.

59. $p^{5/4} = p^{1/4}(\quad)$

60. $y^{-3/4} = y^{1/4}(\quad)$

61. $x^{1/2} + x^{-1/2} = x^{-1/2}(\quad)$

62. $t^{1/3} - t^{-1/3} = t^{-1/3}(\quad)$

63. $a - a^{1/3} = a^{1/3}(\quad)$

64. $k - k^{1/5} = k^{1/5}(\quad)$

65. $x - y = (x^{1/2} + y^{1/2})(\quad)$

66. $a - b = (a^{1/2} - b^{1/2})(\quad)$

67. $8^{2/3}p^{3/4} - 4^{1/2}p^{-1/2} = 2p^{-1/2}(\quad)$

68. $9^{3/2}s^{-1/3} - 9^{1/2}s^{4/3} = 3s^{-1/3}(\quad)$

69. $(x - y) = (x^{1/3} - y^{1/3})(\quad)$

70. $(a + b) = (a^{1/3} + b^{1/3})(\quad)$

Simplify each of the following quantities. Express the answer using only positive exponents.

71. $\dfrac{x^5 y^0 z^{1/4}}{x^{1/3} y^{1/2} z^{2/3}}$

72. $\dfrac{a^{2/3}b^{-1/2}c^{1/4}}{a^{2/3}b^{1/2}c^{-3/4}}$

73. $\dfrac{-25^{3/2}r^{3/3}s^2 t^{1/3}}{5^2 r^{1/4} s^{1/3} t^{1/4}}$

74. $\dfrac{(9u)^{1/2}(v)^{1/3}(25w)^{3/2}}{(8u)^{1/3}(9v)^{3/2}(125w)^{1/3}}$

75. $[(x^{-2/3}y^{1/4}z^{1/5})^{-1/2}]^{-3/2}$

76. $\{[(m^{-1/4})^{1/3}(9p)^{-3/2}(27q)^{1/3}]^{-2}\}^{1/3}$

77. $\dfrac{9^{-2}a^{-1/3}b^{4/3}}{(4a^{2/3}b^{-4/3})^{1/2}}$

78. $\left[\dfrac{16x^{3/4}y^{-3/5}}{24x^0 y^{3/5}}\right]^{-1/3}$

79. $\left(\dfrac{r^{1/2}s^{-2}}{t^{1/2}}\right)^2 \cdot \left(\dfrac{r^{-1/2}s^0}{t^{-4}}\right)^{-2}$

80. $\left(\dfrac{mn^{1/2}}{p^{1/2}q^{3/2}}\right)^{-4} \cdot \left(\dfrac{m^2 n^2}{p^2 q^4}\right)^{3/2}$

81. $(x^{-1/4}y^{-1/3})^{-1} \cdot (x^{-3/2})^2 \cdot (y^{-5/3})^{-2}$

82. $(a^{1/4}b^{-3/8})^{-1} \cdot (a^{-3/4})^{-4} \cdot (b^{-2/3})^4$

Simplify. Assume m and n are positive rational numbers.

83. $(y^{n+1})^2$

84. $(x^{m-1})^3$

85. $(t^{2m})^{3/2}$

86. $\left(\dfrac{s^{3m}}{s^m}\right)^{1/3}$

87. $\left(\dfrac{x^{5n-1}}{x^{3n-2}}\right)^{1/2}$

88. $\left(\dfrac{y^{2n-3}}{y^{n-1}}\right)^{1/3}$

89. $\left(\dfrac{x^{2m}}{y^{2n}}\right)^{1/3} \cdot \left(\dfrac{y^{3n}}{x^{3m}}\right)^{1/2}$

90. $\left(\dfrac{a^{2n}}{b^{-3/2}}\right)^{1/3} \cdot \left(\dfrac{a^{2/3}}{b^{1/4}}\right)^n$

Use a calculator to evaluate each of the following numbers; rounding each answer to four decimal places.

91. $5^{1/2}$

92. $6^{1/2}$

93. $3^{1/3}$

94. $4^{1/3}$

95. $3^{-1/2}$

96. $7^{-1/3}$

97. $9^{3/4}$

98. $100^{2/3}$

Write In Your Own Words

99. What is a rational exponent?

100. Are $\sqrt{9}$ and $9^{1/2}$ equivalent? Why?

101. Are $\sqrt[3]{-8}$ and $-8^{1/3}$ equivalent? Why?

102. Are $-8^{1/3}$ and $(-8)^{1/3}$ equivalent? Why?

103. What conditions are required for $(b^{1/n})^n = (b^n)^{1/n} = b^{n/n} = b$ to be true?

104. If $b < 0$ and n is an even number, why is $b^{1/n}$ not a real number?

105. If n is an odd number, is $b^{1/n}$ always a real number? Explain.

106. Is it better to treat $b^{m/n}$ as $(b^m)^{1/n}$ or $(b^{1/n})^m$? Why?

8.2 Radicals and Rational Exponents

HISTORICAL COMMENT There is some question as to the origin of the radical symbol, although it seems to have some connection with either the capital letter R or its lower-case form. The radical sign as we use it today appears to have been standardized in the latter part of the seventeenth century. Some of the symbols used between 1484 and 1572 to indicate square roots were ℞, ℞², ℞.2ᵃ, ℞□, ℞.q., √, and ⌡. Among symbols used to indicate other roots were ℞³, ℞.3ᵃ, ℞□, ℞.c., ℣, ℣, √³, and ⋎, ⋎⋎.

OBJECTIVES In this section we will learn to
1. identify the parts of a radical expression;
2. interchange between exponential and radical forms; and
3. write radical expressions in their simplest form.

In Section 8.1 we used the symbol $b^{1/n}$ to indicate the *n*th root of *b*. Another symbol that we use to indicate a root of a number is the **radical symbol:**

$$\sqrt{}$$

The symbol $\sqrt{b}$ means *the square root of b.* Since every positive number has two square roots, one positive and one negative, $\sqrt{b}$ will indicate the **principal** (positive) **square root** of *b*.

DEFINITION
Principal Square Root

The **principal square root** of any positive number *b*, written $\sqrt{b}$, is that *positive* number *a* such that $a^2 = b$.

For example, -3 and 3 are both square roots of 9, but $\sqrt{9}$ refers to only one of **them,** $\sqrt{9} = 3$. The other is $-\sqrt{9} = -3$.

EXAMPLE 1   Find each of the following square roots.

(a) $\sqrt{25} = 5$ $5^2 = 25$
(b) $\sqrt{36} = 6$ $6^2 = 36$
(c) $\sqrt{121} = 11$ $11^2 = 121$

When the root of a number is other than a square root, we use an *index* to indicate it. For example, in

$\sqrt[3]{b}$

the index is 3, which indicates that we want to find the *cube root* of *b*.

DEFINITION	The **cube root** of a number b is that number a such that $a^3 = b$.
Cube Root	

The symbol
$$\sqrt[n]{b}$$
means that we want to find the **nth root** of b.

DEFINITION	The **nth root** of a number b is that number a such that $a^n = b$.
nth Root	

Radicals consist of three parts: the **index**, the **radical symbol**, and the **radicand**:

Index ↘   Radical symbol ↙
$\sqrt[n]{b}$ ← Radicand

It is not necessary to show the index when the root we are taking is a square root.
$$\sqrt[2]{b} = \sqrt{b}$$

There is a direct relation between radical notation and rational exponents.

$\sqrt{4} = 4^{1/2} = 2$ $\qquad\qquad \sqrt{4} = \sqrt[2]{4}$
$\sqrt[3]{-27} = (-27)^{1/3} = -3$
$\sqrt[5]{32} = 32^{1/5} = 2$
$\sqrt[4]{a^3} = (a^3)^{1/4} = a^{3/4}$
$(\sqrt[7]{x})^9 = (x^{1/7})^9 = x^{9/7}$

To Convert from Rational Exponents to Radical Notation

$b^{m/n} = (b^{1/n})^m = (\sqrt[n]{b})^m$

or $\qquad\qquad\qquad\qquad\qquad\qquad\qquad b \geq 0$ if n is even

$b^{m/n} = (b^m)^{1/n} = \sqrt[n]{b^m}$

EXAMPLE 2   Rewrite in radical form.

(a) $3^{1/2} = \sqrt{3}$ The index is not shown for square roots.
(b) $5^{1/4} = \sqrt[4]{5}$
(c) $7^{2/3} = \sqrt[3]{7^2} = \sqrt[3]{49}$
(d) $(8^{1/4})^2 = (\sqrt[4]{8})^2$

Notice in Example 2(d) that $(8^{1/4})^2 = 8^{2/4} = 8^{1/2}$. Thus
$$(8^{1/4})^2 = 8^{1/2} = \sqrt{8}$$

EXAMPLE 3 Rewrite using rational exponents.

(a) $\sqrt{5} = 5^{1/2}$
(b) $\sqrt[3]{3^2} = (3^2)^{1/3} = 3^{2/3}$
(c) $\sqrt[3]{x^3 + y^3} = (x^3 + y^3)^{1/3}$; $(x^3 + y^3)^{1/3}$ cannot be simplified further, since exponent do not distribute over sums.
(d) $\sqrt{x^2 + y^2} = (x^2 + y^2)^{1/2}$
(e) $\sqrt{(x + y)^2}$, where $x + y \geq 0$
$$\sqrt{(x + y)^2} = [(x + y)^2]^{1/2} = (x + y)^{2/2} = x + y$$

EXAMPLE 4 Find the indicated roots.

(a) $\sqrt{36} = \sqrt{6^2} = 6^{2/2} = 6$ $6^2 = 36$
(b) $\sqrt[3]{8} = \sqrt[3]{2^3} = 2^{3/3} = 2$ $2^3 = 8$
(c) $\sqrt[3]{a^6 b^9} = a^{6/3} b^{9/3} = a^2 b^3$ $(a^2 b^3)^3 = a^6 b^9$
(d) $\sqrt[4]{625 x^8} = 5x^2$ $(5x^2)^4 = 625 x^8$
(e) $\sqrt[3]{(a + b)^9} = (a + b)^{9/3} = (a + b)^3$
(f) $\sqrt{x^2 + 4x + 4} = \sqrt{(x + 2)^2} = x + 2$, if $x + 2 \geq 0$

In Examples 3(e) and 4(f), the variables were restricted so that the radicand was a positive quantity. The following table shows what happens when the restriction is not there.

	IF $b = 3$	IF $b = -3$
$\sqrt{b^2} = b$	$\sqrt{3^2} \stackrel{?}{=} 3$ $\sqrt{9} \stackrel{?}{=} 3$ $3 \stackrel{\checkmark}{=} 3$ True	$\sqrt{(-3)^2} \stackrel{?}{=} -3$ $\sqrt{9} \stackrel{?}{=} -3$ $3 = -3$ False

To determine $\sqrt{b^2}$ for any real number b, two cases must be considered:

$\sqrt{b^2} = b$ if b is positive or zero ($b \geq 0$)

and

$\sqrt{b^2} = -b$ if b is negative ($b < 0$)

Since $|b| = b$ if $b \geq 0$ and $|b| = -b$ if $b < 0$, both cases can be summarized as follows.

$$\sqrt{b^2} = |b| \quad \text{for all real numbers } b \quad (b \in R)$$

No difficulty arises when the index is odd. For example,

$$\sqrt[3]{3^3} = 3 \quad \text{and} \quad \sqrt[3]{(-3)^3} = -3$$
$$\sqrt[3]{27} = 3 \quad\quad\quad \sqrt[3]{-27} = -3$$
$$3 \overset{\checkmark}{=} 3 \quad\quad\quad -3 \overset{\checkmark}{=} -3$$

EXAMPLE 5   Find the indicated roots.
(a) $\sqrt{x^2} = |x|$
(b) $\sqrt{(x+2)^2} = |x+2|$
(c) $\sqrt[3]{(x-y)^3} = (x-y)$
(d) $\sqrt[5]{x^5} + \sqrt{x^2} = x + |x|$

Expressions involving radicals can often be simplified. However, simplification requires a theorem involving radicals in which we assume the roots exist.

THEOREM 8.1
Properties of Radicals

If a and b represent nonnegative real numbers, then

$$\sqrt[n]{a \cdot b} = \sqrt[n]{a} \cdot \sqrt[n]{b} \quad \text{or} \quad (a \cdot b)^{1/n} = a^{1/n} \cdot b^{1/n}$$

and

$$\sqrt[n]{\frac{a}{b}} = \frac{\sqrt[n]{a}}{\sqrt[n]{b}} \quad \text{or} \quad \left(\frac{a}{b}\right)^{1/n} = \frac{a^{1/n}}{b^{1/n}}$$

To illustrate Theorem 8.1, consider the following examples.

$$\sqrt{144} = \sqrt{9 \cdot 16} = \sqrt{9} \cdot \sqrt{16} = 3 \cdot 4 = 12 \quad\quad \sqrt{144} = 12$$

$$\sqrt{\frac{144}{9}} = \frac{\sqrt{144}}{\sqrt{9}} = \frac{12}{3} = 4 \quad\quad \sqrt{\frac{144}{9}} = \sqrt{16} = 4$$

Simplest Forms for Radicals

For a radical to be in simplest form, the following conditions must be satisfied:

1. All exponents on factors in the radicand must be less than the index.
2. The radicand cannot contain a fraction.
3. The index and **all** of the exponents in the radicand cannot contain any common factors (assuming all of variables in the radicand represent positive numbers).
4. No fraction can have a radical in the denominator.

EXAMPLE 6   Simplify each of the following radicals. Assume all variables represent positive numbers.

SOLUTION (a) $\sqrt{150}$

This radical is not in simplest form because it contains perfect squares in the radicand. To find the perfect squares, we write the radicand in prime factored form:

$$\sqrt{150} = \sqrt{5 \cdot 5 \cdot 6} = \sqrt{5^2 \cdot 6} \qquad 5^2 \text{ is a perfect square.}$$
$$= 5\sqrt{6} \qquad \sqrt[n]{ab} = \sqrt[n]{a}\sqrt[n]{b}$$

(b) $\sqrt{252x^2}$

This radical is not in simplest form because it contains perfect squares in the radicand.

$$\sqrt{252x^2} = \sqrt{2 \cdot 2 \cdot 3 \cdot 3 \cdot 7 \cdot x^2} \qquad \text{Prime factored form}$$
$$= \sqrt{2^2 \cdot 3^2 \cdot 7 \cdot x^2} \qquad 2^2 \text{ and } 3^2 \text{ are perfect squares.}$$
$$= 2 \cdot 3 \cdot x \cdot \sqrt{7} \qquad x > 0$$
$$= 6x\sqrt{7}$$

(c) $\sqrt[4]{a^7 b^6 c}$

This radical is not in simplest form because some exponents on factors in the radicand are greater than the index.

$$\sqrt[4]{a^7 b^6 c} = \sqrt[4]{a^4 a^3 b^4 b^2 c}$$
$$= \sqrt[4]{a^4} \cdot \sqrt[4]{b^4} \cdot \sqrt[4]{a^3 b^2 c} \qquad \sqrt[n]{ab} = \sqrt[n]{a} \cdot \sqrt[n]{b}$$
$$= ab\sqrt[4]{a^3 b^2 c}$$

(d) $\sqrt[4]{x^2 y^6}$

This radical is not in simplest form because 2 is a factor of the index and all of the exponents in the radicand. To see how to simplify this expression, we will write it using rational exponents.

$$\sqrt[4]{x^2 y^6} = (x^2 y^6)^{1/4} = x^{2/4} y^{6/4} = x^{1/2} y^{3/2} = (xy^3)^{1/2}$$

Thus,

$$\sqrt[4]{x^2 y^6} = \sqrt{xy^3} = y\sqrt{xy}$$

Notice that we could have simplified this radical just by dividing the common factor from the index and the exponents in the radicand.

(e) $\quad \sqrt[3]{(x+y)^7} = \sqrt[3]{[(x+y)^2]^3 (x+y)} \qquad (x+y)^6 = [(x+y)^2]^3$
$\qquad\qquad\qquad = (x+y)^2 \sqrt[3]{(x+y)} \qquad \sqrt[3]{b^3} = b, \text{ where } b = (x+y)^2$

(f) $\sqrt{\dfrac{25}{36}}$

This radical is not in simplest form because the radicand contains a fraction.

$$\sqrt{\dfrac{25}{36}} = \dfrac{\sqrt{25}}{\sqrt{36}} = \dfrac{5}{6}$$

(g) $\sqrt{\dfrac{1}{3}}$

This radical also is not in simplest form because the radicand contains a fraction. In order to remove the fraction, the denominator must be made into a perfect square. This can be accomplished by multiplying both the numerator and the denominator of the radicand by 3:

$$\sqrt{\frac{1}{3}} = \sqrt{\frac{1}{3} \cdot \frac{3}{3}} \qquad \text{Fundamental property of fractions}$$

$$= \sqrt{\frac{3}{3^2}} = \frac{\sqrt{3}}{\sqrt{3^2}} = \frac{\sqrt{3}}{3}$$

(h) $\sqrt[3]{\dfrac{1}{4}}$

This radical is not in simplest form because the radicand contains a fraction.

$$\sqrt[3]{\frac{1}{4}} = \sqrt[3]{\frac{1}{2^2}}$$

To eliminate the fraction from the radical, 2^2 will have to be converted into a perfect cube. We multiply the numerator and denominator of 2^2 by 2:

$$\sqrt[3]{\frac{1}{4}} = \sqrt[3]{\frac{1}{2^2}} = \sqrt[3]{\frac{1}{2^2} \cdot \frac{2}{2}} = \sqrt[3]{\frac{2}{2^3}} = \frac{\sqrt[3]{2}}{\sqrt[3]{2^3}} = \frac{\sqrt[3]{2}}{2}$$

(i) $\sqrt[6]{x^4}$

This radical is not in simplest form because the index, 6, and the exponent, 4, have a common factor of 2.

$$\sqrt[6]{x^4} = (x^4)^{1/6} = x^{4/6} = x^{2/3} = \sqrt[3]{x^2}$$

(j) $\sqrt[3]{0.008} = \sqrt[3]{8 \times 10^{-3}} \qquad \text{Scientific notation}$

$$= \sqrt[3]{8} \cdot \sqrt[3]{10^{-3}}$$

$$= 2 \times 10^{-1} = 0.2$$

> **CAUTION**
>
> Remember that
> $\sqrt{x^2 + y^2} \neq x + y.$
> If the expressions were equal, the following would be a true statement:
> $\sqrt{25} = \sqrt{9 + 16}$
> $= \sqrt{9} + \sqrt{16} = 3 + 4 = 7$
> But we know that $\sqrt{25} = 5$, not 7.

Examples of simplifying expressions with radicals in the denominator will be covered in the Section 8.4.

The equation $y = \sqrt{x^2 + 4}$ defines a function, since each value of x will yield one and only one value of y. What is its domain? To determine this, we need to decide if there are any values of x that cannot be used. Since x^2 is nonnegative for any value of x, the radicand is always nonnegative and its square root will exist. The domain is $(-\infty, \infty)$.

EXAMPLE 7 Let $f(x) = \sqrt{x - 7}$. Find the domain of f and find $f(16)$.

SOLUTION In order for the radical to exist, $x - 7$ must be a nonnegative number. If $x \geq 7$, then $x - 7 \geq 0$. The domain is $[7, \infty)$. Furthermore,

$$f(16) = \sqrt{16 - 7} = \sqrt{9} = 3$$

EXAMPLE 8 Let $g(y) = \sqrt{y^2 + 10y + 25}$. Find the domain of g and find $g(-2)$.

SOLUTION $g(y) = \sqrt{y^2 + 10y + 25} = \sqrt{(y+5)^2}$

Since $(y + 5)^2$ is nonnegative for any value of y, the domain is $(-\infty, \infty)$, and

$$g(-2) = \sqrt{(-2)^2 + 10(-2) + 25} = \sqrt{9} = 3$$

Calculators can be used to find decimal approximations to roots of numbers. The sequence of keys used will vary according to the calculator in use, so consult your owner's manual. The next example illustrates the process on a scientific calculator.

EXAMPLE 9 Use a calculator to find $\sqrt[6]{136}$ to three decimal places.

SOLUTION Most calculators have a key labeled y^x for finding roots of numbers. Since $\sqrt[6]{136} = 136^{1/6}$, we find $\sqrt[6]{136}$ by depressing the following sequence of keys:

$$136 \quad \boxed{y^x} \quad \boxed{(} \quad 1 \quad \boxed{\div} \quad 6 \quad \boxed{)} \quad \boxed{=}$$

The result is

$$2.267722025 \approx 2.2677$$

If your calculator also has a reciprocal key, $\boxed{1/x}$, an alternate order is

$$136 \quad \boxed{y^x} \quad 6 \quad \boxed{1/x} \quad \boxed{=}$$

The result, once again, is

$$2.267722025 \approx 2.2677$$

Do You Remember?

Can you match these?

_____ 1. Radical symbol
_____ 2. Index of the radical $\sqrt{b}$
_____ 3. Radicand of $\sqrt[4]{2a^3}$
_____ 4. Principal square root of 9
_____ 5. Exponential form of $\sqrt[5]{4^3}$
_____ 6. One symbol for the nth root of a
_____ 7. Radical form of $(a^{1/4})^3$

a) 3
b) $4^{5/3}$
c) $4^{3/5}$
d) $2a^3$
e) $\sqrt[n]{a}$
f) $\sqrt{}$
g) 2
h) $\sqrt[3]{a^{1/4}}$
i) $\sqrt[4]{a^3}$

Answers: 1. f 2. g 3. d 4. a 5. c 6. e 7. i

Exercise Set 8.2

Rewrite using radical notation.

1. $6^{1/2}$
2. $3^{1/2}$
3. $11^{1/3}$
4. $17^{1/4}$
5. $x^{2/3}$
6. $y^{3/5}$
7. $p^{-1/3}$
8. $r^{-1/5}$
9. $6t^{-3/7}$
10. $7k^{-2/3}$
11. $(5y)^{1/3}$
12. $(13s)^{2/3}$

Rewrite using rational exponents. Express all answers using positive exponents only.

13. $\sqrt{7}$
14. $\sqrt{15}$
15. $\sqrt[3]{2}$
16. $\sqrt[3]{5}$
17. $\sqrt[3]{x^2}$
18. $\sqrt[5]{k^4}$
19. $\sqrt[7]{y^3}$
20. $\sqrt[7]{s^5}$
21. $\sqrt[3]{5^{-2}}$
22. $\sqrt[3]{3^{-2}}$
23. $\sqrt[3]{5x^2}$
24. $\sqrt[3]{4y^2}$
25. $\sqrt[3]{7^3}$
26. $\sqrt[5]{13^5}$
27. $\sqrt[5]{(3^{-2})(m^3)}$
28. $\sqrt[3]{(7^{-2})(k^2)}$
29. $\sqrt{x+y}$
30. $\sqrt{a-b}$
31. $\sqrt[3]{p^3+q^3}$
32. $\sqrt[3]{x^3-y^3}$

Find the indicated roots. Assume that all variables represent positive real numbers.

33. $\sqrt{4}$
34. $\sqrt{25}$
35. $\sqrt[3]{64}$
36. $\sqrt[3]{27}$
37. $\sqrt{1.44}$
38. $\sqrt{1.69}$
39. $\sqrt[3]{-0.008}$
40. $\sqrt[3]{-0.125}$
41. $\sqrt[4]{16p^4}$
42. $\sqrt[4]{81s^4}$
43. $\sqrt{16t^4}$
44. $\sqrt{36s^6}$
45. $\sqrt[7]{r^6 t^9}$
46. $\sqrt[5]{m^{10} p^{15}}$
47. $\sqrt[5]{3^{-5} x^5 y^{-5}}$
48. $\sqrt[5]{5^{-3} k^6 p^{-6}}$
49. $\sqrt[3]{(x+y)^3}$
50. $\sqrt{(a+b)^2}$
51. $\sqrt{(p+q)^4}$
52. $\sqrt[3]{(m+n)^6}$

Simplify each of the following quantities. Assume that all variables represent positive real numbers and that no denominator is zero.

53. $\sqrt{8}$
54. $\sqrt{32}$
55. $\sqrt{20}$
56. $\sqrt{12}$
57. $\sqrt{\dfrac{27}{16}}$
58. $\sqrt{\dfrac{18}{25}}$
59. $\sqrt{\dfrac{128}{50}}$
60. $\sqrt{\dfrac{243}{12}}$
61. $\sqrt{x^3}$
62. $\sqrt{y^5}$
63. $\sqrt{p^3 q}$
64. $\sqrt{r^5 s}$
65. $\dfrac{1}{2}\sqrt{72}$
66. $\dfrac{1}{3}\sqrt{45}$
67. $\dfrac{1}{5}\sqrt{250}$
68. $\dfrac{5}{4}\sqrt{128}$
69. $\sqrt{\dfrac{225}{400}}$
70. $\sqrt{\dfrac{98}{18}}$
71. $\sqrt{\dfrac{x^3}{y^2}}$
72. $\sqrt{\dfrac{a^5}{b^4}}$
73. $\sqrt{0.04}$
74. $\sqrt{0.09}$
75. $\sqrt{9 \cdot 10^4}$
76. $\sqrt{4 \cdot 10^6}$
77. $\sqrt{63ab^2}$
78. $\sqrt{250xy^3}$
79. $\sqrt[3]{24a^7}$
80. $\sqrt[3]{32n^5}$
81. $\sqrt[4]{32p^9}$
82. $\sqrt[5]{96t^7}$
83. $\sqrt[6]{r^6 s^7 t^{13}}$
84. $\sqrt[4]{x^8 y^7 z^6}$
85. $\sqrt{\dfrac{1}{3^3}}$
86. $\sqrt{\dfrac{1}{2^3}}$
87. $\sqrt[3]{\dfrac{2}{3^5}}$
88. $\sqrt[3]{\dfrac{3}{5^4}}$
89. $\sqrt{\dfrac{x}{12}}$
90. $\sqrt{\dfrac{y}{18}}$
91. $\sqrt[3]{\dfrac{1}{16}}$
92. $\sqrt[3]{\dfrac{1}{36}}$
93. $\sqrt[4]{\dfrac{1}{32}}$
94. $\sqrt[4]{\dfrac{1}{27}}$
95. $\sqrt[3]{\dfrac{2}{100}}$
96. $\sqrt[3]{\dfrac{2}{250}}$
97. $\sqrt{\dfrac{3}{x^3}}$
98. $\sqrt{\dfrac{7}{y^3}}$
99. $\sqrt[3]{\dfrac{x}{a^7}}$
100. $\sqrt[4]{\dfrac{t}{t^6}}$

For each square root function, find the indicated values and the domain.

101. $g(x) = \sqrt{x-4}$; $g(4)$ and $g(13)$
102. $f(x) = \sqrt{3-x}$; $f(2)$ and $f(3)$
103. $f(a) = \sqrt{5+2a}$; $f(2)$ and $f\left(-\dfrac{5}{2}\right)$
104. $h(b) = \sqrt{4b+7}$; $h\left(\dfrac{1}{2}\right)$ and $h\left(-\dfrac{7}{4}\right)$
105. $f(x) = \sqrt{x^2+2x+1}$; $f(0)$ and $f(-1)$
106. $g(y) = \sqrt{y^2+6y+9}$; $g(0)$ and $g(-1)$
107. $h(p) = \sqrt{p^2+4p+4}$; $h(0)$ and $h(-2)$
108. $r(s) = \sqrt{s^2+22s+121}$; $r(0)$ and $r(-11)$

Simplify each of the following radicals. All variables represent real numbers and no denominator is zero. Use absolute value as needed.

109. $\sqrt{x^2}$
110. $\sqrt{v^2}$
111. $\sqrt{25k^2}$
112. $\sqrt{36p^2}$
113. $\sqrt[3]{-108y^5}$
114. $\sqrt[3]{-24w^4}$

115. $\sqrt{(s^2)^{-2}}$ 116. $\sqrt{(t^2)^{-4}}$ 117. $\sqrt{\dfrac{98}{r^2}}$

118. $\sqrt{\dfrac{125}{x^2}}$ 119. $\sqrt{\dfrac{64x}{x^3}}$ 120. $\sqrt{\dfrac{96a^2}{a^4}}$

Use a calculator to evaluate each of the following quantities. Round your answers to three decimal places.

121. $\sqrt[3]{6.283}$ 122. $\sqrt[5]{-8.141}$

123. $\sqrt{16.328}$ 124. $\sqrt{29.079}$

125. $(1.82^{1/2})^{1/3}$ 126. $(2.92^{1/3})^{1/2}$

For Extra Thought

127. The diagonal of a square has length d.
    (a) Find the length of a side s in terms of d.
    (b) Find the area A in terms of d.
    (c) Find s and A if $d = 0.17$ ft.
        Round all answers to two decimal places.

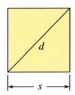

128. The side of an equilateral triangle has length s.
    (a) Find the altitude h in terms of s.
    (b) Find the area A in terms of s.
    (c) Find h and A if $s = 2.2$ in.

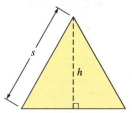

Write In Your Own Words

129. $\sqrt{b}$ is used to indicate the _____ of b.
130. In $\sqrt[n]{b}$, the n is called the _____.
131. Why is $\sqrt{b^2} = |b|$ rather than b? Is $\sqrt{b^2} = b$ true for some real numbers b? For what values of $b \in R$ is $\sqrt{b^2} = b$ false?
132. Does $b^{m/n} = (b^m)^{1/n} = (b^{1/n})^m = \sqrt[n]{b^m}$ for all $b \in R$? Explain.
133. What conditions must be satisfied for a radical expression to be considered to be in **simplest radical form**?
134. State the product rule for radicals in words.
135. State the quotient rule for radicals in words.
136. What is meant by $\sqrt[n]{b} = a$?

8.3 Arithmetic Operations Involving Radicals

HISTORICAL COMMENT   Roots of algebraic polynomials were written in various forms by mathematicians of times past. Before today's notation was commonly accepted, there were many forms in use. Some are given in the following list.

Year	Author	Notation	Modern Interpretation
1202	Fibonacci	radix de 4 et radix de 13	$(\sqrt{4} + \sqrt{13})$
1539	Cardan	℞ V: cu. ℞ 108 p: 10	$\sqrt{\sqrt[3]{108} + 10}$
1572	Bombelli	℞.q.L 20.m.6 p. 1 ⌐	$\sqrt{20 - 6x + x^2}$
1585	Stevin	√ bino 2 + √3	$\sqrt{2 + \sqrt{3}}$
1637	Descartes	$\sqrt{C.\tfrac{1}{2}q + \sqrt{\tfrac{1}{4}qq - \tfrac{1}{27}p^3}}$	$\sqrt[3]{\tfrac{1}{2}q + \sqrt{\tfrac{1}{4}q^2 - \tfrac{1}{27}p^3}}$

8.3 · ARITHMETIC OPERATIONS INVOLVING RADICALS

OBJECTIVES

In this section we will learn to
1. add and subtract radicals; and
2. multiply radicals.

Radicals can be added, subtracted, multiplied, and divided if certain conditions are met. If we are going to add or subtract two or more radicals, they must have the same index and the same radicand. Once this condition is met, we can find the sum or difference by adding or subtracting the coefficients.

EXAMPLE 1 Simplify each of the following.

(a) $\sqrt{6} + 3\sqrt{6} = (1 + 3)\sqrt{6} = 4\sqrt{6}$ Distributive property
(b) $2\sqrt[3]{a} + 4\sqrt[3]{a} - 7\sqrt[3]{a} = (2 + 4 - 7)\sqrt[3]{a} = -\sqrt[3]{a}$
(c) $a\sqrt[7]{b^2} + c\sqrt[7]{b^2} + 3a\sqrt[7]{b^2} = (a + c + 3a)\sqrt[7]{b^2} = (4a + c)\sqrt[7]{b^2}$

When the radicands and indexes are not the same, we can often make them the same by using the properties of radicals.

EXAMPLE 2 Add $\sqrt{32} + \sqrt{50}$.

SOLUTION First we rewrite the two radicals in simplest radical form to see whether the radicands will be the same. If necessary, we can use prime factorization to help us recognize any perfect squares contained in the radicands.

$$\sqrt{32} + \sqrt{50} = \sqrt{2 \cdot 2 \cdot 2 \cdot 2 \cdot 2} + \sqrt{5 \cdot 5 \cdot 2}$$
$$= \sqrt{2^2 \cdot 2^2 \cdot 2} + \sqrt{5^2 \cdot 2}$$
$$= 4\sqrt{2} + 5\sqrt{2} \qquad \sqrt{2^2 \cdot 2^2} = 2 \cdot 2 = 4$$
$$= 9\sqrt{2}$$

To Add Two Radicals
1. If the radicand and index of the radicals are the same, add their coefficients and multiply by the radical.
2. If the radicand and index are *not* the same, use the properties of radicals to simplify each as much as possible, then combine any like terms using Step 1.

In the remainder of this section, we assume all variables represent positive numbers.

EXAMPLE 3 Simplify.

(a) $\sqrt{5} + \sqrt{20} - \sqrt{125} = \sqrt{5} + \sqrt{4 \cdot 5} - \sqrt{25 \cdot 5}$
$= \sqrt{5} + 2\sqrt{5} - 5\sqrt{5}$ Simplify radicands.
$= -2\sqrt{5}$ Combine like terms.

(b) $\sqrt[3]{8x^5} - x\sqrt[3]{27x^2} = \sqrt[3]{2^3 x^3 x^2} - x\sqrt[3]{3^3 x^2}$
$= 2x\sqrt[3]{x^2} - 3x\sqrt[3]{x^2}$ Simplify radicands.
$= -x\sqrt[3]{x^2}$ Combine like terms.

(c) $\dfrac{-3\sqrt{a^3 b^2}}{4} + \dfrac{ab\sqrt{a}}{2} - \dfrac{5ab\sqrt{a}}{3} = \dfrac{-3\sqrt{a^2 \cdot a \cdot b^2}}{4} + \dfrac{ab\sqrt{a}}{2} - \dfrac{5ab\sqrt{a}}{3}$
$= \dfrac{-3ab\sqrt{a}}{4} + \dfrac{ab\sqrt{a}}{2} - \dfrac{5ab\sqrt{a}}{3}$
$= \dfrac{-9ab\sqrt{a} + 6ab\sqrt{a} - 20ab\sqrt{a}}{12}$ The LCD is 12.
$= \dfrac{-23ab\sqrt{a}}{12}$

(d) $\dfrac{-a + \sqrt{c^2 - 4d}}{2} + \dfrac{-a - \sqrt{c^2 - 4d}}{2} = \dfrac{-a + \sqrt{c^2 - 4d} - a - \sqrt{c^2 - 4d}}{2}$
$= \dfrac{-2a}{2} = -a$

(e) $\sqrt{12} + \sqrt{18} = \sqrt{4 \cdot 3} + \sqrt{9 \cdot 2} = 2\sqrt{3} + 3\sqrt{2}$

(f) $\sqrt{2} + \sqrt[3]{2}$ is in simplest form. The terms cannot be combined because the radicals do not have the same index.

To Multiply Two Radicals

To multiply two radicals with the same index, multiply their radicands and retain the index.

$$\sqrt[n]{a}\sqrt[n]{b} = a^{1/n} b^{1/n} = (a \cdot b)^{1/n} = \sqrt[n]{ab}$$

For example,

$$\sqrt{4} \cdot \sqrt{9} = \sqrt{4 \cdot 9} = \sqrt{36} = 6$$

Using this rule, we can multiply or factor expressions involving radicals using the same methods we use for polynomials.

EXAMPLE 4 Find the following products.

(a) $\sqrt{2}\sqrt{8} = \sqrt{2 \cdot 8} = \sqrt{16} = 4$

(b) $3(2 + \sqrt{5}) = 3 \cdot 2 + 3\sqrt{5}$ Distributive property
$= 6 + 3\sqrt{5}$

(c) $\sqrt[3]{5}(\sqrt[3]{3} + \sqrt[3]{2}) = \sqrt[3]{5}\sqrt[3]{3} + \sqrt[3]{5}\sqrt[3]{2}$
$= \sqrt[3]{15} + \sqrt[3]{10}$ $\sqrt[n]{a}\sqrt[n]{b} = \sqrt[n]{ab}$

(d) $(\sqrt{x} + y)(\sqrt{x} - y) = (\sqrt{x})^2 - y^2$ $(a+b)(a-b) = a^2 - b^2$
$= x - y^2$ $(\sqrt{x})^2 = (x^{1/2})^2 = x^{2/2} = x$

(e) $(\sqrt{3} + \sqrt{5})^2 = (\sqrt{3})^2 + 2\sqrt{3}\sqrt{5} + (\sqrt{5})^2$ $(a+b)^2 = a^2 + 2ab + b^2$
$= 3 + 2\sqrt{15} + 5$
$= 8 + 2\sqrt{15}$

(f) $(3\sqrt{2} - \sqrt{3})(\sqrt{6} - \sqrt{2})$
$= 3\sqrt{2}\sqrt{6} - 3(\sqrt{2})^2 - \sqrt{3}\sqrt{6} + \sqrt{3}\sqrt{2}$ FOIL method
$= 3\sqrt{12} - 3 \cdot 2 - \sqrt{18} + \sqrt{6}$
$= 6\sqrt{3} - 6 - 3\sqrt{2} + \sqrt{6}$ $\sqrt{12} = 2\sqrt{3};\ \sqrt{18} = 3\sqrt{2}$

Using the exponential form of a radical is particularly useful when simplifying radicals and finding products of two or more radicals with different indexes.

EXAMPLE 5   Use exponents to carry out the indicated operations.

(a) $\sqrt[4]{9} = \sqrt[4]{3^2} = (3^2)^{1/4} = 3^{2/4} = 3^{1/2} = \sqrt{3}$

(b) $\sqrt{x}\sqrt[3]{x} = x^{1/2}x^{1/3} = x^{(1/2)+(1/3)} = x^{5/6} = \sqrt[6]{x^5}$

(c) $\sqrt[3]{a}(\sqrt[6]{ab} + \sqrt[3]{ab^2}) = a^{1/3}(a^{1/6}b^{1/6} + a^{1/3}b^{2/3})$
$= a^{(1/3)+(1/6)}b^{1/6} + a^{(1/3)+(1/3)}b^{2/3}$
$= a^{3/6}b^{1/6} + a^{2/3}b^{2/3}$
$= (a^3b)^{1/6} + (a^2b^2)^{1/3}$
$= \sqrt[6]{a^3b} + \sqrt[3]{a^2b^2}$

EXAMPLE 6   Simplify $\sqrt[3]{\sqrt{16a^4b^3}}$.

SOLUTION   To simplify this expression, we rewrite it in terms of rational exponents.

$$\sqrt[3]{\sqrt{16a^4b^3}} = [(16a^4b^3)^{1/2}]^{1/3} = (16a^4b^3)^{1/6} = \sqrt[6]{16a^4b^3}$$

EXAMPLE 7   Factor each of the following.

(a) $\sqrt[3]{ab} + 3\sqrt[3]{a}$
(b) $x - y$ as the difference of squares
(c) $x + y$ as the sum of two cubes

SOLUTION   (a) $\sqrt[3]{ab} + 3\sqrt[3]{a} = \sqrt[3]{a}\sqrt[3]{b} + 3\sqrt[3]{a}$
$= \sqrt[3]{a}(\sqrt[3]{b} + 3)$

(b) $x - y = (\sqrt{x})^2 - (\sqrt{y})^2$
$= (\sqrt{x} - \sqrt{y})(\sqrt{x} + \sqrt{y})$ $a^2 - b^2 = (a-b)(a+b)$

(c) Recall that $a^3 + b^3 = (a + b)(a^2 - ab + b^2)$.

$$x + y = (\sqrt[3]{x})^3 + (\sqrt[3]{y})^3$$
$$= (\sqrt[3]{x} + \sqrt[3]{y})[(\sqrt[3]{x})^2 - \sqrt[3]{x}\sqrt[3]{y} + (\sqrt[3]{y})^2]$$
$$= (\sqrt[3]{x} + \sqrt[3]{y})(\sqrt[3]{x^2} - \sqrt[3]{x}\sqrt[3]{y} + \sqrt[3]{y^2}) \qquad (x^{1/3})^2 = (x^2)^{1/3}$$

EXAMPLE 8 Is $\dfrac{-1 + \sqrt{5}}{2}$ a solution to the equation $x^2 + x - 1 = 0$?

SOLUTION

$$\left(\frac{-1 + \sqrt{5}}{2}\right)^2 + \left(\frac{-1 + \sqrt{5}}{2}\right) - 1 \stackrel{?}{=} 0 \qquad \text{Substitute } \frac{-1 + \sqrt{5}}{2} \text{ for } x.$$

$$\frac{1 - 2\sqrt{5} + 5}{4} + \frac{-1 + \sqrt{5}}{2} - 1 \stackrel{?}{=} 0$$

$$\frac{1 - 2\sqrt{5} + 5 - 2 + 2\sqrt{5} - 4}{4} \stackrel{?}{=} 0 \qquad \text{The LCD is 4.}$$

$$\frac{0}{4} \stackrel{\checkmark}{=} 0$$

The equation is satisfied, so $\dfrac{-1 + \sqrt{5}}{2}$ is a solution.

EXAMPLE 9 Write $\sqrt{x\sqrt[3]{x}}$ in simplest radical form. Assume x is positive number.

SOLUTION First we write the expression in exponential form.

$$\sqrt{x\sqrt[3]{x}} = (x \cdot x^{1/3})^{1/2} = x^{1/2}x^{1/6}$$
$$= x^{(1/2)+(1/6)} = x^{4/6}$$
$$= x^{2/3} = \sqrt[3]{x^2}$$

EXAMPLE 10 Write $\sqrt{3xy} \cdot \sqrt[3]{18xy^2}$ as a single radical in simplest form with index 6. Assume x and y represent positive numbers.

SOLUTION Since the indexes are not the same, we must first change the radicals into equivalent expressions with rational exponents.

$$\sqrt{3xy} \cdot \sqrt[3]{18xy^2} = (3xy)^{1/2} \cdot (18xy^2)^{1/3}$$
$$= (3xy)^{3/6} \cdot (18xy^2)^{2/6} \qquad \text{The LCM of } \tfrac{1}{3} \text{ and } \tfrac{1}{2} \text{ is 6.}$$
$$= [(3xy)^3(3^2 \cdot 2xy^2)^2]^{1/6} \qquad 18 = 3^2 \cdot 2$$
$$= [(3^3x^3y^3)(3^4 \cdot 2^2x^2y^4)]^{1/6} \qquad (a \cdot b)^m = a^m \cdot b^m; \ (a^m)^n = a^{mn}$$
$$= (3^7 \cdot 2^2x^5y^7)^{1/6}$$
$$= \sqrt[6]{3^7 \cdot 2^2x^5y^7}$$
$$= \sqrt[6]{3^6 \cdot 3 \cdot 2^2x^5y^6 \cdot y}$$
$$= 3y\sqrt[6]{12x^5y}$$

Exercise Set 8.3

Combine like radicals.

1. $2\sqrt{3} + 5\sqrt{3}$
2. $7\sqrt{2} + 6\sqrt{2}$
3. $\sqrt[3]{5} - 6\sqrt[3]{5}$
4. $11\sqrt[3]{7} - \sqrt[3]{7}$
5. $5x\sqrt{10} - x\sqrt{10}$
6. $8y\sqrt{13} + 2y\sqrt{13}$
7. $\sqrt[4]{3} + 2\sqrt[4]{3} - \sqrt[4]{3}$
8. $\sqrt[4]{2} - 3\sqrt[4]{5} + 3\sqrt[4]{2}$
9. $6x\sqrt{y} - 5x\sqrt{y}$
10. $3m\sqrt{2p} - m\sqrt{p} + \sqrt{2p}$
11. $u\sqrt[3]{3v} - u\sqrt[3]{v} + 2u\sqrt[3]{v}$
12. $7r\sqrt[3]{st} - 5\sqrt[3]{st} - r\sqrt[3]{st}$

Simplify and combine like radicals. Assume all variables represent positive real numbers.

13. $\sqrt{2} + \sqrt{8}$
14. $\sqrt{3} - \sqrt{27}$
15. $3\sqrt[3]{2} + 5\sqrt[3]{16}$
16. $\sqrt[3]{5} - \sqrt[3]{40}$
17. $\sqrt{5} + \sqrt{45} - \sqrt{80} + \sqrt{180}$
18. $\sqrt{3} - \sqrt{12} + 3\sqrt{8} + \sqrt{200}$
19. $\sqrt{75} - \sqrt{125} - 2\sqrt{108} + \sqrt{320}$
20. $\sqrt[5]{-64} + 3\sqrt[5]{2}$
21. $-7\sqrt[4]{162} + 3\sqrt[4]{32}$
22. $\sqrt{27} + \sqrt{288} - \sqrt{48} - \sqrt{162}$
23. $6\sqrt{32} - \sqrt{98} + 3\sqrt{96} - 2\sqrt{54}$
24. $\sqrt{7x^3} + 2x\sqrt{7x}$
25. $-\sqrt{27u^3} + 3u\sqrt{3u}$
26. $-5p\sqrt[3]{8p^5} + 3p^2\sqrt[3]{27p^2}$
27. $-v\sqrt[4]{2v^5} + v^2\sqrt[4]{32v}$
28. $\sqrt{\dfrac{7}{4}} + \dfrac{\sqrt{7}}{2}$
29. $\sqrt{\dfrac{5}{9}} - \dfrac{-2\sqrt{5}}{3}$
30. $7\sqrt{98} - (\sqrt{32} - 2\sqrt{18})$
31. $\sqrt{175} - (\sqrt{28} - \sqrt{63})$
32. $\dfrac{\sqrt{36a^2b^4c}}{2} - \dfrac{b\sqrt{16a^2b^2c}}{3}$
33. $\dfrac{y\sqrt{18xyz}}{5} + \dfrac{\sqrt{32xy^3z}}{4}$
34. $7\sqrt{300a^3} - 2a\sqrt{75a}$
35. $6\sqrt{600u^5} - 2u^2\sqrt{150u}$
36. $-7c\sqrt[3]{375cd^5} + 6d\sqrt[3]{81c^4d^2}$
37. $-5z\sqrt[3]{81y^5z^2} + 4y\sqrt[3]{24y^2z^5}$
38. $-5a\sqrt{75b^3} + \sqrt{18a^2b} + 7b\sqrt{108a^7b} - u\sqrt{50b}$
39. $3\sqrt{24x^3y} - y\sqrt{27x} + 11x\sqrt{54x^2y} - y\sqrt{48x}$
40. $\dfrac{-b + \sqrt{b^2 - 4ac}}{2a} + \dfrac{-b - \sqrt{b^2 - 4ac}}{2a}$
41. $\dfrac{c - \sqrt{c^2 - 4ad}}{2a} + \dfrac{c + \sqrt{c^2 - 4ad}}{2a}$

Simplify.

42. $\dfrac{6 + \sqrt{12}}{2}$
43. $\dfrac{14 + \sqrt{500}}{2}$
44. $\dfrac{30 - 15\sqrt{2}}{5}$
45. $\dfrac{18 - 21\sqrt{7}}{3}$
46. $\dfrac{-16 + 18\sqrt{3}}{-2}$
47. $\dfrac{-91 + 63\sqrt{11}}{-7}$

Multiply and write each expression in simplest radical form. Assume all variables represent positive real numbers.

48. $2(3 + \sqrt{5})$
49. $3(5 - \sqrt{6})$
50. $\sqrt{2}(7 - \sqrt{2})$
51. $\sqrt{11}(3 - \sqrt{11})$
52. $3\sqrt{2}(\sqrt{3} - 2)$
53. $6\sqrt{3}(2\sqrt{2} - \sqrt{3})$
54. $\sqrt{x}(\sqrt{x} + \sqrt{y})$
55. $\sqrt{u}(\sqrt{v} - \sqrt{u})$
56. $\sqrt{ab}(\sqrt{a} - \sqrt{b})$
57. $(\sqrt{2} + \sqrt{3})(\sqrt{2} - \sqrt{3})$
58. $(\sqrt{5} - \sqrt{2})(\sqrt{5} + \sqrt{2})$
59. $\sqrt[3]{6} \cdot \sqrt[3]{36}$
60. $3\sqrt[3]{9} \cdot \sqrt[3]{9}$
61. $2\sqrt[3]{98x} \cdot \sqrt[3]{14x} \cdot \sqrt[3]{2x}$
62. $5\sqrt[4]{8a^3} \cdot 2\sqrt[4]{2a^2} \cdot \sqrt[4]{2^4a^3}$
63. $(\sqrt{2} + \sqrt{5})^2$
64. $(\sqrt{7} - \sqrt{3})^2$
65. $(\sqrt{a+2} - 1)^2$
66. $(\sqrt{c-3} + 1)^2$
67. $(3\sqrt{5} - 2\sqrt{2})(3\sqrt{5} + 2\sqrt{2})$
68. $(7\sqrt{3} - 2\sqrt{6})(7\sqrt{3} + 2\sqrt{6})$
69. $\dfrac{-b + \sqrt{b^2 - 4ac}}{2a} \cdot \dfrac{-b - \sqrt{b^2 - 4ac}}{2a}$
70. $\dfrac{c - \sqrt{c^2 - 4ad}}{2a} \cdot \dfrac{c + \sqrt{c^2 - 4ad}}{2a}$

Factor each of the following as indicated.

71. $3\sqrt{5} + 3 = 3(\quad)$
72. $14\sqrt{2} - 7 = 7(\quad)$
73. $3\sqrt{5} - x\sqrt{5} = \sqrt{5}(\quad)$
74. $2\sqrt{12} + \sqrt{3} = \sqrt{3}(\quad)$
75. $\sqrt[3]{xy} - \sqrt[3]{x} = \sqrt[3]{x}(\quad)$
76. $3\sqrt[3]{81} - 10\sqrt[3]{6} = \sqrt[3]{3}(\quad)$
77. $\sqrt[5]{xy} - \sqrt[5]{xz} = \sqrt[5]{x}(\quad)$
78. $\sqrt[5]{ab} - \sqrt[5]{ac} = \sqrt[5]{a}(\quad)$

Rewrite each expression in simplest radical form. Assume all variables represent positive real numbers.

79. $\sqrt[4]{\sqrt{64}}$
80. $\sqrt[3]{\sqrt{32}}$
81. $\sqrt[4]{x^2} \cdot \sqrt[3]{x^4}$
82. $\sqrt[3]{2x^2} \cdot \sqrt[4]{4x^3}$
83. $\sqrt[5]{ab^2} \cdot \sqrt{ab}$
84. $\sqrt[3]{-2\sqrt{4a^8}}$
85. $\sqrt[6]{a^3}$
86. $\sqrt[9]{x^3 y^6}$
87. $\sqrt[3]{\sqrt[5]{a^{10}}}$
88. $\sqrt[3]{9ab} \cdot \sqrt[4]{4a^3 b}$
89. $\sqrt{2} \cdot \sqrt[3]{3}$
90. $4^2 \cdot \sqrt[3]{4}$
91. $\sqrt{3x^2} \cdot \sqrt[3]{3x^5}$
92. $\sqrt{4b^3} \cdot \sqrt[3]{4b^6}$

Determine whether the given value is a solution to the equation.

93. $x^2 + x - 4 = 0$; $\dfrac{-1 - \sqrt{17}}{2}$
94. $y^2 - 2y - 4 = 0$; $1 + \sqrt{5}$
95. $3a^2 - a - 5 = 0$; $\dfrac{1 - \sqrt{61}}{3}$
96. $5v^2 - 2v - 1 = 0$; $\dfrac{1 - \sqrt{6}}{2}$
97. $p^2 + 9p + 1 = 0$; $\dfrac{-9 + \sqrt{77}}{2}$
98. $c^2 + 7c + 9 = 0$; $\dfrac{-7 + \sqrt{13}}{2}$
99. $5x^2 + 7x - 3 = 0$; $\dfrac{-7 - \sqrt{108}}{12}$
100. $8y^2 + 3y - 11 = 0$; $\dfrac{-3 - \sqrt{359}}{16}$

For Extra Thought

101. Show that the area A of a regular polygon with n sides is equal to one-half the product of the **apothem,** a, (see figure) and the perimeter, P. That is, $A = \frac{1}{2}aP$.

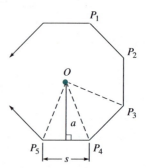

102. Use the formula for the area of a regular polygon from Exercise 101 to find the area of the regular hexagonal roof shown in the figure. Round your answer to four decimal places.

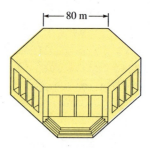

8.4 Division by Radicals; Rationalizing

OBJECTIVES

In this section we will learn to
1. divide by radicals; and
2. simplify expressions with radicals in the denominator by rationalizing the denominator.

In Section 8.2 we saw four conditions that had to be met for an expression involving radicals to be in simplest form. We repeat them here for review, since we will be considering the fourth condition in detail.

Simplest Forms for a Radical

For a radical to be in simplest form, the following conditions must be satisfied:

1. All exponents on factors in the radicand must be less than the index.
2. The radicand cannot contain a fraction.
3. The index and all of the exponents in the radicand cannot contain a common factor (assuming all variables in the radicand represent positive numbers).
4. No fraction can have a radical in the denominator.

The fourth condition is another way of considering division by expressions involving radicals. Recall that

$$\sqrt[n]{\frac{a}{b}} = \frac{\sqrt[n]{a}}{\sqrt[n]{b}}$$

This property can often be used to eliminate a radical from the denominator. *In this section we assume all variables represent positive real numbers.*

EXAMPLE 1 Simplify by eliminating the radical in the denominator.

(a) $\dfrac{6\sqrt[3]{x^7}}{3\sqrt[3]{x^4}} = 2\sqrt[3]{\dfrac{x^7}{x^4}} = 2\sqrt[3]{x^3} = 2x \qquad \dfrac{\sqrt[n]{a}}{\sqrt[n]{b}} = \sqrt[n]{\dfrac{a}{b}}$

(b) $\dfrac{4\sqrt{x}}{8\sqrt{x^3}} = \dfrac{1}{2}\sqrt{\dfrac{x}{x^3}} = \dfrac{1}{2}\sqrt{\dfrac{1}{x^2}} = \dfrac{1}{2x}$

(c) $\dfrac{\sqrt[5]{64x^7y^6}}{\sqrt[5]{2x^2y}} = \sqrt[5]{\dfrac{64x^7y^6}{2x^2y}} = \sqrt[5]{32x^5y^5} = \sqrt[5]{2^5x^5y^5} = 2xy$

In Example 1 we used the second property of radicals to eliminate the radical from the denominator. We can also use the fundamental property of fractions to do this. When we change the denominator of a fraction from an irrational number to a rational number, this process is called **rationalizing the denominator.**

EXAMPLE 2 Simplify $\dfrac{3}{\sqrt{5}}$.

SOLUTION This expression is not in simplest form because it has a radical in the denominator. To eliminate the radical, we must change the irrational number, $\sqrt{5}$, to a rational number. This can be accomplished by multiplying both the numerator and the denominator by $\sqrt{5}$. Notice that

$$\dfrac{\sqrt{5}}{\sqrt{5}} = 1$$

Multiplying an expression by 1 does not change its value.

Irrational → $\dfrac{3}{\sqrt{5}} = \dfrac{3}{\sqrt{5}} \cdot \dfrac{\sqrt{5}}{\sqrt{5}}$ Fundamental property

$= \dfrac{3\sqrt{5}}{\sqrt{5^2}}$ $\sqrt{a} \cdot \sqrt{b} = \sqrt{ab}$

$= \dfrac{3\sqrt{5}}{5}$ ← Rational

EXAMPLE 3 Simplify: $\dfrac{\sqrt{2}}{\sqrt{7}}$

SOLUTION $\dfrac{\sqrt{2}}{\sqrt{7}} = \dfrac{\sqrt{2}}{\sqrt{7}} \cdot \dfrac{\sqrt{7}}{\sqrt{7}} = \dfrac{\sqrt{14}}{7}$

To eliminate a radical from the denominator of a fraction, if necessary we should first simplify it, then multiply the numerator and the denominator by an expression that will make the exponents on the radicand of the denominator equal to the index.

EXAMPLE 4 Simplify: **(a)** $\dfrac{2}{\sqrt[3]{4}}$ **(b)** $\dfrac{2xy}{\sqrt[3]{y}}$ **(c)** $\dfrac{6}{\sqrt[4]{a^5 b^3}}$

SOLUTION **(a)** $\dfrac{2}{\sqrt[3]{4}} = \dfrac{2}{\sqrt[3]{2^2}}$

In order to remove a cube root from the denominator, the radicand must be converted to a perfect cube. The denominator $\sqrt[3]{2^2}$ contains two factors of 2, so we need one additional factor of 2 to make the radicand a perfect cube. We multiply the numerator and the denominator by $\sqrt[3]{2}$.

$\dfrac{2}{\sqrt[3]{2^2}} = \dfrac{2}{\sqrt[3]{2^2}} \cdot \dfrac{\sqrt[3]{2}}{\sqrt[3]{2}}$

$= \dfrac{2\sqrt[3]{2}}{\sqrt[3]{2^3}} = \dfrac{2\sqrt[3]{2}}{2} = \sqrt[3]{2}$

(b) The denominator $\sqrt[3]{y}$ needs two more factors of y to make the radicand y^3, so we multiply the numerator and the denominator by $\sqrt[3]{y^2}$.

$\dfrac{2xy}{\sqrt[3]{y}} = \dfrac{2xy}{\sqrt[3]{y}} \cdot \dfrac{\sqrt[3]{y^2}}{\sqrt[3]{y^2}}$ $\sqrt[3]{y}\,\sqrt[3]{y^2} = \sqrt[3]{y^3}$; the index and exponent are equal.

$= \dfrac{2xy\sqrt[3]{y^2}}{\sqrt[3]{y^3}} = \dfrac{2xy\sqrt[3]{y^2}}{y}$ $\sqrt[3]{y^3} = y$

$= 2x\sqrt[3]{y^2}$ Simplify.

(c) $\dfrac{6}{\sqrt[4]{a^5b^3}} = \dfrac{6}{a\sqrt[4]{ab^3}}$ Simplify first.

$= \dfrac{6}{a\sqrt[4]{ab^3}} \cdot \dfrac{\sqrt[4]{a^3b}}{\sqrt[4]{a^3b}}$

$= \dfrac{6\sqrt[4]{a^3b}}{a\sqrt[4]{a^4b^4}}$ The exponent and index are equal.

$= \dfrac{6\sqrt[4]{a^3b}}{a^2b}$

EXAMPLE 5 Simplify.

(a) $\dfrac{\sqrt{x-y}}{\sqrt{x+y}} = \dfrac{\sqrt{x-y}}{\sqrt{x+y}} \cdot \dfrac{\sqrt{x+y}}{\sqrt{x+y}}$

$= \dfrac{\sqrt{x^2-y^2}}{x+y}$ $\sqrt{x+y} \cdot \sqrt{x+y} = \sqrt{(x+y)^2} = x+y$

(b) $\sqrt[4]{\dfrac{7}{x}} = \sqrt[4]{\dfrac{7}{x} \cdot \dfrac{x^3}{x^3}} = \sqrt[4]{\dfrac{7x^3}{x^4}} = \dfrac{\sqrt[4]{7x^3}}{x}$

(c) $\dfrac{\sqrt{3}}{\sqrt[4]{9}} = \dfrac{\sqrt{3}}{\sqrt[4]{3^2}} = \dfrac{\sqrt{3}}{\sqrt{3}} = 1$ $\sqrt[4]{3^2} = 3^{2/4} = 3^{1/2} = \sqrt{3}$

(d) $\sqrt{3} - \dfrac{2}{\sqrt{3}} = \dfrac{\sqrt{3} \cdot \sqrt{3} - 2}{\sqrt{3}}$ The LCD is $\sqrt{3}$.

$= \dfrac{3-2}{\sqrt{3}} = \dfrac{1}{\sqrt{3}}$

$= \dfrac{1}{\sqrt{3}} \cdot \dfrac{\sqrt{3}}{\sqrt{3}} = \dfrac{\sqrt{3}}{3}$

(e) $\dfrac{\sqrt{2}}{\sqrt{a}} - \dfrac{\sqrt{3}}{\sqrt{2}} + \sqrt{a} = \dfrac{\sqrt{2}\sqrt{2} - \sqrt{3}\sqrt{a} + \sqrt{a}\sqrt{2a}}{\sqrt{2a}}$ The LCD is $\sqrt{2a}$.

$= \dfrac{2 - \sqrt{3a} + a\sqrt{2}}{\sqrt{2a}}$

$= \dfrac{2 - \sqrt{3a} + a\sqrt{2}}{\sqrt{2a}} \cdot \dfrac{\sqrt{2a}}{\sqrt{2a}}$

$= \dfrac{2\sqrt{2a} - a\sqrt{6} + 2a\sqrt{a}}{2a}$

When the denominator of a fraction contains a sum or a difference, more thought is required to choose the expression we need to rationalize it. Recall that

$(a-b)(a+b) = a^2 - b^2$

If $\sqrt{x} + \sqrt{y}$ is multiplied by $\sqrt{x} - \sqrt{y}$, the product is $(\sqrt{x})^2 - (\sqrt{y})^2 = x - y$. So if the irrational number $1 + \sqrt{7}$ is multiplied by $1 - \sqrt{7}$, the product is $1^2 - (\sqrt{7})^2 = 1 - 7 = -6$, a rational number. The expressions $1 - \sqrt{7}$ and $1 + \sqrt{7}$ are said to be **conjugates** of each other.

> **DEFINITION**
>
> **Conjugate of a Binomial**
>
> The **conjugate of a binomial** that contains square roots is a binomial with the same terms but with the opposite sign between them:
>
> $\sqrt{a} + \sqrt{b}$ and $\sqrt{a} - \sqrt{b}$ are conjugates.

EXAMPLE 6 Simplify.

(a) $\dfrac{2}{\sqrt{5} - \sqrt{2}} = \dfrac{2}{\sqrt{5} - \sqrt{2}} \cdot \dfrac{\sqrt{5} + \sqrt{2}}{\sqrt{5} + \sqrt{2}}$ Multiply numerator and denominator by the conjugate, $\sqrt{5} + \sqrt{2}$.

$= \dfrac{2(\sqrt{5} + \sqrt{2})}{5 - 2} = \dfrac{2\sqrt{5} + 2\sqrt{2}}{3}$

(b) $\dfrac{\sqrt{3} + \sqrt{6}}{\sqrt{2} + \sqrt{3}} = \dfrac{\sqrt{3} + \sqrt{6}}{\sqrt{2} + \sqrt{3}} \cdot \dfrac{\sqrt{2} - \sqrt{3}}{\sqrt{2} - \sqrt{3}}$

$= \dfrac{\sqrt{3}\sqrt{2} - \sqrt{3}\sqrt{3} + \sqrt{6}\sqrt{2} - \sqrt{6}\sqrt{3}}{2 - 3}$ FOIL method

$= \dfrac{\sqrt{6} - 3 + \sqrt{12} - \sqrt{18}}{-1}$

$= \dfrac{\sqrt{6} - 3 + 2\sqrt{3} - 3\sqrt{2}}{-1}$

$= -(\sqrt{6} - 3 + 2\sqrt{3} - 3\sqrt{2})$

$= 3 + 3\sqrt{2} - 2\sqrt{3} - \sqrt{6}$

EXAMPLE 7 Simplify: $\dfrac{\sqrt[3]{x^2 y}\sqrt{xy^3}}{\sqrt{x^2 y}}$

SOLUTION One way to simplify an expression of this form is to rewrite each radicand using rational exponents.

$\dfrac{\sqrt[3]{x^2 y}\sqrt{xy^3}}{\sqrt{x^2 y}} = \dfrac{(x^2 y)^{1/3}(xy^3)^{1/2}}{(x^2 y)^{1/2}}$

$= \dfrac{(x^{2/3} y^{1/3})(x^{1/2} y^{3/2})}{xy^{1/2}}$

Now we simplify this using the rules for exponents.

$\dfrac{(x^{2/3} y^{1/3})(x^{1/2} y^{3/2})}{xy^{1/2}} = x^{(2/3)+(1/2)-1} \cdot y^{(1/3)+(3/2)-(1/2)}$

$= x^{1/6} y^{8/6}$ The LCD of the exponents of 6.

$= (xy^8)^{1/6}$

$= \sqrt[6]{xy^8}$

$= \sqrt[6]{x \cdot y^6 \cdot y^2}$

$= y\sqrt[6]{xy^2}$

EXAMPLE 8 Simplify $\dfrac{\sqrt[3]{x} + \sqrt[5]{x^3}}{\sqrt[5]{x^2}}$. Write the result in radical form.

SOLUTION Since this is a sum, the method shown in Example 7 can't be used. To make the radicand in the denominator a perfect fifth power, we multiply both numerator and denominator by $\sqrt[5]{x^3}$.

$$\dfrac{\sqrt[3]{x} + \sqrt[5]{x^3}}{\sqrt[5]{x^2}} = \dfrac{\sqrt[3]{x} + \sqrt[5]{x^3}}{\sqrt[5]{x^2}} \cdot \dfrac{\sqrt[5]{x^3}}{\sqrt[5]{x^3}}$$

$$= \dfrac{x^{1/3} + x^{3/5}}{x^{2/5}} \cdot \dfrac{x^{3/5}}{x^{3/5}} \qquad \text{Exponential form}$$

$$= \dfrac{x^{(1/3)+(3/5)} + x^{(3/5)+(3/5)}}{x^{(2/5)+(3/5)}}$$

$$= \dfrac{x^{14/15} + x^{6/5}}{x}$$

$$= \dfrac{\sqrt[15]{x^{14}} + \sqrt[5]{x^5 \cdot x}}{x}$$

$$= \dfrac{\sqrt[15]{x^{14}} + x\sqrt[5]{x}}{x}$$

In more advanced mathematics classes, rationalizing the numerator of a fraction is a technique that often aids in solving certain types of problems.

EXAMPLE 9 Rationalize the numerator of $\dfrac{2\sqrt{x} - 3\sqrt{y}}{2\sqrt{x} + 3\sqrt{y}}$.

SOLUTION To rationalize the numerator, we multiply both the numerator and the denominator by $2\sqrt{x} + 3\sqrt{y}$:

$$\dfrac{2\sqrt{x} - 3\sqrt{y}}{2\sqrt{x} + 3\sqrt{y}} \cdot \dfrac{2\sqrt{x} + 3\sqrt{y}}{2\sqrt{x} + 3\sqrt{y}}$$

$$= \dfrac{(2\sqrt{x})^2 - (3\sqrt{y})^2}{(2\sqrt{x})^2 + 2(2\sqrt{x})(3\sqrt{y}) + (3\sqrt{y})^2} \qquad \begin{array}{l}(a-b)(a+b) = a^2 - b^2; \\ (a+b)^2 = a^2 + 2ab + b^2\end{array}$$

$$= \dfrac{4x - 9y}{4x + 12\sqrt{xy} + 9y}$$

The next examples illustrate how to work with functions involving radicals.

EXAMPLE 10 Let $f(x) = 3\sqrt{x} + 2$ and $g(x) = 5\sqrt{x} - 2$. Find **(a)** $(f + g)(x)$ and **(b)** $\left(\dfrac{f}{g}\right)(x)$.

SOLUTION **(a)** $(f + g)(x) = 3\sqrt{x} + 2 + 5\sqrt{x} - 2$
$\qquad\qquad\qquad\;\; = 8\sqrt{x}$

(b) $\left(\dfrac{f}{g}\right)(x) = \dfrac{3\sqrt{x}+2}{5\sqrt{x}-2}$

$= \dfrac{3\sqrt{x}+2}{5\sqrt{x}-2} \cdot \dfrac{5\sqrt{x}+2}{5\sqrt{x}+2}$ Rationalize the denominator.

$= \dfrac{15(\sqrt{x})^2 + 6\sqrt{x} + 10\sqrt{x} + 4}{25(\sqrt{x})^2 - 4}$ FOIL method

$= \dfrac{15x + 16\sqrt{x} + 4}{25x - 4}$

EXAMPLE 11   Let $f(a) = \sqrt{a}$ and $g(a) = 2 - \dfrac{1}{\sqrt{a}}$. Find $(f-g)(a)$.

SOLUTION $(f - g)(a) = \sqrt{a} - \left(2 - \dfrac{1}{\sqrt{a}}\right)$

$= \sqrt{a} - \left(2 - \dfrac{1}{\sqrt{a}} \cdot \dfrac{\sqrt{a}}{\sqrt{a}}\right)$ Rationalize the denominator.

$= \sqrt{a} - 2 + \dfrac{\sqrt{a}}{a}$

$= \dfrac{a\sqrt{a} - 2a + \sqrt{a}}{a}$ The LCD is a.

Exercise Set 8.4

Rationalize the denominator and simplify the result if possible. Assume that all variables represent positive real numbers.

1. $\dfrac{1}{\sqrt{2}}$

2. $\dfrac{1}{\sqrt{3}}$

3. $\dfrac{6}{\sqrt{3}}$

4. $\dfrac{4}{\sqrt{2}}$

5. $\dfrac{15}{\sqrt{5}}$

6. $\dfrac{48}{\sqrt{6}}$

7. $\dfrac{-28}{\sqrt{7}}$

8. $\dfrac{-44}{\sqrt{11}}$

9. $\sqrt{\dfrac{1}{3}}$

10. $\sqrt{\dfrac{3}{7}}$

11. $-\sqrt{\dfrac{5}{6}}$

12. $-\sqrt{\dfrac{8}{13}}$

13. $\dfrac{3\sqrt[3]{5}}{\sqrt[3]{3}}$

14. $\dfrac{10\sqrt[3]{2}}{\sqrt[3]{5}}$

15. $\dfrac{-6\sqrt[3]{7}}{5\sqrt[3]{6}}$

16. $\dfrac{-11\sqrt[3]{5}}{10\sqrt[3]{3}}$

17. $\dfrac{x}{\sqrt[4]{x^2}}$

18. $\dfrac{y}{\sqrt[4]{y^2}}$

19. $\dfrac{2a}{\sqrt[3]{a}}$

20. $\dfrac{4p}{\sqrt[3]{2p}}$

21. $\dfrac{x+y}{\sqrt{x+y}}$

22. $\dfrac{s-t}{\sqrt{s-t}}$

23. $\sqrt[3]{\dfrac{3}{r}}$

24. $\sqrt[3]{\dfrac{a}{b}}$

25. $\sqrt[4]{\dfrac{m}{n}}$

26. $\sqrt[4]{\dfrac{5}{k}}$

27. $\dfrac{u}{\sqrt[5]{u}}$

28. $\dfrac{wz}{\sqrt[5]{wz}}$

Given the indicated functions, find their sum, difference or quotient, as specified. Assume that all variables represent positive real numbers.

29. $f(x) = x\sqrt[3]{2}$ and $g(x) = x\sqrt[3]{4}$; $\left(\dfrac{f}{g}\right)(x)$

30. $f(x) = x\sqrt[3]{5}$ and $g(x) = x\sqrt[3]{25}$; $\left(\dfrac{f}{g}\right)(x)$

31. $f(x) = \sqrt[4]{3x}$ and $g(x) = \sqrt[4]{27x^3}$; $\left(\dfrac{f}{g}\right)(x)$

32. $f(a) = \sqrt[4]{2a^2}$ and $g(a) = \sqrt[4]{32a^5}$; $\left(\dfrac{f}{g}\right)(a)$

33. $h(w) = 16\sqrt[5]{w^3}$ and $r(w) = \sqrt[5]{2w^6}$; $\left(\dfrac{h}{r}\right)(w)$

34. $f(k) = 27\sqrt[5]{k^4}$ and $g(k) = \sqrt[5]{27k^7}$; $\left(\dfrac{f}{g}\right)(k)$

35. $f(x) = 3x\sqrt{3}$ and $g(x) = 4x\sqrt{3}$; $(f + g)(x)$

36. $f(x) = \dfrac{3x\sqrt{2}}{2}$ and $g(x) = x\sqrt{2}$; $(f - g)(x)$

37. $f(a) = a\sqrt{5}$ and $g(a) = \dfrac{a}{\sqrt{5}}$; $(f - g)(a)$

38. $f(a) = \sqrt{a}$ and $g(a) = \dfrac{a}{\sqrt{a}}$; $(f - g)(a)$

39. $h(p) = \sqrt{3p}$ and $g(p) = \dfrac{p}{\sqrt{3p}}$; $(h + g)(p)$

40. $r(u) = \sqrt{5u}$ and $s(u) = \dfrac{u}{\sqrt{5u}}$; $(r - s)(u)$

Perform the indicated operations and simplify the result if possible. Assume that all variables represent positive real numbers.

41. $\dfrac{4}{\sqrt{2}} + 2 - \dfrac{3}{\sqrt{3}} - \dfrac{5}{\sqrt{3}}$

42. $\dfrac{3a}{\sqrt{a}} + \dfrac{5}{\sqrt{2}} - 7\sqrt{2} + \dfrac{2a}{\sqrt{a}}$

43. $\dfrac{6w}{\sqrt{2w}} - \dfrac{9\sqrt{3}}{\sqrt{2}} + \dfrac{3w}{5\sqrt{2w}} - \dfrac{3\sqrt{6}}{2}$

44. $\dfrac{7\sqrt{5}}{\sqrt{3}} - \dfrac{2\sqrt{xy}}{\sqrt{x}} + \dfrac{6\sqrt{15}}{2} + \dfrac{3y}{\sqrt{y}}$

45. $\dfrac{x}{\sqrt{x^2+1}} - \dfrac{\sqrt{x^2+1}}{x}$

46. $\dfrac{3a}{\sqrt{a^2-1}} + \dfrac{\sqrt{a^2-1}}{2a}$, $a \ne 1$

Find the quotient and simplify where possible. Assume that all variables represent positive real numbers.

47. $\dfrac{15\sqrt{24} - 6\sqrt{12}}{3\sqrt{3}}$

48. $\dfrac{20\sqrt{10} - 45\sqrt{6}}{6\sqrt{2}}$

49. $\dfrac{6a\sqrt{5} + 8\sqrt{2a^2}}{a\sqrt{2}}$

50. $\dfrac{40t\sqrt{7} - 45\sqrt{15t^2}}{t\sqrt{3}}$

51. $\dfrac{3 + \sqrt{x}}{\sqrt{x}}$

52. $\dfrac{5 - \sqrt{p}}{\sqrt{2p}}$

53. $\dfrac{\sqrt{5w} + \sqrt{6z}}{\sqrt{30wz}}$

54. $\dfrac{\sqrt{7r} - \sqrt{3s}}{\sqrt{21rs}}$

Rationalize the denominators and simplify the result where possible. Assume that all variables represent positive real numbers.

55. $\dfrac{\sqrt{3}}{4 - \sqrt{7}}$

56. $\dfrac{\sqrt{2}}{4 + \sqrt{6}}$

57. $\dfrac{-1}{4\sqrt{3} - \sqrt{5}}$

58. $\dfrac{-1}{2\sqrt{3} - 5\sqrt{2}}$

59. $\dfrac{-2\sqrt{3}}{4\sqrt{3} - 7}$

60. $\dfrac{6\sqrt{5}}{2\sqrt{6} - 5}$

61. $\dfrac{3 - \sqrt{2}}{3 + \sqrt{2}}$

62. $\dfrac{4 + \sqrt{5}}{4 - \sqrt{5}}$

63. $\dfrac{-(3 + \sqrt{5})}{7 - 5\sqrt{2}}$

64. $\dfrac{-(5 + \sqrt{3})}{2\sqrt{5} - 4\sqrt{3}}$

65. $\dfrac{\sqrt{6} - \sqrt{3}}{\sqrt{6} + 2\sqrt{3}}$

66. $\dfrac{\sqrt{2} - \sqrt{6}}{3\sqrt{2} + \sqrt{6}}$

67. $\dfrac{x + y}{\sqrt{x} + \sqrt{y}}$

68. $\dfrac{a - b}{\sqrt{a} - \sqrt{b}}$

69. $\dfrac{3\sqrt{u} - \sqrt{v}}{2\sqrt{u} + \sqrt{v}}$

70. $\dfrac{6\sqrt{k} - 5\sqrt{k}}{3\sqrt{k} + 5\sqrt{k}}$

71. $\dfrac{9\sqrt{xy} - \sqrt{ab}}{9\sqrt{xy} + \sqrt{ab}}$

72. $\dfrac{6\sqrt{2r} - \sqrt{5s}}{\sqrt{2r} + \sqrt{5s}}$

Rationalize each denominator and simplify the result where possible.

73. $\dfrac{a^{1/2} - a^{2/3}}{a^{1/3}}$, $a \ne 0$

74. $\dfrac{x^{3/5} - x^{1/3}}{x^{2/3}}$, $x \ne 0$

75. $\dfrac{x^{1/3}y^{1/3} - x^{1/5}y^{1/5}}{(xy)^{2/5}}$, $x, y \ne 0$

76. $\dfrac{r^{1/7}s^{1/3} - r^{1/5}s^{3/5}}{r^{3/5}s^{2/7}}$, $r, s \ne 0$

77. $\dfrac{(x-y)^{2/3} + (x-y)^{4/3}}{(x-y)^{2/3}}, \quad x \neq y$

78. $\dfrac{(w^2-z^2)^{1/3} - (w^2-z^2)^{5/3}}{(w^2-z^2)^{1/3}}, \quad w \neq \pm z$

Simplify. Assume that all variables represent positive real numbers.

79. $\dfrac{\sqrt{3}}{\sqrt[3]{4}}$

80. $\dfrac{\sqrt{2}}{\sqrt[3]{9}}$

81. $\dfrac{\sqrt[3]{x}}{\sqrt[4]{x}}$

82. $\dfrac{\sqrt{y}}{\sqrt[3]{y}}$

83. $\dfrac{\sqrt[4]{3y^2}}{\sqrt{9y}}$

84. $\dfrac{\sqrt{3^2 a^3 b}}{\sqrt[3]{3^8 ab}}$

85. $\dfrac{\sqrt[3]{x^2 y} \cdot \sqrt[3]{270 x^4 y^2}}{\sqrt[3]{2x^6 y^2}}$

86. $\dfrac{\sqrt[4]{a^3 b^2} \cdot \sqrt[4]{32 a^5 b^3}}{\sqrt[4]{2a^{-4} b^{-3}}}$

REVIEW EXERCISES

Solve each equation by factoring.

87. $a^2 - 4a - 5 = 0$

88. $5t^2 + 4t - 1 = 0$

89. $4w^2 + 12w + 9 = 0$

90. $y^2 - 2y - 3 = 12$

91. $5k^2 - 10k - 15 = 0$

92. $3m^3 - 3m = 0$

93. $\dfrac{1}{2}t^2 + \dfrac{1}{2}t - 3 = 0$

94. $a^2 + \dfrac{1}{3}a - \dfrac{1}{2} = \dfrac{1}{2}a - \dfrac{1}{3}$

95. $3(4 - 2s)^2 + (s + 2)^2 = 37$

96. $\left(\dfrac{1}{2}p - 1\right)^2 + 4 = 4p - 11$

8.5 Solving Equations Involving Radicals

OBJECTIVE In this section we will learn how to solve equations involving radicals.

If an equation involves a radical, it can often be solved by using the **power property of equality.**

Power Property of Equality

If A and B are two algebraic expressions with $A = B$ and if n is a positive integer, then

$A^n = B^n$

When $n = 2$, the property is called the **squaring property.** For example,

$3 = 3$
$3^2 = 3^2$ Squaring property
$9 = 9$

When $n = 3$, the property is called the **cubing property.**

$2 = 2$
$2^3 = 2^3$ Cubing property
$8 = 8$

Using the power property to solve an equation, however, can result in a false solution to the equation. For example, consider what happens when both sides of the equation $x = 2$ are squared:

$x = 2$
$x^2 = 2^2$ Squaring property
$x^2 = 4$

The new equation has two solutions, 2 and -2, while the original equation had only one solution, 2.

$$x^2 = 4$$
$$x^2 - 4 = 0$$
$$(x - 2)(x + 2) = 0$$
$$x - 2 = 0 \quad \text{or} \quad x + 2 = 0 \quad \text{Zero-factor property}$$
$$x = 2 \quad \text{or} \quad x = -2$$

Notice that squaring both sides has introduced the solution $x = -2$, which does not satisfy the original equation. Such a solution is called **extraneous.** The introduction of extraneous solutions when applying the power property of equality necessitates that we *check all solutions* found in this manner.

THEOREM 8.2

When the power property of equality is applied to an equation, the solution set of the original equation will always be a subset of the solution set of the new equation.

Theorem 8.2 states that all, part, or none of the solutions to an equation solved by the power property of equality may be extraneous.

EXAMPLE 1 Solve $\sqrt{x + 2} = 4$.

SOLUTION To eliminate the radical, we square both sides.

$$(\sqrt{x + 2})^2 = 4^2$$
$$x + 2 = 16$$
$$x = 14$$

CHECK $\sqrt{14 + 2} \stackrel{?}{=} 4$
$\sqrt{16} \stackrel{?}{=} 4$
$4 \stackrel{\checkmark}{=} 4$

The solution set is $\{14\}$.

EXAMPLE 2 Solve $\sqrt{x - 2} = 2 - x$.

SOLUTION $\quad(\sqrt{x-2})^2 = (2-x)^2 \quad$ Squaring property

$$x - 2 = 4 - 4x + x^2$$
$$0 = x^2 - 5x + 6$$
$$0 = (x-2)(x-3)$$
$$x - 2 = 0 \quad \text{or} \quad x - 3 = 0$$
$$x = 2 \quad \text{or} \quad x = 3$$

CHECK $\quad \sqrt{2-2} \stackrel{?}{=} 2 - 2 \quad \Big| \quad \sqrt{3-2} \stackrel{?}{=} 2 - 3$
$\qquad\qquad \sqrt{0} \stackrel{?}{=} 0 \quad\quad\quad \Big| \quad\quad \sqrt{1} \stackrel{?}{=} -1$
$\qquad\qquad\quad 0 \stackrel{\checkmark}{=} 0 \quad\quad\quad \Big| \quad\quad\quad 1 \neq -1$

Since only 2 checks, the solution set is {2}. Note that {2} ⊆ {2, 3}, as is asserted by Theorem 8.2.

EXAMPLE 3 Solve $\sqrt[4]{1-x} = 4$.

SOLUTION To eliminate the radical, we raise both sides to the fourth power.

$$(\sqrt[4]{1-x})^4 = 4^4$$
$$1 - x = 256$$
$$-x = 255$$
$$x = -255$$

CHECK $\quad \sqrt[4]{1-(-255)} \stackrel{?}{=} 4$
$\qquad\qquad \sqrt[4]{1 + 255} \stackrel{?}{=} 4$
$\qquad\qquad\quad \sqrt[4]{256} \stackrel{?}{=} 4$
$\qquad\qquad\qquad\quad 4 \stackrel{\checkmark}{=} 4$

The solution set is {−255}.

Some equations involving radicals require that the power property be applied more than once to eliminate the radicals.

EXAMPLE 4 Solve $\sqrt{3x+1} = \sqrt{x} - 1$.

SOLUTION To eliminate the radical on the left side of the equation, we square both sides.

$$(\sqrt{3x+1})^2 = (\sqrt{x} - 1)^2$$
$$3x + 1 = x - 2\sqrt{x} + 1 \qquad (a-b)^2 = a^2 - 2ab + b^2$$

Now we isolate the term involving the radical, $\sqrt{x}$, on one side of the equation and square again.

$$2x = -2\sqrt{x}$$
$$(2x)^2 = (-2\sqrt{x})^2 \qquad \text{Squaring property}$$

Caution

Be careful when squaring expressions involving radicals.

Right

$(\sqrt{x}+5)^2 = (\sqrt{x}+5)(\sqrt{x}+5)$
$\qquad = x + 10\sqrt{x} + 25$

$(2\sqrt{x})^2 = 2^2(\sqrt{x})^2 = 4x$

Wrong

$(\sqrt{x}+5)^2 = x + 25$

$(2\sqrt{x})^2 = 2x$

$$4x^2 = 4x$$
$$4x^2 - 4x = 0$$
$$4x(x-1) = 0$$
$$4x = 0 \quad \text{or} \quad x - 1 = 0$$
$$x = 0 \quad \text{or} \quad x = 1$$

CHECK
$\sqrt{3 \cdot 0 + 1} \stackrel{?}{=} \sqrt{0} - 1 \quad\quad \sqrt{3 \cdot 1 + 1} \stackrel{?}{=} \sqrt{1} - 1$
$\qquad\quad \sqrt{1} \stackrel{?}{=} -1 \quad\quad\quad\quad\quad \sqrt{4} \stackrel{?}{=} 1 - 1$
$\qquad\quad\quad 1 \neq -1 \quad\quad\quad\quad\quad\quad\quad 2 \neq 0$

Since neither solution checks, the solution set is $\emptyset$.

EXAMPLE 5 Solve for y and check:

$$\frac{2}{\sqrt{y+1}} = \sqrt{y+1} + \sqrt{y+2}, \quad y \neq -1$$

SOLUTION We begin by multiplying each side by the LCD.

$$2 = y + 1 + (\sqrt{y+1})(\sqrt{y+2}) \quad \text{The LCD is } \sqrt{y+1}.$$
$$2 = y + 1 + \sqrt{y^2 + 3y + 2}$$

To solve for y, we isolate $\sqrt{y^2 + 3y + 2}$ before squaring.

$$-y + 1 = \sqrt{y^2 + 3y + 2}$$
$$(-y + 1)^2 = (\sqrt{y^2 + 3y + 2})^2 \quad \text{Squaring property}$$
$$y^2 - 2y + 1 = y^2 + 3y + 2$$
$$-5y = 1$$
$$y = -\frac{1}{5}$$

CHECK $\dfrac{2}{\sqrt{y+1}} = \sqrt{y+1} + \sqrt{y+2}$

$$\dfrac{2}{\sqrt{-\frac{1}{5}+1}} \stackrel{?}{=} \sqrt{-\frac{1}{5}+1} + \sqrt{-\frac{1}{5}+2} \quad y = -\frac{1}{5}$$

$$\dfrac{2}{\sqrt{\frac{4}{5}}} \stackrel{?}{=} \sqrt{\frac{4}{5}} + \sqrt{\frac{9}{5}}$$

$$\dfrac{2}{\frac{2}{\sqrt{5}}} \stackrel{?}{=} \dfrac{2}{\sqrt{5}} + \dfrac{3}{\sqrt{5}}$$

$$\frac{2}{\frac{2}{\sqrt{5}} \cdot \frac{\sqrt{5}}{\sqrt{5}}} \stackrel{?}{=} \frac{5}{\sqrt{5}}$$

$$\frac{2}{\frac{2\sqrt{5}}{5}} \stackrel{?}{=} \frac{5}{\sqrt{5}}$$

$$2 \cdot \frac{5}{2\sqrt{5}} \stackrel{?}{=} \frac{5}{\sqrt{5}} \qquad \text{To divide, multiply by the reciprocal of the denominator.}$$

$$\frac{5}{\sqrt{5}} \stackrel{\checkmark}{=} \frac{5}{\sqrt{5}}$$

The solution checks. The solution set is $\left\{-\frac{1}{5}\right\}$.

EXERCISE SET 8.5

Solve each of the following equations. Check each answer. Write $\emptyset$ if no real-number solution exists.

1. $\sqrt{x} = 2$
2. $\sqrt{y} = 3$
3. $\sqrt{a} = -1$
4. $\sqrt{p} = -16$
5. $\sqrt{w+1} = 5$
6. $\sqrt{s-1} = 2$
7. $\sqrt{2t-3} = 1$
8. $\sqrt{3k+1} = 1$
9. $\sqrt{2y-1} = -2$
10. $\sqrt{5m-1} = 2$
11. $\sqrt{3x-1} = 2$
12. $\sqrt{2m+1} = 3$
13. $\sqrt{2u-4} = -4$
14. $\sqrt{4v-5} = -5$
15. $\sqrt{1-3s} = 6$
16. $\sqrt{2-t} = 1$
17. $-\sqrt{3y} = -5$
18. $-\sqrt{2w} = -7$
19. $\sqrt[3]{x} = 2$
20. $\sqrt[3]{t} = 1$
21. $\sqrt[3]{2s} = -1$
22. $\sqrt[3]{5k} = -2$
23. $\sqrt[5]{-a} = 2$
24. $\sqrt[4]{-x} = 2$
25. $\sqrt[4]{2v} = -1$
26. $\sqrt[3]{3y} = -2$
27. $\sqrt[4]{-x} = 3$
28. $\sqrt[4]{1+y} = 3$
29. $-\sqrt[3]{2a-1} = -2$
30. $-\sqrt[3]{3c+2} = -3$
31. $\sqrt{x+1} = \sqrt{3}$
32. $\sqrt{t-2} = \sqrt{3}$
33. $\sqrt{3w-1} = \sqrt{w}$
34. $\sqrt{4s-3} = \sqrt{s}$
35. $\sqrt{7y+3} = \sqrt{2y+5}$
36. $\sqrt{u-4} = \sqrt{2u-17}$
37. $\sqrt{3x+2} = \sqrt{x+4}$
38. $\sqrt{v+4} = -\sqrt{2v+1}$
39. $-\sqrt{3t+5} = \sqrt{t+2}$
40. $\sqrt[3]{k-1} = \sqrt[3]{2k}$
41. $\sqrt[3]{3a-5} = \sqrt[3]{a}$
42. $\sqrt[3]{7w+2} = -\sqrt[3]{2w}$
43. $\sqrt[3]{7m-3} = -\sqrt[3]{m}$
44. $\sqrt{x+5} = \sqrt{x-7}$
45. $\sqrt{9v-2} = \sqrt{9v-1}$
46. $\sqrt{4k-5} = 2\sqrt{k}$
47. $\sqrt{25u+3} = 5\sqrt{u}$
48. $\sqrt{36s-5} = 6\sqrt{s}$
49. $2\sqrt[3]{k} = \sqrt[3]{8k+2}$
50. $\sqrt[3]{16m} = 3\sqrt[3]{2m+3}$
51. $\sqrt[3]{5s+1} = \sqrt[3]{2s+1}$
52. $\sqrt[4]{k-1} = 2\sqrt[4]{k-16}$

Each of the following equations becomes a factorable quadratic equation. Solve and check each answer. Write $\emptyset$ if no real-number solution exists.

53. $u - 5 = \sqrt{u-3}$
54. $\sqrt{p-4} = 4 - p$
55. $m = 1 + \sqrt{m+11}$
56. $4 - k = -\sqrt{k+2}$
57. $\sqrt{a} \cdot \sqrt{a-5} = 6$
58. $\sqrt{v} \cdot \sqrt{v-6} = 4$
59. $\sqrt{x-6} = \dfrac{4}{\sqrt{x}}$
60. $\sqrt{y-9} = \dfrac{6}{\sqrt{y}}$
61. $\sqrt{2y-1} = 2y - 3$
62. $\sqrt{t-1} - 2 = t - 3$
63. $\sqrt{t+2} - 1 = \sqrt{2t-10}$
64. $\sqrt{s} = \sqrt{2s+1} - 1$
65. $\sqrt{p} = p - 6$
66. $\sqrt{y} = y - 12$
67. $\sqrt[3]{v^2 - 1} = 2$
68. $\sqrt[3]{x^2 - 9} = 3$
69. $\sqrt[4]{y^2 - 8} = 1$
70. $\sqrt[4]{m^2 - 19} = 3$

71. $x + 1 = \sqrt{x+4} \cdot \sqrt{x-1}$
72. $y + 3 = \sqrt{y+1} \cdot \sqrt{y+6}$
73. $\sqrt{x+3} - 3\sqrt{x+11} = 2\sqrt{x+18}$
74. $2\sqrt{x+6} = \sqrt{x-1} - \sqrt{x+13}$
75. $\sqrt{y-3} - \dfrac{10}{\sqrt{y-3}} = \sqrt{y+2}$
76. $\sqrt{y+15} - \dfrac{4}{\sqrt{y+7}} = -\sqrt{y+7}$
77. $\sqrt{\sqrt{4-x}} = \sqrt{x+2}$
78. $\sqrt{\sqrt{x^2+11}} = \sqrt{x^2-9}$
79. $\sqrt[3]{\sqrt{3x+1} + \sqrt{3x}} = \sqrt[3]{2}$
80. $\sqrt[3]{\sqrt{x^2+2x+9}} = \sqrt[3]{x+3}$

FOR EXTRA THOUGHT

Round all answers to three decimal points.

81. Find c if $a = \sqrt{6}$ and $b = \sqrt{8}$.

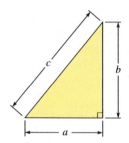

82. Find w if $d = 13$ and $l = \sqrt{12}$.

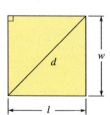

83. Find the height of the ladder on the wall if the base of the ladder is $\sqrt{3}$ m from the wall and the ladder is $2\sqrt{5}$ m long.

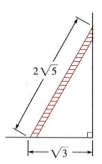

84. The time t in seconds for a pendulum of length l to go through one complete cycle is given by $t = 2\pi\sqrt{l/32}$. Find l if $t = 2$ sec.

REVIEW EXERCISES

Simplify each of the following expressions. Assume all variables represent positive real numbers.

85. $x^{3/4} \cdot x^{1/3}$
86. $y^{7/8} \div y^{2/3}$
87. $3\sqrt{5} + \sqrt{20} - 2\sqrt{50}$
88. $(3\sqrt{2} + 2\sqrt{5})(3\sqrt{2} - 2\sqrt{5})$
89. $\dfrac{12 + 9\sqrt{5}}{3}$
90. $\dfrac{\sqrt{x} - \sqrt{y}}{\sqrt{x} + \sqrt{y}}$
91. $(x^{1/3})^{3/4}$
92. $(0.064)^{1/3}$
93. $2 \div \sqrt[3]{2}$
94. $4 \div \sqrt{6}$
95. $\dfrac{xy}{\sqrt{xy}}$
96. $\dfrac{4x}{\sqrt{x}} + \sqrt{x}$
97. $27^{4/3}$
98. $(x^{1/2} - y^{1/2})^2$
99. $y\sqrt{75} + y\sqrt{27} - y\sqrt{12}$
100. $x\sqrt{98} - x\sqrt{50} + x\sqrt{18}$

8.6 Complex Numbers

HISTORICAL COMMENT In the early development of modern mathematics, mathematicians struggled with the concept of the square root of a negative number. They argued that the square root of a negative number could be neither negative nor positive, since the squares of both are positive numbers. Cardan (1572) called them "fictitious." Girard (1629) called the solutions to equations involving square roots of negative numbers "impossible," but argued that they should be studied further. Descartes (1637) originated the word *imaginaire* (imaginary) to describe these new numbers, and Euler (1748) used the letter i to describe $\sqrt{-1}$. Leibniz (1702) thought of imaginaries as "that wonderful creature of an ideal world, almost an amphibian between things that are and things that are not."

OBJECTIVES In this section we will learn to
1. simplify expressions in which the radicand is negative;
2. find powers of i;
3. identify complex numbers; and
4. add, subtract, multiply, and divide complex numbers.

In earlier sections we considered expressions such as $\sqrt{4}$, $\sqrt[4]{16}$, and the general expression $\sqrt[n]{b}$. For n even, b was restricted to the set of *nonnegative* real numbers. In this section we will consider numbers such as $\sqrt{-1}$. Notice that

$$\sqrt{-1} \neq 1 \quad \text{since} \quad 1^2 = 1 \neq -1$$
$$\sqrt{-1} \neq -1 \quad \text{since} \quad (-1)^2 = 1 \neq -1.$$

Since the square of any *real* number is positive or zero, $\sqrt{-1}$ cannot be a real number. In order to give meaning to the square root of a negative number, mathematicians introduce the **imaginary unit,** i, which is defined as follows.

$$\boxed{i = \sqrt{-1} \quad \text{or} \quad i^2 = -1}$$

Assuming the rules of exponents hold for the number i, we have

$$i = \sqrt{-1} \qquad\qquad i^5 = i^4 \cdot i = 1 \cdot i = i$$
$$i^2 = -1 \qquad\qquad i^6 = (i^2)^3 = (-1)^3 = -1$$
$$i^3 = i^2 \cdot i = -1 \cdot i = -i \qquad i^7 = i^6 \cdot i = -1 \cdot i = -i$$
$$i^4 = (i^2)^2 = (-1)^2 = 1 \qquad i^8 = (i^4)^2 = (1)^2 = 1$$

Notice that only four results occur when i is raised to a positive integer power: i, -1, $-i$, and 1. Since $i^4 = 1$, it is easy to find i^n for any positive integer n.

> *To find the value of i to a positive integer power, divide the integer power by 4 and use the remainder as the power of i for the result.*

Since $i^4 = 1$, if we take i to the remainder power, it is equivalent to i to the original power.

EXAMPLE 1   Find i^{67}.

SOLUTION   We divide the exponent 67 by 4:

$$67 = 4 \cdot 16 + 3$$

The remainder is 3, so

$$i^{67} = (i^4)^{16} \cdot i^3 = (1)^{16} \cdot i^3 = 1 \cdot i^3 = i^3 = -i \qquad i^4 = 1;\ i^3 = -i$$

EXAMPLE 2   Simplify.

(a) $i^{29} = (i^4)^7 \cdot i^1 = (1)^7 \cdot i = i$ The remainder is 1.
(b) $i^{322} = (i^4)^{80} \cdot i^2 = (1)^{80} \cdot i^2 = 1 \cdot (-1) = -1$ The remainder is 2.
(c) $i^{123} = (i^4)^{30} \cdot i^3 = 1^{30} \cdot i^3 = i^3 = -i$ The remainder is 3.
(d) $i^{80} = (i^4)^{20} \cdot i^0 = (1)^{20} \cdot 1 = 1$ The remainder is 0, and $i^0 = 1$.

CAUTION

The property $\sqrt{a}\sqrt{b} = \sqrt{ab}$ *does not apply* if a and b are *both* negative. The radicals must be converted to the i-form first.

RIGHT
$$\sqrt{-9}\sqrt{-4} = 3i \cdot 2i$$
$$= 6i^2$$
$$= -6$$

WRONG
$$\sqrt{-4}\sqrt{-9} = \sqrt{(-4)(-9)}$$
$$= \sqrt{36}$$
$$= 6$$

We can now use properties of radicals to simplify expressions such as $\sqrt{-4}$, $\sqrt{-9}$, or $\sqrt{-45}$. Recall that if a and b are positive or zero, then $\sqrt{ab} = \sqrt{a} \cdot \sqrt{b}$. If we extend this property to cover the situation where there is one negative radicand, then

$$\sqrt{-4} = \sqrt{4(-1)} = \sqrt{4} \cdot \sqrt{-1} = 2i$$
$$\sqrt{-9} = \sqrt{9(-1)} = \sqrt{9} \cdot \sqrt{-1} = 3i$$

This process is generally shortened as follows.

$$\sqrt{-b} = \sqrt{b}\,i, \qquad b > 0$$

For example,

$$\sqrt{-11} = \sqrt{11}\,i = i\sqrt{11}$$
$$\sqrt{-45} = \sqrt{9 \cdot 5(-1)} = \sqrt{9} \cdot \sqrt{5} \cdot \sqrt{-1} = 3\sqrt{5}\,i = 3i\sqrt{5}$$

Numbers such as bi, where b is any real number, are a subset of a larger set called **complex numbers.**

CAUTION

Care must be taken not to write $\sqrt{b}i$ as $\sqrt{bi}$. One way to avoid this in handwritten text is to write it as $i\sqrt{b}$.

A **complex number** is any number of the form $a + bi$, where $a, b \in R$.

A complex number consists of a real and an imaginary part.

$$a + bi$$
Real part is a ↑   ↑—— Imaginary part is b

NOTE The imaginary part is b, not bi.

Thus $2 + 3i$, $4 - 2i$, $5 + 6i$, and $3 - \sqrt{-4} = 3 - 2i$ are all complex numbers. The set of real numbers is a subset of the set of complex numbers, since *real numbers are complex numbers in which the imaginary part is* 0.

$$\text{Real numbers} \begin{cases} 2 = 2 + 0i \\ 3 = 3 + 0i \\ -\dfrac{1}{3} = -\dfrac{1}{3} + 0i \end{cases} \text{Complex form}$$

If the real part of a complex number is zero, the number is said to be a **pure imaginary number.**

$$\text{Pure imaginary numbers} \begin{cases} 2i = 0 + 2i \\ 3i = 0 + 3i \\ -9i = 0 - 9i \end{cases} \text{Complex form}$$

The properties of real numbers stated in Chapter 1 are true for the complex numbers as well as for the real numbers.

> **DEFINITION**
> **Equality of Complex Numbers**
>
> Two complex numbers are equal only when their real parts are equal and their imaginary parts are equal.

This definition is often used to solve equations involving complex numbers.

EXAMPLE 3 Given $2x + yi = 4 + 7i$, solve for x and y.

SOLUTION The real parts are $2x$ and 4; the imaginary parts are y and 7; so

$$2x = 4 \quad \text{and} \quad y = 7$$
$$x = 2 \quad \text{and} \quad y = 7$$

EXAMPLE 4 Solve $(3 + 2i)x + (4 - 2i)y = 2 + 6i$.

SOLUTION First we rewrite the equation so that the real and imaginary parts can be identified.

$$3x + 2xi + 4y - 2yi = 2 + 6i \quad \text{Distributive property}$$
$$(3x + 4y) + (2x - 2y)i = 2 + 6i$$

(a) $3x + 4y = 2$ The real parts are equal.
(b) $2x - 2y = 6$ The imaginary parts are equal.

The system of equations (a) and (b) can be solved by any of the methods of Chapter 4. The reader should verify that the solution is $x = 2$, $y = -1$.

To Add Two Complex Numbers

Treat the two numbers as binomials and add the sum of their real parts to the sum of their imaginary parts:

$$(a + bi) + (c + di) = (a + c) + (bi + di) = (a + c) + (b + d)i$$

EXAMPLE 5 Add: **(a)** $(4 + 6i) + (3 - 2i)$ **(b)** $(8 + 6i) + (3 + 4i) - (5 + 2i)$

SOLUTION **(a)** The real parts are 4 and 3, and the imaginary parts are 6 and -2, so

$$(4 + 6i) + (3 - 2i) = (4 + 3) + (6 - 2)i$$
$$= 7 + 4i$$

(b) First we remove the grouping symbols:

$$8 + 6i + 3 + 4i - 5 - 2i = \overbrace{(8 + 3 - 5)}^{\text{Add the real parts.}} + \overbrace{(6 + 4 - 2)}^{\text{Add the imaginary parts.}}i$$
$$= 6 + 8i$$

To Multiply Two Complex Numbers

Treat the two numbers as binomials and multiply, simplifying any powers of i that may occur:

$$(a + bi)(c + di) = ac + adi + bci + bdi^2$$
$$= (ac - bd) + (ad + bc)i$$

EXAMPLE 6 Multiply $(3 + 5i)(6 + 3i)$.

SOLUTION

$$(3 + 5i)(6 + 3i) = \overset{F}{\underset{\downarrow}{18}} + \overset{O}{\underset{\downarrow}{9i}} + \overset{I}{\underset{\downarrow}{30i}} + \overset{L}{\underset{\downarrow}{15i^2}}$$
$$= 18 + 9i + 30i - 15 \qquad i^2 = -1;\ 15i^2 = -15$$
$$= 3 + 39i$$

EXAMPLE 7 Expand $(2 + i)^3$.

SOLUTION

$$(2 + i)^3 = (2 + i)^2(2 + i)$$
$$= (4 + 4i + i^2)(2 + i) \qquad (a + b)^2 = a^2 + 2ab + b^2$$
$$= (3 + 4i)(2 + i) \qquad i^2 = -1;\ 4 - 1 = 3$$
$$= 6 + 11i + 4i^2 \qquad \text{FOIL method}$$
$$= 2 + 11i \qquad 4i^2 = -4;\ 6 - 4 = 2$$

The process we use to divide two complex numbers is similar to the one we use to rationalize the denominator. The following property is important in using this process.

> The product of two complex conjugates is a **real number:**
> $$(a + bi)(a - bi) = a^2 - b^2i^2 = a^2 - b^2(-1) = a^2 + b^2$$

EXAMPLE 8 Divide $\dfrac{6 + i}{3 + 4i}$. Write the answer in the form $a + bi$.

SOLUTION We multiply the numerator and the denominator by the conjugate of the denominator.

$$\dfrac{6 + i}{3 + 4i} = \dfrac{6 + i}{3 + 4i} \cdot \dfrac{3 - 4i}{3 - 4i} \qquad \text{Fundamental property of fractions}$$

$$= \dfrac{18 - 21i - 4i^2}{3^2 - (4i)^2}$$

$$= \dfrac{18 - 21i - 4(-1)}{9 - 16i^2}$$

$$= \dfrac{18 - 21i + 4}{25} \qquad i^2 = -1$$

$$= \dfrac{22 - 21i}{25} = \dfrac{22}{25} - \dfrac{21}{25}i$$

CHECK To show that the quotient is correct, we need to show that the product of the quotient and the divisor is equal to the dividend. That is,

$$\text{if } \dfrac{a}{b} = c, \quad \text{then} \quad b \cdot c = a$$

$$\left(\dfrac{22}{25} - \dfrac{21}{25}i\right)(3 + 4i) = \dfrac{66}{25} + \dfrac{88}{25}i - \dfrac{63}{25}i - \dfrac{84}{25}i^2$$

$$= \dfrac{66}{25} + \dfrac{88}{25}i - \dfrac{63}{25}i + \dfrac{84}{25} \qquad \dfrac{84}{25}i^2 = \dfrac{84(-1)}{25} = -\dfrac{84}{25}$$

$$= \dfrac{150}{25} + \dfrac{25}{25}i = 6 + i$$

EXAMPLE 9 Divide $\dfrac{\sqrt{5} - \sqrt{-3}}{\sqrt{-2} + \sqrt{3}}$. Write the answer in the form $a + bi$.

SOLUTION First we write $\sqrt{-2}$ as $i\sqrt{2}$ and $\sqrt{-3}$ as $i\sqrt{3}$. This gives us

$$\dfrac{\sqrt{5} - i\sqrt{3}}{\sqrt{3} + i\sqrt{2}}$$

Now we multiply both numerator and denominator by the conjugate of the denominator.

$$\frac{\sqrt{5} - i\sqrt{3}}{\sqrt{3} + i\sqrt{2}} \cdot \frac{\sqrt{3} - i\sqrt{2}}{\sqrt{3} - i\sqrt{2}} = \frac{\sqrt{15} - i\sqrt{10} - 3i + i^2\sqrt{6}}{3 + 2}$$

$$= \frac{\sqrt{15} - i\sqrt{10} - 3i + (-1)\sqrt{6}}{3 + 2} \qquad i^2 = -1$$

$$= \frac{\sqrt{15} - \sqrt{6} - (3 + \sqrt{10})i}{5}$$

$$= \frac{\sqrt{15} - \sqrt{6}}{5} - \frac{(3 + \sqrt{10})}{5}i$$

EXAMPLE 10 Let $f(x) = x^2 + 5x - 4$. Find $f(1 + 2i)$.

SOLUTION
$$f(1 + 2i) = (1 + 2i)^2 + 5(1 + 2i) - 4$$
$$= 1 + 4i + 4i^2 + 5 + 10i - 4$$
$$= 1 + 4i - 4 + 5 + 10i - 4 \qquad i^2 = -1$$
$$= -2 + 14i$$

Exercise Set 8.6

Simplify.

1. i^2
2. $-i^2$
3. $-i^3$
4. i^3
5. i^4
6. $-i^4$
7. $-i^5$
8. i^5
9. i^{20}
10. i^{28}
11. i^{101}
12. i^{97}
13. i^{-8}
14. i^{-4}

Perform the indicated operations and write each answer in the form $a + bi$.

15. $(2 + i) + (3 - 2i)$
16. $(5 - i) + (2 + 7i)$
17. $(6 - 11i) - (2 - i)$
18. $(5 - 3i) - (-1 - 2i)$
19. $(-15 - 2i) - (-3 - 2i)$
20. $(-6 + 5i) - (-6 + 3i)$
21. $(-6 + i) - 2(3 + i)$
22. $-8(1 - i) + 4(3 - 2i)$
23. $i(3 + 4i)$
24. $2i(1 - i)$
25. $(1 + i)(1 - i)$
26. $(3 - i)(3 + i)$
27. $(5 - 7i)(6 - 11i)$
28. $(3 - 13i)(5 + 2i)$
29. $(6 - 5i)^2$
30. $(7 + 2i)^2$
31. $(2 - i)^3$
32. $(1 + i)^3$
33. $\dfrac{5}{i}$
34. $\dfrac{7}{2i}$
35. $\dfrac{4}{1 + i}$
36. $\dfrac{6}{1 - i}$
37. $\dfrac{2 + i}{3 - i}$
38. $\dfrac{5 - i}{2 + i}$
39. $\dfrac{3 + 4i}{5 - 2i}$
40. $\dfrac{6 + 5i}{3 + 7i}$
41. $(2 + 3i)^2 - 5(2 - i) + 4$
42. $(3 - 5i)^2 + 4(7 - 2i) + 6$

Write each of the following radicals in terms of i and simplify.

43. $\sqrt{-4}$
44. $\sqrt{-9}$
45. $\sqrt{-25}$
46. $\sqrt{-16}$
47. $\sqrt{-36}$
48. $\sqrt{-81}$
49. $\sqrt{-32}$
50. $\sqrt{-50}$
51. $\sqrt{-98}$
52. $-2\sqrt{-250}$
53. $-3\sqrt{-200}$
54. $-5 + \sqrt{-2}$
55. $-3 + 5\sqrt{-27}$

56. $-\sqrt{-8} + \sqrt{-18}$
57. $\sqrt{-45} + \sqrt{-20} - \sqrt{-5}$
58. $2 + \sqrt{-5} - 3\sqrt{-80}$
59. $\sqrt{-3} - 5\sqrt{-12} + 2\sqrt{-48}$
60. $\sqrt{-2} - 5\sqrt{-18} + 6\sqrt{-50} - \sqrt{-8}$
61. $7\sqrt{-8} + 5\sqrt{-12} - 6\sqrt{-18} + 2\sqrt{-27}$
62. $11\sqrt{-3} + \sqrt{-75} - \sqrt{-16} + \sqrt{-25}$
63. $\dfrac{1}{\sqrt{-3}}$
64. $\dfrac{1}{\sqrt{-8}}$
65. $\dfrac{3}{\sqrt{-27}}$
66. $\dfrac{2}{1 - \sqrt{-3}}$
67. $\dfrac{3}{2 + \sqrt{-3}}$
68. $\dfrac{5}{7 - \sqrt{-2}}$
69. $\dfrac{1}{1 - \sqrt{-1}}$
70. $\dfrac{1}{3 - \sqrt{-3}}$
71. $\dfrac{6 - 2\sqrt{-50}}{2}$
72. $\dfrac{11 - \sqrt{-2}}{3}$
73. $\dfrac{6 - \sqrt{-7}}{4}$
74. $\dfrac{2 - 2\sqrt{-3}}{2}$

Solve each of the following for x and y.
75. $3x + 5yi = 9 + 10i$
76. $x - 3yi = -5 + 2i$
77. $(3 + 2i)x - (1 - i)y = 4i$
78. $(2 - i)x - (3 - i)y = 7i$
79. $(4x + y - 24)i = 2y + 7 - 3x$
80. $(-3x + 18 - y)i + 17 = 6y - 7x$
81. $3x + (y - 2)i - (3y - 8)i = 5 - 2x$
82. $5 - 4i - 2x + 1 = (3y + 2)i$
83. $9i + 5 - 9ix = 4y + 4iy + 3x$
84. $2ix + 9 - 13i - 9y = -4x - 6iy$

Perform the indicated operations, then simplify the answer and write it in the form a + bi.
85. $\left(-\dfrac{1}{2} - \dfrac{\sqrt{3}}{2}i\right)^3$
86. $\left(-\dfrac{1}{2} + \dfrac{\sqrt{3}}{2}i\right)^3$
87. $\left(\dfrac{1}{\sqrt{2}} - \dfrac{1}{\sqrt{2}}i\right)^4$
88. $\left(\dfrac{1}{\sqrt{2}} + \dfrac{1}{\sqrt{2}}i\right)^4$

Find the indicated value for each polynomial function.
89. $f(a) = 2a^2 - 4a + 3;\ f\left(1 + \dfrac{\sqrt{2}}{2}i\right)$
90. $f(x) = 5x^2 + 2 - 2x;\ f\left(\dfrac{1 - 3i}{5}\right)$
91. $g(y) = 2y^2 + y + 1;\ g\left(\dfrac{-1 + \sqrt{7}i}{4}\right)$
92. $f(p) = p^2 - 3p + 3;\ f\left(\dfrac{3 - \sqrt{2}i}{2}\right)$
93. $g(m) = m^2 + 4m + 7;\ g(2 + \sqrt{3}i)$
94. $r(t) = t^2 - 6t + 19;\ r(5 + \sqrt{10}i)$

WRITE IN YOUR OWN WORDS
95. What is a complex number?
96. What is a pure imaginary number?
97. What is a real number?
98. What does it mean for two complex numbers to be equal?
99. Explain how to add two complex numbers.
100. Explain how to multiply two complex numbers.
101. Explain how to simplify i^{103}.
102. What are conjugate pairs of binomials?

8.7 More Applications

OBJECTIVE To become acquainted with various applications that involve radicals.

Many areas of mathematics use formulas that involve radicals. We will see a few in the following examples.

EXAMPLE 1 What number subtracted from its square root yields -12?

SOLUTION We let n represent the number. Its square root is $\sqrt{n}$. Then

$$\sqrt{n} - n = -12$$
$$\sqrt{n} = n - 12 \qquad \text{Isolate the radical.}$$
$$(\sqrt{n})^2 = (n - 12)^2 \qquad \text{Squaring property}$$
$$n = n^2 - 24n + 144$$
$$0 = n^2 - 25n + 144$$
$$0 = (n - 9)(n - 16) \qquad \text{Factor.}$$
$$n = 9 \quad \text{or} \quad n = 16$$

Now let's check the solution in the words of the problem.

$$\sqrt{16} = 4; \quad 4 - 16 \stackrel{\checkmark}{=} -12$$
$$\sqrt{9} = 3; \quad 3 - 9 \neq -12$$

Therefore the number is 16.

EXAMPLE 2 In statistics, the calculated value of the z-score may be found by the formula

$$z = \frac{P - p}{\sqrt{\dfrac{pq}{n}}},$$

Solve the formula for n.

SOLUTION
$$z^2 = \left(\frac{P - p}{\sqrt{\dfrac{pq}{n}}}\right)^2 \qquad \text{Square both sides to eliminate the radical.}$$

$$= \frac{(P - p)^2}{\left(\sqrt{\dfrac{pq}{n}}\right)^2} \qquad \left(\frac{a}{b}\right)^2 = \frac{a^2}{b^2}$$

$$= \frac{(P - p)^2}{\dfrac{pq}{n}}$$

Now we simplify the complex fraction on the right side of the equation. We multiply numerator and denominator by n:

$$z^2 = \frac{n(P - p)^2}{pq}$$
$$pqz^2 = n(P - p)^2 \qquad \text{Multiply by } pq, \text{ the LCD.}$$
$$\frac{pqz^2}{(P - p)^2} = n \qquad \text{Divide each side by } (P - p)^2.$$

EXAMPLE 3 A point moves over a rectangular coordinate system in such a way that its position (x, y) at any time is determined by the formula $\sqrt{(x + 1)^2 + y} = 4$. Solve the equation for y in terms of x.

SOLUTION

$$\sqrt{(x + 1)^2 + y} = 4$$
$$\left(\sqrt{(x + 1)^2 + y}\right)^2 = 4^2 \qquad \text{Squaring property}$$
$$(x + 1)^2 + y = 16$$
$$y = 16 - (x + 1)^2 \qquad \text{Subtract } (x + 1)^2 \text{ from each side.}$$
$$y = -x^2 - 2x + 15$$

EXAMPLE 4 The formula for finding the radius r of a circle inscribed in a triangle with sides a, b, and c is

$$r = \sqrt{\frac{(s - a)(s - b)(s - c)}{s}}, \qquad \text{where } s = \frac{1}{2}(a + b + c)$$

Find the radius of the circle shown in Figure 8.1.

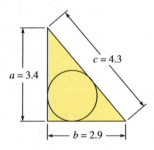

FIGURE 8.1

SOLUTION

$$s = \frac{1}{2}(a + b + c) = \frac{1}{2}(10.6) = 5.3$$

Therefore

$$r = \sqrt{\frac{(5.3 - 3.4)(5.3 - 2.9)(5.3 - 4.3)}{5.3}} = \sqrt{\frac{(1.9)(2.4)(1)}{5.3}}$$

Using a calculator, we find that $r = 0.9$ to the nearest tenth.

Exercise Set 8.7

Solve each of the following applied problems, using a calculator as needed. Round your answer to two decimal places. Let $\pi = 3.1416$.

1. A number exceeds twice its square root by 3. Find the number.

2. The **standard deviation** σ of a theoretical binomial probability distribution is given by $\sigma = \sqrt{npq}$, where n is the number of people, p is the probability of the event occurring, and q is the probability of the event not occurring. Solve for p.

3. In statistics, the number *n* required in a sample to determine an **error** less than or equal to *E* is given by
$$\sqrt{n} = \frac{M \cdot D}{E}$$
Solve for *M*.

4. In statistics, where *n* is one sample size and *N* is a second sample size, and where *s* is the standard deviation of the first sample and *S* is the standard deviation of the second sample, the **pooled estimate** *D* for the standard deviation is given by
$$D = \sqrt{\frac{(n-1)s + (N-1)S}{N+n-2}}$$
Solve for *S*.

5. The area of a regular octagon of side *s* is given by
$$A = \frac{s^2(4 + 2\sqrt{2})}{\sqrt{2}}$$
Find the area *A* of the octagon shown in the figure.

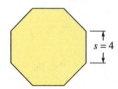

6. The length *s* of one side of a square with diagonal *D* is given by $s = \sqrt{\frac{1}{2}D^2}$. Find the length *s* of a side of the square shown in the figure.

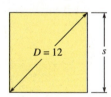

7. The frequency *f* with which a ball on the end of a steel rod oscillates is given by
$$f = \frac{1}{2\pi}\sqrt{\frac{K}{m}}$$

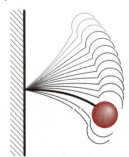

where *m* is the mass of the ball and *K* is a constant of the rod. Find the frequency of oscillation of a ball of mass $m = 0.0625$ when *K* is 102.

8. The lateral surface area *L* of a cone is given by $L = \pi r \sqrt{h^2 + r^2}$, where *h* is the height of the cone and *r* is the radius of the base of the cone. Find the lateral surface area of the given figure.

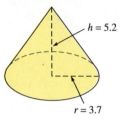

9. The radius *r* of a sphere is given by the equation
$$r = \sqrt{\frac{3V}{4\pi}}$$
where *V* is the volume. Find the radius of a sphere if its volume is $V = 18.23$ cu. in.

10. The volume *V* of a **frustrum** of a pyramid (see figure) is given by
$$V = \frac{h(A + a + \sqrt{Aa})}{3}$$
where *A*, the area of the lower base, is 6.04 in.^{2}; *a*, the area of the upper base, is 4.74 in.^{2}; and *h*, the height, is 3.96 in. Find the volume *V*.

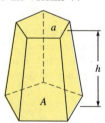

11. The radius *r* of a circle circumscribed about a triangle with sides $a = 12.1$ in., $b = 14.5$ in., and $c = 9.6$ in. is given by
$$r = \frac{abc}{4\sqrt{s(s-a)(s-b)(s-c)}}$$
where $s = \frac{1}{2}(a + b + c)$. Find *r*. (Illustration follows.)

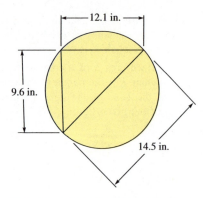

ART TO ACCOMPANY EXERCISE 11

12. Given the equation

$$I^{2/3} = \frac{1}{100} V$$

where I is the current in amperes and V is the voltage in volts, find V in terms of I.

13. In the theory of relativity, the formula

$$m = \frac{m_0}{\sqrt{1 - \frac{v^2}{c^2}}}$$

gives the mass m of an object at a speed v when its mass at rest is m_0 and c is the velocity of light. If $v = 0$, show that $m = m_0$.

14. The velocity v of sound in gasses is given by

$$v = c\sqrt{\frac{T}{M}}$$

Find the velocity of sound in air if $M = 28$, $T = 293$, and $c = 3.41$ m/sec.

15. A string stretched with a tension of 50 newtons produces the note C with a frequency of 256 cps. If f and F are two frequencies, and t and T are the corresponding tensions, then frequency and tension are related by the proportion

$$\frac{f}{F} = \frac{\sqrt{t}}{\sqrt{T}}$$

Find the tension T required to produce a frequency of $F = 512$ cps.

16. The equation of a simple parabola can be written $\sqrt{(x-a)^2 + y^2} = x + a$. Solve this equation for y in terms of the other variables.

17. The path of a projectile shot horizontally from a point y feet above the ground with a velocity of v ft/sec is a parabola with an equation

$$x = \sqrt{\frac{-2v^2 y}{g}}$$

Solve this equation for y in terms of the other variables.

18. The surface area S of a sphere is given by $S = 4\pi r^2$, where r is the radius of the sphere, and the volume is given by $V = \frac{4}{3}\pi r^3$. Solve for V in terms of S and r.

19. The side x of a cube of volume V is given by $x = V^{1/3}$ and the surface area by $S = 6x^2$. Solve for S in terms of V.

20. The area of an equilateral triangle is given by $A = \frac{1}{4}\sqrt{3}\, s^2$. Find the area of an equilateral triangle with side $s = 15$ ft.

21. What is the radius of the largest circle that can be planted with 16 lb of seed if 1 lb of seed plants 50 sq. ft?

22. Find the circumference of a circle circumscribed about a square with a side of $6\sqrt{2}$ cm. See the figure.

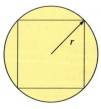

23. A circular lawn $20\sqrt{3}$ ft in diameter has a path 1 ft wide around it. What is the area of the path?

24. A box in the shape of a cube has a surface area of 2,397 cm². Find the length of one side of the cube.

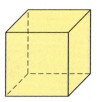

Summary

8.1 Rational Exponents

The properties of exponents as they apply to integer exponents also apply to rational exponents. The expression $b^{1/2}$ means the positive square root of b and $b^{1/3}$ means the cube root of b. In general, $b^{1/n}$ means the nth root of b. If n is even, b must be positive or zero for the expression to represent a real number.

8.2 Radicals and Rational Exponents

Another way to indicate the root of a number is to use a radical symbol. The symbol $\sqrt{b}$ means the **square root** of b; the symbol $\sqrt[3]{b}$ means the **cube root** of b; and the symbol $\sqrt[n]{b}$ means the **nth root** of b. A radical consists of three parts: the **index,** the **radical,** and the **radicand:**

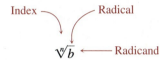

The **principal square root** of a positive number b is written $\sqrt{b}$. It indicates a positive number a such that $a^2 = b$.

The square root of the square of a number is the absolute value of that number:

$$\sqrt{b^2} = |b|$$

Radicals are simplified by using the theorem indicating the multiplication property

$$\sqrt[n]{a \cdot b} = \sqrt[n]{a} \cdot \sqrt[n]{b}$$

and the division property

$$\sqrt[n]{\frac{a}{b}} = \frac{\sqrt[n]{a}}{\sqrt[n]{b}}$$

A radical is in **simplest form** if the following four conditions are satisfied:

1. All exponents on factors in the radicand are less than the index.
2. The radicand does not contain a fraction.
3. The index and all of the exponents in the radicand do not contain a common factor (assuming all of the variables in the radicand represent positive numbers).
4. It is not a fraction with a radical in the denominator.

8.3 Arithmetic Operations Involving Radicals

Radicals with the same index and radicand can be added and subtracted. If the index and radicand are not the same the properties of radicals can often be used to simplify them so that addition and subtraction are possible. Sometimes it is helpful to write the radicals in terms of rational exponents and then use the rules of exponents in the simplification process.

8.4 Division by Radicals; Rationalizing

The fundamental property of fractions is the key to eliminating radicals from the denominator of a fraction. Multiplying both the numerator and the denominator by an appropriate quantity makes the exponents on the radicand equal to the index. This process is called **rationalizing the denominator.** When the denominator is a binomial involving square roots, its **conjugate** is used in the rationalizing process. The conjugate of $\sqrt{a} + \sqrt{b}$ is $\sqrt{a} - \sqrt{b}$.

8.5 Solving Equations Involving Radicals

Equations involving radicals are solved by using the **power property of equality.** When it is used to eliminate a square root, it is called the **squaring property;** when it is used to eliminate a cube root, it is called the **cubing property.**

8.6 Complex Numbers

The number $i = \sqrt{-1}$ is called the **imaginary unit.** When i is raised to positive integer powers, it takes on only four values: i, -1, $-i$, and 1. A number of the form $a + bi$, for real numbers a and b, is called a **complex number.** Addition and multiplication of complex numbers are carried out in the same way that binomials such as $2x + 3y$ and $4x - 7y$ are added and multiplied. Division by complex numbers is accomplished by rationalizing the denominator. If the denominator is a binomial, it is rationalized by multiplying by its **complex conjugate.** The conjugate of $a + bi$ is $a - bi$, and $(a + bi)(a - bi) = a^2 + b^2$, which is a real number.

Cooperative Exercise

EFFECTIVE RATE OF INTEREST

The *effective annual interest rate* is the rate of interest that would be earned if interest were compounded once a year. For example, $1 at 8% compounded quarterly amounts to $1.082 in one year. Thus the total interest earned is $1.082 - $1 = $0.082. This is the same interest that would be earned if $1 were compounded annually at 8.2%, since 8.2% of $1 is $0.082. Thus the *effective annual interest rate* is 8.2%.

Effective rate of interest is often used to compare interest rates that are compounded at different intervals. Many banks and savings institutions advertise various types of accounts together with the *stated* interest rate and the *effective interest rate.* We can determine the effective rate of interest by using the formula

$$E = (1 + r)^n - 1$$

where

E = the effective interest (in decimal form, rounded to the nearest ten thousandth)

n = the number of times the money is compounded in one year

r = the interest rate per compound period (in decimal form)

I. Find and compare the effective rate of interest for each of the following.
   (a) 9% compounded annually
   (b) 9% compounded semiannually
   (c) 9% compounded quarterly
   (d) 9% compounded monthly
   (e) 9% compounded daily (assume 360 days per year)

II. Find the interest you would earn if you were to invest $8,000 at each effective rate of interest from (a) to (e).

III. Would you earn more or less money in interest if you were to invest your money at 9% compounded annually or $8\frac{3}{4}$% compounded quarterly?

Review Exercises

Simplify each of the following expressions. Write all answers with positive exponents.

1. $(0.0081)^{-3/4}$
2. $2^{-1/2} + 3^{-1/2}$
3. $\dfrac{5^{3/4} \cdot 5^{2/3}}{5^{1/3} \cdot 5^{-5/2}}$
4. $\dfrac{x^{-5/2} \cdot x^{5/3}}{x^{7/4}}$
5. $(16a^{-8/9}b^{4/3})^{3/4}$
6. $(a + b)^{-1/2}$
7. $(x + 3)(x - 2)^{-1/2} + (x - 2)^{1/2}$
8. $2a^{-1/2} + 3b^{-1/2}$
9. $(m^{1/2} + n^{1/2})(m^{1/2} - n^{1/2})$
10. $(s^{1/2} + s^{-1/2})^2$

Simplify each of the following expressions. Assume all variables represent positive real numbers.

11. $\sqrt[4]{64t^4}$
12. $\sqrt{400}$
13. $\sqrt[3]{-32}$
14. $\sqrt{-100}$
15. $\sqrt{n^2 + 8n + 16}$
16. $-\sqrt[4]{256k^7}$
17. $(\sqrt{3} + \sqrt{2})(\sqrt{3} - \sqrt{2})$
18. $\sqrt{2} + 5\sqrt{72} + 4\sqrt{50} - \sqrt{128}$
19. $\sqrt{x^6 + x^4}$
20. $(3\sqrt{5} + 4\sqrt{2})(\sqrt{5} - 3\sqrt{2})$
21. $(4\sqrt{3} + 5\sqrt{6})^2$
22. $2\sqrt{x}(7\sqrt{x} + \sqrt{2x} - \sqrt{4x})$
23. $(\sqrt{3} + i)^3$
24. $3\sqrt[3]{x^5y} + \dfrac{x}{y} \cdot \sqrt[3]{125x^2y^4} + \dfrac{3}{y} \cdot \sqrt[3]{-x^8y}$
25. $(\sqrt{a} + \sqrt{a-1})(\sqrt{a} - \sqrt{a-1})$
26. $(\sqrt{x+1} - 1)^2$
27. $\dfrac{\sqrt{6}}{\sqrt{10}}$
28. $\sqrt[3]{\dfrac{4x^2y^7}{9xy^8}}$
29. $\dfrac{\sqrt{3a} - \sqrt{7b}}{\sqrt{21ab}}$
30. $\dfrac{\sqrt{2} - \sqrt{3}}{2\sqrt{3} + \sqrt{2}}$
31. $\dfrac{3k}{\sqrt[3]{2k^2}}$
32. $\dfrac{4a}{\sqrt[3]{2a}}$
33. $\dfrac{5 + 3\sqrt{x}}{5 - 3\sqrt{x}}$
34. $\dfrac{\sqrt{a+b} + \sqrt{a}}{\sqrt{a+b} - \sqrt{a}}$

Solve each of the following equations. Check all solutions, and write φ if no real-number solution exists.

35. $\sqrt{a+1} = 3$
36. $\sqrt{3y+2} = 5$
37. $\sqrt[3]{t+1} = -2$
38. $\sqrt[4]{x} = -1$
39. $\sqrt{s+5} - 2\sqrt{s} = -\sqrt{s-3}$
40. $\sqrt{r+7} = \sqrt{r+2} - \sqrt{4r+17}$
41. $\sqrt{6m-5} - 5 = 0$
42. $\sqrt{x^2 - 5x + 2} = x + 3$
43. $\sqrt{y} \cdot \sqrt{y-7} = 2\sqrt{2}$
44. $\sqrt{2a+3} \cdot \sqrt{3a-1} + 2 = 2$
45. $\dfrac{3}{\sqrt{p+5}} = \sqrt{p+7} + \sqrt{p+5}$
46. $\dfrac{k+2}{\sqrt{k+1}} = \sqrt{k+5}$

Solve for x and y.

47. $2iy - y + 19 = ix + 6x + 12i$
48. $(-3 + 5i)y + 6 + 19i = (4 + 3i)x$

Simplify each of the following complex number expressions.

49. $\sqrt{-9} + \sqrt{-36} - \sqrt{-81}$
50. $-4i(\sqrt{-8} + \sqrt{-50})$
51. $\dfrac{3 + 2i}{2 - i}$
52. $\dfrac{4 - \sqrt{2}i}{3 + \sqrt{2}i}$
53. $(5 - \sqrt{2}i)^2$
54. $(2 - i)(3 - i)(2 + i)$
55. $(3 - \sqrt{2}i)^{-2}$
56. $(\sqrt{3} - i)^3$
57. $(5 - 13i) - (-3 + i)$
58. $(-6 + 7i) + (3 - 8i)$
59. $\dfrac{1 + i}{1 - i}$
60. $\dfrac{5 - \sqrt{-2}}{5 + \sqrt{-2}}$
61. $\sqrt[3]{\sqrt{64}}$
62. $\sqrt{\sqrt[3]{x^6y^9}}$

Chapter Test

Simplify each of the following expressions. Assume all variables represent positive real numbers.

1. $(2x^{1/3})(3x^{1/2})$
2. $a^{1/2}(a^{1/3} + a^{1/2})$
3. $5\sqrt{72}$
4. $\sqrt[3]{-8x^4y^5}$
5. $\sqrt{\dfrac{128}{50}}$
6. $\sqrt{\dfrac{5}{2} + \dfrac{2}{\sqrt{5}}}$
7. $\dfrac{\sqrt{5}}{3\sqrt{2} - \sqrt{5}}$
8. $\dfrac{15\sqrt{20x} - 3\sqrt{18x}}{\sqrt{2}}$
9. $(4\sqrt{3} + \sqrt{5})(4\sqrt{3} - \sqrt{5})$
10. $3\sqrt{20} - \sqrt{125} + 4\sqrt{45}$
11. $\dfrac{3x}{\sqrt[3]{x}}$
12. $\sqrt[4]{\dfrac{3}{4x}}$
13. $8\sqrt{-3} - 2\sqrt{-75} + \sqrt{-16}$
14. $\dfrac{2 - \sqrt{-2}}{\sqrt{-3} + \sqrt{-2}}$

Solve each of the following equations. Check your answers.

15. $\sqrt{3x - 2} = 2$
16. $\sqrt{3x + 1} = x - 1$
17. $\sqrt{y} = y - 6$
18. $\sqrt[3]{y - 8} = 2$

19. $\sqrt{x-2} = -\sqrt{3x-1}$
20. $x - 2yi = 5 + 3i$
21. $\sqrt{3x+8} = \sqrt{7x+12}$
22. $\sqrt{2x+3} = 3 - \sqrt{2x}$
23. $\sqrt[3]{x-2} = 2\sqrt[3]{3x-10}$
24. $\sqrt[4]{y^2+5} = \sqrt[4]{y+3}$

Simplify each of the following expressions.

25. $\dfrac{16^{-2}x^{-1/3}y^{1/2}}{(4x^{2/3}y^{1/3})^{-2}}$

26. $\sqrt[3]{-81x^4y^9z^{17}}$

27. $\sqrt[5]{\sqrt{1024}}$

28. $\sqrt{3} \cdot \sqrt[3]{2}$

29. $\sqrt{\sqrt[3]{x^9}}$

30. $\dfrac{5-2i}{6+i}$

Solve each of the following applied problems.

31. The square root of the sum of three times a number and 2 equals the square root of 23. Find the number.

32. The diagonal of a rectangle is $\sqrt{3x+1}$. If the length of the rectangle is 8 ft and the width is 6 ft, find x.

CHAPTER 9
Quadratic Equations, Functions, and Inequalities

9.1 Solving Polynomial Equations by Factoring and the Square Root Property
9.2 Solving Quadratic Equations by Completing the Square
9.3 Solving Quadratic Equations Using the Quadratic Formula
9.4 Solving Equations Involving Quadratics
9.5 Solving Inequalities Involving Quadratics
9.6 Variation
9.7 More Applications

Application

A group of students charters a bus for a trip to Disneyland for $720. If they can persuade 12 more students to go, each student's cost will decrease by $2. How many students are in the group?

See Section 9.7, Example 2.

One of the most exciting problems of algebra is the algebraic solution of equations. Nearly all mathematicians have at one time or another struggled with this problem. It is simple to find the solutions [roots] of equations of the first- and second-degree [quadratic]. The solutions of third- and fourth-degree equations require more effort but are not difficult. Beyond the fourth degree, it becomes questionable whether they can be solved by algebraic methods.

NIELS HENRIK ABEL (1802–1829)

9.1 Solving Polynomial Equations by Factoring and the Square Root Property

OBJECTIVES In this section we will learn to
1. solve quadratic equations by factoring; and
2. solve quadratic equations by the square root property

In Section 6.6 we solved equations of degree greater than 1 by using the zero-factor theorem. As its name implies, we can use this theorem to find the solution set of an equation that is factorable. The next three examples review this method.

EXAMPLE 1 Solve $6x^3 - 17x^2 + 12x = 0$ by factoring.

SOLUTION

$$6x^3 - 17x^2 + 12x = 0$$
$$x(6x^2 - 17x + 12) = 0 \qquad \text{Remove common factors first.}$$
$$x(3x - 4)(2x - 3) = 0$$
$$x = 0 \quad \text{or} \quad 3x - 4 = 0 \quad \text{or} \quad 2x - 3 = 0 \qquad \text{Zero-factor theorem}$$
$$x = 0 \quad \text{or} \quad x = \frac{4}{3} \quad \text{or} \quad x = \frac{3}{2}$$

We'll check 0 and $\frac{4}{3}$, the check of $\frac{3}{2}$ is left to the reader.

$$6x^3 - 17x^2 + 12x = 0$$

If $x = 0$, then

$$6(0)^3 - 17(0)^2 + 12(0) \stackrel{?}{=} 0$$
$$6 \cdot 0 - 17 \cdot 0 + 12 \cdot 0 \stackrel{?}{=} 0$$
$$0 - 0 + 0 \stackrel{?}{=} 0$$
$$0 \stackrel{\checkmark}{=} 0$$

If $x = \frac{4}{3}$, then

$$6\left(\frac{4}{3}\right)^3 - 17\left(\frac{4}{3}\right)^2 + 12\left(\frac{4}{3}\right) \stackrel{?}{=} 0$$
$$6\left(\frac{64}{27}\right) - 17\left(\frac{16}{9}\right) + 12\left(\frac{4}{3}\right) \stackrel{?}{=} 0$$
$$\frac{128}{9} - \frac{272}{9} + \frac{144}{9} \stackrel{?}{=} 0$$
$$0 \stackrel{\checkmark}{=} 0$$

The solution set is $\left\{0, \dfrac{4}{3}, \dfrac{3}{2}\right\}$.

EXAMPLE 2 Solve $x^4 - 13x^2 + 36 = 0$ by factoring.

SOLUTION

$$x^4 - 13x^2 + 36 = 0$$
$$(x^2 - 4)(x^2 - 9) = 0$$
$$(x - 2)(x + 2)(x - 3)(x + 3) = 0$$

$x - 2 = 0$ or $x + 2 = 0$ or $x - 3 = 0$ or $x + 3 = 0$
$x = 2$ or $x = -2$ or $x = 3$ or $x = -3$

The check is left to the reader. The solution set is $\{-3, -2, 2, 3\}$.

EXAMPLE 3 Solve $x^2 - 9 = 0$ by factoring.

SOLUTION

$$x^2 - 9 = 0$$
$$(x - 3)(x + 3) = 0$$

$x - 3 = 0$ or $x + 3 = 0$
$x = 3$ or $x = -3$

The check is left to the reader. The solution set is $\{-3, 3\}$.

A second method for solving quadratic equations is to use the *square root property*. To illustrate this property, we again consider the equation $x^2 - 9 = 0$, which we solved by factoring in Example 3. This time we rewrite it as follows:

$x^2 - 9 = 0$
$x^2 = 9$ Add 9 to each side of the equation.

We know the solutions to the equation are -3 and 3, both of which are square roots of 9. Thus the solution to the equation can be written

$x = \pm\sqrt{9} = \pm 3$

The Square Root Property

The equation $x^2 = n$ has exactly two solutions,

$x = \sqrt{n}$ and $x = -\sqrt{n}$

The solutions are often written $x = \pm\sqrt{n}$.
The expression $\pm\sqrt{n}$ is read **"the positive or negative square root of n."**

EXAMPLE 4 Solve $x^2 - 45 = 0$ by using the square root property.

SOLUTION
$$x^2 - 45 = 0$$
$$x^2 = 45 \quad \text{Add 45 to each side.}$$
$$x = \pm\sqrt{45} \quad \text{Square root property.}$$
$$x = \pm 3\sqrt{5} \quad \sqrt{45} = \sqrt{9 \cdot 5} = 3\sqrt{5}$$

CHECK
$(3\sqrt{5})^2 \stackrel{?}{=} 45 \quad x = 3\sqrt{5} \quad\quad (-3\sqrt{5})^2 \stackrel{?}{=} 45 \quad x = -3\sqrt{5}$
$9 \cdot 5 \stackrel{?}{=} 45 \quad\quad\quad\quad\quad\quad 9 \cdot 5 \stackrel{?}{=} 45$
$45 \stackrel{\checkmark}{=} 45 \quad\quad\quad\quad\quad\quad 45 \stackrel{\checkmark}{=} 45$

The solution set is $\{-3\sqrt{5}, 3\sqrt{5}\}$.

EXAMPLE 5 Solve $(2a - 1)^2 = -4$ by the square root property.

SOLUTION We let $2a - 1$ play the role of x in the square root property.
$$(2a - 1)^2 = -4$$
$$2a - 1 = \pm\sqrt{-4} \quad \text{Square root property}$$
$$2a - 1 = \pm 2i$$
$$2a = 1 \pm 2i \quad \text{Add 1 to each side.}$$
$$a = \frac{1 \pm 2i}{2} \quad \text{Divide each side by 2.}$$

or

$$a = \frac{1}{2} \pm i \quad \text{Standard form of a complex number}$$

CAUTION

$\frac{1 + 2i}{2}$ cannot be simplified to $1 \pm i$, since that would require that we be able to divide out the 2's. But 2 is not a common factor of *both* terms in the numerator. Only when the numerator and denominator have a common factor can we simplify a rational expression by division.

CHECK
$\left[2\left(\frac{1}{2} + i\right) - 1\right]^2 \stackrel{?}{=} -4 \quad x = \frac{1}{2} + i$
$[1 + 2i - 1]^2 \stackrel{?}{=} -4$
$[2i]^2 \stackrel{?}{=} -4$
$-4 \stackrel{\checkmark}{=} -4 \quad i^2 = -1$

The check of the other solution is left to the reader. The solution set is $\left\{\frac{1}{2} + i, \frac{1}{2} - i\right\}$.

EXAMPLE 6 Solve $y^4 - 64 = 0$.

SOLUTION We begin by factoring the equation into *quadratic factors:*
$$(y^2 - 8)(y^2 + 8) = 0$$
$$y^2 - 8 = 0 \quad \text{or} \quad y^2 + 8 = 0 \quad \text{Zero-factor theorem}$$

Each of these equations can now be solved by the square root property.

$$y^2 - 8 = 0 \qquad\qquad y^2 + 8 = 0$$
$$y^2 = 8 \qquad\qquad y^2 = -8$$
$$y = \pm\sqrt{8} \qquad\qquad y = \pm\sqrt{-8}$$
$$y = \pm 2\sqrt{2} \qquad\qquad y = \pm 2\sqrt{2}\, i$$

Since the fourth powers of $2\sqrt{2}$ and $-2\sqrt{2}$ are the same and the fourth powers of $2\sqrt{2}\, i$ and $-2\sqrt{2}\, i$ are the same, we need to check only two solutions.

CHECK $\qquad\qquad y^4 - 64 = 0$

$$(2\sqrt{2})^4 - 64 \stackrel{?}{=} 0 \qquad x = 2\sqrt{2} \qquad (2\sqrt{2}\, i)^4 - 64 = 0 \qquad x = 2\sqrt{2}\, i$$
$$16(4) - 64 \stackrel{?}{=} 0 \qquad\qquad\qquad\qquad 16(4i^4) - 64 \stackrel{?}{=} 0$$
$$64 - 64 \stackrel{?}{=} 0 \qquad\qquad\qquad\qquad 16(4) - 64 \stackrel{?}{=} 0 \qquad i^4 = 1$$
$$0 \stackrel{\checkmark}{=} 0 \qquad\qquad\qquad\qquad\qquad 64 - 64 \stackrel{?}{=} 0$$
$$\qquad\qquad\qquad\qquad\qquad\qquad\qquad\qquad 0 \stackrel{\checkmark}{=} 0$$

The solution set is $\{-2\sqrt{2}, 2\sqrt{2}, -2\sqrt{2}\, i, 2\sqrt{2}\, i\}$.

EXAMPLE 7 Solve each of the following by the square root property. Simplify all solutions.

(a) $A^3 = 3r^2$ for r \qquad (b) $\dfrac{tx^2}{s} = f$ for x, where $t \neq 0$

SOLUTION (a) $\qquad A^3 = 3r^2$

$$\dfrac{A^3}{3} = r^2 \qquad\text{Divide each side by 3.}$$

$$\pm\sqrt{\dfrac{A^3}{3}} = r \qquad\text{Square root property}$$

$$\pm\sqrt{\dfrac{A^3}{3}\cdot\dfrac{3}{3}} = r \qquad\text{Simplify; eliminate the fraction under the radical.}$$

$$\pm\dfrac{A}{3}\sqrt{3A} = r$$

The solution set is $\left\{-\dfrac{A}{3}\sqrt{3A}, \dfrac{A}{3}\sqrt{3A}\right\}$.

(b) $\dfrac{tx^2}{s} = f$

$$tx^2 = sf \qquad\text{Clear of fractions.}$$

$$x^2 = \dfrac{sf}{t} \qquad\text{Isolate } x^2.$$

$$x = \pm\sqrt{\dfrac{sf}{t}} \qquad\text{Square root property}$$

$$x = \pm \frac{\sqrt{sft}}{t} = \pm \frac{1}{t}\sqrt{sft} \qquad \sqrt{\frac{sf}{t}} = \sqrt{\frac{sft}{t^2}} = \frac{\sqrt{sft}}{t}, \quad t > 0$$

The solution set is $\left\{-\frac{\sqrt{sft}}{t}, \frac{\sqrt{sft}}{t}\right\}$.

Given a solution set, it is not difficult to find the equation that goes with it. Remember, we just reverse the process that we used to find the solutions. For example, to find an equation with a solution set of $\{-2, 5\}$, with no repeated roots, we proceed as follows.

If $\quad x = -2 \quad$ or $\quad x = 5,$

then $\quad x + 2 = 0 \quad$ or $\quad x - 5 = 0.$

This means that

$\quad (x + 2)(x - 5) = 0 \qquad$ Zero-factor theorem

The equation is

$\quad x^2 - 3x - 10 = 0 \qquad$ Multiply.

If we multiply each side of $x^2 - 3x - 10$ by any constant, we generate an equivalent equation. An infinite number of equations exists. We found the simplest one. An equation of degree greater than 2 can be found from its solutions when they are known. The next example shows how we can find an equation with three solutions. Such an equation is called a third-degree or **cubic equation.**

EXAMPLE 8 Find an equation with the solution set $(1, \sqrt{5}, -\sqrt{5})$, with no repeated roots.

SOLUTION
$\quad x = 1 \quad$ or $\quad x = \sqrt{5} \quad$ or $\quad x = -\sqrt{5}$
$\quad x - 1 = 0 \quad$ or $\quad x - \sqrt{5} = 0 \quad$ or $\quad x + \sqrt{5} = 0$
$\quad (x - 1)(x - \sqrt{5})(x + \sqrt{5}) = 0$

Since $x - \sqrt{5}$ and $x + \sqrt{5}$ are conjugates, their product is the difference of two squares. We'll find that product first.

$\quad (x - 1)(x^2 - 5) = 0 \qquad (x - \sqrt{5})(x + \sqrt{5}) = x^2 - 5$
$\quad x^3 - 5x - x^2 + 5 = 0 \qquad$ FOIL method
$\quad x^3 - x^2 - 5x + 5 = 0 \qquad$ Standard form

EXAMPLE 9 Find an equation that has the solution set $\{2, 3 + \sqrt{2}i, 3 - \sqrt{2}i\}$, with no repeated roots.

SOLUTION If $\quad x = 2 \quad$ or $\quad x = 3 + \sqrt{2}i \quad$ or $\quad x = 3 - \sqrt{2}i$

then $\quad x - 2 = 0 \quad$ or $\quad x - 3 - \sqrt{2}i = 0 \quad$ or $\quad x - 3 + \sqrt{2}i = 0$

$\quad (x - 2)(x - 3 - \sqrt{2}i)(x - 3 + \sqrt{2}i) = 0$

Once again, $x - 3 - \sqrt{2}\,i$ and $x - 3 + \sqrt{2}\,i$ are conjugates so it is easier to find their product first.

$$(x - 3 - \sqrt{2}\,i)(x - 3 + \sqrt{2}\,i) = [(x - 3) - \sqrt{2}\,i][(x - 3) + \sqrt{2}\,i]$$
$$= (x - 3)^2 - (\sqrt{2}\,i)^2$$
$$= x^2 - 6x + 9 - 2i^2$$
$$= x^2 - 6x + 9 + 2 \qquad -2i^2 = +2$$
$$= x^2 - 6x + 11$$

$$(x - 2)(x - 3 - \sqrt{2}\,i)(x - 3 + \sqrt{2}\,i) = 0$$
$$(x - 2)(x^2 - 6x + 11) = 0$$
$$x^3 - 8x^2 + 23x - 22 = 0 \qquad \text{Multiply and combine like terms.}$$

EXAMPLE 10 Solve $(y - 3)^2 = 7$ by the square root property.

SOLUTION We think of $y - 3$ as the x in the square root property. Then

$$(y - 3)^2 = 7$$
$$y - 3 = \pm\sqrt{7} \qquad \text{Square root property}$$
$$y = 3 \pm \sqrt{7} \qquad \text{Add 3 to each side.}$$

CHECK $\qquad (y - 3)^2 = 7$

$(3 + \sqrt{7} - 3)^2 \stackrel{?}{=} 7 \qquad y = 3 + \sqrt{7} \qquad \Big| \qquad (3 - \sqrt{7} - 3)^2 \stackrel{?}{=} 7 \qquad y = 3 - \sqrt{7}$
$(\sqrt{7})^2 \stackrel{?}{=} 7 \qquad\qquad\qquad\qquad\quad \Big| \qquad (-\sqrt{7})^2 \stackrel{?}{=} 7$
$7 \stackrel{\checkmark}{=} 7 \qquad\qquad\qquad\qquad\qquad \Big| \qquad 7 \stackrel{\checkmark}{=} 7$

The solution set is $\{3 \pm \sqrt{7}\}$ or $\{3 - \sqrt{7}, 3 + \sqrt{7}\}$.

EXAMPLE 11 Solve $x^2 + 4x + 4 = 3$ by factoring the left-hand side and using the square root property.

SOLUTION
$$x^2 + 4x + 4 = 3$$
$$(x + 2)^2 = 3 \qquad x^2 + 4x + 4 = (x + 2)^2$$
$$x + 2 = \pm\sqrt{3} \qquad \text{Square root property}$$
$$x = -2 \pm \sqrt{3} \qquad \text{Subtract 2 from each side.}$$

We will check one solution and leave the check of the other to the reader.

CHECK $\qquad x^2 + 4x + 4 = 3$

$$(-2 + \sqrt{3})^2 + 4(-2 + \sqrt{3}) + 4 \stackrel{?}{=} 3$$
$$4 - 4\sqrt{3} + 3 - 8 + 4\sqrt{3} + 4 \stackrel{?}{=} 3$$
$$3 \stackrel{\checkmark}{=} 3$$

The solution set is $\{-2 \pm \sqrt{3}\}$ or $\{-2 + \sqrt{3}, -2 - \sqrt{3}\}$.

EXAMPLE 12 Factor the left side of $4x^2 - 20x + 25 = 2$ so that the equation is in the form $(ax - b)^2 = c$ and then solve it by the square root property.

SOLUTION
$$4x^2 - 20x + 25 = 2$$
$$(2x - 5)^2 = 2 \quad \text{Factor the left-hand side.}$$
$$2x - 5 = \pm\sqrt{2} \quad \text{Square root property}$$
$$2x = 5 \pm \sqrt{2} \quad \text{Add 5 to each side.}$$
$$x = \frac{5 \pm \sqrt{2}}{2} \quad \text{Divide each side by 2.}$$

CHECK $\quad 4x^2 - 20x + 25 = 2$

$$4\left(\frac{5 + \sqrt{2}}{2}\right)^2 - 20\left(\frac{5 + \sqrt{2}}{2}\right) + 25 \stackrel{?}{=} 2 \quad x = \frac{5 + \sqrt{2}}{2}$$

$$4\left(\frac{25 + 10\sqrt{2} + 2}{4}\right) - 20\left(\frac{5 + \sqrt{2}}{2}\right) + 25 \stackrel{?}{=} 2 \quad \left(\frac{5 + \sqrt{2}}{2}\right)^2 = \left(\frac{25 + 10\sqrt{2} + 2}{4}\right)$$

$$25 + 10\sqrt{2} + 2 - 50 - 10\sqrt{2} + 25 \stackrel{?}{=} 2 \quad \text{Simplify and multiply.}$$

$$2 \stackrel{\checkmark}{=} 2 \quad \text{Combine like terms.}$$

The check of the other solution is left to the reader. The solution set is

$$\left\{\frac{5 \pm \sqrt{2}}{2}\right\} = \left\{\frac{5}{2} + \frac{\sqrt{2}}{2}, \frac{5}{2} - \frac{\sqrt{2}}{2}\right\}$$

Exercise Set 9.1

Solve each of the following equations by factoring. Remember to remove common factors first.

1. $x^2 - 6x + 8 = 0$
2. $y^2 + y - 2 = 0$
3. $2a^2 + 2 = 5a$
4. $2k^2 - k = 0$
5. $2t^2 - 5t - 3 = 0$
6. $15m^2 - 23m - 28 = 0$
7. $6s^2 - 33s + 45 = 0$
8. $24p^3 - 14p^2 - 20p = 0$
9. $\frac{1}{6}w^2 - \frac{11}{36}w + \frac{1}{12} = 0$
10. $\frac{1}{4}x^3 - \frac{1}{9}x = 0$
11. $15r^2 + 46r + 35 = 0$
12. $q^2 - 81 = 0$
13. $0.015a^2 - 0.019ab + 0.006b^2 = 0$
14. $0.02y^2 + 0.05yz - 0.07z^2 = 0$

Solve each of the following quadratic equations using the square root property.

15. $x^2 = 16$
16. $y^2 = 9$
17. $z^2 = -4$
18. $w^2 = -25$
19. $a^2 - 0.81 = 0$
20. $b^2 - 0.0121 = 0$
21. $v^2 = 10^{-6}$
22. $m^2 = 10^4$
23. $4n^2 - 1 = 0$
24. $16t^2 - 169 = 0$
25. $x^2 = 7$
26. $y^2 = 5$
27. $0.2a^2 = 0.3$
28. $0.03c^2 = 0.11$
29. $50p^2 = 32$
30. $49k^2 = 98$
31. $(2a - 1)^2 = 5$
32. $(3q - 2)^2 = 7$
33. $(4m + 3)^2 = 1$
34. $(7u - 1)^2 = 13$
35. $(2v + 3)^2 + 3 = 0$
36. $(9d - 5)^2 + 7 = 0$
37. $(7x - 5)^2 + 4 = 0$
38. $(y - 11)^2 + 36 = 0$
39. $(x - a)^2 = b^2$
40. $(y - c)^2 = p^2$
41. $81k^2 + 25 = 0$
42. $64q^2 + 225 = 0$

Write a quadratic or cubic equation having the given solution set. Assume no repeated roots.

43. $\{2, 3\}$
44. $\{1, -4\}$
45. $\{0, -1, 5\}$
46. $\{0, 3, -2\}$
47. $\left\{\frac{1}{2}, \frac{1}{3}\right\}$
48. $\left\{-\frac{1}{2}, \frac{1}{5}\right\}$

49. $\{\pm 2\sqrt{2}\}$ 50. $\{\pm 3\sqrt{3}\}$ 51. $\{\pm 3i\}$
52. $\{\pm 2i\}$ 53. $\{2 \pm \sqrt{5}\}$ 54. $\{1 \pm 2\sqrt{3}\}$
55. $\{1 \pm 2i\}$ 56. $\{3 \pm 5i\}$ 57. $\{1 \pm \sqrt{3}\,i\}$
58. $\{4 \pm \sqrt{5}\,i\}$ 59. $\{2a, 3b\}$ 60. $\{6c, -7d\}$
61. $\left\{\dfrac{-b \pm \sqrt{b^2 - 4ac}}{2a}\right\}$ 62. $\left\{\dfrac{-c \pm \sqrt{c^2 - 4ad}}{2a}\right\}$

Factor the left side of each of the following equations into the form $(ax + b)^2 = c$; then solve using the square root property.

63. $a^2 + 2a + 1 = 5$
64. $p^2 - 4p + 4 = 3$
65. $m^2 - 0.6m + 0.09 = -0.18$
66. $k^2 + 1.2k + 0.36 = 0.1$
67. $4x^2 + 12x + 9 = 8$
68. $25y^2 + 60y + 36 = 12$
69. $49s^2 - 42s + 9 = 32$
70. $121w^2 - 44w + 4 = 27$
71. $36t^2 - 108t + 81 = -18$
72. $144r^2 - 168r + 49 = -12$
73. $64u^2 - 112u + 49 = -50$
74. $49q^2 + 98q + 49 = -72$

Solve each of the following equations. Factor first, then solve using the square root property.

75. $x^4 - 81 = 0$
76. $y^4 - 16 = 0$
77. $16a^4 - 64 = 0$
78. $15s^4 - 135 = 0$
79. $p^4 - b^4 = 0,\ b > 0$
80. $t^4 - c^4 = 0,\ c > 0$
81. $m^4 - 5m^2 + 6 = 0$
82. $6r^4 + 13r^2 + 6 = 0$

Solve each of the following for the indicated variable with the indicated restriction.

83. $A = \pi r^2$, for r, where $r > 0$
84. $V = \dfrac{1}{3}\pi r^2 h$, for r, where $r > 0$
85. $S = \dfrac{1}{2}gt^2$, for t, where $t > 0$
86. $a^2 + b^2 = c^2$, for a, where $a > 0$
87. $A = P(1 + r)^2$, for r, where $r > 0$
88. $S = h - r^2$, for r, where $r > 0$

Solve each of the following equations for x.

89. $x^2 - c = 0$
90. $dx^2 = b$
91. $\dfrac{rx^2}{s} = t$
92. $\dfrac{ax^2}{b} = c$
93. $(ax - b)^2 = c$
94. $(cx + d)^2 = d$
95. $(mx + n)^2 = n^2$
96. $(3rx + 2s)^2 = s^2$

For Extra Thought

97. Given the equation $x^2 + 2x - 8 = 0$, explain each step shown.
    (a) $x^2 + 2x = 8$
    (b) $x^2 + 2x + 1 = 8 + 1$
    (c) $x^2 + 2x + 1 = 9$ (d) $(x + 1)^2 = 9$
    (e) $x + 1 = \pm\sqrt{9}$ (f) $x + 1 = \pm 3$
    (g) $x = -1 \pm 3$ (h) $x = 2$ or -4

98. Given the equation $3x^2 - 18x - 21 = 0$, explain each step shown.
    (a) $x^2 - 6x - 7 = 0$
    (b) $x^2 - 6x = 7$
    (c) $x^2 - 6x + 9 = 7 + 9$
    (d) $x^2 - 6x + 9 = 16$ (e) $(x - 3)^2 = 16$
    (f) $x - 3 = \pm\sqrt{16}$ (g) $x - 3 = \pm 4$
    (h) $x = 3 \pm 4$ (i) $x = 7$ or -1

9.2 Solving Quadratic Equations by Completing the Square

OBJECTIVE In this section we will learn to solve quadratic equations by completing the square.

In Section 9.1 we solved equations such as $(x + 2)^2 = 6$ by taking the square root of each side. In this section we will learn to solve equations that are not easily factorable but can be rewritten in the form

$$(ax + b)^2 = c$$

These can then be solved by the square root property. For example, consider the equation

$$x^2 + 6x - 8 = 0$$

Adding 8 to each side of the equation yields

$$x^2 + 6x = 8$$

The left side of the equation is not a perfect-square trinomial, but it can be made into one by adding 9 to each side:

$$x^2 + 6x + 9 = 8 + 9$$
$$(x + 3)^2 = 17$$
$$x + 3 = \pm\sqrt{17} \qquad \text{Square root property}$$
$$x = -3 \pm \sqrt{17} \qquad \text{Subtract 3 from each side.}$$

The solution set is $\{-3 - \sqrt{17}, -3 + \sqrt{17}\}$.

We changed $x^2 + 6x$ into the perfect square $x^2 + 6x + 9$ by using a process called **completing the square**. How did we know that adding 9 would create a perfect-square trinomial? Consider the square of the binomial

$$(x + a)^2 = x^2 + 2ax + a^2$$

How do we know what quantity is necessary to complete the square? *If the coefficient of the second-degree term is* 1, *the necessary quantity is the square of one-half of the coefficient of the first-degree term,* $2ax$, which is $\frac{1}{2}(2a)$, or a.

$$x^2 + 2ax + a^2 = (x + a)^2$$

$\frac{1}{2}$ of $2a$, squared ⤴   ⤴ $\frac{1}{2}$ of $2a$

EXAMPLE 1 Complete the square: (a) $x^2 + 4x$;   (b) $x^2 - 10x$;   (c) $x^2 + 12x$.

SOLUTION In what follows, each quantity in column **II** is the *square* of $\frac{1}{2}$ of the coefficient of x in column **I**. Each quantity in column **III** is $\frac{1}{2}$ the coefficient of x in column **I**.

	I	II	III
(a)	$x^2 + 4x +$	4	$= (x + 2)^2$
(b)	$x^2 - 10x +$	25	$= (x - 5)^2$
(c)	$x^2 + 12x +$	36	$= (x + 6)^2$

In general, to complete the square for the expression $x^2 + Bx$, we add the quantity $\left(\dfrac{B}{2}\right)^2$.

$$x^2 + Bx + \left(\frac{B}{2}\right)^2 = \left(x + \frac{B}{2}\right)^2$$

EXAMPLE 2 Solve $p^2 + 2p + 5 = 0$ by completing the square.

SOLUTION

$p^2 + 2p = -5$ Subtract 5 from each side.

$p^2 + 2p + 1 = -5 + 1$ Complete the square by adding $[\frac{1}{2}(2)]^2 = 1$ to each side.

$(p + 1)^2 = -4$ Write $p^2 + 2p + 1$ as the square of a binomial.

$p + 1 = \pm\sqrt{-4}$ Square root property

$p + 1 = \pm 2i$

$p = -1 \pm 2i$

The check is left to the reader. The solution set is $\{-1 + 2i, -1 - 2i\}$.

EXAMPLE 3 Solve: $3x^2 + 4x - 5 = 0$.

SOLUTION First we write the equation as $3x^2 + 4x = 5$. To complete the square, the lead coefficient must be 1, so we divide each side by 3:

$$3x^2 + 4x = 5$$

$$x^2 + \frac{4}{3}x = \frac{5}{3}$$

Now we complete the square by adding the square of $\frac{1}{2}$ of the coefficient of x to each side: $\frac{1}{2}$ of $\frac{4}{3}$ is $\frac{2}{3}$ and $(\frac{2}{3})^2 = \frac{4}{9}$.

$$x^2 + \frac{4}{3}x + \frac{4}{9} = \frac{5}{3} + \frac{4}{9} \qquad \text{Add } \frac{4}{9} \text{ to each side.}$$

$$\left(x + \frac{2}{3}\right)^2 = \frac{19}{9} \qquad \text{Add the fractions: the LCD is 9.}$$

$$x + \frac{2}{3} = \pm\frac{\sqrt{19}}{3}$$

$$x = -\frac{2}{3} \pm \frac{\sqrt{19}}{3}$$

CHECK $3\left(-\frac{2}{3} - \frac{\sqrt{19}}{3}\right)^2 + 4\left(-\frac{2}{3} - \frac{\sqrt{19}}{3}\right) - 5 \stackrel{?}{=} 0$

$3\left(\frac{4}{9} + \frac{4\sqrt{19}}{9} + \frac{19}{9}\right) - \frac{8}{3} - \frac{4\sqrt{19}}{3} - 5 \stackrel{?}{=} 0$

$\frac{4}{3} + \frac{4\sqrt{19}}{3} + \frac{19}{3} - \frac{8}{3} - \frac{4\sqrt{19}}{3} - \frac{15}{3} \stackrel{?}{=} 0 \qquad 5 = \frac{15}{3}$

$\frac{4 + 19 - 8 - 15}{3} + \frac{4\sqrt{19} - 4\sqrt{19}}{3} \stackrel{?}{=} 0$

$\frac{0}{3} + \frac{0}{3} \stackrel{?}{=} 0$

$0 \stackrel{\checkmark}{=} 0$

The solution set is $\left\{-\dfrac{2}{3} - \dfrac{\sqrt{19}}{3},\ -\dfrac{2}{3} + \dfrac{\sqrt{19}}{3}\right\}$. The check of the other solution is left to the reader.

To Solve a Quadratic Equation by Completing the Square

Step 1 Isolate all terms involving the variable on one side of the equation.

Step 2 If the coefficient of the second-degree term is not 1, divide both sides by that coefficient.

Step 3 Add the square of $\frac{1}{2}$ of the coefficient of the first-degree term to each side of the equation.

Step 4 Express the perfect-square trinomial from Step 3 as the square of a binomial.

Step 5 Use the square root property to solve the resulting equation.

EXAMPLE 4 Solve $3x^2 + 5x + 1 = 0$ by using the five-step process just outlined.

SOLUTION

Step 1 $\quad 3x^2 + 5x = -1$

Step 2 $\quad x^2 + \dfrac{5}{3}x = \dfrac{-1}{3}$

Step 3 $\quad x^2 + \dfrac{5}{3}x + \dfrac{25}{36} = \dfrac{-1}{3} + \dfrac{25}{36} \qquad \dfrac{1}{2} \cdot \dfrac{5}{3} = \dfrac{5}{6};\ \left(\dfrac{5}{6}\right)^2 = \dfrac{25}{36}$

Step 4 $\quad \left(x + \dfrac{5}{6}\right)^2 = \dfrac{-12}{36} + \dfrac{25}{36} = \dfrac{13}{36}$

Step 5 $\quad x + \dfrac{5}{6} = \pm\sqrt{\dfrac{13}{36}} = \pm\dfrac{\sqrt{13}}{6} \qquad \sqrt{36} = 6$

$\quad\quad\quad\quad x = -\dfrac{5}{6} \pm \dfrac{\sqrt{13}}{6} = \dfrac{-5 \pm \sqrt{13}}{6}$

CHECK $\quad 3\left(\dfrac{-5 + \sqrt{13}}{6}\right)^2 + 5\left(\dfrac{-5 + \sqrt{13}}{6}\right) + 1 \stackrel{?}{=} 0 \qquad x = \dfrac{-5 + \sqrt{13}}{6}$

$\quad\quad 3\left(\dfrac{25 - 10\sqrt{13} + 13}{36}\right) + 5\left(\dfrac{-5 + \sqrt{13}}{6}\right) + 1 \stackrel{?}{=} 0$

$\quad\quad \dfrac{25 - 10\sqrt{13} + 13}{12} + \dfrac{-25 + 5\sqrt{13}}{6} + 1 \stackrel{?}{=} 0$

$\quad\quad \dfrac{25 - 10\sqrt{13} + 13 - 50 + 10\sqrt{13} + 12}{12} \stackrel{?}{=} 0$

$\quad\quad \dfrac{0}{12} \stackrel{?}{=} 0$

$\quad\quad 0 \stackrel{\checkmark}{=} 0$

The check of the other solution is similar and is left to the reader. The solution set is $\left\{\dfrac{-5 + \sqrt{13}}{6}, \dfrac{-5 - \sqrt{13}}{6}\right\}$.

Some of the processes we used earlier to solve equations lead to equations that can also be solved by completing the square.

EXAMPLE 5 Solve: $\dfrac{x}{x^2 + 3x + 2} - \dfrac{x + 3}{x + 2} = \dfrac{7}{x + 1}$, $x \neq -1, -2$.

SOLUTION
$$\dfrac{x}{x^2 + 3x + 2} - \dfrac{x + 3}{x + 2} = \dfrac{7}{x + 1}$$
$$\dfrac{x}{(x + 1)(x + 2)} - \dfrac{x + 3}{x + 2} = \dfrac{7}{x + 1}$$

We clear of fractions by multiplying each side by $(x + 2)(x + 1)$, the LCD.

$$(x + 2)(x + 1)\left[\dfrac{x}{(x + 1)(x + 2)} - \dfrac{x + 3}{x + 2}\right] = (x + 2)(x + 1)\left(\dfrac{7}{x + 1}\right)$$

$x - (x + 3)(x + 1) = 7(x + 2)$	Multiply and simplify.
$x - (x^2 + 4x + 3) = 7x + 14$	Multiply.
$x - x^2 - 4x - 3 = 7x + 14$	Remove grouping symbols.
$-x^2 - 3x - 3 = 7x + 14$	Combine like terms on each side.
$-x^2 - 10x = 17$	Isolate x-terms on one side.
$x^2 + 10x = -17$	Multiply each side by -1.

Now we complete the square by adding 25 to each side.

$$x^2 + 10x + 25 = 8$$
$$(x + 5)^2 = 8$$
$$x + 5 = \pm\sqrt{8} = \pm 2\sqrt{2} \qquad \text{Square root property}$$
$$x = -5 \pm 2\sqrt{2}$$

The check is left to the reader. The solution set is $\{-5 - 2\sqrt{2}, -5 + 2\sqrt{2}\}$.

EXAMPLE 6 Solve: $x - \sqrt{x - 1} = 3$.

SOLUTION First we isolate the radical and then square each side.

$$x - 3 = \sqrt{x - 1}$$
$$(x - 3)^2 = (\sqrt{x - 1})^2$$
$$x^2 - 6x + 9 = x - 1$$
$$x^2 - 7x + 10 = 0$$

This equation can be solved either by factoring or completing the square.

FACTORING	COMPLETING THE SQUARE
$x^2 - 7x + 10 = 0$	$x^2 - 7x + 10 = 0$
$(x - 5)(x - 2) = 0$	$x^2 - 7x = -10$
$x - 5 = 0$ or $x - 2 = 0$	$x^2 - 7x + \dfrac{49}{4} = -10 + \dfrac{49}{4}$
$x = 5$ or $x = 2$	$\left(x - \dfrac{7}{2}\right)^2 = \dfrac{9}{4}$
	$x - \dfrac{7}{2} = \pm\dfrac{3}{2}$
	$x = \dfrac{7}{2} \pm \dfrac{3}{2}$
	$x = 5$ or $x = 2$

Factoring is the most efficient method in this example. Since we used squaring to find the solution set, we must check for extraneous solutions.

$$\text{CHECK} \quad 5 - \sqrt{5 - 1} \stackrel{?}{=} 3 \quad \bigg| \quad 2 - \sqrt{2 - 1} \stackrel{?}{=} 3$$
$$5 - \sqrt{4} \stackrel{?}{=} 3 \quad \bigg| \quad 2 - \sqrt{1} \stackrel{?}{=} 3$$
$$5 - 2 \stackrel{?}{=} 3 \quad \bigg| \quad 2 - 1 \stackrel{?}{=} 3$$
$$3 \stackrel{\checkmark}{=} 3 \quad \bigg| \quad 1 \neq 3$$

Only 5 checks. The solution set is {5}.

EXAMPLE 7 Solve the radical function $f(x) = x + \sqrt{x - 4} - 6$.

SOLUTION The solutions to a radical function are those values for which the function is zero.

$$x + \sqrt{x - 4} - 6 = 0$$
$$x - 6 = -\sqrt{x - 4} \qquad \text{Isolate the radical first.}$$
$$(x - 6)^2 = (-\sqrt{x - 4})^2 \qquad \text{Square each side.}$$
$$x^2 - 12x + 36 = x - 4$$
$$x^2 - 13x + 40 = 0$$
$$(x - 8)(x - 5) = 0$$
$$x - 8 = 0 \quad \text{or} \quad x - 5 = 0$$
$$x = 8 \quad \text{or} \quad x = 5$$

$$\text{CHECK} \quad 8 + \sqrt{8 - 4} - 6 \stackrel{?}{=} 0 \quad \bigg| \quad 5 + \sqrt{5 - 4} - 6 \stackrel{?}{=} 0$$
$$8 + \sqrt{4} - 6 \stackrel{?}{=} 0 \quad \bigg| \quad 5 + \sqrt{1} - 6 \stackrel{?}{=} 0$$
$$8 + 2 - 6 \stackrel{?}{=} 0 \quad \bigg| \quad 5 + 1 - 6 \stackrel{?}{=} 0$$
$$4 \neq 0 \quad \bigg| \quad 0 \stackrel{\checkmark}{=} 0$$

Only 5 checks, so the solution set is {5}.

EXPLORING GRAPHICS

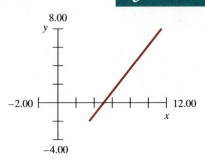

FIGURE 9.1

We can find the solution to Example 7 by using a graphics calculator or a computer. The point where the graph crosses the x-axis is the solution, because $f(x)$ is zero at this point. Figure 9.1 was drawn by a computer. Note that the graph crosses the x-axis where $x = 5$ and that the graph does not extend to the left of $x = 4$. That's because $\sqrt{x - 4}$ is not a real number if $x < 4$.

EXAMPLE 8 Solve the rational function $f(x) = \dfrac{-6}{x + 2} + x + 1$, $x \neq -2$, to find where its graph crosses the x-axis. See the preceding exploring graphics discussion.

SOLUTION The solutions to a rational function are those values that make the function zero.

$$\frac{-6}{x + 2} + x + 1 = 0$$

$$(x + 2)\left(\frac{-6}{x + 2}\right) + (x + 2)(x + 1) = (x + 2) \cdot 0 \quad \text{Multiply each side by } x + 2, \text{ the LCD.}$$

$$-6 + x^2 + 3x + 2 = 0 \quad \text{Simplify.}$$

$$x^2 + 3x - 4 = 0 \quad \text{Combine like terms.}$$

$$(x + 4)(x - 1) = 0$$

$$x + 4 = 0 \quad \text{or} \quad x - 1 = 0$$

$$x = -4 \quad \text{or} \quad x = 1$$

Both solutions check. The graph crosses the x-axis when $x = -4$ or $x = 1$.

EXAMPLE 9 The area of a rectangle is 44 square feet. The length is 7 feet more than the width. Find its perimeter.

SOLUTION We let $x =$ the width and $x + 7 =$ the length. See Figure 9.2. Then

$$x(x + 7) = 44 \quad \text{Area} = \text{length} \cdot \text{width}$$

$$x^2 + 7x - 44 = 0 \quad \text{Simplify.}$$

Since the equation is factorable, we choose this method over the more laborious method of completing the square.

$$(x - 4)(x + 11) = 0$$

$$x - 4 = 0 \quad \text{or} \quad x + 11 = 0$$

$$x = 4 \quad \text{or} \quad x = -11$$

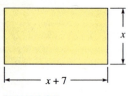

FIGURE 9.2

We reject -11 because a rectangle cannot have negative width. Therefore, the width is $x = 4$ feet and the length is $x + 7 = 4 + 7 = 11$ feet. The perimeter is $P = 2x + 2(x + 7) = 8 + 22 = 30$ feet.

Sometimes we need to use the process of completing the square for tasks other than solving a quadratic equation. This will occur frequently in the next chapter, when we find the center of a circle and the coordinates of the vertex of a parabola. The following is an example of one way it can happen.

EXAMPLE 10   Rewrite $x^2 + 4x + y^2 + 6y = 12$ by completing the squares of binomials.

SOLUTION   To complete the square of $x^2 + 4x$, we must add 4:

$$x^2 + 4x + 4 = (x + 2)^2$$

To complete the square of $y^2 + 6y$, we must add 9:

$$y^2 + 6y + 9 = (y + 3)^2$$

But if we add both 4 and 9 to the left side of the equation, we have to add $4 + 9 = 13$ to the right side.

$$x^2 + 4x + 4 + y^2 + 6y + 9 = 12 + 13$$
$$(x + 2)^2 + (y + 3)^2 = 25$$

Exercise Set 9.2

Complete the square for each of the Exercises 1–8. Then factor it and write it as the square of a binomial.

1. $a^2 + 2a + \underline{}$
2. $b^2 + 4b + \underline{}$
3. $x^2 - 6x + \underline{}$
4. $y^2 - 8y + \underline{}$
5. $r^2 + 10r + \underline{}$
6. $t^2 + 12t + \underline{}$
7. $v^2 - 14v + \underline{}$
8. $u^2 - 16u + \underline{}$

Solve by completing the square. Simplify all solutions.

9. $x^2 - 2x - 3 = 0$
10. $w^2 - 4w - 5 = 0$
11. $a^2 + 8a + 12 = 0$
12. $c^2 + 6c + 5 = 0$
13. $k^2 - k - 2 = 0$
14. $m^2 - m - 6 = 0$
15. $t^2 - 2t + 3 = 0$
16. $w^2 - 4w + 8 = 0$
17. $2s^2 + 3s - 1 = 0$
18. $3r^2 + r - 1 = 0$
19. $3y^2 - 5y + 3 = 0$
20. $x^2 - 9x - 19 = 0$
21. $a^2 + \frac{3}{2}a - \frac{1}{2} = 0$
22. $b^2 + \frac{1}{2}b - 1 = 0$
23. $4w^2 + 8w + 1 = 0$
24. $2c^2 + 5c - 1 = 0$
25. $2v^2 + 7v + 6 = 0$
26. $3t^2 + 4t + 2 = 0$
27. $3a^2 + 7a + 3 = 0$
28. $3q^2 - 8q + 3 = 0$
29. $5m^2 + 7m = 0$
30. $7v^2 - 9v = 0$
31. $t^2 - \frac{4}{3}t + \frac{1}{9} = 0$
32. $s^2 + \frac{7}{15}s - \frac{4}{15} = 0$

Solve by any method.

33. $w^2 + 11w + 30 = 0$
34. $n^2 - 5n - 24 = 0$
35. $(2a - 3)^2 = -11$
36. $(3t + 4)^2 = -1$
37. $x^2 + x + 1 = 0$
38. $y^2 - y + 1 = 0$
39. $p^4 - 625 = 0$
40. $s^4 - 256 = 0$
41. $6c^2 - c - 12 = 0$
42. $35u^2 + 2u - 24 = 0$
43. $3m^2 - 5m - 2 = 0$
44. $60v^2 + v - 10 = 0$

Solve each radical function. Check all solutions.

45. $f(x) = x + \sqrt{x + 5} - 7$
46. $g(p) = 1 + \sqrt{13 - p}$
47. $f(a) = \sqrt{6 + 6a} - 3 - \sqrt{a + 4}$
48. $f(t) = \sqrt{t + 8} + \sqrt{2t - 1} - 4$

Solve each rational function. State any restrictions on the variables.

49. $f(x) = \dfrac{1}{x + 2} - 3x - 1$
50. $g(a) = \dfrac{2a - 1}{3} - \dfrac{1}{a - 1}$
51. $g(v) = \dfrac{5}{v + 1} - 3 - \dfrac{4}{2v - 1}$

52. $f(w) = \dfrac{1}{2w - 5} - 1 + \dfrac{1}{1 - w}$

Write each of the following statements as a quadratic equation and solve it by any method.

53. The sum of twice the square of an integer and three times the integer is 9. Find the integer.
54. The sum of three times the square of an odd integer and twice the odd integer is 33. Find the odd integer.
55. The larger of two numbers exceeds the smaller by 2. Find the two numbers if the sum of their squares is 74.
56. The smaller of two numbers is 4 less than the larger. Find the numbers if the sum of their squares is 58.
57. The sum of the reciprocals of two consecutive integers is $\frac{7}{12}$. Find the integers.
58. A natural number added to its reciprocal is $\frac{122}{11}$. Find the natural number.
59. The length of a rectangle exceeds its width by 8 ft. Find the dimensions if the area is 48 ft².
60. The length of a rectangle is 4 cm shorter than twice the width. Find the dimensions if the area is 96 cm².
61. The base of a triangle is 2 m longer than twice the altitude. Find the base and altitude if the area is 12 m².
62. The second side of a right triangle is 7 units longer than the first side and the third side is 1 unit longer than the second side. Use the Pythagorean theorem to find the lengths of the three sides.

Complete the square in both x and y in Exercises 63–66. Rewrite the left side as the sum of two binomials squared.

63. $x^2 + 2x + y^2 + 4y = 4$
64. $x^2 + 6x + y^2 + 10y = 2$
65. $x^2 - 4x + y^2 - 12y = 7$
66. $x^2 - 14x + y^2 - 6y = 6$

WRITE IN YOUR OWN WORDS

67. Describe how to solve a quadratic equation by factoring.
68. Describe how to solve a quadratic equation by the square root property.
69. Describe how to solve a quadratic equation by completing the square.
70. Why can we conclude that if $(ax + b)(cx + d) = 0$ then $ax + b = 0$ or $cx + d = 0$?
71. Can the method of completing the square be used to solve all quadratic equations? Why or why not?
72. Can the square root property be used to solve all quadratic equations? Why or why not?
73. Can the method of factoring be used to solve all quadratic equations? Why or why not?

9.3 Solving Quadratic Equations Using the Quadratic Formula

OBJECTIVES

In this section we will learn to
1. solve quadratic equations by a formula;
2. use the discriminant to find the number and type of solutions of a quadratic equation; and
3. find the sum and product of the solutions of a quadratic equation without solving it.

In the last section we saw that any quadratic equation can be solved by the method of completing the square. In Section 6.6 we used the equation $ax^2 + bx + c = 0$, $a \neq 0$, to represent any quadratic equation. When we solve this equation by completing the square, the solution set provides a formula that we can use to solve any quadratic equation. We follow the five steps outlined in Section 9.2.

$$ax^2 + bx + c = 0$$

Step 1 $ax^2 + bx = -c$ Isolate the variables on one side of the equation.

Step 2 $\quad x^2 + \dfrac{b}{a}x = \dfrac{-c}{a}$ $\qquad$ Divide by a to make the coefficient of the second-degree term 1.

Step 3 $\quad x^2 + \dfrac{b}{a}x + \dfrac{b^2}{4a^2} = \dfrac{-c}{a} + \dfrac{b^2}{4a^2}$ $\qquad$ Complete the square: $\dfrac{1}{2} \cdot \dfrac{b}{a} = \dfrac{b}{2a}$; $\left(\dfrac{b}{2a}\right)^2 = \dfrac{b^2}{4a^2}$.

Step 4 $\quad \left(x + \dfrac{b}{2a}\right)^2 = \dfrac{b^2 - 4ac}{4a^2}$ $\qquad$ Write as the square of a binomial on the left side; the LCD is $4a^2$ on the right side.

Step 5 $\quad x + \dfrac{b}{2a} = \pm\sqrt{\dfrac{b^2 - 4ac}{4a^2}}$ $\qquad$ Square root property

$\qquad\qquad x + \dfrac{b}{2a} = \pm\dfrac{\sqrt{b^2 - 4ac}}{2a}$ $\qquad$ $\sqrt{4a^2} = 2a$

$\qquad\qquad x = -\dfrac{b}{2a} \pm \dfrac{\sqrt{b^2 - 4ac}}{2a}$ $\qquad$ Subtract $\dfrac{b}{2a}$ from each side.

$\qquad\qquad x = \dfrac{-b \pm \sqrt{b^2 - 4ac}}{2a}$ $\qquad$ The LCD is $2a$.

The solution set is $\left\{\dfrac{-b + \sqrt{b^2 - 4ac}}{2a}, \dfrac{-b - \sqrt{b^2 - 4ac}}{2a}\right\}$.

Rather than write this set in full, we usually just write

$$x = \dfrac{-b \pm \sqrt{b^2 - 4ac}}{2a}, \qquad a \neq 0$$

This equation is called the **quadratic formula**.

THEOREM 9.1

The Quadratic Formula

The solutions of the equation $ax^2 + bx + c = 0$, $a \neq 0$, are given by the formula

$$x = \dfrac{-b \pm \sqrt{b^2 - 4ac}}{2a}$$

The $\pm$ sign is read "plus or minus" rather than positive or negative. The quadratic formula is stated in terms of the variable x, but it is equally true for any other variable. To solve a quadratic equation, we identify the values of a, b, and c and substitute them into the formula.

EXAMPLE 1 Solve $3x^2 + 6x + 1 = 0$ using the quadratic formula.

SOLUTION Note that the equation is already written in standard form. We need only to identify a, b, and c.

$$ax^2 + bx + c = 0$$
$$\downarrow \quad \downarrow \quad \downarrow$$
$$3x^2 + 6x + 1 = 0$$

> **CAUTION**
>
> To identify the values of a, b, and c, first write the equation in standard form:
>
> $ax^2 + bx + c = 0$

Therefore $a = 3$, $b = 6$, and $c = 1$. Now, using the quadratic formula,

$$x = \frac{-b \pm \sqrt{b^2 - 4ac}}{2a}$$

$$= \frac{-6 \pm \sqrt{6^2 - 4 \cdot 3 \cdot 1}}{2 \cdot 3} \quad a = 3, b = 6, c = 1$$

$$= \frac{-6 \pm \sqrt{36 - 12}}{6}$$

$$= \frac{-6 \pm \sqrt{24}}{6}$$

$$= \frac{-6 \pm 2\sqrt{6}}{6} \quad \sqrt{24} = \sqrt{4 \cdot 6} = 2\sqrt{6}$$

$$= \frac{2(-3 \pm \sqrt{6})}{2 \cdot 3} \quad \text{Factor.}$$

$$= \frac{-3 \pm \sqrt{6}}{3} \quad \text{Simplify. } \textit{Only common factors can be divided out.}$$

The solution set is $\left\{\dfrac{-3 + \sqrt{6}}{3}, \dfrac{-3 - \sqrt{6}}{3}\right\}$ or $\left\{-1 + \dfrac{\sqrt{6}}{3}, -1 - \dfrac{\sqrt{6}}{3}\right\}$.

EXAMPLE 2 Solve $4y^2 - 2y - 7 = 0$ using the quadratic formula.

SOLUTION To use the quadratic formula, we replace x with y and think of $4y^2 - 2y - 7 = 0$ as $4y^2 + (-2)y + (-7) = 0$.

$$\begin{array}{ccccc} ay^2 & + & by & + & c & = 0 \\ \downarrow & & \downarrow & & \downarrow & \\ 4y^2 & + & (-2)y & + & (-7) & = 0 \end{array}$$

Therefore $a = 4$, $b = -2$, and $c = -7$, and

$$y = \frac{-b \pm \sqrt{b^2 - 4ac}}{2a}$$

$$= \frac{-(-2) \pm \sqrt{(-2)^2 - 4(4)(-7)}}{2(4)}$$

$$= \frac{2 \pm \sqrt{4 + 112}}{8}$$

$$= \frac{2 \pm \sqrt{116}}{8}$$

$$= \frac{2 \pm \sqrt{4 \cdot 29}}{8}$$

$$= \frac{2 \pm 2\sqrt{29}}{8}$$

$$= \frac{2(1 \pm \sqrt{29})}{2 \cdot 4} \quad \text{Factor the numerator and the denominator.}$$

$$= \frac{1 \pm \sqrt{29}}{4} \qquad \text{Divide out the common factor, 2.}$$

The solution set is $\left\{\dfrac{1 \pm \sqrt{29}}{4}\right\}$ or $\left\{\dfrac{1}{4} + \dfrac{\sqrt{29}}{4}, \dfrac{1}{4} - \dfrac{\sqrt{29}}{4}\right\}$.

The quadratic formula can be used even if an equation is factorable. Since factoring usually requires less effort, it is a good idea first to check to see if the equation can be factored easily.

EXAMPLE 3 Solve: $2x^2 - 4x = -3$.

SOLUTION We think of $2x^2 - 4x = -3$ as $2x^2 + (-4)x + 3 = 0$ to find the values for a, b, and c. Here $a = 2$, $b = -4$, and $c = 3$, so

$$x = \frac{-b \pm \sqrt{b^2 - 4ac}}{2a}$$

$$= \frac{-(-4) \pm \sqrt{(-4)^2 - 4(2)(3)}}{2 \cdot 2}$$

$$= \frac{4 \pm \sqrt{16 - 24}}{4}$$

$$= \frac{4 \pm \sqrt{-8}}{4} = \frac{4 \pm \sqrt{4 \cdot 2}\, i}{4}$$

$$= \frac{4 \pm 2\sqrt{2}\, i}{4} = \frac{2 \pm \sqrt{2}\, i}{2} \qquad \text{Simplify: 2 is a common factor.}$$

The solution set is $\left\{\dfrac{2 \pm \sqrt{2}\, i}{2}\right\}$ or $\left\{1 + \dfrac{\sqrt{2}}{2} i, 1 - \dfrac{\sqrt{2}}{2} i\right\}$.

Notice in Example 3 that when one solution to a quadratic equation with real coefficients is of the form $a + bi$, the other solution is its **conjugate,** $a - bi$.

The quadratic formula also applies to second-degree equations with coefficients that involve nonreal complex numbers. We will consider only one case, in which both a and c are nonreal complex numbers.

EXAMPLE 4 Solve: $2im^2 - 3m + 4i = 0$.

SOLUTION Here $a = 2i$, $b = -3$, and $c = 4i$, so

$$m = \frac{-(-3) \pm \sqrt{(-3)^2 - 4(2i)(4i)}}{2(2i)}$$

$$= \frac{3 \pm \sqrt{9 - 32i^2}}{4i} = \frac{3 \pm \sqrt{9 + 32}}{4i} \qquad i^2 = -1$$

$$= \frac{3 \pm \sqrt{41}}{4i} = \frac{3 \pm \sqrt{41}}{4i} \cdot \frac{i}{i} \qquad \text{Rationalize the denominator;}\ i = \sqrt{-1}.$$

$$= \frac{(3 \pm \sqrt{41})i}{4i^2}$$

$$= \frac{(3 \pm \sqrt{41})i}{-4} \qquad i^2 = -1$$

$$= \frac{(-3 \pm \sqrt{41})i}{4} \qquad \text{Multiply the numerator and denominator by } -1.$$

The solution set is $\left\{\dfrac{(-3 \pm \sqrt{41})i}{4}\right\}$.

When a, b, and c are real numbers, the radicand of the quadratic formula, $b^2 - 4ac$, is called the **discriminant** of the quadratic equation. To see how we use the discriminant, we consider the two solutions to the quadratic equation, which we will designate

$$x_1 = \frac{-b + \sqrt{b^2 - 4ac}}{2a} \quad \text{and} \quad x_2 = \frac{-b - \sqrt{b^2 - 4ac}}{2a}$$

If $b^2 - 4ac = 0$, then

$$x_1 = \frac{-b + 0}{2a} = \frac{-b}{2a} \quad \text{and} \quad x_2 = \frac{-b - 0}{2a} = \frac{-b}{2a} = x_1$$

There is one solution of multiplicity 2 (see Section 6.6). If $b^2 - 4ac > 0$, then the radicand is positive and, although the solutions are real numbers, they are unequal. If $b^2 - 4ac < 0$, the radicand is negative. In this case the solutions are nonreal complex conjugates. *Nonreal* implies that the imaginary part of the complex number is not zero. These observations are outlined in the following table.

The Discriminant of a Quadratic Equation

The discriminant of the quadratic equation $ax^2 + bx + c = 0$, where a, b, and c are real numbers, is given by $b^2 - 4ac$.

Equation	Discriminant Value	Discriminant Result	Nature of Solution(s)
$x^2 + 6x + 9 = 0$	$6^2 - 4(1)(9) = 0$	$b^2 - 4ac = 0$	One real solution of multiplicity 2
$x^2 - 4x + 3 = 0$	$(-4)^2 - 4(1)(3) = 4$	$b^2 - 4ac > 0$	Two real solutions
$x^2 + x + 1 = 0$	$1^2 - 4(1)(1) = -3$	$b^2 - 4ac < 0$	Nonreal complex conjugates

EXAMPLE 5 Determine the nature of the solutions for each of the following equations:

(a) $3x^2 + 4x + 7 = 0$ (b) $4t^2 - 3t - 2 = 0$ (c) $4m^2 - 20m + 25 = 0$

SOLUTION

(a) $b^2 - 4ac = 4^2 - 4(3)(7) = 16 - 84 = -68 < 0$
There are two solutions that are nonreal complex conjugates.

(b) $b^2 - 4ac = (-3)^2 - 4(4)(-2) = 9 + 32 = 41 > 0$
There are two real solutions.

(c) $b^2 - 4ac = (-20)^2 - 4(4)(25) = 400 - 400 = 0$
There is one real solution of multiplicity 2.

EXAMPLE 6 Find all values for K such that to $x^2 + 6x + K = 0$ has two real solutions.

SOLUTION If the solutions are to be real, the discriminant must be positive, or $b^2 - 4ac > 0$. Here $a = 1$, $b = 6$, and $c = K$, so

$$b^2 - 4ac = 6^2 - 4(1)(K) = 36 - 4K$$

Thus, for the discriminant to be positive, $36 - 4K > 0$. Now we solve the inequality.

$$36 - 4K > 0$$
$$-4K > -36 \quad \text{Subtract 36 from each side.}$$
$$K < 9 \quad \text{Divide each side by } -4; \text{ change } > \text{ to } <.$$

EXAMPLE 7 Find all values of K for which the solutions to $Kx^2 - x + 1 = 0$ are nonreal complex conjugates.

SOLUTION If the solutions are to be nonreal complex conjugates, the discriminant must be negative, or $b^2 - 4ac < 0$. Now,

$$b^2 - 4ac = (-1)^2 - 4(K)(1) = 1 - 4K$$

Thus we want $1 - 4K < 0$. Now we solve the inequality for K.

$$1 - 4K < 0$$
$$-4K < -1 \quad \text{Subtract 1 from each side.}$$
$$K > \frac{1}{4} \quad \text{Divide each side by } -4; \text{ change } < \text{ to } >.$$

THEOREM 9.2

The Sum and the Product of the Roots of a Quadratic Equation

If $x_1 = \dfrac{-b + \sqrt{b^2 - 4ac}}{2a}$ and $x_2 = \dfrac{-b - \sqrt{b^2 - 4ac}}{2a}$ are the roots of the quadratic equation $ax^2 + bx + c = 0$, then the sum and the product of the roots can be expressed in terms of the coefficients as follows:

SUM		PRODUCT
$x_1 + x_2 = \dfrac{-b}{a}$	and	$x_1 \cdot x_2 = \dfrac{c}{a}$

We will prove the first equation in the theorem.

PROOF

$$x_1 + x_2 = \frac{-b + \sqrt{b^2 - 4ac}}{2a} + \frac{-b - \sqrt{b^2 - 4ac}}{2a}$$

$$= \frac{-b + \sqrt{b^2 - 4ac} - b - \sqrt{b^2 - 4ac}}{2a}$$

$$= \frac{-2b}{2a} = \frac{-b}{a}$$

The second equation is left as Exercise 97.

The relationship between the sum and the product of the roots of a quadratic equation can be used to check the solutions or to find the equation given the solution set.

EXAMPLE 8 Solve and check: $3x^2 + 6x + 1 = 0$.

SOLUTION From Example 1, the solution set is $\left\{\dfrac{-3 + \sqrt{6}}{3}, \dfrac{-3 - \sqrt{6}}{3}\right\}$. We see that $a = 3$, $b = 6$, and $c = 1$.

CHECK If we let $x_1 = \dfrac{-3 + \sqrt{6}}{3}$ and $x_2 = \dfrac{-3 - \sqrt{6}}{3}$, then

$$x_1 + x_2 = \dfrac{-3 + \sqrt{6}}{3} + \dfrac{-3 - \sqrt{6}}{3}$$

$$= \dfrac{-6}{3} \stackrel{\checkmark}{=} \dfrac{-b}{a}$$

and

$$x_1 \cdot x_2 = \left(\dfrac{-3 + \sqrt{6}}{3}\right)\left(\dfrac{-3 - \sqrt{6}}{3}\right)$$

$$= \dfrac{9 - 6}{9} \qquad (a - b)(a + b) = a^2 - b^2$$

$$= \dfrac{3}{9} = \dfrac{1}{3} \stackrel{\checkmark}{=} \dfrac{c}{a}$$

EXAMPLE 9 Find an equation with solutions $2 + i$ and $2 - i$ by using the sum-and-product concept of Theorem 9.2.

SOLUTION We consider what happens when both sides of the equation $ax^2 + bx + c = 0$ are divided by a:

$$ax^2 + bx + c = 0$$

$$x^2 + \dfrac{b}{a}x + \dfrac{c}{a} = 0$$

$$x^2 + \underbrace{\dfrac{b}{a}}_{\text{The negative of the sum of } x_1 \text{ and } x_2}x + \underbrace{\dfrac{c}{a}}_{\text{The product of } x_1 \text{ and } x_2} = 0$$

Now,

$$x_1 + x_2 = 2 + i + 2 - i = 4 = \dfrac{-b}{a}$$

Thus, $\frac{b}{a} = -4$. Furthermore,

$$x_1 \cdot x_2 = (2 + i)(2 - i) = 4 + 1 = 5 = \frac{c}{a}$$

We now substitute $\frac{b}{a} = -4$ and $\frac{c}{a} = 5$ into $x^2 + \frac{b}{a}x + \frac{c}{a} = 0$. The equation in standard form is $x^2 - 4x + 5 = 0$.

Do You Remember?

Can you match these?

All matchings refer to the equation $ax^2 + bx + c = 0$.

____ 1. The discriminant
____ 2. The quadratic formula
____ 3. The sum of the roots
____ 4. The product of the roots
____ 5. If $b^2 - 4ac = 0$
____ 6. If $b^2 - 4ac > 0$
____ 7. If $b^2 - 4ac < 0$

a) Two real distinct solutions
b) $\frac{c}{a}$
c) $x = \frac{-b \pm \sqrt{b^2 - 4ac}}{2a}$
d) Two nonreal complex conjugate solutions
e) One solution of multiplicity two
f) $b^2 - 4ac$
g) $\frac{-b}{a}$

Answers: 1. f 2. c 3. g 4. b 5. e 6. a 7. d

Exercise Set 9.3

Solve each of the following quadratic equations by using the quadratic formula. Write the equation in standard form first, if necessary.

1. $x^2 - 6x + 8 = 0$
2. $a^2 - a - 2 = 0$
3. $p^2 + 8p + 15 = 0$
4. $y^2 - 7y - 18 = 0$
5. $2w^2 - 7w - 4 = 0$
6. $3s^2 - 10s + 3 = 0$
7. $3k^2 + 3k - 6 = 0$
8. $4x^2 + x - 3 = 0$
9. $2x^2 + 5x - 12 = 0$
10. $v^2 + v + 1 = 0$
11. $y^2 + 3y + 4 = 0$
12. $x^2 - x + 1 = 0$
13. $3a^2 + 2a + 2 = 0$
14. $c^2 - 7c - 2 = 0$
15. $3x^2 + 6 = -19x$
16. $2d^2 - 12 = -5d$
17. $2y^2 + 3y = 2$
18. $t^2 + 2t = -11$
19. $-4y^2 + 3 = 2y$
20. $-5r^2 + 1 = 10r$
21. $c(c - 1) = 5$
22. $x(x - 8) = 6$
23. $x(x + 3) = 1$
24. $12y(3y + 2) = 1$
25. $\frac{x^2}{7} + \frac{x}{4} = -4$
26. $\frac{y^2}{5} + 3 = \frac{y}{4}$
27. $\frac{a^2}{4} + 3a = 5$
28. $\frac{y^2}{2} + 3y = 1$
29. $x^2 + 12 = 0$
30. $y^2 - 16 = 0$

31. $p^2 - 18 = 0$
32. $x^2 + 28 = 0$
33. $x^2 + 5x = 0$
34. $y^2 - 7y = 0$
35. $5a^2 + 3a = 0$
36. $4y^2 - 9y = 0$
37. $(x - 2)^2 = -8$
38. $(2c - 3)^2 = -5$
39. $(a - 2)^2 = 15$
40. $(4p + 1)^2 = 1$
41. $x^2 + 0.6x + 0.1 = 0$
42. $x^2 + 0.1x - 0.1 = 0$
43. $y^2 - y + 0.3 = 0$
44. $y^2 + y + 0.1 = 0$

Use the discriminant to determine the nature of the solution(s) for each of the following equations. **Do not solve.**

45. $x^2 - 3x + 2 = 0$
46. $y^2 - 7y - 8 = 0$
47. $a^2 + a + 1 = 0$
48. $c^2 - c - 3 = 0$
49. $3v^2 - 4v + 2 = 0$
50. $b^2 - 4b + 1 = 0$
51. $4k^2 + 8k + 1 = 0$
52. $2p^2 - p + 2 = 0$
53. $9q^2 - 24q + 16 = 0$
54. $16t^2 + 40t + 25 = 0$
55. $2h^2 + 5h + 4 = 0$
56. $12d^2 + d - 6 = 0$

Find all real values of K for which the solutions to the given equation are as indicated.

57. $a^2 + 6a + K = 0$; two real solutions
58. $p^2 - 7p + K = 0$; two real solutions
59. $p^2 + 8p + K = 0$; one solution
60. $c^2 - 10c + K = 0$; one solution
61. $Km^2 - m - 1 = 0$; complex conjugate solutions
62. $Ky^2 - y - 1 = 0$; complex conjugate solutions
63. $Kd^2 - 2d + 1 = 0$; complex conjugate solutions
64. $Kt^2 + 6t + 5 = 0$; complex conjugate solutions
65. $Kv^2 - 3v + 8 = 0$; two real solutions
66. $Kr^2 - 3r - 2 = 0$; two real solutions

Write a quadratic equation with solutions x_1 and x_2 as given. Use Theorem 9.2.

67. $x_1 = 5, x_2 = -3$
68. $x_1 = -9, x_2 = -6$
69. $x_1 = 0, x_2 = 2$
70. $x_1 = 3, x_2 = 0$
71. $x_1 = 4, x_2 = 5$
72. $x_1 = -1, x_2 = 2$
73. $x_1 = -2, x_2 = -4$
74. $x_1 = 5, x_2 = 7$
75. $x_1 = \dfrac{2+i}{3}, x_2 = \dfrac{2-i}{3}$
76. $x_1 = \dfrac{1+\sqrt{3}}{2}, x_2 = \dfrac{1-\sqrt{3}}{2}$
77. $x_1 = \sqrt{7}, x_2 = -\sqrt{7}$
78. $x_1 = 5 + \sqrt{2}i, x_2 = 5 - \sqrt{2}i$

Solve by any method. Check Exercises 81–84 by Theorem 9.2.

79. $(2x + 5)^2 = -11$
80. $(3a - 7)^2 = -15$
81. $25k^2 - 20k + 4 = 0$
82. $8t^2 - 10t + 3 = 0$
83. $15 - 7v - 4v^2 = 0$
84. $5 - 3z^2 + 2z = 0$
85. $(8x + 19)x = 27$
86. $(24u + 94)u = 25$
87. $36 + \dfrac{60}{s} = \dfrac{-25}{s^2}$
88. $\dfrac{10c + 19}{c} = \dfrac{15}{c^2}$
89. $m - 16 = \dfrac{105}{m}$
90. $25v - 40 + \dfrac{16}{v} = 0$
91. $\dfrac{2x + 11}{2x + 8} = \dfrac{3x - 1}{x - 1}$
92. $\dfrac{2r - 5}{2r + 1} + \dfrac{6}{2r - 3} = \dfrac{7}{4}$
93. $\sqrt{10 - x} - 8 = 2x$
94. $\sqrt{16 - p} + 4 = p$
95. $\sqrt{5b + 1} = 1 + \sqrt{3b}$
96. $6\sqrt{h} - 3 = \sqrt{10h - 1}$
97. Prove the second equation of Theorem 9.2.

For Extra Thought

98. Determine K if the larger root (solution) of the equation $x^2 + Kx - 2 = 0$ is K.

99. The height of a right circular cylinder is twice the radius.
    (a) Express the total surface area in terms of the radius.
    (b) Express the total surface area in terms of the height.

100. A rectangular flower garden has length a and width b. A concrete path of uniform width borders all four sides of the garden. Find the width of the path (in terms of a and b) given that the area of the path equals the area of the garden.

101. The four corners of a square are cut off to form a regular octagon. Find the length of a side of the octagon if the length of a side of the square was 1 yd.

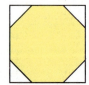

9.4 Solving Equations Involving Quadratics

OBJECTIVES

In this section we will learn to
1. reduce certain equations to quadratic form; and
2. solve equations that can be reduced to quadratic form.

The solutions to many equations that are not quadratic require, in part, the techniques that we used to solve quadratic equations. For example, consider the equation

$$x^3 - 1 = 0$$

In Section 6.4 we saw that to factor $x^3 - 1$, we should view it as the difference of two cubes. Thus

$$x^3 - 1 = (x - 1)(x^2 + x + 1) = 0$$

When we apply the zero-factor theorem, we have

$$x - 1 = 0 \quad \text{or} \quad x^2 + x + 1 = 0$$

The solution to the first equation is $x = 1$. The second equation, $x^2 + x + 1 = 0$, is quadratic, and we can solve it by the quadratic formula:

$$x^2 + x + 1 = 0$$

$$x = \frac{-1 \pm \sqrt{1-4}}{2} = \frac{-1 \pm \sqrt{3}\,i}{2}$$

The solution set is $\left\{1, \frac{-1 + \sqrt{3}\,i}{2}, \frac{-1 - \sqrt{3}\,i}{2}\right\}$. Notice that the equation is of degree three and it has three solutions. The following theorem relates the number of solutions of a polynomial equation to its degree.

THEOREM 9.3   A polynomial equation of degree n has at most n *distinct* solutions.

To say that a solution is *distinct* means it differs from any other solution. For example, a third-degree equation can have one, two, or three distinct solutions. If the equation has only two distinct solutions, one solution must occur twice. If it has only one distinct solution, then the same solution must occur three times, or have multiplicity 3.

EXAMPLE 1   Solve: $x^4 - 3x^2 + 2 = 0$.

SOLUTION   The equation is of degree four. It can have at most four distinct solutions.

$$x^4 - 3x^2 + 2 = 0$$

$$(x^2 - 2)(x^2 - 1) = 0$$

$$x^2 - 2 = 0 \quad \text{or} \quad x^2 - 1 = 0$$

$$x^2 = 2 \quad \text{or} \quad x^2 = 1$$
$$x = \pm\sqrt{2} \quad \text{or} \quad x = \pm 1 \quad \text{Square root property}$$

The solution set is $\{-1, 1, -\sqrt{2}, \sqrt{2}\}$. The check is left to the reader.

EXAMPLE 2 Solve: $p^6 - 64 = 0$.

SOLUTION The equation is of degree six, so it can have at most six distinct solutions. We will factor the equation as the difference of squares.

$$p^6 - 64 = 0$$
$$(p^3 - 8)(p^3 + 8) = 0$$

Each of these can be factored further as the sum or difference of cubes.

$$(p - 2)(p^2 + 2p + 4)(p + 2)(p^2 - 2p + 4) = 0$$
$$p - 2 = 0 \quad \text{or} \quad p^2 + 2p + 4 = 0 \quad \text{or} \quad p + 2 = 0 \quad \text{or} \quad p^2 - 2p + 4 = 0$$

Two solutions are 2 and -2. We can find the other solutions by using the quadratic formula.

$$p^2 + 2p + 4 = 0 \quad \text{or} \quad p^2 - 2p + 4 = 0$$
$$p = \frac{-2 \pm \sqrt{2^2 - 4(1)(4)}}{2} \quad \text{or} \quad p = \frac{-(-2) \pm \sqrt{(-2)^2 - 4(1)(4)}}{2}$$
$$p = \frac{-2 \pm \sqrt{-12}}{2} \quad \text{or} \quad p = \frac{2 \pm \sqrt{-12}}{2}$$
$$p = \frac{-2 \pm 2\sqrt{3}\,i}{2} \quad \text{or} \quad p = \frac{2 \pm 2\sqrt{3}\,i}{2}$$
$$p = -1 \pm \sqrt{3}\,i \quad \text{or} \quad p = 1 \pm \sqrt{3}\,i$$

The solution set is $\{-2, 2, -1 - \sqrt{3}\,i, -1 + \sqrt{3}\,i, 1 - \sqrt{3}\,i, 1 + \sqrt{3}\,i\}$. The check is left to the reader.

EXAMPLE 3 Solve: $x^{1/2} + 4x^{1/4} - 5 = 0$.

SOLUTION To write this equation in quadratic form, we let $m = x^{1/4}$. Then

$$m^2 = (x^{1/4})^2 = x^{1/2}$$

We substitute m^2 for $x^{1/2}$ and m for $x^{1/4}$:

$$x^{1/2} + 4x^{1/4} - 5 = 0$$
$$m^2 + 4m - 5 = 0$$
$$(m - 1)(m + 5) = 0$$
$$m - 1 = 0 \quad \text{or} \quad m + 5 = 0$$
$$m = 1 \quad \text{or} \quad m = -5$$

Now we substitute $x^{1/4}$ back for m:

$$x^{1/4} = 1 \quad \text{or} \quad x^{1/4} = -5$$

$(x^{1/4})^4 = 1^4$ or $(x^{1/4})^4 = (-5)^4$ Power property: raise each side to the fourth power

$x = 1$ or $x = 625$

Since we used the power property, we must check the solutions.

CHECK

$(625)^{1/2} + 4(625)^{1/4} - 5 \stackrel{?}{=} 0$ $x = 625$

$25 + 4(5) - 5 \stackrel{?}{=} 0$

$40 \neq 0$

$1^{1/2} + 4(1)^{1/4} - 5 \stackrel{?}{=} 0$ $x = 1$

$1 + 4(1) - 5 \stackrel{?}{=} 0$

$0 \stackrel{\checkmark}{=} 0$

The solution set is $\{1\}$.

EXAMPLE 4 Solve: $2x^{-2} + 5x^{-1} + 3 = 0$.

SOLUTION We let $m = x^{-1}$; then $m^2 = (x^{-1})^2 = x^{-2}$. The given equation now becomes

$2m^2 + 5m + 3 = 0$ $m = x^{-1}; \ m^2 = x^{-2}$

$(2m + 3)(m + 1) = 0$

$2m + 3 = 0$ or $m + 1 = 0$

$m = \dfrac{-3}{2}$ or $m = -1$

Now we substitute x^{-1} back in for m:

$x^{-1} = \dfrac{1}{x} = \dfrac{-3}{2}$ or $x^{-1} = \dfrac{1}{x} = -1$

We clear of fractions and solve for x.

$2 = -3x$ or $1 = -x$

$x = -\dfrac{2}{3}$ or $x = -1$

CHECK $2x^{-2} + 5x^{-1} + 3 = 0$

$2(-1)^{-2} + 5(-1)^{-1} + 3 \stackrel{?}{=} 0$

$2\left(\dfrac{1}{-1}\right)^2 + 5\left(\dfrac{1}{-1}\right) + 3 \stackrel{?}{=} 0$

$2(-1)^2 + 5(-1) + 3 \stackrel{?}{=} 0$

$2 - 5 + 3 \stackrel{?}{=} 0$

$0 \stackrel{\checkmark}{=} 0$

$2\left(-\dfrac{2}{3}\right)^{-2} + 5\left(-\dfrac{2}{3}\right)^{-1} + 3 \stackrel{?}{=} 0$

$2\left(-\dfrac{3}{2}\right)^2 + 5\left(-\dfrac{3}{2}\right) + 3 \stackrel{?}{=} 0$

$2\left(\dfrac{9}{4}\right) - \dfrac{15}{2} + 3 \stackrel{?}{=} 0$

$\dfrac{9}{2} - \dfrac{15}{2} + \dfrac{6}{2} \stackrel{?}{=} 0$

$\dfrac{9 - 15 + 6}{2} \stackrel{?}{=} 0$

$\dfrac{0}{2} \stackrel{?}{=} 0$

$0 \stackrel{\checkmark}{=} 0$

The solution set is $\left\{-1, -\dfrac{2}{3}\right\}$.

EXAMPLE 5 Solve: $(y - 2)^4 + (y - 2)^2 - 2 = 0$.

SOLUTION This equation is of degree four, so there are at most four distinct solutions. We let $m = (y - 2)^2$; then $m^2 = (y - 2)^4$, and the equation becomes

$$m^2 + m - 2 = 0 \qquad m = (y-2)^2;\ m^2 = (y-2)^4$$
$$(m + 2)(m - 1) = 0$$
$$m + 2 = 0 \quad \text{or} \quad m - 1 = 0$$
$$m = -2 \quad \text{or} \quad m = 1$$
$$(y - 2)^2 = -2 \quad \text{or} \quad (y - 2)^2 = 1 \qquad \text{Substitute } (y-2)^2 \text{ back for } m.$$

We can solve the last two equations by the square root property. The reader should verify that the solution set is $\{1, 3, 2 + \sqrt{2}\,i, 2 - \sqrt{2}\,i\}$.

EXAMPLE 6 Two men working together complete a job in 4 days. If each works alone on the job, the slower worker requires 6 days longer than the faster worker. How long does the faster worker take to do the job alone?

SOLUTION If the faster worker takes d days to do the job alone, then the slower worker takes $d + 6$ days. We summarize in a table.

	NUMBER OF DAYS TO DO THE JOB	FRACTIONAL PART DONE IN ONE DAY
Faster worker	d	$\dfrac{1}{d}$
Slower worker	$d + 6$	$\dfrac{1}{d + 6}$
Together	4	$\dfrac{1}{4}$

The sum of the amount of work each man can do in a day is equal to the amount they can do together in one day.

$$\dfrac{1}{d} + \dfrac{1}{d + 6} = \dfrac{1}{4}$$
$$4(d + 6) + 4d = d(d + 6) \qquad \text{Multiply each side by } 4d(d + 6), \text{ the LCD.}$$
$$4d + 24 + 4d = d^2 + 6d$$
$$d^2 - 2d - 24 = 0$$
$$(d - 6)(d + 4) = 0$$
$$d = 6 \quad \text{or} \quad d = -4$$

We reject -4, since d is the number of *days*. The faster worker takes 6 days alone.

CHECK If the faster worker takes 6 days, the slower worker takes $6 + 6 = 12$ days. The faster worker can do $\frac{1}{6}$ of the job in a day, and the slower worker can do $\frac{1}{12}$ of the job in a day, so together they can do $\frac{1}{6} + \frac{1}{12} = \frac{1}{4}$ of the job in one day. This means that in four days, they can do $4 \cdot \frac{1}{4} = 1$ job working together, as required. The solution checks.

EXAMPLE 7 An industrial plant has a sink for cleaning parts. The sink can be filled by an inlet pipe in 50 minutes, and it drains in 70 minutes when the drain is opened. Suppose a worker is told to fill the sink and doesn't notice that the drain is open. How long will it take the sink to fill up, assuming that the water drains at a constant rate?

SOLUTION We let x denote the number of minutes it takes to fill the sink with both the inlet pipe and the drain open.

	NUMBER OF MINUTES TO FILL OR DRAIN SINK	AMOUNT FILLED OR DRAINED IN ONE MINUTE
Inlet pipe	50	$\frac{1}{50}$
Drain	70	$\frac{1}{70}$
Together	x	$\frac{1}{x}$

The amount filled in one minute with the inlet and drain both open is equal to the amount of water that runs in in one minute minus the amount drained in one minute:

$$\frac{1}{50} - \frac{1}{70} = \frac{1}{x}$$

$7x - 5x = 350$ Multiply each side by $350x$, the LCD.

$2x = 350$

$x = 175$

It will take 175 minutes to fill the sink. The check is left to the reader.

EXAMPLE 8 Solve: $\dfrac{3x + 4}{x^2 - 3x + 2} - \dfrac{x - 5}{x^2 - x - 2} = \dfrac{x}{x^2 - 1}$, $x \neq -1, 1,$ and 2.

SOLUTION We begin by factoring each denominator:

$$\frac{3x + 4}{(x - 2)(x - 1)} - \frac{x - 5}{(x - 2)(x + 1)} = \frac{x}{(x - 1)(x + 1)}$$

Now we multiply each side by the LCD, which is $(x - 2)(x - 1)(x + 1)$.

$(3x + 4)(x + 1) - (x - 5)(x - 1) = x(x - 2)$

$3x^2 + 7x + 4 - x^2 + 6x - 5 = x^2 - 2x$ Remove grouping symbols.

$$2x^2 + 13x - 1 = x^2 - 2x$$
$$x^2 + 15x - 1 = 0$$

When the final equation is solved using the quadratic formula, the solution set is $\left\{\dfrac{-15 - \sqrt{229}}{2}, \dfrac{-15 + \sqrt{229}}{2}\right\}$.

EXAMPLE 9 Solve: $\sqrt{\sqrt{x+1} + \sqrt{x}} = 1$.

SOLUTION We begin by squaring both sides to eliminate the outer radical.

$$\left(\sqrt{\sqrt{x+1} + \sqrt{x}}\right)^2 = 1^2$$
$$\sqrt{x+1} + \sqrt{x} = 1$$
$$\sqrt{x+1} = 1 - \sqrt{x}$$
$$(\sqrt{x+1})^2 = (1 - \sqrt{x})^2 \quad \text{Square each side to eliminate one radical.}$$
$$x + 1 = 1 - 2\sqrt{x} + x$$
$$2\sqrt{x} = 0 \quad \text{Simplify.}$$
$$4x = 0 \quad \text{Square each side}$$
$$x = 0$$

CHECK
$$\sqrt{\sqrt{0+1} + \sqrt{0}} \stackrel{?}{=} 1$$
$$\sqrt{1+0} \stackrel{?}{=} 1$$
$$\sqrt{1} \stackrel{?}{=} 1$$
$$1 \stackrel{\checkmark}{=} 1$$

The solution set is $\{0\}$.

Exercise Set 9.4

Factor and solve.

1. $x^4 - 3x^2 + 2 = 0$
2. $k^4 - 5k^2 + 4 = 0$
3. $p^4 - 6p^2 + 9 = 0$
4. $w^4 - 8w^2 + 16 = 0$
5. $y^4 - 3y^2 - 54 = 0$
6. $m^4 - 13m^2 + 36 = 0$
7. $t^5 + 5t^3 + 6t = 0$
8. $r^5 + r^3 - 56r = 0$
9. $2z^4 - 30z^2 + 112 = 0$
10. $3x^4 + 30x^2 + 63 = 0$
11. $a^3 - 8 = 0$
12. $c^3 - 27 = 0$
13. $s^3 + 1 = 0$
14. $n^3 + 64 = 0$
15. $x^4 + 125x = 0$
16. $y^4 + y = 0$
17. $6k^4 - 6k = 0$
18. $5z^4 - 320z = 0$

Factor as the difference of squares, then solve the resulting equation.

19. $p^6 - 1 = 0$
20. $r^6 - 64 = 0$
21. $3k^6 - 3 = 0$
22. $7m^6 - 7 = 0$
23. $64x^6 - 729 = 0$
24. $3y^7 - 192y = 0$

Use an appropriate substitution to change each of the following equations into a quadratic and solve.

25. $t^{1/3} - 4t^{1/6} + 3 = 0$
26. $y + 2y^{1/2} - 3 = 0$
27. $m^{2/3} - m^{1/3} - 12 = 0$
28. $a^{4/3} + 7a^{2/3} + 12 = 0$
29. $2p^{2/5} + 5p^{1/5} + 2 = 0$
30. $r^{1/2} - 5r^{1/4} + 6 = 0$
31. $(x^2 - x)^2 - 4(x^2 - x) - 12 = 0$
32. $(z^2 + 2z)^2 - (z^2 + 2z) - 6 = 0$
33. $4(t^2 + 1)^2 - 7(t^2 + 1) = 2$
34. $4(m^2 - 1)^2 - 12(m^2 - 1) = -8$
35. $(2r^2 - 1)^2 + 11(2r^2 - 1) = -30$

36. $(k^2 + 2)^2 + 13(k^2 + 2) + 12 = 0$

37. $6q^{-2} - 5q^{-1} - 6 = 0$ 38. $3s^{-2} - 11s^{-1} = 20$

Solve each of the following equations. State any restrictions on the variables.

39. $\dfrac{2}{x - 1} + \dfrac{5}{x - 7} = \dfrac{4}{3x - 1}$

40. $\dfrac{1}{a} + \dfrac{3}{a - 1} - \dfrac{2}{a + 1} = 0$

41. $\dfrac{1}{2} - \dfrac{1}{p - 1} = \dfrac{6}{p^2 - 1}$

42. $\dfrac{m}{m - 2} - 1 + m = \dfrac{1 - m}{2}$

43. $\dfrac{2a + 7}{a^2 - 5a + 4} - \dfrac{a - 8}{a^2 - 4a + 3} = \dfrac{a}{a^2 - 7a + 12}$

44. $\dfrac{k - 7}{k + 4} - \dfrac{k + 3}{k} = \dfrac{3}{k}$

Solve each of the following equations and check all solutions.

45. $\sqrt[3]{x^2 - x - 6} - 2 = 0$ 46. $\sqrt{2z + 1} - \sqrt{z} = 1$

47. $a^2 - 6a - \sqrt{a^2 - 6a} - 3 = 3$ (*Hint:* Use an appropriate substitution.)

48. $\dfrac{4 - x}{\sqrt{x^2 - 8x + 32}} = \dfrac{3}{5}$ 49. $\sqrt{x} - \dfrac{2}{\sqrt{x}} = 1$

50. $\sqrt{\sqrt{x + 16} - \sqrt{x}} = 2$

Rewrite each of the following applied problems as a quadratic equation and solve it.

51. Working together, two cranes can unload a ship in 4 hr. The slower crane working alone requires 6 more hours than the faster crane to unload the ship. How long does it take each crane alone to unload the ship?

52. A roofer and his helper can finish a roofing job in 4 hr. Working alone, the roofer can finish the job 6 hr faster than his helper. How long does it take each of them working alone to finish the job?

53. Computers A and B together can complete a data processing job in 2 hr. Computer A alone can do the job 3 hr faster than computer B. How long does it take each computer to complete the job working alone?

54. Lydia and Myrna can do an advertising layout in 6 days. Working alone, Myrna takes 16 days longer than Lydia to do the layout. How long does it take each person to do the layout by herself?

55. Beverly flies a distance of 600 mi. She can fly the same distance 30 min faster by increasing her average speed by 40 mph. Find Beverly's average speed.

56. Christi travels 40 mi in 45 min and completes her trip of 50 more miles at the same speed. How long does it take her to travel the last 50 mi?

57. Troy travels 36 km down a river and back in 8 hr. If the rate of the boat in still water is 12 km/hr, what is the rate of the current?

58. David and Celeste start at the same time from the same place and travel along roads that are perpendicular to each other. Celeste travels 4 mph faster than David. At the end of 2 hr they are 40 mi apart. How fast is each one traveling?

REVIEW EXERCISES

Solve by any method.

59. $x^2 - 9 = 0$ 60. $y^2 - 16 = 0$
61. $y^2 + 8 = 0$ 62. $x^2 + 12 = 0$
63. $x^2 + x - 12 = 0$ 64. $y^2 + 4y + 3 = 0$
65. $2x^2 - 5x - 3 = 0$ 66. $3x^2 - 2x + 1 = 0$
67. $2y^2 - 4y + 5 = 0$ 68. $4y^2 - 12y + 7 = 0$
69. $x^2 + 4x = -11$ 70. $y^2 + 2 = 2y$
71. $3x - 3 = \dfrac{x + 15}{x}$ 72. $x^2 = \dfrac{2 - x}{3}$
73. $0.2x^2 - 0.5 = -x$ 74. $0.01y^2 = -\dfrac{1}{2}y - 0.2$
75. $x^2 - \dfrac{3}{4}x = \dfrac{1}{2}$ 76. $x\left(x + \dfrac{2}{3}\right) = \dfrac{2}{3}$

𝓕OR 𝓔XTRA 𝓣HOUGHT

77. A rectangle is inscribed in a semicircle of radius 1 m. For what value of x will the area of the rectangle equal 1 m²? See the figure.

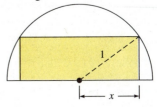

78. A square is inscribed in a circle of radius $\sqrt{3x}$. Find the area of the shaded region outside the square and inside the circle (in terms of x). See the figure.

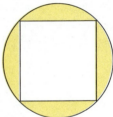

9.5 Solving Inequalities Involving Quadratics

OBJECTIVES In this section we will learn to
1. solve quadratic inequalities;
2. solve inequalities of degree three or greater; and
3. solve inequalities involving rational expressions.

DEFINITION
Quadratic Inequality

A **quadratic inequality** is one that can be written in the form

$$ax^2 + bx + c < 0, \quad a \neq 0$$

or in any form where the inequality symbol is $>$, $\leq$, or $\geq$.

Quadratic inequalities are solved by a variety of methods. One of the easier methods is based on the following observations concerning the product of two numbers C and D:

1. If $C \cdot D > 0$, (here $C \cdot D > 0$ means $C \cdot D$ is positive), *then C and D have the same sign.*
2. If $C \cdot D < 0$, (here $C \cdot D < 0$ means $C \cdot D$ is negative), *then C and D have opposite signs.*

For example, to solve the inequality $x^2 - x - 12 > 0$, first we factor it and write it in the form $C \cdot D > 0$:

$$x^2 - x - 12 > 0$$
$$(x - 4)(x + 3) > 0$$

Observe that the first factor is zero when $x = 4$. Values greater than 4 make the first factor positive while values less than 4 make the first factor negative. The intervals where the first factor is positive or negative can be shown on a number line, as in Figure 9.3. The negative signs to the left of 4 indicate that $x - 4$ is negative when $x < 4$, and the positive signs to the right of 4 indicate that $x - 4$ is positive when $x > 4$. Such a number line is called a **sign graph**.

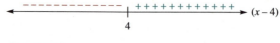

FIGURE 9.3
The signs of $x - 4$

Similarly, $x + 3$ is zero when $x = -3$, positive for $x > -3$, and negative for $x < -3$. The sign graph for $x + 3$ is shown in Figure 9.4. The values where each factor is zero, namely, 4 and -3, are called the **critical numbers** of the inequality.

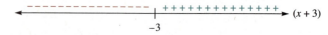

FIGURE 9.4
The signs of $x + 3$

To find the values for which $(x - 4)(x + 3) > 0$, we construct the sign graph for the product of $x - 4$ and $x + 3$ on another line, as shown in Figure 9.5. When the signs of $x - 4$ and $x + 3$ are the same, their product will be positive (> 0), which means that $x^2 - x - 12 > 0$. The product $(x - 4)(x + 3)$ will be positive if x is greater than 4 or if x is less than -3.

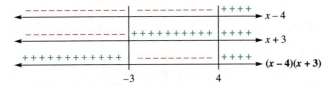

FIGURE 9.5
The signs of $(x - 4)(x + 3)$

The solution set is the union of the regions where the product is positive. The solution set is $(-\infty, -3) \cup (4, \infty)$. In set-builder notation, it is $\{x \mid x < -3$ or $x > 4\}$.

EXAMPLE 1 Solve: $x^2 - 6x + 5 \leq 0$.

SOLUTION

$x^2 - 6x + 5 \leq 0$

$(x - 5)(x - 1) \leq 0$ Factor first.

Now we construct sign graphs showing where $x - 1$ and $x - 5$ are negative or positive. Notice that $x - 5$ will be negative if x is less than 5 and positive if x is greater than 5. Similarly, $x - 1$ will be negative if x is less than 1 and positive if x is greater than 1. See Figure 9.6. The product must be negative or zero. This occurs for values of x between 1 and 5, including both 1 and 5.

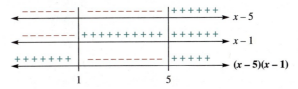

FIGURE 9.6
The signs of $(x - 5)(x - 1)$

The solution set is $[1, 5]$.

EXAMPLE 2 Solve $y(y + 2)(y - 5) > 0$ and graph the solution set.

SOLUTION We construct the sign graphs of each factor in Figure 9.7.

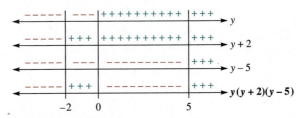

FIGURE 9.7
The signs of $y(y + 2)(y - 5)$

The solution set is the union of the regions where the product is positive. The solution set is $(-2, 0) \cup (5, \infty)$. Its graph is shown in Figure 9.8.

FIGURE 9.8

EXAMPLE 3 Solve: $\dfrac{x-2}{x+2} > 0, \quad x \neq -2$.

SOLUTION An inequality involving a quotient is solved by analyzing the signs of the factors of the numerator and the denominator using a sign graph. The numerator, $x - 2$, is positive when $x > 2$ and negative when $x < 2$. The denominator, $x + 2$, is positive when $x > -2$ and negative when $x < -2$. The quotient is positive when the numerator and the denominator have the same sign and negative when numerator and the denominator have opposite signs. The sign graph is shown in Figure 9.9.

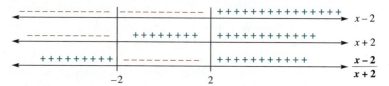

FIGURE 9.9

The signs of $\dfrac{x-2}{x+2}$

Thus $\dfrac{x+2}{x-2} > 0$ when $x < -2$ or $x > 2$. The solution set is $(-\infty, -2) \cup (2, \infty)$.

EXAMPLE 4 Solve: $\dfrac{x+2}{x-3} \leq 2, \quad x \neq 3$.

SOLUTION Our first thought here might be to clear of fractions by multiplying each side by $x - 3$. However, since we don't know whether $x - 3$ is positive or negative, we don't know whether to leave the inequality sign as it is or to reverse its direction. To avoid this difficulty, we proceed in the following way.

$$\dfrac{x+2}{x-3} \leq 2$$

$$\dfrac{x+2}{x-3} - 2 \leq 0 \qquad \text{Subtract 2 from each side.}$$

$$\dfrac{x+2-2(x-3)}{x-3} \leq 0 \qquad \text{Combine the terms on the left side. The LCD is } x - 3.$$

$$\dfrac{x+2-2x+6}{x-3} \leq 0$$

$$\dfrac{8-x}{x-3} \leq 0$$

CAUTION

Never multiply both sides of an inequality by an expression involving variables.

When constructing the sign graph for $8 - x$, note that $8 - x$ is negative when x is greater than 8. The signs to the right of 8 will be negative and those to the left will be positive. See Figure 9.10.

FIGURE 9.10

The solution set is $(-\infty, 3) \cup [8, \infty)$. Note that the square bracket to the left of the 8 indicates that it *is* a member of the solution set, while the parenthesis to the right of the 3 indicates that it is *not* a member of the solution set. If 3 were in the solution set then the denominator of $\frac{8-x}{x-3}$ would be zero. Division by zero is undefined.

EXAMPLE 5 Solve: $\dfrac{2x}{x-2} \leq \dfrac{x}{x-3}$, $x \neq 2, 3$.

SOLUTION

$\dfrac{2x}{x-2} - \dfrac{x}{x-3} \leq 0$ Subtract $\dfrac{x}{x-3}$ from each side.

$\dfrac{2x(x-3) - x(x-2)}{(x-2)(x-3)} \leq 0$

$\dfrac{2x^2 - 6x - x^2 + 2x}{(x-2)(x-3)} \leq 0$ Find the products in the numerator.

$\dfrac{x^2 - 4x}{(x-2)(x-3)} \leq 0$ Combine like terms.

$\dfrac{x(x-4)}{(x-2)(x-3)} \leq 0$ Factor the numerator.

The sign graph is shown in Figure 9.11.

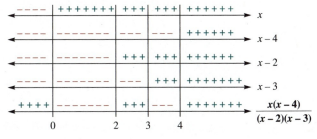

FIGURE 9.11

Remember that x cannot be 2 or 3 because then the denominator would be 0 and division by 0 is undefined. The numerator can be 0, however, so x can be 0 or 4. The solution set is $[0, 2) \cup (3, 4]$, or $\{x \mid 0 \leq x < 2 \text{ or } 3 < x \leq 4\}$.

EXAMPLE 6   Solve $(x - 2)(x^2 - 5x + 6) < 0$ without the use of a sign graph.

SOLUTION
$$(x - 2)(x^2 - 5x + 6) < 0$$
$$(x - 2)(x - 2)(x - 3) < 0 \quad \text{Factor.}$$
$$(x - 2)^2(x - 3) < 0 \quad (x - 2)(x - 2) = (x - 2)^2$$

Since the square of a number is nonnegative, the product $(x - 2)^2(x - 3)$ will be negative only if $x - 3$ is negative, or $x - 3 < 0$. This is equivalent to

$x < 3$ with $x \neq 2$

The solution set is $(-\infty, 2) \cup (2, 3)$.

EXAMPLE 7   Solve: $(x - 2)^2(x - 3) \leq 0$.

SOLUTION   This inequality is almost the same as the one in Example 6; the only difference is the use of $\leq$ instead of $<$. How is the solution set changed? The only change in the solution set is that values of x that make either of the factors zero must be included. These values are 2 and 3, so the solution set is $(-\infty, 3]$.

EXAMPLE 8   A ball projected from ground level vertically upward with an initial velocity of 96 feet/second reaches a height of $h(t) = 96t - 16t^2$ after t seconds.

(a) When will the ball be on the ground?
(b) At what time is the ball more than 80 feet above the ground?

SOLUTION   (a) The ball will be on the ground when $h(t) = 0$, so
$$0 = 96t - 16t^2$$
$$0 = 16t(6 - t)$$
$$16t = 0 \quad \text{or} \quad 6 - t = 0$$
$$t = 0 \quad \text{or} \quad t = 6$$

The ball will be on the ground at the time it is projected and will return to ground level 6 seconds later.

(b) The ball will be more than 80 feet above the ground when $h(t) > 80$. Thus
$$96t - 16t^2 > 80$$
$$-16t^2 + 96t - 80 > 0$$
$$16t^2 - 96t + 80 < 0$$

Dividing each side by 16 and factoring then yields

$$(t - 1)(t - 5) < 0$$

The reader should construct a sign graph to verify that the solution is anywhere in the time interval between 1 and 5 seconds.

Exploring Graphics

When the equation of Example 8 is graphed using a computer, where the vertical axis is h and the horizontal axis is t, it is relatively easy to read the answers to the questions proposed. See Figure 9.12. Notice that the ball is on the ground when $t = 0$ seconds and when $t = 6$ seconds. Notice also that it is 80 feet high when $t = 1$ second on the way up and $t = 5$ seconds on the way down.

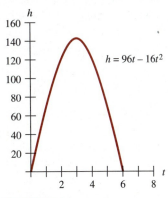

FIGURE 9.12

Exercise Set 9.5

Solve each of the following inequalities and graph the solution set where indicated.

1. $y(y + 1) < 0$
2. $p(p + 2) < 0$
3. $(a + 1)(a + 2) < 0$; graph
4. $(x + 3)(x - 1) < 0$; graph
5. $(w - 2)(w + 3) \leq 0$
6. $(h + 4)(h - 1) \leq 0$
7. $v(v - 3) > 0$; graph
8. $t(t - 4) > 0$; graph
9. $(m + 5)(m - 3) > 0$
10. $(2b - 1)(b + 6) > 0$
11. $(3c - 1)(2c + 3) \geq 0$; graph
12. $(4y + 3)(2y - 5) \geq 0$; graph
13. $(1 - x)(2 + x) \geq 0$
14. $(3 - a)(4 + a) \geq 0$

Factor each of the following inequalities and solve.

15. $h^2 - h - 12 < 0$
16. $r^2 + 4r + 3 < 0$
17. $a^2 + 7a + 6 \leq 0$
18. $x^2 - 5x - 6 \leq 0$
19. $y^2 + 8y + 16 > 0$
20. $t^2 + 6t + 9 > 0$
21. $u^2 - 9u - 22 \geq 0$
22. $p^2 - 7p - 18 \geq 0$
23. $w^2 - 4w + 4 \leq 0$
24. $h^2 - 10h + 25 \leq 0$
25. $r^2 - 9 < 0$
26. $v^2 - 16 < 0$
27. $6b^2 + 7b > 0$
28. $5c^2 - 6c > 0$

Solve each of the following inequalities.

29. $\dfrac{c}{c + 2} < 0$
30. $\dfrac{m}{m - 3} < 0$
31. $\dfrac{1}{a + 4} \leq 1$
32. $\dfrac{1}{w - 2} \leq 1$
33. $\dfrac{t + 3}{t + 2} < 1$
34. $\dfrac{v + 5}{v + 2} < 2$
35. $\dfrac{w + 2}{w + 4} \geq 3$
36. $\dfrac{s + 2}{s + 9} \geq 1$
37. $\dfrac{b - 1}{b - 5} \leq 2$
38. $\dfrac{2v + 6}{v - 1} \geq 5$
39. $\dfrac{4}{h - 3} > \dfrac{-2}{h + 11}$
40. $\dfrac{9}{3y + 4} > \dfrac{1}{2y + 1}$
41. $\dfrac{2x}{x + 2} \leq \dfrac{-2x}{x - 5}$
42. $\dfrac{2p}{p - 2} > \dfrac{3p}{4p + 1}$

Solve each of the following inequalities and graph the answers where indicated.

43. $a(a - 1)(a + 2) \leq 0$
44. $x(x + 1)(x - 2) \leq 0$
45. $(2y - 3)(3y - 1)y \geq 0$; graph
46. $(p - 5)(2p + 1)p \geq 0$; graph

47. $(3w - 4)(w + 1)w < 0$ 48. $(t - 1)(2t + 5)t < 0$
49. $c^2 - 3 < 13$ 50. $-3 < h^2 - 4$
51. $8 > m^2 - 1$; graph 52. $k^2 - 1 < 3$; graph
53. $s^2 - 4 \geq 5$ 54. $x^2 - 9 \geq 7$

Solve each of the following applied problems as indicated.

55. If Lisa works t hours of an 8-hour day, her business shows a profit or loss per hour according to the formula $P(t) = 15t - 3t^2$.
    (a) How many hours during the day is Lisa's profit greater than zero?
    (b) For what values of t does her profit P become less than zero (a loss)?
    (c) How many hours should she work per day if her profit is to be at least $12/hr?

56. The height of a certain projectile above the ground is given by $h(t) = 32t - 16t^2$, where t is the time in seconds.
    (a) For what range of values of t is the projectile off the ground?
    (b) At what time t does the projectile strike the ground?
    (c) For what range of values of t is the projectile less than 16 ft above the ground?

57. A ball is thrown directly upward from ground level at an initial velocity of 40 ft/sec and attains a height of $h(t) = 40t - 16t^2$ after t sec.
    (a) For what range of values of t is the ball above ground level?
    (b) At what time t does the ball strike the ground?
    (c) For what range of values of t is the ball less than 16 ft above the ground?

58. A manufacturer of solar heaters finds that when x units are made and sold, the profit (in thousands of dollars) is given by $P(x) = x^2 - 50x - 5000$, $0 \leq x \leq 200$.
    (a) For what range of values of x will the firm show a loss?
    (b) For what value(s) of x will the firm break even?
    (c) For what range of values of x will the firm show a profit?

59. The tensile strength, S, in pounds per square inch (psi) of a new plastic varies with the temperature according to the formula $S(t) = 500 + 600t - 20t^2$, where t is the temperature in degrees, $0 \leq t < 50$.
    (a) For what temperature range is $S > 4500$ psi?
    (b) For what temperature range is $S \leq 4500$ psi?
    (c) At what temperature, to the nearest degree, is the tensile strength 120 psi?

60. A realtor charges the seller of a house a 7% commission on the selling price. Suppose that the seller wants to clear at least $60,000 after paying the commission and that similar houses are selling for $70,000 or less.
    (a) What is the range of selling prices for the house?
    (b) If the house is sold, what is the possible range for the realtor's commission?

REVIEW EXERCISES

Solve each of the following inequalities and write all solutions using interval notation.

61. $|a - 3| < 0$ 62. $|5t + 1| < 11$
63. $|4x - 1| > 1$ 64. $|3y + 6| \leq 12$
65. $|3a - 1| \leq 8$ 66. $|2v - 6| > 4$
67. $|3y + 2| > -1$ 68. $\left|\frac{1}{3} - h\right| < \frac{2}{3}$
69. $\left|\frac{5 - c}{3}\right| > 4$ 70. $\left|\frac{2x + 1}{3}\right| \leq 5$
71. $\left|\frac{3w - 2}{4}\right| < 1$ 72. $\left|\frac{2k + 1}{3}\right| < 0$

FOR EXTRA THOUGHT

73. A piece of wire x meters long is bent into a square. For which values of x will the area be numerically greater than the perimeter?

74. The height of an isosceles triangle is $a + b$ in. ($a > b > 0$), and the area is less than or equal to $a^2 - b^2$ in.² What is the range of possible values for the base of the triangle?

9.6 Variation

OBJECTIVES

In this section we will learn to
1. write and solve equations involving direct variation;
2. write and solve equations involving inverse variation;
3. write and solve equations involving joint variation; and
4. use variation equations to solve applied problems.

Many jobs in science, mathematics, engineering, and other fields involve working with formulas that express relationships between two or more quantities. The area of a circle varies directly as the square of the radius; the pressure that a gas exerts on its container varies directly with the absolute temperature; the stiffness of a given length of beam varies directly with the product of the width and the square of the depth; the gravitational attraction between two particles varies directly with the product of their masses and inversely with the square of the distance between them.

In this section we will discover how to write and use such formulas.

DEFINITION

Direct Variation

The statement "y **varies directly** with x" means

$$y = kx$$

where k is called the **constant of variation.**

EXAMPLE 1 Translate each equation into words:

(a) $y = kx$ (b) $y = kx^2$ (c) $y = k\sqrt{x}$ (d) $C = \pi d$

SOLUTION
(a) y varies directly with x.
(b) y varies directly with the square of x.
(c) y varies directly with the square root of x.
(d) C varies directly with d. Here the constant of variation is π.

Not all variation is direct. Two additional forms that we will consider are *inverse variation* and *joint variation*.

DEFINITION

Inverse Variation

The statement "y **varies inversely** with x" means

$$y = \frac{k}{x}$$

where k is the **constant of variation.**

EXAMPLE 2   Translate each equation into words:

(a) $y = \dfrac{k}{x^2}$ (b) $y = \dfrac{k}{\sqrt{x}}$

SOLUTION   (a) y varies inversely with the square of x.
(b) y varies inversely with the square root of x.

Note that both direct variation, $y = kx$, and inverse variation, $y = \dfrac{k}{x}$, represent functions. As such they are sometimes written using function notation:

$$y(x) = kx \quad \text{and} \quad y(x) = \dfrac{k}{x}$$

When one variable varies directly with the product of two or more variables, it is said to *vary jointly* with those variables.

> **DEFINITION**
>
> **Joint Variation**
>
> The statement "y **varies jointly** with x and z" means
>
> $y = kxz$
>
> where k is the **constant of variation.**

EXAMPLE 3   Translate each equation into words.

(a) $y = kzt$ (b) $y = \dfrac{kst}{r}$ (c) $y = \dfrac{krt^2}{s^3}$

SOLUTION   (a) y varies jointly with z and t.
(b) y varies jointly with s and t and inversely with r.
(c) y varies jointly with r and the square of t and inversely with the cube of s.

EXAMPLE 4   Name the constant of variation in each of the following common formulas, then translate the statement into words.

(a) The area of a circle is $A = \pi r^2$.
(b) Neglecting air resistance, the distance d an object falls near the earth's surface in a given period of time t is $d = \tfrac{1}{2}gt^2$.

SOLUTION   (a) In $A = \pi r^2$, the constant of variation is π. The area of a circle varies directly with the square of the radius.
(b) In the formula $d = \tfrac{1}{2}gt^2$, the constant of variation is $\tfrac{1}{2}g$. The distance an object falls varies directly with the square of the time it has been falling.

The constant of variation, as the name implies, has the same value for all allowable values of the variables. For this reason it can often be readily determined.

EXAMPLE 5 A quantity y varies directly with x, and $y = 4$ when $x = 2$. Write the variation equation and use the given values of x and y to find k, then rewrite the variation equation using this value of k.

SOLUTION The variation equation is $y = kx$. To find k, we substitute 4 for y and 2 for x.

$$y = kx$$
$$4 = 2k$$
$$k = 2$$

Substituting 2 for k in the original equation, we have $y = 2x$.

EXAMPLE 6 If y varies jointly with x and the square of z, and if $y = 3$ when $x = 4$ and $z = 2$, write the variation equation and use the given values to find k. Rewrite the variation equation using this value of k.

SOLUTION The variation equation is $y = kxz^2$. To find k we substitute 3 for y, 4 for x, and 2 for z.

$$y = kxz^2$$
$$3 = k(4)(2^2)$$
$$k = \frac{3}{16}$$

The equation becomes $y = \frac{3}{16} xz^2$.

EXAMPLE 7 If W varies jointly with the cube of x and the square of y and inversely with the square of z, write the variation equation with the constant of variation, k, and **(a)** find k given that $W = 8$ when $x = 2$, $y = 3$, and $z = 4$; then **(b)** find W when $x = -1$, $y = 4$, and $z = 2$.

SOLUTION The variation equation is $W = k \frac{x^3 y^2}{z^2}$.

(a) $8 = k \frac{(2^3)(3^2)}{4^2}$ $\qquad W = 8, x = 2, y = 3,$ and $z = 4$

$$8 = \frac{9k}{2}$$

$$k = \frac{16}{9}$$

The variation equation is $W = \frac{16}{9} \cdot \frac{x^3 y^2}{z^2}$

(b) $W = \dfrac{16}{9} \cdot \dfrac{(-1)^3(4^2)}{2^2}$ $x = -1, y = 4,$ and $z = 2$

$= \dfrac{-64}{9}$

EXAMPLE 8 According to Hooke's law, the force F necessary to stretch a spring a given distance x varies directly with the distance the spring is stretched. If a 40-kilogram weight stretches the spring 3 centimeters, how far will a 60-kilogram weight stretch it?

SOLUTION The variation equation is

$F(x) = kx$

The force that stretches the spring 3 cm is exerted by the 40-kilogram weight, so $F = 40$ when $x = 3$, or $F(3) = 40$.

$40 = 3k$ Replace $F(x)$ by 40 and x by 3.

$k = \dfrac{40}{3}$

The variation equation is $F(x) = \dfrac{40}{3} x$. If $F(x) = 60$, then

$60 = \dfrac{40}{3} x$

$\dfrac{180}{40} = x$

$x = 4.5$ cm

A 60-kilogram weight would stretch the spring 4.5 centimeters.

In the exercise set that follows, it is not necessary to be familiar with the units of measurement in order to do the work.

Do You Remember?

Can you match these?

____ 1. $y = \dfrac{k}{x}$

____ 2. $y = kx^2$

____ 3. $y = kxz$

____ 4. $y = \dfrac{kx}{z}$

____ 5. $y = kx\sqrt{z}$

a) y varies jointly with x and z.
b) y varies directly with x and inversely with z.
c) y varies inversely with x.
d) y varies jointly with x and the square root of z.
e) y varies directly with the square of x.

Answers: 1. c 2. e 3. a 4. b 5. d

Exercise Set 9.6

Write each variation in the form of an equation.

1. y varies directly with x.
2. R varies directly with s.
3. M varies directly with the square of p.
4. H varies directly with the square of q.
5. V varies directly with the cube of s.
6. R varies directly with the cube of t.
7. y varies inversely with x.
8. A varies inversely with d.
9. S varies directly with the square root of a.
10. H varies directly with the square root of j.
11. V varies jointly with h and the square of r.
12. V varies jointly with L, W, and H.
13. P varies directly with c and inversely with the square root of n.
14. y varies directly with the square of x and inversely with z.

Solve each of the following variation problems.

15. If y varies directly with x and if $y = 3$ when $x = 9$, find y when $x = 12$.
16. If p varies directly with t and if $p = 8$ when $t = 2$, find p when $t = 5$.
17. If A varies directly with the square of s and if $A = 8$ when $s = 2$, find A when $s = 7$.
18. If y varies directly with the square of x and if $y = 3$ when $x = 4$, find y when $x = 16$.
19. If a varies directly with the cube of b and if $a = 6$ when $b = 10$, find a when $b = 2$.
20. If w varies directly with the cube of u and if $w = 12$ when $u = 20$, find w when $u = 2$.
21. If Z varies directly with the square of x and inversely with y, and if $Z = 8$ when $x = 4$ and $y = 3$, find Z when $x = 3$ and $y = 8$.
22. If P varies directly with the cube of m and inversely with the square of n, and if $P = 8$ when $m = 4$ and $n = 6$, find P when $m = 2$ and $n = 12$.
23. If T varies jointly with p and the cube of v and inversely with the square of u and if $T = 24$ when $p = 3$, $v = 2$, and $u = 4$, find T when $p = 4$, $v = 3$, and $u = 2$.
24. If A varies jointly with the square of b, the square of c, and inversely with the cube of d, and if $A = 18$ when $b = 4$, $c = 3$, and $d = 2$, find A when $b = 2$, $c = 6$, and $d = 4$.

25. The intensity of illumination, I, varies inversely with the square of the distance, d, from the light source. If the intensity I is 4 fc (foot-candles) at 15 ft from the source, find the intensity at 10 ft from the source.
26. The distance d an object falls varies directly with the square of the time t (in seconds) that it falls. If the object falls 64 ft in 2 sec, how far can it fall in 10 sec?
27. The resistance R of a conductor varies inversely with the area A of its cross section. If $R = 20$ ohms when $A = 8$ cm^2, find R when $A = 12$ cm^2.
28. The weight W of an object in space varies inversely with the square of its distance d from the center of the earth. If an object weighs 400 lb on the surface of the earth, how much does it weigh 1000 miles from the surface of the earth? Assume the radius of the earth is 4000 miles.
29. The current I in a wire varies directly with the electromotive force E and inversely with the resistance R. In a wire with a resistance of 4 ohms, a current of 30 amp is obtained when the electromotive force is 120 V. Find the current produced when $E = 220$, V and $R = 10$ ohms.
30. The distance d required to stop a midsized car varies directly with the square of its speed, s. If the car stops in 36 ft at a speed of 24 mph, find the distance required to stop it at a speed of 54 mph.
31. The pressure P of a certain gas in a container varies directly with the temperature, T, and inversely with the volume, V. If 300 ft^3 of this gas exert a pressure of 20 psi at a temperature of 500°K, what is the pressure exerted by this gas when the temperature is 400°K and the volume is 500 ft^3?
32. The surface area S of a hollow cylinder varies jointly with the height h and the radius r of the cylinder. If a cylinder with a radius of 5 in. and a height of 7 in. has a surface area of 213.5 in.^{2}, find the surface area when the radius is 8 in. and the height is 11 in.
33. The wind pressure P on a wall varies jointly with the area A of the wall and the square of the velocity v of the wind. If the pressure is 120 lb when the area is 100 ft^2 and the velocity of the wind is 20 mph, find the pressure when the area is 200 ft^2 and the velocity of the wind is 40 mph.
34. Simple interest I earned over a given period of time varies jointly with the principal, p, and the interest rate, r. If \$100.00 at 4% earns \$14.00, how much does \$250.00 at 5% earn in the same period of time?

35. The amount of pollution P entering the atmosphere in a certain area is found to vary directly with the $\frac{2}{3}$ power of N, the number of people in that area. If a population of 8000 people produces 900 tons of pollution per year, find a formula for P in terms of N.

36. Kepler's Third Law states that the time t required for a planet to make one revolution about the sun varies directly with the $\frac{3}{2}$ power of the maximum distance of its orbit from the sun. If the maximum distance of the earth's orbit is 93 million miles and the maximum distance of Mars's orbit is 142 million miles, what is the time required for Mars to make one orbit?

37. The distance traveled in a car moving at a constant rate varies directly with the time. If a car travels 418 mi in 8 hr, how far will the car travel in 13 hr?

38. In Polly's candy recipe, the amount of sugar needed varies directly with the number of people to be served. If $3\frac{1}{3}$ cups of sugar in the recipe will serve 10 people, how many people can be served if 5 cups of sugar are used in the recipe?

39. The resistance R of a wire varies directly with the length l of the wire and the square of the diameter, d. If the resistance of a 12 m length of wire is 8 ohms when the diameter is 0.02 cm, find the resistance of an 18 m piece of wire if the diameter is 0.04 cm.

40. The strength of a rectangular steel beam varies jointly with its width and the square of its depth. The strength of a rectangular beam 1 in. wide and 5 in. deep is 500 psi. What is the strength of this rectangular beam if its width and depth are each doubled?

REVIEW EXERCISES

41. Given $f(x) = \dfrac{3x}{x - 5}$, find:
    (a) $f(0)$; (b) $f(5)$; (c) $f(-2)$; (d) $f(x + h)$.

42. Given $f(x) = 3x - 7$ and $g(x) = \dfrac{x + 7}{3}$, find $f[g(x)]$ and $g[f(x)]$. What can you say about $f(x)$ and $g(x)$?

43. Given $f(x) = -5x + 2$, find $f^{-1}(x)$. Is $f^{-1}(x)$ a function? Graph $f(x)$, $f^{-1}(x)$, and $y = x$ on the same coordinate system. Discuss what you notice about $f(x)$, $f^{-1}(x)$, and the line $y = x$.

44. Given $f(x) = x^2 - 4$ and $g(x) = x - 2$, find:
    (a) $(f + g)(x)$; (b) $(fg)(x)$; (c) $\left(\dfrac{f}{g}\right)(x)$; (d) $f[g(6)]$.

45. Given $f(x) = \dfrac{2x}{4x - 1}$, find its domain.

46. Which of the following are functions?
    (a) $y = \sqrt{x^2 - 4}$; (b) $y = 4$;
    (c) $y = \dfrac{2}{x - 3}$; (d) $x^2 + 4y^2 = 36$;
    (e) $x = \sqrt{y^2 - 9}$; (f) $x - 3y = 4$

47. Given $f(x) = 3x - 5$, find $\dfrac{f(x + h) - f(x)}{h}$.

48. How does a function differ from a relation?

49. Are there any linear relations that are functions? Are there any linear relations that are not functions? Give an example of each, if possible.

50. Are all quadratic relations functions if they are of the type $f(x) = ax^2 + bx + c$, $a \neq 0$? If not, give an example.

9.7 More Applications

OBJECTIVES

In this section we will further explore how to
1. solve applied problems by using quadratic equations; and
2. solve applied problems by using quadratic inequalities.

Applications of quadratic equations and inequalities are too numerous for this book to include an example of each type. In the following examples and exercises, however, we do cover a few important types.

Since quadratic equations often have more than one solution, it is essential that the solutions be checked in the words of the problem. Remember to follow the five steps for solving applied problems we outlined earlier.

EXAMPLE 1 The length of a rectangle exceeds its width by 2 meters. If the area of the rectangle is 224 square meters, find its dimensions.

SOLUTION We let

$w =$ the width

$w + 2 =$ the length

See Figure 9.13. The area of a rectangle is width times length, so

$$w(w + 2) = 224$$
$$w^2 + 2w = 224$$
$$w^2 + 2w - 224 = 0$$
$$(w + 16)(w - 14) = 0$$
$$w + 16 = 0 \quad \text{or} \quad w - 14 = 0$$
$$w = -16 \quad \text{or} \quad w = 14$$

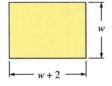

FIGURE 9.13

Since the width of a rectangle cannot be negative, $w = 14$ m and $w + 2 = 16$ m.

CHECK The length, 16 m, is two meters more than the width, 14 m. Since $(16 \text{ m})(14 \text{ m}) = 224 \text{ m}^2$, the area of the rectangle, the solution checks.

EXAMPLE 2 A group of students charters a bus for a trip to Disneyland for $720. If they can persuade 12 more students to go, each student's cost will decrease by $2. How many students are in the group?

SOLUTION We let

$x =$ the number of students in the original group

$x + 12 =$ the number of students in the enlarged group

To determine the cost per student for either the original group or the larger group, we divide the total cost by the number of students going.

Group	Number of Students	Total Cost	Cost per Student
Original	x	$720	$\dfrac{\$720}{x}$
Larger	$x + 12$	$720	$\dfrac{\$720}{x + 12}$

Since the cost per student in the larger group is $2 less per student, we have

$$\frac{720}{x} - \frac{720}{x + 12} = 2$$

$$720(x + 12) - 720x = 2x(x + 12) \qquad \text{Clear of fractions;}$$
$$\text{the LCD is } x(x + 12).$$

$$720x + 8640 - 720x = 2x^2 + 24x \qquad \text{Multiply.}$$
$$8640 = 2x^2 + 24x \qquad \text{Simplify.}$$
$$0 = 2x^2 + 24x - 8640 \qquad \text{Standard form.}$$
$$0 = x^2 + 12x - 4320 \qquad \text{Divide each side by 2.}$$
$$0 = (x - 60)(x + 72) \qquad \text{Factor.}$$

$$x - 60 = 0 \quad \text{or} \quad x + 72 = 0$$
$$x = 60 \quad \text{or} \quad x = -72$$

The number of students cannot be negative, so there are 60 students in the original group.

CHECK If there are 60 students in the group, the cost per student is

$$\frac{\$720}{60}, \quad \text{or } \$12 \text{ each}$$

If there are 12 more students in the group, the cost is

$$\frac{\$720}{72}, \quad \text{or } \$10 \text{ each}$$

This is $2 less than the original cost. The solution checks.

EXAMPLE 3 Two buses, an express and a local, leave Pismo Beach for Los Angeles, a distance of 200 miles. The express makes the trip in one hour less time and travels an average of 10 mph faster. Determine the average rate (speed) of each bus.

SOLUTION We let

$$x = \text{the rate of the local bus}$$
$$x + 10 = \text{the rate of the express bus}$$

The distance traveled by both buses is 200 miles. Since distance = rate · time,

$$\text{time} = \frac{\text{distance}}{\text{rate}}$$

Bus	Rate	Distance	Time
Local	x	200	$\dfrac{200}{x}$
Express	$x + 10$	200	$\dfrac{200}{x+10}$

The local bus takes one hour longer. That means the difference in the times is 1 hour:

$$\frac{200}{x} - \frac{200}{x+10} = 1$$

$$200(x+10) - 200x = x(x+10)$$

$$200x + 2000 - 200x = x^2 + 10x$$

$$0 = x^2 + 10x - 2000$$

$$0 = (x-40)(x+50)$$

$$x - 40 = 0 \quad \text{or} \quad x + 50 = 0$$

$$x = 40 \quad \text{or} \quad x = -50$$

The average speed must be positive, so we reject -50. The local bus averages 40 mph and the express bus goes 10 mph faster, or 50 mph. The check is left to the reader.

EXAMPLE 4 When the inlet pipe and the drain are both left open on a water storage tank, it takes 4 hours to fill the tank. If the drain takes 2 hours longer to empty the full tank than the inlet pipe takes to fill the empty tank when only one is open, how long will it take to fill the empty tank with the drain closed?

SOLUTION We let $x =$ the time it takes the inlet pipe to fill the empty tank with the drain closed. Then $x + 2 =$ the length of time the drain takes to empty the full tank.

	NUMBER OF HOURS TO FILL OR DRAIN	AMOUNT FILLED OR DRAINED IN ONE HOUR
Inlet pipe	x	$\frac{1}{x}$
Drain	$x + 2$	$\frac{1}{x+2}$
Together	4	$\frac{1}{4}$

When the drain is left open, the rate at which it fills in one hour is the difference between the filling and draining rates.

$$\frac{1}{x} - \frac{1}{x+2} = \frac{1}{4}$$

$$4(x+2) - 4x = x(x+2) \quad \text{Multiply each side by } 4x(x+2), \text{ the LCD.}$$

$$4x + 8 - 4x = x^2 + 2x$$

$$0 = x^2 + 2x - 8$$

$$0 = (x+4)(x-2)$$

$$x + 4 = 0 \quad \text{or} \quad x - 2 = 0$$

$$x = -4 \quad \text{or} \quad x = 2$$

We reject -4 because time can't be negative. Therefore it will take 2 hours to fill the tank with the drain closed.

EXAMPLE 5 Arturo has scores of 75, 92, 86, and 71 on his first four of five scheduled tests in biology. If he wants to keep his average score above 80 for the semester, how many points must he get on the last test?

SOLUTION We let x denote the score Arturo must receive on his last test. To find the average of the scores, we find the sum of the scores and divide by the number of scores. That average must be greater than 80.

$$\frac{75 + 92 + 86 + 71 + x}{5} > 80$$

$$\frac{324 + x}{5} > 80$$

$324 + x > 400$ Multiply each side by 5, the LCD.

$x > 76$

Arturo's score must be greater than 76 points.

EXAMPLE 6 A small manufacturing company has found that it can make a profit if the cost of materials is kept under $2000 on any given day. The cost of producing n articles is given by $C(n) = n^2 - 110n + 4800$. How many articles can be produced in a day if the company is to operate profitably?

SOLUTION The cost must be kept under $2000, so

$C < 2000$

$n^2 - 110n + 4800 < 2000$ Substitute $n^2 - 110n + 4800$ for C.

$n^2 - 110n + 2800 < 0$

$(n - 40)(n - 70) < 0$

Now we use a sign graph to determine the interval where $(n - 40)(n - 70)$ is negative; see Figure 9.14.

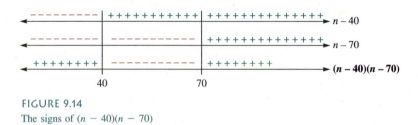

FIGURE 9.14
The signs of $(n - 40)(n - 70)$

We conclude that $(n - 40)(n - 70) < 0$ when $40 < n < 70$. Therefore the company will operate profitably if it manufactures more than 40 but less than 70 articles per day.

Exercise Set 9.7

Solve each of the following applied problems. Some formulas for area and volume are listed on the endsheets.

1. The area of a triangular plate is 20 m². Find the base and height if the base is 6 m longer than the height.
2. Determine the base and altitude of a triangle if the altitude exceeds the base by 3 ft and the area is 14 ft².
3. What are the dimensions of a rectangular piece of metal with length 50 in. more than its width and with area 1875 in.²?
4. A rectangular plot of grass is 10 yd long and 4 yd wide. New strips of grass of equal width are to be planted at one end and one side. Find how wide the strips must be to double the area of the original plot.
5. The hypotenuse of a right triangle is 13 cm long. Find the lengths of the sides of the triangle if one side is 7 cm longer than the other.
6. A rectangular field 250 m long and 180 m wide has a concrete walkway of uniform width as its border. If there are 2616 m² of concrete in the walkway, what is its width?
7. Mel and Nett working together can mow and water a law in 3 hr less time than Mel alone. Nett working alone takes 1 hr longer than Mel. Determine the time it takes them to do the work together.
8. Mindy can review a set of books in 7 hr less time than Jill. Together they can do it in 9 hr less time than Mindy can alone. Find the time they take together to review the books.
9. Rico can row 20 miles upstream and back in 7.5 hr. If his rate of rowing in still water is 6 mph, what is the speed of the current?
10. Celia finds that a trip of 70 mi driven at her normal speed can be decreased by 15 min if she travels 5 mph faster. What is her normal speed?
11. Two fuel lines of different sizes working together can fill a tank in 4 hr. If the larger line alone fills the tank in 4 hr less time than the smaller line, how much time is needed for just the larger line to fill the tank?
12. A gardener sets 180 plants in rows. Each row contains the same number of plants. If there were 40 more plants per row, he would need 6 less rows. How many rows are there?
13. The members of a college chemistry department agree to contribute equal amounts of money to make up a scholarship fund of $280. Then the department hires three new members, resulting in each member's share being reduced by $30. What is the new size of the department?
14. On his first three tests, Dave's scores are 65, 78, and 84. What range of scores on his fourth test will put his average at no less than 70 but less than 80 for the four tests?
15. Sandra's scores on her first four English tests are 68, 76, 83, and 63. What range of scores on a fifth test will give her an average of no less than 70 but less than 80 for the five tests?
16. Plans for a rectangular solar collector call for a height of 1.5 m, but its length has yet to be determined. What is the range of values for this length if the collector will provide 400 watts/m² and it must provide between 2000 and 3500 watts?
17. How far away is the horizon (to the nearest mile) from a plane that is 31,680 ft (6 mi) high? See the figure.

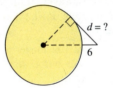

The radius of the earth is 4000 miles

18. How far away is the horizon (to the nearest mile) from a plane that is 8 mi high? See Exercise 17.
19. Given two concentric circles, what should the radius of the inner circle be so that its area is equal to the area between the two circles? (See the figure.) Express your answer in terms of x.

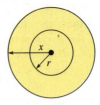

20. A Norman window (a semicircle on a rectangle shown on the next page) is as tall as it is wide, with an area of 16 ft². How wide is the window, to the nearest foot?
21. Matt bought several cartons of large eggs for $11.20. If he had bought extra large eggs at 10¢ more per carton, he would have had two less cartons of eggs for the same amount of money. How much did he pay for a carton of large eggs?

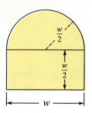

ART TO ACCOMPANY EXERCISE 20

22. Nancy bought $28.00 worth of pizza for a party. If she had bought the pizza at a different pizza parlor, one that charges 50¢ less per pizza, she could have gotten one more pizza for the same amount of money. How many pizzas did she buy?

23. One pipe can fill a tank in 5 hr less time than another. Together, both pipes can fill the tank in 6 hr. How long does it take each pipe alone to fill the tank?

24. A drain takes 3 min longer to empty a sink full of water than it takes for the cold water faucet to fill it when it's empty. If it takes 6 min to fill the empty sink when the drain is accidentally left open, how long does it take to fill the empty sink with the drain closed?

25. If Andy's boat travels 15 mph in still water and it takes him 3 hr more to travel 60 mi up a river than it does to travel 60 mi back down the river, what is the speed of the current?

26. A helicopter takes 1 hr longer to travel 200 mi *against* a 25 mph wind than it does to travel the same distance *with* the wind. What is the speed of the helicopter in still air, to the nearest hundredth?

27. Marilyn contracts to paint some rooms in a house for $336. It takes her 4 hr longer than she expected and so she earns $2/hr less than she anticipated. How long did she think it would take for her to do the job?

28. Marty contracts a yard renovation job for $1200. It takes him 10 hr longer than he had planned, so he earns 50¢/hr less than he expected. How long did he think it would take him to renovate the yard?

29. Janet lives 15 mi from her office. If she drives 15 mph faster than usual, she arrives at the office 10 min earlier than normal. How fast does she usually drive?

30. Jessie rows 5 miles down a river and back in $1\frac{7}{8}$ hours. How fast can he row in still water if the rate of the current is 2 mph?

31. Eunice wants to make a metal box out of a sheet of metal by cutting out square corners and folding up the sides. The box is to have a square base, sides 6 in. high, and a volume of 384 sq. in. What size sheet of metal should she buy?

32. If one pair of opposite sides of a square is doubled and the other pair decreased by 3 ft, the resulting area is 27 ft^2 more than the area of the original square. Find the length of one side of the square.

33. A geology class took a trip costing $288. If there had been 4 more students, it would have cost each student $1 less. How many students are in the class?

34. The ski club went on a ski trip costing $1600. If there had been 8 fewer members on the trip, it would have cost each student $10 more. How many members are in the ski club?

35. A rectangular pool is 20 ft wide and 30 ft long. A strip of grass of uniform width is to be placed around all sides of the pool. If the contractor orders 216 ft^2 of grass sod, how wide will the grass border be?

36. A rectangular house is 25 ft wide and 45 ft across the front. A concrete sidewalk of uniform width and 4 in. deep is to be placed along the two widths and across the front of the house. The contractor orders 89 ft^3 of concrete. How wide is the widewalk?

37. Emilio's revenue, $R(p) = 41 + 26p$, is from selling bags of popcorn at the ballpark at p dollars each. What range of prices can he charge if he wishes to keep his revenue under $80.00?

38. A bullet fired from ground level is at a height $h(t) = 320t - 16t^2$ ft after t sec. During what time interval is the bullet more than 1024 ft above the ground?

39. The cost in cents of producing x radios is given by the function $C(x) = -2x^2 + 300x + 160,000$ for $0 \leq x \leq 700$. How many radios can be produced if the total cost must be kept under $140,000?

40. Maria's revenue is $R(p) = 120 - 10p$ for selling bouquets of flowers each week at a price of p dollars per bouquet. What range of prices can she charge if she wants her weekly revenue to be at least $350.00?

Summary

9.1
Solving Polynomial Equations by Factoring and the Square Root Property

Quadratic equations can be solved by using the **zero-factor theorem,** which we discussed in Section 6.6. After factoring an equation, each factor is set equal to zero. The solutions to the resulting equations form the solution set.

The **square root property** gives another way to solve quadratic equations.

$$\text{If } x^2 = a \quad \text{then} \quad x = \pm\sqrt{a}.$$

Results obtained in this manner may be real numbers or nonreal complex numbers.

9.2
Solving Quadratic Equations by Completing the Square

A third method for solving quadratic equations is called **completing the square.** To use this method, all terms involving the variable are isolated on one side of the equation. Next, a quantity sufficient to make this side a perfect square is added to each side of the equation so that the solution set can be found by the square root property. To complete the square, the coefficient of the second-degree term must be 1. The quantity added to each side is the square of one-half of the coefficient of the first-degree term. In other words, to complete the square of $x^2 + bx$, add

$$\left(\frac{b}{2}\right)^2.$$

9.3
Solving Quadratic Equations Using the Quadratic Formula

When the method of completing the square is applied to the general quadratic equation $ax^2 + bx + c = 0$, $a \neq 0$, the result is called the **quadratic formula:**

$$x = \frac{-b \pm \sqrt{b^2 - 4ac}}{2a}$$

The quadratic formula provides a means of determining the nature of the solutions to a quadratic equation without actually solving the equation. The tool used to do this is the **discriminant** of the equation, $b^2 - 4ac$, which is the radicand of the quadratic formula. When a, b, and c are real numbers,

if $b^2 - 4ac = 0$, there is one solution of multiplicity 2;

If $b^2 - 4ac < 0$, the solutions are nonreal complex conjugates;

if $b^2 - 4ac > 0$, there are two real solutions.

9.4
Solving Equations Involving Quadratics

Many equations can be written in quadratic form after a suitable substitution. Once the substitution has been made, the equation can then be solved by one of the four methods just listed.

If a polynomial is of degree n, it can have at most n distinct solutions.

When r solutions represent the same number, the equation is said to have a **solution of multiplicity r.**

9.5
Solving Inequalities Involving Quadratics

One method of solving quadratic inequalities is the **sign graph.** To use a sign graph, first write the inequality in factored form, then determine the regions in which each factor is positive or negative and indicate these regions on a separate number line for each factor. The region that satisfies the original inequality with respect to all these factors is the solution set.

9.6 Variation

Many relations from science and mathematics can be expressed in terms of variation. Three of the more common types and the symbolism used to represent them are:

Direct: $y = kx$

Inverse: $y = \dfrac{k}{x}$

Joint: $y = kxz$

where k is the **constant of variation.**

9.7 More Applications

To solve an applied problem involving a quadratic form, write a quadratic equation or inequality describing the conditions of the problem. Then find the solutions by one of the preceding methods.

Cooperative Exercise

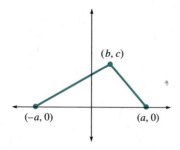

ART TO ACCOMPANY PARTS A–C

PART A

1. Find the coordinates of the point of intersection of the medians.
2. Use some reference to find what this point is called.
3. Is this point useful to us? If so, give an example of its usefulness.

PART B

1. Find the point of intersection of the altitudes.
2. Use some reference to find what this point is called.
3. Is this point useful to us? If so, give an example of its usefulness.

PART C

1. Find the coordinates of the point of intersection of the perpendicular bisectors of the sides.
2. Use some reference to find what this point is called.
3. Is this point useful to us? If so give an example of its usefulness.

Review Exercises

Solve each of the following equations by factoring.

1. $y^2 - 4 = 0$
2. $p^3 - 64p = 0$
3. $12m^3 - 7m^2 - 10m = 0$
4. $30a^2 - 46a - 56 = 0$

Solve each of the following equations by the square root property.

5. $k^2 = 4$
6. $v^2 = 10^{-8}$
7. $64x^2 = 49$
8. $(y - a)^2 = c^2$
9. $121y^2 + 49 = 0$
10. $(3x - 2)^2 + 7 = 0$

Write the quadratic or cubic equation having the given solution set. Assume no repeated roots.

11. $\{1, -1\}$
12. $\left\{-\dfrac{1}{3}, \dfrac{1}{4}\right\}$
13. $\{\pm 2\sqrt{3}\}$
14. $\{0, 2, -3\}$
15. $\{\pm 2i, 3\}$
16. $\{1 \pm i, 0\}$
17. $\{4 \pm 5i\}$

Solve each of the following equations for the indicated variable.

18. $A^3 = \frac{1}{2}bh^2$; for $h > 0$
19. $V = \frac{4}{3}\pi r^3$; for $r > 0$
20. $dx^2 + b = c$; for d

Use the discriminant to determine the nature of the solutions to each of the following equations.

21. $x^2 - 5x - 6 = 0$
22. $y^2 + 2y + 1 = 0$
23. $4m^2 - 2m + 4 = 0$
24. $6a^2 = -7 - 4a$
25. $k^2 - 14 = 0$
26. $(2p - 5)^2 = -3$

Solve for x and y.

27. $(3 - 4i)x - (2 - i)y = 7i$
28. $(x + y - 3)i + 7x + 2y = 5$

Solve by the square root property.

29. $(3a - 5)^2 = 16$
30. $(x - 7)^2 = 6$
31. $(7w - 5)^2 = -8$
32. $(2t + 11)^2 = -27$

Solve by completing the square.

33. $y^2 + 4y + 3 = 0$
34. $3s^2 + 7s - 8 = 0$
35. $6c^2 - c + 1 = 0$
36. $2h^2 + h + 4 = 0$

Solve by the quadratic formula.

37. $x^2 - 5x + 4 = 0$
38. $2w^2 + 3w - 5 = 0$
39. $3k^3 + k^2 + 3k = 0$
40. $3(m - 1)^2 = 5$

Use an appropriate substitution, if needed, to solve each equation.

41. $x^3 - 8 = 0$
42. $t^4 - 6t^2 - 7 = 0$
43. $a^{1/3} - 7a^{1/6} + 6 = 0$
44. $p^{-2} - 11p^{-1} + 30 = 0$
45. $\left(\frac{u^2 + 3}{u}\right)^2 - 6\left(\frac{u^2 + 3}{u}\right) + 8 = 0$
46. $2r^4 - 5r^2 + 2 = 0$

Solve. Write all solutions in interval notation.

47. $(x - 5)(x + 1) < 0$
48. $h^3 + 3h^2 - 4h \geq 0$
49. $\frac{1}{2a + 1} < \frac{9}{3a + 4}$

Solve the following applied problems.

50. A plane regularly flies 2400 mi to Chicago. When there is a headwind of 40 mph, it takes an extra 15 min to get there. What is the speed of the plane in still air?
51. Phil goes 20 mi downstream and back in $11\frac{1}{3}$ hr. If the boat travels $4\frac{1}{4}$ mph in still water, what is the speed of the current?
52. Mary and Betty can finish a certain job in 40 hr. When working alone, Mary takes 39 hr longer than Betty to do the job. How long does it take Betty to do the job working alone?
53. The sum of two natural numbers is 48. The difference of their squares is 36 more than their product. Find the two natural numbers.
54. If P varies jointly with x^2 and y, and if $P = 32$ when $x = 2$ and $y = 4$, find P when $x = 3$ and $y = 5$.

Chapter Test

1. Solve by factoring: $2x^2 - 5x + 2 = 0$.
2. Solve by the square root property: $(3y - 5)^2 = 8$.
3. Solve by completing the square: $y^2 - 4y + 1 = 0$.
4. Solve by the quadratic formula: $4x^2 - x + 5 = 0$.

Solve by any method.

5. $x^4 - 16 = 0$
6. $y(y + 4) = 5$
7. $3p^2 + 6 = -19p$
8. $x^2 + 6x = 0$

Use the discriminant to determine the nature of the solutions in Exercises 9–11. Do not solve.

9. $x^2 + 9x + 8 = 0$
10. $y^2 - 10y + 25 = 0$
11. $p^2 + 2p + 2 = 0$

For Exercises 12–14, find the quadratic equation with the given solution set. Assume no repeated roots.

12. $\{2, 5\}$
13. $\{1 \pm i\}$
14. $\{1 \pm \sqrt{2}\}$

15. Use an appropriate substitution to rewrite the equation $x^{1/2} - 5x^{1/4} + 6 = 0$ as a quadratic equation and then solve it.
16. Solve: $x^2 - 64 = 0$.
17. Solve: $x^2 - 3x - 4 > 0$. Write your answer in interval notation.
18. Solve: $\frac{2x + 3}{x + 2} \leq 1$. Write your answer in interval notation.

19. A man can swim 3 mi up a stream and back in $1\frac{3}{7}$ hr. If the rate of the current is 2 mph, find the rate at which he can swim in still water.

20. The area of a triangular space is 10 ft². Find the dimensions of the base and height if the base is 1 ft more than the height.

21. The floor of a rectangular gazebo is 3 m × 5 m. A concrete sidewalk is poured all around it. If the area formed by the gazebo and sidewalk is now 35 m², find the width of the sidewalk. The sidewalk is of uniform width.

22. The sum of the squares of two natural numbers is 1282. The difference between the two numbers is 8. Find the numbers.

23. A tree was hit by lightning and broken off $\frac{5}{18}$ of the distance from the bottom. The top of the tree hit the

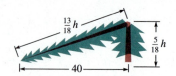

ground 40 ft from the base of the tree. What was the height of the tree?

24. Jeff has a back yard that is 30 ft × 40 ft. He wants to plant flowers in a uniform strip around the yard that will equal exactly half of the area of the yard. How wide a strip will he need to plant?

25. Solve: $\dfrac{4}{x-1} - \dfrac{1}{2x-5} = 1$.

26. If H varies directly with x^2 and inversely with the cube root of y, and if $H = 12$ when $x = 2$ and $y = 8$, find H when $x = 3$ and $y = 27$.

Cumulative Review for Chapters 7–9

1. Simplify: $\dfrac{x^2 - y^2}{x^2 - 2xy + y^2}$.

2. Evaluate $\dfrac{m^2 - 3m - 2}{m^2 - 4}$ for $m = -1$.

In Exercises 3–6, perform the indicated operations and simplify. Assume no denominator is zero.

3. $\dfrac{a^2 - 2ab + b^2}{(c-1)^2} \div \dfrac{a^3 - a^2 b}{ac - a}$

4. $\dfrac{x^2 - 4}{x^3 + 8} \cdot \dfrac{x^3 - 8}{x - 2} \div (x^2 + 2x + 4)$

5. $\dfrac{x}{x-8} - \dfrac{x-2}{x-3}$

6. $\dfrac{\dfrac{x}{x+3} - \dfrac{x}{x-3}}{\dfrac{1}{x-3} - \dfrac{2}{x+3}}$

7. $\dfrac{\dfrac{1}{x} + \dfrac{1}{y}}{\dfrac{1}{y^2} - \dfrac{1}{x^2}}$

8. $\dfrac{\dfrac{x}{x+1} - \dfrac{x}{x-1}}{\dfrac{1}{x-1} - \dfrac{2}{x+1}}$

Solve each of the following equations. State any restrictions on the variables. Check your answers.

9. $x + 3 = \dfrac{x}{3}$

10. $\dfrac{3}{x-4} = \dfrac{4}{x+3}$

11. $\dfrac{1}{x^2 - x - 2} + \dfrac{1}{x+1} = \dfrac{2}{x-2}$

12. $\dfrac{a-1}{a-3} - 3 = a$

In Exercises 13–19, write each of the applied problems as an equation and solve. Check your answers.

13. A gardener can mow a law in 21 min. His helper takes 28 min to mow the same lawn. How long should it take both of them working together?

14. A chemist has 10 ℓ of a 30% alcohol solution. How many liters of pure alcohol should be added to obtain a 40% alcohol solution?

15. If the sides of a square are increased by 2 in., the area is increased by 48 in². Find the length of a side of the original square.

16. Find three consecutive even integers such that one-third of the first is 2 more than the second subtracted from the third.

17. Mrs. Lee invests $12,000 for one year, part at 6.5% and the rest at 5.5%. In one year she earns $700 interest altogether. How much did she invest at each rate?

18. Mr. Johns drove to Newhall at 50 mph and returned at 60 mph. If the total driving time was $5\frac{1}{2}$ hr, how far did he drive?

19. The reciprocal of 2 more than a number is equal to the reciprocal of three times the number. Find the number.

20. Solve: $\sqrt{x+1} = \sqrt{x+4}$.
21. Solve: $5\sqrt{y-3} = 3\sqrt{2y+1}$.
22. Solve: $x - 6 = \sqrt{x^2 + 12}$.
23. Factor: $x^{1/4} - x = x^{1/4}(\quad)$.

Simplify each of the following expressions.

24. $\sqrt{50} - 3\sqrt{2}$
25. $\sqrt{\dfrac{64x}{x^3}}$
26. $\sqrt{48} + \sqrt{162} - \sqrt{288}$
27. $3\sqrt{12} - \sqrt{3}$
28. $3\sqrt[3]{81}$
29. $\dfrac{1}{\sqrt{5}}$
30. $\dfrac{2}{\sqrt[3]{2}}$
31. $\dfrac{12\sqrt{6} - 15\sqrt{6}}{3\sqrt{2}}$
32. $\dfrac{2}{\sqrt{3} - \sqrt{2}}$
33. $\dfrac{\sqrt{5} - \sqrt{2}}{\sqrt{5} + \sqrt{2}}$

34. Solve: $\sqrt[3]{x-1} = \sqrt[3]{3x}$.
35. Solve: $\sqrt{x} = 3 - \sqrt{x+1}$.

Simplify each quantity in Exercises 36–39.

36. $(2 - 3i) - (1 + i)$
37. $(3 + 4i)(3 - 4i)$
38. $\sqrt{-16} + \sqrt{-50} - \sqrt{8}$
39. $\dfrac{1}{2 - i}$

40. Solve by factoring: $x^2 + 2x - 8 = 0$.
41. Solve by completing the square: $y^2 - 6y + 8 = 0$.
42. Solve by the quadratic formula: $x^2 - 6x + 4 = 0$.
43. Solve by any method: $2a^2 - a + 4 = 0$.
44. Solve by completing the square: $3y^2 = 5 - 2y$.

Use the discriminant to determine the nature of the solutions in Exercises 45–48.

45. $x^2 + 2x + 1 = 0$
46. $3y^2 + 5y + 2 = 0$
47. $x^2 + x + 8 = 0$
48. $x^2 + 2x - 1 = 0$

Write the quadratic equation with the given solution set. Assume no repeated roots.

49. $\{2, -3\}$
50. $\left\{\dfrac{1 \pm i}{2}\right\}$
51. $\left\{0, \dfrac{1}{2}\right\}$
52. $\{1 \pm i\}$

Solve each of the following equations.

53. $x^4 - 6x^2 + 8 = 0$
54. $x^4 + x^2 - 2 = 0$
55. $(x + 1)^4 - 5(x + 1)^2 + 4 = 0$
56. $x^{-4} - 10x^{-2} + 9 = 0$

Write each of the following applied problems as an equation and solve. Check your answers.

57. Krystyn rode her bicycle for 17 mi. The first 15 mi were uphill but not too steep. The last 2 mi were uphill and steep, which slowed her down by 3 mph. If it took her 2 hr for the whole trip, find her faster rate.

58. Arturo and Sam together can change a car part in 3 hr. It takes Sam 8 hr longer than Arturo when working alone. How long does it take Arturo working alone?

59. A commuter plane flies between two cities that are 80 mi apart. With a 20 mph tailwind, it arrives 8 min early. Find the speed of the plane in still air.

60. If two pipes, a large one and a small one, are used, a tank can be filled in 4 hr. If only the larger pipe is used, it takes 6 hr less than if only the smaller pipe is used. How long does it take for the larger pipe (alone) to fill the tank?

Solve each of the following inequalities for Exercises 61–64.

61. $\dfrac{x-3}{4} + 2 > \dfrac{x+2}{5}$
62. $6x + 6 < x$
63. $2x - 1 < 2x - 13$
64. $\dfrac{5x-1}{4} \leq \dfrac{3x+2}{2}$

65. If P varies inversely with the cube root of t, and if P is 6 when t is 27, find P when t is 8.

66. The intensity of illumination, I, varies inversely with the square of the distance, d, from the light source. If the intensity is 120 cp (candle power) at a distance of 10 ft, what is the intensity at a distance of 20 ft?

CHAPTER 10
Conic Sections

10.1 Circles
10.2 Parabolas
10.3 Ellipses
10.4 Hyperbolas
10.5 Solving Nonlinear Systems of Equations
10.6 Solving Nonlinear Systems of Inequalities

Application

With a uniformly distributed load, the cable of a suspension bridge hangs in the shape of a parabola according to the formula

$$1000h(x) = 9x^2 + 10{,}000$$

For what value of x does $h(x)$ have its minimum value (the lowest point on the cable) with respect to a level line joining the base of the towers? If the cable is anchored to the tops of the towers, which are 200 feet apart, how tall are the towers?

See Exercise Set 10.2, Exercise 55.

It was Pythagoras who first clearly showed that proofs must follow from assumptions. He was the first to insist that axioms and postulates come first and that close deductive reasoning be used to tie them to any ensuing developments. Of course, he is best remembered for the Pythagorean theorem, which allowed the development of the distance formula, the equations of the conic sections, and many other applications. According to historians, Pythagoras was a "half mystical figure, one-tenth genius and nine-tenths pure fudge."

PYTHAGORAS (about 550 B.C.)

In Chapter 3 we studied the rectangular coordinate system and used it to sketch the graphs of straight lines. In this chapter and the next we will use the same system to graph special equations that are of higher degree than 1. Among the special equations are those with graphs that are circles, parabolas, ellipses, and hyperbolas. The curves represented by these equations have innumerable applications in business and industry.

10.1 Circles

OBJECTIVES

In this section we will learn to
1. find the equation of a circle;
2. identify the center and the radius of a circle from its equation; and
3. sketch the graph of a circle when the center and the radius are known.

The development of the equation of a circle relies heavily on two important formulas from Section 3.2: the distance formula and the formula for determining the midpoint of a line segment.

The Distance and Midpoint Formulas

If (x_1, y_1) and (x_2, y_2) are any two points in a plane, the **distance** d between them is

$$d = \sqrt{(x_2 - x_1)^2 + (y_2 - y_1)^2}$$

and the **midpoint** (x_m, y_m) of the line segment joining the two points is

$$(x_m, y_m) = \left(\frac{x_1 + x_2}{2}, \frac{y_1 + y_2}{2}\right)$$

The following two examples review how these formulas are used.

EXAMPLE 1 Find the distance from $(2, 3)$ to $(6, -7)$.

SOLUTION We let $(x_1, y_1) = (2, 3)$ and $(x_2, y_2) = (6, -7)$. Then

$$d = \sqrt{(x_2 - x_1)^2 + (y_2 - y_1)^2}$$
$$= \sqrt{(6 - 2)^2 + (-7 - 3)^2}$$
$$= \sqrt{4^2 + (-10)^2} = \sqrt{116}$$
$$= 2\sqrt{29} \qquad \sqrt{116} = \sqrt{4 \cdot 29} = 2\sqrt{29}$$

EXAMPLE 2 Find the midpoint of the line segment joining (2, 7) and (4, 9).

SOLUTION We let $(x_1, y_1) = (2, 7)$ and $(x_2, y_2) = (4, 9)$. Then

$$(x_m, y_m) = \left(\frac{x_1 + x_2}{2}, \frac{y_1 + y_2}{2}\right)$$
$$= \left(\frac{2 + 4}{2}, \frac{7 + 9}{2}\right)$$
$$= (3, 8)$$

We now turn our attention to finding the equation of a **circle**.

DEFINITION

Circle

A **circle** is a set of points in the plane that are equidistant from a fixed point, called the **center**. The distance of each point on the circle from the center is called the **radius** of the circle.

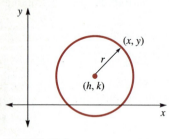

FIGURE 10.1

Let's consider a circle with center at the point (h, k) and with radius r, as in Figure 10.1.

If (x, y) is any point on a circle, the distance from (x, y) to the center, (h, k), is the radius, r. By the distance formula,

$$\sqrt{(x - h)^2 + (y - k)^2} = r \qquad d = \sqrt{(x_2 - x_1)^2 + (y_2 - y_1)^2}, \text{ where}$$
$$x_2 = x, x_1 = h, y_2 = y, y_1 = k, d = r$$

If we square both sides of this expression to remove the radical, the result is the equation of the circle:

$$(\sqrt{(x - h)^2 + (y - k)^2})^2 = r^2$$
$$(x - h)^2 + (y - k)^2 = r^2$$

The Standard Form for the Equation of a Circle

The **equation of a circle** with center at (h, k) and radius r is given by

$$(x - h)^2 + (y - k)^2 = r^2$$

If the center of the circle is at the origin, then $h = k = 0$ and the equation becomes

$$x^2 + y^2 = r^2$$

EXAMPLE 3 Find the equation of the circle with center at (3, 4) and radius 5.

SOLUTION Here $h = 3$, $k = 4$, and $r = 5$, so

$$(x - h)^2 + (y - k)^2 = r^2$$
$$(x - 3)^2 + (y - 4)^2 = 5^2$$
$$(x - 3)^2 + (y - 4)^2 = 25$$

EXAMPLE 4 Find the equation of the circle with center at (−7, 9) and radius $\sqrt{11}$.

SOLUTION Since $h = -7$, $k = 9$, and $r = \sqrt{11}$,

$$(x - h)^2 + (y - k)^2 = r^2$$
$$[x - (-7)]^2 + (y - 9)^2 = (\sqrt{11})^2$$
$$(x + 7)^2 + (y - 9)^2 = 11$$

EXAMPLE 5 Find the equation of the circle that has center at (2, 4) and that is tangent to the *x*-axis. Sketch its graph.

SOLUTION From geometry we know that a circle is tangent to a line if it touches it in exactly one point. Furthermore, the line is perpendicular to the radius at that point. If the circle is tangent to the *x*-axis, its center will lie on a line perpendicular to the *x*-axis and 4 units above it. The graph is shown in Figure 10.2. The radius is 4.

$$(x - h)^2 + (y - k)^2 = r^2$$
$$(x - 2)^2 + (y - 4)^2 = 4^2$$
$$(x - 2)^2 + (y - 4)^2 = 16$$

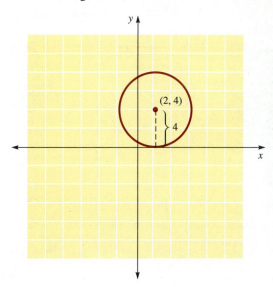

FIGURE 10.2

If we remove the parentheses in the final equation of Example 5 and simplify it, it becomes

$$x^2 - 4x + 4 + y^2 - 8y + 16 = 16$$

or

$$x^2 + y^2 - 4x - 8y + 4 = 0$$

The equation of a circle is often written in this form, which we can generalize as follows.

The General Form for the Equation of a Circle

$$Ax^2 + Ay^2 + Bx + Cy + D = 0$$

For example, in the equation $x^2 + y^2 + 6x - 8y - 24 = 0$, we have $A = 1$, $B = 6$, $C = -8$, and $D = -24$. Furthermore, $x^2 + y^2 + 6x - 8y - 24 = 0$ is a circle with center at $(-3, 4)$ and radius 7. To see this, we complete the square with respect to both x and y.

$$x^2 + y^2 + 6x - 8y - 24 = 0$$
$$(x^2 + 6x) + (y^2 - 8y) = 24$$
$$(x^2 + 6x + 9) + (y^2 - 8y + 16) = 24 + 9 + 16 \quad \text{Add 9 and 16 to each side.}$$
$$(x + 3)^2 + (y - 4)^2 = 49$$
$$[x - (-3)]^2 + (y - 4)^2 = 49$$

Here $h = -3$, $k = 4$, and $r^2 = 49$, giving us $r = 7$, as was desired.

EXAMPLE 6 Find the center and radius of the circle $x^2 + y^2 + 3x - 2y - 1 = 0$.

SOLUTION We complete the square on both x and y.

$$x^2 + y^2 + 3x - 2y - 1 = 0$$
$$(x^2 + 3x) + (y^2 - 2y) = 1$$
$$\left(x^2 + 3x + \frac{9}{4}\right) + (y^2 - 2y + 1) = 1 + \frac{9}{4} + 1 \quad \text{Add } \frac{9}{4} \text{ and 1 to each side}$$
$$\left(x + \frac{3}{2}\right)^2 + (y - 1)^2 = \frac{17}{4}$$
$$\left(x + \frac{3}{2}\right)^2 + (y - 1)^2 = \left(\frac{\sqrt{17}}{2}\right)^2$$

Here $h = -\frac{3}{2}$, $k = 1$, and $r^2 = \frac{17}{4}$, giving us $r = \frac{\sqrt{17}}{2}$. The center is $\left(-\frac{3}{2}, 1\right)$ and the radius is $\frac{\sqrt{17}}{2}$.

If we know the coordinates of both of the endpoints of a line segment forming a diameter of a circle, we can find the equation of the circle once we know the center. The center lies at the midpoint of the diameter.

EXAMPLE 7 Find the equation of the circle with a diameter that has endpoints $(2, 6)$ and $(-4, 10)$.

SOLUTION The center is the midpoint of the diameter, or

$$\left(\frac{2 + (-4)}{2}, \frac{6 + 10}{2}\right) = (-1, 8)$$

The radius r is the distance from either end of the diameter to the center. We use $(2, 6)$.

$$r = \sqrt{[2 - (-1)]^2 + (6 - 8)^2}$$
$$= \sqrt{3^2 + (-2)^2} = \sqrt{9 + 4} = \sqrt{13}$$

The equation is

$$[x - (-1)]^2 + (y - 8)^2 = (\sqrt{13})^2 \quad \text{or} \quad (x + 1)^2 + (y - 8)^2 = 13$$

EXAMPLE 8 Find the equation of the circle that has center at $(3, 4)$ and that passes through the point $(8, 16)$.

SOLUTION If the circle passes through the point $(8, 16)$, then the distance from the center, which is $(3, 4)$, to $(8, 16)$ is the radius of the circle.

$$r = \sqrt{(8 - 3)^2 + (16 - 4)^2}$$
$$= \sqrt{5^2 + 12^2} = \sqrt{169} = 13$$

The equation of the circle is

$$(x - 3)^2 + (y - 4)^2 = 13^2 \quad \text{or} \quad (x - 3)^2 + (y - 4)^2 = 169$$

*E*XPLORING *G*RAPHICS

The graph of the equation in Example 8 can be drawn by a graphics calculator or a computer. In many programs, the equation must be written in a specific form in order to get it done. We will say more about this in Section 10.3.

The graph of $(x - 3)^2 + (y - 4)^2 = 169$ in both Figures 10.3(a) and (b) was drawn by using a computer. Notice in Figure 10.3(a) that the graph does not look like a circle. This is because the computer program assigned different lengths to the "tic" marks on the x- and y-axes. When the units are adjusted to be the same length on both axes, the graph assumes the shape of a circle. [See Figure 10.3(b).]

Graphing with the use of a graphics calculator or a computer requires that the user be familiar with the instruction manual. Each calculator or software package varies in the way it handles graphing.

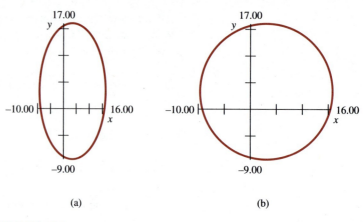

FIGURE 10.3

Do You Remember?

Can you match these?

_____ 1. A circle with center at (3, 4) and radius $r = 5$
_____ 2. The radius of the circle $x^2 + y^2 = 16$
_____ 3. The general form for the equation of a circle
_____ 4. The coordinates of the center of the circle $(x - 2)^2 + (y - 3)^2 = 9$
_____ 5. The distance from the center of a circle to a point on the circle
_____ 6. The standard form for the equation of a circle

a) $(-2, -3)$
b) $(2, 3)$
c) $(x - 3)^2 + (y - 4)^2 = 25$
d) $(x + 3)^2 + (y + 4)^2 = 25$
e) $(x - h)^2 + (y - k)^2 = r^2$
f) the radius
g) 16
h) $Ax^2 + Ay^2 + Bx + Cy + D = 0$
i) 4

Answers: 1. c 2. i 3. h 4. b 5. f 6. e

Exercise Set 10.1

Find the distance between each pair of given points.

1. (3, 5) and (7, 5)
2. (6, 0) and (2, 0)
3. (0, −3) and (0, 7)
4. (8, 1) and (8, −5)
5. (1, 2) and (1, −4)
6. (1, −2) and (3, 4)
7. (9, −7) and (6, −4)
8. (−2, 0) and (−7, 1)
9. (0, 1) and (1, −6)
10. (13, 2) and (5, −4)

Find the coordinates of the midpoint of the line segment having the given endpoints.

11. (1, 3) and (5, 7)
12. (2, −3) and (−4, 3)
13. (4, 9) and (−2, 7)
14. (8, −5) and (−5, 3)
15. (−7, 11) and (5, 4)
16. (−3, 8) and (−2, 5)

Find the coordinates of the center of each circle, find the radius, and graph the circle where indicated.

17. $x^2 + y^2 = 4$; graph
18. $x^2 + y^2 = 1$; graph
19. $x^2 + y^2 = 9$
20. $x^2 + y^2 = 16$
21. $x^2 + (y + 4)^2 = 25$; graph
22. $(x - 3)^2 + y^2 = 36$; graph
23. $(x - 3)^2 + (y - 2)^2 = 25$
24. $(x + 2)^2 + (y + 1)^2 = 1$
25. $(x + 4)^2 + (y - 1)^2 = 16$; graph
26. $(x - 5)^2 + (y + 2)^2 = 25$; graph
27. $(x - 7)^2 + y^2 = 36$
28. $x^2 + (y - 1)^2 = 9$

Write the equation of each circle in standard form,

$$(x - h)^2 + (y - k)^2 = r^2$$

29. Center at (0, 0) and $r = 2$
30. Center at (0, 0) and $r = 3$
31. Center at (1, 2) and $r = 4$
32. Center at (2, 4) and $r = 5$
33. Center at (0, −2) and $r = 6$
34. Center at (5, 0) and $r = 7$
35. Center at (−3, 0) and $r = 3$
36. Center at (0, −3) and $r = 3$
37. Center at (−1, −3) and $r = 5$
38. Center at (−4, −1) and $r = 4$
39. Passes through the origin and has center at (3, −5)
40. Passes through the origin and has center at (−2, 6)
41. Is tangent to the *x*-axis and has center at (−3, −4)
42. Is tangent to the *y*-axis and has center at (3, 2)
43. Consists of the set of all points 5 units from (−3, −5)
44. Consists of the set of all points 6 units from (2, −6)

Determine the coordinates of the center and the radius of each circle by completing the square in each equation.

45. $x^2 + y^2 + 2x - 2y = 0$
46. $x^2 + y^2 - 6x + 8y = 0$
47. $x^2 + y^2 + 8x - 2y + 8 = 0$
48. $x^2 + y^2 + 4x + 6y - 12 = 0$
49. $x^2 + y^2 - 5x + 7y + \dfrac{2}{5} = 0$
50. $x^2 + y^2 - x - y - 1 = 0$
51. $x^2 + y^2 - x + 3y = \dfrac{3}{2}$
52. $x^2 + y^2 - 6x - 10y = -33$
53. $x^2 + y^2 - 8x = 0$
54. $x^2 + y^2 + 18y = -41$
55. $4x^2 + 4y^2 + 20x + 10y = 0$
56. $2x^2 + 2y^2 + 8x + 4y = 0$
57. $3x^2 + 3y^2 + 30x - 18y = 12$
58. $5x^2 + 5y^2 - 10x + 15y = 1$
59. $9x^2 + 9y^2 + 24x + 108y + 232 = 0$
60. $16x^2 + 16y^2 - 128x + 24y - 55 = 0$

For Exercises 61–64, write the equation of the circle satisfied by the given conditions.

61. Passes through the point (−1, 5) and has center (5, −2)
62. Passes through the point (3, 7) and has center (6, 11)
63. A diameter is the line segment with endpoints (5, −1) and (−3, 7).
64. A diameter is the line segment with endpoints (6, 7) and (2, 5)

Write the equation of the circle from each of the following graphs. Leave your answer in standard form.

65.

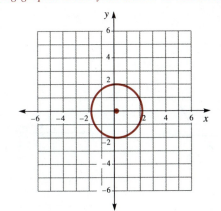

66.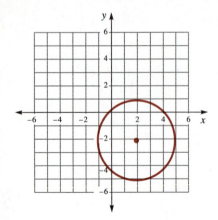

67.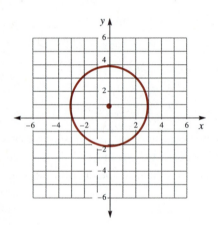

68.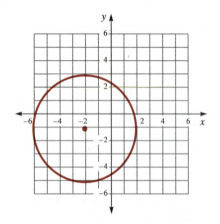

69.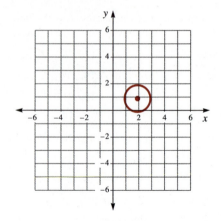

70.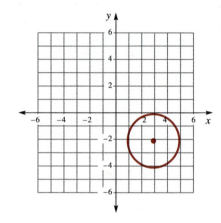

WRITE IN YOUR OWN WORDS

Describe or define:

71. A circle
72. A radius
73. A diameter
74. How to find the coordinates of the midpoint of a line segment
75. How to find the distance between two points
76. Five applications of a circle in our world today; explain why a circle is used in each case

REVIEW EXERCISES

77. Simplify: $\dfrac{1}{\sqrt{3} - \sqrt{2}}$
78. Solve: $\sqrt{x - 5} + 2 = \sqrt{2x - 2}$
79. Solve: $3x^2 - 5x + 1 = 0$

80. Solve: $\dfrac{3x}{4x+1} < \dfrac{2x}{x-2}$

81. Solve: $5x^2 + 4x > 0$

82. The Tretts leave San Luis Obispo traveling north for a distance of 200 mi. The Aldermans leave the same place traveling south for a distance of 200 mi. The Tretts make the trip in 1 hr less time and travel an average of 10 mph faster than the Aldermans. Find the speed of each family.

83. Francesca lives 10 mi from her school. If she drives 10 mph faster than normal, she arrives 12 min earlier than usual. What is her normal speed (rounded to the nearest mph)?

10.2 Parabolas

Parabolas occur naturally. For example, the path that a ball follows when thrown is a parabola. The receiving dish for satellite TV is parabolic in shape, as is the reflector on many flashlights. Mirrors on many telescopes are parabolic, and the cable on a suspension bridge would be parabolic if its load were continuously and uniformly distributed along its length. It actually varies slightly from a parabola because of the weight of the cable itself and the fact that the load is attached to the cable at intervals.

HISTORICAL COMMENT   The Golden Gate Bridge, in San Francisco, California, opened in 1937. Its construction required that the support towers be designed to withstand motion back and forth as well as from side to side. The mathematics necessary to analyze the stresses that could occur required solving *thirty simultaneous equations with six to twenty-one unknowns!* These very difficult calculations, done before computers were available, filled eleven volumes with higher mathematics and allowed the bridge to have soaring open towers, long arcing cables, and a graceful suspended roadbed.

OBJECTIVES   In this section we will learn to
1. find the equation of a parabola;
2. identify the equation of the axis of symmetry of a parabola; and
3. sketch the graphs of parabolas.

The circle of Section 10.1 and the parabola that we will study in this section are two of the curves, called **conic sections,** that result when a cone is cut by a plane. To generate a circle, the plane must cut the cone parallel to its base, as in Figure 10.4(a). A **parabola** results when the plane cuts the cone parallel to a lateral plane, as in Figure 10.4(b).

DEFINITION

Parabola

A **parabola** is the set of all points in a plane equidistant from a fixed point F, called the **focus,** and a fixed line L, called the **directrix.**

The equation of a parabola is derived from its definition. We will do this for a parabola with focus at $F(0, p)$, directrix $y = -p$, and vertex at the origin, as

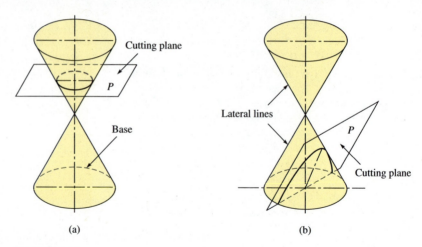

FIGURE 10.4

shown in Figure 10.5. Notice that p is the distance from the vertex to the focus or from the vertex to the directrix. The vertex of a parabola that opens upward is the lowest point on its graph. The vertex of a parabola that opens downward is the highest point on its graph.

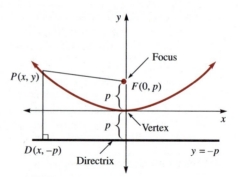

FIGURE 10.5

If the point $P(x, y)$ is any point on the parabola and $D(x, -p)$ is a point directly below P on the directrix, as shown in the Figure 10.5, then by the definition,

distance PF = distance PD

Thus, by the distance formula,

$$\sqrt{(x - 0)^2 + (y - p)^2} = \sqrt{(x - x)^2 + [y - (-p)]^2}$$
$$x^2 + (y - p)^2 = (y + p)^2 \quad \text{Simplify and square each side of the equation.}$$
$$x^2 + y^2 - 2py + p^2 = y^2 + 2py + p^2$$
$$x^2 = 4py$$

This is the equation of a parabola with vertex at $(0, 0)$, focus at $(0, p)$, and directrix $y = -p$.

EXAMPLE 1 Find the distance from the vertex to the focus in each of the following parabolas: (a) $x^2 = 12y$ (b) $5x^2 = 3y$

SOLUTION (a) Here $4p = 12$, so $p = 3$. The distance from the vertex to the focus is 3 units.
(b) First we write the equation as

$$x^2 = \frac{3}{5}y$$

Then $4p = \frac{3}{5}$

$$p = \frac{3}{20}$$

The distance form the vertex to the focus is $\frac{3}{20}$ units.

Notice in Figure 10.5 that the portion of the parabola to the right of the y-axis is **symmetric to** (the mirror image of) the portion to the left of the y-axis. For this reason, the y-axis is called the **axis of symmetry.**

DEFINITION

Axis of Symmetry

The **axis of symmetry** of a parabola is the line drawn through the vertex perpendicular to the directrix.

EXAMPLE 2 Sketch the graph of the parabola $y = x^2$ by constructing a table of values. Find the equation of the axis of symmetry.

SOLUTION The graph is shown in Figure 10.6. The portion of the graph to the right of the

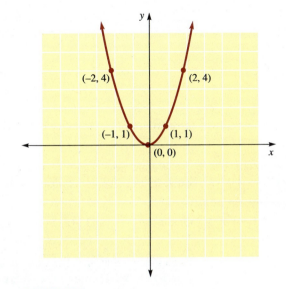

FIGURE 10.6

y-axis is the mirror image of the portion to the left of the *y*-axis, so the axis of symmetry is the *y*-axis itself. Its equation is $x = 0$.

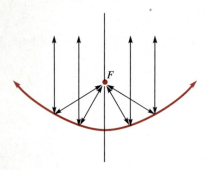

FIGURE 10.7

The *focus* is an important consideration on a parabola. As one application, consider a parabolic reflector. Using physics, it can be demonstrated that rays traveling parallel to the axis of symmetry of a parabola will be reflected to the focus when they strike its surface. (See Figure 10.7.) The reflector on a solar oven concentrates the heat at the focus, and a flashlight has the bulb located at, or near, the focus. On a flashlight, the light from the bulb strikes the surface of the parabolic reflector and is cast forward. Can you think of other examples where parabolic reflectors are commonly used?

Not every parabola has a vertex located at the origin. When a parabola opens up or down, it represents a function and its equation can be written in the form

$$f(x) = ax^2 + bx + c, \quad a, b, c \in R, a \neq 0$$

If $b = c = 0$, then the parabola has its vertex at the origin and its equation can be written

$$f(x) = ax^2$$

When $a > 0$ the parabola opens upward, as in Figure 10.8(a), and its vertex is said to be its **minimum point.** When $a < 0$ the parabola opens downward, as in Figure 10.8(b), and its vertex is said to be its **maximum point.**

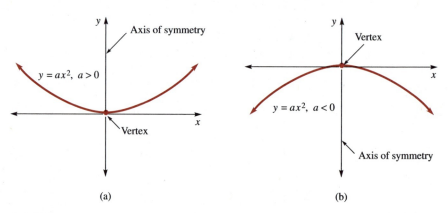

FIGURE 10.8

We are now ready to consider how slight modifications to the equation of a parabola can change its position on the graph and/or its shape.

EXAMPLE 3   Graph $f(x) = x^2$ and $g(x) = x^2 + 1$ on the same set of axes.

SOLUTION   For a given value of *x*, the value of $g(x)$ in $g(x) = x^2 + 1$ will be 1 greater than it is in $f(x) = x^2$. Therefore the vertex of $g(x)$ is 1 unit above the origin. The graphs are shown in Figure 10.9.

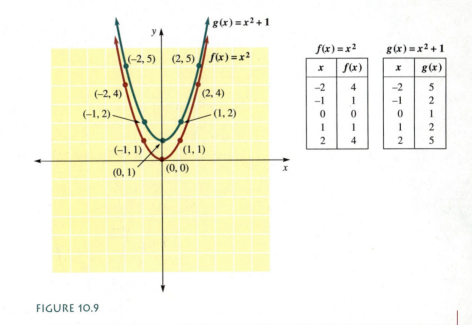

FIGURE 10.9

If we were to change our function g to $g(x) = x^2 - 1$, we would find that the graph has the same shape as $f(x) = x^2$ but its vertex is 1 unit *below* the origin. We say that the graph of $g(x) = x^2 - 1$ is shifted down 1 unit from the graph of $f(x) = x^2$.

EXAMPLE 4 Graph $f(x) = x^2$ and $h(x) = (x + 1)^2$ on the same set of axes. Determine the axis of symmetry of $h(x) = (x + 1)^2$.

SOLUTION Figure 10.10 shows that the shapes of the two graphs are identical; however, the

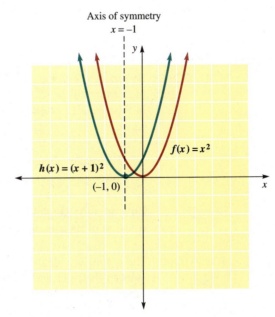

FIGURE 10.10

graph of $h(x) = (x + 1)^2$ is shifted 1 unit to the left of the graph of $f(x) = x^2$. The axis of symmetry of $h(x) = (x + 1)^2$ passes through its vertex, $(-1, 0)$, and has as its equation $x = -1$.

EXAMPLE 5 Sketch the graphs of $f(x) = x^2$ and $g(x) = (x - 1)^2$ on the same axes.

SOLUTION Figure 10.11 shows that the shape of the two parabolas is identical. The graph of $g(x) = (x - 1)^2$ is shifted 1 unit to the right of the graph of $f(x) = x^2$.

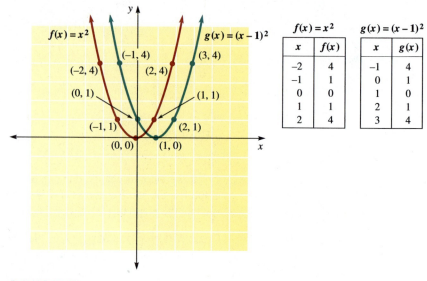

FIGURE 10.11

EXAMPLE 6 Graph: $g(x) = (x + 2)^2 - 3$.

SOLUTION By analogy with the preceding examples, the shape will be the same $f(x) = x^2$, but the vertex will be shifted 2 units to the left and 3 units down. The graph is shown in Figure 10.12 on the next page.

The reader should verify the discussion in Example 6 by constructing a table of values.

EXAMPLE 7 How does the graph of $g(x) = (x - 4)^2 + 2$ compare with $f(x) = x^2$? What is the equation of its axis of symmetry?

SOLUTION The graphs are identical in shape. The vertex of $g(x) = (x - 4)^2 + 2$ is shifted 4 units to the right and 2 units up. Since the axis of symmetry passes through the vertex, its equation is $x = 4$.

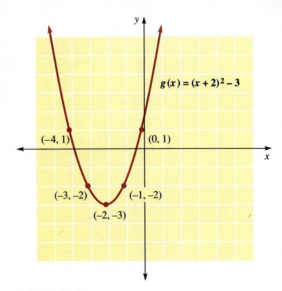

FIGURE 10.12

Summary

With respect to the parabola $y = x^2$, the following observations can be made, where h and k are positive numbers.

Equation	Shift	Vertex	Shape	Opens
$f(x) = x^2 + k$	k units up	$(0, k)$	same	up
$f(x) = x^2 - k$	k units down	$(0, -k)$	same	up
$f(x) = (x - h)^2$	h units to the right	$(h, 0)$	same	up
$f(x) = (x + h)^2$	h units to the left	$(-h, 0)$	same	up
$f(x) = (x - h)^2 + k$	h units to the right and k units up	(h, k)	same	up

When the coefficient a in $f(x) = ax^2$ is different than 1, the parabola will be either wider or narrower. Its vertex remains at the origin.

EXAMPLE 8   Graph $f(x) = x^2$ and $g(x) = \frac{1}{4}x^2$ on the same set of axes by first constructing a table of values.

SOLUTION   The graphs are shown in Figure 10.13 on page 516.

It is not too difficult to determine the vertex of a parabola with a vertical axis of symmetry (one that opens up or down). For example, consider

$$f(x) = ax^2 + bx + c$$

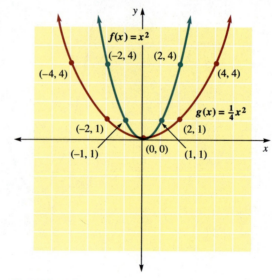

FIGURE 10.13

This can be written

$$f(x) = a\left(x^2 + \frac{b}{a}x\right) + c$$

We now complete the square by adding and subtracting the same amount from the right side of the equation, instead of adding a constant to both sides of the equation, as we did in Section 9.2. This has the net effect of adding zero to the right side.

$$f(x) = a\left(x^2 + \frac{b}{a}x + \frac{b^2}{4a^2}\right) + c - \frac{b^2}{4a}$$

We added $a \cdot \left(\dfrac{b^2}{4a^2}\right) = \dfrac{b^2}{4a}$.   We must subtract $\dfrac{b^2}{4a}$.

$$f(x) = a\left(x + \frac{b}{2a}\right)^2 + \frac{4ac - b^2}{4a} \qquad \text{The LCD of } c \text{ and } \frac{b^2}{4a} \text{ is } 4a.$$

If we let $\dfrac{-b}{2a} = h$ and $\dfrac{4ac - b^2}{4a} = k$, this equation becomes

$$f(x) = a(x - h)^2 + k$$

Notice that when $x = h$,

$$f(x) = a(h - h)^2 + k = k$$

since $a(h - h)^2 = a(0) = 0$.

DEFINITION

The Equation of a Parabola

The equation

$$f(x) = a(x - h)^2 + k$$

represents **a parabola** with vertex at (h, k), opening upward if $a > 0$ and downward if $a < 0$. The axis of symmetry is the line $x = h$. When the equation of a parabola is in the form $y = ax^2 + bx + c$, its vertex is at $\left(\dfrac{-b}{2a}, \dfrac{4ac - b^2}{4a}\right)$.

EXAMPLE 9 Determine the vertex, the axis of symmetry, and the x-intercepts (the points where the graph crosses the x-axis) for $f(x) = x^2 - 6x + 5$. Sketch its graph.

SOLUTION To find the vertex, we complete the square.

$$f(x) = (x^2 - 6x + 9) + 5 - 9 \quad \text{Add and subtract 9.}$$
$$= (x - 3)^2 - 4$$

The vertex is $(3, -4)$. The axis of symmetry is $x = 3$. To determine the x-intercepts, we let $f(x) = 0$.

$$0 = x^2 - 6x + 5$$
$$= (x - 1)(x - 5)$$
$$x = 1 \quad \text{or} \quad x = 5$$

The graph is shown in Figure 10.14.

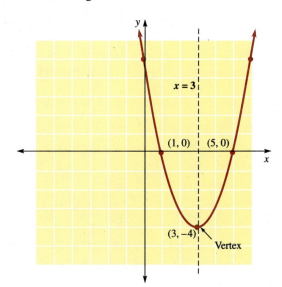

FIGURE 10.14

An alternate way to find the vertex in Example 9 is to use the fact that the x-coordinate of the vertex is $h = \frac{-b}{2a}$.

$$f(x) = ax^2 + bx + c$$
$$= x^2 - 6x + 5$$

To find the x-coordinate of the vertex, we substitute $a = 1$ and $b = -6$:

$$h = \frac{-b}{2a} = \frac{-(-6)}{2(1)} = 3$$

The y-coordinate is then found by substituting 3 for x in the original equation.

$$f(x) = x^2 - 6x + 5$$
$$= (3)^2 - 6(3) + 5 \qquad x = 3$$
$$= 9 - 18 + 5 = -4$$

Thus $k = -4$. The vertex is $(3, -4)$. All that is needed to use this method is to remember that the x-coordinate is $\frac{-b}{2a}$ and that the y-coordinate can be obtained by substituting $\frac{-b}{2a}$ for x in the original equation. Thus the vertex is $\left(\frac{-b}{2a}, f\left(\frac{-b}{2a}\right)\right)$.

EXAMPLE 10 Find the vertex of the parabola $f(x) = 3x^2 + 9x - 4$ by completing the square; determine the direction the parabola opens. Verify the result by calculating $\left(\frac{-b}{2a}, f\left(\frac{-b}{2a}\right)\right)$.

SOLUTION To complete the square, the lead coefficient must be 1. We factor 3 from $3x^2 + 9x$.

$$f(x) = 3x^2 + 9x - 4$$
$$= 3(x^2 + 3x) - 4$$
$$= 3\left(x^2 + 3x + \frac{9}{4}\right) - 4 - \frac{27}{4}$$

We added $3\left(\frac{9}{4}\right) = \frac{27}{4}$. We must subtract $\frac{27}{4}$.

$$f(x) = 3\left(x + \frac{3}{2}\right)^2 - \frac{43}{4} \qquad -4 - \frac{27}{4} = -\frac{43}{4}$$
$$= 3\left[x - \left(-\frac{3}{2}\right)\right]^2 + \left(-\frac{43}{4}\right)$$

The vertex is located at $(h, k) = \left(-\frac{3}{2}, -\frac{43}{4}\right)$. The parabola opens upward, since $a = 3 > 0$.

To verify the results, we need to identify the values of a and b; since $f(x) = 3x^2 + 9x - 4$, we have $a = 3$ and $b = 9$. Therefore

$$\frac{-b}{2a} = \frac{-9}{6} = \frac{-3}{2}$$

$$f\left(\frac{-b}{2a}\right) = 3\left(\frac{-3}{2}\right)^2 + 9\left(\frac{-3}{2}\right) - 4 \qquad$$

$$= 3\left(\frac{9}{4}\right) - \left(\frac{27}{2}\right) - 4 \qquad \text{Substitute } \frac{-3}{2} \text{ for } x.$$

$$= \frac{27}{4} - \frac{54}{4} - \frac{16}{4}$$

$$= -\frac{43}{4}$$

Thus the vertex is $\left(-\frac{3}{2}, \frac{-43}{2}\right)$.

EXAMPLE 11 Find the equation of the parabola shown in Figure 10.15.

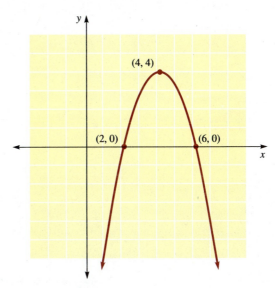

FIGURE 10.15

SOLUTION The vertex (h, k) is the point $(4, 4)$. The x-intercepts are $(2, 0)$ and $(6, 0)$. The equation of a parabola opening up or down can be written

$$f(x) = a(x - h)^2 + k \qquad \text{or} \qquad y = a(x - h)^2 + k$$

Using $y = a(x - h)^2 + k$ and substituting the coordinates of the vertex for h and k yields

$$y = a(x - 4)^2 + 4$$

Since the intercepts are points on the parabola, their coordinates must satisfy its equation, so if we substitute (2, 0) for (x, y), we have

$$0 = a[2 - 4]^2 + 4$$
$$0 = 4a + 4$$
$$-4 = 4a$$
$$-1 = a$$

The equation of the parabola is

$$y = -1(x - 4)^2 + 4 \quad \text{or} \quad f(x) = -1(x - 4)^2 + 4$$

The reader should verify that substituting (6, 0) into $y = a(x - 4)^2 + 4$ also gives $a = -1$.

EXAMPLE 12 Sketch the graph of $20x = y^2$ by constructing a table of values.

SOLUTION The graph is shown in Figure 10.16. Since any vertical line to the right of the y-axis will intersect the graph in more than one point, the equation does not represent a function. The vertical line $x = -1$ is the directrix of the parabola and the dot on the x-axis is the focus.

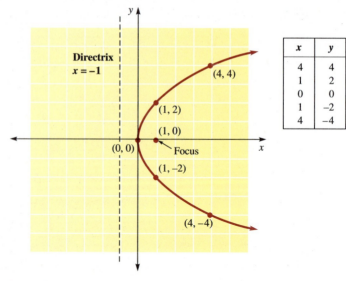

FIGURE 10.16

The equation

$$x = a(y - k)^2 + h$$

represents a parabola with vertex at (h, k), opening to the right if $a > 0$ and to the left if $a < 0$. The axis of symmetry is the line $y = k$.

EXAMPLE 13 Determine the vertex and the axis of symmetry for the parabola whose equation is $y^2 + y + x - 3 = 0$.

SOLUTION In this example the *y*-term is squared, so the axis of symmetry is horizontal.

$$y^2 + y + x - 3 = 0$$
$$x = -y^2 - y + 3 \qquad \text{Solve for } x.$$

Recall that to complete a square, the lead coefficient must be 1.

$$x = -(y^2 + y) + 3 \qquad \text{Factor } -1 \text{ from } y^2 \text{ and } y.$$
$$= -\left(y^2 + y + \frac{1}{4}\right) + 3 + \frac{1}{4} \qquad \text{Complete the square on } y.$$
$$x = -\left(y + \frac{1}{2}\right)^2 + \frac{13}{4}$$

Notice that $-(y^2 + y + \frac{1}{4}) = -y^2 - y - \frac{1}{4}$. This shows that $\frac{1}{4}$ was subtracted. To maintain an equivalent expression, we *added* $\frac{1}{4}$ to the same side of the equation. The vertex is $(\frac{13}{4}, -\frac{1}{2})$. The axis of symmetry is $y = -\frac{1}{2}$. The graph is shown in Figure 10.17.

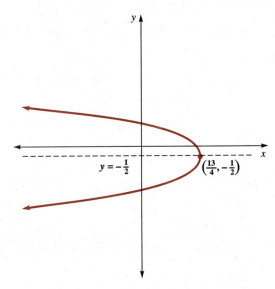

FIGURE 10.17

We could have found the vertex in Example 13 by an alternate method. Since this parabola has a horizontal axis of symmetry, the *y*-coordinate of the vertex is $\frac{-b}{2a}$. We can then find the *x*-coordinate by substituting this value into the original equation for *y* and solving for *x*. The reader is encouraged to verify the coordinates of the vertex by this method.

Exploring Graphics

The graph of $y = x^2 - 6x + 5$, shown in Figure 10.18(a), and the graph of $y = x^2 + 6x - 2$, shown in Figure 10.18(b), were completed using a graphics calculator with a printer attachment. Care must be taken when reading the coordinates of the vertex from the graph. The tic marks on the y-axes represent different values in the two figures.

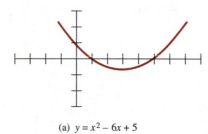

(a) $y = x^2 - 6x + 5$ (b) $y = x^2 + 6x - 2$

FIGURE 10.18

Do You Remember?

Can you match these?

The following refer to the equation $y = a(x - h)^2 + k$.

___ 1. The parabola opens up.
___ 2. The parabola opens down.
___ 3. The vertex is ___.
___ 4. The axis of symmetry is ___.
___ 5. In the form $y = ax^2 + bx + c$, the x-coordinate of the vertex is ___.

a) $\dfrac{-b}{2a}$
b) (h, k)
c) $\dfrac{2a}{b}$
d) $a < 0$
e) $a > 0$
f) $y = k$
g) $x = h$

Answers: 1. e 2. d 3. b 4. g 5. a

Exercise Set 10.2

Sketch the graph of each of the following functions by first constructing a table of values. Determine the vertex and the axis of symmetry.

1. $f(x) = 2x^2$
2. $f(x) = 4x^2$
3. $f(x) = \dfrac{1}{2}x^2$
4. $f(x) = \dfrac{1}{3}x^2$
5. $f(x) = -4x^2$
6. $f(x) = -2x^2$
7. $f(x) = -\dfrac{1}{4}x^2$
8. $f(x) = -\dfrac{1}{2}x^2$

Using $f(x) = x^2$ as a pattern, sketch graphs for Exercises 9–16 and indicate how each parabola is shifted.

9. $f(x) = x^2 + 2$
10. $f(x) = x^2 + 4$
11. $f(x) = x^2 - 3$
12. $f(x) = x^2 - 1$
13. $f(x) = (x - 2)^2$
14. $f(x) = (x - 3)^2$
15. $f(x) = (x + 3)^2 - 4$
16. $f(x) = (x + 2)^2 - 3$

Sketch the graph of each of the following functions, using the coordinates of the vertex and x-intercepts. Determine the axis of symmetry and the direction in which the parabola opens.

17. $f(x) = x^2 - 4$
18. $f(x) = x^2 - 2$
19. $f(x) = x^2 - 1$
20. $f(x) = x^2 - 3$
21. $f(x) = -(x - 1)^2 + 1$
22. $f(x) = (x - 2)^2 - 4$
23. $f(x) = 2(x + 1)^2 - 4$
24. $f(x) = 3(x + 1)^2 - 1$
25. $f(x) = -(x + 2)^2 + 4$
26. $f(x) = -(x + 3)^2 + 1$
27. $f(x) = -2(x - 3)^2 + 2$
28. $f(x) = -2(x - 1)^2 + 4$

Sketch the graph of each of the following parabolas. Determine the axis of symmetry and the direction in which the parabola opens.

29. $x = (y + 2)^2 - 4$
30. $x = (y + 1)^2 - 1$
31. $x = -2(y - 1)^2 + 1$
32. $x = -3(y - 2)^2 + 4$

Rewrite each of the following equations in the form $x = a(y - k)^2 + h$ or $y = a(x - h)^2 + k$; determine the vertex and the axis of symmetry, and graph when indicated.

33. $x^2 + 4x - 2y - 2 = 0$; graph
34. $x^2 - 2x - 3y + 7 = 0$; graph
35. $y^2 - 10y - 3x + 24 = 0$
36. $y^2 + 2y - 4x - 3 = 0$
37. $x^2 - 3x - 3y + 1 = 0$; graph
38. $x^2 - x + 3y + 1 = 0$; graph
39. $y^2 + 6y + \dfrac{1}{2}x + 7 = 0$
40. $y^2 - 6y + 2x + 17 = 0$
41. $2x^2 - 12x - 7y = 10$
42. $x^2 - 2y - 2x = -7$
43. $4y^2 + 16x + 17 = 20y$
44. $y^2 + 2x - 2y = 5$

Determine the equation of the parabola from each graph.

45.

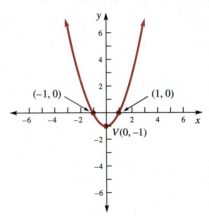

46.

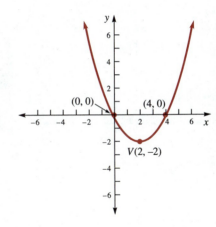

47.

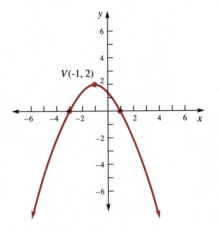

48.

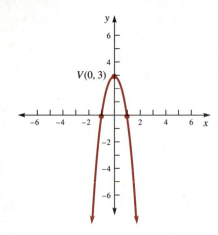

49.

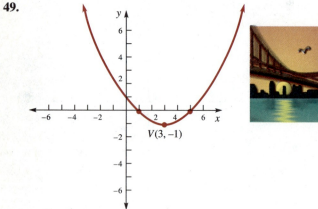

50.

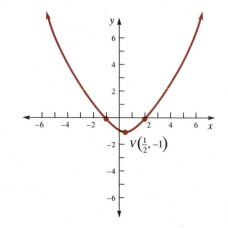

Solve each of the following applied problems as indicated. A calculator may be helpful.

51. A load F supported by a cantilever beam of length x is a function of x by the formula $F(x) = 32 - 2x^2$ lb. What is the load F when $x = 4$?

52. The velocity v of a rocket is a function of the time t from launch time according to the function $v(t) = 100 + 75t - 25t^2$, where v is in feet/second and t is in seconds. Sketch the graph for $t \geq 0$. What is the maximum velocity of the rocket? At what time does the velocity equal zero?

53. The graph of an electric current I forms a parabolic function according to the formula $I(t) = t^2 - 5t + 6$ amps. For what time t (in seconds) is the current equal to zero?

54. The surface area S of a cube forms a parabolic function according to the formula $S(x) = 6x^2$, where x is the length of an edge of the cube. Graph $S(x)$. For what length x does S have its minimum value?

55. With a uniformly distributed load, the cable of a suspension bridge hangs in the shape of a parabola according to the formula $1000\, h(x) = 9x^2 + 10{,}000$. For what value of x does $h(x)$ have its minimum value (the lowest point on the cable) with respect to a level line joining the base of the towers? If the cable is anchored to the tops of the towers (200 ft apart), how tall are the towers?

56. If a company produces n power generators per week, its profit $P(n)$ is given by $P(n) = 110n - n^2 - 1000$. How many generators should the company produce each week to have a maximum profit? What minimum and maximum values of n provide a profit greater than or equal to zero? What values of n provide a negative profit (loss)? Assume $n > 0$.

57. Helen wishes to fence a rectangular area adjacent to her house. One side of the rectangle will be the side of her house. What is the largest area she can fence with 100 ft of fencing?

𝓕OR 𝓔XTRA 𝓣HOUGHT

58. Find the perimeter of the quatrefoil inscribed in a square with a side of 10 cm. See the accompanying figure.

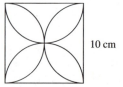

59. Find the length of a belt wrapped around two wheels if their radii are 12 ft and 2 ft and their centers are 20 ft apart. (See the figure.) Round your answer to the nearest foot.

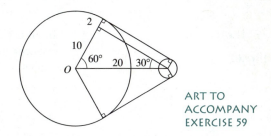

ART TO ACCOMPANY EXERCISE 59

10.3 Ellipses

The ellipse is an important curve in the sciences. In astronomy, the earth and the other planets travel around the sun in elliptical orbits. In the design of machinery like punch presses, the gears are often elliptical. The arches supporting many bridges are elliptical.

HISTORICAL COMMENT   Johannes Kepler (1571–1630) used geometry extensively to establish that the planets of the solar system move in elliptical orbits around the sun. He is thought to be the first to use the term *focus* in conjunction with the ellipse, and the theories he developed concerning the paths of the planets kept mathematicians of his day quite busy establishing their accuracy.

OBJECTIVES   In this section we will learn to
1. recognize an ellipse from its equation; and
2. sketch the graph of an ellipse.

The ellipse, like the circle and the parabola, is a conic section. It results when a plane cuts a cone in the manner depicted in Figure 10.19. An ellipse is easy to draw. Place two pins in a sheet of cardboard and attach one end of a string to each pin, as shown in Figure 10.20. Place a pencil inside the string and pull it taut, at the same time moving it around the pins, as indicated by the arrows.

This process for drawing an **ellipse** is in fact based on its definition.

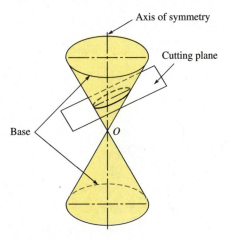

FIGURE 10.19

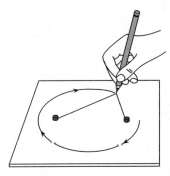

FIGURE 10.20

DEFINITION	An **ellipse** is the set of all points in the plane, the sum of whose distances from two fixed points is a constant. Each fixed point is called a **focus**; we speak of the two points as **foci**.
Ellipse	

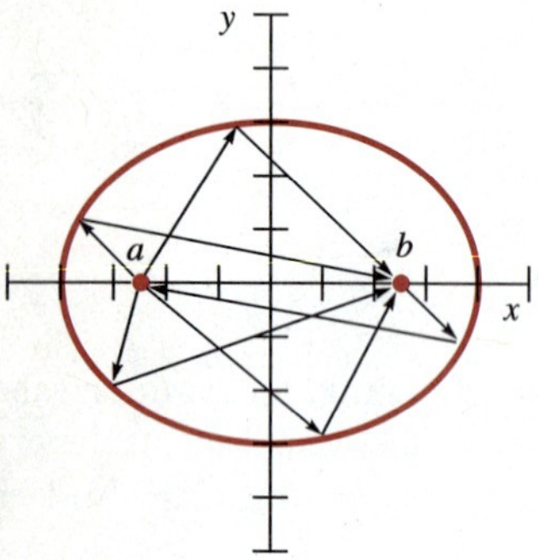

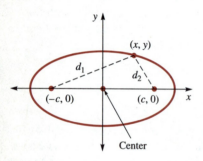

FIGURE 10.21

An interesting and useful property of the ellipse is that any ray emanating from one focus and striking the surface of the ellipse is reflected to the other focus. Where reflection of sound is considered, the property is called the *acoustical property of the ellipse*. The dome on the Capital Building in Washington, D.C., is constructed in this manner. A person standing at one focus, a, can clearly hear a whisper from another person standing at the other focus, b. (See Figure 10.21.) Such structures are called *whispering galleries*.

We will derive the equation of an ellipse with the center at the origin and foci at $(c, 0)$ and $(-c, 0)$. We let (x, y) be any point on the ellipse and $2a$ the constant distance referred to in the definition. (See Figure 10.22.) If we let d_1 and d_2 be the distances of the point (x, y) from the two foci, then

$$d_1 + d_2 = 2a$$

FIGURE 10.22

Both d_1 and d_2 can be found by using the distance formula, as follows:

$$\sqrt{[x - (-c)]^2 + (y - 0)^2} + \sqrt{(x - c)^2 + (y - 0)^2} = 2a$$

To eliminate the radicals, we subtract $\sqrt{(x - c)^2 + y^2}$ from each side and square.

$$(\sqrt{(x + c)^2 + y^2})^2 = (2a - \sqrt{(x - c)^2 + y^2})^2$$
$$x^2 + 2cx + c^2 + y^2 = 4a^2 - 4a\sqrt{(x - c)^2 + y^2} + x^2 - 2cx + c^2 + y^2$$

When like terms are combined, we can rewrite this as

$$cx - a^2 = -a\sqrt{(x - c)^2 + y^2}$$

Squaring each side again gives us

$$c^2x^2 - 2a^2cx + a^4 = a^2x^2 - 2a^2cx + a^2c^2 + a^2y^2$$

which we can write as

$$(a^2 - c^2)x^2 + a^2y^2 = a^2(a^2 - c^2)$$

Finally, we let $a^2 - c^2 = b^2$ and divide both sides by a^2b^2, where $a \neq 0$ and $b \neq 0$.

$$\frac{x^2}{a^2} + \frac{y^2}{b^2} = 1$$

If we let $y = 0$, then

$$\frac{x^2}{a^2} = 1$$

$x^2 = a^2$ Multiply each side by a^2.

$x = \pm a$ Square root property

Two points on the ellipse, therefore, are $(-a, 0)$ and $(a, 0)$. If we let $x = 0$, then

$$\frac{y^2}{b^2} = 1$$

$y^2 = b^2$ Multiply each side by b^2.

$y = \pm b$ Square root property

Thus two additional points on the ellipse are $(0, -b)$ and $(0, b)$. From this we see that $(\pm a, 0)$ and $(0, \pm b)$ are the x- and y-intercepts, respectively, of an ellipse centered at the origin. Figure 10.23 shows the graph with all points labeled.

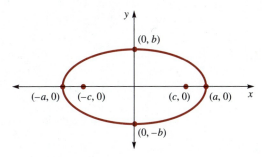

FIGURE 10.23

If $a > b$, the points $(a, 0)$ and $(-a, 0)$ are called the **vertices** of the ellipse. We call the points $(0, b)$ and $(0, -b)$ **co-vertices** for simplicity in referring to them. If $b > a$ then the vertices are $(0, b)$ and $(0, -b)$, while the co-vertices are $(a, 0)$ and $(-a, 0)$.

An ellipse that is longer in the x-direction ($a^2 > b^2$) is often called an **x-ellipse.** If it is longer in the y-direction ($b^2 > a^2$), it is called a **y-ellipse.**

EXAMPLE 1   Sketch the graph of each of the following ellipses:

(a) $x^2 + 4y^2 = 16$ (b) $9x^2 + 4y^2 = 36$

SOLUTION   (a) First we write the equation in the form

$$\frac{x^2}{a^2} + \frac{y^2}{b^2} = 1$$

by dividing each side by 16:

$x^2 + 4y^2 = 16$

$$\frac{x^2}{16} + \frac{y^2}{4} = 1$$

The easiest way to sketch an ellipse centered at the origin is to use the x- and y-intercepts.

$a^2 = 16$ so $a = \pm 4$ and $b^2 = 4$ so $b = \pm 2$

The intercepts are $(-4, 0)$, $(4, 0)$, $(0, -2)$, and $(0, 2)$. The graph of this x-ellipse is shown in Figure 10.24(a).

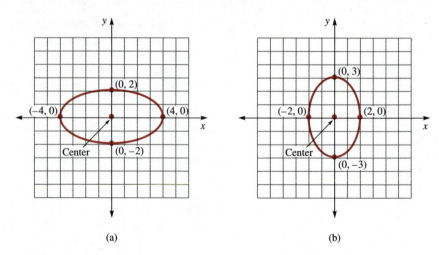

FIGURE 10.24

(b) First we divide each side by 36 to determine the intercepts:

$$9x^2 + 4y^2 = 36$$

$$\frac{x^2}{4} + \frac{y^2}{9} = 1$$

$a^2 = 4$ so $a = \pm 2$ and $b^2 = 9$ so $b = \pm 3$

The intercepts are $(-2, 0)$, $(2, 0)$, $(0, -3)$, and $(0, 3)$. The graph of this y-ellipse is shown in Figure 10.24(b).

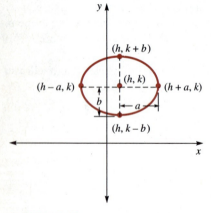

FIGURE 10.25

When the center of an ellipse is shifted from the origin to the point (h, k), its equation becomes

$$\frac{(x - h)^2}{a^2} + \frac{(y - k)^2}{b^2} = 1$$

This is known as the **standard form.** In this form the distance from the center to each vertex is a if $a^2 > b^2$, while the distance from the center to each co-vertex is b. (See Figure 10.25.)

EXAMPLE 2   Sketch the graph of $5(x - 1)^2 + 9(y + 1)^2 = 45$.

SOLUTION   First we write the equation in standard form by dividing each side by 45.

$$5(x - 1)^2 + 9(y + 1)^2 = 45$$

$$\frac{(x - 1)^2}{9} + \frac{(y + 1)^2}{5} = 1 \qquad \text{Standard form. Divide each side by 45.}$$

$a^2 = 9$ so $a = \pm 3$ and $b^2 = 5$ so $b = \pm\sqrt{5}$

Since $a^2 > b^2$, it is an x-ellipse with center at $(h, k) = (1, -1)$. To graph the ellipse, remember that a and b are the distances from the center to the vertices and co-vertices, respectively. The vertices are $(1 - 3, -1) = (-2, -1)$ and

$(1 + 3, -1) = (4, -1)$. In the same manner, the co-vertices are $(1, -1 + \sqrt{5})$ and $(1, -1 - \sqrt{5})$. The graph is shown in Figure 10.26.

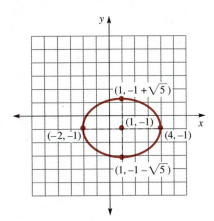

FIGURE 10.26

EXAMPLE 3 Rewrite the equation $9x^2 + 18x + 4y^2 - 8y - 23 = 0$ in standard form and find the center of the ellipse.

SOLUTION We begin by completing the square on x and y.

$$9x^2 + 18x + 4y^2 - 8y - 23 = 0$$
$$9(x^2 + 2x) + 4(y^2 - 2y) = 23$$
$$9(x^2 + 2x + 1) + 4(y^2 - 2y + 1) = 23 + 9 + 4 \quad \text{Add } 9(1) = 9 \text{ and } 4(1) = 4 \text{ to each side.}$$
$$9(x + 1)^2 + 4(y - 1)^2 = 36$$
$$\frac{(x + 1)^2}{4} + \frac{(y - 1)^2}{9} = 1 \quad \text{Divide each side by 36.}$$

The center is at $(-1, 1)$.

Exploring Graphics

EXAMPLE 4 Use a computer or a graphics calculator to sketch the graph of $\frac{x^2}{4} + \frac{y^2}{9} = 1$.

SOLUTION The graph shown in Figure 10.27 was drawn using a computer program written to accommodate conic sections. We have adjusted the distances between the tic marks on the two axes to be the same. The dots on the axes are the foci.

Graphics calculators and many computer programs are designed to sketch the graphs of functions. This means that we have to solve $\frac{x^2}{4} + \frac{y^2}{9} = 1$ for y and write the result in the form of two distinct functions.

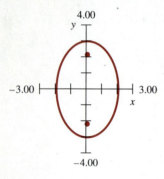

FIGURE 10.27

$$\frac{x^2}{4} + \frac{y^2}{9} = 1$$

$9x^2 + 4y^2 = 36$ Multiply each side by 36, the LCD.

$4y^2 = 36 - 9x^2$ Subtract $9x^2$ from each side.

$2y = \pm\sqrt{36 - 9x^2}$ Square root property

$y = \pm\frac{1}{2}\sqrt{9(4 - x^2)}$

$y = \pm\frac{3}{2}\sqrt{4 - x^2}$

The two equations for y are

I $\quad y = \frac{3}{2}\sqrt{4 - x^2}$ and **II** $\quad y = -\frac{3}{2}\sqrt{4 - x^2}$

Equation **I** represents the upper half of the ellipse, while equation **II** represents the lower half. When both equations were entered into a graphics calculator, the display appeared as in Figure 10.28. The equation represents a y-ellipse even though its graph appears to be an x-ellipse. The reason for the distortion is that the tic marks on the x-axis are farther apart than those on the y-axis.

FIGURE 10.28

EXAMPLE 5 Graph the equation $9x^2 + 18x + 4y^2 - 8y - 23 = 0$ using a computer.

SOLUTION This is the same equation that we rewrote in standard form in Example 3. The center is at $(-1, 1)$, $a = 2$, and $b = 3$. The graph is shown in Figure 10.29. The dots on the vertical axis locate the two foci, which were computed and plotted by the computer. The coordinates of the vertices were found by pointing the cursor at their apparent location. The actual locations of the vertices are $(-1, 4)$, and $(-1, -2)$. Approximations for any point on the ellipse can be obtained in the same manner if the computer program has this feature.

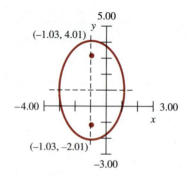

FIGURE 10.29

Do You Remember?

Can you match these?

___ 1. An example of a y-ellipse
___ 2. An example of an x-ellipse
___ 3. Standard form for the equation of an ellipse
___ 4. The center of the ellipse $\dfrac{(x-3)^2}{9} + \dfrac{(y+7)^2}{16} = 1$
___ 5. The distance from the center of the ellipse $\dfrac{x^2}{25} + \dfrac{y^2}{49} = 1$ to either vertex

a) 7
b) $(-3, 7)$
c) $(3, -7)$
d) $\dfrac{(x-h)^2}{a^2} + \dfrac{(y-k)^2}{b^2} = 1$
e) $\dfrac{x^2}{16} + \dfrac{y^2}{4} = 1$
f) 5
g) $\dfrac{x^2}{25} + \dfrac{y^2}{36} = 1$

Answers: 1. g 2. e 3. d 4. c 5. a

Exercise Set 10.3

Determine the coordinates of the center and the vertices of each of the following ellipses. Graph when indicated.

1. $\dfrac{x^2}{4} + \dfrac{y^2}{1} = 1$; graph
2. $\dfrac{x^2}{4} + \dfrac{y^2}{9} = 1$; graph
3. $\dfrac{x^2}{9} + \dfrac{y^2}{4} = 1$
4. $\dfrac{x^2}{25} + \dfrac{y^2}{4} = 1$
5. $\dfrac{x^2}{36} + \dfrac{y^2}{16} = 1$; graph
6. $\dfrac{x^2}{64} + \dfrac{y^2}{49} = 1$; graph
7. $x^2 + 4y^2 = 36$
8. $9x^2 + 4y^2 = 36$
9. $4x^2 + 25y^2 = 100$; graph
10. $25x^2 + y^2 = 25$; graph
11. $16x^2 + 9y^2 = 144$
12. $x^2 + 4y^2 = 16$
13. $\dfrac{(x-1)^2}{9} + \dfrac{(y+2)^2}{4} = 1$; graph
14. $\dfrac{(x+3)^2}{16} + \dfrac{(y-1)^2}{4} = 1$; graph
15. $\dfrac{(x+5)^2}{36} + \dfrac{(y+1)^2}{9} = 1$
16. $\dfrac{(x-2)^2}{1} + \dfrac{(y+5)^2}{9} = 1$
17. $\dfrac{(x-3)^2}{4} + \dfrac{(y-1)^2}{16} = 1$; graph
18. $\dfrac{(x+4)^2}{1} + \dfrac{(y-4)^2}{9} = 1$; graph

Rewrite the equation of each ellipse in standard form. Find the coordinates of the center and the vertices.

19. $x^2 + 4y^2 = 4$
20. $4x^2 + 9y^2 = 36$
21. $36x^2 + y^2 = 36$
22. $25x^2 + 16y^2 = 400$
23. $9x^2 + 16y^2 = 144$
24. $9x^2 + y^2 = 9$

Find the equation of each ellipse from its graph.

25.
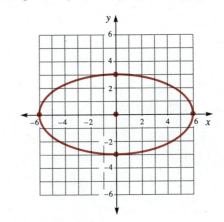

26.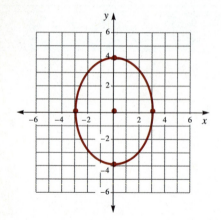

27.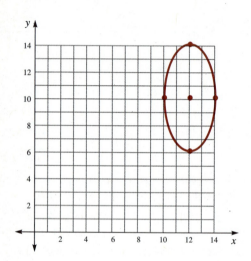

28.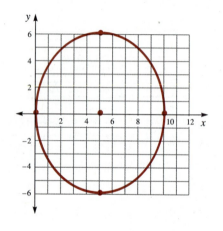

29.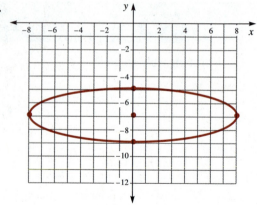

30.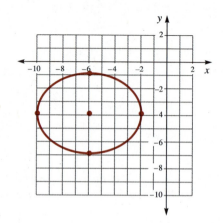

Rewrite each of the following conic sections in standard form. State whether it is a circle, a parabola, or an ellipse. Sketch the graph when indicated.

31. $4x^2 + 25y^2 - 8x + 50y = 71$
32. $25x^2 + 9y^2 + 150x - 36y + 36 = 0$
33. $x^2 + 6x = -2y^2 + 7$
34. $x^2 + y^2 = 6y - 2x - 5$
35. $4y^2 - 24x - 12y = 15$
36. $9x^2 - 8y + 4 = 36x - 4y^2$; graph
37. $x^2 - 4x - y + 8 = 0$; graph
38. $2y^2 - 10y + x + 7 = 0$
39. $3x^2 - 5x + 3y^2 + 7y + \dfrac{19}{6} = 0$

REVIEW EXERCISES

40. Given the equation $x^2 + y^2 = 9$, does it describe a circle, parabola, or ellipse?

41. Solve $x^2 + y^2 = 9$ for y.

Use $y = \sqrt{9 - x^2}$ for Exercises 42–45.

42. Sketch the graph. Is it a conic section? If so, name it.
43. Find the x- and y-intercepts.
44. Complete the ordered pairs $(3, \ )$, $(0, \ )$ and $(-1, \ )$ so that they satisfy the equation.
45. How are the graphs $y = \sqrt{9 - x^2}$ and $x^2 + y^2 = 9$ related?
46. Describe in words how to decide whether the general equation of a conic section, $Ax^2 + Cy^2 + Dx + Ey + F = 0$, $A > 0$, $C \geq 0$, is a circle, a parabola, or an ellipse.

10.4 Hyperbolas

The hyperbola, like the other conic sections, has mathematical and physical properties that are very useful in science and industry. In astronomy, the paths of some comets are hyperbolic in shape. Some telescopes involve both parabolic and hyperbolic mirrors to focus the image; this gives the advantage of a longer focal length in a short telescope. The image is reflected from the *objective* mirror (the parabolic one) to a *secondary* mirror (the hyperbolic one), which then reflects the image back through a hole in the primary mirror to the eyepiece.

HISTORICAL COMMENT   Apollonius of Perga (about 225 B.C.) is said to have done for the study of conic sections what Euclid did for demonstrative geometry. Apollonius's treatment of the conics differed from that of his predecessors. He derived them all from the same cone by passing a plane through the cone at different angles, as we do today. He published eight books on the subject, and all but one are known today.

OBJECTIVES   In this section we will learn to
1. recognize a hyperbola from its equation; and
2. sketch the graph of a hyperbola.

A hyperbola is the conic section formed when a plane cuts both sections (**nappes**) of a cone, forming two separate **branches**, as shown in Figure 10.30. The **hyperbola**, like the ellipse, is defined in terms of the distances of its points from two fixed points.

DEFINITION

Hyperbola

A **hyperbola** is the set of points in the plane, the *difference* of whose distances from two fixed points, is a constant. The fixed points are called the **foci** (singular **focus**).

When a hyperbola is centered at the origin and opens to the right and left, as shown in Figure 10.31, its equation is of the form

$$\frac{x^2}{a^2} - \frac{y^2}{b^2} = 1$$

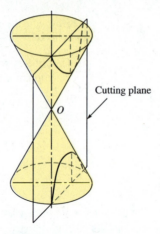

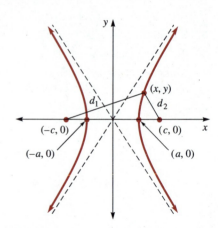

FIGURE 10.30

FIGURE 10.31

The foci are located at $(c, 0)$ and $(-c, 0)$. The points where the graph crosses the x-axis, $(a, 0)$ and $(-a, 0)$ are called the **vertices** of the hyperbola. The constant difference of distances, $d_1 - d_2$, referred to in the definition can be used to derive the equation of the hyperbola, although we shall not do so. The two broken lines shown in Figure 10.31 are called the **asymptotes** of the hyperbola.

> **DEFINITION**
> **Asymptotes of a Hyperbola**
>
> The **asymptotes of a hyperbola** are two lines that always pass through its center and have slope $\pm \dfrac{b}{a}$.

Certain hyperbolas are classified as x- or y-hyperbolas depending on the way that they open. If it opens to the right and the left, as in Figure 10.32(a), the hyperbola is said to be an **x-hyperbola**. If it opens up and down, as in Figure 10.32(b), it is a **y-hyperbola** and the equation is

$$\frac{y^2}{b^2} - \frac{x^2}{a^2} = 1$$

The foci of a y-hyperbola are located at $(0, c)$ and $(0, -c)$ and the vertices at $(0, b)$ and $(0, -b)$.

To sketch a hyperbola, we form a rectangle with sides of width $2a$ and height $2b$ symmetrically located around the center, as in Figures 10.32(a) and (b). We draw the asymptotes through the center and the opposite corners of this rectangle. The hyperbola then passes through its vertices and approaches the asymptotes as x and y get larger and larger.

We can find the equations of the asymptotes by using the fact they they pass through the center and have slope $\pm \dfrac{b}{a}$. When a hyperbola is centered at the origin, the equation of the asymptotes is $y = \pm \dfrac{b}{a} x$. Knowing the equations of the asymptotes is not necessary to sketch their graphs, however. This can be done simply by

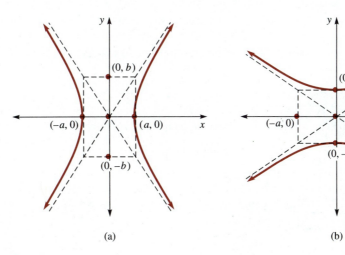

(a) (b)

FIGURE 10.32

extending the diagonals of the rectangle beyond its corners, as pictured in Figure 10.32.

EXAMPLE 1 Identify the hyperbola $\dfrac{x^2}{9} - \dfrac{y^2}{16} = 1$ and sketch its graph.

SOLUTION Since the equation is of the form $\dfrac{x^2}{a^2} - \dfrac{y^2}{b^2} = 1$, it represents an x-hyperbola centered at the origin

$a^2 = 9$ so $a = \pm 3$ and $b^2 = 16$ so $b = \pm 4$

The equation of the asymptotes is $y = \pm \dfrac{4}{3}x$. The graph is shown in Figure 10.33.

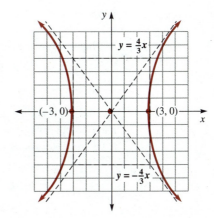

FIGURE 10.33

EXAMPLE 2 Identify the hyperbola $4y^2 - x^2 = 16$. Sketch its graph.

SOLUTION To write the equation in standard form, we divide both sides by 16:

$$\frac{y^2}{4} - \frac{x^2}{16} = 1$$

Since the equation is of the form $\frac{y^2}{b^2} - \frac{x^2}{a^2} = 1$, it represents a *y*-hyperbola centered at the origin.

$a^2 = 16$ so $a = \pm 4$ and $b^2 = 4$ so $b = \pm 2$

The asymptotes are $y = \pm \frac{2}{4}x = \pm \frac{1}{2}x$. The graph is shown in Figure 10.34.

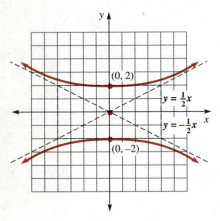

FIGURE 10.34

When the center of a hyperbola is shifted from the origin to (h, k), the **standard form** of the equation becomes

$$\frac{(x-h)^2}{a^2} - \frac{(y-k)^2}{b^2} = 1 \quad \text{or} \quad \frac{(y-k)^2}{b^2} - \frac{(x-h)^2}{a^2} = 1$$

For example,

$$\frac{(x-2)^2}{25} - \frac{(y+3)^2}{9} = 1$$

is an *x*-hyperbola with center at $(2, -3)$, while

$$\frac{(y+7)^2}{9} - \frac{(x+4)^2}{1} = 1$$

is a *y*-hyperbola with center at $(-4, -7)$.

EXAMPLE 3 Write $x^2 + 6x - 2y^2 + 4y - 1 = 0$ in standard form and graph the hyperbola.

SOLUTION We begin by completing the square in both *x* and *y*:

$x^2 + 6x - 2y^2 + 4y - 1 = 0$
$x^2 + 6x - 2y^2 + 4y = 1$ Add 1 to each side.

We must remember that to complete a square, the lead coefficient of the trinomial must be positive. Furthermore, *we must take care when removing the GCF of -2 from the y-terms. All remaining terms in the parentheses will have their signs changed.*

$x^2 + 6x - 2(y^2 - 2y) = 1$
$x^2 + 6x + 9 - 2(y^2 - 2y + 1) = 1 + 9 - 2$ $-2(1) = -2$; add 9 and add -2.
$(x + 3)^2 - 2(y - 1)^2 = 8$
$\frac{(x+3)^2}{8} - \frac{(y-1)^2}{4} = 1$ Divide each side by 8.

The center is at $(h, k) = (-3, 1)$.

$a^2 = 8$ so $a = \pm 2\sqrt{2}$ and $b^2 = 4$ so $b = \pm 2$

The graph is shown in Figure 10.35.

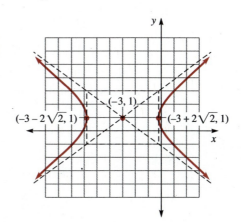

FIGURE 10.35

The Ellipse and the Hyperbola

Equation	Location	Identification
$\dfrac{(x-h)^2}{a^2} + \dfrac{(y-k)^2}{b^2} = 1$	Center at (h, k)	x-ellipse if $a^2 > b^2$ y-ellipse if $b^2 > a^2$
$\dfrac{(x-h)^2}{a^2} - \dfrac{(y-k)^2}{b^2} = 1$	Center at (h, k) Asymptotes with slope $\pm\dfrac{b}{a}$	x-hyperbola
$\dfrac{(y-k)^2}{b^2} - \dfrac{(x-h)^2}{a^2} = 1$	Center at (h, k) Asymptotes with slope $\pm\dfrac{b}{a}$	y-hyperbola

Exploring Graphics

EXAMPLE 4 Sketch the graph of $y^2 + 4y - 2x^2 - 4x + 5 = 0$ using either a computer or a graphics calculator.

SOLUTION Many computer programs designed for graphing conic sections allow their equations to be entered in the form

$$ax^2 + bxy + cy^2 + dx + ey + f = 0$$

which is the general equation of a conic section. In the case of this example, we have $a = -2$, $b = 0$, $c = 1$, $d = -4$, $e = 4$, and $f = 5$. The graph is shown in Figure 10.36(a), where the solid dots are the foci. Their position was computed by the program. To sketch the equation using a graphics calculator, the equation must be written in the form of a function. This can be accomplished by solving for y:

$$y^2 + 4y = 2x^2 + 4x - 5$$
$$y^2 + 4y + 4 = 2x^2 + 4x - 5 + 4 \qquad \text{Complete the square; add 4 to each side.}$$
$$(y + 2)^2 = 2x^2 + 4x - 1$$
$$y + 2 = \pm\sqrt{2x^2 + 4x - 1} \qquad \text{Square root property}$$
$$y = \pm\sqrt{2x^2 + 4x - 1} - 2$$

Both $y = \sqrt{2x^2 + 4x - 1} - 2$ and $y = -\sqrt{2x^2 + 4x - 1} - 2$ are functions and can be plotted on the same screen. The graph as displayed by the calculator is shown in Figure 10.36(b).

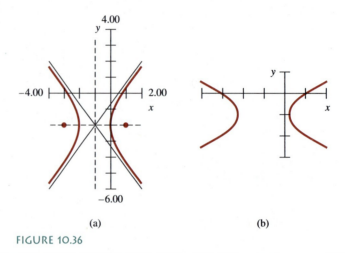

(a)   (b)

FIGURE 10.36

Do You Remember?

Can you match these?

_____ 1. An example of a y-hyperbola

_____ 2. The center of $\dfrac{(x - 4)^2}{16} - \dfrac{(y - 3)^2}{9} = 1$

_____ 3. The slopes of the asymptotes of
$$\dfrac{y^2}{9} - \dfrac{x^2}{16} = 1$$

_____ 4. An example of an x-hyperbola

_____ 5. The vertices of $\dfrac{y^2}{9} - \dfrac{x^2}{16} = 1$

a) $\dfrac{(x - 4)^2}{16} - \dfrac{(y - 3)^2}{9} = 1$

b) $\pm \dfrac{3}{4}$

c) $\pm \dfrac{4}{3}$

d) $\dfrac{y^2}{9} - \dfrac{x^2}{16} = 1$

e) $(3, 4)$

f) $(4, 3)$

g) $(-4, 0), (4, 0)$

h) $(0, -3), (0, 3)$

Answers: 1. d 2. f 3. b 4. a 5. h

Exercise Set 10.4

Graph each hyperbola. Determine the coordinates of the center and the vertices. Graph the asymptotes.

1. $\dfrac{x^2}{4} - \dfrac{y^2}{1} = 1$
2. $\dfrac{x^2}{4} - \dfrac{y^2}{9} = 1$
3. $\dfrac{y^2}{16} - \dfrac{x^2}{9} = 1$
4. $\dfrac{y^2}{4} - \dfrac{x^2}{4} = 1$
5. $\dfrac{(x-2)^2}{4} - \dfrac{(y+1)^2}{4} = 1$
6. $\dfrac{(x+3)^2}{25} - \dfrac{(y-2)^2}{4} = 1$
7. $\dfrac{(y+3)^2}{9} - \dfrac{(x-1)^2}{4} = 1$
8. $\dfrac{(y-2)^2}{16} - \dfrac{(x+5)^2}{16} = 1$
9. $\dfrac{(x+7)^2}{1} - \dfrac{(y-6)^2}{1} = 1$
10. $\dfrac{(x-5)^2}{16} - \dfrac{(y-4)^2}{25} = 1$
11. $\dfrac{(y-4)^2}{1} - \dfrac{x^2}{16} = 1$
12. $\dfrac{(y+7)^2}{25} - \dfrac{x^2}{1} = 1$
13. $\dfrac{x^2}{36} - \dfrac{(y-1)^2}{25} = 1$
14. $\dfrac{x^2}{49} - \dfrac{(y+2)^2}{4} = 1$

Exercises 15–18 are also hyperbolas. Graph these by plotting points. Label the asymptotes.

15. $xy = 9$
16. $xy = 4$
17. $x = \dfrac{16}{y}$
18. $y = \dfrac{25}{x}$

Rewrite each equation in standard form. Graph the hyperbola if indicated.

19. $x^2 - 4y^2 = 4$
20. $9x^2 - 4y^2 = 36$
21. $16y^2 - 9x^2 = 144$; graph
22. $25y^2 - 4x^2 = 100$; graph
23. $(x+1)^2 - (y-2)^2 = 4$
24. $(x-2)^2 - (y+3)^2 = 9$
25. $8(x+3)^2 - 4(y-1)^2 = 32$
26. $(x-1)^2 - 3y^2 = 3$
27. $16(y-3)^2 - 9(x-2)^2 = 144$; graph
28. $2(y-1)^2 - (x+2)^2 = 2$; graph

Name each of the following conic sections.

29. $x^2 + y^2 = 4$
30. $x^2 - y = 0$
31. $3x^2 - y^2 = 108$
32. $3x^2 + y^2 = 108$
33. $5x^2 + 5y^2 = 125$
34. $2x^2 - 3y^2 = 5$
35. $y^2 = 16x$
36. $x^2 + 9y^2 = 36$

Rewrite each of the following equations in standard form: that of a circle, an ellipse, a parabola, or a hyperbola.

37. $4y^2 = x^2 - 3x + 4$
38. $x^2 + 9y^2 - 2x + 36y = -28$
39. $4y^2 - 24x - 12y = 15$
40. $3x^2 + 3y^2 - 5x + 7y + 4 = 0$
41. $x^2 + 6x = -3y^2 - 6$
42. $3x^2 - 4y^2 + 6x + 6y = 0$
43. $x^2 + 4x = y^2 - 4y + 1$
44. $x^2 + 25 = 3y + 10x$
45. $x^2 + y^2 = 6y - 2x - 5$
46. $y^2 - 6x - 3y = 5$
47. $9x^2 - 8y + 31 = 36x - 4y^2$
48. $x^2 + 2x = 4y^2 - 8y + 7$

REVIEW EXERCISES

Solve each of the following systems of equations by the method indicated.

49. $\left.\begin{array}{l} 3x + 2y = 5 \\ 5x - 3y = 7 \end{array}\right\}$ by elimination

50. $\left.\begin{array}{l} x + 2y + z = 5 \\ 2x + y - z = 3 \\ x + y + 2z = 0 \end{array}\right\}$ by elimination

51. $\left.\begin{array}{l} 7x - 5y = 8 \\ 2x + 3y = 1 \end{array}\right\}$ by substitution

52. $\left.\begin{array}{l} 4x - 5y = 1 \\ 2x + y = 9 \end{array}\right\}$ by substitution

53. $\left.\begin{array}{l} x + y - z = -3 \\ 3x + y + 2z = -2 \\ 2x - y - 3z = -4 \end{array}\right\}$ by elimination

54. $\left.\begin{array}{l} 5x + 2y - 3z = -3 \\ 2x - 3y + 2z = 9 \\ x + y + 4z = 8 \end{array}\right\}$ by elimination

WRITE IN YOUR OWN WORDS

Define and describe each of the following.

55. A hyperbola
56. An ellipse
57. A parabola

58. The asymptotes of a hyperbola

59. A conic section

60. Give an application of **(a)** a parabola; **(b)** an ellipse; and **(c)** a hyperbola.

10.5 Solving Nonlinear Systems of Equations

OBJECTIVES In this section we will learn to
1. solve systems of nonlinear equations by the elimination method; and
2. solve systems of nonlinear equations by the substitution method.

In Section 4.1 we saw that the solution to a consistent system of linear equations was given by the coordinates of the point(s) of intersection of their graphs. In this section we consider some of the methods we can use to solve a system of equations in which one or both of the equations are of degree 2. Such a system is called a **nonlinear system of equations.**

EXAMPLE 1 Find the solution set to the following system. Graph the system.

$$\text{I} \quad x^2 + y^2 = 25$$
$$\text{II} \quad y + 2x = 10$$

SOLUTION To solve this system, we use substitution. We begin by solving equation **II** for y:

$$\text{II} \quad y + 2x = 10$$
$$y = 10 - 2x$$

Now we substitute this value for y into equation **I**:

$$x^2 + (10 - 2x)^2 = 25$$
$$x^2 + 100 - 40x + 4x^2 = 25$$
$$5x^2 - 40x + 75 = 0 \qquad \text{Collect like terms.}$$
$$x^2 - 8x + 15 = 0 \qquad \text{Divide each side by 5.}$$
$$(x - 5)(x - 3) = 0$$
$$x - 5 = 0 \quad \text{or} \quad x - 3 = 0$$
$$x = 5 \quad \text{or} \quad x = 3$$

To find y, we substitute $x = 5$ and $x = 3$ into equation **II**:

$$\begin{array}{c|c} y + 2x = 10 & y + 2x = 10 \\ y + 2 \cdot 5 = 10 & y + 2 \cdot 3 = 10 \\ y = 0 & y = 4 \end{array}$$

The solution set consists of ordered pairs, (5, 0) and (3, 4). To sketch the graph, we note that equation **I** is a circle with center at the origin and radius 5. Equation **II** is a straight line. The graphs are shown in Figure 10.37.

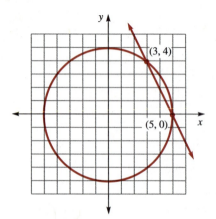

FIGURE 10.37

From Figure 10.37 we can see that the maximum number of intersections of a line and a circle is 2. The following discussion shows how it is easy to be misled into believing a point is a solution to a system when in fact it isn't. In Example 1, we substituted $x = 3$ into equation **II**, $y = 2x + 5$, to find that one point of intersection was (3, 4). If we had substituted $x = 3$ into equation **I**, $x^2 + y^2 = 25$, we would have found two different values for y, namely, 4 and -4.

$$x^2 + y^2 = 25$$
$$9 + y^2 = 25$$
$$y^2 = 16$$
$$y = \pm 4 \quad \text{Square root property}$$

It appears that (3, 4) and (3, -4) are both solutions. Notice, however, that although (3, -4) lies on the circle, it does not lie on the line. Thus (3, -4) is an **extraneous solution**. A rule of thumb to follow is **when solving a nonlinear system of equations, the second variable should be found, whenever possible, using the lower-degree equation.**

*C*AUTION

When solving nonlinear systems of equations, always check the solutions to determine if any of them are extraneous.

When the form of each equation in a system is recognizable, it is possible to estimate the maximum number of solutions before solving. For example, a circle and an ellipse could intersect in as many as four points or not intersect at all. Some possibilities are shown in Figure 10.38.

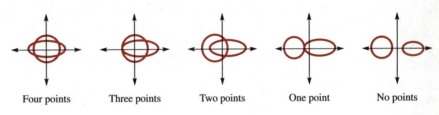

Four points   Three points   Two points   One point   No points

FIGURE 10.38

We can also use the elimination method to solve some nonlinear systems.

EXAMPLE 2   Solve the following system by elimination and graph the two equations, showing their intersection point(s).

$$\text{I} \quad x^2 + y^2 = 7$$
$$\text{II} \quad 3x^2 - 2y^2 = 6$$

SOLUTION   Equation **I** is a circle and equation **II** is a hyperbola. To find the points of intersection by the elimination method, we multiply both sides of equation **I** by 2 and add the result to equation **II**.

$$\begin{array}{llll} \text{I} & x^2 + y^2 = 7 & \rightarrow & \text{III} \quad 2x^2 + 2y^2 = 14 \\ \text{II} & 3x^2 - 2y^2 = 6 & \rightarrow & \text{II} \quad \underline{3x^2 - 2y^2 = 6} \\ & & & \quad\quad 5x^2 \quad\quad\quad = 20 \quad \text{Add III to II.} \\ & & & \quad\quad\quad x^2 = 4 \\ & & & \quad\quad\quad x = \pm 2 \end{array}$$

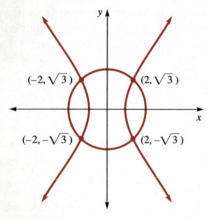

FIGURE 10.39

We can now substitute $x = 2$ and $x = -2$ back into equation **I** as follows:

$$x^2 + y^2 = 7 \qquad\qquad x^2 + y^2 = 7$$
$$2^2 + y^2 = 7 \quad x = 2 \qquad (-2)^2 + y^2 = 7 \quad x = -2$$
$$y^2 = 3 \qquad\qquad y^2 = 3$$
$$y = \pm\sqrt{3} \qquad\qquad y = \pm\sqrt{3}$$

The ordered pairs in the solution set are as follows: $(2, -\sqrt{3})$, $(2, \sqrt{3})$, $(-2, -\sqrt{3})$, and $(-2, \sqrt{3})$. These four solutions to the system represent the four points of intersection on the graph. Their graphs are shown in Figure 10.39.

The solutions were obtained from equation **I**. We can verify them by substituting $x = \pm 2$ into equation **II**.

$$3x^2 - 2y^2 = 6$$
$$3(\pm 2)^2 - 2y^2 = 6 \quad x = 2 \text{ or } x = -2$$
$$3 \cdot 4 - 2y^2 = 6$$
$$-2y^2 = -6$$
$$y^2 = 3 \quad \text{so} \quad y = \pm\sqrt{3}$$

EXAMPLE 3 Solve the system

I $\quad x^2 - xy - 2y^2 = 0$
II $\quad x^2 - y + y^2 = 1$

SOLUTION We can solve these equations by factoring equation **I** to get x in terms of y.

$$x^2 - xy - 2y^2 = 0$$
$$(x - 2y)(x + y) = 0$$
$$x - 2y = 0 \quad \text{or} \quad x + y = 0$$
$$x = 2y \quad \text{or} \quad x = -y$$

Now we substitute these values for x into equation **II**, beginning with $x = -y$,

II $\qquad x^2 - y + y^2 = 1$
$\qquad\qquad (-y)^2 - y + y^2 = 1 \quad x = -y$
$\qquad\qquad 2y^2 - y - 1 = 0$
$\qquad\qquad (2y + 1)(y - 1) = 0$
$\qquad\qquad 2y + 1 = 0 \quad \text{or} \quad y - 1 = 0$
$\qquad\qquad y = -\dfrac{1}{2} \quad \text{or} \quad y = 1$

Since $x = -y$, we get $x = \frac{1}{2}$ or $x = -1$. Two solutions are $\left(\dfrac{1}{2}, -\dfrac{1}{2}\right)$ and $(-1, 1)$.

When $x = 2y$,

$$x^2 - y + y^2 = 1$$
$$(2y)^2 - y + y^2 = 1 \quad x = 2y$$
$$5y^2 - y - 1 = 0$$

Since this equation doesn't factor, we'll use the quadratic formula to determine its solution set.

$$y = \frac{1 \pm \sqrt{1 + 20}}{10} = \frac{1 \pm \sqrt{21}}{10}$$

Knowing that $x = 2y$, we can now determine the complete solution set:

$$\left\{\left(\frac{1}{2}, -\frac{1}{2}\right), (-1, 1), \left(\frac{1 + \sqrt{21}}{5}, \frac{1 + \sqrt{21}}{10}\right), \left(\frac{1 - \sqrt{21}}{5}, \frac{1 - \sqrt{21}}{10}\right)\right\}$$

The check is left to the reader. Since the final ordered pairs were obtained from equation **II**, it's necessary to show that they satisfy equation **I**.

EXAMPLE 4 Find two natural numbers with a product of 40 and with squares that add to 89.

SOLUTION We let x = the first natural number
y = the second natural number

Then

$xy = 40$ Their product is 40.
$x^2 + y^2 = 89$ The sum of their squares is 89.

The methods used to solve Example 3 can be used to find the numbers. The solution is $x = 5$ and $y = 8$. Since we are seeking *natural* numbers, the other solution to the equation, $x = -5$, $y = -8$, must be rejected.

Exploring Graphics

EXAMPLE 5 Find all real-number solutions to the following system and graph the system.

I $2x^2 - y^2 = 1$
II $xy = 1$

SOLUTION We solve equation **II** for y and substitute the result into equation **I**:

$$xy = 1$$
$$y = \frac{1}{x}$$

I $2x^2 - y^2 = 1$
$$2x^2 - \left(\frac{1}{x}\right)^2 = 1 \qquad y = \frac{1}{x}$$
$$2x^2 - \frac{1}{x^2} = 1$$
$$2x^4 - 1 = x^2 \qquad \text{Clear of fractions.}$$
$$2x^4 - x^2 - 1 = 0$$

$$(2x^2 + 1)(x^2 - 1) = 0 \quad \text{Factor.}$$

$$2x^2 + 1 = 0 \quad \text{or} \quad x^2 - 1 = 0$$

$$x^2 = -\frac{1}{2} \quad \text{or} \quad x^2 = 1$$

$$x = \pm \frac{\sqrt{2}}{2}i \quad \text{or} \quad x = \pm 1 \quad \text{Square root property}$$

The only real-number values for x are 1 and -1. Since $y = \frac{1}{x}$, we get $y = 1$ or $y = -1$. The two solutions are $(1, 1)$ and $(-1, -1)$. The solution set is $\{(1, 1), (-1, -1)\}$.

The graphs in Figure 10.40 (the solid curve and the dotted curve) were drawn using a computer program that graphs conic sections without having to write their equations in standard form first. If we want to use a graphics calculator, we first have to solve the equations for y before we can enter them. The equations we would use in a graphics calculator are

$$y = \sqrt{2x^2 - 1}, \quad y = -\sqrt{2x^2 - 1}, \quad \text{and} \quad y = \frac{1}{x}$$

The graphs of the two equations are hyperbolas. The graph of $y = \frac{1}{x}$ has the x- and y-axes as its asymptotes. When we point the cursor at the apparent points of intersection, an approximation of their coordinates appears on the graph.

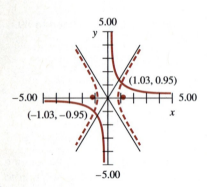

FIGURE 10.40

Exercise Set 10.5

Solve each of the following systems of equations by the substitution method.

1. $x^2 + y^2 = 9$
$x = 3$

2. $x^2 + y^2 = 16$
$y = -4$

3. $y^2 = 4x$
$x = -y$

4. $x^2 + 4y^2 = 32$
$x + 2y = 0$

5. $3y^2 - 2x^2 = 25$
$x + y = 0$

6. $x^2 = 2y$
$y = x - 3$

7. $x^2 + 3y^2 = 28$
$x = 2y$

8. $xy = 6$
$x - y = 1$

9. $x^2 + y^2 + 2x + 4y = 5$
$x = 4y$

10. $y^2 + x^2 + 2x = 3$
$x - y = 1$

11. $(x - 6)^2 = 3(y + 3)$
$x - y = 3$

12. $x^2 + y^2 + 3x - 4y = 10$
$3x + 2y = -7$

Solve each of the following systems of equations by the elimination method.

13. $x^2 + y^2 = 9$
$x^2 - y^2 = 9$

14. $2x^2 + 2y^2 = 3$
$3x^2 - 2y^2 = 2$

15. $4x^2 + y^2 = 4$
$x^2 - y^2 = 1$

16. $x^2 + y^2 = 10$
$16x^2 + y^2 = 25$

17. $x^2 + y^2 = 16$
$y^2 = 4 - x$

18. $x^2 + y^2 = 9$
$x^2 + 4y^2 = 36$

19. $x^2 - 2y^2 = 1$
$x^2 + 4y^2 = 25$

20. $2x^2 + y^2 = 24$
$-x^2 + y^2 = 12$

21. $2x^2 + 3y^2 = 4$
$4x^2 + 2y^2 = 8$

22. $x^2 + 4y^2 = 17$
$-2x^2 + 3y^2 = 10$

23. $x^2 - y^2 = 2$
$-x^2 + y^2 = 0$

24. $x^2 + y^2 = 20$
$x^2 - y = 0$

Solve each of the following systems of equations by any method.

25. $x^2 - 4x + y^2 - 4y = 1$
    $x^2 - 4x + y = -5$
26. $x^2 + y^2 + 2x - y = 14$
    $x^2 + y^2 + x - y = 9$
27. $x^2 - xy - y^2 = 1$
    $x^2 + xy + y^2 = 1$
28. $2x^2 + xy - 4y^2 = -12$
    $x^2 - 2y^2 = -4$
29. $x^2 + 2xy + y^2 = 9$
    $x^2 + 2xy - 8y^2 = 0$
30. $2x^2 + 3xy + 3y^2 = 3$
    $2x^2 + 3xy - 3y^2 = -3$
31. $x^2 - xy = 54$
    $ xy - y^2 = 18$
32. $x^2 + 3xy + 2y^2 = 6$
    $ -3xy - 3y^2 = -6$

Write each of the following applied problems as a system of equations (in two variables) and solve it.

33. Find two numbers with a sum of $\frac{65}{8}$ and a product of 1.
34. Find two numbers with a difference of 1 and a product of 1.
35. Find the dimensions of a rectangle that has an area of 100 m² and a perimeter of 58 m.
36. Find the dimensions of a rectangle that has an area of 128 ft² and a perimeter of 132 ft.
37. Find two natural numbers with a product of 63 and a difference of 2.
38. Find two natural numbers with a product of 42 and squares that add to 85.
39. Find two numbers with a product of 72 and reciprocals that add to $\frac{1}{4}$.
40. Find two numbers such that the product of their reciprocals is $\frac{1}{9}$ and the sum of their reciprocals is $\frac{5}{6}$.
41. The product of two numbers is 48 and the difference of their reciprocals is $\frac{1}{24}$. Find the numbers.
42. Find two numbers such that the sum of six times the smaller one and four times the larger one is 3, and the sum of their squares is $\frac{5}{18}$.

REVIEW EXERCISES

Solve each system of inequalities by graphing.

43. $x + y \leq 5$
    $x - y > 2$
44. $x - 5y \leq 3$
    $x + y \leq 7$
45. $x \leq 1$
    $y \leq 3$
    $y + x \geq 0$
46. $x \geq -2$
    $y \geq -2$
    $y < x + 1$

Solve each system of equations by the indicated method.

47. $\left. \begin{array}{l} 9x + 2y = 1 \\ 5x + y = 0 \end{array} \right\}$ Cramer's rule

48. $\left. \begin{array}{l} 3x + 2y = 5 \\ 2x + y = 3 \end{array} \right\}$ graphing

49. $\left. \begin{array}{l} 4x - y = 7 \\ 3x + 3y = 9 \end{array} \right\}$ elimination

50. $\left. \begin{array}{l} 2x - y = 0 \\ 3x + 4y = 11 \end{array} \right\}$ substitution

51. $\left. \begin{array}{l} x + 3y = 7 \\ 3x - y = 5 \end{array} \right\}$ matrices

52. $\left. \begin{array}{l} x + y + z = 3 \\ x - y + z = 2 \\ 2x + 2y - z = 3 \end{array} \right\}$ Cramer's rule

53. $\left. \begin{array}{l} x = y + z + 2 \\ y = x - z + 3 \\ z = x + y + 4 \end{array} \right\}$ elimination

54. $\left. \begin{array}{l} \dfrac{2}{3}x + \dfrac{3}{4}y = 1 \\ 2x + 3y = \dfrac{7}{3} \end{array} \right\}$ any method

10.6 Solving Nonlinear Systems of Inequalities

OBJECTIVE In this section we will learn to solve systems of nonlinear inequalities by graphing.

A **nonlinear inequality** is an inequality such that one or more of the terms is of degree 2 or greater. For example,

$$x^2 + y^2 \leq 4$$
$$x^2 + 2y^2 - x + y > 3$$
$$x - y^2 \geq 0$$

are three examples of nonlinear inequalities.

The solution to a nonlinear inequality is a region of the rectangular coordinate system. Consider the four graphs shown in Figures 10.41(a) through (d). Figure 10.41(a) is the graph of $x^2 + y^2 \leq 4$. To determine what region is the solution set (shaded), a test point that does not lie on the *boundary* of the region is used in the same manner as for linear inequalities. For example, if the test point is (0, 0), then

$$x^2 + y^2 \leq 4$$
$$0^2 + 0^2 \stackrel{?}{\leq} 4 \qquad x = 0, y = 0$$
$$0 \stackrel{\checkmark}{\leq} 4$$

Since $0 \leq 4$ is a true statement, the region containing this test point (the interior of the circle) and its boundary form the solution set.

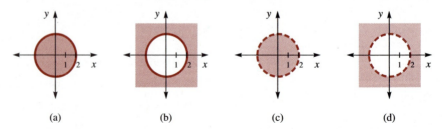

(a) (b) (c) (d)

FIGURE 10.41

Figure 10.4(b) is the graph of $x^2 + y^2 \geq 4$. In this case the previous test point does not satisfy the inequality since

$$x^2 + y^2 \geq 4$$
$$0^2 + 0^2 \stackrel{?}{\geq} 4$$
$$0 \not\geq 4$$

The test point is in the interior of the circle, while the region satisfying the inequality is outside the circle, including its boundary.

Figures 10.41(c) and (d) are the graphs of $x^2 + y^2 < 4$ and $x^2 + y^2 > 4$, respectively. In each case the circle itself is not part of the solution set. This is represented as a broken line.

A linear inequality in two variables divides the plane into two regions called *half-planes*. In the same manner, a nonlinear inequality divides the plane into two regions.

> The solution to a nonlinear system of inequalities is that portion of the plane that lies in the intersection of the graphs.

EXAMPLE 1   Solve the system $x^2 + y^2 < 16$ and $x + y \leq -2$ by graphing.

SOLUTION   The equation $x^2 + y^2 = 16$ is a circle with radius 4 centered at the origin. The circle itself does not belong to the solution set and must be shown as a broken line. Since the test point (0, 0) satisfies the inequality $x^2 + y^2 < 16$,

$$0^2 + 0^2 \stackrel{\checkmark}{<} 16$$

all points in the interior of the circle belong to the solution set. We shade the interior of the circle lightly to indicate this. (See Figure 10.42.)

Next we graph the line $x + y = -2$, showing it as a solid boundary of the solution set. The inequality $x + y \leq -2$ is not satisfied by the test point (0, 0), since

$$0 + 0 \not\leq -2$$

To indicate this, we shade the region on the opposite side of the line from (0, 0). The solution set is that region that satisfies both inequalities (contains double shading). The graph is shown in Figure 10.42.

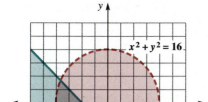

FIGURE 10.42

EXAMPLE 2   Solve the system

$$y + x^2 \leq 1$$
$$x^2 + 4y^2 \geq 16$$

SOLUTION   The graph of $y + x^2 = 1$, or $y = -x^2 + 1$, is a parabola opening downward with its vertex at (0, 1); see Figure 10.43. Although (0, 0) is the easiest test point to use to determine a solution set, we will use (1, 2) to illustrate that the choice for a test point is not limited to (0, 0). If we use (1, 2) as a test point in $y + x^2 \leq 1$, then we have

$$1 + 2^2 \stackrel{?}{\leq} 1$$
$$5 \not\leq 1$$

Thus (1, 2) is not a part of the solution set. Since (1, 2) lies outside of the parabola, the graph for the first inequality consists of the interior of the parabola together with its boundary.

The equation $x^2 + 4y^2 = 16$ is an ellipse. In standard form, it becomes

$$\frac{x^2}{16} + \frac{y^2}{4} = 1 \qquad \text{Divide each side by 16.}$$

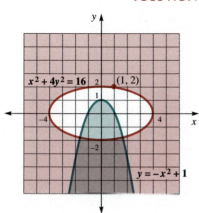

FIGURE 10.43

The center is at the origin, the vertices at $(-4, 0)$ and $(4, 0)$, and the co-vertices at $(0, -2)$ and $(0, 2)$. If we use $(1, 2)$ as a test point in $x^2 + 4y^2 \geq 16$, we find that

$$1^2 + 4 \cdot 2^2 \stackrel{?}{\geq} 16$$
$$1 + 16 \stackrel{?}{\geq} 16$$
$$17 \stackrel{\checkmark}{\geq} 16$$

Thus, $(1, 2)$ is a part of the solution set. The graph of $x^2 + 4y^2 \geq 16$ consists of the region outside of the ellipse, with the ellipse as the boundary. The solution to the given system is that region in Figure 10.43 that has double shading, including the boundary.

In Examples 1 and 2 we considered systems involving just two inequalities. A system may involve any number of inequalities

EXAMPLE 3   Solve the system

$$y \geq x^2 + 4x + 4$$
$$x^2 + y^2 \leq 16$$
$$|y| < 4$$

SOLUTION   The equation $y = x^2 + 4x + 4$ can be written as $y = (x + 2)^2$, which is a parabola opening upward with vertex at $(-2, 0)$. The test point $(0, 0)$ does not satisfy the inequality and so is not in the solution set. The graph of the solution set is the solid boundary of the parabola together with its interior. (See Figure 10.44.) The graph of $x^2 + y^2 \leq 16$ is the interior of the circle $x^2 + y^2 = 16$ together with its solid boundary. To graph $|y| < 4$, recall that this means $-4 < y < 4$. The graph consists of all points that lie between the lines $y = 4$ and $y = -4$. The lines are *not* a part of the solution set but they do serve as a boundary. The region that has the triple shading in Figure 10.44 shows the solution set.

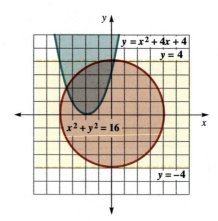

FIGURE 10.44

EXPLORING GRAPHICS

EXAMPLE 4 Use a computer to sketch the solution set for the inequalities of Example 3.

SOLUTION The graphs of the equations that represent the boundaries are shown in Figures 10.45(a) and (b). Notice that the computer plots all graphs in the same mode, either all solid or all broken. If we want to display the correct solution set after the computer sketches the boundaries, we will have to modify by hand the way the lines are drawn. We determined the final shading in Example 3.

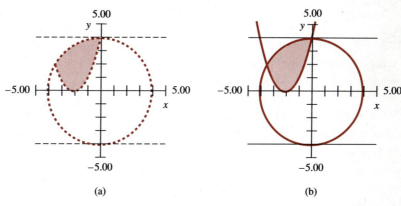

FIGURE 10.45

EXERCISE SET 10.6

Solve each system of inequalities by graphing.

1. $x^2 + y^2 < 25$
   $x + y < 5$
2. $y > x^2 + 2x + 1$
   $y \leq x + 3$
3. $x^2 + y^2 \leq 4$
   $x - y \leq \frac{1}{2}x + 2$
4. $x^2 + y^2 \leq 16$
   $x + y \geq 4$
5. $x + y < 4$
   $xy \geq 1$
   (shade only in Quadrant I)
6. $x^2 + y^2 + 2x - 6y + 1 \leq 0$
   $x + y \leq 4$
7. $y \geq x^2 - 1$
   $y \leq x$
8. $x^2 + y^2 \leq 36$
   $x \geq 3$
9. $x^2 - y^2 < 25$
   $x + y \leq 4$
10. $x^2 - 4 \leq y$
    $x^2 + y^2 \leq 25$
11. $9x^2 + 25y^2 \leq 225$
    $x^2 + y^2 \geq 16$
12. $16x^2 + 25y^2 \leq 400$
    $x^2 - y^2 \geq 4$
13. $x^2 - 6y + y^2 \leq -5$
    $9x^2 + 16y^2 < 144$
14. $y > x^2 - 4$
    $y - x < 2$
15. $y - x^2 > -1$
    $y + 2x < -1$
16. $x^2 - y^2 \geq 4$
    $x^2 + y^2 \geq 25$
17. $4x^2 + y^2 \leq 16$
    $x^2 + 4y^2 \geq 16$
18. $x^2 + y^2 - 6x - 8y \leq 0$
    $xy < 12$
19. $x^2 + y^2 - 10x + 21 \geq 0$
    $x^2 + y^2 - 10x + 2y + 1 \leq 0$
20. $x^2 + y^2 \leq 9$
    $16x^2 - 4y^2 \leq 64$
21. $x^2 - y^2 \leq -4$
    $x^2 + y^2 \leq 10$
22. $y < -x^2 + 1$
    $y \geq x^2 - 2$

23. $y \leq 4 - x^2$
    $y \geq x^2 - 4$
    $|x| < 1$

24. $x^2 + y^2 < 16$
    $x^2 + y^2 > 4$
    $|x| > 1$

25. $x^2 + y^2 \leq 36$
    $x^2 + y^2 \geq 16$
    $|x| > 3$

26. $x^2 + y^2 \leq 25$
    $x \geq 1$
    $y < 4$

27. $x^2 + y^2 \leq 9$
    $|x| < 2$
    $|y| < 2$

28. $y \leq 4 - x^2$
    $y > x^2 - 4$
    $|x| < 1$

29. $x^2 + y \leq 5$
    $x^2 + y^2 < 25$
    $|x| < 2$

30. $y \geq x^2 - 4x + 4$
    $x^2 + y^2 < 16$
    $|y| < 3$

Write In Your Own Words

31. What is a linear inequality? Give an example.

32. What is a *system* of linear inequalities? Give an example.

33. What is a linear equation? Give an example.

34. What is a *system* of linear equations? Give an example.

35. What is a nonlinear inequality? Give an example.

36. What is a *system* of nonlinear inequalities? Give an example.

37. What is a nonlinear equation? Give an example.

38. What is a *system* of nonlinear equations? Give an example.

39. Choose which of the following is a description of the solution set of the system
    $$x^2 + y^2 < 16$$
    $$x > 1$$

    (a) All points lie outside the circle and above the line.
    (b) All points lie inside the circle and above the line.
    (c) All points lie outside the circle and to the right of the line.
    (d) All points lie inside the circle and to the right of the line.

40. Choose which of the following is a description of the solution set of the system
    $$x^2 - y^2 < 4$$
    $$y < 2$$

    (a) All points lie inside the hyperbola and below the line.
    (b) All points lie outside the hyperbola and below the line.
    (c) All points lie inside the hyperbola and to the left of the line.
    (d) All points lie outside the hyperbola and to the left of the line.

For Extra Thought

41. Find a range of ski lengths that is greater than or equal to 50 cm to fit inside a closet measuring 50 cm by 80 cm by 2 m. Round your range of answers to the nearest centimeter.

42. Equilateral triangle *ABC* has a 24-ft side. Find the area of the shaded region rounded to one decimal point. *Note:* The three arcs touch the sides *AB*, *AC*, and *BC* at their midpoints.

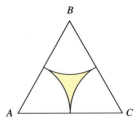

43. A piece of wire $6x$ cm long is bent into an equilateral triangle. For which values of x (rounded to one decimal point) will the area be numerically less than the perimeter?

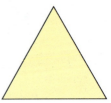

Summary

10.1 Circles

When the coordinates of two points (x_1, y_1) and (x_2, y_2) in a plane are known, the distance between them can be found by using the **distance formula**:
$$d = \sqrt{(x_2 - x_1)^2 + (y_2 - y_1)^2}$$

The distance formula can be used to find the equation of a circle. A circle with center at (h, k) and radius r has equation

$$(x - h)^2 + (y - k)^2 = r^2$$

The **midpoint** of the line segment joining (x_1, y_1) and (x_2, y_2) is found by averaging the coordinates of the endpoints. This yields the formula

$$(x_m, y_m) = \left(\frac{x_1 + x_2}{2}, \frac{y_1 + y_2}{2}\right)$$

10.2–10.4
Parabolas, Ellipses, and Hyperbolas

A **conic section** is the curve that results when a circular cone is cut by a plane. The conic sections studied in this chapter are the circle, the parabola, the ellipse, and the hyperbola. Their **standard form for the equations** are as follows:

Circle $\quad (x - h)^2 + (y - k)^2 = r^2 \quad$ Center at (h, k) and radius r

Parabola $\quad \left.\begin{array}{l} f(x) = a(x - h)^2 + k \\ \text{or} \quad y = a(x - h)^2 + k \\ \text{and} \quad x = a(y - k)^2 + h \end{array}\right\} \quad$ Vertex at (h, k)

Ellipse $\quad \dfrac{(x - h)^2}{a^2} + \dfrac{(y - k)^2}{b^2} = 1 \quad$ Center at (h, k)

Hyperbola $\quad \left.\begin{array}{l} \dfrac{(x - h)^2}{a^2} - \dfrac{(y - k)^2}{b^2} = 1 \\ \dfrac{(y - k)^2}{b^2} - \dfrac{(x - h)^2}{a^2} = 1 \end{array}\right\} \quad$ Center at (h, k)

Parabolas
Parabolas open upward or downward, or to the right or to the left, depending on whether a is positive or negative. A parabola that opens up or down represents a function, while one that opens to the right or to the left does not.

Ellipses
An ellipse is called an **x-ellipse** if $a^2 > b^2$ and a **y-ellipse** if $b^2 > a^2$.

Hyperbolas
A hyperbola is called an **x-hyperbola** if its branches open to the right and left and a **y-hyperbola** if its branches open up and down. The asymptotes of a hyperbola are the two straight lines that pass through its center and have slope $\pm\dfrac{b}{a}$.

10.5
Solving Nonlinear Systems of Equations

A **nonlinear system of equations** is one in which one or more of the equations has degree 2 or greater. Such systems are solved using a variety of methods, including substitution, elimination, and factoring with substitution. The number of solutions to such a system will vary depending on the degree of the equations.

10.6
Solving Nonlinear Systems of Inequalities

A **solution set to a nonlinear system of inequalities** is found by sketching the graph of the solution set of each inequality on the same coordinate system. That portion of the plane, if any, that contains shading from every inequality in the system forms the solution set of the system.

Cooperative Exercise

The city of Cedarvale owns an older metal building that is in the shape of a semi-ellipsoid. A youth group approaches the city council about using the building for various youth activities. The city council members decide that this is an excellent idea provided the costs are analyzed and certain conditions are met. One of the conditions is that the youth group must make some calculations to help the council determine the cost of renovating the building. The council has requested the following list of figures.

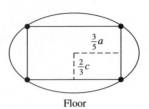

Floor

I. Approximate the volume (in cubic feet) of the building. (See the accompanying figure.) The council needs to calculate the costs of heat and air conditioning, which are a function of volume. The volume of an ellipsoid is approximately $\frac{4}{3}\pi abc$.

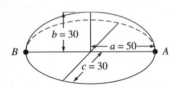

II. (a) Approximate the total area of the floor of the building. Find the cost to install hardwood flooring in the rectangular part (see figure) if a local contractor will furnish and install the hardwood floor for $9.25/ft² excluding tax. The area of an ellipse is approximately πab or πac or πbc, where a, b, and c are the same as in Part I.
(b) Approximate the area of the floor remaining outside the rectangular part. Find the cost (excluding tax) of having commercial carpet installed in this area at a cost of $12.75/yd². Because of the elliptical shape of the floor, there will be a waste of approximately 18% of the carpeting. See the figure for Part II(a).

III. (a) Referring to the figure in Part I, a fluorescent light strip is to be installed along the top (inside) of the building, from 8 ft inside of point A to 8 ft inside of point B. Find the length of the light strip.
(b) Because the building needs additional electrical outlets, a conduit must be installed all around the inside of the building, along the floor line. Find the length of the conduit.

IV. OPTIONAL
Approximate the inside surface area of the building (in square feet) so that the council can decide whether or not to install some kind of insulating material for temperature and sound control. *Note:* You will need to develop an algebraic formula to approximate the surface area. *Hint:* The surface area of a sphere is $4\pi r^2$, where r is the radius of the sphere.

Review Exercises

Identify each of the following conic sections and graph when indicated.

1. $x^2 + y^2 = 4$; graph
2. $y = x^2$; graph
3. $x = y^2$
4. $2x^2 + 2y^2 = 18$
5. $x^2 + 4y^2 = 4$; graph
6. $9x^2 + 2y^2 = 18$
7. $x^2 + 2y^2 = 2$
8. $25x^2 + 4y^2 = 100$
9. $x^2 - y^2 = 16$; graph
10. $y^2 - x^2 = 16$
11. $xy = 4$; graph by plotting points
12. $5x^2 + 5y^2 = 45$
13. $9x^2 - 16y^2 = 144$
14. $25x^2 + 9y^2 = 225$
15. $x^2 - y + 4x - 12 = 0$
16. $y^2 - 2x + 2y + 1 = 0$

17. Two vertices of an ellipse are $(-3, 2)$ and $(5, -3)$. Find the distance between the vertices.
18. The coordinates of the endpoints of a diameter of a circle are $(6, 1)$ and $(2, 3)$. Find the coordinates of the center.
19. The coordinates of the endpoints of a diameter of a circle are $(5, -6)$ and $(-7, 5)$. Find the equation of the circle.
20. Find the coordinates of the center and the length of the radius for the circle $(x - 2)^2 + (y + 1)^2 = 16$.
21. Determine the coordinates of the center and the length of the radius for the circle $x^2 + y^2 + 2x - 2y = 2$.

22. Write the equation of the circle that has center at $(3, -1)$ and that passes through the point $(5, 2)$.

Identify each of the following conic sections and graph when indicated.

23. $y = -x^2$
24. $y = x^2 + 2x + 3$; graph
25. $3y^2 = 4x$
26. $x^2 - y^2 = 4$; graph
27. $2x^2 + 2y^2 = 32$; graph
28. $5x^2 - 4y^2 = 20$
29. $x^2 - y^2 = -4x - 4y + 1$
30. $x^2 + 2y^2 + 6x - 8y = 0$; graph
31. $3x^2 - 5x + 7y = -3y^2 + 1$
32. $x - 3 = 2y^2 + 4y + 2$

Solve each of the following systems of equations.

33. $x^2 + y^2 = 6$
    $x = y$
34. $x^2 - y^2 = 2x - 1$
    $y = x$
35. $4x^2 + 4y^2 = 6$
    $6x^2 - 4y^2 = 4$
36. $5x^2 + y^2 = 10$
    $x^2 - y^2 = 2$
37. $x^2 + y^2 = 1$
    $x^2 - y^2 = -7$
38. $2x^2 - xy + y^2 = 8$
    $x^2 - y^2 = 0$
39. $x^2 - 4x = y - 3$
    $x^2 = 3x + y - 2$
40. $x^2 - 2xy + y^2 = 0$
    $x^2 + xy + y^2 = 3$

Solve each of the following systems of inequalities by graphing.

41. $x^2 + y^2 < 36$
    $x + y < 4$
42. $x^2 + y^2 \le 24$
    $|x| < 4$
43. $x^2 + 36y^2 \le 72$
    $x^2 + y^2 \ge 16$
44. $y \le -x^2 + 4$
    $y \ge x^2 - 1$
45. $y > -x^2 - 2x - 1$
    $y < x + 2$
46. $y \ge |x + 1| - 2$
    $y \le -|x + 1| + 3$

Chapter Test

Name and graph each of the following conic sections.

1. $x^2 + y^2 = 25$
2. $x^2 - y^2 = 64$
3. $y = (x - 3)^2 + 4$
4. $4x^2 + y^2 = 4$
5. $x = \frac{1}{2}(y + 2)^2 - 3$
6. $(x + 2)^2 + (y - 1)^2 = 9$
7. $\frac{(x + 3)^2}{4} + \frac{(y + 1)^2}{9} = 1$
8. $\frac{(x - 1)^2}{16} - \frac{(y - 3)^2}{9} = 1$

Change each of the following equations into standard form and name the conic it describes.

9. $x^2 + y^2 + 4x + 6y = 3$
10. $y = x^2 + 6x + 5$
11. $x^2 - 2x - y^2 + 2y = 4$
12. $x^2 + 8x + 4y^2 + 8y = -4$

Solve the following systems of equations.

13. $x^2 + y^2 = 5$
    $x^2 - y^2 = 13$
14. $x^2 - y^2 = 12$
    $x = 2y$

Solve the systems of inequalities in Exercises 15–17 by graphing.

15. $x^2 + y^2 \le 9$
    $x^2 - y^2 \ge 4$
16. $4x^2 + y^2 \le 16$
    $y \ge x + 2$
17. $x^2 - y^2 \le 4$
    $x \ge 1$
    $y \le 2$

18. Graph $xy = 4$ by plotting points.

19. Given the two points $(1, 2)$ and $(5, 3)$: **(a)** find the coordinates of the midpoint; **(b)** find the distance from the point $(5, 3)$ to the midpoint; and **(c)** write the equation of the circle (in standard form) with the given points as the endpoints of its diameter.

20. A pool is to be made in the form of the ellipse pictured in the accompanying figure. Write the equation of the ellipse (in standard form).

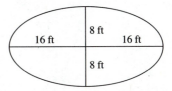

CHAPTER 11
Exponential and Logarithmic Functions

11.1 Exponential Functions
11.2 Logarithmic Functions
11.3 Properties of Logarithms
11.4 Solving Exponential and Logarithmic Equations
11.5 More Applications

Application

When light passes through a relatively clear body of water, the light intensity I is reduced according to the formula $I(d) = I_0 e^{-kd}$, where I_0 is the light intensity at the surface, d is the distance in feet below the surface of the water, and k is the *coefficient of extinction*. Find the depth of Lake Nacimiento, which has $k = 0.0485$, at which the light is reduced to 1% of that at the surface.

See Exercise Set 11.5, Exercise 19.

A scientist worthy of the name, especially a mathematician, experiences in his work the same excitement as an artist; his pleasure is as great and even greater when he makes a new discovery or comes up with a new idea.

HENRI POINCARÉ (1854–1912)

In Chapter 5 we developed the laws of exponents and in Chapter 8 we extended them to cover rational exponents. In this chapter we will introduce exponential functions and a few of their wide variety of applications. Exponential functions are used in physics to describe rates of radioactive decay; they are used in the biological sciences to describe the growth of bacterial cultures and populations; and they are used in finance to calculate compound interest, to name just a few examples.

11.1 Exponential Functions

OBJECTIVES

In this section we will learn to
1. identify and graph exponential functions; and
2. solve exponential equations.

In Chapter 8 we learned to work with exponential expressions in which the base was a real number and the exponent was a rational number. Recall that $b^{m/n} = \sqrt[n]{b^m} = (\sqrt[n]{b})^m$, provided that $\sqrt[n]{b}$ exists. For example,

$$9^{1/2} = \sqrt{9} = 3$$

$$16^{3/4} = (\sqrt[4]{16})^3 = 2^3 = 8$$

and

$$27^{-1/3} = \frac{1}{\sqrt[3]{27}} = \frac{1}{3}$$

However, $(-4)^{1/2}$ does not exist in the system of real numbers.

We now ask the question, "What is meant by b^m, where b is a positive real number and m is irrational?" The complete answer requires concepts taught in calculus and is beyond the scope of this book. It is possible, however, to gain some insight by considering an expression such as $3^{\sqrt{2}}$. Recall that $\sqrt{2}$ is irrational and as such is a nonterminating, nonrepeating decimal. An approximation is

$$\sqrt{2} = 1.4142136$$

The values $3^1 = 3$, $3^{1.4} = 3^{14/10}$, $3^{1.41} = 3^{141/100}$, $3^{1.414} = 3^{1414/1000}$, ..., can be computed, because the exponents are rational numbers. If the process were to continue in the same manner forever, the results would approach $3^{\sqrt{2}}$ so closely that any difference would be insignificant.

We can define **exponential functions** as follows.

DEFINITION

Exponential Function

A function of the form $f(x) = b^x$, $b > 0$, $b \neq 1$, is an **exponential function** with domain equal to the set of real numbers and with range equal to the set of positive real numbers.

555

For example, $f(x) = 3^{2x}$, $f(x) = 5^{-x/2}$, and $f(x) = 2^{-x}$ are all exponential functions.

The behavior of exponential functions is best observed by considering their graphs.

EXAMPLE 1 Sketch the graph of $f(x) = 2^x$.

SOLUTION We first construct a table of values for some rational values of x. Assume from the discussion above that a smooth curve will pass through these points as well as all others corresponding to irrational values of x. The graph is shown in Figure 11.1. Notice that for each unit change in x, the value of $f(x)$ grows to twice its previous value. When x changes from 2 to 3, $f(x)$ changes from 4 to 8. The function $f(x) = 2^x$ is an example of **exponential growth**.

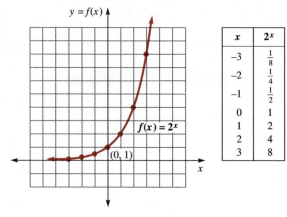

FIGURE 11.1

EXAMPLE 2 Sketch the graph of $f(x) = 2^{-x}$.

SOLUTION The function $f(x) = 2^{-x}$ is the same as $f(x) = \dfrac{1}{2^x} = \left(\dfrac{1}{2}\right)^x$. We begin by constructing

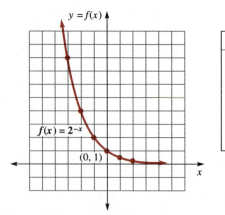

FIGURE 11.2

a table of values. The graph is shown in Figure 11.2. Notice in this case that $f(x)$ *decreases* as x increases. This is an example of **exponential decay.**

Notice in Examples 1 and 2 that both graphs cross the y-axis where $y = 1$.

EXPLORING GRAPHICS

EXAMPLE 3 Figure 11.3 shows the graphs of the exponential functions

(1) $f(x) = \left(\dfrac{1}{3}\right)^x$

(2) $f(x) = 3^x$

(3) $f(x) = (0.1)^x$

(4) $f(x) = -2^x$

(5) $f(x) = -\left(\dfrac{1}{2}\right)^x$

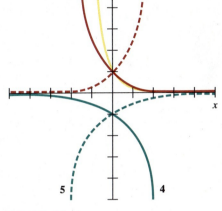

FIGURE 11.3

plotted on the same set of axes by a computer. Notice that they all cross the y-axis where $y = 1$ or $y = -1$. Recall that in the case of $-b^x$, the exponent applies only to b and not to the negative sign before it.

EXAMPLE 4 Graph $f(x) = 2^x + 2$ and $g(x) = 2^{|x|}$ on the same coordinate system.

SOLUTION To graph each equation, we construct a table of values, plot the points in the table,

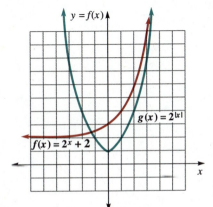

FIGURE 11.4

x	$2^x + 2$
-3	$\frac{17}{8}$
-2	$\frac{9}{4}$
-1	$\frac{5}{2}$
0	3
1	4
2	6
3	10

x	$2^{|x|}$
-3	8
-2	4
-1	2
0	1
1	2
2	4
3	8

and then join them with a smooth curve. The graphs are shown in Figure 11.4. Notice that the graph of $f(x) = 2^x + 2$ is the graph of $f(x) = 2^x$ (Figure 11.1) shifted 2 units up. Can you describe how the graph of $g(x) = 2^{|x|}$ relates to $f(x) = 2^x$?

As we mentioned in the chapter introduction, the decay of radioactive materials and the growth of bacterial cultures can be represented by exponential functions. Some exponential functions that result in a continuous change by increasing in size (growth) or decreasing in size (decay) have a special irrational base, the number e, where $e \approx 2.7182818284590$. In fact, because e is irrational, its decimal form neither terminates nor repeats. Scientific and graphics calculators have a special key, $\boxed{e^x}$, for calculating powers of e. Such calculations are necessary to solve many problems that arise in science, finance, and engineering, as well as in many other fields.

EXAMPLE 5 A particular type of radioactive material decays according to the equation $A(t) = 200e^{-0.05t}$, where $A(t)$ is in grams and t represents time in years.

(a) How much is initially present?
(b) How much is present when $t = 10$?
(c) Sketch the decay graph.

SOLUTION (a) The initial amount occurs when $t = 0$, so $A(0) = 200e^0 = 200 \cdot 1 = 200$ g.
(b) When $t = 10$ years, $A(10) = 200e^{-0.05(10)} = 200e^{-0.5}$. The value of $e^{-0.5}$ can be found with a calculator using either an $\boxed{e^x}$ or a $\boxed{y^x}$ key. To do this on a calculator with an $\boxed{e^x}$ key, enter 0.5 and press the $\boxed{\pm}$ key to change the sign to negative. Now press the $\boxed{e^x}$ key. The result is approximately 0.60653. Thus $A(10) = 200(0.60653) = 121.3$ g to the nearest tenth of a gram.

If a calculator does not have an $\boxed{e^x}$ key, the $\boxed{y^x}$ key can be used with y approximating e with a value of 2.71828. To do this, enter 2.71828 and press the $\boxed{y^x}$ key. Enter 0.5 and press the $\boxed{\pm}$ key, then press the $\boxed{=}$ key. The result may differ slightly from that calculated using the $\boxed{e^x}$ key, since 2.71828 is an approximation for e.

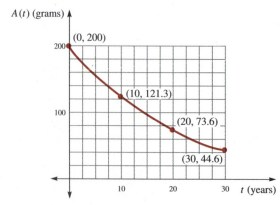

FIGURE 11.5

(c) To sketch the decay graph, we find several ordered pairs satisfying the equation by assigning values to t and computing the value of $A(t)$. The graph is shown in Figure 11.5. From the graph we can see that one-half the original amount has decayed after about 15 years. In Section 11.4 we will learn how to calculate the half-life of a radioactive substance.

$\mathcal{E}$XPLORING $\mathcal{G}$RAPHICS

EXAMPLE 6 Figure 11.6 shows the graphs of five exponential equations plotted by a computer on the same set of axes. Each one has the same basic shape as $y = 2^x$. Find the equation of each curve.

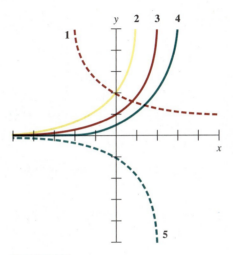

FIGURE 11.6

SOLUTION Recall from the discussion of parabolas how slight modifications in a formula can shift the graph of an equation to the right, to the left, up, or down. The same concepts apply here.

Curve **3** is the graph of $f(x) = 2^x$. Curve **4** is the graph of $f(x) = 2^x$ shifted one unit to the right, so its equation is $f(x) = 2^{x-1}$. Curve **2** is the graph of $f(x) = 2^x$ shifted one unit to the left, so its equation is $f(x) = 2^{x+1}$. Curve **1** is the graph of $f(x) = 2^{-x}$ shifted one unit up; its equation is $f(x) = 2^{-x} + 1$. Curve **5** is the reflection of $f(x) = 2^x$, so its equation is $f(x) = -2^x$.

In practice it is not uncommon for the need to arise to solve an equation where the variable is an exponent. For example, to solve for x if $2^x = 32$, we can proceed as follows:

$2^x = 32$

$2^x = 2^5$ 32 is 2^5.

Now, since the base, 2, is the same on both sides of the equation and since the two quantities are equal, the exponents must also be equal. Thus,

$$x = 5$$

The result of this example leads us to the following theorem.

> **THEOREM 11.1**
>
> If $b > 0$, $b \neq 1$, and m and n are real numbers, then
>
> $$b^m = b^n$$
>
> if and only if $m = n$.

EXAMPLE 7 Solve for y if $2^y = 128$.

SOLUTION We first need to express 128 as a power of 2, if possible.

$$2^y = 128$$
$$2^y = 2^7 \qquad 128 = 2^7$$
$$y = 7 \qquad \text{Theorem 11.1: } b^m = b^n \text{ implies } m = n.$$

The solution set is $\{7\}$.

EXAMPLE 8 Solve for x if $3^{2x} = \dfrac{1}{27}$.

SOLUTION Since $27 = 3^3$, we have

$$3^{2x} = \frac{1}{3^3} \qquad \frac{1}{27} = \frac{1}{3^3}$$
$$3^{2x} = 3^{-3} \qquad \text{Express each side in terms of the same base, 3.}$$
$$2x = -3 \qquad \text{Theorem 11.1: } b^m = b^n \text{ implies } m = n.$$
$$x = -\frac{3}{2}$$

The solution set is $\left\{-\dfrac{3}{2}\right\}$.

EXAMPLE 9 Solve $8^{x+3} = 64^x$ for x.

SOLUTION Both 8 and 64 are powers of 8, so we write

$$8^{x+3} = (8^2)^x = 8^{2x}$$
$$x + 3 = 2x \qquad \text{Theorem 11.1: } b^m = b^n \text{ implies } m = n.$$
$$x = 3$$

The solution set is $\{3\}$.

A second approach to solving this example is to write both 8 and 64 as powers of 2:

$$8^{x+3} = 64^x$$
$$(2^3)^{x+3} = (2^6)^x \qquad 8 = 2^3 \text{ and } 64 = 2^6$$
$$2^{3x+9} = 2^{6x}$$
$$3x + 9 = 6x \qquad b^m = b^n \text{ implies } m = n.$$
$$9 = 3x$$
$$3 = x$$

When funds are invested at an interest rate that is **compounded annually** (credited once a year) for a given number of years, they grow to an amount A given by the formula

$$A = p(1 + r)^t$$

where p is the principal (the amount originally invested), r is the rate of interest earned, and t is the time in years. To see how this formula is developed, consider a principal p invested at a rate r for a one-year period. The interest it would earn during that year would be $I = prt = pr$, since $t = 1$. The interest is added to the principal at the end of the first year, making the new balance

$$A_1 = p + pr = p(1 + r)$$

The new principal $A_1 = p(1 + r)$ earns interest at the rate r during the second year, so the balance at the end of the second year is

$$A_2 = p(1 + r) + p(1 + r)r = p(1 + r)(1 + r) = p(1 + r)^2$$

The pattern of multiplying the previous year's principal by $1 + r$ yields the following amounts.

End of Year	Amount after Each Compounding
0	$A = p$
1	$A = p(1 + r)$
2	$A = p(1 + r)^2$
3	$A = p(1 + r)^3$
⋮	⋮
n	$A = p(1 + r)^n$

For example, an investment of $1000 at 8% interest for a period of two years will grow to

$$A = 1000(1 + 0.08)^2$$
$$= 1000(1.08)^2$$
$$= \$1166.40$$

The value of $(1.08)^2$ can be found on a calculator using either the $\boxed{x^2}$ or the $\boxed{y^x}$ key. To use the $\boxed{x^2}$ key, first enter 1.08 and then press the $\boxed{x^2}$ key. The result will read 1.1664. To use the $\boxed{y^x}$ key, first enter 1.08 then press the $\boxed{y^x}$ key. Now enter 2 and press the $\boxed{=}$ key. The result is also 1.1664. Thus $A = 1000(1.1664) = \$1166.40$. Note that $(1.08)^2$ need not be completed separately, as indicated by the following sequence of key strokes:

$$A = 1.08 \;\; \boxed{x^2} \;\; \boxed{\times} \;\; 1000 \;\; \boxed{=} \;\; \$1166.40$$

When interest is compounded more than once a year, the annual interest rate r must be divided by the number of times n it is compounded and paid per year. For example, if interest is paid four times a year, then one-fourth of the annual rate is paid each interest period and the number of interest periods is four times as great.

> **The General Compound Interest Formula**
>
> The value of an investment A when p dollars are invested at $r\%$ with interest paid n times each year for t years is given by
>
> $$A = p\left(1 + \frac{r}{n}\right)^{nt}$$

Daily interest is often compounded using what is called a *banker's year*, which contains 360 days.

EXAMPLE 10 Find the amount of money that will be on deposit at the end of 9 years if $1300 is invested at 9% compounded monthly.

SOLUTION The interest is compounded monthly, so there are 12 payment periods each year. Thus, by the general formula,

$$A = 1300\left(1 + \frac{0.09}{12}\right)^{12 \cdot 9} \qquad n = 12, r = 0.09, \text{ and } t = 9$$

$$= 1300(1 + 0.0075)^{108}$$

To determine the amount using a calculator, we use the following sequence of keystrokes:

$$A = 1300 \;\; \boxed{\times} \;\; 1.0075 \;\; \boxed{y^x} \;\; 108 = \$2913.46$$

EXAMPLE 11 Find the amount of money on deposit at the end of 6 years if $1200 is initially deposited at 5% compounded daily.

SOLUTION The interest is compounded daily, so there are 360 payment periods each year.

$$A = 1200\left(1 + \frac{0.05}{360}\right)^{360 \cdot 6}$$

$$= 1200(1.000138889)^{2160}$$

$$= \$1619.80$$

When interest is **compounded continuously,** the amount A is given by the formula

$$A = pe^{rt}$$

Notice that we use the base e for continuous compounding.

EXAMPLE 12 Find the amount of money on deposit at the end of 4 years if $1500 is invested at 4.25% compounded continuously.

SOLUTION $A = pe^{rt} = 1500e^{0.0425(4)}$ $p = \$1500, r = 0.0425, t = 4$

To use a scientific calculator, we press the following sequence of keys:

$A = 0.0425$ ☒ 4 ☐ e^x ☒ 1500 ☐ $\$1777.96$

Do You Remember?

Can you match these?

___ 1. A possible graph for $f(x) = 3^x$
___ 2. A possible graph for $f(x) = 2^{-x}$
___ 3. The compound interest formula compounded n times per year
___ 4. The compound interest formula compounded annually
___ 5. The compound interest formula compounded continuously

a) $A = p\left(1 + \dfrac{r}{n}\right)^{nt}$

b) $A = pe^{rt}$

c) $A = p(1 + r)^t$

d)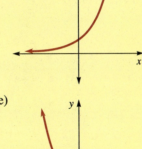

e)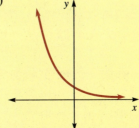

Answers: 1. d 2. e 3. a 4. c 5. b

Exercise Set 11.1

Complete the ordered pairs or graph when indicated.

1. $f(x) = 2^x$; (0,), (1,), (2,)
2. $f(x) = 3^x$; (−2,), (0,), (2,)
3. $f(x) = \left(\dfrac{1}{2}\right)^x$; graph
4. $f(x) = \left(\dfrac{1}{3}\right)^x$; graph
5. $f(x) = 3^{2x}$; (−2,), (0,), (2,)
6. $f(x) = 2^{3x}$; (−1,), (0,), (1,)
7. $f(x) = 2^{x-1}$; (−2,), (−1,), (1,)
8. $f(x) = 2^{-x+1}$; (−1,), (0,), (1,)
9. $f(x) = 5^{-x}$; graph
10. $f(x) = 2^{-3x}$; graph
11. $f(x) = (0.5)^{-x}$; (0,), (1,)

12. $f(x) = (0.25)^{-x}$; $(-1, \quad)$, $(0, \quad)$, $(1, \quad)$
13. $f(x) = 3^x + 1$; graph
14. $f(x) = 2^x - 1$; graph
15. $f(x) = \left(\dfrac{2}{3}\right)^x$; $(-2, \quad)$, $(-1, \quad)$, $(0, \quad)$
16. $f(x) = \left(\dfrac{3}{4}\right)^x$; $(0, \quad)$, $(2, \quad)$, $(4, \quad)$
17. $f(x) = 2^{3-2x}$; $(0, \quad)$, $(1, \quad)$, $(2, \quad)$
18. $f(x) = 3^{2-3x}$; $(0, \quad)$, $(1, \quad)$, $(2, \quad)$
19. $f(x) = 1^x$; $(0, \quad)$, $(1, \quad)$, $(2, \quad)$
20. $f(x) = 1^{2x}$; $(-2, \quad)$, $(0, \quad)$, $(2, \quad)$
21. $f(x) = (0.1)^x$; graph
22. $f(x) = (0.1)^{1+x}$; graph
23. $f(x) = \left(\dfrac{3}{2}\right)\left(\dfrac{1}{3}\right)^{-x}$; $(0, \quad)$, $(1, \quad)$
24. $f(x) = 3(2)^{-x}$; $(-1, \quad)$, $(0, \quad)$, $(1, \quad)$
25. $f(x) = 3 - 2(2)^{x-2}$; $(2, \quad)$, $(4, \quad)$
26. $f(x) = 1 - 2(2)^{x-4}$; $(3, \quad)$, $(4, \quad)$, $(5, \quad)$
27. $f(x) = 3^x$; graph
28. $f(x) = 3^{-x}$; graph

Solve each of the following exponential equations.

29. $2^x = 4$
30. $3^x = 9$
31. $4^x = 64$
32. $5^x = 125$
33. $3^{-x} = \dfrac{1}{27}$
34. $2^{-x} = \dfrac{1}{16}$
35. $2^{3x} = 8$
36. $5^{3x} = 625$
37. $6^{-2x} = \dfrac{1}{216}$
38. $4^{-3x} = \dfrac{1}{16}$
39. $4^x = 8$
40. $9^x = 27$
41. $\left(\dfrac{1}{2}\right)^{3x+1} = 64$
42. $\left(\dfrac{1}{3}\right)^{2x+3} = 81$
43. $(216)^{x+2} = \dfrac{1}{36}$
44. $(81)^{4x-3} = \dfrac{1}{27}$
45. $\left(\dfrac{2}{5}\right)^x = \dfrac{125}{8}$
46. $\left(\dfrac{3}{4}\right)^{-x} = \dfrac{16}{9}$
47. $4^{3x-1} = 256$
48. $9^{4x-3} = 243$

Use a calculator to find the amount of money A accumulated for each of the following investments. Daily compounding is based on a 360-day year.

49. $1000 invested for 2 years at 8% compounded annually
50. $1000 invested for 5 years at 7% compounded annually
51. $500 invested for 12 years at 10% compounded semi-annually
52. $300 invested for 7 years at 11% compounded semi-annually
53. $1250 invested for 5 years at 12% compounded quarterly
54. $2000 invested for 3 years at 8% compounded quarterly
55. $1000 invested for 4 years at 12% compounded monthly
56. $1100 invested for 8 years at 9% compounded monthly
57. $1500 invested for 6 years at 11% compounded daily
58. $1750 invested for 2 years at 13% compounded daily
59. $1000 invested for 11 years at 8.5% compounded continuously
60. $2000 invested for 6 years at 7.5% compounded continuously

Solve each of the following applied problems as indicated. Round to four decimal places as needed.

61. The number of bacteria $n(t)$ in a culture after t hours is given by $n(t) = 200e^{0.25t}$.
    (a) How many bacteria are in the culture at $t = 0$ hours?
    (b) How many bacteria are in the culture at $t = 5$ hours?

62. The amount of radioactive substance that remains after t hours is given by $A(t) = 500e^{-0.05t}$.
    (a) Determine the amount of the substance at $t = 0$ hr.
    (b) Determine the amount at $t = 4$ hr.
    (c) Determine the amount at $t = 100$ hr.

63. The number of grams $N(t)$ of potassium present in a solution t hours after a chemical reaction begins is given by $N(t) = 400e^{-0.055t}$.
    (a) How much potassium remains at $t = 0$ hr?
    (b) How much remains at $t = 3$ hr?
    (c) How much remains at $t = 7$ hr?

64. The world population is approximately equal to $P(t) = (5 \cdot 10^9)e^{0.02t}$, where t is in years.
    (a) Find the population when $t = 0$ yr.
    (b) Find the population when $t = 10$ yr.
    (c) Find the population when $t = 30$ yr.

WRITE IN YOUR OWN WORDS

65. What is an exponential decay function? Give an example of an exponential decay function.
66. What is an exponential growth function? Give an example of an exponential growth function.
67. In an exponential function $f(x) = b^x$, we restrict b by stating that $b > 0$, $b \neq 1$. What does the function become if $b = 1$? If $b = 0$? What happens to the function if $b < 0$? Try these on your calculator and explain what happens.

68. How do the graphs of $f(x) = 2^{-x}$ and $g(x) = (\frac{1}{2})^x$ compare? Explain why.

69. Which of the compound interest formulas does your bank use? How much more interest would you earn if you deposited $1000 at 8% interest compounded continuously compared to compounded daily?

70. Why is one exponential function called a "growth" function and another called a "decay" function? Describe what you notice about the graphs of these two types of exponential functions.

For Extra Thought

71. Triangle ABC is inscribed in a semicircle and semicircles are drawn on its legs. Show that the area of the large semicircle is equal to the sum of the areas of the two smaller semicircles. Show that the area of the triangle is equal to the sum of the areas of the shaded regions.

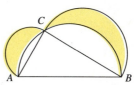

72. Find the area of the pictured ice skating rink. The measure of arc AC is 90°.

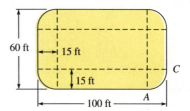

11.2 Logarithmic Functions

HISTORICAL COMMENT Until the early part of the seventeenth century, long laborious computations had to be carried out using just arithmetic. Much of the work that was underway at that time would have been subjected to long delays had it not been for the invention of logarithms. John Napier (1550–1617), who developed logarithms to the base e, and Henry Briggs (1561–1631), who developed logarithms to the base 10, were the early pioneers in this field. Napier's work appeared in 1614 under the title, "A Description of the Wonderful Law of Logarithms."

The need for logarithms as a computational tool has been displaced by the introduction of the calculator and the computer. They remain extremely important in the sciences, however, for solving exponential equations, as well as for multitude of other reasons.

OBJECTIVES In this section we will learn
1. the definition of a logarithm;
2. how to sketch graphs of logarithms; and
3. the relationship between logarithms and exponential functions.

We begin by defining a **logarithm**. Notice in the following definition that *a logarithm is nothing more than an exponent* with certain special properties.

Definition
Logarithm

The logarithm to the base b of a number x is the power to which the base b must be raised to produce x. In symbols this becomes

$$\log_b x = y \quad \text{if and only if} \quad b^y = x, \quad \text{where } b > 0, b \neq 1$$

This definition can be read two ways:

1. $\log_b x = y$ if and only if $b^y = x$
2. $b^y = x$ if and only if $\log_b x = y$

$y = \log_b x$ is read, "y is the logarithm to the base b of x."

The various parts of a logarithm are identified as follows:

The logarithm of the number x → $\log_b x = y$ ← is the power
to which the base b must be raised ↑   ↑ to produce x.

To work effectively with logarithms, it is often necessary to change them to exponential form. For example,

$\log_2 8 = y$ can be written $2^y = 8$

In the same manner, we can use the definition of a logarithm to change from exponential to logarithmic form:

$3^5 = t$ can be written $\log_3 t = 5$

In the examples that follow, the symbol "⇔" is interpreted to mean "if and only if."

EXAMPLE 1 Write each expression in exponential form:

(a) $\log_2 8 = 3$ (b) $\log_3 81 = 4$ (c) $\log_{10} 1000 = 3$

SOLUTION (a) We recall that $\log_b x = y$ if and only if $b^y = x$. Thus

$\log_2 8 = 3 \Leftrightarrow 2^3 = 8$

(b) $\log_3 81 = 4 \Leftrightarrow 3^4 = 81$
(c) $\log_{10} 1000 = 3 \Leftrightarrow 10^3 = 1000$

EXAMPLE 2 Write each expression in logarithmic form:

(a) $3^2 = 9$ (b) $2^5 = 32$ (c) $10^2 = 100$

SOLUTION (a) We recall that $b^y = x$ if and only if $\log_b x = y$. Thus

$3^2 = 9 \Leftrightarrow \log_3 9 = 2$

(b) $2^5 = 32 \Leftrightarrow \log_2 32 = 5$
(c) $10^2 = 100 \Leftrightarrow \log_{10} 100 = 2$

EXAMPLE 3 Solve each of the following logarithmic equations for x, y, or b:

(a) $\log_b 64 = 2$ (b) $\log_{3/2} \dfrac{27}{8} = y$ (c) $\log_4 x = 3$

SOLUTION (a) First we write the equation in exponential form:

$$b^2 = 64$$
$$b = 8 \quad \text{By definition } b > 0; \text{ the base must be a positive number.}$$

The solution set is $\{8\}$.

(b) First we write the equation in exponential form:

$$\left(\frac{3}{2}\right)^y = \frac{27}{8}$$
$$\left(\frac{3}{2}\right)^y = \left(\frac{3}{2}\right)^3$$
$$y = 3$$

The solution set is $\{3\}$.

(c) First we write the equation in exponential form.

$$4^3 = x$$
$$64 = x$$

The solution set is $\{64\}$.

Certain logarithms remain constant for any base. For example,

$\log_b b = 1$ since $b^1 = 1$ and $\log_b 1 = 0$ since $b^0 = 1$

In words this becomes:

---

1. The logarithm of any positive number b to the base b is 1.

$$\log_b b = 1$$

2. The logarithm of 1 to any base is 0.

$$\log_b 1 = 0$$

EXAMPLE 4 Find the logarithm of each number.

(a) $\log_4 4 = 1$
(b) $\log_7 7 = 1$
(c) $\log_5 5 = 1$
(d) $\log_8 8 = 1$

EXAMPLE 5 Simplify: (a) $\log_3(\log_8 8)$ (b) $\log_4[\log_4(\log_4 16)]$

SOLUTION (a) $\log_3(\log_8 8) = \log_3 1$ $\log_8 8 = 1$
$= 0$ $\log_3 1 = 0$

(b) We begin by noting that $\log_4 16 = 2$ since $4^2 = 16$. Then

$\log_4[\log_4(\log_4 16)] = \log_4[\log_4(2)]$ $\log_4 16 = y \Leftrightarrow 4^y = 16$ so $y = 2$

$= \log_4 \dfrac{1}{2}$ $\log_4 2 = y \Leftrightarrow 4^y = 2$ so $y = \dfrac{1}{2}$

$= -\dfrac{1}{2}$ $\log_4 \dfrac{1}{2} = y \Leftrightarrow 4^y = \dfrac{1}{2}$ so $y = -\dfrac{1}{2}$

In Section 11.1 we considered the exponential function $y = f(x) = b^x$. The inverse of $y = b^x$, which we obtain by interchanging x and y, is the function $x = b^y$. Every value of x corresponds to one and only one value of y. When the inverse, $x = b^y$, is written in logarithmic form, it becomes

$y = \log_b x$

Recall from the definition of a logarithm that $b \neq 1$. If $b = 1$, then $x = b^y$ becomes $x = 1^y = 1$. But $x = 1^y$ is the vertical line $x = 1$, which is not a function.

> Exponential and logarithmic functions are inverses of each other.

The graphs of the two functions $y = \log_b x$ and $y = b^x$ are shown in Figure 11.7. Notice that the domain of the logarithmic function is $(0, \infty)$ and its range is $(-\infty, \infty)$.

> The logarithm of a negative number or zero does not exist.

In order for a function to have an inverse that is also a function, it must be *one-to-one*. This means that each x-value corresponds to only one y-value and that each y-value corresponds to only one x-value. One way to determine whether a function is a **one-to-one function** is to draw its graph.

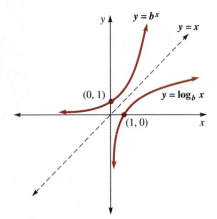

FIGURE 11.7

DEFINITION

One-to-One Function

A function is **one-to-one** if every horizontal line intersects its graph in at most one point.

EXAMPLE 6 Sketch the graph of $y = \log_2 x$. Is it one-to-one?

SOLUTION First we write the equation in exponential form:

$y = \log_2 x \Leftrightarrow 2^y = x$

Next we assign values to y and compute the corresponding values for x in order to construct a table of values. The graph is shown in Figure 11.8. Since any horizontal line will intersect the graph in at most one point, the function is one-to-one.

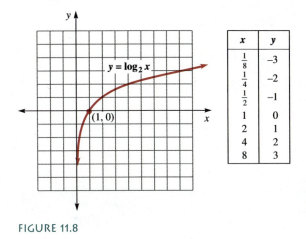

FIGURE 11.8

There is an interesting relationship between a one-to-one function and its inverse, and it is expressed in terms of the composition of functions:

If $f(x)$ and $g(x)$ are inverses, then $f[(g(x)] = g[f(x)] = x$.

EXAMPLE 7 Show that $f(x) = \log_2 x$ and $g(x) = 2^x$ are inverses by showing that $f[g(x)] = g[f(x)] = x$.

SOLUTION First we will show that $f[g(x)] = x$.

$f[g(x)] = \log_2(2^x) \Leftrightarrow 2^{f[g(x)]} = 2^x$ $y = \log_b x \Leftrightarrow b^y = x$
$f[g(x)] = x$ If $b^m = b^n$, then $m = n$.

We must also show that $g[f(x)] = x$.

$g[f(x)] = 2^{\log_2 x} \Leftrightarrow \log_2 g[f(x)] = \log_2 x$ $x = b^y \Leftrightarrow \log_b x = y$
$g[f(x)] = x$ If the logarithms of two numbers are equal, then the numbers are equal. (We will prove this later.)

Thus $f[(g(x)] = g[f(x)] = x$. The functions are inverses.

EXPLORING GRAPHICS

The graphs of $y = \log_{10} x$, $y = \log_2 x$, and $y = \log_e x$ are shown in Figure 11.9; they were drawn by a computer. Notice that they all cross the x-axis where $x = 1$.

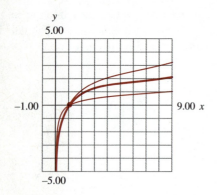

FIGURE 11.9

Can you identify which graph belongs to each equation? Also notice that x is always positive in the graph of a logarithm.

Do You Remember?

Can you match these?

___ 1. The exponential form of $\log_c d = m$
___ 2. The logarithmic form of $s^t = r$
___ 3. $\log_b b = ?$
___ 4. The logarithm of 1 to any base
___ 5. Numbers that do not have logarithms

a) negative numbers or zero
b) 0
c) 1
d) $\log_s r = t$
e) $\log_t r = s$
f) $\log_r t = s$
g) $d^m = c$
h) $c^m = d$
i) $m^c = d$

Answers: 1. h 2. d 3. c 4. b 5. a

Exercise Set 11.2

Write each of the following exponential equations in logarithmic form.

1. $3^3 = 27$
2. $4^2 = 16$
3. $2^5 = 32$
4. $5^3 = 125$
5. $10^0 = 1$
6. $5^0 = 1$
7. $9^{1/2} = 3$
8. $36^{1/2} = 6$
9. $10^{-1} = 0.1$
10. $7^{-1} = \dfrac{1}{7}$
11. $3^{-4} = \dfrac{1}{81}$
12. $2^{-3} = \dfrac{1}{8}$
13. $8^{2/3} = 4$
14. $16^{3/4} = 8$
15. $81^{-3/4} = \dfrac{1}{27}$
16. $125^{-2/3} = 0.04$
17. $8^1 = 8$
18. $10^1 = 10$
19. $\left(\dfrac{3}{4}\right)^2 = \dfrac{9}{16}$
20. $\left(\dfrac{2}{3}\right)^3 = \dfrac{8}{27}$
21. $(0.3)^3 = 0.027$
22. $(0.15)^2 = 0.0225$
23. $0.001^{1/3} = 0.1$
24. $0.008^{1/3} = 0.2$

Write each of the following logarithmic equations in exponential form.

25. $\log_6 36 = 2$
26. $\log_9 81 = 2$
27. $\log_9 9 = 1$
28. $\log_3 3 = 1$
29. $\log_6 1 = 0$
30. $\log_5 1 = 0$

31. $\log_{16} 8 = \dfrac{3}{4}$
32. $\log_{27} 9 = \dfrac{2}{3}$
33. $\log_{10}(0.001) = -3$
34. $\log_{10}(0.01) = -2$
35. $\log_3 \left(\dfrac{1}{9}\right) = -2$
36. $\log_2 \left(\dfrac{1}{8}\right) = -3$
37. $\log_{1/5} 25 = -2$
38. $\log_{1/3} 81 = -4$
39. $\log_{1/2} 32 = -5$
40. $\log_{2/3} \left(\dfrac{27}{8}\right) = -3$

Solve for the variable in each of the following equations.

41. $\log_2 x = 2$
42. $\log_3 x = 2$
43. $\log_y(0.001) = -3$
44. $\log_y(0.01) = -2$
45. $\log_{1/3} 9 = k$
46. $\log_{25}\left(\dfrac{1}{5}\right) = m$
47. $\log_b(1000) = \dfrac{3}{2}$
48. $\log_t 8 = \dfrac{3}{2}$
49. $\log_{2/3} c = -2$
50. $\log_{1/4} w = -3$
51. $\log_2 \left(\dfrac{1}{8}\right) = p$
52. $\log_5 \left(\dfrac{1}{5}\right) = v$

Simplify in Exercises 53–60.

53. $\log_4(\log_5 5)$
54. $\log_7(\log_8 8)$
55. $\log_9[\log_3(\log_3 27)]$
56. $\log_4[\log_2(\log_2 16)]$
57. $\log_{10}(\log_{10} 10)$
58. $\log_2(\log_7 49)$
59. $\log_b(\log_a a^b); \; a, b > 0$
60. $\log_a(\log_b b^a); \; a, b > 0$

Given $f(x) = \log_{10} x$ and $g(x) = 10^x$, answer the following questions:

61. What is the domain of f?
62. What is the range of f?
63. Graph $f(x)$, $g(x)$, and $h(x) = x$ on the same coordinate system. What do you conclude about f and g?
64. Find $f^{-1}(x)$.
65. What is the domain of $f^{-1}(x)$?
66. What is the range of $f^{-1}(x)$?
67. What do you conclude about $f^{-1}(x)$ and $g(x)$?
68. Find $f[g(x)]$ and $g[f(x)]$. What do you conclude about the relationship of $f[g(x)]$ and $g[f(x)]$? Is $f(x)$ a one-to-one function? Is $g(x)$ a one-to-one function?
69. Find $f[g(1)]$, $f[g(2)]$, and $f[g(3)]$. What is the value of $f[g(a)]$, for all a in the domain of f and g?
70. Find $g[f(1)]$, $g[f(2)]$, and $g[f(3)]$. What is the value of $g[f(a)]$, for all a in the domain of f and g?

REVIEW EXERCISES

71. Determine the domain and range of the function $f = \{(x, y) \,|\, y = 1^x\}$. Is it a one-to-one function?
72. Determine the domain and range of the function $f = \{(x, y) \,|\, y = (\tfrac{1}{2})^x\}$. Is it a one-to-one function?
73. Is $f(x) = 3^x$ a function? What is its domain and range? Is it a one-to-one function?
74. The population of the town of Corona is given by the function $f(t) = 800e^{0.06t}$, where t is the time in years.
    (a) What is the initial population of Corona?
    (b) What will the population be in 5 years?
    (c) What will the population be in 10 years?
    (d) Graph the function. Is it a growth or a decay function?
75. The value of a certain computer is given by the function $f(t) = 2200e^{-0.2t}$, where t is the time in years.
    (a) What is the initial value of the computer?
    (b) What is the value of the computer in 1 year?
    (c) What is the value of the computer in 3 years?
    (d) Graph the function. Is it a growth or a decay function?
76. Is $f(x) = 2^{-x}$ a function? What is its domain and range? Is it a one-to-one function?
77. How are logarithmic and exponential functions related? Illustrate your answer.
78. Why is the logarithm of 1 to any base equal to zero?
79. Why doesn't the logarithm of a negative number exist? Try $\log_{10}(-5)$ on your calculator. What happens? Why?
80. Why does $b^{\log_b x} = x$? Is this true for all x? For what values of x is it true? Why?
81. Why does the logarithm of any positive natural number to the same base equal 1? Would it be true if the number were zero?

11.3 Properties of Logarithms

OBJECTIVES

In this section we will learn to
1. use the properties of logarithms to simplify logarithmic expressions; and
2. use the properties of logarithms to solve logarithmic equations.

Before calculators and computers were available, long calculations such as

$$\sqrt[4]{\frac{(3.265)^6(24.28)^7}{(0.036)^4(25.41)^3}}$$

were carried out by using logarithms. The scientific calculator has eliminated this need; however, the theorems that enabled mathematicians to make these calculations are still very useful in higher mathematics.

THEOREM 11.2

Properties of Logarithms

Let x and y be real numbers or algebraic expressions such that $x > 0$ and $y > 0$. If b is a positive real number such that $b \neq 1$ and if k is any real number, then

IN SYMBOLS	IN WORDS
(a) $\log_b xy = \log_b x + \log_b y$	The logarithm of a product is the sum of the logarithms of the factors.
(b) $\log_b \frac{x}{y} = \log_b x - \log_b y$	The logarithm of a quotient is the difference of the logarithms of the numerator and the denominator.
(c) $\log_b x^k = k \log_b x$	The logarithm of a number raised to a power is the power times the logarithm of the number.

Before proving of any of the properties of logarithms it must be emphasized that **the properties of logarithms in Theorem 11.2 do not apply to the following expressions:**

$$\log_b(x+y), \quad (\log_b x)\cdot(\log_b y), \quad \log_b(x-y), \quad (\log_b x)^y, \quad \text{and} \quad \frac{\log_b x}{\log_b y}$$

We will prove Theorem 11.2(a) and leave the remaining parts to be proven as Exercises 71 and 72. The proofs of the theorems are based on the relationship between logarithms and exponential expressions.

PROOF $\quad \log_b xy = \log_b x + \log_b y$

We let $\log_b x = M$ and $\log_b y = N$. From the definition of a logarithm, we have

$$x = b^M \quad \text{and} \quad y = b^N$$

Multiplying the left-hand sides, x and y, and the right-hand sides, b^M and b^N, gives us

$$x \cdot y = b^M \cdot b^N = b^{M+N} \qquad a^m \cdot a^n = a^{m+n}$$

We can write this in logarithmic form as

$$\log_b xy = M + N$$

Now when we substitute $\log_b x$ for M and $\log_b y$ for N, we get

$$\log_b xy = \log_b x + \log_b y$$

This completes the proof.

In the examples that follow and throughout the rest of the text, the base will not be indicated when it is 10.

DEFINITION
Common Logarithms

Base 10 logarithms are called **common logarithms**. We write

$$\log_{10} x = \log x$$

EXAMPLE 1 Express each of the following expressions as algebraic sums of logarithms:

(a) $\log_b rst$ (b) $\log x^2 y$ (c) $\log \dfrac{st}{r}$

SOLUTION

(a) $\log_b rst = \log_b r + \log_b s + \log_b t \qquad \log_b xy = \log_b x + \log_b y$

(b) $\log x^2 y = \log x^2 + \log y \qquad \log_b xy = \log_b x + \log_b y$
    $= 2 \log x + \log y \qquad \log_b x^k = k \log_b x$

(c) $\log \dfrac{st}{r} = \log st - \log r \qquad \log_b \dfrac{x}{y} = \log_b x - \log_b y$

$\phantom{(c) \log \dfrac{st}{r}} = \log s + \log t - \log r \qquad \log_b xy = \log_b x + \log_b y$

EXAMPLE 2 Write $\log \sqrt[3]{\dfrac{ab}{c^2}}$ as an algebraic sum of logarithms.

SOLUTION

$\log \sqrt[3]{\dfrac{ab}{c^2}} = \log \left(\dfrac{ab}{c^2}\right)^{1/3}$

$\phantom{\log \sqrt[3]{\dfrac{ab}{c^2}}} = \dfrac{1}{3} \log \dfrac{ab}{c^2} \qquad \log_b x^k = k \log_b x$

$\phantom{\log \sqrt[3]{\dfrac{ab}{c^2}}} = \dfrac{1}{3}(\log ab - \log c^2) \qquad \log_b \dfrac{x}{y} = \log_b x - \log_b y$

$\phantom{\log \sqrt[3]{\dfrac{ab}{c^2}}} = \dfrac{1}{3}(\log a + \log b - 2 \log c) \qquad \begin{array}{l}\log_b xy = \log_b x + \log_b y; \\ \log_b x^k = k \log_b x\end{array}$

Just as we used Theorem 11.2 to write a logarithm as an algebraic sum, we can use it to write the sum or difference of logarithms as a single logarithm with a coefficient of 1.

EXAMPLE 3 Write $2 \log_e x + 5 \log_e y - \log_e z$ as a single logarithm with a coefficient of 1.

SOLUTION
$$2 \log_e x + 5 \log_e y - \log_e z = \log_e x^2 + \log_e y^5 - \log_e z \quad\quad k \log_b x = \log_b x^k$$
$$= \log_e x^2 y^5 - \log_e z \quad\quad \log_b x + \log_b y = \log_b xy$$
$$= \log_e \frac{x^2 y^5}{z} \quad\quad \log_b x - \log_b y = \log_b \frac{x}{y}$$

Suppose that we are given the values $\log 2 = 0.3010$, $\log 3 = 0.4771$, and $\log 5 = 0.6990$. These values can be used to find the logarithm of any number that can be expressed in terms of 2, 3, and 5. The next examples illustrate how this is done.

EXAMPLE 4 Find $\log 75$ using the preceding information.

SOLUTION
$$\log 75 = \log(3 \cdot 5^2)$$
$$= \log 3 + \log 5^2$$
$$= \log 3 + 2 \log 5$$
$$= 0.4771 + 2(0.6990)$$
$$= 1.8751$$

EXAMPLE 5 Find $\log 18\sqrt{10}$ using the preceding information.

SOLUTION
$$\log 18\sqrt{10} = \log(2 \cdot 3^2 \cdot 10^{1/2})$$
$$= \log 2 + \log 3^2 + \log 10^{1/2}$$
$$= \log 2 + 2 \log 3 + \frac{1}{2} \log 10$$
$$= 0.3010 + 2(0.4771) + \frac{1}{2}(1) \quad\quad \log 10 = \log_{10} 10 = 1$$
$$= 0.3010 + 0.9542 + 0.5$$
$$= 1.7552$$

EXAMPLE 6 Use the properties of logarithms to verify that
$$\frac{1}{2} \log 36 - \frac{1}{3} \log 8 + \frac{3}{5} \log 32 = \log 24$$

SOLUTION We can rewrite the given expression
$$\log 36^{1/2} - \log 8^{1/3} + \log 32^{3/5} = \log 6 - \log 2 + \log 8 \quad\quad 36^{1/2} = 6; \ 8^{1/3} = 2;$$
$$\quad\quad 32^{3/5} = (32^{1/5})^3 = 2^3 = 8$$
$$= (\log 6 + \log 8) - \log 2 \quad\quad \text{Only } \log 2 \text{ is subtracted.}$$

$$= \log(6 \cdot 8) - \log 2$$
$$= \log \frac{6 \cdot 8}{2}$$
$$= \log 24$$

EXAMPLE 7 Solve for x in $\log_3 x + \log_3(x + 8) = 2$.

SOLUTION First we combine terms using Theorem 11.2:

$$\log_3 x + \log_3(x + 8) = 2$$
$$\log_3[x(x + 8)] = 2 \qquad \log_b x + \log_b y = \log_b xy$$

Now we use the definition of a logarithm to rewrite this equation as an exponential equation.

$$x(x + 8) = 3^2$$
$$x^2 + 8x = 9$$
$$x^2 + 8x - 9 = 0$$
$$(x - 1)(x + 9) = 0$$
$$x - 1 = 0 \quad \text{or} \quad x + 9 = 0$$
$$x = 1 \quad \text{or} \quad x = -9$$

If $x = -9$ then $\log_3 x$ is the logarithm of a negative number and does not exist. Thus -9 is not in the solution set. The solution set is $\{1\}$.

> **CAUTION**
>
> $\log x - \log y + \log z$ **is not equal to** $\log x - (\log y + \log z)$;
>
> $\log x - \log y + \log z =$
> $(\log x + \log z) - \log y$

> **CAUTION**
>
> When solving an equation involving logarithms, check all solutions to insure that none creates a logarithm of a negative number or zero. Logarithms of negative numbers and zero do not exist.

In problem solving, equations sometimes occur that require the use of one or both of the following relationships to complete the solution.

Relationships Between Logarithms and Numbers

1. If $\log_b M = \log_b N$ then $M = N$.
2. If $M = N$ then $\log_b M = \log_b N$.

We used the first relationship to establish the results of Example 7 in Section 11.2. We are now in a position to prove it.

PROOF

$\log_b M = \log_b N$	
$\log_b M - \log_b N = 0$	Subtract $\log_b N$ from each side.
$\log_b \dfrac{M}{N} = 0$	Theorem 11.2
$\dfrac{M}{N} = b^0$	Definition of a logarithm
$\dfrac{M}{N} = 1$	
$M = N$	Multiply each side by N.

EXAMPLE 8   Solve for x if $\log_b x = \log_b 2 + \log_b y$, where $x > 0$ and $y > 0$.

SOLUTION   We begin by simplifying the right side.

$$\log_b x = \log_b 2 + \log_b y = \log_b 2y \qquad \log_b x + \log_b y = \log_b xy$$

Since the logarithms of the two numbers are equal, the numbers themselves are equal:

$$x = 2y$$

The solution set is $\{2y\}$.

EXAMPLE 9   Find $\log_3 2$ given that $\log 2 = 0.3010$ and $\log 3 = 0.4771$.

SOLUTION   Since we are given logarithms to the base 10, we convert base 3 to base 10 as follows. We let $x = \log_3 2$; then

$$3^x = 2 \qquad \text{Definition of a logarithm}$$

Now we take the logarithm of each side to the base 10:

$$\log 3^x = \log 2 \qquad \text{If } M = N, \text{ then } \log M = \log N.$$
$$x \log 3 = \log 2 \qquad \log x^y = y \log x$$
$$x = \frac{\log 2}{\log 3} \qquad \text{Divide each side by log 3.}$$
$$x = \frac{0.3010}{0.4771} \qquad \text{See caution.}$$
$$= 0.6309$$

Caution

$\dfrac{\log_b x}{\log_b y}$ is not equal to

$\log_b x - \log_b y$;

$\log_b \dfrac{x}{y} = \log_b x - \log_b y$

Do You Remember?

Can you match these?

_____ 1. $\log_b xy$
_____ 2. $\log_b x^k$
_____ 3. $\log_b \dfrac{x}{y}$
_____ 4. $\log_b b$
_____ 5. $\log_b 1$
_____ 6. $\log_b(-2)$

a) Does not exist.
b) 1
c) $\log_b x + \log_b y$
d) 0
e) $k \log_b x$
f) $\log_b x - \log_b y$

Answers: 1. c 2. e 3. f 4. b 5. d 6. a

Exercise Set 11.3

Write each of the following expressions as an algebraic sum of logarithms.

1. $\log uvw$
2. $\log rst$
3. $\log_b \dfrac{xy}{z}$
4. $\log_b \dfrac{pq}{n}$
5. $\log_4 \dfrac{a^2}{b^3}$
6. $\log_6 \dfrac{k^3}{m^2}$
7. $\log_e \sqrt{\dfrac{x+y}{x-y}}$
8. $\log_e \sqrt{(x-y)^2(y-z)}$
9. $\log \left(\dfrac{xy^{2/3}}{z^2}\right)^{3/2}$
10. $\log \left(\dfrac{x^2}{x^{1/3}z^{1/2}}\right)^{1/2}$
11. $\log_e y\sqrt[5]{x}$
12. $\log_e x^3\sqrt[4]{y}$
13. $\log_b \sqrt[3]{\dfrac{x^3 y}{z^5}}$
14. $\log_b \sqrt[5]{\dfrac{x^4 y^2}{z^7}}$
15. $\log_e \dfrac{\sqrt{x^3}}{\sqrt[3]{y^4}}$
16. $\log_e \dfrac{\sqrt[3]{a^2}}{\sqrt[7]{b^5}}$
17. $\log \sqrt[4]{a^2 b^{-3/4} c^{1/3}}$
18. $\log \sqrt[3]{xy^{2/3}z^{-1/4}}$
19. $\log 2\pi\sqrt{\dfrac{L}{g}}$
20. $\log \pi \sqrt{\dfrac{2L}{R^2}}$
21. $\log \sqrt{(s-a)(s-b)(s-c)}$
22. $\log \sqrt{x^2(x-a)^3}$

Express each of the following quantities as a single logarithm with a coefficient of 1.

23. $\log a + \log b + \log c$
24. $\log x + \log y + 2 \log z$
25. $\log_e r - \log_e s + \log_e t$
26. $\log_e m + \log_e n - \log_e p$
27. $2 \log_e b + 3 \log_e c$
28. $4 \log_2 s - 2 \log_2 t$
29. $\dfrac{1}{2}\log 12 - \dfrac{1}{2}\log 9 + \dfrac{1}{3}\log 27$
30. $\dfrac{1}{4}\log 8 + \dfrac{1}{4}\log 2 + \dfrac{1}{3}\log 8$
31. $\dfrac{1}{2}(\log a + 4 \log b - 2 \log c)$
32. $\dfrac{1}{3}(\log x + 6 \log y - 3 \log z)$
33. $\dfrac{3}{2}\log_e x - 4 \log_e y$
34. $\dfrac{2}{3}\log_e a - 5 \log_e b$

Use the properties of logarithms to verify that the left side of each equation is equal to the right side.

35. $\log 2 - \log 3 + \log 5 = \log \dfrac{10}{3}$
36. $3 \log 2 - 4 \log 3 = \log \dfrac{8}{81}$
37. $\dfrac{1}{2}\log 25 - \dfrac{1}{2}\log 64 + \dfrac{2}{3}\log 27 = \log \dfrac{45}{8}$
38. $\log 5 - 1 = \log \dfrac{1}{2}$
39. $\log 10 - 2 = \log 0.1$
40. $2 \log 3 + 4 \log 2 - 3 = \log 0.144$
41. $\log 2x^3 - \log 2\sqrt{x^3} - \log \sqrt{x} = \log x$
42. $2 \log b - \log b + \log \sqrt{c} + \log \sqrt{c^5} = \log bc^3$
43. $\dfrac{2}{3}\log x + \dfrac{1}{3}\log y + \dfrac{5}{3}\log y - \dfrac{5}{3}\log x = \log \dfrac{y^2}{x}$
44. $\log \dfrac{9}{12} - \log 38 + 4 \log 2 + \log \dfrac{76}{12} + \log \dfrac{8}{4} = 2 \log 2$
45. $\log \dfrac{16}{27} - \log \dfrac{2}{3} - \log \dfrac{8}{9} = 0$
46. $\log \dfrac{5}{36} - \log \dfrac{2}{27} + \log \dfrac{8}{15} = 0$
47. $\log \dfrac{54}{35} + \log \dfrac{55}{4} - \log \dfrac{18}{70} - \log \dfrac{165}{4} = \log 2$
48. $\log \dfrac{9}{4} + \log \dfrac{6}{45} - \log \dfrac{15}{10} = -\log 5$

Solve for the indicated variable in each of the following equations.

49. $\log_2 x = y + c$; for x
50. $\log a = 2 \log b$; for a
51. $\log_e I = \log_e a - T$; for I
52. $2 \log x + 3 \log y = 4 \log z - 2$; for y
53. $\log(x + 3) = \log x + \log 3$; for x
54. $\log_2 8 + \log_2 x = 4$; for x
55. $\log_2 x + \log_2(x - 2) = 3$; for x
56. $\log_3(2x - 5) - 2 = \log_3(3x - 2)$; for x

Given $\log 2 = 0.3010$, $\log 3 = 0.4771$, and $\log 5 = 0.6990$, use the properties of logarithms to evaluate Exercises 57–70. Use a calculator and round to four decimal places as needed.

57. $\log 8$
58. $\log 30$
59. $\log 12$

60. $\log 100$
61. $\log 16\sqrt{2}$
62. $\log 27\sqrt{3}$
63. $\log 0.25$
64. $\log \sqrt[3]{6}$
65. $\log_3 2$
66. $\log_2 5$
67. $\log_5 8$
68. $\log_5 18$
69. $\log_3 16$
70. $\log_3 125$

71. Prove that $\log_b \dfrac{x}{y} = \log_b x - \log_b y$.

72. Prove that $\log_b x^k = k \log_b x$.

Write in Your Own Words

73. What are "common logarithms"? Who is considered the father of common logarithms?

74. Why do you think 10 is used as the base of common logarithms?

75. Are there other bases used for logarithms? If so, what are they?

76. State in words:

$$\log ab = \log a + \log b$$

$$\log \frac{a}{b} = \log a - \log b$$

$$\log a^k = k \log a$$

77. Is there any reason we should study logarithms now that we have calculators?

78. If you were asked to solve for x in $3^x = 7$, how would you approach it? Explain each step you would use. Could you approximate x by graphing? Could you use a calculator? Could you use logarithms?

79. In the equation $\log_b xy = \log_b x + \log_b y$, why are x, y, and b restricted to being positive numbers and $b \neq 1$?

11.4 Solving Exponential and Logarithmic Equations

OBJECTIVES

In this section we will learn to
1. solve equations involving logarithms; and
2. solve exponential equations involving logarithms.

Many of the problems in this section require the use of a calculator capable of displaying the logarithm of a number and of raising a number to a power. Although we will use the equal sign in the examples, it should be noted that results obtained in this way are only approximations.

Sometimes equations are stated in terms of logarithms, and we must use the properties of logarithms to find the solution set. The next few examples expand on the technique that was introduced in Example 8 of Section 11.3.

EXAMPLE 1 Solve: $\log(x - 2) + \log(x - 3) = \log(x + 13)$.

SOLUTION We begin by applying Theorem 11.2:

$$\log(x - 2) + \log(x - 3) = \log(x + 13)$$
$$\log[(x - 2)(x - 3)] = \log(x + 13) \qquad \log_b x + \log_b y = \log_b xy$$

Now we use the fact that if $\log M = \log N$, then $M = N$.

$$(x - 2)(x - 3) = x + 13$$
$$x^2 - 5x + 6 = x + 13$$
$$x^2 - 6x - 7 = 0$$
$$(x - 7)(x + 1) = 0$$
$$x - 7 = 0 \quad \text{or} \quad x + 1 = 0$$
$$x = 7 \quad \text{or} \quad x = -1$$

We reject -1 since $\log(x - 2) = \log(-3)$ when $x = -1$. The logarithm of a negative number does not exist. The solution set is $\{7\}$.

The number e, which was introduced in Section 11.1, is encountered so frequently in mathematics as a base of logarithms that a special symbol, **ln**, is used to designate logarithms to the base e.

DEFINITION	The symbol "ln x" is read **"the natural logarithm of x."** We write
The Natural Logarithm	$\log_e x = \ln x$

EXAMPLE 2 Solve: $\ln(x - 2) = 5$. Round the answer to three decimal places.

SOLUTION The base of the logarithm, ln, is e, so by definition of a logarithm we have

$$x - 2 = e^5$$
$$x = e^5 + 2$$

To obtain a decimal approximation for the result, a scientific calculator with an $\boxed{e^x}$ key or a $\boxed{y^x}$ key can be used. The order in which the entries are made can vary with the calculator. Several possibilities are as follows:

5 $\boxed{e^x}$ $\boxed{+}$ 2 $\boxed{=}$ 150.41316

Some graphics calculators require this order:

$\boxed{e^x}$ 5 $\boxed{+}$ 2 $\boxed{=}$ 150.41316

Calculators using RPN logic differ from either of the preceding key sequences, using the order:

5 $\boxed{\text{ENTER}}$ $\boxed{e^x}$ 2 $\boxed{+}$

The display then reads 150.41316. The owner's manual for each calculator provides instructions and should be consulted. Refer to Section 11.1 for instructions on the use of the $\boxed{y^x}$ key. The solution set is $\{150.413\}$ rounded to three decimal places.

EXAMPLE 3 Solve: $\log(x^2 - 1) = 1 + \log 2 + \log(x - 1)$.

SOLUTION We begin by simplifying the right side:

$$\log(x^2 - 1) = 1 + \log 2 + \log(x - 1) = 1 + \log[2(x - 1)]$$

Therefore

$$\log(x^2 - 1) - \log(2x - 2) = 1$$
$$\log \frac{x^2 - 1}{2x - 2} = 1$$

Now we change this last equation to exponential form.

$$\frac{x^2 - 1}{2x - 2} = 10^1 = 10$$

$$x^2 - 1 = 20x - 20 \qquad \text{The LCD is } 2x - 2.$$

$$x^2 - 20x + 19 = 0$$

$$(x - 19)(x - 1) = 0$$

$$x = 19 \quad \text{or} \quad x = 1$$

We reject 1, because $\log(x^2 - 1) = \log 0$ when $x = 1$ and $\log 0$ does not exist. The solution set is $\{19\}$.

EXAMPLE 4 Solve $3^x = 8$, rounding the answer to three decimal places.

SOLUTION Since 3 and 8 can't both be expressed as an integer power of the same base, we begin by taking the logarithm of each side.

$$\log 3^x = \log 8 \qquad \text{If } M = N, \text{ then } \log_b M = \log_b N.$$

$$x \log 3 = \log 8$$

$$x = \frac{\log 8}{\log 3}$$

To use a calculator to find $\log 8$, we enter 8 and press the key marked $\boxed{\log}$. This gives us $\log 8 = 0.903089987$. Similarly, $\log 3 = 0.477121254$. Thus

$$x = \frac{0.903089987}{0.477121254} = 1.893$$

Using the following sequence of keys will yield the result directly:

$$8 \; \boxed{\log} \; \boxed{\div} \; 3 \; \boxed{\log} \; \boxed{=} \; 1.892789261 \approx 1.893$$

The solution set is $\{1.893\}$.

EXAMPLE 5 Solve $15^{2x+1} = 20^{x+2}$, rounding the answer to three decimal places.

SOLUTION We begin by taking the logarithm of each side:

$$\log 15^{2x+1} = \log 20^{x+2} \qquad \text{If } M = N, \text{ then } \log_b M = \log_b N.$$

$$(2x + 1)\log 15 = (x + 2)\log 20 \qquad \log_b x^k = k \log_b x$$

$$2x \log 15 + \log 15 = x \log 20 + 2 \log 20 \qquad \text{Remove grouping symbols.}$$

$$2x \log 15 - x \log 20 = 2 \log 20 - \log 15 \qquad \text{Isolate the variable.}$$

$$(2 \log 15 - \log 20)x = 2 \log 20 - \log 15$$

Now we divide each side by the coefficient of x:

$$x = \frac{2 \log 20 - \log 15}{2 \log 15 - \log 20}$$

We use a calculator to find the logarithms and carry out the arithmetic calculations. If each part is calculated separately, we have

$$x = \frac{2.6021 - 1.1761}{2.3522 - 1.3010} = 1.3565$$

The solution set is {1.357}. See if you can get the final result on your calculator without doing each calculation separately.

EXAMPLE 6 A radioactive substance decays according to the formula $A(t) = A_0 e^{-0.05t}$, where t is measured in years and A_0 is the initial amount present. What is its **half-life** rounded to the nearest tenth of a year, where the half-life is the time it takes for one-half of the substance to decay?

SOLUTION If A_0 is the original amount, then $\frac{1}{2} A_0$ is the amount that remains at half-life.

$$A(t) = A_0 e^{-0.05t}$$

$$\frac{1}{2} A_0 = A_0 e^{-0.05t} \qquad \text{At the half-life, one-half of the substance remains, so } A = \frac{1}{2} A_0.$$

$$\frac{1}{2} = e^{-0.05t} \qquad \text{Divide each side by } A_0.$$

Since we are considering a power of e, we take the natural logarithm of each side. Recall that the base for ln is e, so that $\ln e = 1$.

$$\ln \frac{1}{2} = \ln e^{-0.05t}$$

$$\ln \frac{1}{2} = -0.05t \ln e$$

$$\ln 0.5 = -0.05t \qquad \ln e = 1$$

$$\frac{\ln 0.5}{-0.05} = t$$

To find ln 0.5, we enter 0.5 and press the $\boxed{\ln}$ key. This gives us $\ln 0.5 \approx 0.693147$. Thus

$$\frac{-0.693147}{-0.05} = t$$

$$13.9 = t$$

The half-life is approximately 13.9 years.

Exercise Set 11.4

Solve each of the following logarithmic equations.

1. $\log(x + 2) = 3$
2. $\log(x + 1) = 1$
3. $\log(8y + 40) = \log 1000$
4. $\log(5y - 40) = \log 100$
5. $\ln(y + 3) = 4$
6. $\ln(x - 2) = 5$
7. $\ln(2x - 5) = \ln 3$
8. $\ln(6y + 7) = \ln 8$
9. $\ln t + \ln(2t - 5) = \ln 3$
10. $\ln(k + 2) + \ln(k - 1) = \ln 6$
11. $\log(y + 3) - \log(2y - 5) = -3$
12. $\log(4x + 8) - 2 = -\log 3x$
13. $\log_5 x - \log_5(3x + 2) = -\log_5 4$
14. $\log_2(a - 1) + \log_2 3 = \log_2(a + 5)$

15. $\log y = 1 + \log(y - 1)$
16. $\log(2p^2 + 3) = 1 + \log(p + 1)$
17. $\log_3(x - 3) + \log_3 x = \log_3 4$
18. $\log 100y + \log(y - 9) = \log 1000$
19. $\log s - \log 8 = \log 10$
20. $\log_8 x + \log_8 5 = \log_8 64$
21. $\log_3 3^x = \log_3 27$
22. $\log_2 2^p = \log_2 16$
23. $\log(u + 4)^2 - \log(18u + 7) = \log 1$
24. $\log(m + 6)^2 - \log(9m + 14) = \log 2$

Solve each of the following exponential equations. Round your answers to four decimal places if needed. If no real-number solution exists, write $\varnothing$.

25. $4^x = 7$
26. $3^y = 25$
27. $5^{2d} = 4$
28. $10^{3v-2} = 37$
29. $3^{2+p} = 5^p$
30. $3^{w+1} = 4^{w-1}$
31. $10^{3a} = 57$
32. $10^{2k} = 65$
33. $e^{3x} = 21$
34. $e^{x+2} = 15$
35. $e^{0.32x} = 632$
36. $e^{-1.4x} = 13$
37. $500e^{-0.12x} = 123$
38. $100e^{0.25x} = 48$
39. $9^x = 7^{x+2}$
40. $20^{2x-1} = 30^{3x-2}$
41. $1000^{x^2+2x} = 100^{x-2}$
42. $7^{a^2+2a+1} = 1$
43. $7^{\log x} = 12$
44. $3^{\log x} = 5$
45. $10^{3 \log x} = 8$
46. $e^{5 \ln x} = 6$
47. $e^{x^2} = 6$
48. $e^{x^2} = e^{-x}$
49. $2^{x^2} = 4^{x-2}$
50. $5^{x^2+x} = 25^{x+1}$
51. $5^{a^2+a} = 25$
52. $7^{t^2+t} = 1$

Solve each of the following applied problems. Round your answers to four decimal places if needed.

53. The amount of radioactive substance present after t days is given by the formula $A(t) = 400e^{-kt}$. If $A(3) = 430$, find k.

54. The amount of radioactive substance after t weeks is given by the formula $A(t) = A_0 e^{-0.04t}$, where A_0 is the initial amount. How many weeks does it take for $A(t)$ to decay to one-half its initial amount?

55. The number of items per day that a worker can produce after t full days of practice is given by the formula $P(t) = 400(1 - e^{-t})$. How many full days of practice will it take until this worker can produce at least 300 items/day?

56. The population P of a city t years from now is given by the formula $P(t) = 30{,}000e^{0.05t}$. In how many years will the population be 70,000?

57. If you invest $2000 in the bank at 8.5% interest compounded continuously, how long will it take for the $2000 to double? Round your answer to the nearest whole year.

58. The concentration of a pain medicine in a person's system decreases according to the formula $f(t) = 3e^{-0.12t}$, where t is in hours. Find how long it takes the concentration to be one-half of its original amount.

REVIEW EXERCISES

59. Graph: $f(x) = 5^x$.
60. Graph: $f(x) = \left(\dfrac{1}{3}\right)^x$.
61. Graph: $f(x) = \log x$.
62. Graph: $f(x) = \ln x$.
63. Graph: $f(x) = \log_3 5$.
64. Graph: $f(x) = \log_2 5$.

For Extra Thought

65. The cube in the picture has one-inch edges. How long is the diagonal AB?

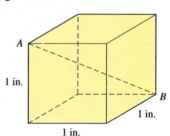

66. A right circular cone has a cross section that is an isosceles right triangle with a base of $3\sqrt{2}$ cm. What is the cone's surface area? Round your answer to two decimal places.

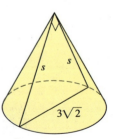

11.5 More Applications

OBJECTIVE In this section we will learn to use logarithms to solve applied problems.

Using logarithms to solve applied problems requires knowing both how to find the logarithm of a number and how to find a number if its logarithm is known. For example, using a calculator,

$$\log 18 = 1.2552725 \quad \text{to seven decimal places}$$

In exponential form, this logarithm is

$$10^{1.2552725} = 18$$

If we had been asked to find x, where

$$\log x = 1.2552725$$

then

$$x = 10^{1.2552725}$$

Using a calculator with a $\boxed{10^x}$ key, we proceed as follows:

Enter 1.2552725.
Press the $\boxed{10^x}$ key.
The result is $17.99999979 \approx 18$.

For calculators with a $\boxed{y^x}$ key, the steps are as follows:

Enter 10.
Press $\boxed{y^x}$.
Enter 1.2552725.
Press $\boxed{=}$.
The result is $17.99999979 \approx 18$.

EXAMPLE 1 Find the value of the variable in each of the following:

(a) $\log x = -1.235$ (b) $\log x = 3.714$ (c) $\ln x = 0.3041$

SOLUTION (a) First we write the equation in exponential form:

$$10^{-1.235} = x$$

We then enter -1.235 in the calculator and press the $\boxed{10^x}$ key. The result is 0.05821032.

(b) We enter 3.714 and press the $\boxed{10^x}$ key. The result is 5176.0683.

(c) The base of ln is e, so in exponential form the equation becomes $e^{0.3041} = x$. We enter 0.3041 and press the $\boxed{e^x}$ key; the result is 1.13554046.
The $\boxed{y^x}$ key can be used in place of the $\boxed{e^x}$ key by letting $y = 2.71828$.

EXAMPLE 2 An amount p invested at $r\%$ compounded annually for t years will have a final value of $A(t) = p(1 + r)^t$. How many years will it take for p dollars invested at 8% to double?

SOLUTION When the investment doubles, $A(t) = 2p$. Thus,

$$2p = p(1 + 0.08)^t$$
$$2 = (1.08)^t \quad \text{Simplify by dividing each side by } p.$$
$$\log 2 = \log(1.08)^t$$
$$\log 2 = t \log(1.08) \quad \log_b x^k = k \log_b x$$
$$\frac{\log 2}{\log (1.08)} = t \quad \text{Isolate the } t.$$

We use a calculator to find t with the following keystrokes:

2 [log] [÷] 1.08 [log] [=] 9.00646 ≈ 9.01

It will take approximately 9 years.

EXAMPLE 3 How much money must be invested today at 9% compounded annually to have a value of $10,000 in 8 years?

SOLUTION We need to find p when $A(t) = \$10{,}000$, $r = 0.09$, and $t = 8$.

$$10{,}000 = p(1 + 0.09)^8 = p(1.09)^8$$
$$p = \frac{10{,}000}{(1.09)^8} = \$5018.66 \quad \text{Use a calculator.}$$

Although it is not as efficient in this case, we can also determine the value of p by using logarithms.

$$\log p = \log \frac{10{,}000}{(1.09)^8} \quad \text{If } M = N, \text{ then } \log_b M = \log_b N.$$
$$\log p = \log 10{,}000 - 8 \log 1.09 \quad \text{Theorem 11.2}$$
$$\log p = 3.700588016 \quad \text{Use a calculator.}$$
$$p = 10^{3.700588016}$$
$$= \$5018.66$$

So if $5018.69 is invested today at 9% interest compounded annually, it will be worth approximately $10,000 in 8 years.

EXAMPLE 4 Certain quantities involving optics are governed by the equation

$$L = a + b \log D$$

Find D if $L = 14$, $a = 7.3$, and $b = 3.5$.

SOLUTION Substitution yields

$$14 = 7.3 + 3.5 \log D$$
$$14 - 7.3 = 3.5 \log D$$
$$\frac{6.7}{3.5} = \log D$$
$$1.91429 \approx \log D$$
$$D \approx 10^{1.91429} \approx 82.1$$

EXAMPLE 5 A bacterial culture grows according to the law $P = P_0 e^{kt}$, where P_0 is the initial population. If there were 1000 organisms present initially and 3500 present when $t = 3$ hours, determine k and find how many organisms will be present when $t = 5$ hours.

SOLUTION To find k, we substitute 3500 for P, 1000 for P_0, and 3 for t:

$3500 = 1000 e^{3k}$	
$3.5 = e^{3k}$	Divide each side by 1000.
$\ln 3.5 = \ln e^{3k}$	If $M = N$, then $\log_b M = \log_b N$.
$\ln 3.5 = 3k \ln e$	$\ln x^k = k \ln x$
$1.25276 \approx 3k$	$\ln e = 1$
$0.41759 \approx k$	

The equation for growth becomes $P = P_0 e^{0.41759 t}$.

When $t = 5$ and $P_0 = 1000$, we have

$$P = 1000 e^{0.41759(5)}$$

Evaluating this with a calculator with an $\boxed{e^x}$ key, we get

$$P = 8068.346773 \approx 8068$$

Exercise Set 11.5

Use logarithms (as needed) to solve each of the following applied problems. Round your answers to three decimal places if needed.

1. A body falling in a vacuum falls s ft in t sec,

    where $t = \sqrt{\dfrac{2s}{g}}$ and $g = 32$ ft/sec^2

    is the gravitational constant. Find t when $s = 1653$ ft.

2. The period T of a pendulum of length L is given by

    $T = 2\pi \sqrt{\dfrac{L}{g}},$ where $g = 981.0$ cm/sec^2.

    If $L = 281.3$ cm, find T. Use $\pi = 3.1416$. See the figure.

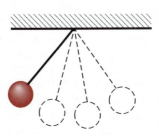

3. A person without a parachute falls with a velocity of v ft/sec, with $v = w(1 - e^{-32t/w})$, where w is the weight of the person in pounds and t is the falling time in seconds. Find the velocity of a 200-lb person who has been falling for 10 sec.

4. The current I (in amps) of an inductive electrical circuit is given by the formula $I = \dfrac{E}{R}(1 - e^{-Rt/L})$, where E is the voltage, R is the resistance, t is the time in seconds, and L is the inductance. If $E = 500$, $R = 50$, and $L = 0.5$, how many seconds will it take for the current to equal 4.8 amps?

5. A radioactive substance decays according to the formula $A(t) = A_0 e^{-0.00041t}$, where t is in years. How long will it take for the amount $A(t)$ to be equal to one-half the initial amount A_0?

6. A colony of bacteria with an initial size of 1000 increases its population by 10% each day. After x days its population is approximated by the formula $P(x) = 1000 e^{0.1x}$. How long will it take for the colony to double in size?

7. If an initial investment of p dollars earns interest at an interest rate r compounded annually, the value after t years is given by $A(t) = p(1 + r)^t$. At an annual rate of 10%, how long will it take an initial investment of \$3000 to accumulate to an amount of at least \$6000?

8. A radioactive substance has a half-life of 138 days. How long will it take for 90% of a sample of this substance to decay, given that $A(t) = A_0 e^{-0.005t}$. (See Exercise 5.)

9. Radioactive carbon, ^{14}C, is used by geologists to estimate the age of fossils. If an organism contains an amount A_0 of ^{14}C when it dies, then the amount A of ^{14}C remaining t years later is given by $A(t) = A_0 e^{-0.000124t}$. If a fossil contains 60% of the amount of ^{14}C that was present when the organism died, what is the approximate age of the fossil?

10. The growth equation of a certain city is approximated by $A(t) = A_0 e^{0.0282t}$, where the initial population A_0 in 1981 was 935,000 ($t = 0$). Calculate the population for 1998. *Note: t is in years.*

11. The number of bacteria in a colony grows from 100 to 400 in 16 hr. Using $A(t) = A_0 e^{kt}$, where A_0 is the initial population and A is the number in t hours, determine k and find the number of bacteria present after 20 hr.

12. An initial amount A_0 of money invested at 8% compounded continuously grows to an amount A according to the formula $A(t) = A_0 e^{0.08t}$, where t is in years. How long will it take for the initial investment to double?

13. The expected population P of a city with initial population P_0 is given by the formula $P(t) = P_0 3^{kt}$, where t is in years. If the initial population is 16,000 at $t = 0$ and 20,000 at $t = 10$, (a) determine k; (b) find the expected population for $P(40)$.

14. The atmospheric pressure P varies approximately according to the formula $P(d) = 30(10)^{-0.09d}$, where d is the distance in miles above sea level. Find the atmospheric pressure at the top of Mount Whitney, which is 14,496 ft above sea level.

15. Use the formula in Exercise 14 to determine the atmospheric pressure at the top of Mount Everest, 29,028 ft above sea level.

16. The difference in the intensity level of two sounds with intensities I and i is defined by $L = 10 \log \left(\dfrac{I}{i}\right)$ decibels. Find the intensity level L in decibels (dB) of the sound from an electric pump if I is 175.6 times greater than i.

17. Find the intensity level L (in decibels) of each of the following sounds if
$$L = 10 \log \left(\dfrac{I}{i}\right), \quad \text{where } i = 10^{-16} \text{ watts/cm}^2$$
   (a) A whisper, at $I = 10^{-13}$ watts/cm^2
   (b) Normal conversation, at $I = 3.16 \times 10^{-10}$ watts/cm^2
   (c) A loud motorcycle, at $I = 10^{-8}$ watts/cm^2

18. The voltage V across a certain capacitor in a circuit is given by $V(t) = 80(1 - e^{-0.04t})$, where t is in seconds. How much time t is required for $V(t)$ to equal 50 volts?

19. When light passes through a relatively clear body of water, the light intensity I is reduced according to the formula $I(d) = I_0 e^{-kd}$, where I_0 is the light intensity at the surface, d is the distance in feet below the surface of the water, and k is called the *coefficient of extinction*. Find the depth in Lake Nacimiento, which has $k = 0.0485$, at which the light is reduced to 1% of that at the surface.

20. The learning curve in typing is given by the formula $N(n) = 80(1 - e^{-0.07n})$ where N is the number of words typed per minute after n weeks of study. About how many weeks will it take an individual to learn to type 50 words per minute?

21. The limiting magnitude L of an optical telescope with lens diameter D inches is given by the formula $L(D) = 8.8 + 5.1 \log D$. Find the diameter of a telescope with a limiting magnitude of 20.6.

22. An investment of \$2000 is made in a one-year account that pays 7.25% interest compounded continuously. What is the effective rate of yield for this investment (rounded to two decimal places)? Use $A = pe^{rt}$.

23. An investment of \$1300 is made in a one-year account that pays 8.8% interest compounded continuously. What is the effective rate of yield for this investment (rounded to two decimal places)? Use $A = pe^{rt}$.

24. A new car that originally costs $20,120 has a depreciated value of $16,800 after one year. Find the value of the car when it is two years old, using the formula $V(t) = V_0 e^{kt}$. Here V is the value at any time, V_0 is the initial value, and t is the time in years.

25. The sales N (in units) of a hair dryer, after spending d dollars in advertising, is given by the formula $N(d) = 100(1 + e^{0.0012d})$.

    (a) Find the amount spent on advertising if 3000 units are sold.

    (b) How many units will be sold if the company spends $800 on advertising (rounded to the nearest whole number)?

26. A new motorcycle sells for $6170; one year later it has a depreciated value of $5000. Find the value of the motorcycle when it is three years old by using the formula $V(t) = V_0 e^{kt}$.

27. A radioactive isotope has a half-life of 18,400 yr. If originally there are 100 g of this isotope, how much will remain after 9000 yr (rounded to four decimal places)? Use $A(t) = A_0 e^{-kt}$.

28. A $1600 investment grows to $1800 when compounded continuously for two years. Find the rate of interest correct to two decimal places. How long will it take for the original investment to double at this interest rate? Round your answer to two decimal places.

Summary

11.1 Exponential Functions

An **exponential function** takes the form $f(x) = b^x$, where $b > 0$ and $b \neq 1$. The **domain** of an exponential function is the set of real numbers and the **range** is the set of **positive** real numbers. The graph of an exponential function of the form $f(x) = b^x$ always crosses the y-axis where $y = 1$. If both sides of an exponential equation can be expressed in terms of the same base,

$$b^m = b^n,$$

the solution can be found by setting $m = n$.

11.2 Logarithmic Functions

The **logarithmic function** $y = \log_b x$ is equivalent to the exponential equation $b^y = x$. The logarithmic function $\log_b x = y$ is the inverse of the exponential function $y = b^x$. From the definition it can be seen that *a logarithm is an exponent.* Some useful observations concerning logarithms to the base b, $b \neq 1$, $b > 0$, are:

1. The logarithm of any positive number b to the base b is 1:

   $$\log_b b = 1$$

2. The logarithm of 1 to any base is 0:

   $$\log_b 1 = 0$$

3. The logarithm of a negative number or zero does not exist.

There are two forms of logarithms in common use: the natural logarithm, with base e, and the common logarithm, with base 10.

11.3 Properties of Logarithms

Historically, logarithms were used to carry out complex numerical computations. Computers and calculators have taken over this role, but the properties involved are still very useful in higher mathematics. The basic properties are:

1. $\log_b xy = \log_b x + \log_b y$

2. $\log_b \dfrac{x}{y} = \log_b x - \log_b y$

3. $\log_b x^k = k \log_b x$

Logarithms to the base 10 are often written without indicating the base:

$$\log_{10} x = \log x$$

The following relationship exists between numbers and their logarithms:

1. If $\log_b M = \log_b N$ then $M = N$.
2. If $M = N$ then $\log_b M = \log_b N$.

11.4 Solving Exponential and Logarithmic Equations

The expression $\ln x$ is read, "**the natural logarithm of x**" and has a base of e. Solving a logarithmic equation requires using the properties of logarithms. Solutions to logarithmic equations should always be checked to ensure that none requires taking the logarithm of a negative number or zero.

Cooperative Exercise

Dean and Mary have chosen a house they would like to purchase at a price of $78,000.00. They are looking at three loan options. Based upon their income, they can make monthly loan payments of no more than $700.00 per month.

LOAN A
City Bank will provide a 20-year loan at 8.75% annual interest rate. They require a 10% down payment and three points to make the loan.

1. How much will the down payment be?
2. What will they pay in points?
3. What will their monthly payment be?
4. How much interest will they have paid at the end of 20 years?
5. How much will they have paid (total) for the house in 20 years?

LOAN B
Community Savings and Loan will provide a 20-year loan at 7.9% for the first 5 years (at payments of $550.00 per month), 8.6% for the next 5 years (at payments of $600 per month), and 9.3% for the last 10 years (payments to be calculated). They require a 7% down payment and one and one-half points to make the loan.

1. How much will the down payment be?
2. What will they pay in points?
3. What will their monthly payments be for the last ten years?
4. How much interest will they have paid at the end of 20 years?
5. How much will they have paid (total) for the house in 20 years?

LOAN C
The local Credit Union will provide a loan at 8% (but only for 15 years). They require a 10% down payment and one and one-half points to make the loan.

1. How much will the down payment be?
2. What will they pay in points?
3. What will their monthly payments be?
4. How much interest will they have paid at the end of 15 years?
5. How much will they have paid (total) for the house in 15 years?

Use the information about loans A, B, and C to answer each of the following questions.

1. What are some positive and negative things about loan A?
2. What are some positive and negative things about loan B?
3. What are some positive and negative things about loan C?
4. Which of the three loans would you consider to be the best choice for Dean and Mary? Give reasons for your choice.

Review Exercises

Solve or graph as indicated.

1. Solve: $2^x = 16$.
2. Solve: $5^{2x} = 125$.
3. Graph: $f(x) = 2^x$.
4. Graph: $f(x) = 2^{|x|}$.
5. Graph: $f(x) = \left(\dfrac{1}{4}\right)^x$
6. Graph: $f(x) = 3^{-x}$.
7. Solve: $7^{3x-2} = 343$.
8. Solve: $2^{-5x+1} = 128$.

Write each equation in logarithmic form.

9. $2^3 = 8$
10. $3^5 = 243$
11. $\left(\dfrac{2}{3}\right)^2 = \dfrac{4}{9}$
12. $(0.12)^2 = 0.0144$

Write each equation in exponential form.

13. $\log_7 49 = 2$
14. $\log_8 1 = 0$
15. $\log_{1/2} 64 = -6$
16. $\log_{1/6} 36 = -2$

Solve for the variable in each of the following equations.

17. $\log_5 x = 5$
18. $\log_{1/2} 4 = k$
19. $\log_{3/4} c = -3$
20. $\log_9 \dfrac{1}{9} = w$

Express each of the following expressions as a logarithmic sum or difference.

21. $\log xy\sqrt{z}$
22. $\log \dfrac{x^{1/3} y^{3/4}}{z^{1/2}}$
23. $\log_b \dfrac{a^4}{b^3}$
24. $\log_b \dfrac{a^{1/3}}{b^{-3/4}}$

Express each of the following quantities as a single logarithm with a coefficient of 1.

25. $2 \log 3 - 5 \log 2 + \log 6$
26. $\log_b x - \log_b y + 3 \log_b z$
27. $\dfrac{1}{2} \log 16 + \log 9 - \dfrac{1}{3} \log 125$
28. $\dfrac{1}{2}(\log x - 5 \log y + 2 \log z)$

Solve each of the following equations. Round your answer to four decimal places if needed.

29. $\log 6x - \log(2 + x) = 1$
30. $\log_6 x + \log_6 5 = \log_6 30$
31. $7^x = 4$
32. $10^{3x} = 25$
33. $7^x = 5^{x+1}$
34. $e^{x-3} = 17$

Solve each of the following applied problems.

35. The amount A of radioactive substance present after t days is given by $A(t) = 600e^{-kt}$. If $A(8) = 300$, find k. Find the amount present after 16 days.

36. The intensity of sound is given by
$$L(I) = 10 \log \left[\dfrac{I}{10^{-16}}\right]$$
where I is the power of sound in watts/cm². Find $L(2.82 \times 10^{-8})$.

37. The amount of money A accumulated at compound interest is given by $A = p\left(1 + \dfrac{r}{n}\right)^{nt}$, where n is the number of times the money is compounded per year, t is the number of years, r is the annual interest rate, and p is the initial amount invested. Find A if $p = \$3000$, $r = 9\%$, $n = 12$, and $t = 3$ yr.

38. The force of the wind on the side of a house trailer varies jointly with the area of the side and the square of the velocity of the wind. There is a force of 2 lb/ft² when the velocity of the wind is 20 mph and the area is 240 ft. Find the force of the wind with a velocity of 40 mph on the side of a trailer with the same area.

39. The amount A of money accumulated with continuously compounded interest is given by $A = pe^{rt}$. Find the amount A if $p = \$2000$, $r = 7\%$, and $t = 7$ yr.

40. The force with which the earth attracts an object that is above its surface varies inversely with the square of the distance of the object from the center of the earth. An object on the surface of the earth (3960 mi from the center) is attracted with a force of 180 lb. What is the force of attraction of an object that is 4000 mi from the center of the earth?

Chapter Test

1. Graph: $f(x) = \left(\frac{1}{3}\right)^x$.

2. Graph: $f(x) = \left(\frac{1}{2}\right)^{-x}$.

3. Graph: $f(x) = 3^{2x}$.

4. Solve: $2^x = 32$.

5. Solve: $5^{2x-1} = 25$.

6. Write $2^7 = 128$ in logarithmic form.

7. Write $\log_3 x = y$ in exponential form.

8. Express $\log x^2 y \sqrt[3]{z}$ as an algebraic sum of logarithms.

9. Express $2 \log 5 + 3 \log 2 - \log 50$ as a single logarithm with a coefficient of 1.

Solve Exercises 10–20, rounding your answers to two decimal places as needed.

10. $\log x + \log(x - 3) = \log 4$

11. $5^x = 4$

12. $e^x = 15$

13. $2^{3x} = 5^{x-1}$

14. $\log 5x + 2 = 3 \log 10$

15. Find the amount of money you will have if you invest $1000 compounded continuously at 9% for 6 years.

16. When an annuity is paid out, the amount of money that remains is given by $A(t) = A_0 e^{-0.05t}$. Find how long it takes for one-half of the money to be paid out, where t is in years.

17. The atmospheric pressure P varies according to the formula $P(d) = 30(10)^{-0.09d}$, where d is the distance in miles above sea level. Find the atmospheric pressure on an airplane flying at an altitude of 31,680 ft above sea level.

18. The population of a city grows from 12,000 to 18,000 in 10 years. Use $P(t) = P_0 e^{kt}$ to find how long it takes (rounded to the nearest year) for the population to double.

19. How much money will be in an account at the end of four years if $1500 is invested at 8.5% interest compounded quarterly?

20. The output of the radioactive power supply for a certain machine is given by $f(t) = 40e^{-0.003t}$, where t is in hours and $f(t)$ is in watts. What will be the output of the radioactive power supply at the end of 10 hr (rounded to three decimal places)?

CHAPTER 12
Sequences, Series, and the Binomial Theorem

12.1 Sequences, Series, and Summation Notation
12.2 Arithmetic Sequences and Series
12.3 Geometric Sequences and Series
12.4 The Binomial Theorem

Application

A motorcycle loses 10% of its value each year. Write a sequence that shows its value at the end of each of the first three years if its original cost was $3000. What is the general term of the sequence?

See Section 12.1, Example 11.

I have always studied mathematics because I enjoy it, not to become a well known person. I find real joy in reading about what others have already discovered and think how I would have approached it. Sometimes I cannot think of another or better way and so rejoice in what they have already done. If I am lucky enough to be able to add something to their work, I give them all of the credit, believing they would have done even better had they been in my place.
— MARQUIS PIERRE SIMON DE LAPLACE (1749–1827)

12.1 Sequences, Series, and Summation Notation

OBJECTIVES

In this section we will learn to
1. recognize a sequence;
2. differentiate between a sequence and a series; and
3. use summation notation.

The sequences and series that we will introduce in this chapter are a small part of an extensive branch of mathematics. They have many applications in finance and engineering, as well as other fields of science. Logarithms of numbers can be found by using a series. We begin with the definition of a **sequence**.

DEFINITION

Sequence

A **sequence** is a function with domain equal to the set of positive integers. The numbers that make up a sequence are called its **terms**. When a sequence is written in the form

$$a_1, a_2, a_3, a_4, \ldots, a_n, \ldots$$

the *subscript* indicates where the term appears in the sequence. For example, a_1 is the first term, a_3 is the third term, and a_n is the nth term, or **general term**.

For example, the sequence 2, 4, 6, 8, ... is defined by the function

$$f(x) = 2x,$$

since $f(1) = 2 \cdot 1 = 2$, $f(2) = 2 \cdot 2 = 4$, $f(3) = 2 \cdot 3 = 6$, $f(4) = 2 \cdot 4 = 8$, and so on. Similarly, the sequence 1, 4, 9, 16, ... is defined by the function $g(x) = x^2$.

A sequence can be generated by a formula that describes the nth term as a function of n.

EXAMPLE 1 What are the terms of the sequence with nth term $a_n = 3n$?

SOLUTION Since the domain is the set of positive integers, we let $n = 1, 2, 3, 4, \ldots$.

When $n = 1$, $a_1 = 3 \cdot 1 = 3$.

When $n = 2$, $a_2 = 3 \cdot 2 = 6$.
When $n = 3$, $a_3 = 3 \cdot 3 = 9$.

The terms of the sequence are therefore 3, 6, 9, 12, ..., $3n$, ...

EXAMPLE 2 What are the terms of the sequence with nth term $a_n = 5 + 2n$?

SOLUTION When $n = 1$, $a_1 = 5 + 2 \cdot 1 = 5 + 2 = 7$.
When $n = 2$, $a_2 = 5 + 2 \cdot 2 = 9$.
When $n = 3$, $a_3 = 5 + 2 \cdot 3 = 11$.

The terms of the sequence are 7, 9, 11, 13, ..., $5 + 2n$, ...

Sequences such as those in Examples 1 and 2 are said to be **infinite,** since no last value of n is specified. The sequence with nth term $a_n = 2n - 1$ for $n = 1, 2, 3, 4$ is a **finite** sequence, since it is limited to the following four terms:

$a_1 = 2 \cdot 1 - 1 = 1$
$a_2 = 2 \cdot 2 - 1 = 3$
$a_3 = 2 \cdot 3 - 1 = 5$
$a_4 = 2 \cdot 4 - 1 = 7$

EXAMPLE 3 Find the first four terms of the sequences with nth terms **(a)** $a_n = \dfrac{n+1}{n+2}$ and **(b)** $a_n = x^{2n}$.

SOLUTION **(a)** $a_1 = \dfrac{1+1}{1+2} = \dfrac{2}{3}$ $n = 1$

$a_2 = \dfrac{2+1}{2+2} = \dfrac{3}{4}$ $n = 2$

$a_3 = \dfrac{3+1}{3+2} = \dfrac{4}{5}$ $n = 3$

$a_4 = \dfrac{4+1}{4+2} = \dfrac{5}{6}$ $n = 4$

The first four terms are $\dfrac{2}{3}, \dfrac{3}{4}, \dfrac{4}{5},$ and $\dfrac{5}{6}$.

(b) $a_1 = x^{2 \cdot 1} = x^2$
$a_2 = x^{2 \cdot 2} = x^4$
$a_3 = x^{2 \cdot 3} = x^6$
$a_4 = x^{2 \cdot 4} = x^8$

The first four terms are x^2, x^4, x^6, and x^8.

EXAMPLE 4 Find the general term of the following sequences:

(a) 3, 5, 7, 9, ...   (b) $-\dfrac{1}{2}, \dfrac{3}{4}, -\dfrac{5}{8}, \dfrac{7}{16}, \ldots$

SOLUTION (a) The sequence 3, 5, 7, 9, ... is the set of odd integers beginning with 3. Since the domain is $\{1, 2, 3, 4, \ldots\}$, we must express the sequence in terms of values of $n \in \{1, 2, 3, 4, \ldots\}$. This can involve a certain amount of trial and error. Let's try $n + 2$:

TRIAL	n	TERM	RESULT	CONCLUSION
$n + 2$	1	$1 + 2$	3	Right
	2	$2 + 2$	4	Wrong

The second term must be 5. An odd integer is 1 more or less than an even integer. Let's try again with that in mind:

TRIAL	n	TERM	RESULT	CONCLUSION
$2n + 1$	1	$2 \cdot 1 + 1$	3	Right
	2	$2 \cdot 2 + 1$	5	Right

An inspection for $n = 3, 4$, and 5 confirms that $a_n = 2n + 1$.

(b) First we notice that the signs of the terms alternate from negative to positive. Suppose we represent the sign for any term as $(-1)^n$. When n is an odd integer, $(-1)^n$ is negative, and when n is an even integer, $(-1)^n$ is positive. The numerators of the fractions are the odd integers beginning with 1. Since any odd integer is 1 less than the next even integer, we can represent the numerators as $2n - 1$. Since the denominators are powers of 2, we can represent them as 2^n. Thus we can write the general term as

$$a_n = (-1)^n \dfrac{2n - 1}{2^n}$$

EXAMPLE 5 A sequence has a first term of 9, and each term that follows is 2 greater than the previous term. Find the first eight terms.

SOLUTION The first term is 9 so the second term must be 11, since each term increases by 2. Now, since the second term is 11, the third term is 2 greater, or 13. The remaining terms will each increase by 2; the first eight terms are

9, 11, 13, 15, 17, 19, 21, 23

We say that a sequence like the one in Example 5 is defined **recursively.** Many sequences are described in terms of a **recursion formula,** which relates any term a_{n+1} to the term that precedes it, a_n. Such a definition is called a **recursive definition.** For example, if

$$a_{n+1} = a_n + 2$$

then

$$a_2 = a_1 + 2, \quad a_3 = a_2 + 2, \quad a_4 = a_3 + 2, \ldots, \quad a_{26} = a_{25} + 2$$

To find the terms of a sequence using a recursion formula, at least one term of the sequence must be known.

EXAMPLE 6 Find the first five terms of the sequence in which $a_1 = 4$ and $a_{n+1} = a_n + 3$.

SOLUTION According to the formula, each term is 3 greater than the preceding term, so

$$a_1 = 4$$
$$a_2 = a_1 + 3 = 4 + 3 = 7$$
$$a_3 = a_2 + 3 = 7 + 3 = 10$$
$$a_4 = a_3 + 3 = 10 + 3 = 13$$
$$a_5 = a_4 + 3 = 13 + 3 = 16$$

The first five terms are 4, 7, 10, 13, and 16.

When the terms of a sequence are added, the result is called a *series.* For example, 2, 5, 8, 11, ... is a *sequence* of terms, while their sum, $2 + 5 + 8 + 11 + \ldots$, is a **series.**

DEFINITION

Series

A **series** is the indicated sum of a finite or infinite sequence of ordered terms.

A series can be written in compact form called **sigma** or **summation notation.** We use the Greek letter Σ (sigma) to indicate addition.

DEFINITION

Sigma or Summation Notation

A sum of terms $a_1, a_2, \ldots, a_n$ can be notated using **sigma** or **summation notation**:

$$a_1 + a_2 + a_3 + a_4 + \cdots + a_n = \sum_{i=1}^{n} a_i$$

The subscript i that ranges from 1 to n is called the **index of summation.**

EXAMPLE 7 Evaluate the series $\sum_{i=1}^{5} 2^i$.

SOLUTION The series is obtained by replacing i successively with the positive integers from 1 to 5, as indicated by the index of summation, and adding the resulting terms.

$$\sum_{i=1}^{5} 2^i = 2^1 + 2^2 + 2^3 + 2^4 + 2^5$$
$$= 2 + 4 + 8 + 16 + 32$$
$$= 62$$

NOTE An index of summation can begin with any integer.

EXAMPLE 8 Evaluate $\sum_{i=3}^{5} (i^2 - 2)$.

SOLUTION
$$\sum_{i=3}^{5} (i^2 - 2) = (3^2 - 2) + (4^2 - 2) + (5^2 - 2)$$
$$= 7 + 14 + 23 = 44$$

EXAMPLE 9 Write the series $\frac{1}{3} + \frac{3}{5} + \frac{5}{7} + \frac{7}{9}$ using summation notation.

SOLUTION The numerators are the first four odd integers beginning with 1, while the denominators are the four odd integers beginning with 3. Recalling that an odd integer is 1 more or less than an even integer and an even integer takes the form $2i$, where i is an integer, we can describe the numerators as $2i - 1$ and the denominators as $2i + 1$. That makes the general term of the sequence $\frac{2i - 1}{2i + 1}$. Thus,

$$\frac{1}{3} + \frac{3}{5} + \frac{5}{7} + \frac{7}{9} = \sum_{i=1}^{4} \frac{2i - 1}{2i + 1}$$

EXAMPLE 10 Evaluate $\sum_{i=0}^{2} (-1)^i (4i - 1)$.

SOLUTION
$$\sum_{i=0}^{2} (-1)^i (4i - 1) = (-1)^0 (4 \cdot 0 - 1) + (-1)^1 (4 \cdot 1 - 1) + (-1)^2 (4 \cdot 2 - 1)$$
$$= (1)(-1) + (-1)(3) + (1)(7)$$
$$= -1 - 3 + 7$$
$$= 3$$

EXAMPLE 11 A motorcycle loses 10% of its value each year. Write a sequence that shows its value at the end of each of the first three years if its original cost was $3000. What is the general term of the sequence?

SOLUTION At the end of the first year, its value is
$$a_1 = 3000 - 0.10(3000) = \$2700$$
At the end of the second year, its value is
$$a_2 = 2700 - 0.10(2700) = \$2430$$
At the end of the third year, its value is
$$a_3 = 2430 - 0.10(2430) = \$2187$$
The first three terms of the sequence are
$$\$2700, \$2430, \text{ and } \$2187$$
We will now find the general term. If a_n is the motorcycle's value at the end of n years and if a_0 was its original value, then
$$a_1 = a_0 - 0.10a_0 = a_0(1 - 0.10) = 0.90a_0 \quad a_0 \text{ is a common factor.}$$
$$a_2 = 0.90a_0 - 0.10(0.90a_0) = 0.90a_0(1 - 0.10) = 0.90a_0(0.90) = (0.90)^2 a_0$$
$$a_3 = (0.90)^2 a_0 - 0.10(0.90)^2 a_0 = (0.90)^2 a_0(1 - 0.10) = (0.90)^2 a_0(0.90)$$
$$= (0.90)^3 a_0$$

In other words the value of the motorcycle at the end of n years is a_n and is given by the formula
$$a_n = (0.90)^n a_0 = (0.90)^n 3000$$

Do You Remember?

Can you match these?

___ 1. The domain of a sequence
___ 2. The sum of a finite or an infinite sequence
___ 3. $\sum_{i=1}^{n} a_i$
___ 4. $a_{n+1} = a_n + b$
___ 5. Σ

a) A recursive formula
b) Sigma
c) The positive integers
d) A series
e) Summation notation

Answers: 1. c 2. d 3. e 4. a 5. b

Exercise Set 12.1

Find the first five terms of each sequence with the given general term.

1. $a_n = n$
2. $a_n = n + 1$
3. $a_n = 2n + 3$
4. $a_n = 2n$
5. $a_n = n^3$
6. $a_n = n^2$
7. $a_n = \dfrac{n-1}{n+1}$
8. $a_n = \dfrac{n+3}{n+2}$

9. $a_n = (-2)^{n+1}$
10. $a_n = (-3)^{n-1}$
11. $a_n = \dfrac{n}{n^2 + 1}$
12. $a_n = \dfrac{n^2 - 1}{n}$
13. $a_n = 3^{-n}$
14. $a_n = 2^{-n}$
15. $a_n = n^2 + n^{-2}$
16. $a_n = n^2 + n^{-1}$
17. $a_n = 3n(-x)^n$
18. $a_n = 2n(-x)^{n+1}$
19. $a_n = (nx)^{n+1}$
20. $a_n = (nx)^n$

Find the general term for each of the following sequences.

21. $2, 3, 4, 5, \ldots$
22. $3, 4, 5, 6, \ldots$
23. $1, 4, 9, 16, \ldots$
24. $1, 8, 27, 64, \ldots$
25. $3, 9, 27, 81, \ldots$
26. $2, 4, 8, 16, \ldots$
27. $-1, 1, -1, 1, \ldots$
28. $-10, -20, -30, -40, \ldots$
29. $\dfrac{1}{2}, \dfrac{2}{3}, \dfrac{3}{4}, \dfrac{4}{5}, \ldots$
30. $\dfrac{1}{3}, \dfrac{1}{9}, \dfrac{1}{27}, \dfrac{1}{81}, \ldots$
31. $4, 8, 12, 16, \ldots$
32. $4, 7, 10, 13, \ldots$
33. $-1, \dfrac{4}{3}, -\dfrac{3}{2}, \dfrac{8}{5}, \ldots$
34. $-\dfrac{1}{2}, \dfrac{4}{3}, -\dfrac{9}{4}, \dfrac{16}{5}, \ldots$
35. $x^2, x^3, x^4, x^5, \ldots$
36. $x, x^4, x^9, x^{16}, \ldots$
37. $\dfrac{\sqrt{x}}{3}, \dfrac{\sqrt[3]{x}}{9}, \dfrac{\sqrt[4]{x}}{27}, \dfrac{\sqrt[5]{x}}{81}, \ldots$
38. $-\dfrac{x^2}{2}, \dfrac{x^3}{4}, -\dfrac{x^4}{8}, \dfrac{x^5}{16}, \ldots$

Evaluate each of the following series.

39. $\sum_{i=1}^{4} i$
40. $\sum_{i=1}^{3} (i + 2)$
41. $\sum_{i=2}^{3} (2i - 1)$
42. $\sum_{i=3}^{4} (5i + 2)$
43. $\sum_{i=1}^{3} (-1)^i$
44. $\sum_{i=1}^{4} (i - 2)(i + 3)$
45. $\sum_{i=2}^{6} i(-1)^i$
46. $\sum_{i=3}^{8} (-2)^i$
47. $\sum_{i=2}^{4} \dfrac{i+1}{i-1}$
48. $\sum_{i=1}^{5} \dfrac{2i-1}{i+2}$
49. $\sum_{i=1}^{7} i^2$
50. $\sum_{i=3}^{7} (i^2 - 1)$
51. $\sum_{i=2}^{3} (-1)^i (3i - 1)$
52. $\sum_{i=1}^{3} (-1)^{i+1}(i^2 - 1)$

Write each of the following sums using sigma notation.

53. $3 + 6 + 9 + 12$
54. $2 + 4 + 6 + 8 + 10$
55. $1 + 4 + 9 + 16 + 25$
56. $16 + 25 + 36 + 49$
57. $\dfrac{1}{2} + \dfrac{2}{3} + \dfrac{3}{4} + \dfrac{4}{5} + \dfrac{5}{6}$
58. $\dfrac{2}{5} + \dfrac{4}{6} + \dfrac{6}{7} + \dfrac{8}{8}$
59. $-\dfrac{1}{3} + \dfrac{2}{4} - \dfrac{3}{5} + \dfrac{4}{6} - \dfrac{5}{7}$
60. $\dfrac{5}{2} - \dfrac{7}{3} + \dfrac{9}{4}$
61. $7 - 11 + 15 - 19$
62. $2 - 5 + 10 - 17$
63. $x - \dfrac{1}{2}x^3 + \dfrac{1}{3}x^5 - \dfrac{1}{4}x^7$
64. $\dfrac{1}{2} - \dfrac{1}{4}x^2 + \dfrac{1}{6}x^4 - \dfrac{1}{8}x^6$

Find the first six terms of each sequence with the given recursive definition.

65. $a_1 = 2;\ a_{n+1} = 3a_n$
66. $a_1 = 1;\ a_{n+1} = a_n + 2$
67. $a_1 = -1;\ a_{n+1} = 4a_n + 5$
68. $a_1 = -2;\ a_{n+1} = 5a_n + 3$

WRITE IN YOUR OWN WORDS

69. What is a sequence?
70. What is a series?
71. What is a recursive definition of a sequence?
72. What does the symbol Σ mean?
73. What is the subscript on a_i for?
74. What is a finite sequence? Give an example of one.
75. What is an infinite sequence? Give an example of one.
76. Give an example of an application of a sequence or a series.
77. Describe what is meant by the general term of a sequence.

12.2 Arithmetic Sequences and Series

OBJECTIVES

In this section we will learn to
1. recognize arithmetic sequences; and
2. use formulas to find the nth term and the sum of an arithmetic sequence.

We begin by considering the definition of an **arithmetic sequence**.

CHAPTER 12 · SEQUENCES, SERIES, AND THE BINOMIAL THEOREM

DEFINITION **Arithmetic Sequence**	An **arithmetic sequence** is one in which the difference between any two consecutive terms is a constant.

If $a_1, a_2, a_3, \ldots, a_n$ are terms of a finite arithmetic sequence and the constant (difference) is d, then $a_2 = a_1 + d$, $a_3 = a_2 + d$, and $a_4 = a_3 + d$. An arithmetic sequence can be defined recursively by the formula

$$a_{n+1} = a_n + d$$

EXAMPLE 1 Determine which of the following finite sequences is arithmetic. If it is, identify the common difference.

(a) $3, 5, 7, 9, \ldots, 2n + 1$
This is an arithmetic sequence with a common difference of 2.

(b) $5, 2, -1, -4, \ldots, (8 - 3n)$
This is an arithmetic sequence with common difference of -3.

(c) $1^2, 2^2, 3^2, 4^2, \ldots, n^2$
This is *not* an arithmetic sequence. There is no common difference between consecutive terms.

(d) $\dfrac{1}{2}, \dfrac{2}{3}, \dfrac{5}{6}, \ldots, \dfrac{n+2}{6}$

If there is a common difference, it is not as obvious, as in (a) and (b). To find a common difference, we must subtract any two successive terms:

$$\frac{2}{3} - \frac{1}{2} = \frac{4}{6} - \frac{3}{6} = \frac{1}{6} \quad \text{or} \quad \frac{5}{6} - \frac{2}{3} = \frac{5}{6} - \frac{4}{6} = \frac{1}{6}$$

The sequence is arithmetic with a common difference of $\dfrac{1}{6}$.

To find an expression for the nth term of a given arithmetic sequence, we must consider the pattern that is formed when d is added to a term to get the next term.

Term Number	Term	Number of Times d Is Added
1	a_1	0
2	$a_2 = a_1 + d$	1
3	$a_3 = a_2 + d = a_1 + 2d$	2
4	$a_4 = a_3 + d = a_1 + 3d$	3
$\vdots$	$\vdots$	$\vdots$
n	$a_n = a_1 + (n-1)d$	$n - 1$

> **nth Term of an Arithmetic Sequence**
>
> The **nth term of an arithmetic sequence**, a_n, is given by the formula
>
> $$a_n = a_1 + (n-1)d$$
>
> where a_1 is the first term and d is the common difference.

EXAMPLE 2 Find the ninth term of the arithmetic sequence 3, 7, 11, ...

SOLUTION In the ninth term, n is 9 and a_n becomes a_9. Therefore

$a_n = a_1 + (n-1)d$
$a_9 = 3 + (9-1)(4)$ $a_1 = 3, d = 4, n = 9$
$ = 3 + 32 = 35$

EXAMPLE 3 The fifteenth term of an arithmetic sequence is 27 and the common difference is 2. Find the first term.

SOLUTION We substitute 27 for a_n, 15 for n, and 2 for d. Then

$a_n = a_1 + (n-1)d$
$27 = a_1 + (15-1)(2)$
$27 = a_1 + 28$
$-1 = a_1$

EXAMPLE 4 The seventh term of an arithmetic sequence is 20, and the eighteenth term is 53. Find the first term and the common difference.

SOLUTION We use the formula $a_n = a_1 + (n-1)d$ for $n = 7$ and $n = 18$. The seventh term is 20, so

$\phantom{\text{I }}20 = a_1 + (7-1)d$ $a_7 = 20, n = 7$
I $20 = a_1 + 6d$

The eighteenth term is 53, so

$\phantom{\text{II }}53 = a_1 + (18-1)d$ $a_{18} = 53, n = 18$
II $53 = a_1 + 17d$

We now have two equations, **I** and **II**, in two variables, which we can solve by the elimination method.

I $20 = a_1 + 6d$
II $\underline{53 = a_1 + 17d}$
III $33 = 11d$ **III** = **II** + (−1)**I**
$\phantom{\text{III }}3 = d$

The common difference is 3. Substituting $d = 3$ into equation **I** yields

$$a_1 + 6 \cdot 3 = 20$$
$$a_1 + 18 = 20$$
$$a_1 = 2$$

The first term is 2.

Occasionally we need to find the sum of the terms of an arithmetic sequence. If the sequence involves only a few terms, finding the sum is a relatively easy process. If the sequence contains many terms, however, finding the sum would be time-consuming if it were not for the fact that a formula exists for just this purpose. To get an idea of how the formula works, consider the sum of the finite arithmetic sequence

$$1, 3, 5, 7, 9$$

The sum is

$$S_5 = 1 + 3 + 5 + 7 + 9$$

This can also be written

$$S_5 = 9 + 7 + 5 + 3 + 1$$

When we add these two series term-by-term, an interesting pattern emerges:

$$S_5 = 1 + 3 + 5 + 7 + 9$$
$$S_5 = 9 + 7 + 5 + 3 + 1$$
$$2S_5 = 10 + 10 + 10 + 10 + 10 \qquad \text{Add the corresponding terms.}$$
$$ = 5(10) = 5(1 + 9) \qquad \text{Rewrite 10 as } 1 + 9.$$
$$S_5 = \frac{5}{2}(1 + 9) \qquad \text{Divide each side by 2.}$$

Notice that there are five terms in the sequence and that the sum can be written as the product of one-half the number of terms (5) times the sum of the first term (1) and last term (9). This observation is confirmed in the following theorem.

Theorem 12.1

The Sum of an Arithmetic Sequence

The sum S_n of the first n terms of an arithmetic sequence is given by the formula

$$S_n = \frac{n}{2}(a_1 + a_n)$$

To derive this formula, we proceed as follows. We let $a_1, a_2, a_3, \ldots, a_n$ be the first n terms of an arithmetic sequence. The sum of the sequence is

$$S_n = a_1 + a_2 + a_3 + \cdots + a_n$$

Since the common difference between any two terms is d, we can write the series that represents the sum as

I $\quad S_n = a_1 + (a_1 + d) + (a_1 + 2d) + \cdots + a_n$

We can also write this series starting with the last term; then each term will be d *less* than the preceding term.

II $\quad S_n = a_n + (a_n - d) + (a_n - 2d) + \cdots + a_1$

When equations **I** and **II** are added term-by-term, we get

$$\begin{aligned}\textbf{I} \quad S_n &= a_1 + (a_1 + d) + (a_1 + 2d) + \cdots + a_n \\ \textbf{II} \quad S_n &= a_n + (a_n - d) + (a_n - 2d) + \cdots + a_1 \\ \hline 2S_n &= (a_1 + a_n) + (a_1 + a_n) + (a_1 + a_n) + \cdots + (a_1 + a_n)\end{aligned}$$

Since the expression $(a_1 + a_n)$ appears once for each of the n terms, the right side of the equation can be written $n(a_1 + a_n)$. Thus

$$2S_n = n(a_1 + a_n) \quad \text{or} \quad S_n = \frac{n}{2}(a_1 + a_n)$$

EXAMPLE 5 Find the sum of the sequence 1, 2, 3, 4, ..., 72.

SOLUTION
$$S_n = \frac{n}{2}(a_1 + a_n)$$
$$S_{72} = \frac{72}{2}(1 + 72)$$
$$= 36(73) = 2628$$

The formula $S_n = \frac{n}{2}(a_1 + a_n)$ can also be written in an alternate form that does not require that the nth term be known in order to find the sum of the terms.

$$S_n = \frac{n}{2}(a_1 + a_n)$$
$$= \frac{n}{2}[a_1 + a_1 + (n-1)d] \qquad a_n = a_1 + (n-1)d$$
$$= \frac{n}{2}[2a_1 + (n-1)d]$$

The Sum of an Arithmetic Sequence: Alternate Formula

$$S_n = \frac{n}{2}[2a_1 + (n-1)d]$$

EXAMPLE 6 Given that $a_n = 4 + 3n$, find S_{16}.

SOLUTION We begin by finding a_1.

$$a_n = 4 + 3n$$

$$a_1 = 4 + 3 \cdot 1 \qquad n = 1$$
$$= 7$$

Now we use the alternate formula. The difference is 3.

$$S_n = \frac{n}{2}[2a_1 + (n-1)d]$$

$$S_{16} = \frac{16}{2}[2(7) + (16-1)(3)] \qquad a_1 = 7, d = 3, n = 16$$

$$= 8(59) = 472$$

EXAMPLE 7 Find the number n of terms of a finite arithmetic sequence if $a_1 = 5$, $a_n = 32$, and $S_n = 185$.

SOLUTION
$$S_n = \frac{n}{2}(a_1 + a_n)$$

$$185 = \frac{n}{2}(5 + 32)$$

$$370 = 37n \qquad \text{Clear of fractions and simplify.}$$
$$10 = n$$

EXAMPLE 8 Find the indicated sum: $\sum_{i=1}^{18}(-2i + 6)$.

SOLUTION The given expression represents the sum of an arithmetic sequence. The index i from 1 to 18 shows that there are 18 terms, so $n = 18$. To find the sum, we need to find a_1 and a_{18}.

$$a_1 = -2(1) + 6 = 4$$
$$a_{18} = -2(18) + 6 = -30$$
$$S_n = \frac{n}{2}(a_1 + a_n)$$
$$S_{18} = \frac{18}{2}[4 + (-30)]$$
$$= 9(-26) = -234$$

EXAMPLE 9 Show that the sum of the first n terms of $5 + 9 + 13 + \cdots$ is $2n^2 + 3n$.

SOLUTION The first term is 5 and the difference is 4. We can use the alternate formula to express the sum in terms of n.

$$S_n = \frac{n}{2}[2a_1 + (n-1)d]$$

$$= \frac{n}{2}[2(5) + (n-1)4] \qquad a_1 = 5, d = 4$$

$$= \frac{n}{2}[4n + 6] = 2n^2 + 3n$$

EXAMPLE 10 Find the sum of all the even three-digit numbers between 1 and 1000.

SOLUTION The first even three-digit number is 100, and the last one less than 1000 is 998. Even numbers differ by 2, so $d = 2$. To determine n, we use the formula

$$a_n = a_1 + (n-1)d$$
$$998 = 100 + (n-1)2 \qquad a_1 = 100, a_n = 998, d = 2$$
$$998 = 98 + 2n$$
$$900 = 2n$$
$$450 = n$$

So there are 450 three-digit even numbers between 1 and 100. Their sum is

$$S_n = \frac{n}{2}(a_1 + a_n)$$
$$S_{450} = \frac{450}{2}(100 + 998)$$
$$= 225(1098)$$
$$= 247{,}050$$

Exercise Set 12.2

Determine which of the following finite sequences are arithmetic. Find the common difference for each arithmetic sequence.

1. 1, 2, 3, 4, 5
2. 2, 4, 6, 8, 10
3. 3, 6, 9, 12, 15
4. 4, 8, 12, 16, 20
5. 1, 4, 7, 11
6. 2, 5, 8, 12
7. 20, 15, 10, 5
8. 30, 24, 18, 12
9. 2, 4, 8, 16
10. 1, 3, 6, 9
11. $x + 3y, 2x + 2y, 3x + y$
12. $x - 2y, 2x - y, 3x$
13. $-7, -14, -21, -28$
14. $1, 0, -1, -2$

The following are arithmetic sequences or series. Find the indicated unknowns.

15. $a_1 = -3, d = 5; a_{10}; a_8$
16. $a_1 = -1, d = 3; a_{18}; a_5$
17. $a_1 = 3, d = 6; a_{13}; a_6$
18. $a_1 = 4, d = 2; a_7; a_{14}$
19. $a_1 = 4, d = 3; a_7; S_{10}$
20. $a_1 = 1, d = 3, a_n = 37; a_3; S_n$
21. $a_{15} = 45, S_{15} = 255; a_1; d$
22. $a_{24} = 163, S_{24} = 1980; a_1; d$
23. $a_1 = 82, d = -13, a_n = -9; n; S_n$
24. $a_1 = 46, d = -9, a_n = -125; n; S_n$
25. $a_9 = -12, a_{13} = 3; a_1; S_6$
26. $a_1 = 24, a_{24} = -22; S_{24}; a_4$
27. $a_n = 9n - 3; S_{20}; S_{24}$
28. $a_n = 4 + 3n; S_{15}; S_{20}$
29. $a_1 = 36, a_3 = 66, S_n = 234; n; d$
30. $a_1 = 12, a_n = 38, S_n = 1325; n; d$
31. $a_1 = 12, S_{28} = 1092; d; a_{28}$

32. $a_1 = 71$, $S_{20} = -100$; d; a_{20}
33. $a_1 = 4$, $d = 0.4$, $S_n = 30$; n; a_n
34. $a_1 = 6$, $d = -\dfrac{5}{3}$, $S_n = -15$; n; a_n
35. $d = -4$, $a_n = -56$, $S_n = -420$; a_1; n
36. $d = 3$, $a_n = 1$, $S_n = -39$; a_1; n

Find each of the following sums.

37. $\displaystyle\sum_{i=1}^{10} (3i - 1)$
38. $\displaystyle\sum_{i=1}^{7} (2i - 1)$
39. $\displaystyle\sum_{i=5}^{12} (2i - 5)$
40. $\displaystyle\sum_{i=13}^{20} (5i - 1)$
41. $\displaystyle\sum_{i=12}^{18} \dfrac{2i - 1}{3}$
42. $\displaystyle\sum_{i=12}^{16} \dfrac{8 - 2i}{3}$

Solve or show, as indicated.

43. Find the sum of the first 50 natural numbers.
44. Find the sum of all the odd integers from 1 through 101.
45. Find the sum of all positive two-digit integers.
46. Find the sum of all integer multiples of 8 between 9 and 199.
47. Find the sum of all integer multiples of 7 between 8 and 103.
48. Show that the sum of the first n natural numbers is $\dfrac{n(n + 1)}{2}$.
49. Show that the sum of the first n terms of the series $3 + 5 + 7 + 9 + \cdots$ is $2n + n^2$.
50. Show that the sum of the first n terms of the series $10 + 14 + 18 + \cdots$ is $8n + 2n^2$.
51. Show that the sum of the first n even natural numbers is $n + n^2$.
52. Marie deposits $100 in a savings account on her daughter's first birthday and $200 on her second birthday. If she continues this pattern through her daughter's eighteenth birthday, how much money will she have deposited in the account by then?
53. Cynthia starts a new job with an annual salary of $16,500 in 1995. If she receives a $900 raise each year, what will her annual salary be in the year 2006?

REVIEW EXERCISES

54. Solve: $\log_2(x - 3) = 4$.
55. Graph: $f(x) = \log_2 x$.
56. Solve: $3^x = 5^{x-1}$.
57. How much interest will be earned if $7200 is invested for 2 years at a 9.13% interest rate compounded continuously?
58. Solve: $\log(x - 3) + \log x = 1$.
59. A certain radioactive material decays according to the formula $f(t) = 20e^{kt}$, where t is in years and 20 g is the initial amount.
    (a) Find k if the amount after 10 years is 19 g.
    (b) Find the amount remaining after 200 years.
60. How long will it take for an amount of money p to double if it is invested at 7.5% compounded continuously? (Round your answer to the nearest whole year.)

12.3 Geometric Sequences and Series

HISTORICAL COMMENT The early Egyptians were interested in geometry for both practical and recreational purposes. The Rhind Papyrus (c. 1650 B.C.), which was copied by the scribe Ahmes from an earlier work, contains examples of geometric sequences. An example of a purely recreational problem from early Egyptian times states that there are seven houses each of which contains seven cats each of which eats seven mice; each mouse has eaten seven ears of spelt, each of which would have produced seven hekats of grain. The problem is to find the total of all of these. Variations of this problem appear throughout history and are frequently found in recreational mathematics publications today.

12.3 · GEOMETRIC SEQUENCES AND SERIES

OBJECTIVES In this section we will learn to
1. recognize geometric sequences; and
2. use formulas to find the nth term and the sum of a geometric sequence.

In the preceding section we studied the concept of an arithmetic sequence, in which there is a common difference between any two successive terms. The sequence 1, 2, 4, 8, 16, . . . is not arithmetic because there is not a common difference between any two terms. However, any one of the terms can be obtained from the one preceding it by multipying by 2. Sequences of this type are called *geometric sequences*.

> **DEFINITION**
> **Geometric Sequence**
>
> A **geometric sequence** is one in which each term after the first is obtained by multiplying the preceding one by a constant r, called the **common ratio.**

For example, the sequence $3, 1, \frac{1}{3}, \frac{1}{9}, \ldots$ is geometric. To obtain any term other than the first one, we multiply the preceding term by $\frac{1}{3}$.

If a_1 is the first term of a geometric sequence, then the second term is $a_1 r$, where r is the common ratio. In the same manner, the third term is obtained by multiplying the second term by r, so the third term $a_3 = a_1 r^2$. Similarly, the fourth term is $a_4 = a_1 r^3$. From this we see that we can define a geometric sequence recursively as

$$a_{n+1} = a_n r$$

When we know two successive terms of a geometric sequence, we can find the common ratio by solving the recursion formula for r. To do this, we divide each side by a_n:

$$r = \frac{a_{n+1}}{a_n}$$

For example, in the geometric sequence

$$2, \frac{4}{3}, \frac{8}{9}, \frac{16}{27}, \ldots$$

we can find the common ratio by taking the ratio of any two consecutive terms. Let's use the second and third terms:

$$r = \frac{a_{n+1}}{a_n} = \frac{\frac{8}{9}}{\frac{4}{3}} = \frac{8}{9} \cdot \frac{3}{4} = \frac{2}{3}$$

EXAMPLE 1 Find the common ratio for each of the following geometric sequences.

(a) 1, 3, 9, 27,

Since $\frac{3}{1} = 3$, $\frac{9}{3} = 3$, and $\frac{27}{9} = 3$, the common ratio is 3.

(b) 8, 4, 2, 1, ...

Since $\frac{4}{8} = \frac{1}{2}$, $\frac{2}{4} = \frac{1}{2}$, and $\frac{1}{2} = \frac{1}{2}$, the common ratio is $\frac{1}{2}$.

(c) $\frac{1}{3}, -\frac{2}{15}, \frac{4}{75}, -\frac{8}{375}, \ldots$

The ratio is $\frac{a_{n+1}}{a_n} = -\frac{2}{5}$.

We can find the *n*th term of a geometric sequence by observing the pattern of terms.

Term Number	Term	Number of Times r Is Multiplied
1	a_1	0
2	$a_1 r$	1
3	$a_1 r^2$	2
4	$a_1 r^3$	3
⋮	⋮	⋮
n	$a_1 r^{n-1}$	$n - 1$

nth Term of a Geometric Sequence

The **n**th term of a geometric sequence, a_n, is given by

$$a_n = a_1 r^{n-1}$$

where a_1 is the first term and r is the common ratio.

EXAMPLE 2 Find r and a_n for the sequence 3, 6, 12, ...

SOLUTION

$r = \dfrac{a_{n+1}}{a_n} = \dfrac{6}{3} = 2$

$a_n = a_1 r^{n-1} = 3(2)^{n-1}$ $a_1 = 3; r = 2$

EXAMPLE 3 The first term of a geometric sequence is 3 and the sixth term is 729. Find r.

SOLUTION

$a_n = a_1 r^{n-1}$

$729 = 3r^{6-1}$ $a_6 = 729, n = 6, a_1 = 3$

$243 = r^5$ Divide each side by 3.

$3^5 = r^5$ $243 = 3^5$

$3 = r$ Take the fifth root of each side.

We can find a formula for the sum of the terms in a finite geometric sequence of length n in a manner analogous to how we found the sum of the terms of a finite arithmetic sequence. We assume $r \neq 1$, since otherwise $a_1 = a_2 = \cdots = a_n$. If S_n represents the sum of the terms of the sequence, then

I $S_n = a_1 + a_1 r + a_1 r^2 + \cdots + a_1 r^{n-1}$

Next we multiply both sides of equation **I** by r to get

II $rS_n = a_1 r + a_1 r^2 + a_1 r^3 + \cdots + a_1 r^{n-1} + a_1 r^n$

Note in equation **II** that the term preceding $a_1 r^n$ is a $a_1 r^{n-1}$ since the exponent on r decreases by 1 on each term as we move from right to left. Now if we subtract equation **II** from equation **I**, all of the terms on the right side are eliminated, except a_1 and $a_1 r^n$:

$$\begin{aligned}
\textbf{I} \quad & S_n = a_1 + a_1 r + a_1 r^2 + \cdots + a_1 r^{n-1} \\
\textbf{II} \quad & rS_n = a_1 r + a_1 r^2 + a_1 r^3 + \cdots + a_1 r^{n-1} + a_1 r^n \\
& S_n - rS_n = a_1 - a_1 r^n \qquad \textbf{I} - \textbf{II} \\
& S_n(1-r) = a_1(1 - r^n) \qquad \text{Factor.} \\
& S_n = \frac{a_1(1-r^n)}{1-r} \qquad \text{Divide each side by } 1-r;\ r \neq 1.
\end{aligned}$$

THEOREM 12.2

The Sum of a Geometric Sequence

The sum S_n of the first n terms of a geometric sequence is given by the formula

$$S_n = \frac{a_1(1-r^n)}{1-r}$$

where a_1 is the first term and r is the common ratio ($r \neq 1$). The indicated sum of a geometric sequence is called a **geometric series.**

EXAMPLE 4 Find the sum of a geometric sequence of five terms with $a_1 = 2$ and $r = \dfrac{1}{3}$.

SOLUTION

$$S_n = \frac{a_1(1-r^n)}{1-r}$$

$$S_5 = \frac{2\left[1 - \left(\dfrac{1}{3}\right)^5\right]}{1 - \dfrac{1}{3}} = \frac{2\left(1 - \dfrac{1}{243}\right)}{\dfrac{2}{3}} \qquad a_1 = 2, r = \dfrac{1}{3}, n = 5$$

$$= 2\left(\frac{242}{243}\right) \cdot \frac{3}{2} = \frac{242}{81}$$

EXAMPLE 5 The sixth term of a geometric sequence is $\frac{1}{2}$, and also $r = \frac{1}{2}$. Find a_1 and S_9.

SOLUTION

$$a_n = a_1 r^{n-1}$$

$$\frac{1}{2} = a_1 \left(\frac{1}{2}\right)^{6-1} \qquad a_6 = \frac{1}{2}, r = \frac{1}{2}, n = 6$$

$$\frac{1}{2} = a_1 \left(\frac{1}{2}\right)^5$$

$$16 = a_1 \qquad \text{Multiply each side by 32 and reduce.}$$

Now we use the value of a_1 in the formula for S_n to find S_9.

$$S_n = \frac{a_1(1 - r^n)}{1 - r}$$

$$S_9 = \frac{16\left[1 - \left(\frac{1}{2}\right)^9\right]}{1 - \frac{1}{2}} = \frac{16\left(1 - \frac{1}{512}\right)}{\frac{1}{2}} \qquad a_1 = 16, r = \frac{1}{2}, n = 9$$

$$= \frac{16\left(\frac{511}{512}\right)}{\frac{1}{2}} = 32\left(\frac{511}{512}\right)$$

$$= \frac{511}{16}$$

EXAMPLE 6 Find n, the number of terms, in a geometric sequence for which $a_1 = 2$, $r = -2$, and $S_n = -42$.

SOLUTION

$$S_n = \frac{a_1(1 - r^n)}{1 - r}$$

$$-42 = \frac{2[1 - (-2)^n]}{1 - (-2)} = \frac{2[1 - (-2)^n]}{3} \qquad S_n = -42, a_1 = 2, r = -2.$$

We multiply each side by $\frac{3}{2}$ to isolate the expression containing n.

$$-63 = 1 - (-2)^n$$
$$-64 = -(-2)^n \qquad \text{Subtract 1 from each side.}$$
$$64 = (-2)^n \qquad \text{Multiply each side by } -1.$$
$$(-2)^6 = (-2)^n \qquad 64 = (-2)^6$$
$$6 = n \qquad \text{If } a^m = a^n, \text{ then } m = n.$$

If $-1 < r < 1$, we can find the sum of a geometric sequence containing an infinite number of terms. For example, the sum of the terms of the infinite sequence

$\frac{1}{2}, \frac{1}{4}, \frac{1}{8}, \frac{1}{16}, \ldots$ is 1. To see this, we think of a number line one unit long, where each point is halfway between the preceding point and 1, as shown in Figure 12.1.

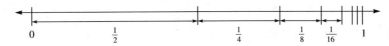

FIGURE 12.1

Even though there are an infinite number of subdivisions of the line from 0 to 1, the sum of its parts can never exceed 1. Thus,

$$\frac{1}{2} + \frac{1}{4} + \frac{1}{8} + \frac{1}{16} + \frac{1}{32} + \cdots = 1$$

It should be pointed out that the sum will be 1 only if we include an *infinite* number of terms. If we terminate the series after even billions of terms, the sum will be very *close* to 1 but not quite there.

We can find a formula for the sum of an infinite number of terms of a geometric sequence if we require that the common ratio r lies between -1 and 1, that is, $-1 < r < 1$. To see why this makes it possible, let's consider what happens as the exponent of $r^n = (0.01)^n$ increases.

$(0.01)^1 = 0.01$
$(0.01)^2 = 0.0001$
$(0.01)^3 = 0.000001$
$(0.01)^4 = 0.00000001$

We can make the value of $(0.01)^n$ as small as we want by letting n become sufficiently large. In the formula

$$S_n = \frac{a_1(1 - r^n)}{1 - r}$$

r^n will approach zero if $-1 < r < 1$ and n grows very large. If S_∞ represents the sum of an infinite number of terms, then

$$S_\infty = \frac{a_1(1 - 0)}{1 - r} = \frac{a_1}{1 - r}$$

Sum of an Infinite Geometric Sequence

The **sum S_∞ of an infinite geometric sequence** is given by the formula

$$S_\infty = \frac{a_1}{1 - r}$$

where $-1 < r < 1$ and the subscript ∞ means that there is an infinite number of terms.

If $r > 1$ or $r < -1$, then r^n does not approach zero as n becomes very large. The infinite sum for those values of r does not exist.

EXAMPLE 7 Find the sum of the infinite geometric series $4 + 2 + 1 + \frac{1}{2} + \cdots$.

SOLUTION Here $r = \frac{1}{2}$, which means that the formula for the sum of an infinite sequence applies.

$$S_\infty = \frac{a_1}{1 - r}$$

$$= \frac{4}{1 - \frac{1}{2}} = \frac{4}{\frac{1}{2}} \qquad a_1 = 4, r = \frac{1}{2}$$

$$= 4 \cdot 2 = 8$$

The sum of the series is 8.

EXAMPLE 8 Write $0.\overline{2} = 0.222\ldots$ as a rational number.

SOLUTION The number $0.222\ldots$ can be thought of as an infinite sum as follows:

$$0.222\ldots = 0.2 + 0.02 + 0.002 + 0.0002 + \cdots$$

This is in fact an infinite geometric sequence with $a_1 = 0.2$ and $r = 0.1$, so

$$S_\infty = \frac{a_1}{1 - r}$$

$$= \frac{0.2}{1 - 0.1} = \frac{0.2}{0.9} \qquad a_1 = 0.2, r = 0.1$$

$$= \frac{2}{9}$$

The sum of the series is $\frac{2}{9}$, so we can write $0.222\ldots = \frac{2}{9}$.

EXAMPLE 9 Evaluate $\sum_{i=1}^{4} \left(\frac{1}{3}\right)^i$ by using the formula for the sum of a geometric sequence.

SOLUTION First let's write out the sum:

$$\sum_{i=1}^{4} \left(\frac{1}{3}\right)^i = \frac{1}{3} + \frac{1}{9} + \frac{1}{27} + \frac{1}{81}$$

This is a finite geometric series with $n = 4$, $a_1 = \frac{1}{3}$, and $r = \frac{1}{3}$, so

$$S_n = \frac{a_1(1 - r^n)}{1 - r}$$

$$S_4 = \frac{\frac{1}{3}\left[1 - \left(\frac{1}{3}\right)^4\right]}{1 - \frac{1}{3}} \qquad a_1 = \frac{1}{3}, r = \frac{1}{3}, n = 4$$

$$= \frac{\frac{1}{3}\left(1 - \frac{1}{81}\right)}{\frac{2}{3}} = \frac{\frac{1}{3}\left(\frac{80}{81}\right)}{\frac{2}{3}}$$

$$= \frac{1}{3}\left(\frac{80}{81}\right) \cdot \frac{3}{2} = \frac{40}{81}$$

The sum of the series is $\frac{40}{81}$.

EXAMPLE 10 Evaluate $\sum_{i=1}^{\infty} \left(\frac{1}{3}\right)^i$.

SOLUTION Again $r = \frac{1}{3}$ and $a_1 = \frac{1}{3}$, but this time the series is infinite.

$$S_\infty = \frac{a_1}{1 - r}$$

$$= \frac{\frac{1}{3}}{1 - \frac{1}{3}} = \frac{\frac{1}{3}}{\frac{2}{3}}$$

$$= \frac{1}{3} \cdot \frac{3}{2} = \frac{1}{2}$$

The sum is $\frac{1}{2}$.

Solution to the Historical Problem

EXAMPLE 11 There are seven houses, each of which contains seven cats, each of which eats seven mice; each mouse has eaten seven ears of spelt, each of which would have produced seven hekats of grain. The problem is to find the total of all of these.

SOLUTION There are 7 houses.
Each house has seven cats, so there are $7 \cdot 7 = 7^2$ cats.
Each cat eats seven mice, so there were $7^2 \cdot 7 = 7^3$ mice.
Each mouse has eaten seven ears of spelt, so $7^3 \cdot 7 = 7^4$ ears of spelt have been eaten.
Each ear would have produced 7 hekats of grain, so there would have been $7^4 \cdot 7 = 7^5$ hekats.
The total of all these is the geometric sequence $7 + 7^2 + 7^3 + 7^4 + 7^5$, or

$$S_5 = \frac{7(1 - 7^5)}{1 - 7} = 19{,}607$$

12.4 · THE BINOMIAL THEOREM

34. $\sum_{i=1}^{\infty} 14\left(\frac{5}{7}\right)^{i-1}$ — wait, illegible

36. $2 - 2 + 2 - 2 + \cdots$

37. $\sum_{i=1}^{\infty} 2\left(\frac{3}{4}\right)^{i-1}$ 38. $\sum_{i=1}^{\infty} 2\left(-\frac{3}{4}\right)^{i-1}$ 39. $\sum_{i=1}^{\infty} 2\left(\frac{7}{5}\right)^{i-1}$

40. $\sum_{i=1}^{\infty}\left(\frac{1}{3}\right)^{i}$ 41. $\sum_{i=1}^{\infty}\left(\frac{3}{7}\right)^{i}$ 42. $\sum_{i=1}^{\infty}\left(\frac{5}{8}\right)^{i}$

Write each of the following decimal numbers as a rational number.

43. $0.333\ldots$ 44. $0.111\ldots$
45. $0.666\ldots$ 46. $0.999\ldots$
47. $0.212121\ldots$ 48. $0.474747\ldots$
49. $0.123123123\ldots$ 50. $0.102102102\ldots$

Solve each of the following applied problems. Round to four decimal places as needed.

51. A rubber ball is dropped from a height of 3 ft. Each time it strikes the ground it bounces back up two-thirds of the distance it fell. Find the total distance traveled by the ball before coming to rest.

52. From what height should the ball of Exercise 51 be dropped so that the total distance traveled is 25 ft?

53. An object suspended from a spring moves up and down. It each second it moves half as far as it did the previous second. If the object moves 8 cm the first second, how far will it move in 10 sec?

54. The tip of a pendulum moves through an arc of 14 in. in its first swing. If each successive arc is then nine-tenths as long as the previous swing, find the distance the pendulum travels in 6 swings.

55. A point on a pendulum travels 8 cm on the first swing and 5% less on each successive swing. Find the distance traveled by the point in 8 swings.

56. The city of Morro Bay had a population of 10,000 in 1995. If the population increases by 2%/yr, what will its population be in the year 2005?

FOR EXTRA THOUGHT

57. A man standing 16 ft from an 18-ft streetlight casts a shadow 8 ft long. How tall is the man?

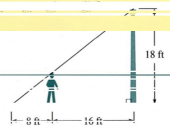

58. The midpoints of the sides of two squares are joined as shown. Find the ratio of the areas of squares $WXYZ$ and $ABCD$.

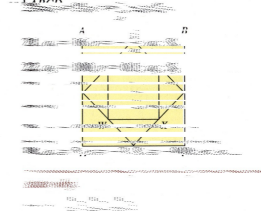

12.4 The Binomial Theorem

OBJECTIVES   In this section we will learn to
1. use factorial notation; and
2. expand binomials by use of the binomial theorem.

Suppose we found it necessary to expand $(a + b)^8$. One way is to find the product using $(a + b)$ as a factor eight times. This method is far too time-consuming for any large exponent, however, and we can replace it with a formula called the *binomial theorem*. Before we can consider the formula, we first need to learn a new notation called **n factorial,** written "$n!$".

DEFINITION	*n* **factorial,** symbolized by $n!$, is the product of the first n positive integers:
n **Factorial**	$$n! = n(n-1)(n-2)\cdots 3 \cdot 2 \cdot 1$$

EXAMPLE 1 Evaluate each factorial.

(a) $4! = 4 \cdot 3 \cdot 2 \cdot 1 = 24$
(b) $6! = 6 \cdot 5 \cdot 4 \cdot 3 \cdot 2 \cdot 1 = 720$

EXAMPLE 2 Evaluate $5 \cdot 4!$.

SOLUTION $5 \cdot 4! = 5 \cdot 4 \cdot 3 \cdot 2 \cdot 1 = 120$

Example 2 illustrates that factorials can be defined recursively. Thus,

$$n! = n(n-1)! \qquad \text{where } n \geq 2$$

EXAMPLE 3 Evaluate each of the following: (a) $\dfrac{7!}{3!}$ (b) $\dfrac{(n+1)!}{(n-1)!}$

SOLUTION (a) $\dfrac{7!}{3!} = \dfrac{7 \cdot 6 \cdot 5 \cdot 4 \cdot 3!}{3!}$

$\qquad\qquad = 7 \cdot 6 \cdot 5 \cdot 4 \qquad$ Divide out $3!$.

$\qquad\qquad = 840$

(b) $\dfrac{(n+1)!}{(n-1)!} = \dfrac{(n+1)(n)(n-1)!}{(n-1)!}$

$\qquad\qquad\quad = (n+1)(n) \qquad$ Divide out $(n-1)!$

Another new notation that will prove useful for the binomial theorem is $\binom{n}{r}$.

$$\binom{n}{r} = \frac{n!}{r!(n-r)!}, \qquad \text{where } n \geq r \geq 0, n \text{ and } r \text{ integers}$$

EXAMPLE 4 Evaluate the following: (a) $\binom{6}{2}$ (b) $\binom{7}{3}$

SOLUTION (a) $\binom{6}{2} = \dfrac{6!}{2!(6-2)!} = \dfrac{6!}{2!4!}$

$= \dfrac{6 \cdot 5 \cdot 4!}{2 \cdot 1 \cdot 4!}$ Write 6! recursively.

$= \dfrac{30}{2}$ Divide out 4!.

$= 15$

(b) $\binom{7}{3} = \dfrac{7!}{3!(7-3)!} = \dfrac{7!}{3!4!}$

$= \dfrac{7 \cdot 6 \cdot 5 \cdot 4!}{3 \cdot 2 \cdot 1 \cdot 4!}$ Write 7! recursively.

$= \dfrac{7 \cdot 6 \cdot 5}{3 \cdot 2 \cdot 1}$ Divide out 4!.

$= 35$

EXAMPLE 5   Write $6 \cdot 5 \cdot 4$ in factorial notation.

SOLUTION   We know that

$$6! = 6 \cdot 5 \cdot 4 \cdot 3 \cdot 2 \cdot 1 \quad \text{and} \quad 3! = 3 \cdot 2 \cdot 1$$

thus

$$6 \cdot 5 \cdot 4 = \dfrac{6 \cdot 5 \cdot 4 \cdot 3 \cdot 2 \cdot 1}{3 \cdot 2 \cdot 1} = \dfrac{6!}{3!}$$

What do 0! and 1! mean? If we recall the recursive formula

$$n! = n(n-1)!$$

then

$$1! = 1(1-1)! = 1 \cdot 0!$$

Since $1! = 1$ by definition, and $1 \cdot 0! = 1!$, then 0! must also be defined to be 1 for consistency.

$$1! = 0! = 1$$

The next two equations now follow:

$$\binom{n}{0} = \dfrac{n!}{0!(n-0)!} = \dfrac{n!}{1 \cdot n!} = 1$$

$$\binom{n}{n} = \dfrac{n!}{n!(n-n)!} = \dfrac{n!}{n!0!} = 1$$

We are now ready to consider the *binomial expansion*. We begin by looking at various powers of $a + b$.

$(a + b)^0 = 1$

$(a + b)^1 = a + b$

$(a + b)^2 = a^2 + 2ab + b^2$

$(a + b)^3 = a^3 + 3a^2b + 3ab^2 + b^3$

$(a + b)^4 = a^4 + 4a^3b + 6a^2b^2 + 4ab^3 + b^4$

$(a + b)^5 = a^5 + 5a^4b + 10a^3b^2 + 10a^2b^3 + 5ab^4 + b^5$

We can make several observations about each of these expansions that apply to the expansion of $(a + b)^n$ in general:

1. The exponents on the first and last terms of the expansion are the same as the power of the binomial.
2. The sum of the exponents on any term of the expansion is the same as the power of the binomial.
3. As we read from left to right, the exponents on a decrease by 1, while the exponents on b increase by 1.
4. The coefficients of any term can be expressed in the form $\binom{n}{r}$, where r is 1 less than the number of the term; so $r = 0$ on the first term, $r = 1$ on the second term, and so on.

For example, for $n = 4$,

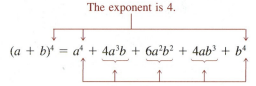

The sum of the exponents on each term is 4 and the exponents on a decrease by 1 as the exponents on b increase by 1.

When we apply these observations to $(a + b)^n$, the **binomial theorem** is the result.

The Binomial Theorem

$$(a + b)^n = \binom{n}{0}a^n + \binom{n}{1}a^{n-1}b + \binom{n}{2}a^{n-2}b^2 + \cdots + \binom{n}{n-1}ab^{n-1} + \binom{n}{n}b^n$$

n is a positive integer.

If we write each coefficient of the binomial expansion in factorial notation and simplify, an alternate form results.

The Binomial Theorem: Alternate Form

$$(a + b)^n = a^n + na^{n-1}b + \frac{n(n-1)}{2!}a^{n-2}b^2 + \cdots + nab^{n-1} + b^n$$

EXAMPLE 6 Verify the binomial theorem for $(a + b)^5$ by comparing it to the general theorem itself.

SOLUTION
$$(a + b)^5 = \binom{5}{0}a^5 + \binom{5}{1}a^4b + \binom{5}{2}a^3b^2 + \binom{5}{3}a^2b^3 + \binom{5}{4}ab^4 + \binom{5}{5}b^5$$

$$= a^5 + \frac{5!}{1!4!}a^4b + \frac{5!}{2!3!}a^3b^2 + \frac{5!}{3!2!}a^2b^3 + \frac{5!}{4!1!}ab^4 + b^5$$

When we evaluate each factorial, we have
$$(a + b)^5 = a^5 + 5a^4b + 10a^3b^2 + 10a^2b^3 + 5ab^4 + b^5$$

EXAMPLE 7 Use the alternate form of the binomial theorem to expand $(x - 3y)^4$.

SOLUTION We think of $(x - 3y)^4$ as $[x + (-3y)]^4$. Then if $(a + b)^4 = [x + (-3y)]^4$, we have $a = x$ and $b = -3y$. Therefore

$$[x + (-3y)]^4 = x^4 + 4x^3(-3y) + \frac{4 \cdot 3}{2!}x^2(-3y)^2 + \frac{4 \cdot 3 \cdot 2}{3!}x(-3y)^3 + (-3y)^4$$

$$= x^4 - 12x^3y + 54x^2y^2 - 108xy^3 + 81y^4$$

It's possible to find a particular term without carrying out the entire binomial expansion. Notice that

$$\binom{n}{2}a^{n-2}b^2 \quad \text{represents the third term in the binomial expansion}$$

and

$$\binom{n}{3}a^{n-3}b^3 \quad \text{represents the fourth term}$$

The following formula gives us a way to find any term in the binomial expansion when we know r, the number of the term.

rth Term in the Binomial Expansion

The **rth term of the binomial expansion** is given by the formula
$$\binom{n}{r - 1} a^{n-(r-1)}b^{r-1}$$

EXAMPLE 8 Find the fifth term of the expansion of $(c^2 + 2d)^8$.

SOLUTION Here $r = 5$, $n = 8$, $a = c^2$, and $b = 2d$, so

$$\binom{8}{5 - 1}(c^2)^{8-(5-1)}(2d)^4 = \binom{8}{4}(c^2)^4(2d)^4$$

$$= \frac{8!}{4!4!} c^8(16d^4)$$
$$= 1120c^8d^4$$

EXAMPLE 9 Find the term containing x^6 in the expansion of $(x^2 + y^3)^7$.

SOLUTION Here $a = x^2$, $b = y^3$, and $n = 7$. Notice that $x^6 = (x^2)^3 = a^3$. Solving for r yields $r = 5$. To see this, notice once again that in the binomial expansion,

$$(a + b)^n = \binom{n}{0}a^n + \binom{n}{1}a^{n-1}b + \binom{n}{2}a^{n-2}b^2 + \cdots + \binom{n}{n}b^n$$

the exponent on a decreases by 1 on each successive term. Since the expansion in this problem involves a seventh-degree polynomial, the exponent 3 occurs on the fifth term, as follows:

Term	1	2	3	4	5	
Exponent	a^7	a^6	a^5	a^4	a^3	$\leftarrow a^3 = (x^2)^3 = x^6$

We use the formula for the rth term with $r = 5$:

$$\binom{n}{r-1}a^{n-(r-1)}b^{r-1} = \binom{7}{4}(x^2)^{7-(5-1)}(y^3)^4$$
$$= \frac{7!}{4!3!}(x^2)^3(y^3)^4$$
$$= 35x^6y^{12}$$

Do You Remember?

Can you match these?

___ 1. The third term of $(a + b)^n$

___ 2. $\frac{6!}{3!}$

___ 3. $\binom{8}{6}$

___ 4. The formula for the rth term in the binomial expansion

___ 5. The fourth term of $(a + b)^n$

a) 120

b) $\binom{n}{r-1}a^{n-(r-1)}b^{r-1}$

c) 56

d) 28

e) $\frac{n(n-1)(n-2)}{3!}a^{n-3}b^3$

f) $\binom{n}{2}a^{n-2}b^2$

Answers: 1. f 2. a 3. d 4. b 5. e

Exercise Set 12.4

Evaluate each of the following quantities.

1. $3!$
2. $5!$
3. $10!$
4. $8!$
5. $\dfrac{20!}{19!}$
6. $\dfrac{18!}{17!}$
7. $\dfrac{6!}{4!2!}$
8. $\dfrac{7!}{5!2!}$
9. $\dfrac{5!}{0!(5-0)!}$
10. $\dfrac{9!}{0!(9-0)!}$
11. $\dfrac{11!}{8!(11-8)!}$
12. $\dfrac{12!}{7!(12-7)!}$
13. $\binom{6}{5}$
14. $\binom{4}{3}$
15. $\binom{5}{1}$
16. $\binom{3}{1}$
17. $\binom{7}{7}$
18. $\binom{2}{2}$
19. $\binom{5}{0}$
20. $\binom{4}{0}$

Write each of the following quantities as the quotient of two factorials.

21. 8
22. 6
23. $5 \cdot 6 \cdot 7$
24. $7 \cdot 8 \cdot 9$
25. $12 \cdot 11 \cdot 10 \cdot 9$
26. $8 \cdot 7 \cdot 6 \cdot 5$
27. $3 \cdot 2 \cdot 1$
28. $4 \cdot 3 \cdot 2 \cdot 1$

Write each of the following quotients without factorial notation; simplify.

29. $\dfrac{n!}{(n-1)!}$
30. $\dfrac{(n+1)!}{n!}$
31. $\dfrac{(n+3)!}{(n-1)!}$
32. $\dfrac{n!}{(n-3)!}$

Use the binomial theorem to expand each of the following powers.

33. $(x + y)^3$
34. $(u + v)^4$
35. $(a + 2b)^5$
36. $(h + 3k)^6$
37. $(u - v)^7$
38. $(x - y)^5$
39. $(2x - y)^4$
40. $(3u - 2v)^6$
41. $(3h^2 - 5k)^4$
42. $(3r^2 - 4s)^5$

Find the indicated term in each expansion and simplify.

43. $(x^2 - y)^8$; third
44. $(t + 3)^7$; fourth
45. $(r + 2s)^9$; sixth
46. $(3x - 2y)^7$; fifth
47. $(2p^2 - q^2)^7$; fifth
48. $(u^3 - 3v^2)^6$; third
49. $\left(\dfrac{x}{2} - 3\right)^{12}$; fourth
50. $\left(\dfrac{p}{2} - \dfrac{1}{3}\right)^{14}$; seventh

Approximate the following powers by using the binomial expansion. Use only the first four terms. A calculator may be helpful. Do not round the results.

51. $(1.01)^8$ *Hint:* $(1.01)^8 = (1 + 0.01)^8$
52. $(0.99)^7$
53. $(0.98)^{12}$
54. $(2.01)^5$

For Extra Thought

Pascal's triangle is another way to find the coefficients in a binomial expansion.

Pascal's triangle

55. Can you see a pattern? If so, write the next line of numbers.
56. Use Pascals triangle to expand $(a + b)^8$.
57. Add the numbers in each row; look for a pattern. Come up with a generalization for finding the sum of the numbers in the *n*th row.
58. Write two other patterns that you can see in Pascal's triangle. (There are many patterns.)

Summary

12.1 Sequences, Series, and Summation Notation

A **sequence** is a function with domain equal to the set of positive integers. The numbers that make up a sequence are called its **terms**. When a sequence is written in the form

$$a_1, a_2, a_3, \ldots, a_n, \ldots$$

the subscripts indicate the number of the term. If no last value of n is specified, the sequence is **infinite.** When a last term is specified, the sequence is **finite.** A sequence can be generated by a formula that describes the nth term in the sequence as a function of n, or it can be defined by a **recursion formula,** which relates any term a_{n+1} to the preceding term, a_n.

A **series** is the indicated sum of a finite or infinite sequence. A finite series $a_1 + a_2 + a_3 + \cdots + a_n$ can be written using **summation notation** as follows:

$$\sum_{i=1}^{n} a_i$$

where i is the index of summation and the number of terms is n.

12.2–12.3 Arithmetic and Geometric Sequences and Series

Two important types of sequences are **arithmetic sequences** and **geometric sequences.** Any two successive terms of an arithmetic sequence differ by a constant called the **common difference.** In a geometric sequence, any two successive terms have the same ratio, that is,

$$\frac{a_{n+1}}{a_n} = r, \qquad n \geq 1$$

where r is a constant called the **common ratio.**

The following recursion formulas describe arithmetic and geometric sequences.

Arithmetic	Geometric
$a_{n+1} = a_n + d$	$a_{n+1} = a_n r$

The nth term and sum of a sequence can be found by using the following formulas.

Type of Sequence	nth Term	Sum of n Terms	Abbreviations Used
Arithmetic	$a_n = a_1 + (n-1)d$	$S_n = \dfrac{n}{2}(a_1 + a_n)$ $S_n = \dfrac{n}{2}[2a_1 + (n-1)d]$	a_1 = first term n = term number d = common difference
Finite geometric	$a_n = a_1 r^{n-1}$	$S_n = \dfrac{a_1(1-r^n)}{1-r}$	a_1 = first term n = term number r = common ratio
Infinite geometric	$a_n = a_1 r^{n-1}$	$S_\infty = \dfrac{a_1}{1-r},\ -1 < r < 1$	a_1 = first term n = term number r = common ratio

12.4 The Binomial Theorem

Factorial notation, symbolized by $n!$, is used to indicate the product of the first n positive integers:

$$n! = n(n-1)(n-2)(n-3) \cdots 3 \cdot 2 \cdot 1$$

When defined recursively, $n! = n(n-1)!$.

The **binomial theorem** is a formula for evaluating binomials raised to positive integer powers. There are two common forms:

$$(a + b)^n = a^n + na^{n-1}b + \frac{n(n-1)}{2!}a^{n-2}b^2 + \cdots + nab^{n-1} + b^n$$

and

$$(a + b)^n = \binom{n}{0}a^n + \binom{n}{1}a^{n-1}b + \binom{n}{2}a^{n-2}b^2 + \cdots + \binom{n}{n-1}ab^{n-1} + \binom{n}{n}b^n$$

where

$$\binom{n}{r} = \frac{n!}{r!(n-r)!}, \quad \binom{n}{0} = 1, \quad \binom{n}{n} = 1$$

and $0! = 1! = 1$.

The **rth term of a binomial expansion** is given by the formula

$$\binom{n}{r-1}a^{n-(r-1)}b^{r-1}$$

COOPERATIVE EXERCISE

PART 1
Find the center of a circle passing through the points $A(3, 7)$, $B(7, 5)$, and $C(0, 6)$ by finding the coordinates of the point of intersection of the perpendicular bisectors of the line segments joining adjacent pairs of points A, B, and C. Find the length of the radius and write the standard form for the equation of a circle, $(x - h)^2 + (y - k)^2 = r^2$. Graph the circle.

PART II
Find the equation of a parabola passing through the points $A(3, 7)$, $B(7, 5)$, and $C(0, 6)$. Use the general form for the equation of a parabola, $y = ax^2 + bx + c$. *Hint:* Use systems of equations. Graph the parabola.

PART III
Use the information obtained in Parts I and II to:

(a) Find the y-coordinates of the point on the circle and the one on the parabola when $x = 3$. Are the y-coordinates the same? Did you expect them to be? Why or why not?

(b) Find the positive x-intercepts of the circle and the parabola. Are the x-intercepts the same? Did you expect them to be? Why or why not?

(c) Find the coordinates of the vertex of the parabola. Does this point lie on the circle? Did you expect it to? Why or why not?

REVIEW EXERCISES

Write the first four terms of each sequence having the given formula for a_n.

1. $a_n = n^2$
2. $a_n = (-2)^{n+1}$
3. $a_n = (-1)^n(2n + 1)$
4. $a_n = (-1)^{n+1}x^n$

Find the indicated unknowns for each of the following sequences.

5. $a_1 = 1, d = 4; a_8; a_{11}$
6. $a_1 = 2, d = -3; a_6; a_9$
7. $a_9 = 18, S_9 = 80; a_1; d$
8. $a_n = 2 + 5n; S_{12}; S_{18}$
9. $a_1 = 5, r = 1; a_6; S_6$
10. $a_1 = 64, r = \frac{1}{2}; a_6; S_6$

11. $r = -3, a_1 = 5; a_6; S_6$

12. $a_3 = 125, r = \dfrac{1}{5}; a_1; S_4$

In Exercises 13–16, write each sum in expanded form and find the total.

13. $\sum_{i=1}^{10} (i - 1)$

14. $\sum_{i=1}^{4} (2)^i$

15. $\sum_{i=1}^{5} (-1)^i$

16. $\sum_{i=1}^{8} (3i - 2)$

17. Find the sum of all the even integers between 5 and 59.

18. Find three numbers that form an arithmetic sequence if their sum is 33 and their product is 1320.

Evaluate each of the following quantities.

19. $7!$

20. $\dfrac{35!}{33!}$

21. $\binom{8}{6}$

22. $\dfrac{n!}{(n-2)!}$

23. $\dfrac{13!}{9!(13-9)!}$

24. $\binom{5}{0}$

Use the binomial theorem to expand each of the following powers.

25. $(a - b)^5$

26. $(3x - 2y)^6$

27. $(2p^2 - 3q^2)^4$

28. $(u - v)^8$

Find the indicated term in each expansion.

29. $(x - 2y)^{10}$; fourth

30. $(x - y)^{14}$; middle term

31. $(2a - b)^{11}$; fifth

32. $\left(\dfrac{p}{2} + \dfrac{1}{3}\right)^{15}$; seventh

Solve each of the following applied problems. Round your answers to four decimal places, if needed.

33. At the end of each year, a certain motorcycle is worth 91% of its value at the beginning of that year. If the motorcycle cost $3800 new, what is its value at the end of three years?

34. Twyeffort Plastering had a gross income of $56,000 during its first year of operation. If the income increases by $11,000 each succeeding year, find the total income during the first 6 years of operation.

35. A ball is dropped from a height of 6 ft. Each time the ball strikes the ground, it rebounds seven-eighths of the distance it just fell. Find the total distance the ball travels before coming to rest.

36. The tip of a pendulum travels 11 in. on the first swing and 8% less on each successive swing. Find the distance traveled by the tip of the pendulum in 12 swings.

37. With a calculator, approximate the value of $(1.03)^{12}$ using the first four terms of the binomial expansion. Do not round off.

Chapter Test

Write the first five terms of each of the following sequences.

1. $a_n = n^2 - 3$

2. $a_n = 3^{n-1}$

3. $a_n = (-1)^n (2n - 1)$

4. $a_1 = 2; a_{n+1} = a_n + 3$

Find the sum.

5. $\sum_{i=1}^{5} (i + 3)$

6. $\sum_{i=1}^{6} (-1)^{i+1} 2^i$

In Exercises 7–9, find the indicated unknowns for each of the following sequences.

7. $a_1 = 5, d = 3; a_5; S_5$

8. $a_1 = 4, r = 2; a_4; S_4$

9. $a_1 = 64, r = \dfrac{1}{2}; S_\infty$

10. Evaluate: $\binom{6}{3}$.

11. Evaluate: $\dfrac{8!}{3!5!}$.

12. Evaluate: $\binom{7}{0}$.

13. Expand $(a + b)^6$.

14. Expand $(x - y)^5$.

15. Find the fifth term of $(2x - y)^9$.

16. Change $0.111\ldots$ into common fraction form.

17. Change $0.161616\ldots$ into common fraction form.

18. Find the sum of all the multiples of 7 between 13 and 99.

19. A ball is dropped from a height of 8 m. Each time it strikes the ground, it rebounds 60% of the distance it fell. Find the total distance the ball travels before coming to rest.

20. The tip of the pendulum on a grandfather clock travels through an arc of 30 cm on its first swing. If it travels 10% less on each successive swing, how far will it travel in 8 swings? Round your answer to four decimal places.

Cumulative Review for Chapters 10–12

For each conic section in Exercises 1–6, (a) put it in standard form, (b) name it, (c) sketch the graph, and (d) find the coordinates of the center or the vertex.

1. $x^2 + y^2 - 2x = 3$
2. $x^2 + y^2 + x - y = \dfrac{1}{2}$
3. $4x^2 + y^2 + 4y - 24x = -24$
4. $x^2 - y^2 + 2y - 2x = 4$
5. $f(x) = -2x^2 + 4x - 3$
6. $f(x) = -2x^2 + 8x - 5$

Use the given graph to write the equation of the conic section; name it.

7.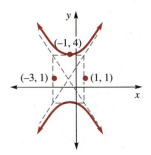

8.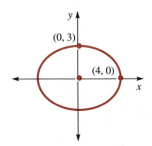

9.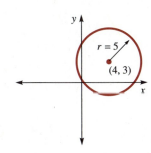

10.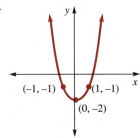

Solve each system of nonlinear equations.

11. $x^2 + y^2 = 4$
    $x^2 + 4y^2 = 4$

12. $x^2 + y^2 = 13$
    $2x + y = -1$

13. $y = 3x^2 - 4x + 5$
    $y = 3x^2 + 2x + 1$

For Exercises 14 and 15, solve each system of nonlinear inequalities by graphing.

14. $x^2 + y^2 \le 25$
    $x^2 - y^2 \ge 9$

15. $25x^2 + 4y^2 \le 100$
    $x^2 - y^2 \le -4$

16. Graph $f(x) = \log_{10} x$. Find the domain and the range.

17. Graph $f(x) = -\left(\dfrac{1}{2}\right)^x$. Find the domain and the range.

18. Solve: $\log_3 x + \log_3 2 = 2$.

19. Solve: $\log_2 16 = \log_2(x - 1)$.

20. Given $P(x) = 14.7e^{-0.00004x}$, find $P(2640)$.

21. Given $P(t) = P_0 e^{0.05t}$ and $P_0 = 50{,}000$, find $P(30)$.

22. For $f(x) = b^x$, find $f^{-1}(x)$.

23. When an initial amount of money, A_0, is invested at 6% compounded continuously, it grows to an amount A according to the formula $A(t) = A_0 e^{0.06t}$, where t is time in years. How long will it take the initial amount to double (rounded to the nearest whole year)?

24. Find the first four terms of the sequence with general term $a_n = \dfrac{2}{n}$.

25. Find the first three terms of the sequence with $a_1 = 12$ and $a_n = \dfrac{a_{n-1}}{n}$, $n \ge 2$.

26. Find the general term a_n for the following sequence: 1, 3, 5, 7, 9, ...

27. Find $\displaystyle\sum_{i=1}^{4}(-3)^i$.

28. Find S_{31} for the arithmetic sequence with $a_1 = 100$ and $d = -5$.
29. Find a_n for the sequence 1, 2, 4, 8, ...
30. Find the sum of the first five terms of the geometric sequence with $a_1 = 4$ and $r = -\dfrac{1}{2}$.
31. Find $\sum\limits_{i=1}^{\infty} 2 \cdot \left(\dfrac{1}{2}\right)^{i-1}$.

Use the binomial theorem for Exercises 32–35.

32. Expand $(x + 2)^3$.
33. Find the fifth term of $(x + 1)^6$.
34. Find the eighth term of $(y - 1)^{10}$.
35. Expand $(x^2 - y)^5$.

Appendices

Appendix A: Graphics Calculators   A-2
Appendix B: Tables   A-14
Appendix C: Answers   A-20

APPENDIX A
GRAPHICS CALCULATORS

Handheld calculators are so commonplace today that it is difficult to find a student in any mathematics class that doesn't own either a graphics calculator or a scientific calculator. Your instructor will set the tone concerning their use in this course.

The TI-81 and TI-82:   A Brief Introduction

A graphics calculator can perform all of the operations that are found on a scientific calculator as well as to graph certain functions. They can also handle many advanced functions and can be programmed to perform repetitive tasks. The keyboard of the TI-81 is shown in Figure A1 and that of the TI-82 is shown in Figure A2. Notice that the keyboards are similar in appearance and that many of the keys are labeled the same.

The **display screen** at the top of the calculator is where any entries for calculations and graphing is displayed.

The **graphing keys** are located immediately below the display screen. These five keys serve to define and manipulate the graphs of functions.

The **2nd key** (blue) is used to access the functions displayed above the other keys on the keyboard. To access them, first press the 2nd key and then the desired key.

The **ALPHA key** allows access to letters such as A, B, and C that appear above other keys.

The **insert,** INS , and **delete,** DEL , **keys** are used for editing previous entries. When the insert key is pressed, a blinking underline appears where the insertion will be made. When the delete key is pressed, the character under the curser is deleted. On the T1-82, the delete and insert functions share the same key.

The **arrow keys,** blue keys on the TI-81 and purple on the TI-82, have arrowheads on their surfaces; these are used to move the cursor around in the display for editing purposes. Depressing an arrow key moves the cursor left or right on the screen.

The **advanced function keys** are MATH , MATRX , PRGM , and VARS . These are used to perform operations that we will not discuss in this appendix.

The **numeric keypad** is comprised of the twelve keys at the bottom half of

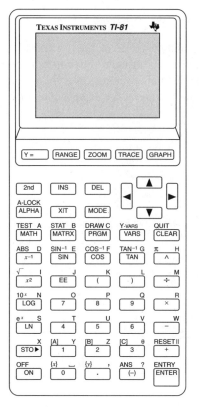

FIGURE A1
The TI-81

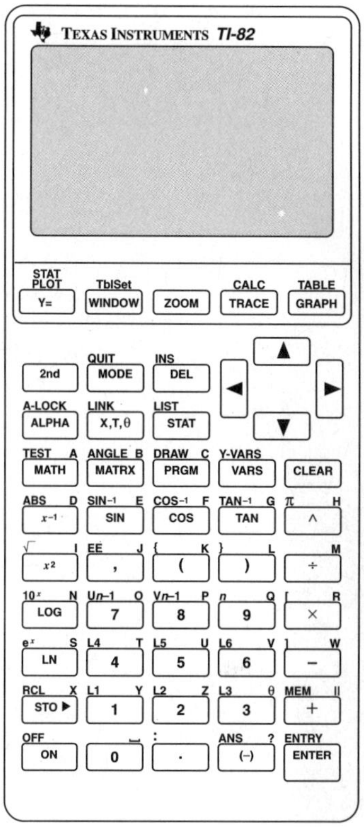

FIGURE A2
The TI-82

the calculator. It is used to enter positive and negative numbers and the decimal point.

The **operation keys** are the four blue keys (TI-81) or purple keys (TI-82) marked $\boxed{\div}$, $\boxed{\times}$, $\boxed{-}$, and $\boxed{+}$ and located on the right side of the calculator. They are used to divide, multiply, subtract, and add, respectively.

The **ENTER key** is located in the lower right corner of the keypad on both calculators. It is used to execute the calculations that have been typed in.

Getting Started

Turn your calculator on by pressing the $\boxed{\text{ON}}$ key located in the lower left corner of the keypad. The **cursor** (a blinking square) will appear on the screen. If the cursor is difficult to see, you can change the contrast by pressing and releasing the $\boxed{\text{2nd}}$ function key and then holding the down-arrow key, $\boxed{\triangledown}$, to lighten the background, or the up-arrow key, $\boxed{\triangle}$, to darken it. Often all that is necessary to improve the visibility of the screen is to tilt it slightly so that the light strikes it at a different angle.

Arithmetic Calculations

The calculation mode of the TI-81 and TI-82 graphics calculators operates in a slightly different manner from a four-function calculator or from a scientific calculator. One distinct advantage over many calculators is that everything you enter will show and remain on the screen. This makes entries easy to correct when an error is made.

EXAMPLE 1 Calculate $15 - 4 \cdot 3$.

SOLUTION We type

15 $\boxed{-}$ 4 $\boxed{\times}$ 3

and press the $\boxed{\text{ENTER}}$ key. The answer is 3. The screen is shown in Figure A3

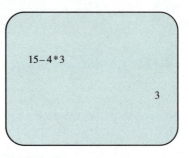

FIGURE A3

Both calculators follow the normal order of operations. We must be careful not to leave anything out when entering a calculation.

EXAMPLE 2 Calculate $5(11 - 7) + 12$.

SOLUTION We type

5 $\boxed{(}$ 11 $\boxed{-}$ 7 $\boxed{)}$ $\boxed{+}$ 12

and press the $\boxed{\text{ENTER}}$ key. The result is 32. The screen is shown in Figure A4.

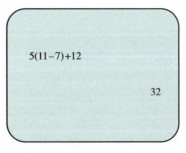

FIGURE A4

Editing Calculations

The TI-81 and TI-82 have an advantage over many calculators. *If an error is discovered in an entered calculation, it can be changed quickly if done before the* ENTER *key is pressed.*

EXAMPLE 3 Evaluate $x(x + 3 + 6x) - 4$ for $x = 15$.

SOLUTION We type incorrectly

15　(　15　+　3　+　6　)　15　−　4

but do not press the ENTER key. The calculation is entered incorrectly, as shown on the screen in Figure A5(a). The second parenthesis is in the wrong place and the multiplication sign is missing. To correct this, we use the left-arrow key, ◁, to move the cursor over the second parenthesis and press the × key to replace it. Then we move the cursor to the position where the parenthesis belongs by using the right-arrow key, ▷, and use the INS key followed by the) key to insert it. (The insert key on the TI-82 is accessed by first pressing the 2nd key and then the DEL key. Now we press the ENTER key to complete the calculation. The corrected screen and answer are shown in Figure A5(b).

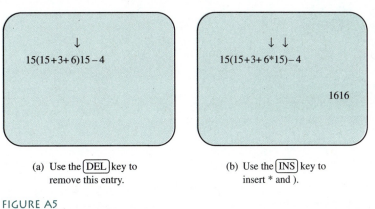

(a) Use the DEL key to remove this entry.

(b) Use the INS key to insert * and).

FIGURE A5

Note: *If you do not notice the error until you have pressed* ENTER, *it is still possible to make the correction without entering the entire expression again.* By pressing the up-arrow key, △, on the TI-81, we will cause the last expression entered on the screen to reappear. We can now make the corrections. A second method of recapturing the last expression works on both the TI-81 and TI-82: we press the 2nd key followed by the **ENTRY key,** located above the ENTER key on the lower right corner of the keyboard.

Clearing the Screen

To clear the entire screen on the TI-81, we use the CLEAR **key.** To clear a single character, we position the cursor over the character and press the DEL key. Pressing the CLEAR key on the TI-82 clears the current computation if it has not been completed but it leaves other calculations on the screen. Pressing the $\boxed{\text{CLEAR}}$ key a second time clears the entire screen.

Computations Involving Negative Numbers and Exponents

To type a negative number, we press the $\boxed{(-)}$ key in the last row and then the desired number.

EXAMPLE 4 Calculate $225 - (-412)$.

SOLUTION We type

225 $\boxed{-}$ $\boxed{(}$ $\boxed{(-)}$ 412 $\boxed{)}$

and press the $\boxed{\text{ENTER}}$ key. The screen is shown in Figure A6.

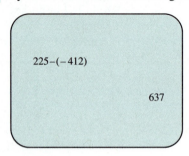

FIGURE A6

The square of a number can be indicated two ways on the TI-81 and TI-82. The first way is to enter the number and then press the $\boxed{x^2}$ key. The second method is to enter the desired number followed by the $\boxed{\wedge}$ key and the number 2. The easiest way to enter powers higher than 2 is to use the $\boxed{\wedge}$ key followed by the desired power.

EXAMPLE 5 Calculate $15^2 - (-3)^4$.

SOLUTION We type

15 $\boxed{x^2}$ $\boxed{-}$ $\boxed{(}$ $\boxed{(-)}$ 3 $\boxed{)}$ $\boxed{\wedge}$ 4

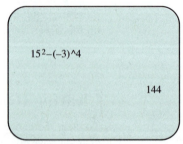

FIGURE A7

and then press the ENTER key. The answer is 144. The screen is shown in Figure A7.

EXAMPLE 6 Calculate $3^5 - 4^2(17 + 9(15 - 2))$.

SOLUTION We type

3   ^   5   −   4 x^2 (17   +   9   (15   −   2   )   )

and press the ENTER key. The answer is -1901. The screen is shown in Figure A8.

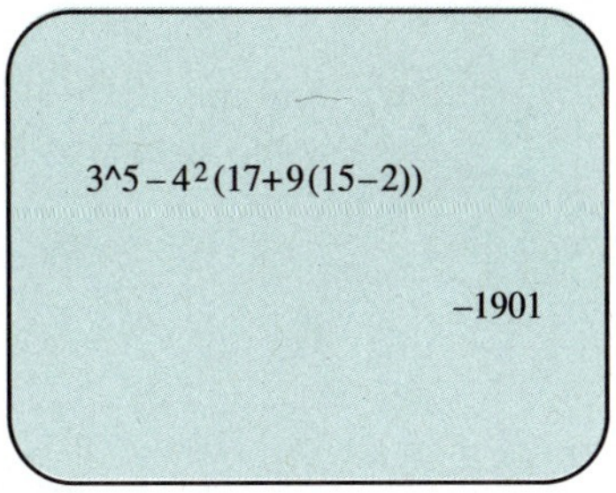

```
3^5–4²(17+9(15–2))
                        –1901
```

FIGURE A8

The ANS Key

When a computation has been completed, the result is stored in a location that can be accessed by pressing the 2nd key followed by the **ANS** and **ENTER** keys. Supposes we complete a calculation and then inadvertently clear the screen before writing the answer down. By typing

2nd   ANS   ENTER

"Ans" and the result of the previous computation will appear on the screen.

EXAMPLE 7 Calculate $2 + 3^2 + 9^4$. Clear the screen and then recover the answer.

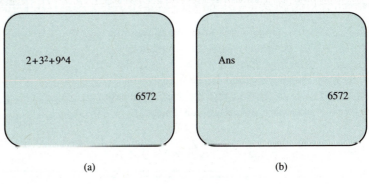

(a)   (b)

FIGURE A9

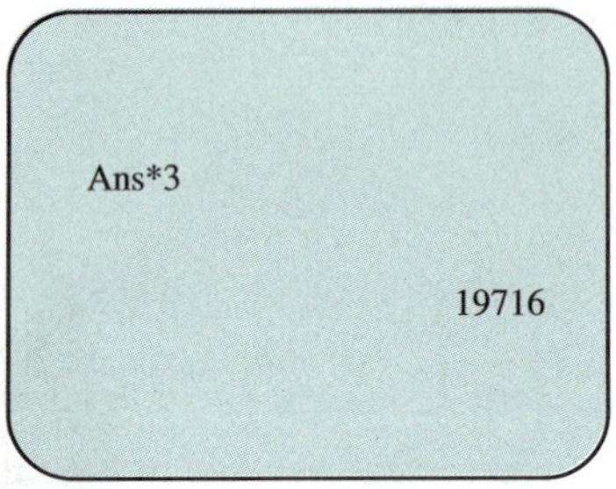

SOLUTION We type

2 $\boxed{+}$ 3 $\boxed{x^2}$ $\boxed{+}$ 9 $\boxed{\wedge}$ 4 $\boxed{\text{ENTER}}$

The answer is 6572; the screen is shown in Figure A9(a).

To clear the screen we press $\boxed{\text{CLEAR}}$. Then to recover the answer, we press $\boxed{\text{2nd}}$ $\boxed{\text{ANS}}$ $\boxed{\text{ENTER}}$.

The screen is shown in Figure A9(b)

```
Ans*3
         19716
```

FIGURE A10

EXAMPLE 8 Multiply the result of Example 7 by 3.
If the answer is still on the screen, we just type

$\boxed{\times}$ 3 $\boxed{\text{ENTER}}$

The answer is 19716. The screen is shown in Figure A10.

Do You Remember?

Can you match these?

_____ 1. A single key that allows you to recall the last computation
_____ 2. The key that allows you to insert a character into a calculation
_____ 3. The key that causes a calculation to be completed
_____ 4. The key used to indicate that a number is negative
_____ 5. The key used to enter an exponent other than 2
_____ 6. The key used to delete a character in a computation
_____ 7. A combination of keys that allows you to recall the previous calculation
_____ 8. The keys that allow you to move the cursor about the screen
_____ 9. The keystrokes that are used to recover the result of the last calculation

a) $\boxed{\text{DEL}}$ $\boxed{\text{INS}}$
b) $\boxed{\text{ENTER}}$
c) $\boxed{\wedge}$
d) $\boxed{\triangle}$
e) $\boxed{\text{DEL}}$
f) The arrows
g) $\boxed{(-)}$
h) $\boxed{\text{2nd}}$ $\boxed{\text{ENTER}}$
i) $\boxed{\text{2nd}}$ $\boxed{\text{ANS}}$ $\boxed{\text{ENTER}}$
j) $\boxed{\text{INS}}$

Answers: 1. d 2. j 3. b 4. g 5. c 6. e 7. h 8. f 9. i

Exercise Set A1

1. Carry out the following computations:
   (a) $3 \cdot 8 - \dfrac{12.6}{3}$; (b) $15(13.2 - 4.56) + 11$;
   (c) $(-3.2)^5 + (2.3)^2$.
2. Evaluate $3x^2 - 2x + 4$ for $x = 25$.
3. Type in $24 + 12 \div 4 + 3$.
   (a) Press the [ENTER] key and record the result.
   (b) Use the up-arrow key, [△] (TI-81), or [2nd] [(−)] [ENTER] to redisplay the computation. Change it to $24 + 12/(4 + 3)$. Press the [ENTER] key and record the result.
   (c) Use the [2nd] key and the [ENTRY] key to redisplay the computation. Change it to $4 + 22/(4 + 3)$ by replacing 12 by 22. Press the [ENTER] key and record the result.

Solutions

1. (a) 19.8   (b) 140.6   (c) -330.25432
2. 1829
3. (a) 30   (b) 25.71428571   (c) 7.142857143

Graphing with the TI-81 and TI-82

The graphics capabilities of both of these calculators is so vast that the information that follows is just enough to get you started. It does not necessarily present the most efficient methods of graphing and is intended only to get you interested enough to explore the remaining possibilities discussed in the instruction manual that comes with the calculator.

The five graphing keys on the TI-81 and the TI-82 are the [Y=] key, the [RANGE] key (TI-81) or the [WINDOW] key (TI-82), the [ZOOM] key, the [TRACE] key, and the [GRAPH] key. These keys appear in the first row, just below the screen. Their functions are as follows:

[Y=] key	When you press Y=, an edit screen appears that allows you to enter up to four functions on the TI-81 and up to eight on the TI-82. To enter an equation such as $y = x + 4$, you would press the [X|T] key (TI-81) or the [X,T,θ] key (TI-82) to obtain the x, or you can also press the [ALPHA] key followed by X. The other characters are obtained in the usual manner.
[RANGE] key (TI-81) or [WINDOW] key (TI-82)	When you press the [RANGE] key, an edit screen appears so you can define the lengths of the x- and y-axes that will appear in the display, as well as the distances between marks on the axes. The lengths of the axes and the scales on the two axes do not have to be the same.
[ZOOM] key	When you press the [ZOOM] key, a menu of instructions appears that allows you to change the dimensions of the viewing rectangle.

TRACE key	When you press the TRACE key, you can move the cursor along the curve by using the ◁ and ▷ keys to display the *x*- and *y*-coordinates of its location.
GRAPH key	When you press the GRAPH key, the functions that were entered will be graphed on the screen.

To follow the next few examples, make sure your calculator is in **rectangular mode.** On the TI-81, press the MODE key. At the bottom of the screen, the word Rect should be shaded. If it isn't, use the arrow keys to move the blinking cursor over Rect and press the ENTER key. On the T1-82, press the WINDOW key and use the ▷ key to move the shading to **FORMAT.** If the word **RectGC** is not shaded, use the arrow keys to move the blinking cursor to it and press the ENTER key to select it

EXAMPLE 1 Graph $y = 2x + 4$ and determine where it crosses the *x*-axis by using the TRACE key. Remember that X|T and X,T,θ both result in an X appearing on the screen.

SOLUTION Press the Y= key and type

2 X|T + 4

by the $Y_1=$ on the screen. Next press the GRAPH key. The screen shown in Figure A11 should look similar to yours, since your RANGE settings may vary. *Do not clear the screen at this point.* Now press the TRACE key and use the arrow keys to move the **trace** (a small blinking square) along the graph until the *y*-coordinate shown at the bottom of the screen is as close to zero as possible. The *x*-coordinate is approximately −2.

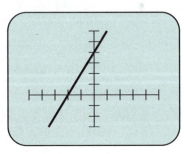

FIGURE A11

EXAMPLE 2 Sketch the graph of $y = 2x + 4$ for $-5 \leq x \leq 5$, $-3 \leq y \leq 6$.

SOLUTION This is the equation of Example 1, which should still be in your calculator. If you cleared the equation, re-enter it now. To set $-5 \leq x \leq 5$ and $-3 \leq y \leq 6$, press the RANGE key (TI-81) or the WINDOW key (TI-82). Then use the arrow keys to move the cursor around the screen and enter Xmin = −5, Xmax = 5, Ymin = −3, and Ymax = 6. Press the GRAPH key. The screen is shown in Figure A12.

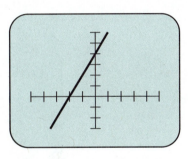

FIGURE A12

To exit the graphing mode and return to the home screen, press $\boxed{\text{2nd}}$ $\boxed{\text{CLEAR}}$.

The TI-81 is designed to allow up to four graphs to appear on the screen at one time; the TI-82 will allow up to eight. The next example shows how this feature can be used to find where two graphs intersect.

EXAMPLE 3 Graph $y = x + 1$ and $y = -2x + 3$ on the same screen, $-5 \leq x \leq 5$, $-5 \leq y \leq 5$. Use the $\boxed{\text{TRACE}}$ key to determine the point where they intersect.

SOLUTION Press the $\boxed{\text{Y=}}$ key and set $Y_1 = X + 1$. Next use the $\boxed{\triangledown}$ key to move to $Y_2=$ and set $Y_2 = -2X + 3$. We press the $\boxed{\text{RANGE}}$ (TI-81) or $\boxed{\text{WINDOW}}$ (TI-82) key and the screen appears as Figure A13(a). To set the maximum and minimum values for x and y, delete whatever entries are on the screen and enter -5 for both Xmin and Ymin and 5 for both Xmax and Ymax, as required. Now set Xscl = 1 and Yscl = 1. The new screen is shown in Figure A13(b). *Note:* Xres = 1 does not appear in WINDOW on the TI-82.

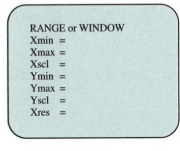

(a)   (b)

FIGURE A13

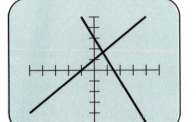

FIGURE A14

When you press the $\boxed{\text{GRAPH}}$ key, the new screen shown in Figure A14 appears. Press the $\boxed{\text{TRACE}}$ key and use the arrow keys to move the tracer to the apparent point of intersection. The value of x and y at this point appear at the bottom of the screen. You can get a better approximation by magnifying (zooming in on) the point of intersection. To do this, press the $\boxed{\text{ZOOM}}$ key to access the menus on the screen, as shown in Figure A15.

You can select **Zoom In** by using the arrow keys or you can type the number of Zoom In, 2, and press the $\boxed{\text{ENTER}}$ key. A new screen will appear with the

A-12 APPENDIX A · GRAPHICS CALCULATORS

```
        TI-81                           TI-82
  ZOOM                           ZOOM
  1: Box                         1: ZBox
  2: Zoom In                     2: Zoom In
  3: Zoom Out                    3: Zoom Out
  4: Set Factors                 4: ZDecimal
  5: Square                      5: ZSquare
  6: Standard                    6: ZStandard
  7↓ Trig                        7↓ ZTrig
```

FIGURE A15

tracer at the center. The view is improved by a factor of 4 in this manner. You can repeat the zoom operation as many times as you want, moving the tracer to the apparent intersection point after each zoom. After a few zooms, you can read the intersection as $x = 0.66118421$ and $y = 1.6617063$. (Yours may vary slightly). The actual intersection is $x = \frac{2}{3} \approx 0.666667$ and $y = \frac{5}{3} \approx 1.6666667$. You're getting close.

When using the ZOOM feature, the RANGE screen is automatically reset to reflect the changes in the maximum and minimum settings. You can press the $\boxed{\text{RANGE}}$ or $\boxed{\text{WINDOW}}$ key to observe the new settings; they're shown in Figure A16. They may vary slightly, depending on where you set the trace to approximate the intersection and on the number of times you zoomed in on it.

```
  RANGE
  Xmax = .3585526315
  Xmin = .9835526314
  Xscl = 1
  Ymin = 1.374007937
  Ymax = 1.999007937
  Yscl = 1
  Xres = 1
```

FIGURE A16

EXAMPLE 4   Graph $y = x^2 - 2x - 3$; $-6 \leq x \leq 6$, $-6 \leq y \leq 6$. Where does it cross the x-axis?

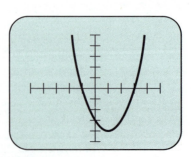

FIGURE A17

SOLUTION   Begin by setting the max and min values to 6 and −6 on the RANGE or WINDOW screen. Press the Y= key and enter the function as $Y_1 = X^2 − 2X − 3$; then press the GRAPH key. The screen is shown in Figure A17.

The graph appears to cross the x-axis at −1 and 3. The result can be checked in the original equation.

Do You Remember?

Can you match these?

_____ 1. The key used to draw the graph of an equation
_____ 2. The key used to enter an equation for graphing
_____ 3. The key used to magnify a portion of a graph
_____ 4. The key used to set the dimensions of the graph on the screen
_____ 5. The sequence of keys used to quit the graphing feature and return to the calculation screen

a) 2nd CLEAR
b) ZOOM
c) Y=
d) RANGE
e) GRAPH

Answers: 1. e 2. c 3. b 4. d 5. a

Exercise Set A2

1. Graph $y = 2x − 3$. Use the ZOOM feature to verify that it crosses the x-axis at $x = 1.5$.
2. Graph $y = x + 5$ and $y = −x + 2$ on the same screen. Where do they appear to intersect after three zooms?

SOLUTIONS
1. Verify
2. $x = −1.501645$, $y = 3.499504$; answers will vary slightly.

APPENDIX B
TABLES

Table 1 Squares and Square Roots

n	n^2	$\sqrt{n}$	$\sqrt{10n}$	n	n^2	$\sqrt{n}$	$\sqrt{10n}$
1	1	1.000	3.162	51	2601	7.141	22.583
2	4	1.414	4.472	52	2704	7.211	22.804
3	9	1.732	5.477	53	2809	7.280	23.022
4	16	2.000	6.325	54	2916	7.348	23.238
5	25	2.236	7.071	55	3025	7.416	23.452
6	36	2.449	7.746	56	3136	7.483	23.664
7	49	2.646	8.367	57	3249	7.550	23.875
8	64	2.828	8.944	58	3364	7.616	24.083
9	81	3.000	9.487	59	3481	7.681	24.290
10	100	3.162	10.000	60	3600	7.746	24.495
11	121	3.317	10.488	61	3721	7.810	24.698
12	144	3.464	10.954	62	3844	7.874	24.900
13	169	3.606	11.402	63	3969	7.937	25.100
14	196	3.742	11.832	64	4096	8.000	25.298
15	225	3.873	12.247	65	4225	8.062	25.495
16	256	4.000	12.649	66	4356	8.124	25.690
17	289	4.123	13.038	67	4489	8.185	25.884
18	324	4.243	13.416	68	4624	8.246	26.077
19	361	4.359	13.784	69	4761	8.307	26.268
20	400	4.472	14.142	70	4900	8.367	26.458
21	441	4.583	14.491	71	5041	8.426	26.646
22	484	4.690	14.832	72	5184	8.485	26.833
23	529	4.796	15.166	73	5329	8.544	27.019
24	576	4.899	15.492	74	5476	8.602	27.203
25	625	5.000	15.811	75	5625	8.660	27.386
26	676	5.099	16.125	76	5776	8.718	27.568
27	729	5.196	16.432	77	5929	8.775	27.749
28	784	5.292	16.733	78	6084	8.832	27.928
29	841	5.385	17.029	79	6241	8.888	28.107
30	900	5.477	17.321	80	6400	8.944	28.284
31	961	5.568	17.607	81	6561	9.000	28.460
32	1024	5.657	17.889	82	6724	9.055	28.636
33	1089	5.745	18.166	83	6889	9.110	28.810
34	1156	5.831	18.439	84	7056	9.165	28.983
35	1225	5.916	18.708	85	7225	9.220	29.155
36	1296	6.000	18.974	86	7396	9.274	29.326
37	1369	6.083	19.235	87	7569	9.327	29.496
38	1444	6.164	19.494	88	7744	9.381	29.665
39	1521	6.245	19.748	89	7921	9.434	29.833
40	1600	6.325	20.000	90	8100	9.487	30.000
41	1681	6.403	20.248	91	8281	9.539	30.166
42	1764	6.481	20.494	92	8464	9.592	30.332
43	1849	6.557	20.736	93	8649	9.644	30.496
44	1936	6.633	20.976	94	8836	9.695	30.659
45	2025	6.708	21.213	95	9025	9.747	30.822
46	2116	6.782	21.448	96	9216	9.798	30.984
47	2209	6.856	21.679	97	9409	9.849	31.145
48	2304	6.928	21.909	98	9604	9.899	31.305
49	2401	7.000	22.136	99	9801	9.950	31.464
50	2500	7.071	22.361	100	10000	10.000	31.623

Table 2 Common Logarithms

n	0	1	2	3	4	5	6	7	8	9
1.0	.0000	.0043	.0086	.0128	.0170	.0212	.0253	.0294	.0334	.0374
1.1	.0414	.0453	.0492	.0531	.0569	.0607	.0645	.0682	.0719	.0755
1.2	.0792	.0828	.0864	.0899	.0934	.0969	.1004	.1038	.1072	.1106
1.3	.1139	.1173	.1206	.1239	.1271	.1303	.1335	.1367	.1399	.1430
1.4	.1461	.1492	.1523	.1553	.1584	.1614	.1644	.1673	.1703	.1732
1.5	.1761	.1790	.1818	.1847	.1875	.1903	.1931	.1959	.1987	.2014
1.6	.2041	.2068	.2095	.2122	.2148	.2175	.2201	.2227	.2253	.2279
1.7	.2304	.2330	.2355	.2380	.2405	.2430	.2455	.2480	.2504	.2529
1.8	.2553	.2577	.2601	.2625	.2648	.2672	.2695	.2718	.2742	.2765
1.9	.2788	.2810	.2833	.2856	.2878	.2900	.2923	.2945	.2967	.2989
2.0	.3010	.3032	.3054	.3075	.3096	.3118	.3139	.3160	.3181	.3201
2.1	.3222	.3243	.3263	.3284	.3304	.3324	.3345	.3365	.3385	.3404
2.2	.3424	.3444	.3464	.3483	.3502	.3522	.3541	.3560	.3579	.3598
2.3	.3617	.3636	.3655	.3674	.3692	.3711	.3729	.3747	.3766	.3784
2.4	.3802	.3820	.3838	.3856	.3874	.3892	.3909	.3927	.3945	.3962
2.5	.3979	.3997	.4014	.4031	.4048	.4065	.4082	.4099	.4116	.4133
2.6	.4150	.4166	.4183	.4200	.4216	.4232	.4249	.4265	.4281	.4298
2.7	.4314	.4330	.4346	.4362	.4378	.4393	.4409	.4425	.4440	.4456
2.8	.4472	.4487	.4502	.4518	.4533	.4548	.4564	.4579	.4594	.4609
2.9	.4624	.4639	.4654	.4669	.4683	.4698	.4713	.4728	.4742	.4757
3.0	.4771	.4786	.4800	.4814	.4829	.4843	.4857	.4871	.4886	.4900
3.1	.4914	.4928	.4942	.4955	.4969	.4983	.4997	.5011	.5024	.5038
3.2	.5051	.5065	.5079	.5092	.5105	.5119	.5132	.5145	.5159	.5172
3.3	.5185	.5198	.5211	.5224	.5237	.5250	.5263	.5276	.5289	.5302
3.4	.5315	.5328	.5340	.5353	.5366	.5378	.5391	.5403	.5416	.5428
3.5	.5441	.5453	.5465	.5478	.5490	.5502	.5514	.5527	.5539	.5551
3.6	.5563	.5575	.5587	.5599	.5611	.5623	.5635	.5647	.5658	.5670
3.7	.5682	.5694	.5705	.5717	.5729	.5740	.5752	.5763	.5775	.5786
3.8	.5798	.5809	.5821	.5832	.5843	.5855	.5866	.5877	.5888	.5899
3.9	.5911	.5922	.5933	.5944	.5955	.5966	.5977	.5988	.5999	.6010
4.0	.6021	.6031	.6042	.6053	.6064	.6075	.6085	.6096	.6107	.6117
4.1	.6128	.6138	.6149	.6160	.6170	.6180	.6191	.6201	.6212	.6222
4.2	.6232	.6243	.6253	.6263	.6274	.6284	.6294	.6304	.6314	.6325
4.3	.6335	.6345	.6355	.6365	.6375	.6385	.6395	.6405	.6415	.6425
4.4	.6435	.6444	.6454	.6464	.6474	.6484	.6493	.6503	.6513	.6522
4.5	.6532	.6542	.6551	.6561	.6571	.6580	.6590	.6599	.6609	.6618
4.6	.6628	.6637	.6646	.6656	.6665	.6675	.6684	.6693	.6702	.6712
4.7	.6721	.6730	.6739	.6749	.6758	.6767	.6776	.6785	.6794	.6803
4.8	.6812	.6821	.6830	.6839	.6848	.6857	.6866	.6875	.6884	.6893
4.9	.6902	.6911	.6920	.6928	.6937	.6946	.6955	.6964	.6972	.6981
5.0	.6990	.6998	.7007	.7016	.7024	.7033	.7042	.7050	.7059	.7067
5.1	.7076	.7084	.7093	.7101	.7110	.7118	.7126	.7135	.7143	.7152
5.2	.7160	.7168	.7177	.7185	.7193	.7202	.7210	.7218	.7226	.7235
5.3	.7243	.7251	.7259	.7267	.7275	.7284	.7292	.7300	.7308	.7316
5.4	.7324	.7332	.7340	.7348	.7356	.7364	.7372	.7380	.7388	.7396
n	0	1	2	3	4	5	6	7	8	9

Continued

Table 2 (*continued*)

n	0	1	2	3	4	5	6	7	8	9
5.5	.7404	.7412	.7419	.7427	.7435	.7443	.7451	.7459	.7466	.7474
5.6	.7482	.7490	.7497	.7505	.7513	.7520	.7528	.7536	.7543	.7551
5.7	.7559	.7566	.7574	.7582	.7589	.7597	.7604	.7612	.7619	.7627
5.8	.7634	.7642	.7649	.7657	.7664	.7672	.7679	.7686	.7694	.7701
5.9	.7709	.7716	.7723	.7731	.7738	.7745	.7752	.7760	.7767	.7774
6.0	.7782	.7789	.7796	.7803	.7810	.7818	.7825	.7832	.7839	.7846
6.1	.7853	.7860	.7868	.7875	.7882	.7889	.7896	.7903	.7910	.7917
6.2	.7924	.7931	.7938	.7945	.7952	.7959	.7966	.7973	.7980	.7987
6.3	.7993	.8000	.8007	.8014	.8021	.8028	.8035	.8041	.8048	.8055
6.4	.8062	.8069	.8075	.8082	.8089	.8096	.8102	.8109	.8116	.8122
6.5	.8129	.8136	.8142	.8149	.8156	.8162	.8169	.8176	.8182	.8189
6.6	.8195	.8202	.8209	.8215	.8222	.8228	.8235	.8241	.8248	.8254
6.7	.8261	.8267	.8274	.8280	.8287	.8293	.8299	.8306	.8312	.8319
6.8	.8325	.8331	.8338	.8344	.8351	.8357	.8363	.8370	.8376	.8382
6.9	.8388	.8395	.8401	.8407	.8414	.8420	.8426	.8432	.8439	.8445
7.0	.8451	.8457	.8463	.8470	.8476	.8482	.8488	.8494	.8500	.8506
7.1	.8513	.8519	.8525	.8531	.8537	.8543	.8549	.8555	.8561	.8567
7.2	.8573	.8579	.8585	.8591	.8597	.8603	.8609	.8615	.8621	.8627
7.3	.8633	.8639	.8645	.8651	.8657	.8663	.8669	.8675	.8681	.8686
7.4	.8692	.8698	.8704	.8710	.8716	.8722	.8727	.8733	.8739	.8745
7.5	.8751	.8756	.8762	.8768	.8774	.8779	.8785	.8791	.8797	.8802
7.6	.8808	.8814	.8820	.8825	.8831	.8837	.8842	.8848	.8854	.8859
7.7	.8865	.8871	.8876	.8882	.8887	.8893	.8899	.8904	.8910	.8915
7.8	.8921	.8927	.8932	.8938	.8943	.8949	.8954	.8960	.8965	.8971
7.9	.8976	.8982	.8987	.8993	.8998	.9004	.9009	.9015	.9020	.9025
8.0	.9031	.9036	.9042	.9047	.9053	.9058	.9063	.9069	.9074	.9079
8.1	.9085	.9090	.9096	.9101	.9106	.9112	.9117	.9122	.9128	.9133
8.2	.9138	.9143	.9149	.9154	.9159	.9165	.9170	.9175	.9180	.9186
8.3	.9191	.9196	.9201	.9206	.9212	.9217	.9222	.9227	.9232	.9238
8.4	.9243	.9248	.9253	.9258	.9263	.9269	.9274	.9279	.9284	.9289
8.5	.9294	.9299	.9304	.9309	.9315	.9320	.9325	.9330	.9335	.9340
8.6	.9345	.9350	.9355	.9360	.9365	.9370	.9375	.9380	.9385	.9390
8.7	.9395	.9400	.9405	.9410	.9415	.9420	.9425	.9430	.9435	.9440
8.8	.9445	.9450	.9455	.9460	.9465	.9469	.9474	.9479	.9484	.9489
8.9	.9494	.9499	.9504	.9509	.9513	.9518	.9523	.9528	.9533	.9538
9.0	.9542	.9547	.9552	.9557	.9562	.9566	.9571	.9576	.9581	.9586
9.1	.9590	.9595	.9600	.9605	.9609	.9614	.9619	.9624	.9628	.9633
9.2	.9638	.9643	.9647	.9652	.9657	.9661	.9666	.9671	.9675	.9680
9.3	.9685	.9689	.9694	.9699	.9703	.9708	.9713	.9717	.9722	.9727
9.4	.9731	.9736	.9741	.9745	.9750	.9754	.9759	.9763	.9768	.9773
9.5	.9777	.9782	.9786	.9791	.9795	.9800	.9805	.9809	.9814	.9818
9.6	.9823	.9827	.9832	.9836	.9841	.9845	.9850	.9854	.9859	.9863
9.7	.9868	.9872	.9877	.9881	.9886	.9890	.9894	.9899	.9903	.9908
9.8	.9912	.9917	.9921	.9926	.9930	.9934	.9939	.9943	.9948	.9952
9.9	.9956	.9961	.9965	.9969	.9974	.9978	.9983	.9987	.9991	.9996
n	0	1	2	3	4	5	6	7	8	9

Table 3 Natural Logarithms

x	$\ln x$	x	$\ln x$	x	$\ln x$
0.0		4.5	1.5041	9.0	2.1972
0.1	−2.3026	4.6	1.5261	9.1	2.2083
0.2	−1.6094	4.7	1.5476	9.2	2.2192
0.3	−1.2040	4.8	1.5686	9.3	2.2300
0.4	−0.9163	4.9	1.5892	9.4	2.2407
0.5	−0.6931	5.0	1.6094	9.5	2.2513
0.6	−0.5108	5.1	1.6292	9.6	2.2618
0.7	−0.3567	5.2	1.6487	9.7	2.2721
0.8	−0.2231	5.3	1.6677	9.8	2.2824
0.9	−0.1054	5.4	1.6864	9.9	2.2925
1.0	0.0000	5.5	1.7047	10	2.3026
1.1	0.0953	5.6	1.7228	11	2.3979
1.2	0.1823	5.7	1.7405	12	2.4849
1.3	0.2624	5.8	1.7579	13	2.5649
1.4	0.3365	5.9	1.7750	14	2.6391
1.5	0.4055	6.0	1.7918	15	2.7081
1.6	0.4700	6.1	1.8083	16	2.7726
1.7	0.5306	6.2	1.8245	17	2.8332
1.8	0.5878	6.3	1.8405	18	2.8904
1.9	0.6419	6.4	1.8563	19	2.9444
2.0	0.6931	6.5	1.8718	20	2.9957
2.1	0.7419	6.6	1.8871	25	3.2189
2.2	0.7885	6.7	1.9021	30	3.4012
2.3	0.8329	6.8	1.9169	35	3.5553
2.4	0.8755	6.9	1.9315	40	3.6889
2.5	0.9163	7.0	1.9459	45	3.8067
2.6	0.9555	7.1	1.9601	50	3.9120
2.7	0.9933	7.2	1.9741	55	4.0073
2.8	1.0296	7.3	1.9879	60	4.0943
2.9	1.0647	7.4	2.0015	65	4.1744
3.0	1.0986	7.5	2.0149	70	4.2485
3.1	1.1314	7.6	2.0281	75	4.3175
3.2	1.1632	7.7	2.0412	80	4.3820
3.3	1.1939	7.8	2.0541	85	4.4427
3.4	1.2238	7.9	2.0669	90	4.4998
3.5	1.2528	8.0	2.0794	95	4.5539
3.6	1.2809	8.1	2.0919	100	4.6052
3.7	1.3083	8.2	2.1041		
3.8	1.3350	8.3	2.1163		
3.9	1.3610	8.4	2.1281		
4.0	1.3863	8.5	2.1401		
4.1	1.4110	8.6	2.1518		
4.2	1.4351	8.7	2.1633		
4.3	1.4586	8.8	2.1748		
4.4	1.4816	8.9	2.1861		

Table 4 Powers of e

x	e^x	e^{-x}	x	e^x	e^{-x}
0.00	1.00000	1.00000	1.60	4.95302	0.20189
0.01	1.01005	0.99004	1.70	5.47394	0.18268
0.02	1.02020	0.98019	1.80	6.04964	0.16529
0.03	1.03045	0.97044	1.90	6.68589	0.14956
0.04	1.04081	0.96078	2.00	7.38905	0.13533
0.05	1.05127	0.95122			
0.06	1.06183	0.94176	2.10	8.16616	0.12245
0.07	1.07250	0.93239	2.20	9.02500	0.11080
0.08	1.08328	0.92311	2.30	9.97417	0.10025
0.09	1.09417	0.91393	2.40	11.02316	0.09071
0.10	1.10517	0.90483	2.50	12.18248	0.08208
			2.60	13.46372	0.07427
0.11	1.11628	0.89583	2.70	14.87971	0.06720
0.12	1.12750	0.88692	2.80	16.44463	0.06081
0.13	1.13883	0.87810	2.90	18.17412	0.05502
0.14	1.15027	0.86936	3.00	20.08551	0.04978
0.15	1.16183	0.86071			
0.16	1.17351	0.85214	3.50	33.11545	0.03020
0.17	1.18530	0.84366	4.00	54.59815	0.01832
0.18	1.19722	0.83527	4.50	90.01713	0.01111
0.19	1.20925	0.82696			
			5.00	148.41316	0.00674
0.20	1.22140	0.81873	5.50	224.69193	0.00409
0.30	1.34985	0.74081			
0.40	1.49182	0.67032	6.00	403.42879	0.00248
0.50	1.64872	0.60653	6.50	665.14163	0.00150
0.60	1.82211	0.54881			
0.70	2.01375	0.49658	7.00	1096.63316	0.00091
0.80	2.22554	0.44932	7.50	1808.04241	0.00055
0.90	2.45960	0.40656	8.00	2980.95799	0.00034
1.00	2.71828	0.36787	8.50	4914.76884	0.00020
1.10	3.00416	0.33287			
1.20	3.32011	0.30119	9.00	8130.08392	0.00012
1.30	3.66929	0.27253	9.50	13359.72683	0.00007
1.40	4.05519	0.24659			
1.50	4.48168	0.22313	10.00	22026.46579	0.00005

APPENDIX C

ANSWERS

This appendix provides answers to all odd-numbered problems within the exercise sets, and all of the answers for the chapter reviews, cumulative reviews, and chapter tests. Complete solutions to these problems are available in the Student's Solutions Manual.

CHAPTER 1

Exercise Set 1.1
1. False **3.** True **5.** True **7.** False **9.** True
11. False **13.** True **15.** 1 **17.** $-18, 0, 1$
19. $-\sqrt{3}, \sqrt{8}$ **21.** 1 **23.** -18 **25.** {a, d} **27.** $\emptyset$
29. {4} **31.** {1, 2, 3, 4} **33.** {1, 2, 3, 4, 5} **35.** True
37. False **39.** True **41.** True **43.** True **45.** True
47. True **49.** {0, 1, 2, 3, 4, ...} **51.** {2, 4, 6, 8} = B
53. {0, 2, 4, 6, 8} **55.** {0} **57.** infinite set
59. infinite set **61.** {Canada, Mexico, United States}
63. {Iowa, Utah, Ohio} **65.** {November}
67. $\{x \mid x \text{ is a month of the year}\}$ **69.** $\{x \mid x \in W\}$
71. $\{x \mid x \text{ is a month of the year starting with N}\}$
73. {V, P}, {V, S}, {V, L}, {P, S}, {P, L}, {S, L}
75. {V, P, S, L} **77.** $\not\subseteq$ **79.** $\in$ **81.** $\subseteq$ **83.** $\not\subseteq$
85. 0.75 **87.** $0.27\overline{27}$, rational **89.** $0.18\overline{18}$, rational
91. no **93.** no **95.** no **97.** no
99. I have a set of wrenches.
101. He studies in the student union building.
103. I only have a finite amount of time to study.

Exercise Set 1.2
1. < **3.** > **5.** < **7.** > **9.** > **11.** > **13.** >
15. > **17.** -8 **19.** 5 **21.** $-a$ **23.** a **25.** $-b$
27. $-|x|$ **29.** $|x|$ **31.** $|-a|$ **33.** $7 < 12$
35. $x + 1 > 0$ **37.** $m + 2 \geq 0$ **39.** $3y - 2 \leq 0$
41. $9x + 7 = -9$ **43.** 9 **45.** 0 **47.** $(7 \cdot 8)y$
49. $5 + b$ **51.** p **53.** $2a + 6$ **55.** $3(t + 2)$
57. $13r - 52t$ **59.** Commutative property of addition
61. Additive identity property
63. Multiplicative identity property
65. Reflexive property
67. Associative property of multiplication

69. Multiplicative inverse property **71.** $(2 + 3)a = 5a$
73. $(11 - 5)m = 6m$ **75.** $(7 + 1)x = 8x$ **77.** $3 \neq 8$
79. $6 + 0 = 6$ **81.** $3 \in N$ **83.** $6(m + 2) = 6m + 12$
85. $13 - 4 \neq 4 - 13$ **87.** $\in$ **89.** $\subseteq$ **91.** $\in$
93. $\frac{1}{5}(5) = 1$ Inverse property

$[7 + (-7)] = 0$ Additive inverse property

$67 + 1 + 0 + 33 = 67 + 33 + 1$ Commutative property; Identity property

$ = (67 + 33) + 1$ Associative property

$ = 100 + 1$

$ = 101$

95. $[(-58) + 58] + (-13) + \frac{1}{2}(29 - 3)$ Commutative property; associative property

$ = [0] + (-13) + \frac{1}{2}(26)$ Additive inverse property

$ = 0 + (-13) + 13$

$ = 0 + [(-13) + 13]$ Associative property

$ = 0 + 0$ Additive inverse property

$ = 0$ Additive identity property

A-20

Exercise Set 1.3
1. 6   3. -32 5. 0   7. -23 9. 29   11. -18
13. -6 15. $\dfrac{15}{11}$ 17. $-\dfrac{3}{7}$ 19. $-\dfrac{11}{8}$ 21. -3
23. 3   25. $-9p$ 27. $-5a + 20b$ 29. 20,602 ft
31. -120 33. 70   35. -546 37. -3 39. 32
41. 0   43. $-\dfrac{15}{8}$ 45. -3 47. 0   49. 137
51. 0   53. 10   55. -13 57. $\dfrac{44}{1225}$ 59. -2
61. undefined   63. -80 65. 5,600   67. $\dfrac{25}{3}$
69. undefined   71. $-\dfrac{9}{5}$ 73. 2.376   75. -30
77. 0   79. 2   81. $\dfrac{29}{10}$
83. commutative property of addition
85. associative property of addition
87. distributive property   89. transitive property
91. True   93. True   95. False   97. False
99. False
101. $\{3 - 16 \cdot 3\}(25)$ Exponents first
    $= \{3 - 48\}(25)$ Multiplication next
    $= \{-45\}(25)$ Addition inside grouping symbol
    $= -1125$ Multiplication
103. $|-7| = 7$, which is greater than -7
105. Reduce first; multiply numerator; multiply denominator.
107. Invert the divisor and multiply.
109. A set is a *subset* of another set but not an *element* of another set.

Review Exercises
1. set   2. natural   3. {i, r, a, t, o, n, l}   4. **W** and **N**
5. $\{-2, 0, 2, 4, 6, 8\}$ 6. intersection   7. no   8. subset
9. is less than   10. left   11. additive inverses
12. integers, **J**   13. rational   14. irrational   15. $0.27\overline{27}$
16. closure   17. commutative   18. distributive
19. symmetric   20. $-\dfrac{57}{70}$ 21. $-\dfrac{5}{2}$ 22. $-\dfrac{416}{153}$
23. $-\dfrac{21}{4}$ 24. $\dfrac{361}{45}$ 25. $-\dfrac{61}{36}$ 26. $-\dfrac{52}{45}$ 27. -15
28. -4 29. -45 30. -13 31. 4   32. negative
33. 4   34. undefined   35. $-\dfrac{7}{3}$ 36. $\dfrac{1}{20}$
37. $-19x + 13y$ 38. 5   39. $-x$ 40. -9 41. 6
42. $-\dfrac{3}{4}$ 43. -0.64 44. $-13x$ 45. -24 46. $\dfrac{25}{4}$
47. 0   48. $\dfrac{16}{17}$ 49. undefined   50. 0

Chapter Test
1. $\{1, 2, 3, 4\}$ 2. The set of all students ten feet tall
3. $0.55\overline{5}$ 4. $\{1, 3\}$ 5. $\dfrac{35}{32}$ 6. $\dfrac{11}{30}$ 7. 3   8. -14
9. distributive property
10. associative property of addition
11. multiplicative inverse property
12. additive inverse property
13. associative property of multiplication
14. reflexive property   15. transitive property
16. *N* (natural numbers)   17. *Q* (rational numbers)
18. *J* (integers)   19. *Q* (rational numbers)
20. *W* (whole numbers)   21. *W* (whole numbers)
22. *W* (whole numbers)   23. $\{x \mid x \in \{J, a, n, u, r, y\}\}$
24. $8 \div 2 = 4$ but $2 \div 8 = \dfrac{1}{4}$
25. addition and multiplication   26. -24
27. multiplication   28. *H* (irrational numbers)
29. $\varnothing$ (empty set)   30. additive identity
31. -1 32. 4   33. $-4.1\overline{36}$ 34. $\dfrac{35}{8}$ 35. $\dfrac{1}{2}$
36. -1836 37. $-\dfrac{35}{2}$ 38. $\dfrac{1297}{300}$ 39. $\dfrac{49}{9}$ 40. -120

CHAPTER 2

Exercise Set 2.1
1. $\{8\}$ 3. $\{-5\}$ 5. $\{3\}$ 7. $\left\{\dfrac{3}{2}\right\}$ 9. $\left\{-\dfrac{11}{3}\right\}$
11. $\{-4\}$ 13. $\{-4\}$ 15. $\left\{\dfrac{2}{3}\right\}$ 17. $\left\{\dfrac{3}{2}\right\}$ 19. $\{0\}$
21. $\{10\}$ 23. conditional; $\{-1\}$ 25. identity; $\{a \mid a \in R\}$
27. conditional; $\{6\}$ 29. identity; $\{y \mid y \in R\}$ 31. $\{10\}$
33. $\left\{-\dfrac{3}{5}\right\}$ 35. $\left\{\dfrac{3}{4}\right\}$ 37. $\{-1\}$ 39. $\left\{-\dfrac{1}{3}\right\}$
41. $\left\{\dfrac{17}{9}\right\}$ 43. $\{1\}$ 45. $\{8\}$ 47. $\left\{-\dfrac{15}{2}\right\}$ 49. $\left\{\dfrac{15}{14}\right\}$
51. $\left\{-\dfrac{45}{14}\right\}$ 53. $\left\{\dfrac{3}{8}\right\}$ 55. $\left\{\dfrac{7}{16}\right\}$ 57. $\{-2\}$
59. $\{15\}$ 61. $\left\{\dfrac{10}{3}\right\}$ 63. $\left\{-\dfrac{17}{10}\right\}$ 65. $\left\{\dfrac{15}{2}\right\}$ 67. $\{0\}$
69. $\left\{-\dfrac{7}{41}\right\}$ 71. $\left\{-\dfrac{9}{26}\right\}$ 73. F; $5 \in N$ 75. True
77. True, $x = 2$ 79. False, $\{1, 2\} \cup \{2, 3\} = \{1, 2, 3\}$
81. True   83. False, additive identity property   85. True
87. True   89. True
91. Two equations with the same solution set; $x + 2 = 6$ is equivalent to $x = 4$.

93. An equation with a graph that is a straight line; $3x + 2y = 5$
95. Multiplying both sides of an equation by the same number gives an equivalent equation. If $x = 4$ then $2x = 8$.
97. An identity is an equation that is true for any real number. For example, $x + 5 = x + 5$.

Exercise Set 2.2

1. Let h be the cost of a hotdog; $6h$
3. Let g be the cost of a gallon of gas; $10g$
5. Let x be the number of dimes; $10x$
7. Let x be the number; $3(x + 7)$
9. Let a be Mario's age; $3a$
11. Let x be the number; $(x \div 4) - 8$
13. Let x be the number; $(x + 50) \div 2$
15. Let x be the third score; $(86 + 92 + x) \div 3$
17. Let x be the first odd integer and $x + 2$ the next consecutive odd integer; $x + (x + 2)$
19. Let x be first multiple of 4 and $x + 4$ the next consecutive multiple of 4; $x + (x + 4)$
21. Let L be Luis's grade; $L + 10$
23. Let x be the first integer and $x + 4$ the second integer; $x + (x + 4)$
25. $3h - 26$ 27. $\dfrac{y}{11}$ 29. $\dfrac{1}{4}a + 8$ 31. $6r - 51$
33. Let x be the integer and $2x + 1$ the other integer; $x + (2x + 1) = 35$
35. Let x be the number; $x + 8 = 10$
37. Let n be the number; $11 + 3n = 24$
39. Let $x, x + 2,$ and $x + 4$ be the three consecutive odd numbers; $x + (x + 2) + (x + 4) = 21$
41. Let $x, x + 1,$ and $x + 2$ be the three consecutive natural numbers; $x + (x + 1) + (x + 2) - 30 = 18$
43. Let x be the first number and $x + 6$ the other number; $x + (x + 6) = 22$; $x = 8$; $x + 6 = 14$
45. Let x be an integral multiple of 4 and $x + 4$ the next consecutive integral multiple of 4; $x + (x + 4) = 28$; $x = 12, x + 4 = 16$
47. Let x be the number; $3(x - 6) = 3$; $x = 7$
49. Let x be one number and $6x + 1$ the other number; $x + (6x + 1) = 22$; $x = 3$; $6x + 1 = 19$
51. Let x be the length of a side; $3x = 81$; $x = 27$ ft
53. Let x be the measure of an angle; $4x = 360, x = 90°$
55. Let x be the width; $6x = 90$; $x = 15$ cm (width); $2x = 30$ cm (length)
57. Let $x =$ short side, $2x =$ long side; $6x = 30$; $x = 5$ cm; $2x = 10$ cm
59. $100°$ 61. 17 m 63. $30°, 150°$ 65. 4 cm
67. $x = 12.5$ in.; $x - 5 = 7.5$ in.
69. $x = 12$ cm; $x + 6 = 18$ cm; $x + 3 = 15$ cm
71. $x = 13$ m; $x + 2 = 15$ m 73. 3 75. $-\dfrac{3}{16}$
77. $13x + 32$ 79. -3 81. 24 83. 2 85. 6 87. $\dfrac{9}{10}$
89. Opposite sides are parallel and equal; adjacent angles are supplementary.

Exercise Set 2.3

1. $\dfrac{I}{pt}$ 3. $\dfrac{C}{2\pi}$ 5. $\dfrac{c - ax}{b}$ 7. $\dfrac{S - 2lh}{2l + 2h}$ 9. $\dfrac{S}{2\pi r}$
11. $\dfrac{2A}{h} - b$ 13. $\dfrac{2A}{h}$ 15. $\dfrac{2A}{b}$ 17. $\dfrac{A - P}{Pr}$ 19. $\dfrac{V}{\pi r^2}$
21. $\dfrac{T_2 P_1 V_1}{P_2 V_2}$ 23. $\dfrac{100B}{C}$ 25. $\dfrac{R_1 R_2}{R_1 + R_2}$ 27. $\dfrac{5}{11}D + 15$
29. $\dfrac{f_0}{M}$ 31. $\dfrac{5}{9}(F - 32)$ 33. $\dfrac{S - A}{DS}$ 35. $\dfrac{y^2}{2x}$
37. $\dfrac{3V + \pi h^3}{3\pi h^2}$ 39. $\dfrac{6V - HS_0 - 4HS_1}{H}$ 41. $\dfrac{C^2}{4\pi}$ 43. $\dfrac{P^2}{16}$
45. $r = 2$ 47. $h = 16$ 49. $W = 4$ 51. $b = 95$
53. $R = \dfrac{40}{3}$ 55. $f = 15$ 57. $x = 2$ 59. $H = \dfrac{49}{8}$
61. $119.5°F$ 63. 65.7 65. 56 ft 67. 37.7 ft³
69. $67, 69$ 71. $-4, -9$ 73. $L = 25$ km, $W = 15$ km
75. 18 oz
77. A formula is a literal (in letters) translation of a sentence.
79. $C = 1.0725(18D) = 19.305D$

Exercise Set 2.4

1. $\{4, -4\}$ 3. $\emptyset$ 5. $\{169, -169\}$ 7. $\emptyset$
9. $\left\{-\dfrac{7}{36}, \dfrac{7}{36}\right\}$ 11. $\left\{-\dfrac{1}{4}, \dfrac{1}{4}\right\}$ 13. $\{7, -11\}$ 15. $\emptyset$
17. $\{-2, 8\}$ 19. $\{6, 0\}$ 21. $\left\{-\dfrac{1}{2}\right\}$ 23. $\emptyset$
25. $\{14, -10\}$ 27. $\left\{-\dfrac{27}{2}, \dfrac{33}{2}\right\}$ 29. $\left\{-\dfrac{2}{5}, \dfrac{8}{5}\right\}$
31. $\{-12, 3\}$ 33. $\left\{-\dfrac{3}{2}, \dfrac{3}{4}\right\}$ 35. $\left\{-\dfrac{7}{6}\right\}$
37. $\{y \mid y \in R\}$ 39. $\left\{\dfrac{3}{2}\right\}$ 41. $\{p \mid p \le 0\}$
43. $\left\{-\dfrac{8}{3}, \dfrac{10}{9}\right\}$ 45. $\emptyset$ 47. $\{p \mid p \ge 4\}$
49. $\{y \mid y \in R\}$ 51. $\left\{-3, -\dfrac{1}{5}\right\}$ 53. $\left\{\dfrac{3}{4}\right\}$ 55. $\left\{\dfrac{11}{2}\right\}$
57. $\left\{\dfrac{6}{5}, 4\right\}$ 59. False 61. True 63. True 65. True
67. $\{-3, 3\}$ 69. $\{0.0, -1.0\}$ 71. $\{0.9, 1.3\}$
73. $|x| = 8$; $\{-8, 8\}$ 75. $\left|\dfrac{2x - 3}{4}\right| = 0$; $\left\{\dfrac{3}{2}\right\}$
77. $\left|\dfrac{3x}{4} + 1\right| = -10$; $\emptyset$ 79. $|x| = 4$ 81. $|x| = 3$
83. $x = \dfrac{(x_2 - x_1)(y - y_1)}{y_2 - y_1} + x_1$ 85. $h = \dfrac{S - 2\pi r^2}{2\pi r}$

87. The absolute value of x equals p, or x is a member of the set of all numbers p units from the origin.
89. Because $|x|$ is always greater than or equal to zero; if $x < 0$ then $|x| = -x$.

Exercise Set 2.5

1. $(-\infty, -7)$
3. $[0, \infty)$
5. $(-\infty, 2]$
7. $[-9, 4]$
9. $[3, 5)$
11. $(-\infty, 4)$
13. $\{x \mid x \leq 4, x \in N\}$
15. $\left[\dfrac{8}{3}, \infty\right)$
17. $\left(-\dfrac{1}{2}, \infty\right)$
19. $\{x \mid x \geq 1, x \in J\}$
21. $\left[-\dfrac{5}{3}, \infty\right)$
23. $(-3, 4]$
25. $\{x \mid -3 \leq x \leq 3, x \in J\}$
27. $\{x \mid -2 < x < 3\}$
29. $\left[\dfrac{13}{5}, \infty\right)$
31. $\left(-\infty, -\dfrac{1}{5}\right)$
33. $\{x \mid x \geq 12, x \in J\}$
35. $\left\{x \mid x \geq \dfrac{13}{8}, x \in W\right\}$
37. $\left\{x \mid x < \dfrac{21}{4}\right\}$
39. $\left(-\infty, \dfrac{38}{27}\right)$
41. $(2, 5)$
43. $\{y \mid y \in R\}$
45. $[-5, 4)$
47. $(-\infty, 27]$ **49.** $(-15, 21)$ **51.** $x \leq -12$
53. $x \geq 9$ **55.** $x \geq 69$ **57.** $n = 5$ **59.** $[72, 92)$
61. $94 \leq x < 124$ **63.** $\{-2, 4\}$ **65.** $\left\{-\dfrac{14}{3}, 0\right\}$
67. $\left\{\dfrac{13}{6}\right\}$
69. An inequality is a mathematical statement relating two or more quantities that are not equal
71. Two or more inequalities are equivalent if they have the same solution set
73. The endpoints are included in a closed interval, whereas they're not in an open interval.

Exercise Set 2.6

1. $-3 < x < 3$ **3.** $-4 \leq x \leq 4$ **5.** $y < -3$ or $y > 3$
7. $-8 \leq x - 5 \leq 8$ **9.** $2x - 1 \geq 4$ or $2x - 1 \leq -4$
11. $|x| < 2$
13. $|y| > 3$ **15.** $|x + 1| \leq 6$ **17.** $|5x - 1| > 3$
19. $|y| < 3$ **21.** $|y - 3| \leq 5$ **23.** $|x + 3| > 5$
25. $|6y - 2| > 4$
27. $(-2, 2)$
29. $\{x \mid x \geq 5$ or $x \leq -5, x \in J\}$
31. $\{0\}$ **33.** No solution
35. all real numbers
37. $\{y \mid y \in J\}$
39. $[1, 4]$
41. all real numbers
43. $(7, 9)$

45. No solution

47. $\left(-\infty, \frac{1}{7}\right] \cup \left[\frac{3}{7}, \infty\right)$

49. No solution

51. $\left(-\infty, \frac{5}{3}\right] \cup \left[\frac{11}{3}, \infty\right)$

53. $\left[-\frac{4}{3}, 2\right]$

55. No solution   **57.** $x \geq \frac{c-b}{a}$ or $x \leq \frac{-c-b}{a}$

59. $x > \frac{2bc}{a}$ or $x < 0$

61. $\{x \mid x < 8, x \in N\}$ $\{1, 2, 3, 4, 5, 6, 7\}$

63. $|x| > 5$; $x > 5$ or $x < -5$

65. $|x| \leq 10$; $-10 \leq x \leq 10$

67. $|x + 3| - 5 \geq 0$; $x \geq 2$ or $x \leq -8$

69. The price is less than or equal to $3.50.

71. The price is more than $14.00.   **73.** $\left\{-1, \frac{5}{3}\right\}$

75. $\left\{\frac{4}{15}, \frac{1}{10}\right\}$ **77.** $\left(-\infty, -\frac{2}{5}\right]$ **79.** $\left(\frac{7}{11}, \infty\right)$

81. $\left[\frac{5}{2}, \infty\right)$

83. By definition of absolute value, $|x| < c$ is the set of all numbers between $-c$ and c.

85. The solution set of $|x| < 8$ is the set of all numbers *less* than 8 units from the origin, while the solution set of $|x| > 8$ is the set of all numbers *more* than 8 units from the origin.

87. $|-3 + (-7)| \stackrel{?}{\geq} |-3| + |-7|$
$|-10| \stackrel{?}{\geq} 3 + 7$
$10 \stackrel{\checkmark}{\geq} 10$

Exercise Set 2.7
1. 55, 56, 57   **3.** 14   **5.** 25   **7.** 12, 15, 18   **9.** 82
11. 80°   **13.** $L = 21$ m, $W = 12$ m   **15.** 12 ft
17. $W_2 = 15$ lb   **19.** $2\frac{1}{4}$ ft from the 540-lb weight
21. $150   **23.** 78 at $3, 22 at $4.50
25. $16,250 at 15%; $8,750 at 5%   **27.** $64
29. 24 min at 10 mph, 12 min at 15 mph
31. 50 mph, 55 mph   **33.** 64 mph   **35.** 170 miles
37. 32 l **39.** 2 qt   **41.** 10 gal
43. 25 lb of the $5 coffee and 50 lb of the $6.50 coffee

Review Exercises
1. $11x$ **2.** $5y$ **3.** subtraction   **4.** addition   **5.** $\{7\}$
6. If both sides of an equation are multiplied by the same nonzero number, the result is an equivalent equation.
7. If both sides of an equation are divided by the same nonzero number, the result is an equivalent equation.
8. $\{19\}$ **9.** $\{6\}$ **10.** $\{-5\}$ **11.** $x \in R$ **12.** identity
13. conditional   **14.** $\{p, -p\}$, where $p > 0$
15. $\{x \mid x \geq 1, x \in W\}$ **16.** $0.\overline{285714}$
17. $\{x \mid x \geq 1, x \in W\}$ **18.** $\{x \mid -2 \leq x \leq 3, x \in J\}$
19. associative   **20.** inverses   **21.** inverses   **22.** 1
23. $ac = bc$ **24.** $a + (-b)$ **25.** $\left\{\frac{13}{3}\right\}$ **26.** $\{-4, 4\}$
27. $[0, 6]$ **28.** $\emptyset$ **29.** negative   **30.** $\frac{x - 21}{7}$
31. $2(x - 6)$ **32.** $x \neq -4, 0,$ or 2 **33.** $x = 12$
34. $y = \frac{7}{3}$ **35.** $x = -1, -\frac{5}{3}$
36. (number line from -2 to 2)
37. $m = \pm 6$ **38.** $a = -1, \frac{2}{3}$ **39.** $y = -1, 5$
40. $(-3, 3)$
41. $x = \frac{4}{3}, x \neq 0$
42. $\left[-\frac{8}{3}, \infty\right)$
43. $\left\{x \mid x \leq \frac{11}{5}, x \in J\right\}$
44. $\left[\frac{11}{9}, \infty\right)$
45. $\left[-\frac{4}{3}, 2\right]$
46. $\{x \mid x \in N\}$
47. $x = \frac{y - b}{m}$ **48.** $t = \frac{I}{pr}$ **49.** $r = \frac{A - p}{pt}$
50. $a = f$ **51.** $\frac{3x - 2}{8}$ **52.** $\frac{7 - 5y}{4}$ **53.** $|x| \leq 5$
54. $|y| \geq 5$
55. $x + (x + 2) + (x + 4) = 114$; 36, 38, 40
56. $5x + 25(30 - x) = 450$; 15 nickels, 15 quarters
57. $3.75x + 4.00(40 - x) = 153.00$; $x = 28$ lb at $3.75/lb; 12 lb at $4/lb
58. $x = 12\frac{1}{2}$ ft, $2x - 5 = 20$ ft

59. 3 m, 3 m, and 5 m   **60.** 5 hr   **61.** 575 km
62. 10 hr   **63.** $2\frac{5}{11}$ hr   **64.** 150 km

Chapter Test

1. $16x - 2$ **2.** $\{-25, 25\}$ **3.** $\left(-\infty, \frac{1}{9}\right) \cup (1, \infty)$
4. $\left\{\frac{9}{5}\right\}$ **5.** $\left\{-\frac{4}{3}, 2\right\}$ **6.** $\frac{A}{1 + rt}$ **7.** 270 miles
8. $\left\{-\frac{7}{4}\right\}$ **9.** $\left\{x \mid x \geq -\frac{1}{4}, x \in R\right\}$ or $\left[-\frac{1}{4}, \infty\right)$
10. $w = 12$ in., $l = 24$ in.   **11.** At least 94 points
12. $\left[-\frac{3}{7}, -\frac{1}{7}\right]$ **13.** 15, 16   **14.** 6   **15.** 12 m
16. $\left\{\frac{61}{6}\right\}$ **17.** $\left\{-\frac{3}{7}\right\}$ **18.** 65, 67
19. 13 km × 21 km   **20.** $28/yd²   **21.** 4 qt
22. $3000 at 7%; $4000 at 9%   **23.** $20,000
24. 36 mph   **25.** 50 min (after James leaves).

CHAPTER 3

Exercise Set 3.1

1.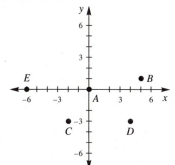

3. I   **5.** II   **7.** II   **9.** III   **11.** III   **13.** I
15.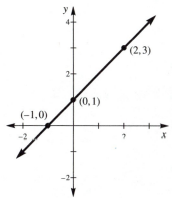

17.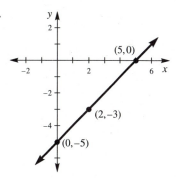

19.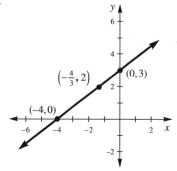

21.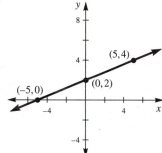

23.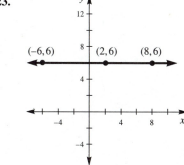

25.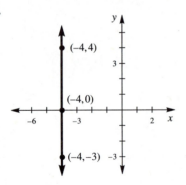

27.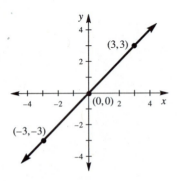

29.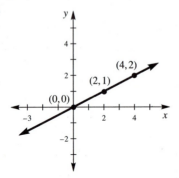

31. (a), (b), (c) **33.** (a), (c), (d)

35.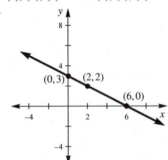

37.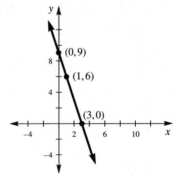

39.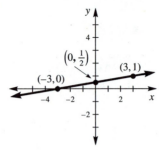

41.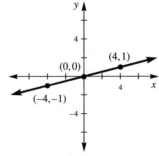

43.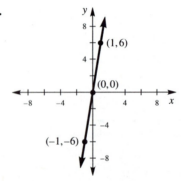

45.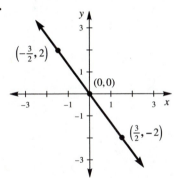

47.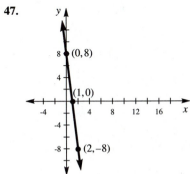

49.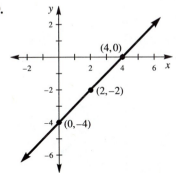

51.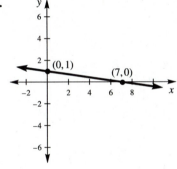

53. $x = 3y$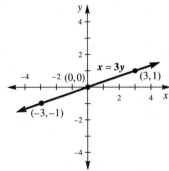

55. $2x = 5y - 5$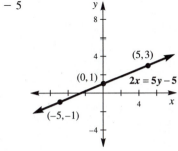

57. $\dfrac{y}{2} = x + 1$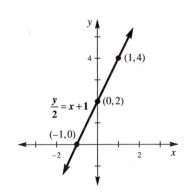

59.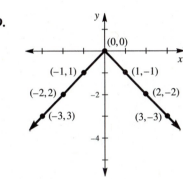

61.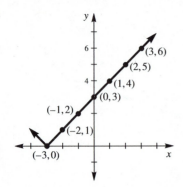

63. (a) $90; (b) 30; (c) 40; (d) $90; (e) $180

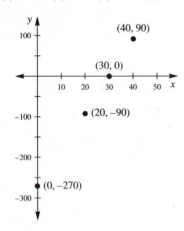

65. The Cartesian coordinate system consists of two perpendicular lines marked off in units of measure.

67. "Coordinates" is used because *together* they represent one point.

Exercise Set 3.2

1. $\frac{1}{3}$ **3.** -1 **5.** $-\frac{1}{4}$ **7.** 0 **9.** undefined **11.** $\frac{3}{4}$

13. -1 **15.** $\frac{d-b}{c-a}$ **17.** $\frac{b_2 - b_1}{a_2 - a_1}$

19. 0, horizontal

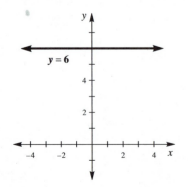

21. undefined, vertical

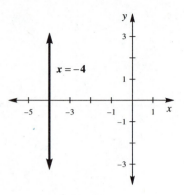

23. 1, rises

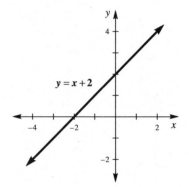

25. -1, falls

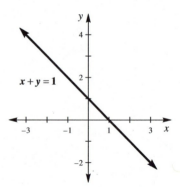

27. 1, rises

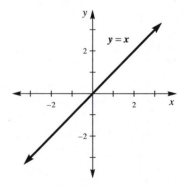

29. no **31.** no **33.** yes **35.** 1
37. undefined **39.** $-\dfrac{7}{8}$ **41.** $-\dfrac{1}{2}$ **43.** $\dfrac{1}{2}$ **45.** 5
47. 10 **49.** 13 **51.** $\sqrt{(b_2 - b_1)^2 + (a_2 - a_1)^2}$
53. $\left(\dfrac{3}{2}, 7\right)$ **55.** $(-4, 0)$ **57.** $\left(\dfrac{m+s}{2}, \dfrac{n+t}{2}\right)$
59. $\left(-\dfrac{1}{12}, \dfrac{1}{12}\right)$
61. parallel

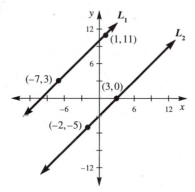

63. perpendicular

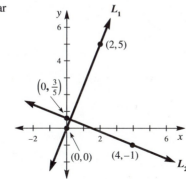

65. neither

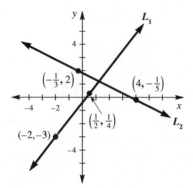

67. perpendicular

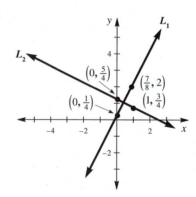

69. $m_{AB} = 1$, $m_{DC} = 1$

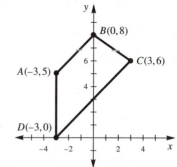

71. $m_{BC} = -\dfrac{4}{3}$, $m_{CD} = \dfrac{3}{4}$

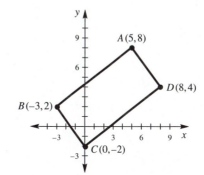

73. $y = \dfrac{59}{7}$

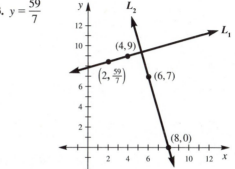

75. parallel: $m = \frac{4}{3}$; perpendicular: $m = -\frac{3}{4}$

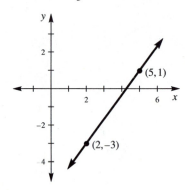

77. parallel: $m = \frac{7}{5}$; perpendicular: $m = -\frac{5}{7}$

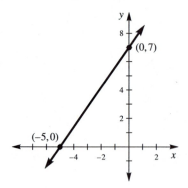

79. Slopes are negative reciprocals; lines are perpendicular.

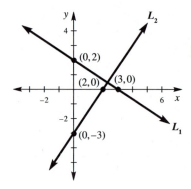

81. Slope is rise over run.
83. Congruent triangles are two triangles with the same shape and size.
85. A vertical line has undefined slope because division by zero is not defined.

Exercise Set 3.3
1. $x - y = -2$ **3.** $3x + y = 11$ **5.** $x - 2y = 3$
7. $2x + 5y = 10$ **9.** $y = 4$ **11.** $x = 2$
13. $3x - y = -13$ **15.** $x - y = 4$ **17.** $y = 2$
19. $y = 2$ **21.** $2x + y = 8$ **23.** $5x + 6y = -8$
25. $y = 4$ **27.** $x = -5$ **29.** $x - 12y = -1$
31. $5x - 6y = -1$ **33.** $y = 2x + 5$
35. $y = -\frac{1}{3}x + \frac{1}{2}$ **37.** $y = -5$ **39.** $x = 3$
41. $x - 4y = -4$ **43.** $2x - y = 0$ **45.** $x - 2y = -5$
47. $3x + 5y = 10$ **49.** $4x + 3y = 9$ **51.** $x - 2y = 3$
53. $2x + 5y = -5$ **55.** $x = 0$ **57.** $y = \frac{1}{2}x$
59. $x = 3$ **61.** $y = \frac{1}{3}x + 4$ **63.** $y = -\frac{3}{8}x + \frac{5}{2}$
65. $y - 2x = 0$ **67.** $x - y = -5$
$x + 2y = 5$ $x - y = 3$
69.

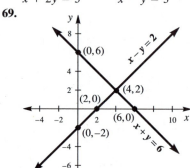

71.

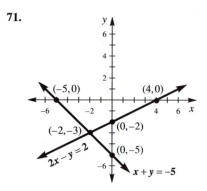

73. The opposite sides of a rectangle are parallel. It would save time to write the equations using the slope and one vertex for each line.
75. Yes: the slope–intercept form could be eliminated. We keep both forms because sometimes one is easier to use than the other.
77. The line $y = mx + b$ is easier to graph because we have the slope (m) and the y-intercept (b). To graph $ax + by = c$, solve for $y = -\frac{ax}{b} + \frac{c}{b}$, then plot

the point $\left(0, \dfrac{c}{b}\right)$ and use the slope $-\dfrac{a}{b}$ to locate a second point on the line.

Exercise Set 3.4
1. yes, D{−1, 1, 2, 3}; R{2, 3, 7, 8}
3. yes, D{−4, 0, 2, 6}; R{−8, 16}
5. no, D{2, 3, 6}; R{−7, −5, 7}
7. no, D{2, 3, 4}; R{0, 1, 3, 4}
9. $D(-\infty, \infty); R(-\infty, \infty)$ 11. $D(-\infty, \infty); R(-\infty, \infty)$
13. $D(-\infty, \infty); R[0, \infty)$ 15. $D(-\infty, \infty); R(-\infty, \infty)$
17. $D(-\infty, \infty); R\{-7\}$ 19. $D(-\infty, \infty); R[0, \infty)$
21. $D(-\infty, \infty); R(-\infty, 0]$ 23. Yes, $D\{x \mid x \neq 0\}$
25. yes, $D\{x \mid x \neq 1\}$ 27. yes, $D\{x \mid x \neq -3\}$
29. no, D{4} 31. yes, $D\{x \mid x \neq 5\}$
33. $D[-4, 2]; R[-1, 3]$
35. $D(-\infty, -6] \cup [0, \infty); R(-\infty, \infty)$
37. $D[-3, -1]; R[0, 8]$ 39. $D(-\infty, \infty); R(-\infty, 1]$
41. D{2}; $R(-\infty, \infty)$ 43. $D[-2, 3]; R[1, 5]$
45. $f(1) = -2; f(-1) = -4; f(3) = 0$
47. $h(0) = -1; h(-1) = 0; h(1) = 0$
49. $g(4) = 12; g(-4) = 20; g(0) = 16$
51. $f(0) = -8; f(2) = 6; f(-1) = -6$
53. $f\left(\dfrac{-b}{2a}\right) = \dfrac{-b^2 + 4ac}{4a}; f\left(\dfrac{b}{2a}\right) = \dfrac{3b^2 + 4ac}{4a}$

55.
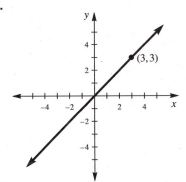

57.
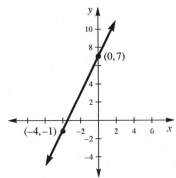

59.
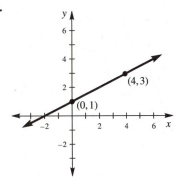

61. yes 63. no 65. yes 67. no
69. (a) 0 (b) $15,000 (c) 7.5 yr
71. (a) $P(n) = 50n - 100$
    (b) $n = 2$
    (c) $n > 2$
    (d) $0 \leq n < 2$
    (e) As many as possible

73. $2x + y = 2$ 75. $\left(\dfrac{3}{2}, 4\right)$ 77. 10

Exercise Set 3.5
1. −1 3. −4 5. −18 7. −1 9. 6 11. 3
13. −4 15. $a + 3$ 17. $x + 1$ 19. $x + 2$
21. (a) {(1, 0), (2, 1), (3, 2), (4, 3)}
    (b) D{1, 2, 3, 4}, R{0, 1, 2, 3}
23. (a) {(4, −2), (1, −1), (0, 0), (1, 1), (4, 2)}
    (b) D{0, 1, 4}, R{−2, −1, 0, 1, 2}
25. (a) {(1, 1), (2, 4), (3, 9), (4, 16)}
    (b) D{1, 2, 3, 4}, R{1, 4, 9, 16}
27. (a) {(−5, −3), (1, −1), (7, 1), (13, 2)}
    (b) D{−5, 1, 7, 13}, R{−3, −1, 1, 2}

29.
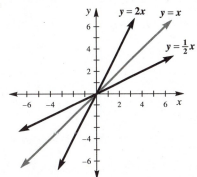

31.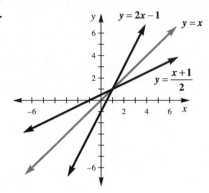

33.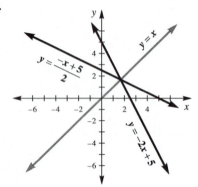

35.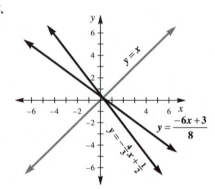

9. $A(4, 4)$ **10.** -4 **11.** $\sqrt{(x_2 - x_1)^2 + (y_2 - y_1)^2}$

12. $\left(\dfrac{x_1 + x_2}{2}, \dfrac{y_1 + y_2}{2}\right)$ **13.** $y = mx + b$

14. $y - y_1 = m(x - x_1)$

15. (a) -10 (b) 5 (c) 2 (d) -10 (e) $\dfrac{22}{5}$

16.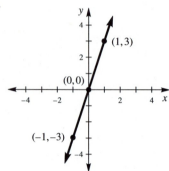

17.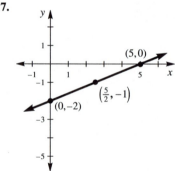

18.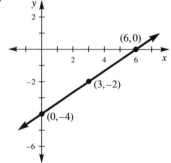

37. (a) 2 (b) $(-\infty, \infty)$ **39.** $(-\infty, 0) \cup (0, \infty)$ **41.** 2
43. 0 **45.** 0 **47.** $y = \dfrac{2}{3}x + \dfrac{8}{3}$ **49.** $y = -7x + 10$
51. $y = -\dfrac{1}{2}x$ **53.** $y = 5$ **55.** $y = 0$
57. $y = \dfrac{2}{3}x - \dfrac{1}{3}$ **59.** $y = -\dfrac{1}{2}x + \dfrac{5}{2}$ **61.** 5

Review Exercises

1. zero **2.** $m = \dfrac{y_2 - y_1}{x_2 - x_1}$ **3.** parallel **4.** undefined
5. perpendicular **6.** $-\dfrac{1}{5}$ **7.** $\dfrac{1}{2}$ **8.** parallel

19.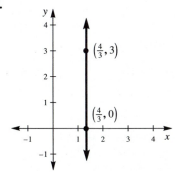

20. $3x + 2y = 6$ **21.** $2x - y = -1$
22. $4x + 5y = 33$ **23.** $x + y = 2$ **24.** $x + 2y = 7$
25. $y = 0$ **26.** $x = 4$ **27.** $3x - y = 1$
28. $x + y = 0$ **29.** $2x + 3y = \dfrac{13}{12}$ **30.** $x + y = 0$
31. $y = -1$ **32.** $y = \dfrac{1}{2}$ **33.** $x = 0$ **34.** $y = -x$
35. $y = x$ **36.** $x = -4$ **37.** $y = -3$ **38.** $y = \dfrac{3}{2}x + 3$
39. $y = -\dfrac{4}{3}x - 4$ **40.** $y = -\dfrac{2}{5}x + 3$ **41.** $y = \dfrac{1}{5}x - 3$
42. $y = 5x + 19$ **43.** (a) $\left(\dfrac{3}{2}, \dfrac{5}{2}\right)$ (b) $d = \sqrt{34}$
44. (a) $(2, 3)$ (b) $d = \sqrt{8} = 2\sqrt{2}$
45. (a) $\left(\dfrac{7}{2}, \dfrac{11}{2}\right)$ (b) $d = \sqrt{18} = 3\sqrt{2}$
46. (a) $\left(-\dfrac{11}{2}, -\dfrac{9}{2}\right)$ (b) $d = \sqrt{82}$
47. (a) $\left(\dfrac{13}{24}, \dfrac{31}{20}\right)$
 (b) $d = \sqrt{\dfrac{25}{144} + \dfrac{361}{100}} = \sqrt{\dfrac{13321}{3600}} = \sqrt{\dfrac{13321}{60}}$
48. (a) $\left(-\dfrac{11}{28}, -\dfrac{13}{30}\right)$ (b) $d = \sqrt{\dfrac{9}{196} + \dfrac{49}{225}} = \sqrt{\dfrac{11629}{44100}}$
49. $y = -x$

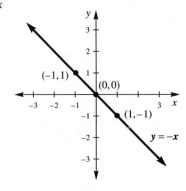

50. $y = 4x$

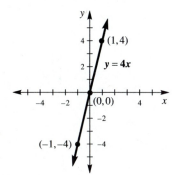

51. $y = x + 2$

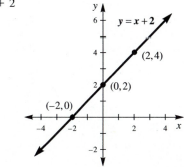

52. $2y = x - 5$

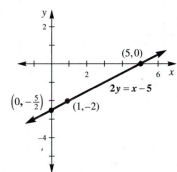

53. $y = x + 14$

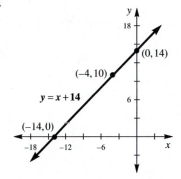

54. $y = 10x - 5$

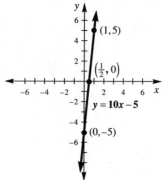

55. $4y = 7x + 35$

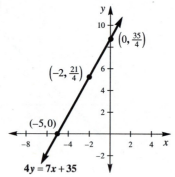

56. $y = 3x - 1$

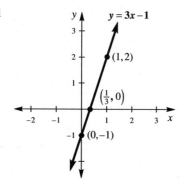

57. perpendicular **58.** parallel **59.** neither **60.** neither
61. $x = 1$ **62.** $y = 5$
63. $f(0) = 0$; $f(-2) = -10$; $f(3) = 15$; $\dfrac{f(x+h) - f(x)}{h} = 5$
64. No: when $x = 0$, $y = 1$ or 2.
65. $D\{x \mid x = 4\}$, $R\{y \mid y \in R\}$, no
66. $f(g(3)) = 1$; $g(f(0)) = -2$; $(f + g)(1) = 0$;
$(f \cdot g)(-2) = -27$; $\left(\dfrac{f}{g}\right)(0) = -\dfrac{5}{7}$
67. $f^{-1}(x) = \dfrac{1}{4}(x + 9)$

68.

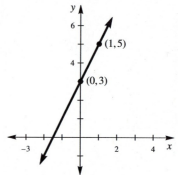

69. $D\{x \mid x \in R\}, R\left\{x \mid x \neq \dfrac{3}{2}\right\}$

Chapter Test
1. $m = \dfrac{4}{3}$ **2.** $d = 10$ **3.** $(6, 5)$ **4.** $y = \dfrac{4}{3}x - 3$
5. $(0, -3), \left(\dfrac{9}{4}, 0\right)$

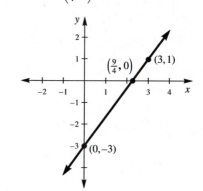

6. $2x - 3y = -6$ **7.** $y = \dfrac{1}{2}x + 5$ **8.** $y = 4$ **9.** $x = 6$
10. $y = -\dfrac{80}{63}x + \dfrac{41}{63}$ **11.** $y = \dfrac{1}{2}x$
12. $m = -\dfrac{2}{3}$, y-intercept, 3)

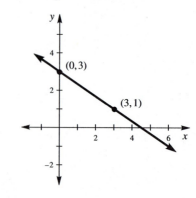

13. $y = 3x - 4$
14. Yes, because the slopes of AB & BC are equal.
15. perpendicular   16. $y = \frac{3}{5}x - 3$
17. $y = 4x - 3$

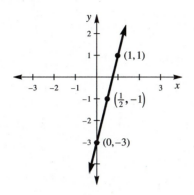

18. $x + y = 5$; negative slope   19. $5x - 7y = 18$
20. $y = \frac{1}{2}x + \frac{1}{2}$
21.

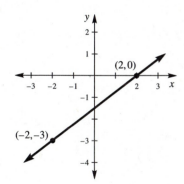

22.

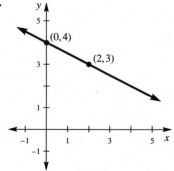

23.

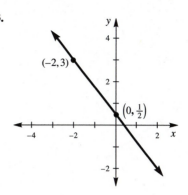

24. $f^{-1}(x) = \frac{x+2}{3}$ 25. $D\{x \mid x \in R\}; R\{y \mid y \in R\}$
26. $D\{x \mid x \in R\}; R\left\{x \mid x = \frac{1}{4}\right\}$
27. $f(g(1)) = 4; (f+g)(2) = -1; (f-g)(-1) = 7$
28. $g(0) = 2, g(1) = 0, g(2) = 0$ 29. yes

Cumulative Review Chapters 1–3
1. $\{1, 2, 3, 4\}$ 2. $\{0, 1, 2, 3\}$ 3. $\{-3, -2, -1, 0, 1, 2\}$
4. {Monday}   5. $\{1, 2, 3\}$ 6. $\{2\}$
7. A = {natural numbers}   8. $\{1, 2, 3, 4, 6\}$ 9. True
10. True   11. False   12. True   13. False   14. True
15. True   16. True   17. True   18. True   19. False
20. 9   21. 6   22. 4   23. 1   24. 0   25. undefined
26.

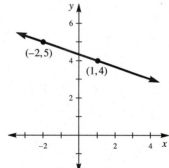

27.

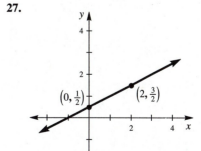

28.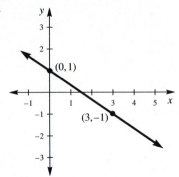

29.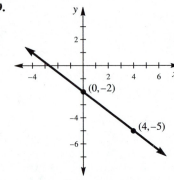

30. -8 31. 3 32. $\dfrac{16}{9}$ 33. $\{7\}$ 34. $\{-2\}$

35. $\left\{-\dfrac{7}{6}\right\}$ 36. $\left\{-\dfrac{9}{2}\right\}$ 37. $\{21\}$ 38. $\{15\}$

39. $\left\{-\dfrac{66}{13}\right\}$ 40. $\{5\}$ 41. $\left\{-\dfrac{1}{6}\right\}$ 42. $\left\{\dfrac{4}{3}\right\}$

43. $\left\{-\dfrac{1}{12}\right\}$ 44. $13, 14$ 45. -14 46. $16x$ ft

47. $15d$ in. 48. $25, 32$ 49. $0, 5$ 50. $8\dfrac{1}{4}$ m $\times$ $11\dfrac{1}{4}$ m

51. 5 cm 52. $24, 28$ 53. $h = \dfrac{2A}{b}$ 54. $\dfrac{y-b}{m}$

55. $\dfrac{R_1 R_2}{R_2 - R_1}$ 56. $\dfrac{A-p}{pt}$ 57. $\dfrac{c-ax}{b}$ 58. $\dfrac{9}{5}C + 32$

59. $\{\pm 5\}$ 60. $\varnothing$ 61. $\left\{-\dfrac{5}{4}\right\}$ 62. $\{-6, 0\}$ 63. $\left\{\dfrac{2}{3}\right\}$

64. $\left\{-\dfrac{1}{2}, -\dfrac{11}{10}\right\}$ 65. $(-\infty, 2]$ 66. $\left[\dfrac{16}{7}, \infty\right)$

67. $(2, 6)$ 68. $(-\infty, 3] \cup (4, \infty)$ 69. $[21, \infty)$

70. $\left(-\dfrac{4}{3}, 2\right)$ 71. 82 points 72. $-2 < m < 2$

73. $a \geq \dfrac{9}{2}$ or $a \leq \dfrac{5}{2}$ 74. $x = 0$ 75. $y \in \mathbf{R}$

76. $p \geq \dfrac{3}{2}$ or $p \leq 0$ 77. $a \geq -3$ or $a \leq -5$

78. (a) $\left(\dfrac{3}{2}, \dfrac{7}{2}\right)$ (b) 1 (c) $x - y = -2$

79. (a) $\left(\dfrac{1}{2}, \dfrac{3}{2}\right)$ (b) -1 (c) $x + y = 2$

80. (a) $(1, 4)$ (b) 0 (c) $y = 4$

81. (a) $\left(\dfrac{3}{8}, \dfrac{3}{8}\right)$ (b) -1 (c) $4x + 4y = 3$

82. (a) $\left(3, \dfrac{9}{4}\right)$ (b) undefined (c) $x = 3$

83. (a) $(2, -1)$ (b) -1 (c) $x + y = 1$

84. $y = \dfrac{3}{2}x + 3$ 85. $y = 5x - 5$ 86. $y = -\dfrac{1}{2}x - 2$

87. $y = -3$ 88. $y = 3x$ 89. $y = \dfrac{3}{2}x - \dfrac{1}{2}$ 90. $y = 1$

91. yes 92. $D\{x \mid x \in R\}$; $\{7\}$
93. $D\{x \mid x \in R\}$; $R\{x \mid x \neq 2\}$
94. $f(0) = -7$; $f(-2) = -13$; $f(a) = 3a - 7$; $f(a + 2) = 3a - 1$
96. yes; $D\{x \mid x \in R\}, R\{y \mid y \in R\}$ 97. yes 98. no
99. yes 100. no
101. $g(0) = 0$; $g(1) = -2$; $g(a) = 3a^2 - 5a$
102. $f(g(2)) = 3$; $g(f(0)) = -7$; $(f + g)(3) = 10$;
$(f - g)(0) = 3$; $(f \cdot g)(-1) = 21$; $\left(\dfrac{f}{g}\right)(4) = \dfrac{7}{8}$

CHAPTER 4

Exercise Set 4.1

1. $\{(6, 4)\}$

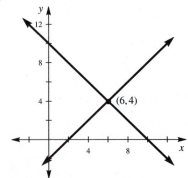

3. {(3, 0)}

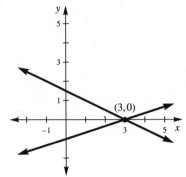

5. {(2, −1)}

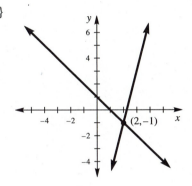

7. {(2, −1)}   **9.** $\left\{\left(\frac{5}{2}, -\frac{1}{4}\right)\right\}$ **11.** $\left\{\left(-\frac{3}{4}, 2\right)\right\}$ **13.** {(3, 2)}

15. inconsistent   **17.** {(6, 4)}   **19.** $\left\{\left(\frac{54}{11}, -\frac{27}{11}\right)\right\}$

21. {(1, 2)}   **23.** {(1, 1)}   **25.** {(3, 2)}   **27.** dependent
29. {(2, 1)}   **31.** {(3, 1)}   **33.** inconsistent

35. $\left\{\left(\frac{22}{3}, 4\right)\right\}$ **37.** $\left\{\left(\frac{2}{5}, -2\right)\right\}$ **39.** $\left\{\left(\frac{3}{2}, -3\right)\right\}$

41. $\left\{\left(\frac{43}{74}, \frac{43}{58}\right)\right\}$ **43.** {(18, 4)}   **45.** {(3, 7)}

47. 32, 42   **49.** 12.5 cm × 7.5 cm   **51.** $1.50
53. $6500 at 6%, $3500 at 7%   **55.** 5 l of each
57. $3\frac{1}{3}$ lb peanuts, $6\frac{2}{3}$ lb cereal   **59.** $a = 0$, $b = 4$
61. $a = 4$, $b = -\frac{2}{3}$ **63.** $a = 3$, $b = 2$

65. independent: the lines meet at one point; dependent: they are the same line; inconsistent: the lines are parallel.
67. The point of intersection is sometimes difficult or impossible to determine.

69.

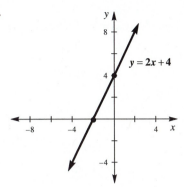

71.

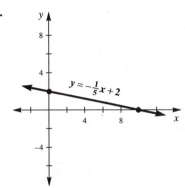

73.

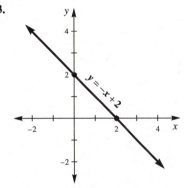

75. $y = 3x + 5$ **77.** $5x + 2y = 17$ **79.** $y = \frac{1}{4}x - 1$

Exercise Set 4.2

1. {(3, −1, 5)}   **3.** {(1, 2, −1)}   **5.** {(1, 2, 3)}
7. {(2, 3, 4)}   **9.** ∅   **11.** {(−3, 0, −9)}
13. $\left\{\left(\frac{335}{29}, -\frac{146}{87}, -\frac{31}{29}\right)\right\}$ **15.** {(2, 0, 3)}
17. $\left\{\left(0, -\frac{6}{5}, \frac{7}{5}\right)\right\}$ **19.** {(1, 1, 1)}   **21.** {(1, 1, 1)}
23. {(1, 1, −1)}   **25.** {(0, 1, 1, 1)}
27. 17 nickels, 32 dimes, 6 quarters
29. $10,000 at 5%, $5000 at 6%, $10,000 at 8%

31. *A* is fastest: it takes 120 minutes.
33. The point (ordered triple) where the three planes meet
35. A plane
37. $f(0) = -1$; $f(1) = 2$; $f(-2) = -7$; $f(b) = 3b - 1$
39. $f(g(0)) = 2$; $(f + g)(2) = 8$; $(f \cdot g)(-1) = -12$
41. yes: D$\{x \mid x \in R\}$; R$\{x \mid x \neq \frac{4}{3}\}$

Exercise Set 4.3

1. $\begin{bmatrix} 4 & -6 & -8 \\ 1 & -3 & 2 \end{bmatrix}$ **3.** $\begin{bmatrix} -7 & 9 & 18 \\ 4 & -6 & -8 \end{bmatrix}$

5. $\begin{bmatrix} 1 & -3 & 2 \\ 8 & -18 & 0 \end{bmatrix}$ **7.** $\begin{bmatrix} 2 & 1 & 4 & 1 \\ 1 & -1 & 3 & 0 \\ -1 & 2 & 1 & -3 \end{bmatrix}$

9. $\begin{bmatrix} 2 & -2 & 6 & 0 \\ 2 & 1 & 4 & 1 \\ -1 & 2 & 1 & -3 \end{bmatrix}$ **11.** $\begin{bmatrix} 1 & -1 & 3 & 0 \\ 2 & 1 & 4 & 1 \\ 0 & 1 & 4 & -3 \end{bmatrix}$

13. $\begin{bmatrix} 1 & -1 & 3 & 0 \\ 0 & 3 & -2 & 1 \\ -1 & 2 & 1 & -3 \end{bmatrix}$ **15.** $\begin{bmatrix} 2 & -1 & | & 0 \\ -1 & 1 & | & 1 \end{bmatrix}$

17. $\begin{bmatrix} 1 & 1 & 1 & | & 3 \\ 2 & 1 & -1 & | & 2 \\ 1 & 1 & -2 & | & 5 \end{bmatrix}$ **19.** $x + 2y = 5$; $3x + y = 4$

21. $x + z = 2$; $x - y + z = 3$; $z = 4$
23. $\{(3, -2)\}$, independent   **25.** inconsistent
27. $\{(4, -1)\}$ **29.** $\{(\frac{1}{2}, 1)\}$ **31.** $\{(2, -1, 4)\}$
33. $\{(4, -4, 3)\}$ **35.** $\{(2, -3)\}$ **37.** $\{1, 2\}$
39. $\{(-\frac{5}{7}, -\frac{2}{7}, -\frac{3}{7}, \frac{2}{7})\}$ **41.** $\{(-0.87, -0.85)\}$
43. $\{(-0.41, 0.46)\}$ **45.** $\{(\frac{3b}{5a}, -\frac{b}{5})\}$
47. $\{(\frac{6a + 2b}{a^2 + b^2}, \frac{6b - 2a}{a^2 + b^2})\}$ **49.** $a = \frac{1}{2}, b = 3$
51. $a = \frac{1}{2}, b = 2, c = 1$

Exercise Set 4.4

1. 1   **3.** 0   **5.** 312   **7.** 42   **9.** $\{30\}$ **11.** $r = 4$
13. $x = 1$ **15.** $\{(4, 3)\}$ **17.** $\{(\frac{11}{7}, \frac{12}{7})\}$ **19.** $\{(6, 4)\}$
21. $\{(-1, 2, 1)\}$ **23.** $\{(1, 1, 1)\}$ **25.** $\{(2, -3, 2)\}$
27. inconsistent   **29.** $\{(\frac{1}{2}, 0, 1)\}$ **31.** $\{(\frac{4}{3}, -4)\}$
33. $\{(\frac{64}{33}, \frac{-160}{9})\}$ **35.** $x \cdot 0 - y \cdot 0 = 0$

37. $akd - kcb = k(ad - cb) = k\begin{vmatrix} a & b \\ c & d \end{vmatrix}$

39. Expand about column 1 and simplify.
41. $3x - 2y - 1 = 0$ is the equation; $(1, 1)$ and $(-3, -5)$ both satisfy.
43. crawl stroke: 10 min; breast stroke: 5 min
45. Sherry: \$4.50; Jason: \$1.50; Lisa: \$6.00
47. $\{(-\frac{3}{7}, \frac{1}{14})\}$ **49.** $\{(-\frac{16}{11}, \frac{69}{44}, \frac{71}{22})\}$

Exercise Set 4.5

1.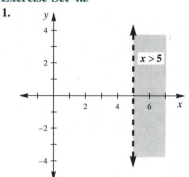

3.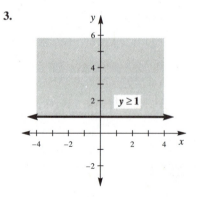

5.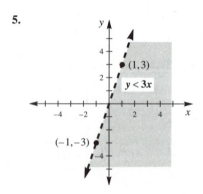

7.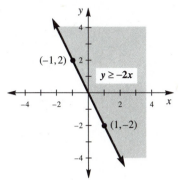

9.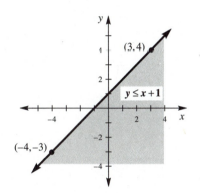

11.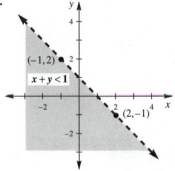

13.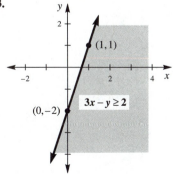

15.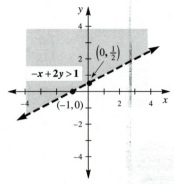

17.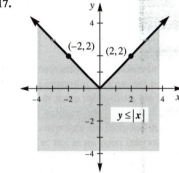

19.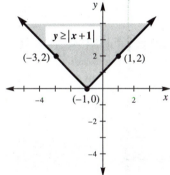

21.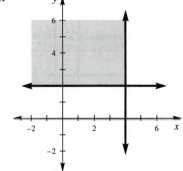

23.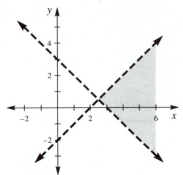

25.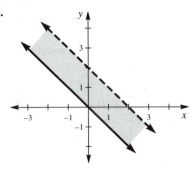

27.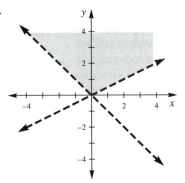

29.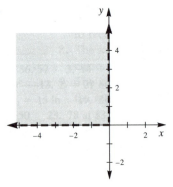

31.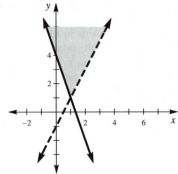

33.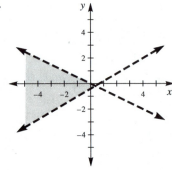

35.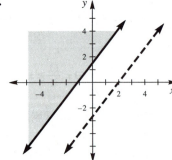

37.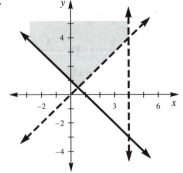

39.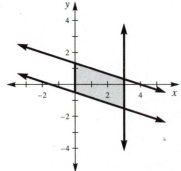

41.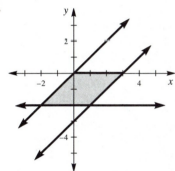

43.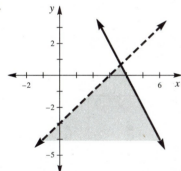

45.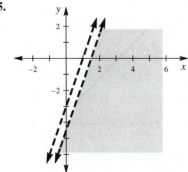

47. yes **49.** yes **51.** no **53.** $y \leq x$ **55.** $2x + 3y \geq 6$
57. $y \geq 4$
59. $x + y \geq 4$; (1, 4), (4, 1), (5, 3)

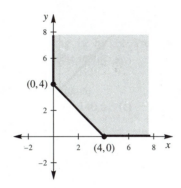

61. $5x + 8y \leq 240$, $x \geq 0$, $y \geq 0$

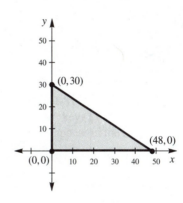

63.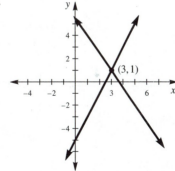

65. $\left\{\left(\dfrac{14}{11}, \dfrac{12}{11}\right)\right\}$ **67.** $\{(1, 1, 1)\}$

Exercise Set 4.6

1.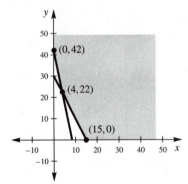

3.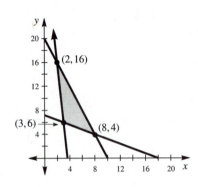

5. Maximum: $P(0, 8) = 16$ **7.** Maximum: $C(5, 0) = 25$
9. $P(0, 5) = 15$
11. Making 432 of Bar A and 504 of Bar B yields maximum receipts of \$401.76.
13. Mixing 6 oz of type A and 10 oz of type B yields a maximum profit of 68¢/packet.
15. Taking 3 of Tablet M and 2 of Tablet P minimizes cost at 70¢/day.
17. 17,802 ft^3

Exercise Set 4.7

1. $\frac{3}{2}, -\frac{17}{2}$ **3.** 17, 19

5. speed of boat $= 17\frac{1}{2}$ mph; speed of current $= 2\frac{1}{2}$ mph

7. 55 km/hr and 45 km/hr
9. large cola: 90¢; small cola: 60¢   **11.** \$2000
13. $\frac{3}{4}$ l of 40%; $1\frac{1}{4}$ l of 80%
15. 4 lb of peanuts; 6 lb of pecans
17. dictionary: $\frac{45}{26}$ in. thick; encyclopedia: $\frac{7}{13}$ in. thick
19. \$2500 for the first company; \$4500 for the second company

21. $2x + 3y \leq 240, x > 0, y \geq 0$

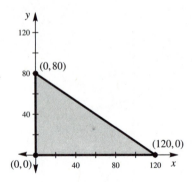

23. $M + P \leq 6, M - P \geq 1, M \geq 0, P \geq 0$

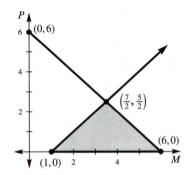

25. (a) $p = \$20, q = 500$
(b)

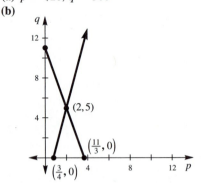

27. (a) $p = \$4, q = 4$
(b) no, they are equal
(c) yes; 12 units

29. \$5000 at 6%; \$5000 at 5%; \$5000 in business

31. P = number of days producing paperbacks; H = number of days producing hardbacks (a)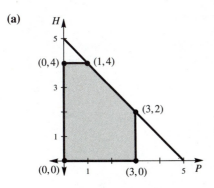

(b) $A(3, 0)$, $B(3, 2)$, $C(1, 4)$, $D(0, 4)$, $E(0, 0)$
(c) A: \$1200; B: \$2600; C: \$3200; D: \$2800; $E = \$0$
(d) vertex B: 1 day producing the paperback, 4 days producing the hardback

Review Exercises

1. $\left\{\left(\dfrac{21}{11}, \dfrac{8}{11}\right)\right\}$ 2. dependent 3. inconsistent
4. independent
5.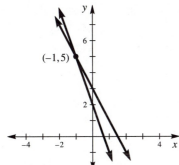
6. $\left\{\left(\dfrac{5}{6}, \dfrac{7}{3}\right)\right\}$ 7. $\{(2, 1)\}$ 8. $\left\{\left(\dfrac{13}{6}, -\dfrac{17}{6}\right)\right\}$
9.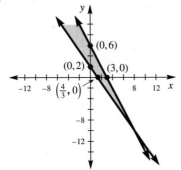
10. $\{(0, 10)\}$

11.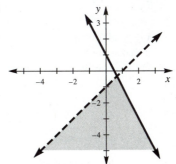

12.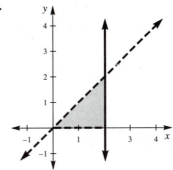

13. $\{(2, -1, -2)\}$ 14. $\left\{\left(\dfrac{52}{29}, -\dfrac{4}{29}\right)\right\}$
15.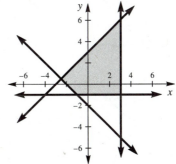

16. $\left\{\left(\dfrac{21}{2}, -84\right)\right\}$ 17. $\begin{bmatrix} 3 & -2 & | & 1 \\ 2 & -5 & | & -3 \end{bmatrix}$
18. $3x + y = 4$
    $2x + 3y = 5$
19. $\left\{\left(\dfrac{5}{3}, -\dfrac{2}{3}\right)\right\}$ 20. $\{(1, -1)\}$ 21. 18 22. $\{(1, 2, 1)\}$
23. $\{(1, 0)\}$ 24. 3 25. $\{(3, -1, 1)\}$ 26. $\left\{\left(\dfrac{7}{5}, 2, \dfrac{9}{5}\right)\right\}$
27. $\{1\}$ 28. 0 29. yes: add $3R_1$ to R_2 30. not correct
31. yes: add $3R_1$ to R_2 32. yes: add R_2 to R_3
33. $\{(-5, 2)\}$

34. Evaluate both sides to get
$bc^2 + ab^2 + a^2c - a^2b - ac^2 - b^2c$.
35. oats at $5.50/bushel; rye at $4.00/bushel
36. Hot dogs are 90 cents; shakes are 80 cents.
37. $9.50 for a shirt; $1.80 for a pair of socks
38. 5 gal of 6%; 5 gal of 2%   39. 295 mph and 175 mph
40. 7 quarters, 11 dimes
41. $800 at 5%, $1000 at 8%; $1200 at 6%
42. 2 mph   43. 165 mph   44. $60\frac{5}{11}$ mph
45. 10 hr for Company A; 15 hr for Company B
46. 5 beginning and 5 experienced instructors
47. 10 units of model A, 12 units of model B
48. 10 units of model x, 2 units of model y; P = $56

Chapter Test
1.
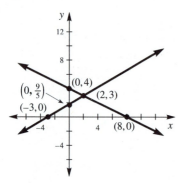
2. {(1, 2)}   3. {(2, 3)}
4.
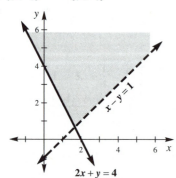
5. {(3, 3, −1)}   6. $1000 at 8%; $2000 at 6%
7. 25, −9   8. {(1, −2)}   9. {(1, 1, 1)}   10. $\left\{-\frac{5}{2}\right\}$
11. {−7}   13. $\left\{\left(\frac{1}{3}, 2\right)\right\}$ 14. $\left\{\left(2, \frac{1}{2}, -1\right)\right\}$
15. {(2, 1)}   16. $\left(\frac{1}{3}, \frac{2}{3}, 0\right)$
17. 12.5 l of 20%; 12.5 l of 30%

18. Current goes at 10 mph; boat goes at 20 mph.
19. $l = 13$ in., $w = 5$ in.
20. cashews: $3.50/lb; peanuts: $2.50/lb
21. first pound: $2.10; each additional pound: $0.70
22. Angelina; 84 items

CHAPTER 5

Exercise Set 5.1
1. 8   3. $\frac{1}{10}$ 5. 1   7. −32   9. 100   11. $-\frac{1}{5}$
13. $\frac{1}{9}$ 15. 0.04   17. $\frac{4}{9}$ 19. $\frac{25}{81}$ 21. 16   23. 10
25. 1   27. $\frac{9}{125}$ 29. $\frac{1}{10}$ 31. 1.8×10^4
33. 8.0×10^{-2} 35. 4.3×10^7 37. 8.1×10^{-9}
39. 8,000   41. 0.011   43. 21   45. 0.0317   47. 7,800
49. 0.000192   51. 120   53. 770   55. 100,000
57. 0.07   59. 0.12   61. 50   63. a^9 65. t 67. p^8
69. $\frac{1}{c^{15}}$ 71. s^4 73. $\frac{1}{c^{12}}$ 75. s^2t 77. $\frac{8x^9}{3y}$ 79. $\frac{8}{x^2}$
81. 1   83. $\frac{1}{p^2}$ 85. $\frac{4b^3}{3a^5}$ 87. $\frac{m^9n^2}{8}$ 89. $4p^8q^2$
91. x^{n+m} 93. a^{5n-1} 95. p^2 97. t^{m-n} 99. $a + b$
101. 1   103. 2.0×10^3 hr   105. 2.112×10^{15} km
107. 29,688.12 ft³   109. $\left(\frac{x}{y}\right)^{-m} = \frac{1}{\left(\frac{x}{y}\right)^m} = \left(\frac{y}{x}\right)^m$
111. $(3 + 4)^2 \neq 3^2 + 4^2$ because $49 \neq 25$

Exercise Set 5.2
1. yes   3. no   5. yes   7. no   9. yes   11. no
13. (a) $5p^2 + p$ (b) 2nd degree   (c) 5   (d) binomial
15. (a) $23r$ (b) 1st degree   (c) 23   (d) monomial
17. (a) $-q^5 - 9q^3 + 5q^0$ (b) 5th degree   (c) −1   (d) trinomial
19. (a) $2x^3 + 5x^2 - 9x$ (b) 3rd degree   (c) 2   (d) trinomial
21. $7a$ 23. $3y$ 25. $-6m^2 - 3m$ 27. $12ab^2$
29. $-\frac{1}{4}t$ 31. $8x^2$ 33. $-b$ 35. $-3t + 3$
37. $-b^2 + 3b - 1$ 39. $3a + 5$ 41. $-4k - 17$
43. $-2x^2y - 3xy^2 - 4$ 45. $-6r^2 + 16r - 5$
47. $-14x^2y + 6xy^2 + 18xy$ 49. $2x + 4y$
51. $-7x + 5y + 8$ 53. $6m + 1$ 55. $21s - 14t$
57. $24r - 30s$ 59. $9x^2 - 15xy + 6y^2 + 3$
61. $5x^2 + x + 1$ 63. $4x^{3m} + 2x^{2m} + 9x - 5$
65. $3x^{3a} - 7x^a - 19x - 4$
67. $x^{5b} + 6x^{4b} + 3x^{3b} - 7x^{2b} - 6x - 4$
69. $-6.06x^2 + 1.142x - 17.849$ 71. −0.4392

73. $P(x) = 60x - 50$; $P(50) = \$2950$
75. $P(x) = x^2 + 6x - 20$; $P(10) = 140$;
$C(10) = 160$; $R(10) = 300$
77. $\left[\left(\dfrac{11}{18}, \dfrac{19}{18}\right)\right]$
79.

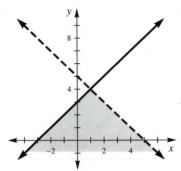

81. $-3xy^5$ **83.** $\dfrac{1}{x^2 y^{16}}$

Exercise Set 5.3
1. $6x^5$ **3.** $56m^6$ **5.** $-p^9 q^{12}$ **7.** t^7 **9.** $49a^9$
11. $2y^3 - 6y^2$ **13.** $-15t^4 - 30t^3 - 45t^2$
15. $-3s^4 + 27s$ **17.** $a^4 b + 7a^3 b^2 - a^2 b^3$
19. $-3x^5 + 19x^4 - 57x^3$ **21.** $a^{3n} + 3a^{2n} - 4a^n$
23. $u^3 + 2u^2 v + 2uv^2 + v^3$
25. $6x^5 + 3x^4 y - 2x^3 y^2 + 2x^2 y^3 - y^5$
27. $x^{3n} + x^n y^{2n} - 2y^{3n}$ **29.** $x^2 + 2x + 1$
31. $4p^2 - 12p + 9$ **33.** $r^2 - s^2$ **35.** $9k^2 - 25$
37. $4h^4 - 36h^2 + 81$ **39.** $16y^4 - 9$
41. $x^3 + 6x^2 + 12x + 8$ **43.** $x^3 - 6x^2 + 12x - 8$
45. $a^2 + 2ab + b^2 - 16$
47. $25p^2 + 10pq + q^2 - 50p - 10q + 25$
49. $x^2 + 2x - 3$; -3 **51.** $k^2 + 11k + 28$
55. $28y^2 + 36y - 40$ **57.** $56r^2 + 2rs - 30s^2$
59. $21x^4 + 32x^2 y^2 - 5y^4$ **61.** $66m^4 + 23m^2 n - 63n^2$
63. $10p^3 - 41p^2 + 31p - 6$
65. $24a^3 - 71a^2 b + 51ab^2 - 10b^3$
67. $q^5 - 3q^4 r + 2q^3 r^2 + 4q^2 r^3 - 3qr^4 - r^5$ **69.** $a^{3n} + b^{3n}$
71. $2a^4 - 5a^3 + 5a^2 + 11a - 10$
73. $6x^4 - 13x^3 y - 41x^2 y^2 - 27xy^3 - 5y^4$ **75.** $-12x$
77. $-12x^n y^n$ **79.** $4u^3 + 4u^2 v - 5uv^2 - 3v^3$
81. $p^{3m+5} - 2p^{3m+6} q^{m+1} + p^{2m+5} q^3 - p^m q^{m+3} + 2p^{m+1} q^{2m+4} - q^{m+6}$
83. $8x^2 + 10x + 3$
85. $9x^2 + 15x + \dfrac{13}{2} + \pi\left(4x^2 + 6x + \dfrac{9}{4}\right)$
87. $14x^2 + 20x + 7 + \pi\left(4x^2 + 6x + \dfrac{9}{4}\right)$ **89.** 39.48 yd^2
91. $\dfrac{1}{x^8}$ **93.** $\dfrac{1}{x^4}$ **95.** $\dfrac{-1}{3}$ **97.** $\dfrac{-a}{4}$ **99.** $-b^2 - 2b + 4$

Exercise Set 5.4
1. $a + 2$ **3.** $-3t + 4$ **5.** $2t^2 - 3t + 1$
7. $8x^2 + 4x - 1$ **9.** $-x - 3 + \dfrac{4}{x}$ **11.** $-9a + 11 - \dfrac{3}{2a^2}$
13. $6ab^2 + 4b$ **15.** $a^{4m} - a^{2m}$ **17.** $-14t^{12n} + 9t^{8n}$
19. $-7a^7 b^4 c^5 + 4a^4 b^5 c^{12}$ **21.** $\dfrac{15b^8 c^7 d}{11} - \dfrac{5b^5 c^7 d^5}{11} - \dfrac{5}{22}$
23. $x + 1$ **25.** $a + 3$ **27.** $t + 3$ **29.** $x^2 + 5x - 6$
31. $v - 2$ **33.** $5a^2 + 9a + \dfrac{1}{7} + \dfrac{-\frac{9}{7}}{7a - 2}$
35. $4x^2 + 2x + 1$; 3 **37.** $3s^2 + \dfrac{3}{5}s + 1 + \dfrac{-\frac{3}{5}s - 2}{5s^2 + 1}$
39. $x^2 - 3x + 5$ **41.** $8x - 5y + \dfrac{8xy^2 - 5y^3}{3x^2 - 7xy + y^2}$
43. $x^n + y^n$ **45.** $x^n - y^n$ **47.** $x^{2n} - x^n y^n + y^{2n}$
49. $Q = x + 4$; $R = 1$ **51.** $Q = \dfrac{5}{4}x + \dfrac{33}{16}$; $R = -\dfrac{111}{16}$
53. $5x - 1$ **55.** $5x^2 + 4x - 1$ **57.** $h(x) = 2x - 1$
59. $a(x) = x - 7$ **61.** $5x^3 + 15x^2 - x - 1$
63. $9m^2 + 12mn + 4n^2$ **65.** $121a^2 - 36b^2$
67. $y^4 - 10y^3 + 25y^2 - 49$ **69.** $21x^2 + xy - 36y^2$
71. $8x^3 - 12x^2 y + 6xy^2 - y^3$
73.

75.

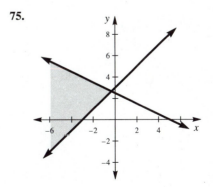

77. $\left\{\left(\frac{17}{11}, \frac{7}{11}\right)\right\}$ **79.** $\{(2, 1)\}$

81. The divisor is what we are dividing by; the dividend is what we are dividing into.

83. When the new dividend is of lower degree than the divisor

Exercise Set 5.5

1. $x - 5$ **3.** $a^2 + 6a + 5 + \frac{15}{a-2}$ **5.** $3t^2 - 4t + 4 + \frac{4}{t+1}$

7. $x^4 + 2x^3 + 4x^2 + 8x + 16$ **9.** $2y^3 + y^2 + 3y + 3$

11. $r^3 + r^2 + 5r + 3$

13. $a^5 + a^4 + a^3 + a^2 + a + 1 + \frac{-2}{a-1}$

15. $2t^3 + 3t^2 + 14t + 70 + \frac{357}{t-5}$

17. $5y^3 - 10y^2 + 2y - 10 + \frac{23}{y+2}$

19. $2m^3 - 6m^2 + 15m - 40 + \frac{113}{m+3}$

21. $4x^2 - 5x + 3$ **23.** $5x^2 + 3x - 8 + \frac{2}{x-3}$

25. $-3x^3 + 9x^2 - 5x + 11 + \frac{27}{x-3}$

27. $-2x^3 - 8x^2 + 5x + 1 + \frac{-7}{x+7}$ **29.** $3x^2 - 2x + 6$

31. $x + \frac{1}{3}$ **33.** $x^4 + \frac{1}{2}x^2 + \frac{5}{4}$ **35.** $7x^2 + \frac{21}{2}x + \frac{65}{4} + \frac{\frac{155}{8}}{x-\frac{3}{2}}$

37. $\frac{1}{2}x + \frac{5}{2}$ **39.** $x + 8 + \frac{6}{6x-5}$

41. $-20x^3y - 7x^2y + 20xy$ **43.** $-29x^2 - 57x + 330$

45. $9a^2 + 12ab + 4b^2 - c^2$ **47.** $9x^4 - 12x^2y + 4y^2$

49. $x^{2m+4} + 4x^{m+5}y^{3m-2} - x^{m-1}y^{2m-1} - 4y^{5m-3}$

51. $9x^4 - 12x^3y^2 + 4x^2y^4$

53.

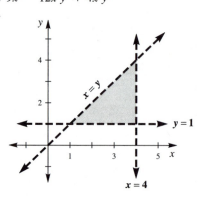

55. $r^2 = \frac{x^2 + 3x + 2}{\pi}$ **57.** $\pi = \frac{3}{4}(x^2 + 2x + 1)$

59. It's a shortcut method.

61. They're not needed for this method

Review Exercises

1. $\frac{7}{2}$ **2.** 2   **3.** $-\frac{25}{9}$ **4.** 35   **5.** $\frac{10,000}{9}$ **6.** $\frac{20,449}{81}$

7. x^9 **8.** $\frac{1}{y}$ **9.** x^{m+n} **10.** -1 **11.** $\frac{2x^2}{y}$ **12.** x^{2n+2}

13. $14x$ **14.** $\frac{y^3}{64}$ **15.** $10x - 4$ **16.** $-x - 4$ **17.** $\frac{x^4z^8}{9y^6}$

18. $\frac{9}{2x^8y^6}$ **19.** $\frac{y^2}{4x^{12}}$ **20.** $15x^4$ **21.** $9x^3 - 9x^2$

22. $a^2 - 4a + 4$ **23.** $6x^2 - 23x + 21$

24. $p^3 - 3p^2 + 3p - 1$ **25.** $49x^2 - 4y^2$

26. $4x^4 - 16x^2y^2 - 33y^4$ **27.** $5x^2 - 10x$

28. $4x^2 - 4x - 8$ **29.** $2p^5 + p^3 + 10p^2 - 6p - 15$

30. $2 + \frac{4}{a}$ **31.** $3t^2 - 4t + 1$ **32.** $-x + 3$

33. $x - 9 + \frac{23}{x+2}$ **34.** $m^2 + 3m + 9$ **35.** $16a^2 + 4a + 1$

36. $x^2 - 7x + 13 + \frac{-12x - 30}{x^2 + 2x + 3}$ **37.** $p^3 + 11p - 5$

38. $3x^3 + 9x^2 + 20x + 60 + \frac{189}{x-3}$

39. $\frac{1}{2}x^2 - \frac{7}{4}x - \frac{5}{8} + \frac{-\frac{117}{16}}{x-\frac{1}{2}}$ **40.** $17x + 16$

41. (a) $P(x) = 4x + 16$ (b) $A(x) = x^2 + 6x + 5$

42. (a) $A(x) = \frac{1}{2}(x^2 + 5x)$ (b) $P(x) = 3x + 11$

43. (a) $P(x) = 4x + 16 + \pi(x + 3)$

(b) $A(x) = 2x^2 + 16x + 30 + \frac{1}{2}\pi(x^2 + 6x + 9)$

(c) $A(x) = \frac{1}{2}\pi(x^2 + 6x + 9)$

44. (a) $A_c(x) = \pi x^2$ (b) $A_r(x) = 8x^2 - 2\pi x^2$

(c) $P_c(x) = 2\pi x$ (d) $P_r(x) = 12x$

Chapter Test

1. $30x^3$ **2.** $-5x^2 + 7x - 2$ **3.** $9x^2 - 12x + 4$

4. $49x^2 - 9$ **5.** $x + 13 + \frac{32}{x-3}$ **6.** $x - \frac{14}{5} + \frac{\frac{16}{5}}{5x-1}$

7. $20m^2 - 3m - 56$ **8.** $6y^4 - 19y^2 - 7$

9. $a^{2m+1} + a^{2m} + 2a^m$ **10.** $m^3 - n^3$

11. (a) $-x^2 + 3$ (b) second   (c) binomial

12. (a) $x^3 + 2x^2 + x$ (b) third   (c) trinomial

13. (a) 0   (b) zero   (c) monomial

14. (a) $15y^3 - 11y^2 + 1$ (b) third (c) trinomial
15. $-p - 16$ 16. $-x^2 + x + 3$ 17. $9y - 7x$
18. $x^5 - 4x - 5$ 19. $\dfrac{3x}{10y}$ 20. $-r^2$ 21. 15
22. 6×10^8 23. $7x^2 - 8x + 32 + \dfrac{-32x + 125}{x^2 - 4}$
24. $x^2 + \dfrac{1}{3} + \dfrac{-\frac{29}{6}}{x - \frac{1}{2}}$ 25. $-2x^3y - 2xy^2 - 9xy - 8y$
26. $8x^3 - 12x^2 + 6x - 1$ 27. $9x^2 - 6x - 3$ 28. $x - 1$
29. $x + 2$ 30. $\dfrac{1}{2}(3x^2 + 4x - 7)$

Chapter 6

Exercise Set 6.1

1. $x + 2$ 3. $2m - 3$ 5. $7s + 1$ 7. $y - 3$
9. $x^2 + 2x - 3$ 11. $3t^2 - 4t + 2$ 13. $2a^2 - 10a + 1$
15. $3p^2 + 16p - 10$ 17. $4s^2 - 6s + 1$
19. $8r^2 - 21r + 14$ 21. $2(x + 5)$ 23. $11(y - 5)$
25. $5a(3a - 10)$ 27. $8(1 - 2t)$ 29. $3(a + 3b - 2)$
31. $xy(x - y)$ 33. $6mp(2m - p + 3)$ 35. $t(a + b - c)$
37. prime 39. $3r^2s^3(5r^3 - 8s)$ 41. $a(5 + 7a - 9a^2)$
43. $7x^2(4x^2 - 8x - 7)$ 45. $2r^4s^3(4r - 3s^2 - 10)$
47. $16xyz(3 - 4xz - 5y)$ 49. $9a(7a^2c - 9ad + 8cd)$
51. $(x + 1)(x + 2)$ 53. $(2a + 1)(a - 1)$
55. $(a + c)(m + n)$ 57. $(x - y)^2$
59. $(t - 1)(t^2 - t - 1)$ 61. $(m - 5)(4 - m)$
63. $\left(\dfrac{1}{2} - a\right)\left(\dfrac{1}{3} + a\right)$ 65. $(x - 1)(x - 2)$
67. $(2a + 3)(2a + 5)$
69. $(a - b)^3(a^2 + b)^2[(a - b)(a^2 + b) - 3]$
71. $a^m(a + 1)$ 73. $x^{2m}(x - 1)$ 75. $y(y^{m-1} + 1)$
77. $r^{2m+2}(r - 1)$ 79. $(c + d)(x + y)$
81. $2(a - 2)(a + b)$ 83. $(r + 2s)(m + 2n)$
85. $(x - y)(a - b)$ 87. $(2ab - 5c)(2xy - 3z)$
89. $(x - 5)(x + 2)$ 91. $(5x - 1)(x + 3)$
93. $(7y - 1)(4y - 3)$ 95. $(4x + 1)(2x + 1)$
97. $3(5x + 2)(2x - 5)$ 99. $x^2 - y^2$ 101. $-3x - y + 4$
103. $9m^2 + 12mn + 4n^2$ 105. $a^3 - b^3$
107. $x^4 + 2x^3 + 4x^2 + 8x + 16$ 109. $a^{2m} + 2a^m a^n + a^{2n}$
111. $-6x^3 + 26x^2 + 20x$ 113. $2x^3 - 18x^2 + 54x - 54$
115. No: it is $3(x + 2)(2x + 1)$. 117. yes

Exercise Set 6.2

1. $(x + 2)(x + 1)$ 3. $(m + 6)(m + 1)$ 5. $(y - 1)^2$
7. $(p - 3)(p - 2)$ 9. $(x - 3)(x + 2)$
11. $(t + 3)(t - 1)$ 13. $(y - 10)(y + 2)$
15. $(m + 6)(m - 5)$ 17. $(r + 8a)(r - 2a)$
19. $(a - 12b)(a + 5b)$ 21. $2(x + 6)(x + 4)$
23. prime 25. $5(p - 1)(p - 8)$
27. $(5 + s)(2 - s)$ or $-(s - 2)(s + 5)$
29. $x(x - 8)(x + 4)$ 31. $4a(a - 5b)(a + b)$
33. $\left(r + \dfrac{1}{5}\right)\left(r + \dfrac{1}{3}\right)$ 35. $\left(k + \dfrac{1}{3}\right)\left(k - \dfrac{1}{2}\right)$ 37. prime
39. $(ab - 2)^2$ 41. $-3(m + 4)(m - 3)$
43. $-7a\left(a^2 + \dfrac{1}{2}a - \dfrac{1}{16}\right)$ 45. $-(x - 6)(x - 2)$
47. $3st(t + 9s)(t + 2s)$ 49. $5pq(p + q)^2$
51. $(x^2 + 4)(x^2 + 2)$ 53. $(a^2 + 4)^2$
55. $2(m^2 + 5)(m^2 + 4)$ 57. $-5(t^2 + 6)(t^2 - 5)$
59. $-k\left(k - \dfrac{1}{6}\right)\left(k - \dfrac{1}{5}\right)$ 61. $(x + 3)(x - 2)(x + 1)$
63. $(a - 3)^2(a + 2)$ 65. $(t + 7)^3$ 67. $\left(k - \dfrac{1}{3}\right)^3$
69. $(x^m + 6)(x^m + 2)$ 71. $(y^m - 0.2)(y^m - 0.1)$
73. $c^m(c + 5)(c + 6)$ 75. $s^2(s^m - 2)(s^m + 1)$
77. $k^3(k^m - 0.7)(k^m - 0.3)$ 79. $0.65y(0.9y - 0.11)$
81. $-0.93t(0.11t + 0.9)$ 83. $3x(2x - 1)$
85. $x(x + 3)(x + 1)$ 87. $3x^2 + 4x - 2$
89. $x^2 + 3x + 4 + \dfrac{7}{x - 2}$ 91. x^{15} 93. x^{6n+2} 95. $81x^4$
97. Factor: $25x^2 - 20x + 4 = (5x - 2)^2$; so $s(x) = 5x - 2$ is the length of a side. Therefore $s(2) = 8$.
99. Factor: $\dfrac{1}{3}\left(5\pi x^2 - \dfrac{5}{2}\pi x + 5\pi\right)$
$= \dfrac{1}{3}\pi\left(x^2 - \dfrac{1}{2}x + 1\right) \cdot 5 = \dfrac{1}{3}\pi\left[\sqrt{x^2 - \dfrac{1}{2}x + 1}\right]^2 \cdot 5$, so
$r(x) = \sqrt{x^2 - \dfrac{1}{2}x + 1}$ is the radius. Therefore $r(4) = \sqrt{15}$.

Exercise Set 6.3

1. $(2a + 1)(a + 1)$ 3. $(3x + 1)(x + 2)$
5. $(2s - 1)(s + 4)$ 7. $2(3r + 4)(2r - 1)$
9. $(4m - 5)(2m + 1)$ 11. $2(4x + 17)(x - 1)$
13. $(2a + 5)^2$ 15. $4y^3(2y - 5)(y + 5)$
17. $(3c + 2d)(3c - 10d)$ 19. $(4 + 3k)(3 - 2k)$
21. $(2 + 3t)(1 - 4t)$ 23. $(3ab - 10)(2ab + 1)$
25. $-2x^2(k + 3a)(k + 4a)$ 27. $(5y + 8)(3y + 5)$
29. $-2(8x - 3y)(2x + y)$ 31. $(ab + 3cd)(ab + 2cd)$
33. no 35. no 37. yes 39. yes 41. yes
43. $2(x + 2)^2(x - 2)^2$
45. $3(a + 3b)(a - 3b)(a + b)(a - b)$
47. $(c + 2d)(c - d)(c^2 - 2cd + 4d^2)(c^2 + cd + d^2)$
49. $(3x + 2y)(3x - 2y)(x + 3y)(x - 3y)$
51. $2(7c + d)(7c - d)(c + d)(c - d)$
53. $\left(\dfrac{2}{3}y + \dfrac{1}{2}\right)\left(\dfrac{2}{3}y + \dfrac{1}{3}\right)$ 55. $\left(\dfrac{3}{5}m + \dfrac{1}{3}n\right)\left(\dfrac{1}{2}m - \dfrac{1}{4}n\right)$
57. $(3x^m - 2y^n)(x^m + 4y^n)$ 59. $(4c^m + 9d^n)(3c^m - d^n)$

61. $(x + 5)(x + 2)$ **63.** $[6(m + n) + 5][4(m + n) + 3]$
65. $2(3x + 2y)(4a + 13b)$ **67.** $(2x - 1)(2x - y)$
69. $a(a - b)$ **71.** $(y - x)(72x - 1)$ **73.** $x(x - 2)(x + 1)$
75. $[3(x - y) + 5(a + b)][2(x - y) - 7(a + b)]$
77. $63(2y - 1)$ **79.** $(x - 1)^2$ **81.** $(x + 7)(x + 4)$
83. $(y - 3)(y - 1)$ **85.** $(x - 7)(x - 2)$ **87.** $(y + 2z)^2$
89. $(x + 9y)(x - 3y)$ **91.** $(x + y)(x + 29y)$
93. $(b + 9)(b + 2)$
95. $-3(x^4 - 4x^2 - 5)$: factor out -3.
$-3(x^2 - 5)(x^2 + 1)$: factor the trinomial.
97. $2(3y - 5)^2(4x - 1)[(3y - 5)(4x - 1) - 2]$: common factor
99. $ax^2 + bx + c$: If there are two numbers m and n such that $m \cdot n = a \cdot c$ and $m + n = b$, rewrite it as $ax^2 + mx + nx + c$. It then can be regrouped and factored.

Exercise Set 6.4
1. $(x + 1)(x - 1)$ **3.** $2(a + 5)(a - 5)$
5. $(xy + 4z)(xy - 4z)$
7. $(11p + 12q)(11p - 12q)$
9. $(a^2 + b^2)(a + b)(a - b)$ **11.** $3(r^m + s^m)(r^m - s^m)$
13. $b^3 \left(\frac{1}{4}a + \frac{1}{3}b\right)\left(\frac{1}{4}a - \frac{1}{3}b\right)$
15. $ab(0.8a + 0.7b)(0.8a - 0.7b)$ **17.** $4ab$ **19.** $-24mn$
21. $(a^4 + b^4)(a^2 + b^2)(a + b)(a - b)$
23. $(x - y)(x^2 + xy + y^2)$ **25.** $(p + q)(p^2 - pq + q^2)$
27. $(3a - 2b)(9a^2 + 6ab + 4b^2)$
29. $(xy + wz)(x^2y^2 - wxyz + w^2z^2)$
31. $(a^m - b^m)(a^{2m} + a^m b^m + b^{2m})$
33. $(2x - 1)(x^2 - x + 1)$ **35.** $2(3y^2 + 6y + 4)$
37. $(0.2x + 0.1y)(0.04x^2 - 0.02xy + 0.01y^2)$
39. $2xy(2x - 3y)(4x^2 + 6xy + 9y^2)$
41. $2b(3a^2 + b^2)$ **43.** $\left(\frac{1}{3}s - \frac{1}{2}t\right)\left(\frac{1}{9}s^2 + \frac{1}{6}st + \frac{1}{4}t^2\right)$
45. $(a + 2)^2$ **47.** $(x - 1)^2$ **49.** $(2x + 3y)^2$
51. $3p(p - q)^2$ **53.** $(c^m + 7d^m)^2$ **55.** $(y + 3)^2(y - 3)^2$
57. $(a + 3)^2$ **59.** $3m(m^2 - 5)^2$ **61.** $(x - y - 2a + 2b)^2$
63. $\left(\frac{1}{4}w + \frac{1}{5}z\right)^2$ **65.** $(a - b + 1)(a + b + 1)$
67. $(y + 3x + 3)(y - 3x + 3)$
69. $(s - 2t - 4)(s + 2t + 4)$ **71.** $(3b - 7)(2a - c)$
73. $(2m + 3n)(4m^2 - 6mn + 9n^2 + 1)$
75. $2(4x - y)(16x^2 + 4xy + y^2 - 1)$
77. $3(x + y + 5)(x - y + 1)$ **79.** $(a + b - c)(x - y)$
81. $(c - 2d)(3ab - 5c^2)$ **83.** $5(s + t)(s - t)(x + y)(x - y)$
85. $(3a - x)(b + c)(b - c)$
87. $4(2y + z)(2y - z)(3a + 2b)(3a - 2b)$
89. $x(x + y + 1)(x - y - 1)$
91. $(3x^2 + 3y - 2)(3x^2 - 3y + 2)$
93. $(x - y)(x^2 + xy + y^2 - 1)$
95. $(x + 3)^2(x - 3)$ **97.** $(x + 1)^2(x^2 - x + 1)$
99. $(4x + y)(16x^2 - 4xy + y^2 - 1)$
101. $(2x - y)[4x^2 + 2xy + y^2 + 1]$

103. $h(1) = 0$; $h(2) = 0$; $h\left(\frac{3}{2}\right) = 4$, so $h(1)$ and $h(2)$ are back at starting position, while $h\left(\frac{3}{2}\right)$ is 4 ft above $h(1) = h(2) = 0$.

Exercise Set 6.5
1. $9(2x + 3)$ **3.** $0.8(3 - 2x)$ **5.** $5xy(3x + 7y)$
7. $0.3x(2x + 1)(2x - 1)$ **9.** $4(2x + 3)(x - 1)$
11. $(x + 5)(3 + y)$ **13.** $(x + 6)(x + 7)$
15. $(3x + 2)(3x + 4)$ **17.** $(0.3x + 0.4)(0.3x - 0.4)$
19. $3y(2y + 5)(2y - 5)$ **21.** $\frac{1}{2}(a + 1)^2$ **23.** $(3x - 2)^2$
25. $x(2x - 1)^2$ **27.** $(a - 4)^2(a^2 - 4a - 4)$
29. $(a - b)(x - y)$ **31.** $3(x - 2)(x^2 + 2x + 4)$
33. $(1 + 0.2x)(1 - 0.2x + 0.04x^2)$
35. $(x^2 + 1)(x + 1)(x - 1)$ **37.** $3(y + 4)(y^2 - 4y + 16)$
39. $2(3a + 2b)(9a^2 - 6ab + 4b^2)$
41. $(x^2 + 2)(x + 1)(x - 1)$ **43.** $0.1(y + 3)(y + 6)$
45. $-ax^2(x - 13)(x + 7)$ **47.** $-3xy(x - 3)(x - 8)$
49. $0.1(2y + 3)(y - 2)$ **51.** $(3x + 7)(2x - 1)$
53. $5(x^2 + 5x - 3)$ **55.** $(3 - 5x)(1 + 2x)$
57. $2y(6 - x)(1 - 6x)$ **59.** $\left(x + \frac{5}{3}\right)\left(x - \frac{1}{3}\right)$
61. $\frac{1}{6}(2p + 3)(p - 1)$ **63.** $\frac{1}{18}(4a - 3)(a - 3)$
65. $(x^2 + 2)(x + 1)$ **67.** $2(4x - 3y)(3a - 5b)$
69. $(x + y + z)(x + y - z)$
71. $(x + 3 + 5y)(x + 3 - 5y)$ **73.** $(x + 3)^2(x - 3)$
75. $(x + y + 2)(x + y - 1)$
77. $(2x + y + 5)(2x + y - 3)$
79. $(3m - 6n - 4)(m - 2n + 3)$
81. $2(x^2 + 9)(x + 3)(x - 3)$ **83.** $(3x^2 - 1)(x^2 + 5)$
85. $(x + 2)(x - 2)(x^2 - 2x + 4)(x^2 + 2x + 4)$
87. $(x - 3y - 3)(x^2 + 3xy + 3x + 9y^2 + 18y + 9)$
89. $4(m - p + 4)(m^2 + mp - 4m + p^2 - 8p + 16)$
91. $(x - 2)(x^2 + 3x + 4)$ **93.** $x(x + 1)(x - 1)$
95. $(x - y)(x^2 + xy + y^2 - 6)$ **97.** $6(4a - 3b)(5x + 2y)$
99. $(2x - 1)(4x^2 + 2x + 1)(x + 1)(x^2 - x + 1)$
101. $(x - 1)^4(2y + 1)^2$ **103.** $(x - 2z + y)(x - 2z - y)$
105. $(3x - 5y + 4)(3x - 5y - 4)$
107. Factor: $x^3 - 3x^2 + 3x - 1 = (x - 1)^3$, so $s(x) = x - 1$. Therefore $V(1) = 0$ and $s(1) = 0$.
109. $A(x) = 8x = \frac{1}{2} \cdot 16 \cdot x$, so $b = 16$ and $h(x) = x$. Therefore $A(12) = 96$ and $h(12) = 12$.

Exercise Set 6.6
1. $\{-3, 0\}$ **3.** $\{1, 2\}$ **5.** $\{-3, 3\}$ **7.** $\{3, 7\}$
9. $\{-4, 12\}$ **11.** $\{9\}$ **13.** $\left\{-\frac{3}{4}, 7\right\}$ **15.** $\left\{-\frac{8}{5}, \frac{3}{2}\right\}$
17. $\{-2, 3\}$ **19.** $\{-18, -2, 0\}$ **21.** $\left\{-\frac{1}{8}, 0, \frac{1}{2}\right\}$

23. $\{-\frac{1}{4}, 0, \frac{2}{3}\}$ **25.** $\{1, 2, 5\}$ **27.** $\{-1, 1, \frac{4}{3}\}$
29. $\{-\frac{1}{4}, \frac{1}{3}, \frac{1}{2}\}$ **31.** $\{\frac{1}{3}\}$ **33.** $\{-\frac{1}{4}, \frac{1}{2}\}$ **35.** $\{\frac{2}{3}, \frac{3}{2}\}$
37. $x^2 - 5x + 6 = 0$ **39.** $x^2 + 3x = 0$
41. $x^3 - x^2 - 2x = 0$ **43.** $x^2 - \frac{9}{2}x - \frac{5}{2} = 0$
45. $x^3 + \frac{1}{21}x^2 - \frac{2}{21}x = 0$ **47.** $\{-\frac{1}{2}\}$ **49.** $\{-1, \frac{13}{15}\}$
51. $\{-\frac{2}{3}, \frac{5}{2}\}$ **53.** $\{-5, 3\}$ **55.** $\{-1\}$ **57.** $\{-1, 2\}$
59. $\{4, 6\}$ **61.** 11, 13   **63.** 30 m × 70 m   **65.** 9, 12
67. $b = \frac{9}{2}$ cm, $h = 12$ cm   **69.** 6, 8, 10

Exercise Set 6.7
1. $\frac{1}{3}$ **3.** 3   **5.** −12   **7.** 9, 11   **9.** 5 in. × 12 in.
11. 5 yd   **13.** 1000 miles   **15.** 10 ft × 20 ft
17. 20 ft × 50 ft   **19.** 3, 4, 5   **21.** 16 in.
23. 18 in. × 36 in.   **25.** 14 cm × 28 cm   **27.** $b = 3$ in.
29. $R = 10$ m   **31.** 2 sec   **33.** 4 sec
35. 5 sec after release   **37.** either 3 items or 16 items

Review Exercises
1. $5(x - 2)$ **2.** $3x(x^2 - 4x - 1)$ **3.** $4xy(4x + 7y)$
4. $(x + 1)(x + 2)$ **5.** $2x(2y - 3)(8 - 5x)$
6. $2a(2b + 3c - 9d)$ **7.** $13xy^2z(7xy - 5z - 3xz^2)$
8. $y^{2m}(y^{m+1} - 1)$ **9.** $(s + 8)(s - 8)$
10. $2(a - 2)(a^2 + 2a + 4)$ **11.** $6(x - 5)(x - 1)$
12. $(4m - 3)^2$ **13.** $3rs^2(r - 4s)(r + 3s)$
14. $(x^2 + 4y^2)(x + 2y)(x - 2y)$ **15.** $(b - 3)^3$
16. $(k^m - 4r^m)(k^m - r^m)$ **17.** $3\left(y + \frac{1}{2}\right)\left(y + \frac{1}{4}\right)$
18. $(s^m - 2t^m)(s^{2m} + 2s^m t^m + 4t^{2m})$ **19.** $(5a + 6b)^2$
20. $(x + 3)(x + y)$ **21.** $(a + 2b + c)(a + 2b - c)$
22. $(k + 2)(3k - 2m)$ **23.** $(6 - x)(1 - 2x)$
24. $-7(y - 6)(y + 1)$ **25.** $(3k + m + 3)(3k - m - 3)$
26. $(2a - b)(2a + b)(4a^2 + 2ab + b^2)(4a^2 - 2ab + b^2)$
27. $(x - 1)^2(6a + b)$ **28.** $\left(\frac{1}{3}t - \frac{1}{4}\right)^2$
29. $(5s - 1)(7s^2 - s + 1)$ **30.** $-4ab$
31. $n^2(0.7m - 1.1n)(0.7m + 1.1n)$
32. $(x^4 + y^4)(x^2 + y^2)(x + y)(x - y)$
33. $\left(\frac{1}{9}a^2 + \frac{1}{4}b^2\right)\left(\frac{1}{3}a + \frac{1}{2}b\right)\left(\frac{1}{3}a - \frac{1}{2}b\right)$
34. $-(2x - 5)(10x - 19)$ **35.** $2(3y - 2w)(x + z)$
36. $10a(a - 2b)^2$
37. $(s^2 - 6)(s^2 - 3)$
38. $2m(m^2 - 5)^2$ **39.** $(a - b + c)(z - t)$
40. $(a - 2)(a^2 + 2a + 3)$ **41.** $xy(y + 3x)(4y - 3x)$
42. $2(3k - 8m)(k + 2m)$

43. $(a + b)(a - 3b)(a^2 - ab + b^2)(a^2 + 3ab + 9b^2)$
44. $(6x^m + y^n)(3x^m - 4y^n)$ **45.** $\{-2, -1\}$
46. $\{-2, 0, \frac{4}{3}\}$ **47.** $\{-2, 8\}$ **48.** $\{-\frac{5}{2}, \frac{1}{3}\}$
49. $\{-\frac{1}{2}, \frac{7}{2}\}$ **50.** $\{-4, -3, 0\}$ **51.** $x^2 - x - 2 = 0$
52. $x^3 + 2x^2 - 3x = 0$ **53.** $x^2 - \frac{1}{6}x - \frac{1}{6} = 0$
54. $x^3 - \frac{34}{5}x^2 - \frac{7}{5}x = 0$ **55.** $(x + 2)(x - 2)$
56. $2(x + 1)(x - 1)$ **57.** $(x - 1)(x^2 + x + 1)$
58. $(x + 3)(x^2 - 3x + 9)$ **59.** 8, 9 or −8, −9   **60.** 25 ft
61. 2 ft   **62.** 75 miles

Chapter Test
1. $3x(x + 1)$ **2.** $-5(x - 2y)(1 + 2z)$ **3.** $(x + 2)(x + 1)$
4. $(m - 4)^2(m + 2)$ **5.** $(3x - 2y + m)(3x - 2y - m)$
6. $3x^2y^3(5x^3 - 8y)$ **7.** $(x - 12y)(x + 5y)$
8. $(m - p)(y + 2)^2$ **9.** $(4y - 3x)(16y^2 + 12xy + 9x^2 + 1)$
10. $2(x - 6y)^2$ **11.** $(2a - x^2)(b - c)$
12. $(4x + 1)(2x - 3)$ **13.** $\{\frac{5}{2}, 8\}$ **14.** $\{1, 4\}$
15. $\{-2, 2\}$ **16.** $\{-1, 0, 2\}$ **17.** $\{-\frac{5}{2}, 3\}$
18. $\{-\frac{1}{3}, 0, 2\}$ **19.** $\{-4, -3, 3\}$ **20.** $\{0, 9\}$
21. $\{-\frac{7}{3}, 0, 4\}$ **22.** $\{-4, \frac{1}{2}\}$ **23.** $\{-2, 2\}$
24. $2x^3 - 5x^2 - 3x = 0$ **25.** $x^3 + x^2 - 4x - 4 = 0$
26. 6 in.   **27.** 12 cm
28. boat: 30 mph; current: 10 mph

Cumulative Review Chapters 4–6
1. $\{(5, 1)\}$ **2.** $\{(2, 2)\}$ **3.** $\left\{\left(\frac{10}{3}, \frac{4}{3}\right)\right\}$ **4.** $\{(0, 0)\}$
5. $\{(1, 1)\}$ **6.** $\{(1, 1, 1)\}$ **7.** Dependent   **8.** $\{(8, 10)\}$
9. $\{(1, 1, -1)\}$ **10.** $\{(3, 3)\}$ **11.** $\{(2, 1, 2)\}$
12.

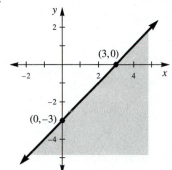

13.

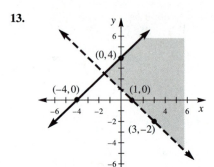

14. $\frac{47}{3}, \frac{25}{3}$ **15.** 32, 34 **16.** plane: 180 mph, wind: 20 mph
17. 6.125 lb peanuts, 1.875 lb pecans
18. $p = 2, q = 12$ belts
19. $4500 at 5%, $4500 at 7%, $1000 at 8%
20. 15 at 8 points, 5 at 12 points, 180 points total **21.** x^{a-b}
22. x^{a+b} **23.** x^{3m} **24.** $-\frac{16y}{3}$ **25.** $3888y$ **26.** 3×10^3
27. 7×10^1 **28.** $7x^2 - 2x - 8$, 2nd degree, trinomial
29. $-\frac{2}{3}y^2 + \frac{3}{2}y$, 2nd degree, binomial **30.** $7x - 3y$
31. $7a^2 + ab$ **32.** $-a^2 + 4a - 4$ **33.** $2m^2 + m - 9$
34. $2x^2 + 6x - 10$ **35.** $t^3 + t^2 + t$ **36. (a)** 36, **(b)** 2
37. $x^2 - 16$ **38.** $4x^2 + 4x + 1$ **39.** $6t^2 - t - 12$
40. $a^{3n} - 3a^{2n} + 4a^n$ **41.** $36x^7$ **42.** $t^{2n} - w^{2n-2}$
43. $-18 - \frac{24}{a}$ **44.** $9r^2 - 8r + 6 - \frac{1}{r}$ **45.** $x + 5$
46. $9t^2 + 3t + 1$ **47.** $2a^2 + 4a + 3 + \frac{-15}{a+2}$
48. $x + \frac{-\frac{1}{6}}{x - \frac{1}{3}}$ **49.** $3(2n - 1)$ **50.** $15xy(3x - y + 2)$
51. $(p - 4)^2$ **52.** $(a + b)(x + y)$ **53.** $(2b - a)(3y - 2x)$
54. $(x - y)(x + y)(-2y)$ **55.** $(y - 1)^3$
56. $(x - 5)(x - 1)$ **57.** $(a + 7b)(a - 6b)$
58. $(4 + t)(2 - t)$ **59.** $(x + 3)(x - 3)(x + 2)(x - 2)$
60. $2ab(3b - a)^2$ **61.** $\left(r + \frac{1}{3}\right)\left(r + \frac{1}{4}\right)$
62. $(x^n + 6)(x^n + 3)$ **63.** $5(1 - 3a - 2a^2)$
64. $3(2y + 3x)(10y - 3x)$
65. $\frac{1}{75}(5x^2 - 13xy + 10y^2)$ or $\frac{1}{15}\left(x^2 - \frac{13}{5}xy + 2y^2\right)$
66. $(x + 3)(8x + 5)$ **67.** $(x + 5)(x - 5)$
68. $(x - 3)(x^2 + 3x + 9)$ **69.** $(6a - 5b)^2$
70. $2(x - 1)(x^2 - 2x + 4)$
71. $y(0.6x + 0.7y)(0.6x - 0.7y)$
72. $(a + 2)(a + 1)(a^2 - a + 1)$ **73.** $(x - y)(e + f - a)$
74. $(x + 1 + 2y)(x + 1 - 2y)$ **75.** $\{\pm 2\}$ **76.** $\{-3, 4\}$
77. $\left\{\pm 2, \frac{4}{3}\right\}$ **78.** $\{0, \pm 3\}$ **79.** $\{6\}$ **80.** $\left\{-\frac{7}{5}\right\}$
81. $x^2 - 7x = 0$ **82.** $x^2 - x - 6 = 0$
83. $x^3 + x^2 - 2x = 0$ **84.** 7, 12 **85.** $h = 8, b = 5$
86. 11 in. **87.** 3 sec **88.** 1 ft

CHAPTER 7

Exercise Set 7.1
1. $\frac{1}{3}$ **3.** $\frac{3}{4z}$ **5.** -1 **7.** $m + n$ **9.** $\frac{a}{4}$ **11.** $\frac{a}{a+4}$
13. $\frac{3r}{13a^2}$ **15.** $\frac{xy}{x-y}$ **17.** 1 **19.** $\frac{m-4}{m+2}$ **21.** equivalent
23. equivalent **25.** equivalent **27.** $x \neq 0$
29. $b \neq 0, -\frac{3}{2}$ **31.** $p \neq -1, 4$ **33.** $x \neq 3, y \neq 1$
35. $a \neq b$ **37.** $x \neq 3, y \neq 1$ **39.** $30a^3b^2; a, b \neq 0$
41. $3(a + 1); a \neq -3, -1$ **43.** $-5p(p^2 + q^2); p \neq \pm q$
45. $m^2 - 16; m \neq \pm 4$ **47.** $(y + 1)(3x + 2); y \neq 1, x \neq -\frac{2}{3}$
49. $p - q; p \neq q$ **51.** $\frac{6}{13}$ **53.** undefined **55.** 0
57. $\frac{1-y}{3y^2(x-y)}$ **59.** $x - y$ **61.** $\frac{2-a}{a-b}$ **63.** $\frac{a+3}{a-3}$
65. $\frac{x^2 + y^2}{x^2 - y^2}$ **67.** $\frac{x^2 - xy + y^2}{x - y}$ **69.** $\frac{2a - 5}{a}$
71. They have no common factors.
73. Multiplying both numerator and denominator by the same nonzero number produces an equivalent equation.
75. 0 and $\frac{3}{2}$, because these numbers make the denominator zero.

Exercise Set 7.2
1. $\frac{x}{2}$ **3.** $6ap$ **5.** $\frac{x-y}{2}$ **7.** $4p$ **9.** 4 **11.** $3n^2$
13. 1 **15.** $\frac{y+2}{2y}$ **17.** $\frac{a+1}{a+2}$ **19.** $\frac{t-1}{t-3}$ **21.** $\frac{a+1}{a+2}$
23. 1 **25.** $\frac{3nx^3}{2y^2}$ **27.** $\frac{w(y-1)}{w-x}$
29. $\frac{-(r^2 + rs + s^2)(r + s)^2}{r^2 + s^2}$ **31.** $\frac{2a+b}{a+b}$ **33.** 3
35. $\frac{(m+2)(x+y)}{(2x+y)^2}$ **37.** $\frac{5q}{2}$ **39.** $\frac{-7(x-2)}{3(x-3)(x+2)(x-8)}$
41. $\frac{k-1}{k^2-k+1}$ **43.** $(x-3)(x+2)$ **45.** $-b(a+b)$
47. $\frac{w^3z^3}{5a^4b^5}$ **49.** $\frac{3x+5}{x-1}$ **51.** $\frac{x-1}{x+2}$ **53.** $\frac{p-3}{p-1}$
55. (a) $-\frac{5}{2}, 0$ **(b)** $\{x \mid x \neq 5\}$ **57.** $\frac{3}{t-1}$ **59.** $x^3 - 1$

61. (a) $(t + 1)^2$ (b) 1   **63.** $\{y|y \neq 0, 2\}$
65. $\{-8, -1\}$ **67.** $\left\{-\dfrac{1}{5}\right\}$ **69.** $\{-2, 3\}$
71. $\{-2, -1, 0\}$ **73.** $\left\{-\dfrac{4}{5}, \dfrac{2}{3}\right\}$ **75.** $\left\{-\dfrac{1}{3}, \dfrac{3}{2}\right\}$
77. $\left\{-\dfrac{5}{2}, -\dfrac{1}{3}, 0\right\}$
79. (a) $x = 12$ ft, $x + 4 = 16$ ft, $x + 8 = 20$ ft
   (b) 48 ft   (c) $144   (d) $38.40
81. Invert the divisor and multiply.
83. Invert the second and third expressions, factor, reduce, and multiply. For example, $\dfrac{3x}{(x-2)(x-2)} \div \dfrac{x}{(x-2)(x-1)} \div \dfrac{x-1}{x-2} = \dfrac{3x}{(x-2)(x-2)} \cdot \dfrac{(x-2)(x-1)}{x} \cdot \dfrac{x-2}{x-1} = 3$

Exercise Set 7.3
1. $\dfrac{1}{x}$ **3.** $\dfrac{1}{y}$ **5.** $\dfrac{5a}{b}$ **7.** $7b$ **9.** 1   **11.** $\dfrac{1}{a-b}$ **13.** 2
15. 2   **17.** 1   **19.** $\dfrac{1}{x-2}$ **21.** $\dfrac{1}{r+t}$ **23.** $\dfrac{1}{m^2+m+1}$
25. $-\dfrac{1}{10a}$ **27.** $\dfrac{12+7x}{12x^2}$ **29.** $\dfrac{3a-7}{30a}$
31. $\dfrac{-6m+3}{(m+3)(m-4)}$ **33.** 0   **35.** $\dfrac{5}{m+2}$ **37.** $\dfrac{2}{x^2-1}$
39. $\dfrac{4k^2+5k+6}{k(2k+3)}$ **41.** $\dfrac{-2x}{x^2-4}$ **43.** $\dfrac{c^2-21c}{(c-4)(c-5)(c+3)}$
45. $\dfrac{6s+5}{(4s-1)(3s-4)}$ **47.** $\dfrac{ab^2-a}{b}$ **49.** $\dfrac{m^2}{m-1}$
51. $\dfrac{-k^2}{k^2+2k+1}$ **53.** $\dfrac{a^2+ab+a-v}{u+v}$ **55.** $5m^2+5m$
57. $\dfrac{2x}{(x+1)^2}$ **59.** $\dfrac{11a^2+10a+3}{12a^2(a+1)^2}$ **61.** $\dfrac{11k^2-6k-5}{3k(k^2-1)}$
63. $\dfrac{x^{3n}+2x^{2n}+2x^n}{(x^n+1)(x^n-1)(x^{2n}+x^n+1)}$
65. 3   **67.** $\dfrac{-(x^2+xy+y^2)}{2}$ **69.** abx^2 **71.** $\dfrac{x^4-x^2y^2+y^4}{x^2+y^2}$
73. $\dfrac{(x+3)(x-3)(x-2)(x^2+4)^2}{(x+4)(x-4)(x+1)(x^2+2)(x^2+3x+1)}$
75. It stands for least common denominator.
77. Change to equivalent fractions with the same denominator, then add the numerators.

Exercise Set 7.4
1. 2   **3.** $\dfrac{c}{a}$ **5.** $\dfrac{14}{9}$ **7.** $\dfrac{2}{x}$ **9.** $\dfrac{3}{4}$ **11.** $\dfrac{k}{k+1}$
13. $\dfrac{ab}{b-a}$ **15.** $\dfrac{a-1}{a^3}$ **17.** $12b$ **19.** $\dfrac{x}{x-1}$ **21.** $\dfrac{v}{2}$
23. 1   **25.** 1   **27.** 1   **29.** 2   **31.** $\dfrac{7x+1}{x+1}$

33. $(2t+3)(t-2)$ **35.** $\dfrac{3b(2b-3)}{(b-3)(b-1)}$ **37.** $3-2p$
39. $2m-9$ **41.** $\dfrac{s^2-17}{s^2-6s-27}$ **43.** $\dfrac{a+1}{a}$ **45.** $\dfrac{1-2p}{2p+1}$
47. $\dfrac{w+5}{w+4}$ **49.** $\dfrac{-(x+y)}{2(x-y)}$ **51.** 12   **53.** $\dfrac{x+y}{y-x}$ **55.** $\dfrac{(rs)^3}{s-r}$
57. $\dfrac{1}{p+1}$ **59.** $\dfrac{r-2s}{s+5r}$ **61.** $\dfrac{1}{2}$ **63.** $\dfrac{(a+3)(a-4)}{3(a^2+a-1)}$
65. $\left\{\dfrac{11}{2}\right\}$ **67.** $\left\{\left(\dfrac{14}{5}, -\dfrac{3}{5}\right)\right\}$ **69.** $\left[-\dfrac{2}{3}, 2\right]$ **71.** $\dfrac{2}{x-1}$
73. $\dfrac{2x^2-4x+3}{(x-1)(2x-3)}$

Exercise Set 7.5
1. $\{-8\}$ **3.** $\{56\}$ **5.** $\{6\}$ **7.** $\{4\}$ **9.** $\{2\}$
11. $\{9\}$; $r \neq 0$ **13.** $\left\{\dfrac{1}{7}\right\}$; $y \neq 0$ **15.** $\{-4\}$; $x \neq -3$
17. $\left\{\dfrac{8}{3}\right\}$; $s \neq -\dfrac{2}{3}$ **19.** $\varnothing$; $m \neq 2, 4$ **21.** $\left\{\dfrac{9}{11}\right\}$; $x \neq 0$
23. $\varnothing$, $x \neq \pm 4$ **25.** $\varnothing$; $x \neq \dfrac{1}{2}$ **27.** $\left\{\dfrac{1}{6}\right\}$; $x \neq 5$
29. $\varnothing$; $x \neq 3$ **31.** $\{-2\}$; $x \neq \pm 1$ **33.** $\varnothing$; $x \neq 0$
35. $\{-4, 3\}$; $y \neq -1$ **37.** $\{-3\}$; $a \neq -1, 2$
39. $p(1+rt)$ **41.** $-\dfrac{a}{5}$ **43.** $\dfrac{16t^2+s}{t}$ **45.** $\dfrac{pF}{F-p}$
47. $\dfrac{ab}{5-3a}$ **49.** $y = \dfrac{x^3-3}{3x^3-2}$ **51.** $y = \dfrac{-x^2+2x-2}{x^2-2x-1}$
53. $\varnothing$; $a \neq \pm 3$ **55.** $\{-13\}$; $r \neq -5, 1, 9$
57. $\{2\}$; $k \neq -\dfrac{3}{5}, \dfrac{5}{4}, 6$ **59.** $x + \dfrac{1}{x}$ **61.** $\dfrac{600}{x}$ miles
63. $\dfrac{126}{N}$ **65.** A fraction involving polynomials
67. Factoring and reducing
69. A fraction divided by a fraction

Exercise Set 7.6
1. -7 or -1 **3.** 9   **5.** $\dfrac{13}{19}$ **7.** $\dfrac{7}{2}$ **9.** $-\dfrac{1}{2}$ **11.** $5\dfrac{1}{2}$ hr
13. $1\dfrac{1}{2}$ hr; 90 miles   **15.** Penny: 50 mph; Bill: 70 mph
17. 10 mph   **19.** 3 mph   **21.** $6\dfrac{3}{7}$ hr   **23.** 12 hr
25. 30 hr   **27.** 36 min   **29.** $11\dfrac{1}{4}$ hr   **31.** 8 ft, 12 ft
33. 2040   **35.** $9000   **37.** $x = \dfrac{7}{5}$; $x + 1 = \dfrac{12}{5}$; $4x - 2 = \dfrac{18}{5}$
39. 20   **41.** $\dfrac{3}{10}$

Review Exercises
1. rational   **2.** relatively prime   **3.** $\dfrac{A \cdot C}{B \cdot C}$ **4.** $\dfrac{x}{x+4}$

5. $\dfrac{-1}{a+2}$ 6. $-\dfrac{7}{2}$ 7. $x^2 - x$ 8. $2pq + 2p^2$
9. 4 or -3 10. $p - 6$ 11. $\dfrac{5}{3}$ 12. $\dfrac{(a+3)(b+3)}{a-3}$
13. $\dfrac{(m-3)(n-2)}{n+2}$ 14. $-9q - 7$ 15. 0 16. $\dfrac{y^3z}{2x^2}$
17. $\dfrac{4s^2 - 2r^2}{r+s}$ 18. $\dfrac{p(p-4)}{p^2 - 2p + 1}$ 19. $\dfrac{5x^2 - 16x + 4}{x}$
20. $\dfrac{5}{x^2-4}$ 21. $\dfrac{x(2x+1)}{2(x+1)}$ 22. $\dfrac{3(p^2-1)}{5}$ 23. $1 - 2y$
24. $-\dfrac{1}{2}$ 25. $\dfrac{t-1}{5(2t-1)}$ 26. $\dfrac{y-x}{y+x}$ 27. $\dfrac{-a}{b}$ 28. $\dfrac{x+y}{a-b}$
29. $\{-30\}$ 30. $\{-1\}$ 31. $\left\{\dfrac{11}{3}\right\}$ 32. $\left\{\dfrac{1}{5}\right\}$ 33. $\left\{\dfrac{192}{125}\right\}$
34. $\varnothing$ 35. $\{-6, 3\}$ 36. $\{-1, 9\}$ 37. $\{1\}$ 38. $\left\{-\dfrac{9}{2}\right\}$
39. $\left\{\dfrac{12}{7}\right\}$ 40. $\{-7\}$ 41. -1 42. $\dfrac{9}{13}$
43. 3.5 hr; 224 miles 44. $2\dfrac{11}{12}$ min 45. 60 mph
46. $10\dfrac{10}{13}$ hr 47. $y = 2a$ 48. $y = \dfrac{2x - ax}{5x - 3}$

Chapter Test
1. $\dfrac{4}{5}$ 2. $\dfrac{x+2}{x-2}$ 3. $\dfrac{6}{x+3}$ 4. $\dfrac{x+y}{x(x-y)}$
5. $\dfrac{9x+2}{(x-2)(x+2)}$ 6. $\dfrac{1}{x+2}$ 7. $\dfrac{1}{x-1}$ 8. $\{-1\}$; $x \neq 2, 3$
9. $\{2\}$; $x \neq -2$ 10. $\left\{\dfrac{6}{11}\right\}$; $x \neq 0, -3$
11. $\left\{-\dfrac{39}{16}\right\}$; $x \neq -\dfrac{5}{2}$ 12. $\left\{-\dfrac{2}{5}\right\}$; $a \neq \pm 2, -3$
13. $\left\{\dfrac{3}{4}\right\}$; $y \neq 0, 1$ 14. $\{x \mid x \in R\}$, $x \neq 0, 6$ 15. $\dfrac{y^2+3}{y-1}$
16. 5
17. $2\dfrac{6}{11}$ hr 18. 8 mph 19. $4\dfrac{2}{7}$ hr 20. $6000
21. 20 hr 22. $\dfrac{2x}{3x+4}$; $x \neq 0, -2$

CHAPTER 8

Exercise Set 8.1
1. 5 3. 3 5. $\dfrac{1}{2}$ 7. 8 9. -3 11. 9 13. 8
15. 3 17. $\dfrac{4}{9}$ 19. 2 21. $\dfrac{8}{343}$ 23. $\dfrac{5}{2}$
25. not a real number 27. $p^{4/5}$ 29. $10t$ 31. a^2
33. $18m^{5/6}$ 35. $\dfrac{3}{k^{13/6}}$ 37. $a^{1/2}$ 39. $2r$ 41. 1
43. $\dfrac{y}{2^{1/3}}$ 45. $\dfrac{xy^2}{3}$ 47. $\dfrac{2^{2/3}p}{q^{5/3}}$ 49. $x + x^{3/4}$ 51. $k - 1$
53. $\dfrac{15}{t^{5/12}} - 6t^{5/6}$ 55. $x - y$ 57. $a + 2a^{1/2}b^{1/4} + b^{1/2}$
59. p 61. $x + 1$ 63. $a^{2/3} - 1$ 65. $x^{1/2} - y^{1/2}$
67. $2p^{5/4} - 1$ 69. $x^{2/3} + x^{1/3}y^{1/3} + y^{2/3}$ 71. $\dfrac{x^{14/3}}{y^{1/2}z^{5/12}}$
73. $-5r^{7/20}s^{5/3}t^{1/12}$ 75. $\dfrac{y^{3/16}z^{3/20}}{x^{1/2}}$ 77. $\dfrac{b^2}{162a^{2/3}}$ 79. $\dfrac{r^2}{s^4 t^9}$
81. $\dfrac{y^{11/13}}{x^{11/4}}$ 83. y^{2n+2} 85. t^{3m} 87. $x^{n+(1/2)}$ 89. $\dfrac{y^{5n/6}}{x^{5m/6}}$
91. 2.2361 93. 1.4422 95. 0.5774 97. 5.1962
99. It is an exponent that is a rational number.
101. Yes: $\sqrt[3]{-8} = \sqrt[3]{(-2)^3} = -2 = -(8)^{1/3} = -8^{1/3}$
103. b must be positive or zero when n is even.
105. Yes, since an odd root of a negative number is a negative real number.

Exercise Set 8.2
1. $\sqrt{6}$ 3. $\sqrt[3]{11}$ 5. $\sqrt[3]{x^2}$ 7. $\dfrac{1}{\sqrt[3]{p}}$ 9. $\dfrac{6}{\sqrt[7]{t^3}}$ 11. $\sqrt[3]{5y}$
13. $7^{1/2}$ 15. $2^{1/3}$ 17. $x^{2/3}$ 19. $y^{3/7}$ 21. $\dfrac{1}{5^{2/3}}$
23. $5^{1/3}x^{2/3}$ 25. 7 27. $\dfrac{m^{3/5}}{3^{2/5}}$ 29. $(x+y)^{1/2}$
31. $(p^3 + q^3)^{1/3}$ 33. 2 35. 4 37. 1.2 39. -0.2
41. $2p$ 43. $4t^2$ 45. r^2t^3 47. $\dfrac{x}{3y}$ 49. $x + y$
51. $(p + q)^2$ 53. $2\sqrt{2}$ 55. $2\sqrt{5}$ 57. $\dfrac{3\sqrt{3}}{4}$ 59. $\dfrac{8}{5}$
61. $x\sqrt{x}$ 63. $p\sqrt{pq}$ 65. $3\sqrt{2}$ 67. $\sqrt{10}$ 69. $\dfrac{3}{4}$
71. $\dfrac{x\sqrt{x}}{y}$ 73. 0.2 75. 3×10^2 77. $3b\sqrt{7a}$
79. $2a^2\sqrt[3]{3a}$ 81. $2p^2\sqrt[6]{2p}$ 83. $rst^2\sqrt[6]{st}$ 85. $\dfrac{\sqrt{3}}{9}$
87. $\dfrac{\sqrt[3]{6}}{9}$ 89. $\dfrac{\sqrt{3x}}{6}$ 91. $\dfrac{\sqrt[3]{4}}{4}$ 93. $\dfrac{\sqrt[3]{8}}{4}$ 95. $\dfrac{\sqrt[3]{20}}{10}$
97. $\dfrac{\sqrt{3x}}{x^2}$ 99. $\dfrac{\sqrt[3]{a^2 x}}{a^3}$ 101. $0, 3$; $x \geq 4$
103. $3, 0$; $x \geq -\dfrac{5}{2}$ 105. $1, 0$; $x \in R$ 107. $2, 0$; $P \in R$
109. $|x|$ 111. $5|k|$ 113. $-3y\sqrt[3]{4y^2}$ 115. $\dfrac{1}{s^2}$
117. $\dfrac{7\sqrt{2}}{|r|}$ 119. $\dfrac{8}{|x|}$ 121. 1.845 123. 4.041
125. 1.105 127. (a) $s = \dfrac{d}{\sqrt{2}}$ (b) $A = \left(\dfrac{d}{\sqrt{2}}\right)^2 = \dfrac{1}{2}d^2$
(c) $s = 0.12$ ft, $A = 0.01$ ft^2 129. square root

131. If $b \geq 0$ then $\sqrt{b^2} = b$; if $b < 0$ then $\sqrt{b^2} = |b|$. But $\sqrt{b^2} = b$ is false if $b < 0$ since a square root is always positive.
133. There are no nth powers inside the radical, where n is the index.
135. The nth root of a divided by the nth root of b is equal to the nth root of the quotient of a and b.

Exercise Set 8.3
1. $7\sqrt{3}$ 3. $-5\sqrt[3]{5}$ 5. $4x\sqrt{10}$ 7. $2\sqrt[4]{3}$ 9. $x\sqrt{y}$
11. $u\sqrt[3]{3v} + u\sqrt[3]{v}$ 13. $3\sqrt{2}$ 15. $13\sqrt[3]{2}$ 17. $6\sqrt{5}$
19. $-7\sqrt{3} + 3\sqrt{5}$ 21. $-15\sqrt[4]{2}$ 23. $17\sqrt{2} + 6\sqrt{6}$
25. 0 27. $v^2\sqrt[4]{2v}$ 29. $\sqrt{5}$ 31. $6\sqrt{7}$ 33. $\frac{8y}{5}\sqrt{2xyz}$
35. $50u^2\sqrt{6u}$ 37. $-7yz\sqrt[3]{3y^2z^2}$
39. $6x\sqrt{6xy} - 7y\sqrt{3x} + 33x^2\sqrt{6y}$ 41. $\frac{c}{a}$ 43. $7 + 5\sqrt{5}$
45. $6 - 7\sqrt{7}$ 47. $13 - 9\sqrt{11}$ 49. $15 - 3\sqrt{6}$
51. $3\sqrt{11} - 11$ 53. $12\sqrt{6} - 18$ 55. $\sqrt{uv} - u$
57. -1 59. 6 61. $28x$ 63. $7 + 2\sqrt{10}$
65. $a + 3 - 2\sqrt{a+2}$ 67. 37 69. $\frac{c}{a}$ 71. $\sqrt{5} + 1$
73. $3 - x$ 75. $\sqrt[3]{y} - 1$ 77. $\sqrt[5]{y} - \sqrt[5]{z}$ 79. $\sqrt[4]{8}$
81. $x\sqrt[6]{x^5}$ 83. $\sqrt[10]{a^7b^9}$ 85. $\sqrt{a}$ 87. $\sqrt[3]{a^2}$ 89. $\sqrt[6]{72}$
91. $x^2\sqrt[6]{243x^4}$ 93. yes 95. no 97. yes 99. no
101. The area of $\triangle OP_4P_5$ is $\frac{1}{2}sa$. Since there are n of these sides, there are n equal triangles. The perimeter p is the sum of the n sides, or ns, and the area of the regular polygon with n sides of perimeter p is
$A = n\left(\frac{1}{2}sa\right) = \frac{1}{2}ap$.

Exercise Set 8.4
1. $\frac{\sqrt{2}}{2}$ 3. $2\sqrt{3}$ 5. $3\sqrt{5}$ 7. $-4\sqrt{7}$ 9. $\frac{\sqrt{3}}{3}$
11. $\frac{-\sqrt{30}}{6}$ 13. $\sqrt[3]{45}$ 15. $\frac{-\sqrt[3]{252}}{5}$ 17. $\sqrt{x}$
19. $2\sqrt[3]{a^2}$ 21. $\sqrt{x+y}$ 23. $\frac{\sqrt[3]{3r^2}}{r}$ 25. $\frac{\sqrt[4]{mn^3}}{n}$
27. $\sqrt[5]{u^4}$ 29. $\frac{\sqrt[3]{4}}{2}$ 31. $\frac{\sqrt{3x}}{3x}$ 33. $\frac{8\sqrt[5]{16w^2}}{w}$
35. $7x\sqrt{3}$ 37. $\frac{4}{5}a\sqrt{5}$ 39. $\frac{4}{3}\sqrt{3p}$
41. $\frac{6\sqrt{2} + 6 - 8\sqrt{3}}{3}$ 43. $\frac{33\sqrt{2w} - 60\sqrt{6}}{10}$
45. $\frac{-\sqrt{x^2+1}}{x(x^2+1)}$ 47. $10\sqrt{2} - 4$ 49. $3\sqrt{10} + 8$
51. $\frac{3\sqrt{x} + x}{x}$ 53. $\frac{5w\sqrt{6z} + 6z\sqrt{5w}}{30wz}$ 55. $\frac{4\sqrt{3} + \sqrt{21}}{9}$

57. $\frac{-(4\sqrt{3} + \sqrt{5})}{43}$ 59. $24 + 14\sqrt{3}$ 61. $\frac{11 - 6\sqrt{2}}{7}$
63. $21 + 7\sqrt{5} + 15\sqrt{2} + 5\sqrt{10}$ 65. $\frac{3\sqrt{2} - 4}{2}$
67. $\frac{x\sqrt{x} - x\sqrt{y} + y\sqrt{x} - y\sqrt{y}}{x - y}$ 69. $\frac{6u - 5\sqrt{uv} + v}{4u - v}$
71. $\frac{81xy - 18\sqrt{abxy} + ab}{81xy - ab}$ 73. $a^{1/6} - a^{1/3}$
75. $\frac{(xy)^{14/15} - (xy)^{4/5}}{xy}$ 77. $1 + (x - y)^{2/3}$ 79. $\frac{\sqrt[6]{108}}{2}$
81. $\sqrt[12]{x}$ 83. $\frac{\sqrt[4]{3}}{3}$ 85. $3\sqrt[3]{5y}$ 87. $\{-1, 5\}$ 89. $\left\{-\frac{3}{2}\right\}$
91. $\{-1, 3\}$ 93. $\{-3, 2\}$ 95. $\left\{\frac{5}{13}, 3\right\}$

Exercise Set 8.5
1. $\{4\}$ 3. $\emptyset$ 5. $\{24\}$ 7. $\{2\}$ 9. $\emptyset$ 11. $\{3\}$
13. $\{0\}$ 15. $\left\{-\frac{35}{3}\right\}$ 17. $\left\{\frac{25}{3}\right\}$ 19. $\{8\}$ 21. $\left\{-\frac{1}{2}\right\}$
23. $\{-32\}$ 25. $\emptyset$ 27. $\{-81\}$ 29. $\left\{\frac{9}{2}\right\}$ 31. $\{2\}$
33. $\left\{\frac{1}{2}\right\}$ 35. $\left\{\frac{2}{5}\right\}$ 37. $\{1\}$ 39. $\emptyset$ 41. $\left\{\frac{5}{2}\right\}$
43. $\left\{\frac{3}{8}\right\}$ 45. $\emptyset$ 47. $\emptyset$ 49. $\emptyset$ 51. $\{0\}$ 53. $\{4\}$
55. $\{5\}$ 57. $\{9\}$ 59. $\{8\}$ 61. $\left\{\frac{5}{2}\right\}$ 63. $\{7\}$ 65. $\{9\}$
67. $\{-3, 3\}$ 69. $\{-3, 3\}$ 71. $\{5\}$ 73. $\emptyset$ 75. $\emptyset$
77. $\{0\}$ 79. $\left\{\frac{3}{16}\right\}$ 81. 3.742 83. 4.123 m 85. $x^{13/12}$
87. $5\sqrt{5} - 10\sqrt{2}$ 89. $4 + 3\sqrt{5}$ 91. $x^{1/4}$ 93. $\sqrt[3]{4}$
95. $\sqrt{xy}$ 97. 81 99. $6y\sqrt{3}$

Exercise Set 8.6
1. -1 3. i 5. 1 7. $-i$ 9. 1 11. i 13. 1
15. $5 - i$ 17. $4 - 10i$ 19. -12 21. $-12 - i$
23. $-4 + 3i$ 25. 2 27. $-47 - 97i$ 29. $11 - 60i$
31. $2 - 11i$ 33. $-5i$ 35. $2 - 2i$ 37. $\frac{1}{2} + \frac{1}{2}i$
39. $\frac{7}{29} + \frac{26}{29}i$ 41. $-11 + 17i$ 43. $2i$ 45. $5i$
47. $6i$ 49. $4\sqrt{2}i$ 51. $7\sqrt{2}i$ 53. $-30\sqrt{2}i$
55. $-3 + 15\sqrt{3}i$ 57. $4\sqrt{5}i$ 59. $-\sqrt{3}i$
61. $-4\sqrt{2}i + 16\sqrt{3}i$ 63. $\frac{-\sqrt{3}i}{3}$ 65. $\frac{-\sqrt{3}i}{3}$
67. $\frac{6}{7} - \frac{3\sqrt{3}i}{7}$ 69. $\frac{1+i}{2}$ 71. $3 - 5\sqrt{2}i$
73. $\frac{6 - \sqrt{7}i}{4}$ 75. $x = 3, y = 2$ 77. $x = \frac{4}{5}, y = \frac{12}{5}$

79. $x = 5, y = 4$ 81. $x = 1, y = 3$ 83. $x = \frac{2}{3}, y = \frac{3}{4}$
85. 1   87. -1 89. 0   91. 0   93. $16 + 8\sqrt{3}i$
95. A number of the form $a + bi$, where $i = \sqrt{-1}$
97. A number of the form $a + bi$, where $b = 0$
99. Add their real parts and then add their imaginary parts.
101. $i^{103} = (i^4)^{25} \cdot i^3 = 1 \cdot i^3 = i^3 = -i$. Use the fact that $i^4 = 1$.

Exercise Set 8.7

1. 9   3. $M = \frac{E\sqrt{n}}{D}$ 5. $A = 77.25$ 7. $f = 6.43$
9. $r = 2.09$ in.   11. $r = 7.30$ in.   15. $T = 200$ newtons
17. $y = \frac{-x^2 g}{2v^2}$ 19. $s = 6v^{2/3}$ 21. 15.96 ft
23. $(20\sqrt{3} + 1)\pi \approx 111.97$ ft^2

Review Exercises

1. $\frac{1}{(0.3)^3}$ or $\frac{1}{0.027}$ 2. $\frac{3^{1/2} + 2^{1/2}}{6^{1/2}}$ 3. $5^{43/12}$ 4. $\frac{1}{x^{31/12}}$
5. $\frac{8b}{a^{2/3}}$ 6. $\frac{1}{(a+b)^{1/2}}$ 7. $\frac{2x+1}{(x-2)^{1/2}}$ 8. $\frac{2b^{1/2} + 3a^{1/2}}{a^{1/2}b^{1/2}}$
9. $m - n$ 10. $\frac{(s+1)^2}{s}$ 11. $2t\sqrt{2}$ 12. 20
13. $-2\sqrt[3]{4}$ 14. $10i$ 15. $n + 4$ 16. $-4k\sqrt[4]{k^3}$
17. 1   18. $43\sqrt{2}$ 19. $x^2\sqrt{x^2+1}$ 20. $-9 - 5\sqrt{10}$
21. $198 + 120\sqrt{2}$ 22. $10x + 2\sqrt{2}x$ 23. $8i$
24. $\left(8x - \frac{3x^2}{y}\right)\sqrt[3]{x^2 y}$ 25. 1   26. $x + 2 - 2\sqrt{x+1}$
27. $\frac{\sqrt{15}}{5}$ 28. $\frac{1}{3y}\sqrt[3]{12xy^2}$ 29. $\frac{3a\sqrt{7b} - 7b\sqrt{3a}}{21ab}$
30. $\frac{3\sqrt{6} - 8}{10}$ 31. $\frac{3}{2}\sqrt[3]{4k}$ 32. $2\sqrt[3]{4a^2}$
33. $\frac{25 + 30\sqrt{x} + 9x}{25 - 9x}$ 34. $\frac{2a + b + 2\sqrt{a(a+b)}}{b}$ 35. $\{8\}$
36. $\left\{\frac{23}{3}\right\}$ 37. $\{-9\}$ 38. $\varnothing$ 39. $\{4\}$ 40. $\varnothing$
41. $\{5\}$ 42. $\left\{-\frac{7}{11}\right\}$ 43. $\{8\}$ 44. $\left\{\frac{1}{3}\right\}$ 45. $\left\{-\frac{31}{8}\right\}$
46. $\left\{-\frac{1}{2}\right\}$ 47. $x = 2, y = 7$ 48. $x = 3, y = -2$
49. 0   50. $28\sqrt{2}$ 51. $\frac{4+7i}{5}$ 52. $\frac{10-7\sqrt{2}i}{11}$
53. $23 - 10\sqrt{2}i$ 54. $15 - 5i$ 55. $\frac{7 + 6\sqrt{2}i}{121}$ 56. $-8i$
57. $8 - 14i$ 58. $-3 - i$ 59. i 60. $\frac{23 - 10\sqrt{2}i}{27}$
61. 2   62. $xy^{3/2}$

Chapter Test

1. $6x^{5/6}$ 2. $a^{5/6} + a$ 3. $30\sqrt{2}$ 4. $-2xy\sqrt[3]{xy^2}$ 5. $\frac{8}{5}$

6. $\frac{5\sqrt{10} + 4\sqrt{5}}{10}$ 7. $\frac{5 + 3\sqrt{10}}{13}$ 8. $15\sqrt{10x} - 9\sqrt{x}$
9. 43   10. $13\sqrt{5}$ 11. $3\sqrt[3]{x^2}$ 12. $\frac{\sqrt[4]{12x^3}}{2x}$
13. $(4 - 2\sqrt{3})i$ 14. $(2 - \sqrt{6}) + (2\sqrt{2} - 2\sqrt{3})i$
15. $\{2\}$ 16. $\{5\}$ 17. $\{9\}$ 18. $\{16\}$ 19. $\varnothing$
20. $x = 5, y = -\frac{3}{2}$ 21. $\{-1\}$ 22. $\left\{\frac{1}{2}\right\}$ 23. $\left\{\frac{78}{23}\right\}$
24. $\varnothing$ 25. $\frac{xy^{7/6}}{16}$ 26. $-3xy^3z^5\sqrt[3]{3xz^2}$ 27. 2
28. $\sqrt[6]{108}$ 29. $x^{3/2}$ or $x\sqrt{x}$ 30. $\frac{28 - 17i}{37}$ 31. 7
32. 33

CHAPTER 9

Exercise Set 9.1

1. $\{2, 4\}$ 3. $\left\{\frac{1}{2}, 2\right\}$ 5. $\left\{-\frac{1}{2}, 3\right\}$ 7. $\left\{\frac{5}{2}, 3\right\}$
9. $\left\{\frac{1}{3}, \frac{3}{2}\right\}$ 11. $\left\{-\frac{7}{5}, -\frac{5}{3}\right\}$ 13. $\left\{\frac{3b}{5}, \frac{2b}{3}\right\}$ 15. $\{\pm 4\}$
17. $\{\pm 2i\}$ 19. $\{\pm 0.9\}$ 21. $\left\{\pm \frac{1}{1000}\right\}$ 23. $\left\{\pm \frac{1}{2}\right\}$
25. $\{\pm \sqrt{7}\}$ 27. $\left\{\pm \frac{\sqrt{6}}{2}\right\}$ 29. $\left\{\pm \frac{4}{5}\right\}$ 31. $\left\{\frac{1 \pm \sqrt{5}}{2}\right\}$
33. $\left\{-1, -\frac{1}{2}\right\}$ 35. $\left\{\frac{-3 \pm \sqrt{3}i}{2}\right\}$ 37. $\left\{\frac{5 \pm 2i}{7}\right\}$
39. $\{a \pm |b|\}$ 41. $\left\{\pm \frac{5}{9}i\right\}$ 43. $x^2 - 5x + 6 = 0$
45. $x^3 - 4x^2 - 5x = 0$ 47. $x^2 - \frac{5}{6}x + \frac{1}{6} = 0$
49. $x^2 - 8 = 0$ 51. $x^2 + 9 = 0$ 53. $x^2 - 4x - 1 = 0$
55. $x^2 - 2x + 5 = 0$ 57. $x^2 - 2x + 4 = 0$
59. $x^2 - (2a + 3b)x + 6ab = 0$ 61. $ax^2 + bx + c = 0$
63. $\{-1 \pm \sqrt{5}\}$ 65. $\{0.3 \pm 3\sqrt{0.02}i\}$ 67. $\left\{\frac{-3 \pm 2\sqrt{2}}{2}\right\}$
69. $\left\{\frac{3 \pm 4\sqrt{2}}{7}\right\}$ 71. $\left\{\frac{3 \pm \sqrt{2}i}{2}\right\}$ 73. $\left\{\frac{7 \pm 5\sqrt{2}i}{8}\right\}$
75. $\{\pm 3, \pm 3i\}$ 77. $\{\pm \sqrt{2}, \pm \sqrt{2}i\}$ 79. $\{\pm b, \pm bi\}$
81. $\{\pm \sqrt{2}, \pm \sqrt{3}\}$ 83. $\frac{\sqrt{A\pi}}{\pi}$ 85. $\frac{\sqrt{2Sg}}{g}$ 87. $\frac{\sqrt{AP}}{P} - 1$
89. $\pm \sqrt{c}$ 91. $\frac{\pm \sqrt{rst}}{r}$ 93. $\frac{b \pm \sqrt{c}}{r}$ 95. $\frac{-n \pm |n|}{m}$
97. (a) Add 8 to both sides.
    (b) Add 1 to both sides.
    (c) Combine like terms.
    (d) Factor the left side.
    (e) Take the square root of both sides.

(f) Simplify.
(g) Add -1 to both sides.
(h) Combine like terms.

Exercise Set 9.2
1. $1, (a+1)^2$ **3.** $9, (x-3)^2$ **5.** $25, (r+5)^2$
7. $49, (v-7)^2$ **9.** $\{-1, 3\}$ **11.** $\{-6, -2\}$ **13.** $\{-1, 2\}$
15. $\{1 \pm \sqrt{2}i\}$ **17.** $\left\{\dfrac{-3 \pm \sqrt{17}}{4}\right\}$ **19.** $\left\{\dfrac{5 \pm \sqrt{11}i}{6}\right\}$
21. $\left\{\dfrac{-3 \pm \sqrt{17}}{4}\right\}$ **23.** $\left\{\dfrac{-2 \pm \sqrt{3}}{2}\right\}$ **25.** $\left\{-2, -\dfrac{3}{2}\right\}$
27. $\left\{\dfrac{-7 \pm \sqrt{13}}{6}\right\}$ **29.** $\left\{-\dfrac{7}{5}, 0\right\}$ **31.** $\left\{\dfrac{2 \pm \sqrt{3}}{3}\right\}$
33. $\{-6, -5\}$ **35.** $\left\{\dfrac{3 \pm \sqrt{11}i}{2}\right\}$ **37.** $\left\{\dfrac{-1 \pm \sqrt{3}i}{2}\right\}$
39. $\{\pm 5, \pm 5i\}$ **41.** $\left\{-\dfrac{4}{3}, \dfrac{3}{2}\right\}$ **43.** $\left\{-\dfrac{1}{3}, 2\right\}$ **45.** $\{4\}$
47. $\{5\}$ **49.** $\left\{\dfrac{-7 \pm \sqrt{61}}{6}\right\}; x \neq -2$
51. $\left\{\dfrac{1 \pm \sqrt{15}i}{4}\right\}; v \neq -1, \dfrac{1}{2}$ **53.** -3
55. 7, 5 or $-5, -7$ **57.** 3, 4 **59.** 4 ft $\times$ 12 ft
61. altitude: 3 m; base: 8 m **63.** $(x+1)^2 + (y+2)^2 = 9$
65. $(x-2)^2 + (y-6)^2 = 47$
67. Write it in standard form; factor the left side; set each factor equal to zero; solve the two resulting equations.
69. Arrange it so that all terms containing the variable are on one side (in standard form) and the constant is on the other side. Make the leading coefficient a 1. Add the square of one-half of the coefficient of the linear term to both sides. Factor the left side and use the square root method to find the solutions.
71. Yes, because it was derived using the general quadratic equation.
73. No. The quadratic equation must be factorable.

Exercise Set 9.3
1. $\{2, 4\}$ **3.** $\{-5, -3\}$ **5.** $\left\{-\dfrac{1}{2}, 4\right\}$ **7.** $\{-2, 1\}$
9. $\left\{-4, \dfrac{3}{2}\right\}$ **11.** $\left\{\dfrac{-3 \pm \sqrt{7}i}{2}\right\}$ **13.** $\left\{\dfrac{-1 \pm \sqrt{5}i}{3}\right\}$
15. $\left\{6, -\dfrac{1}{3}\right\}$ **17.** $\left\{-2, \dfrac{1}{2}\right\}$ **19.** $\left\{\dfrac{-1 \pm \sqrt{13}}{4}\right\}$
21. $\left\{\dfrac{1 \pm \sqrt{21}}{2}\right\}$ **23.** $\left\{\dfrac{-3 \pm \sqrt{13}}{2}\right\}$ **25.** $\left\{\dfrac{-7 \pm \sqrt{1743}i}{8}\right\}$
27. $\{-6 \pm 2\sqrt{14}\}$ **29.** $\{\pm 2\sqrt{3}i\}$ **31.** $\{\pm 3\sqrt{2}\}$
33. $\{-5, 0\}$ **35.** $\left\{-\dfrac{3}{5}, 0\right\}$ **37.** $\{2 \pm 2\sqrt{2}i\}$

39. $\{2 \pm \sqrt{15}\}$ **41.** $\{-0.3 \pm 0.1i\}$ **43.** $\left\{\dfrac{1 \pm \sqrt{0.2}i}{2}\right\}$
45. two real solutions **47.** complex conjugates
49. complex conjugates **51.** two real solutions
53. one real solution of multiplicity 2
55. complex conjugates **57.** $K < 9$ **59.** $K = 16$
61. $K < -\dfrac{1}{4}$ **63.** $K > 1$ **65.** $K < \dfrac{9}{32}$
67. $x^2 - 2x - 15 = 0$ **69.** $x^2 - 2x = 0$
71. $x^2 - 9x + 20 = 0$ **73.** $x^2 + 6x + 8 = 0$
75. $x^2 - \dfrac{4}{3}x + \dfrac{5}{9} = 0$ **77.** $x^2 - 7 = 0$ **79.** $\left\{\dfrac{-5 \pm \sqrt{11}i}{2}\right\}$
81. $\left\{\dfrac{2}{5}\right\}$ **83.** $\left\{-3, \dfrac{5}{4}\right\}$ **85.** $\left\{-\dfrac{27}{8}, 1\right\}$ **87.** $\left\{-\dfrac{5}{6}\right\}$
89. $\{-5, 21\}$ **91.** $\left\{-3, -\dfrac{1}{4}\right\}; x \neq -4, 1$ **93.** $\left\{-\dfrac{9}{4}\right\}$
95. $\{0, 3\}$ **99.** (a) $S = 6\pi r^2$ (b) $S = \dfrac{3}{2}\pi h^2$
101. $-1 + \sqrt{2}$

Exercise Set 9.4
1. $\{\pm 1, \pm \sqrt{2}\}$ **3.** $\{\pm \sqrt{3}\}$ **5.** $\{\pm 3, \pm \sqrt{6}i\}$
7. $\{0, \pm \sqrt{3}i, \pm \sqrt{2}i\}$ **9.** $\{\pm 2\sqrt{2}, \pm \sqrt{7}\}$
11. $\{2, -1 \pm \sqrt{3}i\}$ **13.** $\left\{-1, \dfrac{1 \pm \sqrt{3}i}{2}\right\}$
15. $\left\{0, -5, \dfrac{5 \pm 5\sqrt{3}i}{2}\right\}$ **17.** $\left\{0, 1, \dfrac{-1 \pm \sqrt{3}i}{2}\right\}$
19. $\left\{\pm 1, \dfrac{-1 \pm \sqrt{3}i}{2}, \dfrac{1 \pm \sqrt{3}i}{2}\right\}$
21. $\left\{\pm 1, \dfrac{-1 \pm \sqrt{3}i}{2}, \dfrac{1 \pm \sqrt{3}i}{2}\right\}$
23. $\left\{\pm \dfrac{3}{2}, \dfrac{-3 \pm 3\sqrt{3}i}{4}, \dfrac{3 \pm 3\sqrt{3}i}{4}\right\}$ **25.** $\{1, 729\}$
27. $\{-27, 64\}$ **29.** $\left\{-32, -\dfrac{1}{32}\right\}$ **31.** $\left\{-2, 3, \dfrac{1 \pm \sqrt{7}i}{2}\right\}$
33. $\left\{\pm \dfrac{\sqrt{5}i}{2}, \pm 1\right\}$ **35.** $\left\{\pm \sqrt{2}i, \pm \dfrac{\sqrt{10}i}{2}\right\}$ **37.** $\left\{-\dfrac{3}{2}, \dfrac{2}{3}\right\}$
39. $\left\{\dfrac{16 \pm \sqrt{409}}{17}\right\}; x \neq 1, 7, \dfrac{1}{3}$ **41.** $\{-3, 5\}; p \neq \pm 1$
43. $\left\{\dfrac{53}{14}\right\}; a \neq 1, 3, 4$ **45.** $\left\{\dfrac{1 \pm \sqrt{57}}{2}\right\}$ **47.** $\{3 \pm 3\sqrt{2}\}$
49. $\{4\}$ **51.** 6 hr, 12 hr **53.** A: 3 hr; B: 6 hr
55. 200 mph **57.** 6 km/hr **59.** $\{\pm 3\}$ **61.** $\{\pm 2\sqrt{2}i\}$
63. $\{-4, 3\}$ **65.** $\left\{3, -\dfrac{1}{2}\right\}$ **67.** $\left\{1 \pm \dfrac{\sqrt{6}}{2}i\right\}$
69. $\{-2 \pm \sqrt{7}i\}$ **71.** $\left\{-\dfrac{5}{3}, 3\right\}$ **73.** $\left\{\dfrac{-5 \pm \sqrt{35}}{2}\right\}$

75. $\left\{\dfrac{3 \pm \sqrt{41}}{8}\right\}$ **77.** $\dfrac{\sqrt{2}}{2}$

Exercise Set 9.5
1. $(-1, 0)$
3. ←——(——)——|——→ $-2\ -1\ 0$
5. $[-3, 2]$
7. ←——)——|——|——(——→ $0\ \ \ 3$
9. $(-\infty, -5) \cup (3, \infty)$
11. ←——]——|——[——→ $-\frac{3}{2}\ \ 0\ \frac{1}{3}$
13. $[-2, 1]$ **15.** $(-3, 4)$ **17.** $[-6, -1]$
19. $(-\infty, -4) \cup (-4, \infty)$ **21.** $(-\infty, -2] \cup [11, \infty)$
23. $\{2\}$ **25.** $(-3, 3)$ **27.** $\left(-\infty, -\dfrac{7}{6}\right) \cup (0, \infty)$
29. $(-2, 0)$ **31.** $(-\infty, -4) \cup [-3, \infty)$
33. $(-\infty, -2)$ **35.** $[-5, -4)$ **37.** $(-\infty, 5) \cup [9, \infty)$
39. $\left(-11, -\dfrac{19}{3}\right) \cup (3, \infty)$ **41.** $(-2, 0] \cup \left[\dfrac{3}{2}, 5\right]$
43. $(-\infty, -2] \cup [0, 1]$
45. ←——]——[——→ $0\ \frac{1}{3}\ \ \frac{3}{2}$
47. $(-\infty, -1) \cup \left(0, \dfrac{4}{3}\right)$ **49.** $(-4, 4)$
51. ←——(——|——|——|——)——→ $-3\ -1\ \ 1\ \ 3$
53. $(-\infty, -3] \cup [3, \infty)$
55. (a) $(0, 5)$ (b) $(5, 8]$ (c) $[1, 4]$
57. (a) $\left(0, \dfrac{5}{2}\right)$ (b) $\dfrac{5}{2}$ sec   (c) $\left(0, \dfrac{1}{2}\right) \cup \left(2, \dfrac{5}{2}\right]$
59. (a) $(10, 20)$ (b) $[0, 10] \cup [20, 50)$ (c) $31°$
61. $\varnothing$ **63.** $(-\infty, 0) \cup \left(\dfrac{1}{2}, \infty\right)$ **65.** $\left[-\dfrac{7}{3}, 3\right]$
67. $(-\infty, \infty)$ **69.** $(-\infty, -7) \cup (17, \infty)$ **71.** $\left(-\dfrac{2}{3}, 2\right)$
73. $x > 16$

Exercise Set 9.6
1. $y = kx$ **3.** $M = kp^2$ **5.** $V = ks^2$ **7.** $y = \dfrac{k}{x}$
9. $S = k\sqrt{a}$ **11.** $V = khr^2$ **13.** $P = \dfrac{kC}{\sqrt{n}}$ **15.** 4

17. 98   **19.** $\dfrac{6}{125}$ **21.** $\dfrac{27}{16}$ **23.** 432   **25.** 9 foot-candles
27. $\dfrac{40}{3}$ ohms   **29.** 22 amps   **31.** $\dfrac{48}{5}$ lb/ft²   **33.** 960 lb
35. $P = \dfrac{9}{4}N^{2/3}$ **37.** 679.25 miles   **39.** 48 ohms
41. (a) 0   (b) undefined   (c) $\dfrac{6}{7}$ (d) $\dfrac{3x + 3h}{x + h - 5}$
43.

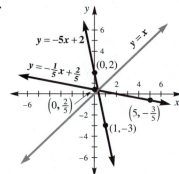

45. $\left\{x \in R, x \neq \dfrac{1}{4}\right\}$ **47.** 3

Exercise Set 9.7
1. base: 10 m; height: 4 m   **3.** 25 in. × 75 in.
5. 5 cm and 12 cm   **7.** $2\sqrt{3}$ hr   **9.** 2 mph
11. $(2 + 2\sqrt{5})$ hr   **13.** 7 members   **15.** [60, 110)
17. 219 miles   **19.** $\dfrac{x\sqrt{2}}{2}$ **21.** 70¢   **23.** 10 hr, 15 hr
25. 5 mph   **27.** 24 hr   **29.** 30 mph   **31.** 20 in. × 20 in.
33. 32 students   **35.** 2 ft   **37.** $p < \$1.50$
39. $200 < x \leq 700$

Review Exercises
1. $\{2, -2\}$ **2.** $\{-8, 0, 8\}$ **3.** $\left\{-\dfrac{2}{3}, 0, \dfrac{5}{4}\right\}$ **4.** $\left\{-\dfrac{4}{5}, \dfrac{7}{3}\right\}$
5. $\{\pm 2\}$ **6.** $\left\{\pm\dfrac{1}{10^4}\right\}$ **7.** $\left\{\pm\dfrac{7}{8}\right\}$ **8.** $\{a \pm c\}$
9. $\left\{\pm\dfrac{7}{11}i\right\}$ **10.** $\left\{\dfrac{2 \pm \sqrt{7}i}{3}\right\}$ **11.** $x^2 - 1 = 0$
12. $12x^2 + x - 1 = 0$ **13.** $x^2 - 12 = 0$
14. $x^3 + x^2 - 6x = 0$ **15.** $x^3 - 3x^2 + 4x - 12 = 0$
16. $x^3 - 2x^2 + 2x = 0$ **17.** $x^2 - 8x + 41 = 0$
18. $\dfrac{A}{b}\sqrt{2Ab}$ **19.** $\dfrac{\sqrt[3]{6\pi^2 V}}{2\pi}$ **20.** $\dfrac{c-b}{x^2}$
21. two real solutions   **22.** one real solution of multiplicity 2
23. complex conjugates   **24.** complex conjugates
25. two real solutions   **26.** complex conjugates
27. $x = -\dfrac{14}{5}, y = -\dfrac{21}{5}$ **28.** $x = -\dfrac{1}{5}, y = \dfrac{16}{5}$ **29.** $\left\{\dfrac{1}{3}, 3\right\}$

CHAPTER 10 · EXERCISE SET 10.1

30. $\{7 \pm \sqrt{6}\}$ 31. $\left\{\dfrac{5 \pm 2\sqrt{2}i}{7}\right\}$ 32. $\left\{\dfrac{-11 \pm 3\sqrt{3}i}{2}\right\}$
33. $\{-3, -1\}$ 34. $\left\{\dfrac{-7 \pm \sqrt{145}}{6}\right\}$ 35. $\left\{\dfrac{1 \pm \sqrt{23}i}{12}\right\}$
36. $\left\{\dfrac{-1 \pm \sqrt{31}i}{4}\right\}$ 37. $\{1, 4\}$ 38. $\left\{-\dfrac{5}{2}, 1\right\}$
39. $\left\{0, \dfrac{-1 \pm \sqrt{35}i}{6}\right\}$ 40. $\left\{\dfrac{3 \pm \sqrt{15}}{3}\right\}$
41. $\{2, -1 \pm \sqrt{3}i\}$ 42. $\{\pm\sqrt{7}, \pm i\}$ 43. $\{6^6, 1\}$
44. $\left\{\dfrac{1}{5}, \dfrac{1}{6}\right\}$ 45. $\{1, 3, 1 \pm \sqrt{2}i\}$ 46. $\left\{\pm\dfrac{\sqrt{2}}{2}, \pm\sqrt{2}\right\}$
47. $(-1, 5)$ 48. $[-4, 0] \cup [1, \infty)$
49. $\left(-\dfrac{4}{3}, -\dfrac{1}{2}\right) \cup \left(-\dfrac{1}{3}, \infty\right)$ 50. -640 mph
51. 1.75 mph 52. 65 hr 53. $18, 30$ 54. 900

Chapter Test

1. $\left\{\dfrac{1}{2}, 2\right\}$ 2. $\left\{\dfrac{5 \pm 2\sqrt{2}}{3}\right\}$ 3. $\{2 \pm \sqrt{3}\}$
4. $\left\{\dfrac{1 \pm \sqrt{79}i}{8}\right\}$ 5. $\{\pm 2, \pm 2i\}$ 6. $\{-5, 1\}$
7. $\left\{-6, -\dfrac{1}{3}\right\}$ 8. $\{-6, 0\}$ 9. two distinct real roots
10. one real root of multiplicity 2 11. complex conjugates
12. $x^2 - 7x + 10 = 0$ 13. $x^2 - 2x + 2 = 0$
14. $x^2 - 2x - 1 = 0$ 15. $\{16, 81\}$
16. $\{4, -2 \pm 2\sqrt{3}i\}$ 17. $(-\infty, -1) \cup (4, \infty)$
18. $(-2, -1]$ 19. 5 mph 20. $b = 5$ ft, $h = 4$ ft
21. 1 m 22. $21, 29$ 23. 60 ft 24. ≈ 3.86 ft
25. $x = 3$ or 4 26. $H = 18$

Cumulative Review Chapters 7–9

1. $\dfrac{x + y}{x - y}$ 2. $-\dfrac{2}{3}$ 3. $\dfrac{a - b}{a(c - 1)}$ 4. $\dfrac{x - 2}{x^2 - 2x + 4}$
5. $\dfrac{7x - 16}{(x - 3)(x - 8)}$ 6. $\dfrac{6x}{x - 9}$ 7. $\dfrac{xy}{x - y}$ 8. $\dfrac{2x}{x - 3}$
9. $\left\{-\dfrac{9}{2}\right\}$ 10. $\{25\}; x \neq -3, 4$ 11. $\{-3\}; x \neq -1, 2$
12. $\left(\dfrac{1 \pm \sqrt{33}}{2}\right); a \neq 3$ 13. 12 min 14. $1\dfrac{2}{3}$ liters
15. 11 in. 16. $12, 14, 16$
17. $\$4000$ at 6.5%, $\$8000$ at 5.5% 18. 300 miles
19. 1 20. $\emptyset$ 21. $\{12\}$ 22. $\emptyset$ 23. $1 - x^{3/4}$
24. $2\sqrt{2}$ 25. $\dfrac{8}{|x|}$ 26. $4\sqrt{3} - 3\sqrt{2}$ 27. $5\sqrt{3}$
28. $9\sqrt[3]{3}$ 29. $\dfrac{\sqrt{5}}{5}$ 30. $\sqrt[3]{4}$ 31. $-\sqrt{3}$

32. $2(\sqrt{3} + \sqrt{2})$ 33. $\dfrac{7 - 2\sqrt{10}}{3}$ 34. $\left\{-\dfrac{1}{2}\right\}$
35. $\left\{\dfrac{16}{9}\right\}$ 36. $1 - 4i$ 37. 25 38. $-2\sqrt{2} + (4 + 5\sqrt{2})i$
39. $\dfrac{2 + i}{5}$ 40. $\{-4, 2\}$ 41. $\{2, 4\}$ 42. $\{3 \pm \sqrt{5}\}$
43. $\left\{\dfrac{1 \pm \sqrt{31}i}{4}\right\}$ 44. $\left\{-\dfrac{5}{3}, 1\right\}$
45. one real root of multiplicity 2 46. two distinct real roots
47. complex conjugates 48. two distinct real roots
49. $x^2 - x - 6 = 0$ 50. $2x^2 - 2x + 1 = 0$
51. $2x^2 - x = 0$ 52. $x^2 - 2x + 2 = 0$ 53. $\{\pm 2, \pm\sqrt{2}\}$
54. $\{\pm 1, \pm\sqrt{2}i\}$ 55. $\{-3, -2, 0, 1\}$ 56. $\left\{\pm\dfrac{1}{3}, \pm 1\right\}$
57. 9 mph 58. 4 hr 59. 100 mph 60. 6 hr
61. $(-17, \infty)$ 62. $\left(-\infty, -\dfrac{6}{5}\right)$ 63. $\emptyset$ 64. $[-5, \infty)$
65. 9 66. 30 cp

CHAPTER 10

Exercise Set 10.1

1. 4 3. 10 5. 6 7. $3\sqrt{2}$ 9. $5\sqrt{2}$ 11. $(3, 5)$
13. $(1, 8)$ 15. $\left(-1, \dfrac{15}{2}\right)$
17.

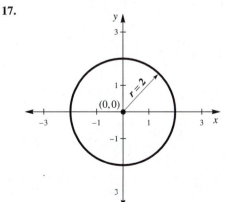

19. $(0, 0), r = 3$

21.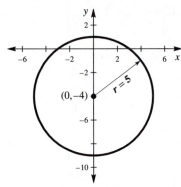

23. $(3, 2)$, $r = 5$

25.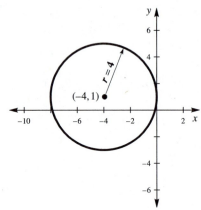

27. $(7, 0)$, $r = 6$ **29.** $x^2 + y^2 = 4$
31. $(x - 1)^2 + (y - 2)^2 = 16$ **33.** $x^2 + (y + 2)^2 = 36$
35. $(x + 3)^2 + y^2 = 9$ **37.** $(x + 1)^2 + (y + 3)^2 = 25$
39. $(x - 3)^2 + (y + 5)^2 = 34$
41. $(x + 3)^2 + (y + 4)^2 = 16$
43. $(x + 3)^2 + (y + 5)^2 = 25$
45. $(-1, 1)$, $r = \sqrt{2}$ **47.** $(-4, 1)$, $r = 3$
49. $\left(\frac{5}{2}, -\frac{7}{2}\right)$, $r = \frac{\sqrt{1810}}{10}$ **51.** $\left(\frac{1}{2}, -\frac{3}{2}\right)$, $r = 2$
53. $(4, 0)$, $r = 4$ **55.** $\left(-\frac{5}{2}, -\frac{5}{4}\right)$, $\frac{5\sqrt{5}}{4}$
57. $(-5, 3)$, $r = \sqrt{38}$ **59.** $\left(-\frac{4}{3}, -6\right)$, $r = 2\sqrt{3}$
61. $(x - 5)^2 + (y + 2)^2 = 85$
63. $(x - 1)^2 + (y - 3)^2 = 32$
65. $x^2 + y^2 = 4$ **67.** $x^2 + (y - 1)^2 = 9$
69. $(x - 2)^2 + (y - 1)^2 = 1$
71. The set of all points in a plane equidistant from a fixed point (the center)
73. A line through the center of a circle that joins two points on the circle

75. It is the square root of the sum of the squares of the difference of the x-coordinates and the difference of the y-coordinates.

77. $\sqrt{3} + \sqrt{2}$ **79.** $\left\{\frac{5 \pm \sqrt{13}}{6}\right\}$

81. $\left(-\infty, -\frac{4}{5}\right) \cup (0, \infty)$ **83.** 18 mph

Exercise Set 10.2
1. vertex: $(0, 0)$; axis: $x = 0$

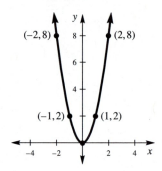

3. vertex: $(0, 0)$; axis: $x = 0$

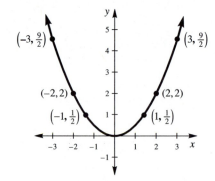

5. vertex: $(0, 0)$; axis $x = 0$

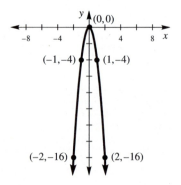

7. vertex: (0, 0); axis: $x = 0$

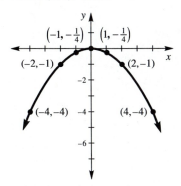

9. Shift up 2 units.

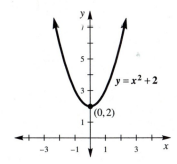

11. Shift down 3 units.

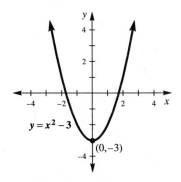

13. Shift right 2 units.

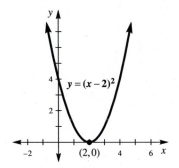

15. Shift left 3 units and down 4 units.

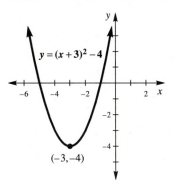

17. $x = 0$; opens upward

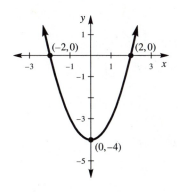

19. $x = 0$; opens upward

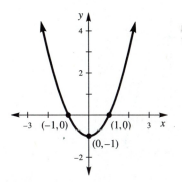

21. $x = 1$; opens downward

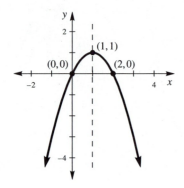

23. $x = -1$, opens upward

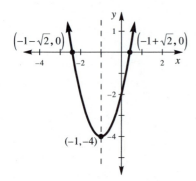

25. $x = -2$; opens downward

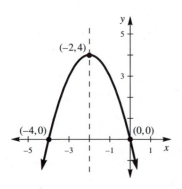

27. $x = 3$; opens downward

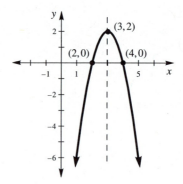

29. $y = -2$; opens to the right

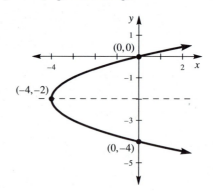

31. $y = 1$; opens to the left

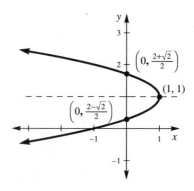

33. $y = \frac{1}{2}(x + 2)^2 - 3$

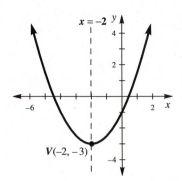

35. $x = \frac{1}{3}(y - 5)^2 - \frac{1}{3}$; $V\left(-\frac{1}{3}, 5\right)$; $y = 5$

37. $y = \frac{1}{3}\left(x - \frac{3}{2}\right)^2 - \frac{5}{12}$

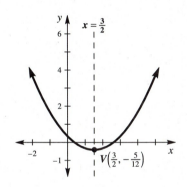

39. $x = -2(y + 3)^2 + 4$; $V(4, -3)$; $y = -3$

41. $y = \frac{2}{7}(x - 3)^2 - 4$; $V(3, -4)$; $x = 3$

43. $x = -\frac{1}{4}\left(y - \frac{5}{2}\right)^2 + \frac{1}{2}$; $V\left(\frac{1}{2}, \frac{5}{2}\right)$; $y = \frac{5}{2}$

45. $y = x^2 - 1$ **47.** $y = -\frac{1}{2}(x + 1)^2 + 2$

49. $y = \frac{1}{4}(x - 3)^2 - 1$ **51.** 0 **53.** 2 sec or 3 sec

55. $x = 0$ ft for minimum; 100 ft **57.** 1250 ft² **59.** 89 ft

Exercise Set 10.3

1.

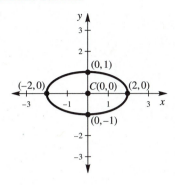

3. center: (0, 0); vertices: (±3, 0), (0, ±2)

5.

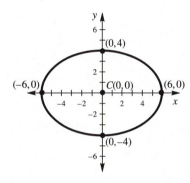

7. center: (0, 0); vertices: (±6, 0), (0, ±3)

9.

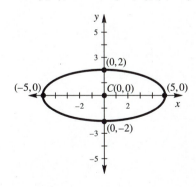

11. center: (0, 0); vertices: (±3, 0), (0, ±4)

13.
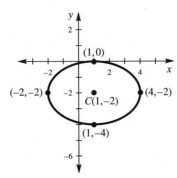

15. center: $(-5, -1)$; vertices: $(-11, -1), (1, -1), (-5, 2), (-5, -4)$

17.
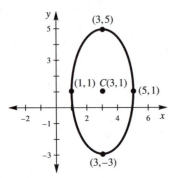

19. $\dfrac{x^2}{4} + \dfrac{y^2}{1} = 1$; center: $(0, 0)$; vertices: $(\pm 2, 0), (0, \pm 1)$

21. $\dfrac{x^2}{1} + \dfrac{y^2}{36} = 1$; center: $(0, 0)$; vertices: $(\pm 1, 0), (0, \pm 6)$

23. $\dfrac{x^2}{16} + \dfrac{y^2}{9} = 1$; center: $(0, 0)$; vertices: $(\pm 4, 0), (0, \pm 3)$

25. $\dfrac{x^2}{36} + \dfrac{y^2}{9} = 1$ **27.** $\dfrac{(x-12)^2}{4} + \dfrac{(y-10)^2}{16} = 1$

29. $\dfrac{x^2}{64} + \dfrac{(y+7)^2}{4} = 1$ **31.** $\dfrac{(x-1)^2}{25} + \dfrac{(y+1)^2}{4} = 1$; ellipse

33. $\dfrac{(x+3)^2}{16} + \dfrac{y^2}{8} = 1$; circle

35. $x = \dfrac{1}{6}\left(y - \dfrac{3}{2}\right)^2 - 1$; parabola

37. $y = (x-2)^2 + 4$; parabola

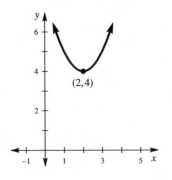

39. $\left(x - \dfrac{5}{6}\right)^2 + \left(y + \dfrac{7}{6}\right)^2 = 1$; circle   **41.** $y = \pm\sqrt{9 - x^2}$

43. x-intercepts: $(\pm 3, 0)$; y-intercept: $(0, 3)$

45. They are the same for $y \geq 0$, but the first one is not defined for $y < 0$.

Exercise Set 10.4

1. center: $(0, 0)$; vertices: $(\pm 2, 0)$

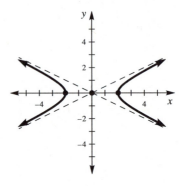

3. center: $(0, 0)$; vertices: $(0, \pm 4)$

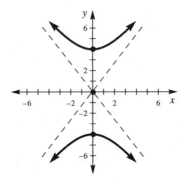

5. center: $(2, -1)$; vertices $(4, -1), (0, -1)$

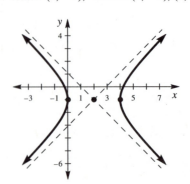

7. center: (1, −3); vertices (1, 0), (1, −6)

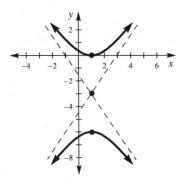

9. center: (−7, 6); vertices: (−6, 6), (−8, 6)

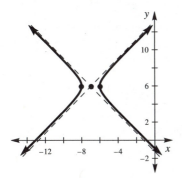

11. center: (0, 4); vertices: (0, 3), (0, 5)

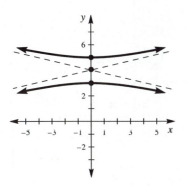

13. center: (0, 1); vertices: (6, 1), (−6, 1)

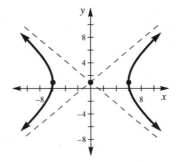

15. center: (0, 0); $x = 0$, $y = 0$

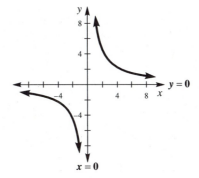

17. center: (0, 0); $x = 0$, $y = 0$

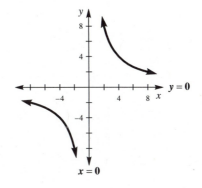

19. $\dfrac{x^2}{4} - \dfrac{y^2}{1} = 1$

21. $\dfrac{y^2}{9} - \dfrac{x^2}{16} = 1$

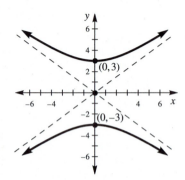

23. $\dfrac{(x+1)^2}{4} - \dfrac{(y-2)^2}{4} = 1$

25. $\dfrac{(x+3)^2}{4} - \dfrac{(y-1)^2}{8} = 1$

27. $\dfrac{(y-3)^2}{9} - \dfrac{(x-2)^2}{16} = 1$ 29. circle 31. hyperbola

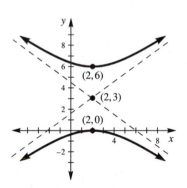

33. circle 35. parabola 37. $\dfrac{y^2}{\frac{7}{16}} - \dfrac{\left(x - \frac{3}{2}\right)^2}{\frac{7}{4}} = 1$

39. $x = \dfrac{1}{6}\left(y - \dfrac{3}{2}\right)^2 - 1$ 41. $\dfrac{(x+3)^2}{3} + \dfrac{y^2}{1} = 1$

43. $\dfrac{(x+2)^2}{1} - \dfrac{(y-2)^2}{1} = 1$ 45. $(x+1)^2 + (y-3)^2 = 5$

47. $\dfrac{(x-2)^2}{1} + \dfrac{(y-1)^2}{\frac{9}{4}} = 1$ 49. $\left\{\left(\dfrac{29}{19}, \dfrac{4}{19}\right)\right\}$

51. $\left\{\left(\dfrac{29}{31}, -\dfrac{9}{31}\right)\right\}$ 53. $\{(-1, -1, 1)\}$

55. A hyperbola is the set of all points in the plane the difference of whose distances from two fixed points (the foci) is a constant.

57. A parabola is the set of all points in the plane equidistant from a point (the focus) and a line (the directrix).

59. A figure formed by cutting a cone by a plane.

Exercise Set 10.5

1. $\{(3, 0)\}$ 3. $\{(0, 0), (4, -4)\}$ 5. $\{(5, -5), (-5, 5)\}$
7. $\{(4, 2), (-4, -2)\}$ 9. $\left\{(-4, -1), \left(\dfrac{20}{17}, \dfrac{5}{17}\right)\right\}$
11. $\{(3, 0), (12, 9)\}$ 13. $\{(\pm 3, 0)\}$ 15. $\{(\pm 1, 0)\}$
17. $\{(4, 0), (-3, \pm\sqrt{7})\}$ 19. $\{(\pm 3, 2), (\pm 3, -2)\}$
21. $\{(\pm\sqrt{2}, 0)\}$ 23. $\varnothing$ 25. $\{(2, -1)\}$
27. $\{(\pm 1, 0), (1, -1), (-1, 1)\}$
29. $\{(2, 1), (-2, -1), (-4, 1), (4, -1)\}$
31. $\{(9, 3), (-9, -3)\}$ 33. $\dfrac{1}{8}, 8$ 35. 25 m × 4 m
37. 9, 7 39. 6, 12 41. 6, 8 or −6, −8

43.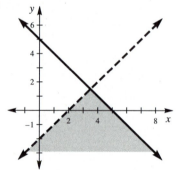

45.

47. $(-1, 5)$ 49. $(2, 1)$ 51. $\left(\dfrac{11}{5}, \dfrac{8}{5}\right)$ 53. $\varnothing$

Exercise Set 10.6

1.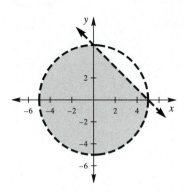

3.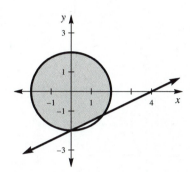

5.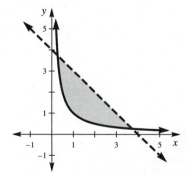

7.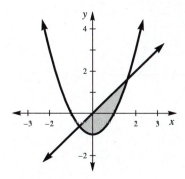

9.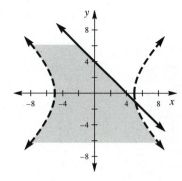

11.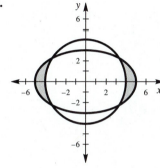

13.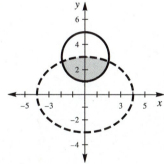

15.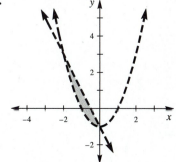

17.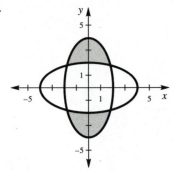

19.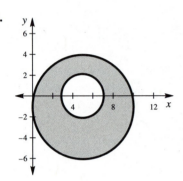

21.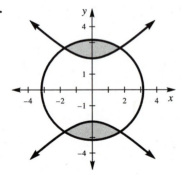

23.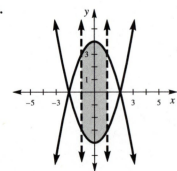

25.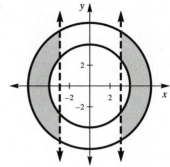

27.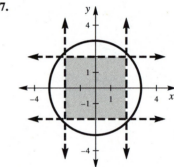

29.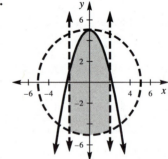

31. $x + y \leq 4$; see text. **33.** $x + y = 5$; see text.
35. $x^2 + y < 3$; see text. **37.** $x^2 + 4y^2 = 36$; see text.
39. (d) **41.** $9 \leq$ length $\leq 10\sqrt{489} \approx 221$ cm
43. $x < 6.9$

Review Exercises
1. circle

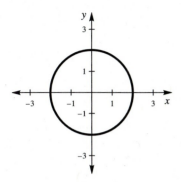

2. parabola

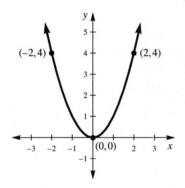

3. parabola
4. circle
5. ellipse

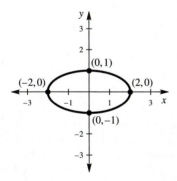

6. ellipse
7. ellipse
8. ellipse

9. hyperbola

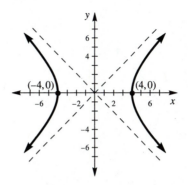

10. hyperbola
11. hyperbola

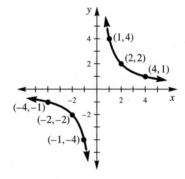

12. circle   **13.** hyperbola
14. ellipse
15. parabola
16. parabola   **17.** $\sqrt{89}$ **18.** $(4, 2)$
19. $(x + 1)^2 + \left(y + \dfrac{1}{2}\right)^2 = \dfrac{265}{4}$
20. center: $(2, -1)$; radius $= 4$
21. center: $(-1, 1)$; radius $= 2$
22. $(x - 3)^2 + (y + 1)^2 = 13$
23. parabola
24. parabola

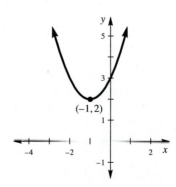

25. parabola
26. hyperbola

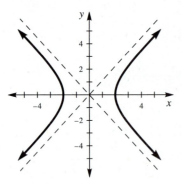

27. circle

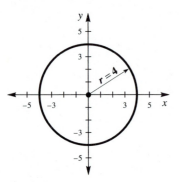

28. hyperbola
29. hyperbola
30. $\dfrac{(x+3)^2}{17} + \dfrac{(y-2)^2}{\frac{17}{2}} = 1$; ellipse

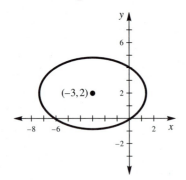

31. circle
32. parabola
33. $\{(\sqrt{3}, \sqrt{3}), (-\sqrt{3}, -\sqrt{3})\}$ 34. $\left\{\left(\dfrac{1}{2}, \dfrac{1}{2}\right)\right\}$
35. $\left\{\left(1, \pm\dfrac{\sqrt{2}}{2}\right), \left(-1, \pm\dfrac{\sqrt{2}}{2}\right)\right\}$ 36. $\{(\pm\sqrt{2}, 0)\}$

37. $\varnothing$ 38. $\{(2, 2), (-2, -2), (\sqrt{2}, -\sqrt{2}), (-\sqrt{2}, \sqrt{2})\}$
39. $\{(1, 0)\}$ 40. $\{(1, 1), (-1, -1)\}$
41.

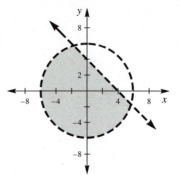

42.

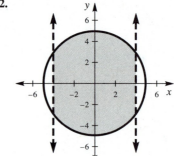

43.

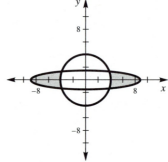

44.

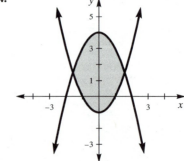

45.

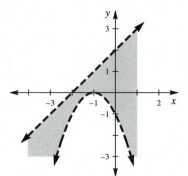

46.

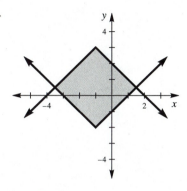

Chapter Test
1. circle

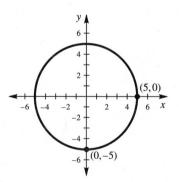

2. hyperbola

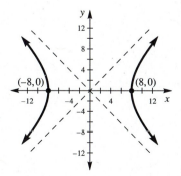

3. parabola

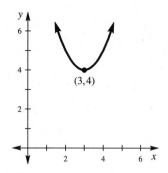

4. ellipse

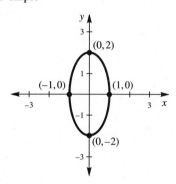

5. parabola

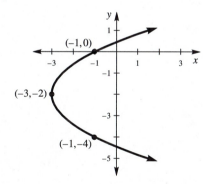

6. circle

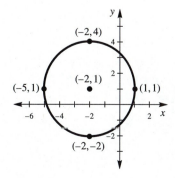

7. ellipse

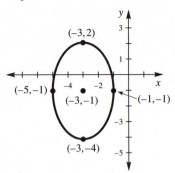

8. hyperbola

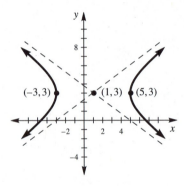

9. $(x + 2)^2 + (y + 3)^2 = 16$; circle
10. $y = (x + 3)^2 - 4$; parabola
11. $\dfrac{(x - 1)^2}{4} - \dfrac{(y - 1)^2}{4} = 1$; hyperbola
12. $\dfrac{(x + 4)^2}{16} + \dfrac{(y + 1)^2}{4} = 1$; ellipse
13. ∅   **14.** $\{(4, 2), (-4, -2)\}$
15.

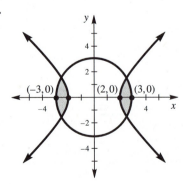

16.

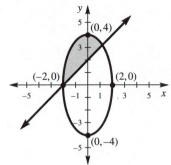

17.

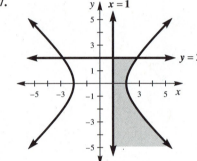

18.

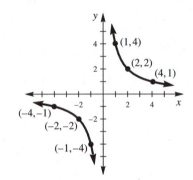

19. (a) $\left(3, \dfrac{5}{2}\right)$ (b) $\dfrac{\sqrt{17}}{2}$ (c) $(x - 3)^2 + \left(y - \dfrac{5}{2}\right)^2 = \dfrac{17}{4}$
20. $\dfrac{x^2}{256} + \dfrac{y^2}{64} = 1$

Chapter 11

Exercise Set 11.1
1. (0, 1), (1, 2), (2, 4)

3.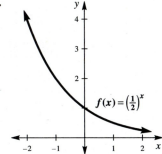

5. $\left(-2, \frac{1}{81}\right), (0, 1), (2, 81)$ 7. $\left(-2, \frac{1}{8}\right), \left(-1, \frac{1}{4}\right), (1, 1)$

9.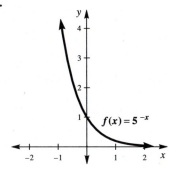

11. (0, 1), (1, 2)

13.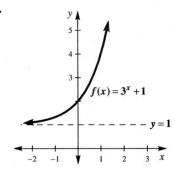

15. $\left(-2, \frac{9}{4}\right), \left(1, \frac{3}{2}\right), (0, 1)$ 17. $(0, 8), (1, 2), \left(2, \frac{1}{2}\right)$

19. (0, 1), (1, 1), (2, 1)

21.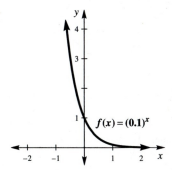

23. $\left(0, \frac{3}{2}\right), \left(1, \frac{9}{2}\right)$ 25. $(2, 1), (4, -5)$

27.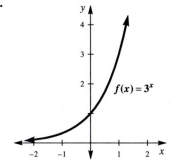

29. {2}   31. {3}   33. {3}   35. {1}   37. $\left\{\frac{3}{2}\right\}$

39. $\left\{\frac{3}{2}\right\}$ 41. $\left\{-\frac{7}{3}\right\}$ 43. $\left\{-\frac{8}{3}\right\}$ 45. {−3}

47. $\left\{\frac{5}{3}\right\}$ 49. $1166.40   51. $1612.55   53. $2257.64

55. $1612.23   57. $2901.90   59. $2547.21

61. (a) 200   (b) 698

63. (a) 400 g   (b) 339.1575 g   (c) 272.1803 g

65. An exponential decay function is one in which the slope is always negative; $f(x) = 2^{-x}$

67. If $b = 1$ it becomes a horizontal line. If $b = 0$ it is the x-axis. If $b < 0$ it is an alternating function (one that goes up and down).

69. Answer varies. Only 1¢ more in a year.

71. The area of the larger of the two small semicircles is $\frac{1}{8}\pi BC^2$. The area of smallest semicircle is $\frac{1}{8}\pi AC^2$. The area of triangle ABC is $\frac{1}{2}(BC)(AC)$. The area of the semicircle through ACB is $\frac{1}{8}\pi AB^2$. The sum of the areas

of the shaded regions is

$$\frac{1}{2}(BC)(AC) + \frac{1}{8}\pi BC^2 + \frac{1}{8}\pi AC^2 - \frac{1}{8}\pi AB^2$$

$$= \frac{1}{2}(BC)(AC) + \frac{1}{8}\pi(BC^2 + AC^2 - AB^2)$$

By the Pythagorean theorem, $AB^2 = BC^2 + AC^2$, so the sum of the shaded areas reduces to $\frac{1}{2}(BC)(AC)$, which is the area of the triangle.

Exercise Set 11.2
1. $\log_3 27 = 3$ **3.** $\log_2 32 = 5$ **5.** $\log_{10} 1 = 0$
7. $\log_9 3 = \frac{1}{2}$ **9.** $\log_{10} \frac{1}{10} = -1$ **11.** $\log_3 \frac{1}{81} = -4$
13. $\log_8 4 = \frac{2}{3}$ **15.** $\log_{81} \frac{1}{27} = \frac{-3}{4}$ **17.** $\log_8 8 = 1$
19. $\log_{3/4} \frac{9}{16} = 2$ **21.** $\log_{0.3} 0.027 = 3$
23. $\log_{0.001} 0.1 = \frac{1}{3}$ **25.** $6^2 = 36$ **27.** $9^1 = 9$
29. $6^0 = 1$ **31.** $16^{3/4} = 8$ **33.** $10^{-3} = 0.001$
35. $3^{-2} = \frac{1}{9}$ **37.** $\left(\frac{1}{5}\right)^{-2} = 25$ **39.** $\left(\frac{1}{2}\right)^{-5} = 32$
41. {4} **43.** {10} **45.** {−2} **47.** {100} **49.** $\left\{\frac{9}{4}\right\}$
51. {−3} **53.** 0 **55.** 0 **57.** 0 **59.** 1
61. $\{x \in R \mid x > 0\}$ **63.** f and g are inverses.
65. $D\{x \mid x \in R\}$ **67.** $f^{-1}(x) = g(x)$ **69.** 1, 2, 3, a
71. $D\{x \mid x \in R\}$; $R\{y \mid y = 1\}$; not one-to-one
73. yes: domain: R; range: $\{y \in R \mid y \geq 0\}$; it is one-to-one.
75. (a) $2200 (b) $1801.21 (c) $1207.39
  (d) a decay function

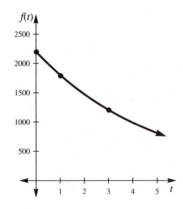

77. They are inverse functions.
79. The base ($b > 0$) raised to any power is never negative. An error occurs on the calculator when $b < 0$.
81. $\log_b b = 1$ because $b^1 = b$ for all allowable b; no

Exercise Set 11.3
1. $\log u + \log v + \log w$ **3.** $\log_b x + \log_b y - \log_b z$
5. $2 \log_4 a - 3 \log_4 b$ **7.** $\frac{1}{2}[\log_e(x+y) - \log_e(x-y)]$
9. $\frac{3}{2}\log x + \log y - 3 \log z$ **11.** $\log_e y + \frac{1}{5}\log_e x$
13. $\log_b x + \frac{1}{3}\log_b y - \frac{5}{3}\log_b z$ **15.** $\frac{3}{2}\log_e x - \frac{4}{3}\log_e y$
17. $\frac{1}{2}\log a - \frac{3}{16}\log b + \frac{1}{12}\log c$
19. $\log 2 + \log \pi + \frac{1}{2}\log L - \frac{1}{2}\log g$
21. $\frac{1}{2}[\log(s-a) + \log(s-b) + \log(s-c)]$ **23.** $\log abc$
25. $\log_e \frac{rt}{s}$ **27.** $\log_e b^2 c^3$ **29.** $\log 2\sqrt{3}$ **31.** $\log \frac{b^2}{c}\sqrt{a}$
33. $\log_e \frac{x^{3/2}}{y^4}$ **49.** $\{2^{y+c}\}$ **51.** $\{ae^{-T}\}$ **53.** $\left\{\frac{3}{2}\right\}$ **55.** {4}
57. 0.9030 **59.** 1.0791 **61.** 1.3545 **63.** −0.6020
65. 0.6309 **67.** 1.2918 **69.** 2.5236
73. logarithms to the base 10; developed by Henry Briggs
75. Yes: one example is base e (natural logarithms), which is used extensively.
77. Yes: their properties are used in many areas.
79. Because logarithms of zero or negative numbers do not exist.

Exercise Set 11.4
1. {998} **3.** {120} **5.** $\{e^4 - 3\}$ **7.** {4} **9.** {3}
11. ∅ **13.** {2} **15.** $\left\{\frac{10}{9}\right\}$ **17.** {4} **19.** {80}
21. {3} **23.** {1, 9} **25.** {1.4037} **27.** {0.4307}
29. {4.3013} **31.** {0.5853} **33.** {1.0148} **35.** {20.1528}
37. {11.6869} **39.** {15.4859} **41.** ∅ **43.** {18.9230}
45. {2} **47.** {±1.3386} **49.** ∅ **51.** {−2, 1}
53. −0.0241 **55.** 2 days **57.** 8 yr
59. $f(x) = 5^x$

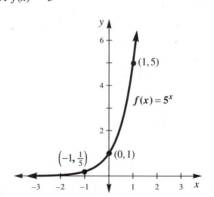

61. $f(x) = \log x$

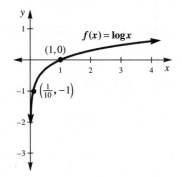

63. $f(x) = \log_3 5 = 1.465$

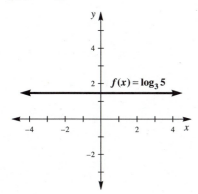

65. $\sqrt{3}$

Exercise Set 11.5
1. 10.164 sec **3.** 159.621 ft/sec **5.** 1690.603 yr
7. 7.273 yr **9.** 4119.561 yr **11.** 0.087, 565
13. (a) $k = 0.020$ **(b)** 38,531 **15.** 9.601
17. (a) 30 dB **(b)** 64.997 dB **(c)** 80 dB
19. 94.952 ft **21.** 205.933 in. **23.** 9.2%
25. (a) $2806.08 **(b)** 361 units **27.** 71.2454 g

Review Exercises
1. $x = 4$ **2.** $x = \dfrac{3}{2}$

3. $f(x) = 2^x$

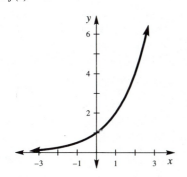

4. $f(x) = 2^{|x|}$

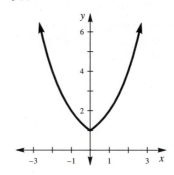

5. $f(x) = \left(\dfrac{1}{4}\right)^x$

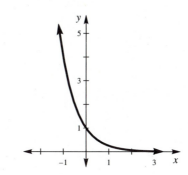

6. $f(x) = 3^{-x}$

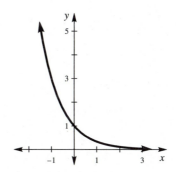

7. $\left\{\dfrac{5}{3}\right\}$ **8.** $\left\{-\dfrac{6}{5}\right\}$ **9.** $\log_2 8 = 3$ **10.** $\log_3 243 = 5$
11. $\log_{2/3} \dfrac{4}{9} = 2$ **12.** $\log_{0.12} 0.0144 = 2$ **13.** $7^2 = 49$
14. $8^0 = 1$ **15.** $\left(\dfrac{1}{2}\right)^{-6} = 64$ **16.** $\left(\dfrac{1}{6}\right)^{-2} = 36$
17. $\{3125\}$ **18.** $\{-2\}$ **19.** $\left\{\dfrac{64}{27}\right\}$ **20.** $\{-1\}$
21. $\log x + \log y + \dfrac{1}{2}\log z$ **22.** $\dfrac{1}{3}\log x + \dfrac{3}{4}\log y - \dfrac{1}{2}\log z$
23. $4\log_e a - 3\log_e b$ **24.** $\dfrac{1}{3}\log_b a + \dfrac{3}{4}$ **25.** $\log \dfrac{27}{16}$

26. $\log_e \dfrac{xz^3}{y}$ **27.** $\log \dfrac{36}{5}$ **28.** $\log \sqrt{\dfrac{xz^2}{y^5}}$ **29.** $\varnothing$
30. $\{6\}$ **31.** $\{0.7124\}$ **32.** $\{0.4660\}$ **33.** $\{4.7833\}$
34. $\{5.8332\}$ **35.** $k = 0.0866;\ 150.1042$ **36.** 84.5025 dB
37. $\$3925.94$ **38.** 8 lbs/ft^2 **39.** $\$3264.63$
40. 176.42 lb

Chapter Test
1. $f(x) = \left(\dfrac{1}{3}\right)^x$

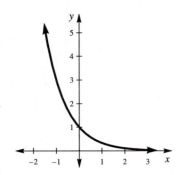

2. $f(x) = \left(\dfrac{1}{2}\right)^{-x}$

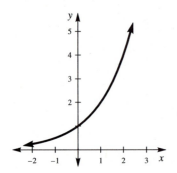

3. $f(x) = 3^{2x}$

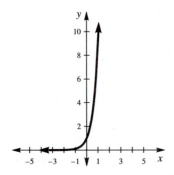

4. $\{5\}$ **5.** $\left\{\dfrac{3}{2}\right\}$ **6.** $\log_2 128 = 7$ **7.** $x = 3^y$
8. $2 \log x + \log y + \dfrac{1}{3} \log z$ **9.** $\log 4$ **10.** $\{4\}$
11. $\{0.86\}$ **12.** $\{2.71\}$ **13.** $\{-3.42\}$ **14.** $\{2\}$
15. $\$1716.01$ **16.** 13.86 yr **17.** $P = 8.65$
18. 17 yr **19.** $\$2099.93$ **20.** $f(10) = 38.818$ watts

CHAPTER 12

Exercise Set 12.1
1. $1, 2, 3, 4, 5$ **3.** $5, 7, 9, 11, 13$ **5.** $1, 8, 27, 64, 125$
7. $0, \dfrac{1}{3}, \dfrac{1}{2}, \dfrac{3}{5}, \dfrac{2}{3}$ **9.** $4, -8, 16, -32, 64$
11. $\dfrac{1}{2}, \dfrac{2}{5}, \dfrac{3}{10}, \dfrac{4}{17}, \dfrac{5}{26}$ **13.** $\dfrac{1}{3}, \dfrac{1}{9}, \dfrac{1}{27}, \dfrac{1}{81}, \dfrac{1}{243}$
15. $2, \dfrac{17}{4}, \dfrac{82}{9}, \dfrac{257}{16}, \dfrac{626}{25}$ **17.** $-3x, 6x^2, -9x^3, 12x^4, -15x^5$
19. $x^2, 8x^3, 81x^4, 1024x^5, 15625x^6$ **21.** $a_n = n + 1$
23. $a_n = n^2$ **25.** $a_n = 3^n$ **27.** $a_n = (-1)^n$
29. $a_n = \dfrac{n}{n+1}$ **31.** $a_n = 4n$ **33.** $a_n = (-1)^n \dfrac{2n}{n+1}$
35. $a_n = x^{n+1}$ **37.** $a_n = \dfrac{\sqrt[n+1]{x}}{3^n}$ **39.** 10 **41.** 8 **43.** -1
45. 4 **47.** $\dfrac{20}{3}$ **49.** 140 **51.** -3 **53.** $\sum_{i=1}^{4} 3i$
55. $\sum_{i=1}^{5} i^2$ **57.** $\sum_{i=1}^{5} \dfrac{i}{i+1}$ **59.** $\sum_{i=1}^{5} (-1)^i \dfrac{i}{i+2}$
61. $\sum_{i=1}^{4} (-1)^{i+1}(4i+3)$ **63.** $\sum_{i=1}^{4} (-1)^{i+1} \dfrac{x^{2i-1}}{i}$
65. $2, 6, 18, 54, 162, 486$ **67.** $-1, 1, 9, 41, 169, 681$
69. A sequence is a function with domain equal to the set of positive integers.
71. A formula that defines a sequence by relating each term to its preceding term.
73. To indicate which term it is in the sequence.
75. An infinite sequence has no last term, for example, $1, 2, 3, 4, \ldots$
77. The general term of a sequence represents every term in the sequence with a formula; for example, $a_n = n$ represents the set of natural numbers.

Exercise Set 12.2
1. yes; 1 **3.** yes; 3 **5.** no **7.** yes; -5 **9.** no
11. yes; $x - y$ **13.** yes; -7 **15.** $a_{10} = 42;\ a_8 = 32$
17. $a_{13} = 75;\ a_6 = 33$ **19.** $a_7 = 22;\ S_{10} = 175$
21. $a_1 = -11;\ d = 4$ **23.** $n = 8;\ S_n = S_8 = 292$
25. $a_1 = -42;\ S_6 = -\dfrac{783}{4}$ **27.** $S_{20} = 1830;\ S_{24} = 2628$

29. $n = 4$; $d = 15$ **31.** $d = 2$; $a_{28} = 66$
33. $n = 6$; $a_n = 6$
35. $n = 14$; $a_1 = -4$ or $n = 15$; $a_1 = 0$ **37.** 155
39. 96   **41.** $\dfrac{203}{3}$ **43.** 1275   **45.** 4905   **47.** 728
53. $26,400
55. $f(x) = \log_2 x$

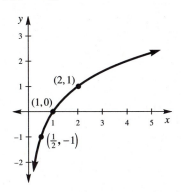

57. $1442.41   **59.** (a) $k = -0.0051$ (b) 7.1697

Exercise Set 12.3
1. geometric; $r = 2$; $a_n = 2^{n-1}$
3. arithmetic; $d = 5$; $a_n = 5n$
5. geometric; $r = \dfrac{1}{2}$; $a_n = \left(\dfrac{1}{2}\right)^{n+1}$
7. arithmetic; $d = \dfrac{1}{2}$; $a_n = \dfrac{x+n-1}{2}$
9. neither   **11.** neither   **13.** geometric; $r = \dfrac{1}{4}$; $a_n = \dfrac{1}{2^{2n-1}}$
15. geometric; $r = -\dfrac{1}{5}$; $a_n = \dfrac{1}{3}\left(-\dfrac{1}{5}\right)^{n-1}$
17. $a_4 = -189$; $S_4 = -140$ **19.** $a_7 = 1$; $S_7 = 43$
21. $a_1 = -1024$; $S_4 = -816$
23. $a_1 = 3^8 = 6561$; $S_\infty = \dfrac{19683}{4}$
25. $a_1 = (1.02)^5$; $a_2 = (1.02)^6$ **27.** $n = 7$; $a_7 = 192$
29. $\dfrac{3}{4}$ **31.** 16   **33.** Does not exist: $|r| = 2$.
35. Does not exist: $|r| = 1$.   **37.** $-\dfrac{25275}{65536}$ **39.** 1
41. $\dfrac{3}{4}$ **43.** $\dfrac{1}{3}$ **45.** $\dfrac{2}{3}$ **47.** $\dfrac{7}{33}$ **49.** $\dfrac{41}{333}$
51. 15 ft   **53.** $\dfrac{1023}{64} = 15.9844$ cm   **55.** 40.3895 cm
57. $h = 6$ ft

Exercise Set 12.4
1. 6   **3.** 3,628,800   **5.** 20   **7.** 15   **9.** 1   **11.** 165
13. 6   **15.** 5   **17.** 1   **19.** 1   **21.** $\dfrac{8!}{7!}$ **23.** $\dfrac{7!}{4!}$

25. $\dfrac{12!}{8!}$ **27.** $\dfrac{3!}{1!}$ or $\dfrac{3!}{0!}$ **29.** n
31. $n(n+1)(n+2)(n+3)$ **33.** $x^3 + 3x^2y + 3xy^2 + y^3$
35. $a^5 + 10a^4b + 40a^3b^2 + 80a^2b^3 + 80ab^4 + 32b^5$
37. $u^7 - 7u^6v + 21u^5v^2 - 35u^4v^3 + 35u^3v^4 - 21u^2v^5 + 7uv^6 - v^7$
39. $16x^4 - 32x^3y + 24x^2y^2 - 8xy^3 + y^4$
41. $81h^8 - 540h^6k + 1350h^4k^2 - 1500h^2k^3 + 625k^4$
43. $28x^{12}y^2$ **45.** $4032r^4s^5$ **47.** $280p^6q^8$
49. $-\dfrac{1485}{128}x^9$ **51.** 1.082856   **53.** 0.78464
55. 1   5   10   10   5   1   **57.** 2^{n-1}

Review Exercises
1. 1, 4, 9, 16   **2.** 4, -8, 16, -32
3. $-3, 5, -7, 9$ **4.** $x, -x^2, x^3, -x^4$
5. $a_8 = 29$; $a_{11} = 41$ **6.** $a_6 = -13$; $a_9 = -22$
7. $a_1 = -\dfrac{2}{9}$; $d = \dfrac{41}{18}$ **8.** $S_{12} = 414$; $S_{18} = 891$
9. $a_6 = 5$; $S_6 = 30$ **10.** $a_6 = 2$; $S_6 = 126$
11. $S_6 = -910$; $a_6 = -1215$
12. $a_1 = 5^5 = 3125$; $S_4 = 3900$ **13.** 45   **14.** 30
15. -1 **16.** 92   **17.** 864   **18.** 10, 11, 12   **19.** 5040
20. 1190   **21.** 28   **22.** $n(n-1)$ **23.** 715   **24.** 1
25. $a^5 - 5a^4b + 10a^3b^2 - 10a^2b^3 + 5ab^4 - b^5$
26. $729x^6 - 2916x^5y + 4860x^4y^2 - 4320x^3y^3 + 2160x^2y^4 - 576xy^5 + 64y^6$
27. $16p^8 - 96p^6q^2 + 216p^4q^4 - 216p^2q^6 + 81q^8$
28. $u^8 - 8u^7v + 28u^6v^2 - 56u^5v^3 + 70u^4v^4 - 56u^3v^5 + 28u^2v^6 - 8uv^7 + v^8$
29. $-960x^7y^3$ **30.** $-3432x^7y^7$ **31.** $42240a^7b^4$
32. $\dfrac{5005}{373248}p^9$ **33.** $2863.57   **34.** $501,000
35. 90 ft   **36.** 86.9459 in.   **37.** 1.42534

Chapter Test
1. $-2, 1, 6, 13, 22$ **2.** 1, 3, 9, 27, 81
3. $-1, 3, -5, 7, -9$ **4.** 2, 5, 8, 11, 14   **5.** 30
6. -42 **7.** $a_5 = 17$; $S_5 = 55$ **8.** $a_4 = 32$; $S_4 = 60$
9. $S_\infty = 128$ **10.** 20   **11.** 56   **12.** 1
13. $a^6 + 6a^5b + 15a^4b^2 + 20a^3b^3 + 15a^2b^4 + 6ab^5 + b^6$
14. $x^5 - 5x^4y + 10x^3y^2 - 10x^2y^3 + 5xy^4 - y^5$
15. $4032x^5y^4$ **16.** $\dfrac{1}{9}$ **17.** $\dfrac{16}{99}$ **18.** 728   **19.** 32 m
20. 170.8598 cm

Cumulative Review Chapters 10–12

1. (a) $(x - 1)^2 + y^2 = 4$ (b) circle
   (c)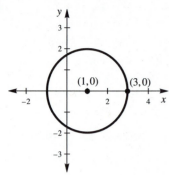
   (d) center: $(1, 0)$; $r = 2$

2. (a) $\left(x + \dfrac{1}{2}\right)^2 + \left(y - \dfrac{1}{2}\right)^2 = 1$ (b) circle
   (c)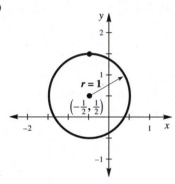
   (d) center: $\left(-\dfrac{1}{2}, \dfrac{1}{2}\right)$; $r = 1$

3. (a) $\dfrac{(x - 3)^2}{4} + \dfrac{(y + 2)^2}{16} = 1$ (b) ellipse
   (c)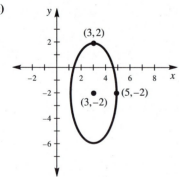
   (d) center: $(3, -2)$

4. (a) $\dfrac{(x - 1)^2}{4} - \dfrac{(y - 1)^2}{4} = 1$ (b) hyperbola
   (c)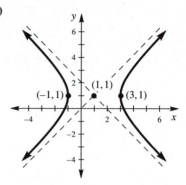
   (d) center: $(1, 1)$

5. (a) $f(x) = -2(x - 1)^2 - 1$ (b) parabola
   (c)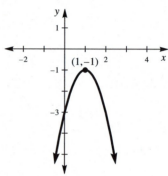
   (d) vertex: $(1, -1)$

6. (a) $f(x) = -2(x - 2)^2 + 3$ (b) parabola
   (c)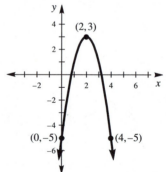
   (d) vertex: $(2, 3)$

7. $\dfrac{(y - 1)^2}{9} - \dfrac{(x + 1)^2}{4} = 1$; hyperbola

8. $\dfrac{x^2}{16} + \dfrac{y^2}{9} = 1$; ellipse

9. $x^2 + y^2 - 8x - 6y = 0$; circle

10. $f(x) = x^2 - 2$; parabola

11. $\{(\pm 2, 0)\}$ 12. $\left\{(-2, 3), \left(\dfrac{6}{5}, -\dfrac{17}{5}\right)\right\}$ 13. $\left\{\left(\dfrac{2}{3}, \dfrac{11}{3}\right)\right\}$

14.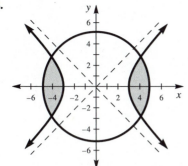

15.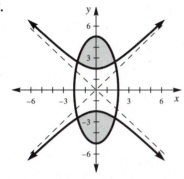

16. D$\{x \mid x > 0\}$; R$\{y \mid y \in R\}$

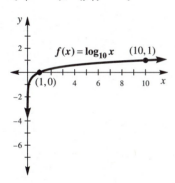

17. D$\{x \mid x \in R\}$; R$\{y \mid y < 0\}$

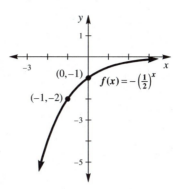

18. $\left\{\dfrac{9}{2}\right\}$ **19.** $\{17\}$ **20.** 13.2   **21.** 224,084

22. $f^{-1}(x) = \log_b x$ **23.** 12 yr   **24.** 2, 1, $\dfrac{2}{3}, \dfrac{1}{2}$

25. 12, 6, 2   **26.** $a_n = 2n - 1$ **27.** 60   **28.** 775

29. $a_n = 2^{n-1}$ **30.** $\dfrac{11}{4}$ **31.** 4   **32.** $x^3 + 6x^2 + 12x + 8$

33. $15x^2$ **34.** $-120y^3$

35. $x^{10} - 5x^8 y + 10x^6 y^2 - 10x^4 y^3 + 5x^2 y^4 - y^5$

Index

Abel, Niels Henrik, 445
Abscissa, 105, 155
Absolute value
   explanation of, 15–16, 31, 62–63, 79
   linear equations involving, 62–66, 96
   linear inequalities involving, 79–83, 97
ac test, 311–312, 340
Addition
   with fractions, 22
   with functions, 147–148
   with polynomials, 260–261
   with radicals, 409–410
   with rational expressions, 359–360, 388
   with signed numbers, 23, 31–32
   symbol for, 14
   of two complex numbers, 431
Addition property
   of equality, 38, 173
   of inequality, 68–69
Additive identity, 17
Additive inverse, 7, 14, 17
Algebra of functions, 148, 156
Algebraic expressions
   common, 47–48
   greatest common factor of, 296–298
   translating verbal statements into, 48–50
Algebraic term, 259, 290
Apollonius of Perga, 533
Applied problems
   distance-rate-time, 91–92
   explanation of, 49–50, 96
   factoring used to solve, 334–338, 341
   fulcrum, 90–91
   historical comments on, 92–93, 334
   linear equations and inequalities used to solve, 231–237

logarithms used to solve, 583–585
mixture, 88–90
quadratic equations and quadratic inequalities to solve, 488–492
rational equations used to solve, 378–385, 388
steps used to solve, 85–87, 231
translating verbal statements into algebraic symbols in, 47–51
Arithmetic sequences
   explanation of, 600
   nth term of, 600–602, 622
   sum of, 602–605
Associative properties, 17, 31
Asymptotes, 534
Augmented matrix, 196–198
Average speed, 58
Axis of abscissa, 105
Axis of ordinates, 105
Axis of symmetry, 511

Back substitution, 199–200, 240
Bhaskara, 334
Binomial theorem, 615–620, 622–623
Binomials
   conjugate of, 418
   division of polynomials by, 276–279
   explanation of, 259, 290
   product of two, 267
   square of, 318
Boundary line, 216
Braces, 3, 30
Branches, 533
Briggs, Henry, 565

Calculators. *See* Graphics calculators; Scientific calculators
Cantor, George, 3
Cardano, Girolamo, 12, 408, 428
Cartesian coordinate system. *See* Rectangular coordinate system

Center, of circle, 502
Circles
   area of, 54
   development of equation of, 501–502
   explanation of, 502, 550–551
   general form for equation of, 504–505
   graphics calculators and computers used to draw graphs of, 505–506
   standard form for equation of, 502–504, 551
Closed intervals, 69
Closure, 16, 17, 31
Coefficient, 302
Coefficient matrix, 196
Collinear, 119
Common denominators, 22. *See also* Denominators
Common fractions, 21
Common logarithms
   explanation of, 573
   table of, A–16 – A–17
Common ratio, 607
Commutative properties, 16, 17, 31
Completing the square
   explanation of, 453–454
   quadratic equations solved by, 454–457, 495
   uses for, 459
Complex conjugates, 432, 440
Complex numbers
   adding two, 431
   dividing two, 432–433
   equality of, 430
   explanation of, 428–430, 440
   historical comment on, 428
   multiplying two, 431–432
Complex rational expressions, 366–370, 388
Compound inequality, 74

I-1

Compound interest formula, 562
Computers
   to draw graphs of functions, 141, 142
   for exponential functions, 559–563
   to find solutions to systems of equations in two variables, 180–181
   to graph conic sections, 529–530, 537–538
   to solve quadratic equations, 458
Conditional equations, 42, 96
Conic sections. *See also specific conic sections*
   explanation of, 509, 551
   graphics calculators and computers to graph, 537–538
Conjugates
   of binomials, 418
   complex, 432, 440
   explanation of, 417, 439, 463
Constant of variation, 483–485
Constants, 4
Constraints, 224, 241
Coordinates, 6, 104–105
Counting numbers, 3
Cramer's rule, 210–214, 241
Critical numbers, 476
Cube roots, 401, 439
Cubic equations, 449
Cubing property, 422

Dantzig, George, 224
de Fermat, Pierre, 127
de Laplace, Pierre Simon, 593
de Lineriis, Johannes, 21
Decimal notation, 255–256
Degree, of polynomials, 259, 290, 469–470
Demand, 236, 237
Denominators
   common, 22
   least common, 22, 31, 360–364, 373–376
   of rational expressions, 347–351. *See also* Rational expressions
   rationalizing, 415–417, 439
Descartes, René, 103, 104, 247, 408, 428
Determinants
   explanation of, 205, 241
   historical comment on, 205
   used to solve systems of equations, 205–210

Difference of two cubes, 317–318, 341
Difference of two squares
   explanation of, 268, 341
   factoring and, 315–317
Dimensions, of matrix, 195, 240
Diophantus, 47
Direct variation, 483, 484
Directrix, 509
Discriminant, 464–465, 495
Distance formula, 121, 155, 501–502, 550
Distance-rate-time problems. *See also* Applied problems
   historical comment on, 378
   rational expressions used to solve, 380–382
   steps to solving, 91–92
Distributive properties, 17, 31, 39
Division
   dividing polynomials by synthetic, 284–288, 291
   explanation of, 26
   with fractions, 21, 32
   with functions, 147–150
   with polynomials, 274–281, 286–288, 291
   with rational expressions, 355–356
   with signed numbers, 25
   symbol for, 14
   of two complex numbers, 432–433
   by zero, 26
Division property
   of equality, 40–41
   of inequality, 73–74
Domain
   of exponential function, 555, 587
   of relation, 136, 156
Double negative property, 14, 31

Elements
   explanation of, 3, 30
   of matrix, 195
   notation for, 3–4
Elimination method
   Gauss-Jordan, 200–203, 241
   to solve systems of linear equations, 173–176, 178, 179
   to solve systems of nonlinear equations, 541
   by using graphics calculators or computers, 180–181
Ellipses
   accoustical property of, 526
   deriving equation of, 526–527

   explanation of, 525–526, 537, 551
   graphics calculators and computers to graph, 529–530
   historical comment on, 525
   sketching graphs of, 527–529
Empty sets, 3, 4, 30
Entries, 195
Equality
   addition property of, 38, 173
   division property of, 40–41
   multiplication property of, 40, 174
   power property of, 422–425
   properties of, 16, 31, 38
   subtraction property of, 38–39
   symbols of, 13
Equations. *See also* Linear equations; Literal equations; Nonlinear equations; Quadratic equations
   conditional, 42, 96
   cubic, 449
   equivalent, 38, 96
   involving radicals, 422–426, 440
   involving variation, 483–486
   reduced to quadratic form, 469–474, 495
   second-degree, in one variable, 326
   solution or root of, 37
   to solve geometry problems, 51
   to solve number problems, 50
Equilibrium price, 236
Equivalent equations, 38, 96, 173
Equivalent fractions, 25
Equivalent inequalities, 68
Euler, Leonard, 135, 295, 428
Exponential decay, 557
Exponential form, 247
Exponential functions
   explanation of, 555, 587
   graphing, 556–558
   relationship between logarithms and, 568–569
   solutions to, 558–559
   using computers and scientific calculators for, 559–563
Exponential growth, 556
Exponents. *See also* Rational exponents
   explanation of, 26, 247–248, 290
   historical comment on, 247
   negative, 250
   power rules for, 253–254, 290
   product rule for, 248–250
   quotient rule for, 251–252

symbol for, 14
zero, 251
Extraneous solutions, 541

Factorability, 312–314, 340
Factorial notation, 616–617, 622
Factors/factoring. *See also* Greatest common factor (GCF)
  applied problems solved by, 334–338, 341
  in exponents, 247
  by grouping, 298–300, 319–320, 341
  nonlinear equations solved by, 326–332, 341
  of polynomials, 296–298, 315–320, 322–324, 341
  quadratic equations solved by, 445–446, 457, 495
  of type I trinomials, 302–306
  of type II trinomials, 307–314
Fibonacci, Leonardo, 6, 183, 408
Finite sets, 4, 30
Focus
  of ellipse, 526
  of hyperbola, 533
  of parabola, 509, 512
FOIL method, 267–268, 291
Fraction form, 25
Fractions
  addition and subtraction with, 22
  division with, 21, 32
  equivalent, 25
  explanation of, 31
  fundamental property of, 21, 31
  historical comment about, 21
  multiplication with, 21
  simplification of, 22
Fulcrum problems, 90–91
Function notation, 138, 156
Functions. *See also* Linear functions
  addition, subtraction, multiplication, and division of, 147–150
  composition of, 150–151, 156
  explanation of, 137–138, 156
  inverse of, 151–153
  linear, 139

Gauss, Carl Friedrich, 205, 247
Gauss-Jordan elimination method, 200–203, 241
General form, 128, 132, 156
General term, 593

Geometric sequences
  explanation of, 607–608
  historical comment on, 606
  nth term of, 608–609, 621
  sum of, 609–613
Geometric series, 609
Geometry problems, 51, 87–88
Graphics calculators
  to draw graphs of functions, 141, 142
  to evaluate polynomials, 270
  to find solutions to systems of equations in two variables, 180–181
  to graph conic sections, 522, 529–530, 537–538
  logarithms on, 579, 580, 583–585
  to solve quadratic equations, 458
  TI-81 and TI-82, A–2—A–13
Graphs/graphing
  calculators and computers used for, 141, 142, 180–181, 481
  linear inequalities in two variables, 215–220
  of lines, 109
  as points on number lines, 7
  of relations, 136–137, 141
  to solve systems of linear equations, 169–173, 180
  used to solve linear programming problems, 229
Greatest common factor (GCF)
  of algebraic expressions, 296–300
  in factoring, 340
  of integers, 295–296
Grouping
  factoring by, 298–300, 319–320, 341
  symbols used for, 27, 32

Half-open intervals, 69
Half-plane, 216–219
Harriot, Thomas, 68
Hilbert, David, 167
Hippocrates, 54
Horizontal axis, 105
Horizontal lines, 107, 132, 156
Hyperbolas
  asymptotes of, 534–537
  explanation of, 533–534, 551
  historical comment on, 533

Identity, 42, 96
Identity properties, 17, 31

Imaginary units, 428, 440
Index
  of radicals, 401, 439
  of summation, 596, 597
Inequalities. *See also* Linear inequalities; Quadratic inequalities
  addition and subtraction properties of, 68–69
  compound, 74
  division property of, 73–74
  equivalent, 68
  multiplication property of, 72
  nonlinear systems of, 546–549
  phrases concerning, 75
  symbols of, 13
Infinite sets, 4, 30
Infinity, 69
Integers
  explanation of, 30
  greatest common factor of, 295–296
  set of, 7–8
Intersection, 5, 30
Interval notation, 69
Intervals, 69–71
Inverse properties, 17, 31
Inverse variation, 483, 484
Irrational numbers, 8–9, 31

Kepler, Johannes, 525
Kowa, Seki, 205

Lagrange, Joseph Louis, 347
Leading coefficient, 302
Least common denominator (LCD)
  explanation of, 22, 31
  of rational expressions, 360–364, 373–376, 388
Leibniz, Gottfried Wilhelm, 247, 265, 428
Like terms, 39
Line segments, 121–123, 156
Linear equation systems
  Cramer's rule to solve, 210–214
  determinants to solve, 205–210
  elimination method to solve, 173–176, 178, 179
  explanation of, 167, 240
  graphing method to solve, 169–173, 180
  methods to solve, 168–169, 240
  number of solutions to, 171
  substitution method to solve, 176–181
  in three or more variables, 183–193
  using matrices, 195–203

Linear equations
  explanation of, 37, 104, 132, 155
  historical comment on, 167
  involving rational expressions, 373–376, 388
  in one variable
    involving absolute value, 62–66, 96
    methods to solve, 37–45, 96
    steps to solve, 43
    to solve applied problems, 231–237
  in three or more variables
    consistent systems of, 186–188, 240
    dependent systems of, 188–193
    explanation of, 184, 240
    graph of, 185–186
    historical comment on, 183
    inconsistent systems of, 188, 240
    steps to solve, 184–185, 193
  in two variables. See also Rectangular coordinate system
    example of, 103–104
Linear functions. See also Functions
  explanation of, 139, 156
  steps to finding inverse of, 152–153
Linear inequalities. See also Inequalities
  in one variable
    explanation of, 68, 96
    historical comment on, 68
    solutions for, 68–76, 96
    using absolute value, 79–83, 96
  systems of, 220–221, 241
  in two variables
    graphs of, 215–220, 241
    solutions to, 216
    used to solve applied problems, 231–237
Linear programming
  application of, 224–229
  explanation of, 224, 241
  graphing used to solve problems in, 229
  historical comment on, 224
Lines
  finding slope of, 113–117
  graph of, 109
  parallel, 118
  perpendicular, 118–119
  writing equations of, 128–133, 156
Listing method, to describe sets, 4, 30
Literal equations
  explanation of, 54–55, 96

historical comment on, 54
methods for solving, 55–60
Logarithmic functions, 565–569, 587
Logarithms
  common, 573
  explanation of, 565–567
  historical comment on, 565
  natural, 579
  properties of, 572–576, 587–588
  relationship between exponential functions and, 568–569
  sketching graphs of, 568
  solving equations involving, 578–581, 588
  tables of, A–16 – A–18

Mathematical sentences, 49–50, 96. See also Applied problems
Matrix
  augmented, 196
  coefficient, 196
  explanation of, 195, 240
  square, 196, 205, 241
Members, 3, 30
Midpoint
  formula for, 501–502
  of line segment, 121–123, 156, 551
Mixture problems, 88–90
Monomials
  division of polynomials by, 274–276, 291
  explanation of, 259, 290
Multiplication
  with fractions, 21
  with functions, 147–150
  with polynomials, 265–270, 290–291, 334–335
  with radicals, 410–412
  with rational expressions, 353–355
  with signed numbers, 24–25, 32
  symbol for, 14, 265
  of two complex numbers, 431–433
Multiplication property
  of equality, 40, 174
  of inequality, 72
  of zero, 18–19, 31
Multiplicative identity, 17
Multiplicative inverse, 17
Multiplicity, 495

n factorial, 615–616
Napier, John, 565
Nappes, 533

Natural logarithms
  explanation of, 579
  table of, A–18
Natural numbers, 3, 7, 30
Negative exponents, 250, 290
Negative infinity, 69
Negative numbers
  explanation of, 6
  historical comment on, 6
Newton, Issac, 103
Noether, Emmy, 167
Nonlinear equations
  factoring used to solve, 327–332, 341
  historical comment on, 326
Nonlinear systems of equations
  explanation of, 540
  solutions to, 541–544, 551
Nonlinear systems of inequalities
  computers used to sketch solution set for, 549
  explanation of, 546
  solutions to, 546–548, 551
nth root
  explanation of, 394, 401, 439
  finding, 395–396
Null sets, 4. See also Empty sets
Number lines, 6–7
Number problems, 50, 379–380
Number system, 6
Numbers. See also Real numbers
  complex, 428–433
  positive and negative, 6
  pure imaginary, 430
  rational, 8, 30

Objective function, 226, 241
One-to-one function, 568–569
Ordered pairs
  explanation of, 103–104, 155
  on rectangular coordinate system, 106–108
  relation as set of, 136
Ordered quadruples, 193
Ordered triples, 184
Ordinate, 105, 155
Origin, 104
Orseme, Nicole, 393
Oughtred, William, 68

Parabolas
  axis of symmetry of, 511–512
  equation of, 509–511, 517, 551
  explanation of, 509, 515, 551

focus of, 509, 512
graphics calculator used to graph, 522
graphs of, 511–521
historical comment on, 509
Parallel lines, 118, 155
Perfect-square trinomials, 311, 318, 341
Perpendicular lines, 118–119, 155
Planes, 185–186, 240
Poincaré, Henri, 555
Point-slope form, 128–129, 132, 156
Polynomials
addition with, 260–261, 290
classifying, 259–260
degree of, 259, 290, 469–470
division with, 274–281, 291
explanation of, 258–259, 290
factoring, 296–298, 315–320, 322–324, 341
graphics calculators for evaluating, 270
historical comments on, 265, 408
multiplication with, 265–270, 290–291, 334–335
prime, 305
relatively prime, 350, 388
subtraction with, 261–263, 290
Positive numbers, 6
Power property of equality, 422–425, 471
Power rules for exponents, 253–254, 290
Powers of e table, A–19 – A–20
Prime polynomials, 305
Principal square roots, 400, 439
Product rule for exponents, 248–250, 290
Proportions, 384
Pure imaginary numbers, 430
Pythagoras, 501
Pythagorean theorem
early application of, 334
explanation of, 120–121, 335, 341, 501
used to solve applied problems, 335–336

Quadrants, 104, 155
Quadratic equations
applied problems solved using, 488–492
completing the square to solve, 452–457, 495

discriminant of, 464–465, 495
explanation of, 326
factoring used to solve, 445–446, 457, 495
quadratic formula used to solve, 460–464, 495
square root property to solve, 446–451
sum and product of roots of, 465–467
zero-factor theorem to solve, 326–327, 341
Quadratic formula
explanation of, 460–461, 495
quadratic equations solved by using, 461–464
Quadratic inequalities
applied problems solved using, 488–492
explanation of, 476
steps to solve, 476–480, 495
Quotient rule for exponents, 251–252, 290

Radical notation, 401–403
Radicals
addition and subtraction with, 409–410
applications involving, 434–436
arithmetic operations involving, 408–412, 439
division by, 414–420, 439
explanation of, 401, 439
multiplication with, 410–412
properties of, 403
simplest form for, 403–406, 439
solving equations involving, 422–426, 440
symbol for, 400
Radicand, 401, 439
Radius, 502
Raising to powers, 26
Range
of exponential function, 555, 587
of relation, 136, 156
Rational exponents. See also Exponents
converting to radical notation from, 401–403
explanation of, 393–396, 439
historical comment on, 393
rules for, 396–398
Rational expressions
addition with, 359–360, 388
building, 351

division with, 355–356, 388
explanation of, 347–349, 387
fundamental principle of, 349–350, 387–388
least common denominator of, 360–364, 373–376, 388
multiplication with, 353–355, 388
simplifying, 349–350, 366–370, 388
solving equations involving, 373–376, 388
subtraction with, 360, 388
used to solve applied problems, 378–385
Rational numbers, 8, 30
Rationalizing the denominator, 415–420, 439
Ratio-proportion problems. See also Applied problems
rational expressions used to solve, 384–385
Ratios
common, 607
explanation of, 384
Real numbers
historical comment on, 12
operations with, 21–28, 31–32
product of complex conjugates as, 432
properties of, 13–19, 31
set of, 9, 30–31
Recorde, Robert, 62
Rectangular coordinate system
explanation of, 104, 155
historical comment on, 103, 104
locating points on, 104–106
Recursion formula, 617, 622
Reflexive property, 16, 31
Relations
graph of, 136–137, 141
inverse of, 151–152, 156
as set of ordered pairs, 136, 156
Relatively prime polynomials, 350, 388
Rhind Papyrus, 606
Root
of equation, 37, 96
of multiplicity, 329
symbol for, 14, 439
Row operations, 196, 240
Row transformation theorem, 196, 240
rth term in binomial expansion, 619–620, 623
Rule method. See Set-builder notation

Scientific calculators
  used for exponential functions, 563
  used to find decimal approximations to roots of numbers, 406
Scientific notation
  converted to decimal notation, 255–256
  explanation of, 254–255, 290
Second-degree equation in one variable, 326
Sequences
  explanation of, 593–596, 621
  infinite and finite, 594
  series *vs.*, 596
Series, 596, 621
Set-builder notation, 4, 30
Sets
  explanation of, 3, 30
  of integers, 7–8
  of irrational numbers, 8–9
  methods of describing, 4
  of nature numbers, 7
  notation for, 3–4
  operations using, 5
  of rational numbers, 8
  of real numbers, 9
  of whole numbers, 7
Sigma, 596
Sign graph, 476, 495
Signed numbers
  explanation of, 22–23
  operations with, 23–25, 31–32
Simplex method, 229
Slope
  explanation of, 113, 114, 155
  positive and negative, 116
  zero, 117
Slope-intercept form, 129–132, 156
Solution of multiplicity, 495
Solution set
  of equivalent equations, 173
  explanation of, 38, 96
  of linear inequality in two variables, 216
Solutions, extraneous, 423
Square matrix, 196, 205, 241
Square root property
  explanation of, 446
  solving quadratic equations using, 447–451, 495

Square roots
  principal, 400, 439
  symbol for, 439
  table of, A–15 – A–16
Squaring property, 422, 423
Standard form
  ellipse and, 528–529, 551
  hyperbola and, 536, 551
  parabola and, 551
  polynomials in, 259–260, 290
Subsets, 5, 30
Substitution method
  back, 199–200
  to solve systems of linear equations, 176–181
  to solve systems of nonlinear equations, 542–543
  by using graphics calculators or computers, 181
Substitution property, 16, 31
Subtraction
  with fractions, 22
  with functions, 147–150
  with radicals, 409–410
  with rational expressions, 360
  with signed numbers, 24, 32
  symbol for, 14
Subtraction property
  of equality, 38–39
  of inequality, 68–69
Sum of two cubes, 317, 341
Summation notation, 596, 597, 621
Supply, 236, 237
Symbols
  for comparison, 13
  for grouping, 27, 32
  for operations, 14
  for radical, 400
Symmetric property, 16, 31
Symmetry, 511–512
Synthetic division
  dividing polynomials by, 286–288
  explanation of, 284–286, 291

TEMATH (Tools for Exploring Mathematics), 141
Theorems, 18
TI-81 and TI-82 graphics calculators, A–2 – A–13
Transitive property, 16, 31

Trinomials
  determining signs of binomial factors of, 310
  division of polynomials by, 279–281
  explanation of, 259, 290
  factoring of type I, 302–306, 340
  factoring of type II, 307–314, 340
  perfect-square, 311, 318, 341

Union, of sets, 5

Variables, 4
Variation
  constant of, 483–485
  direct, 384, 483
  explanation of, 496
  inverse, 483, 484
Verbal problems. *See* Applied problems
Vertical axis, 105
Vertical line test, 141, 156
Vertical lines, 107, 132, 156
Vertices
  of ellipse, 527
  explanation of, 226
  of hyperbola, 534
  of parabola, 511, 512, 515–516
Viéte, François, 37, 167

Whole numbers, 7, 30
Word problems. *See* Applied problems
Work problems, 378, 382–383. *See also* Applied problems

x-axis, 104, 105
x-intercept, 106–107

y-axis, 104, 105
y-intercept, 106–107

Zero
  division by, 26
  multiplication property of, 18–19, 31
Zero exponent, 251, 290
Zero-factor theorem
  explanation of, 326–327, 341
  generalized, 329–330

Properties of Radicals

$$\sqrt[n]{ab} = \sqrt[n]{a} \cdot \sqrt[n]{b} \quad \text{or} \quad (ab)^{1/n} = a^{1/n}b^{1/n}$$

$$\sqrt[n]{\frac{a}{b}} = \frac{\sqrt[n]{a}}{\sqrt[n]{b}} \quad \text{or} \quad \left(\frac{a}{b}\right)^{1/n} = \frac{a^{1/n}}{b^{1/n}}, \text{ where } a \text{ and } b \text{ are not both negative}$$

$b^{1/n} = \sqrt[n]{b}$, where $b \geq 0$ when n is even

$\sqrt[n]{b^n} = b$ if n is odd; $\sqrt[n]{b^n} = |b|$ if n is even

Complex Numbers

$$i = \sqrt{-1}; \quad i^2 = -1$$

If $a, b \in R$, then:

$a + bi = c + di$ if and only if $a = c$ and $b = d$

$(a + bi) + (c + di) = (a + c) + (b + d)i$

$(a + bi)(c + di) = (ac - bd) + (ad + bc)i$

$a + bi$ and $a - bi$ are **complex conjugates.**

$(a + bi)(a - bi) = a^2 + b^2$

The Quadratic Formula; The Discriminant

The quadratic formula $x = \dfrac{-b \pm \sqrt{b^2 - 4ac}}{2a}$;   the **discriminant** is $b^2 - 4ac$.

If $b^2 - 4ac > 0$, there are two distinct real solutions.

If $b^2 - 4ac = 0$, there is one real solution of multiplicity 2.

If $b^2 - 4ac < 0$, the two solutions are nonreal complex conjugates.

If x_1 and x_2 are solutions of $ax^2 + bx + c = 0$, then

$$x_1 + x_2 = \frac{-b}{a} \quad \text{and} \quad x_1 x_2 = \frac{c}{a}$$

The Slope of a Line

$$m = \frac{y_2 - y_1}{x_2 - x_1}$$

The Midpoint of a Line Segment

$$(x_m, y_m) = \left(\frac{x_1 + x_2}{2}, \frac{y_1 + y_2}{2}\right)$$

Linear Equations

General form	$ax + by = c$
Point–slope form	$y - y_1 = m(x - x_1)$
Slope–intercept form	$y = mx + b$
Horizontal line	$y = $ constant
Vertical line	$x = $ constant

Conic Sections

Circle $(x - h)^2 + (y - k)^2 = r^2$; center at (h, k), radius r

Parabola $y = a(x - h)^2 + k$; vertex at (h, k); opens up if $a > 0$, down if $a < 0$

Ellipse $\dfrac{(x - h)^2}{a^2} + \dfrac{(y - k)^2}{b^2} = 1$; center at (h, k)

x-ellipse if $a > b$; y-ellipse if $b > a$

Hyperbola $\dfrac{(x - h)^2}{a^2} - \dfrac{(y - k)^2}{b^2} = 1$; x-hyperbola; center at (h, k)

$\dfrac{(y - k)^2}{b^2} - \dfrac{(x - h)^2}{a^2} = 1$; y-hyperbola; center at (h, k)

Variation

Direct $y = kx$ — y varies directly as x

Inverse $y = \dfrac{k}{x}$ — y varies inversely as x

Joint $y = kxz$ — y varies jointly with x and z

yard: yd

Logarithms

$y = \log_b x$ if and only if $b^y = x$, where $b > 0$, $b \neq 1$

$\log_b b = 1$; $\log_b 1 = 0$; $b^{\log_b x} = x$

Properties of Logarithms

$\log_b xy = \log_b x + \log_b y$

$\log_b \dfrac{x}{y} = \log_b x - \log_b y$

$\log_b x^k = k \log_b x$

The Distance Formula

$$d = \sqrt{(x_2 - x_1)^2 + (y_2 - y_1)^2}$$

Arithmetic Sequences and Series

$a_1, a_1 + d, a_1 + 2d, \ldots, a_1 + (n - 1)d, \ldots$ is an **arithmetic sequence**, where a_1 is the first term, d is the **common difference**, and n is the number of the term

The **nth term:** $a_n = a_1 + (n - 1)d$

The **sum:** $S_n = \dfrac{n}{2}(a_1 + a_n) = \dfrac{n}{2}[2a_1 + (n - 1)d]$